AF737433

STUDIES IN NEOTROPICAL MAMMALOGY
Essays in Honor of Philip Hershkovitz

PHILIP HERSHKOVITZ

FIELDIANA

Zoology

NEW SERIES, NO. 39

STUDIES IN NEOTROPICAL MAMMALOGY
Essays in Honor of Philip Hershkovitz

Edited by
Bruce D. Patterson

Division of Mammals
Field Museum of Natural History
Chicago, Illinois 60605-2496

Robert M. Timm

Division of Mammals
Field Museum of Natural History

Present address:
Museum of Natural History
Department of Systematics and Ecology
University of Kansas
Lawrence, Kansas 66045

Accepted for publication July 30, 1985
December 31, 1987
Publication 1382

PUBLISHED BY FIELD MUSEUM OF NATURAL HISTORY

FIELDIANA: ZOOLOGY
New Series, No. 39

Studies in Neotropical Mammalogy:
Essays in Honor of Philip Hershkovitz
Bruce D. Patterson and Robert M. Timm, Editors

© 1987 Field Museum of Natural History
Library of Congress Catalog Card Number: 87-82549
ISSN 0015-0754
PRINTED IN THE UNITED STATES OF AMERICA

Table of Contents

Preface

In the early 1980s, we discussed the possibility of a testimonial volume for Philip Hershkovitz with Larry Marshall, then with the Department of Geology, Field Museum. As the senior mammalogist at Field Museum and a student of South American mammalogy for almost half a century, Hershkovitz had generously provided invaluable advice and assistance to each of us in the early stages of our careers. We felt a *Festschrift* in his honor might repay a portion of our debts to him and, at the same time, serve as an independent, lasting tribute to his life-work.

In the entire history of Field Museum, only three testimonial volumes had been produced in honor of museum scientists. Each recognized the contributions of men who were both preeminent scientists and museum administrators: Wilfred H. Osgood, Chief Curator of Zoology, 1921–1941; Karl P. Schmidt, Chief Curator of Zoology, 1941–1956; and Rainer Zangerl, Chief Curator and Chairman of Geology, 1962–1974. Although Hershkovitz has never served in an upper-level administrative capacity, his contributions to the museum through distinguished and continuing research clearly qualified him for this honor.

However, plans for a testimonial volume in *Fieldiana: Zoology* did not materialize until November 1983. By that time, Marshall had assumed a new position at the University of Arizona and Hershkovitz had just celebrated his 74th birthday. Given realistic editing and publication schedules, we were faced with the prospect of producing the volume nearly midway between traditionally celebrated anniversary dates. Nevertheless, such timing is somehow fitting: Hershkovitz the man is both extemporaneous and unconventional.

Another notable departure from the *Festschrift* tradition is evident from the table of contents: Hershkovitz himself is a contributor! On many occasions Hershkovitz had lamented the lack of a historical review of South American mammalogy. During the present information explosion, scientists are hard-pressed to keep up with current developments of direct relevance to their research; much less are they afforded the occasion to amble through historical records in Latin, German, French, Spanish, and Portuguese, even though these records are full of interesting and relevant information. As a result of his 50 years in the discipline, Hershkovitz may be unique in his broad knowledge of both historical literature and current research on Neotropical mammals. The editors therefore prevailed upon him to write such a historical survey to complement and enhance this volume. We convinced him that, by assembling a historical analysis of the subject, he would provide a tremendous service to younger workers.

Other contributions to the volume came from friends and colleagues of Hershkovitz. All share an interest in the distribution, taxonomy, and natural history of Neotropical mammals, and each one was inspired to honor Hershkovitz with their contribution. Each of the contributions focuses on those fields of Neotropical mammalogy to which Hershkovitz has contributed most significantly.

We owe thanks to numerous persons connected with this volume. First and foremost, Tanisse Bezin, Managing Editor of Field Museum Press, deserves recognition. Her keen eye for grammar and style eliminated numerous editorial inconsistencies forwarded by the volume editors. Graham Harles, Field Museum Press copy editor, provided skillful editing and proofreading. The Scientific Editor for *Fieldiana,* Timothy Plowman, endured countless interruptions during production of this volume and served as corresponding editor for our own papers. Translations of abstracts into Spanish and Portuguese were kindly provided by Myriam Ibarra (an Ecuadorean ichthyologist) and Debra Moskovits (a Brazilian ecologist), who offered these as their own tributes to Hershkovitz. Assistance in assembling the indices was provided by Mary Anne Rogers.

Finally, we are enormously indebted to a dedicated body of reviewers, who critically evaluated papers in this volume. Their constructive advice and recommendations made editorial tasks far lighter. The editors gratefully thank: W. T. Atyeo, P. V. August, K. Benirschke, W. A. Clemens, J. A. Davis, W. B. Davis, M. R. Dawson, M. D. Engstrom, J. Fooden, G. L. Forman, M. H. Gallardo, A. L. Gardner, H. H. Genoways, W. E. Glanz, M. S. Hafner, D. Hunsaker II, R. J. Izor, J. A. W. Kirsch, K. F. Koopman, M. A. Mares, R. E. Martin, T. J. McCarthy, G. G. Musser, P. Myers, J. L. Patton, O. P. Pearson, R. H. Pine, W. B. Quay, L. Radinsky, O. J. Reichman, D. S. Rogers, R. W. Thorington, Jr., W. D. Turnbull, J. H. Wahlert, S. D. Webb, C. Wemmer, J. O. Whitaker, Jr., D. E. Wilson, R. G. Wolff, and A. E. Wood, in addition to anonymous reviewers of our own papers.

B. D. Patterson
R. M. Timm
Chicago, Illinois

A Biographical Sketch of Philip Hershkovitz, with a Complete Scientific Bibliography

Bruce D. Patterson

Philip Hershkovitz was born October 12, 1909, in Pittsburgh, Pennsylvania, to Aba Hershkovitz and Bertha Halpern. He was the second of four children and their only son. He attended primary and secondary schools in Pittsburgh, graduating from Schenley High School in February 1927. In 1929 he enrolled at the University of Pittsburgh where he majored in zoology, serving as an Undergraduate Assistant in that department during 1930–1931. Having exhausted Pittsburgh's course offerings in zoology and seeking to pursue a career in mammalogy, he was advised to transfer to another school with an expanded curriculum (Harvard University, University of Michigan, or University of California, Berkeley). In his junior year (1931), he transferred to the University of Michigan at Ann Arbor because of its proximity to his home. There he became an Undergraduate Assistant in the Museum of Zoology, working under the supervision of Professor and Curator Lee R. Dice during 1931–1932. He supplemented the meager earnings of this position with taxidermy jobs, which supported him during the early years of the Great Depression.

His first fieldwork was undertaken during the summer of 1932. He went to the San Marcos region of Texas to collect blind cave salamanders (*Typhlomolge rathbuni*) for Professor Uhlenhuth of the University of Maryland Medical School. Having a principal interest in mammals, he wanted to collect small mammals in areas surrounding the caves, but Dice could spare no traps for him and told him to purchase some in Texas.

While hitchhiking from Ann Arbor to Texas, Hershkovitz stopped to visit friends in Chicago. There, a chance visit to Field Museum of Natural History secured him the traps and supplies he needed and seemingly set the course of his later career. Colin Sanborn, then Curator of Mammals during Wilfred Osgood's tenure as Chief Curator of Zoology (1921–1941), befriended Hershkovitz and loaned him the necessary supplies. As a consequence, the mammals that Hershkovitz collected in Texas that first of many field seasons were deposited in the Field Museum collections. He now maintains that his chance visit to Field Museum in 1932 indelibly fixed that institution as the place at which to pursue his career goals.

Hershkovitz's formal education was delayed by the worsening economic situation during 1933. No longer able to afford tuition, he sought advice on subsistence during the Depression, and was told that Ecuador and Paraguay were undoubtedly the least expensive countries in this hemisphere in which to live. Transportation costs decided the issue, and in 1933 he set sail via the Grace Line from New York to Guayaquil, Ecuador for the whopping sum of $600, one-way.

He stayed in Ecuador until 1937. During this period, he mastered Spanish and learned how to live off the land in the Neotropics. His boots disintegrated after six months' time and thereafter he went barefoot. He assembled a fine collection of Ecuadorean mammals for the Museum of Zoology, University of Michigan, supporting his activities in part by selling horses bought on the Peruvian frontier.

He then returned to the University of Michigan where he again enrolled as an undergraduate, graduating in 1938 with a Bachelor of Science degree. By this time, Dice had moved from the Museum of Zoology to the Laboratory of Vertebrate Genetics, and William H. Burt had assumed the curatorship in the Museum. Hershkovitz spent the years 1938–1941 as a graduate student enrolled at the University of Michigan, working on his Ecuadorean collection under Burt's direction. From 1939–1941, he was supported in this work by a Graduate Assistantship. In 1940 he received his Master of Science degree and immediately entered the doctoral program.

Two years before the expected completion of his doctoral program, the Curator of Reptiles and Amphibians at the Museum of Zoology, Helen Gage, told Hershkovitz about the Walter Rathbone Bacon Travelling Scholarship of the United States National Museum. This program was customarily reserved for postdoctoral support, but Mrs. Gage strongly urged him to apply immediately. Thus encouraged, Hershkovitz submitted a brief proposal for work in the Santa Marta region of northern Colombia; his compliance with Mrs. Gage's wishes in this matter was so perfunctory that he failed to include a map of the proposed itinerary. But Remington Kellogg at the National Museum had long wished to obtain a Bacon Scholar for the Mammal Division and asked Hershkovitz to send the omitted material. Much to his surprise, Hershkovitz was awarded the scholarship and left Ann Arbor immediately for Washington. He spent two months there studying the then very poor collection of Neotropical mammals. Afterward he spent two years in Colombia (1941–1943) collecting mammals, other vertebrates, and ectoparasites. The resulting collection was the National Museum's first large and representative Neotropical mammal accession.

In 1943 Hershkovitz's work was interrupted by World War II, and he returned to Ann Arbor to enlist in the Armed Services. He was assigned to the Office of Strategic Services (OSS) and served from 1943–1946 in the European Theater. While serving in France, he met Anne Marie Pierrette, whom he married in 1946. The two returned to the United States, where in 1946 and 1947 he continued his Bacon Scholarship studies of Colombian mammals in Washington. The first of three children (Francine, Michael, and Mark) was born in 1947.

About this time, he was contacted regarding the opening of a curatorial position at Field Museum in Chicago, an opportunity he eagerly hailed for several reasons: (1) The comprehensive collections of Neotropical mammals at Field Museum would be a tremendous resource for what he had already decided would be his life's work; (2) he had the highest regard for W. H. Osgood, who as a principal authority on South American mammals would be a great personal resource on which to draw; (3) the press of family responsibilities made continuation of his graduate studies untenable; and (4) aspirations to a curatorial position had been the *raison d'etre* of his graduate program; a curatorial position made the graduate degree superfluous. Thus he jumped at the offer of employment at Field Museum, knowing full well that it marked the end of his graduate program at Michigan. Like many similar institutions, Michigan had a final year-in-residence degree requirement. Unfortunately, Osgood died in June of 1947, and what might have been a remarkably productive apprenticeship under Osgood never came to pass.

Upon his arrival at Field Museum, Hershkovitz found an uncurated backlog of some four or five years of accessions. Nevertheless, he wasted little time in returning to the field, prompted in part by postwar housing shortages in Chicago. (One can almost hear him now, telling the Museum's Director Clifford Gregg that the nearest affordable housing was in Bogotá!) In 1948 he and his family moved to Colombia where he resumed his inventory of the mammals of that country. He remained in Colombia until the press of curatorial duties and a gently delivered ultimatum from Sanborn finally recalled him to Chicago in 1952.

The collections he made in Colombia, first for the National Museum, then for Field Museum, were to be the heart of all his subsequent research. But unlike others studying the mammal faunas of specific geographical regions, Hershkovitz found it unsatisfying to assess the systematics of Colombian mammals without following them across national boundaries. Studies of a species or species group in Colombia led him to evaluate its context within genera, families, and even orders; and the remarkable diversity of Colombia's mammal fauna led him into most major groups and most Neotropical subregions. In the course of his career, he has published dozens of generic, tribal, and familial revisions, covering all 12 orders of Neotropical mammals. Few spatial and temporal boundaries have withstood the onslaught of his studies of Neotropical mammals. As examples one can point to the cosmopolitan *Catalog of Living Whales* (1966)—after all, most cetaceans do occur in South American waters—and studies of Oligocene and later fossils (1974, 1982).

One senses that the Department of Zoology during Hershkovitz's early years was a stimulating, harmonious one. Chief Curator Karl P. Schmidt took an almost paternal interest in junior staff and served as a confidant on the most personal of matters. In addition to Colin Sanborn, who was most considerate of his junior curator's interests and talents, Hershkovitz shared mammalogical problems and topics with Dwight Davis, Curator of Anatomy, and Bryan Patterson, Curator of Vertebrate Paleontology. During the early and mid-1950s, Hershkovitz established a vigorous and productive research program and participated in all aspects of departmental affairs.

However, upon Schmidt's retirement in 1957, Austin S. Rand became Chief Curator of Zoology, and neither Rand nor Hershkovitz did much to disguise their antipathy for one another. Over the ensuing years, Hershkovitz increasingly detached himself from museum operations, culminating with Joseph Moore's appointment as Curator of Mammals in 1961, and Hershkovitz's appointment that year to Research Curator. No one before or since has held this title at Field Museum. Hershkovitz formally retired in 1971, although his work has continued unabated as Curator Emeritus. During his career, he assisted countless students in mammal projects, but has served on only a single graduate committee, that of Jack Fooden, now himself a renowned biologist and primate specialist at Field Museum.

Few scientists can claim the independence in research that is indicated in Hershkovitz's bibliography. Of his approximately 300 scientific, popular, and encyclopedia articles, only three represent collaborative efforts. The first, with William P. Harris, an important benefactor of the Museum of Zoology of the University of Michigan, was suggested by Burt in recognition of Harris's interests in squirrels and in token repayment for his patronage of the museum. The second, with Paul Rode, came about one afternoon in the Muséum National d'Histoire Naturelle in Paris when Hershkovitz offhandedly suggested that designating a lectotype might solve a nomenclatural problem that Rode had encountered in his research. Rode insisted that Hershkovitz share authorship on the resulting paper. Later, after further study in the United States, Hershkovitz arrived at a contrary opinion and wrote a paper, with Rode as coauthor, correcting their earlier one.

Independent thought is also exemplified by the sometimes heated debates in which Hershkovitz has participated over the years. His published reviews and the discussion sections of many of his papers record his clearly enunciated views on such topics as the role of penial morphology in rodent taxonomy, the age and derivation of the South American fauna, panbiogeography, evolution of pelage coloration, and the systematic position of certain species (e.g., *Dolichocebus*). While such firmly held views brand him as something other than conciliatory or diplomatic, they accurately reflect his abiding passion and zest in science. Unfortunately, some acerbic exchanges had the effect of stifling the scientific dialogue to which they were offered (e.g., penial morphology).

Hershkovitz has focused his research on Neotropical mammals, their origin, evolution, dispersal, classification, nomenclature, and systematics. Specialists in these fields are well aware of his impact. However, he is perhaps most widely known for his work on three general topics of Neotropical mammalogy: faunal origins, metachromism, and New World monkeys. It would be folly to attempt to review all of his research, and more definitive appraisals on selected topics can be found scattered throughout this volume. However, some comments on these general issues seem in order.

As late as his revision of phyllotine rodents (1962), Hershkovitz adhered to traditional notions of the derivation of certain South American taxa, notably "cricetid" rodents, from North and Middle American stocks. This hypothesis of origins has been advocated most articulately by George G. Simpson, Bryan Patterson, and Rosendo Pascual, and more recently by Larry G. Marshall and S. David Webb. However, in the early 1960s, Hershkovitz was approached by Rupert Wenzel, Curator of Insects at Field Museum, who questioned him on the evidence for Plio-Pleistocene origins of the Neotropical cricetids. Wenzel's studies of the ectoparasites of Panamanian mammals suggested much earlier, South American origins. His interest piqued, Hershkovitz reviewed available evidence, synthesizing continental drift (which was then becoming established in geological circles) and neontological studies of mammals (especially those of Hooper and Musser, which showed a relatively sharp dichotomy between simple and complex penis-types of cricetids). He concluded that continental drift permitted a much greater role for paleotropical stocks in South American faunal origins than was allowed by the Simpsonian school, which in turn pointed to a much greater time period for independent evolution. Interestingly, and perhaps even characteristically, Hershkovitz concluded that South American rodents were not only not derived from North American stocks, but instead gave rise to them. These views were published in 1966, 1969, and 1972.

Hershkovitz's theory of metachromism, or deterministic evolution of pelage coloration through the loss of one or the other or both classes of hair pigments (eumelanins and phaeomelanins), was first published in 1968. Since then he has used it repeatedly in describing geographic variation in platyrrhine monkeys (e.g., 1977). However, the origins of this concept stem from his earlier work on the *Sciurus granatensis* group in northern Colombia where populations of squirrels thoroughly isolated from one another show similar progressions of pelage patterns. Few workers other than Neotropical primatologists (and not all of these) have accepted his interpretations, although the theory is potentially applicable to a variety of other, mostly diurnal taxa showing pelage pattern variations. While Timothy Lawlor detailed some theoretical misgivings with the theory in a 1969 paper in *Evolution* (rebutted by Hershkovitz in 1970), to my knowledge it has not been substantially refuted. The theory is eminently testable: refutation would simply entail showing that pelage pattern variation of taxa arranged by metachromism is not congruent with well-established phyletic patterns.

Finally, some explanation seems warranted for Hershkovitz's current devotion to primates. Indeed, many recent workers unschooled in mammalian systematics think of him as a primatologist. Nothing could be further from the truth, as he hastens to point out. He had published several articles on primates in the course of working up his Colombian collections, but gave these taxa no special attention until the 1960s. Then government funding for primate studies soared, largely because of interest in biomedical applications, especially for the complex and taxonomically confused Callitrichidae. For almost 20 years, Hershkovitz has focused first on the Callitrichidae and Callimiconidae, now on the Cebidae. His slower progress through these groups is attributable to the vast body of current knowledge about them; his 1977 and subsequent works serve as model syntheses of skin and skull morphology with biochemistry, karyology, ethology, serology, and epidemiology. By his own estimation, monkeys do not culminate his studies of Neotropical mammals, but rather represent a large and complex group to be covered in his attempt to treat all South American mammals. After seven years of work on Volume II of his primate monograph, he has nearly completed generic revisions of cebids lacking prehensile tails and is beginning comparative studies of their organ systems. In 1984 he submitted another grant proposal for this work, totaling one-half million dollars in direct costs. His is not a modest work; it has been described by Pine (1982; Vol. 6, Spec. Publ. Ser., Pymatuning Lab. Ecol.) as "the most heroically monumental revisionary monograph ever devoted to a Neotropical group."

In 1984, Hershkovitz turned 75 years old. The 14 years he spent in the field in South America have served him well, for he seems younger than many men 15 years his junior. His tireless energy is best indicated by his habitual use of stairs rather than elevators (even his two divisional offices are three floors apart), a continuing program of fieldwork (most recently in Brazil during 1986 and 1987), and a museum workday that extends from 9 a.m. to 6 p.m., uninterrupted by coffee breaks or even lunch. Visitors to his home, now within walking distance of the Museum, know of his office there which relieves the chronic insomnia of advancing years. He is an outstanding cook, a genial host, a trusted and valued friend, and an awesomely productive scientist.

Publications of Philip Hershkovitz

1938

1. A new caecilian from Ecuador. Occasional Papers, Museum of Zoology, University of Michigan, 370:1–3.
2. Two new squirrels from Ecuador. Occasional Papers, Museum of Zoology, University of Michigan, 391:1–6 (with W. P. Harris).
3. A review of the rabbits of the *andinus* group and their distribution in Ecuador. Occasional Papers, Museum of Zoology, University of Michigan, 393:1–15.

1940

4. Four new oryzomyine rodents from Ecuador. Journal of Mammalogy, 21:78–84.
5. Notes on the distribution of the akodont rodent, *Akodon mollis,* in Ecuador with a description of a new race. Occasional Papers, Museum of Zoology, University of Michigan, 418:1–3.
6. A new spiny mouse of the genus *Neacomys* from eastern Ecuador. Occasional Papers, Museum of Zoology, University of Michigan, 419:1–4.

1941

7. The South American harvest mice of the genus *Reithrodontomys.* Occasional Papers, Museum of Zoology, University of Michigan, 441:1–7.

1944

8. A systematic review of the Neotropical water rats of the genus *Nectomys* (Cricetinae). Miscellaneous Publications, Museum of Zoology, University of Michigan, 58:1–88.

1945

9. Désignation d'un lectotype de *Callithrix penicillatus* (E. Geoffroy). Bulletin du Museum National d'Histoire Naturelle, Paris 17(3):221–222 (with P. Rode).

1947

10. A correction. Journal of Mammalogy, 28(1): 68 (with P. Rode).
11. Mammals of northern Colombia. Preliminary report no. 1: Squirrels (Sciuridae). Proceedings of the United States National Museum, 97:1–46.

1948

12. Mammals of northern Colombia. Preliminary report no. 2: Spiny rats (Echimyidae), with supplemental notes on related forms. Proceedings of the United States National Museum, 97:125–140.
13. Mammals of northern Colombia. Preliminary report no. 3: Water rats (genus *Nectomys*), with supplemental notes on related forms. Proceedings of the United States National Museum, 98:49–56.
14. The technical name of the Virginia deer with a list of the South American forms. Proceedings of the Biological Society of Washington, 61:41–48.
15. Names of mammals dated from Frisch, 1775, and Zimmermann, 1777. Journal of Mammalogy, 29(3):272–277.

1949

16. Technical names for the fallow deer and Virginia deer. Journal of Mammalogy, 30(1): 94.
17. Generic names of the four-eyed pouch opossum and the woolly opossum (Didelphidae). Proceedings of the Biological Society of Washington, 62:11–12.
18. Technical names of the African muishond (genus *Zorilla*) and the Colombian hog-nosed skunk (genus *Conepatus*). Proceedings of the Biological Society of Washington, 62:13–16.
19. Mammals of northern Colombia. Preliminary report no. 4: Monkeys (Primates), with taxonomic revisions of some forms. Proceedings of the United States National Museum, 98:323–427.
20. Mammals of northern Colombia. Preliminary report no. 5: Bats (Chiroptera). Proceedings of the United States National Museum, 99:429–454.
21. Status of names credited to Oken, 1816. Journal of Mammalogy, 30(3):289–301.
22. Tapirs: Strange mammals native to Asia and tropical America from Mexico south. Chicago Natural History Museum Bulletin, 20(9):6–7.

1950

23. Mammals of northern Colombia. Preliminary report no. 6: Rabbits (Leporidae), with notes on the classification and distribution of the South American forms. Proceedings of the United States National Museum, 100: 327–375.

1951

24. Mammals from British Honduras, Mexico, Jamaica and Haiti. Fieldiana: Zoology, 31(47):547–569.

1953

25. *Zorilla* I. Geoffroy and *Spilogale* Gray, generic names for African and American polecats, respectively. Journal of Mammalogy, 34(3):378–382.
26. Four years on a zoological expedition in Colombia. Chicago Natural History Museum Bulletin, 24(1):3–4.
27. The reindeer—Important to man in fact and fancy. Chicago Natural History Museum Bulletin, 24(12):3–4.

1954

28. Mammals of northern Colombia, Preliminary report no. 7: Tapirs (genus *Tapirus*), with a systematic review of American species. Proceedings of the United States National Museum, 103:465–496.
29. What the groundhog undergoes to make a "holiday." Chicago Natural History Museum Bulletin, 25(2):3–4.
30. Who's a cow? Chicago Natural History Museum Bulletin, 25(7):5.

31. Some ecological aspects of natural versus artificial rehabilitation of a water basin area in Bogotá, Colombia. Boletin del Instituto de La Salle, Bogotá, 41(193/194):80–83.

1955

32. South American marsh rats genus *Holochilus* with a summary of sigmodont rodents. Fieldiana: Zoology, 37:639–673.
33. [Review] Mammals, a guide to familiar American species. Chicago Natural History Museum Bulletin, 26(7):7.
34. Notes on American monkeys of the genus *Cebus*. Journal of Mammalogy, 36:449–452.
35. Status of the generic name *Zorilla* (Mammalia): Nomenclature by rule or by caprice. Proceedings of the Biological Society of Washington, 68:185–192.
36. On the cheek pouches of the tropical American paca, *Agouti paca* (Linnaeus, 1766). Saügetierkundliche Mitteilungen, 3(2):67–70.
37. Know your rabbits. Sports Afield, 134(6): 36–41, 88.

1956

38. Comments on *Galerella* Gray, *Herpestes* Illiger, *Leucomitra* Howell, *Icticyon* Lund, *Lutreola* Wagner, *Oryctogale* Merriam, *Paracynictis* Pocock. Opinion 384 International Commission of Zoological Nomenclature, 12(5):71–190 (in text).
39. Critical remarks on the status of names in Oken's "Lehrbuch." Opinion 417, International Commission on Zoological Nomenclature, 14(1):33–35.

1957

40. Comments on *Canis dingo* Meyer. Opinion 451, International Commission on Zoological Nomenclature, 15(17):335–336.
41. Comments on the validation of *Muntiacus* Rafinesque. Opinion 460, International Commission on Zoological Nomenclature, 15(26):467–468.
42. Comments on the generic name *Mormoops* Leach. Opinion 462, International Commission on Zoological Nomenclature, 16(1): 8–9.
43. Comments on *Sciurus gambianus*. Opinion 464, International Commission on Zoological Nomenclature, 16(3):36–39.
44. Comments on the validation of *silvestris* Schreber, 1777 [*Felis* (*catus*) *silvestris*]. Opinion 465, International Commission on Zoological Nomenclature, 16(4):49.
45. Comments on the validation of the name *Phacochoerus* Cuvier. Opinion 466, International Commission on Zoological Nomenclature, 16(5):67–68.
46. Comments on the validation of the name *Odobenus* Brisson. Opinion 467, International Commission on Zoological Nomenclature, 16(6):84–85.
47. The systematic position of the marmoset, *Simia leonina* Humboldt (Primates). Proceedings of the Biological Society of Washington, 70:17–20.
48. The type locality of *Bison bison* Linnaeus. Proceedings of the Biological Society of Washington, 70:31–32.
49. A synopsis of the wild dogs of Colombia. Novedades Colombianas Museo de Historia Naturale Universidad del Cauca (Popayán), no. 3:157–161.
50. On the possible occurrence of the spectacled bear, *Tremarctos ornatus* (F. Cuvier, 1825), in Panama. Säugetierkundliche Mitteilungen, 5(3):122–123.

1958

51. [Review] Biological investigations in the Selva Lacandona, Chiapas, Mexico. Quarterly Review of Biology, 33(1):67.
52. [Review] Mammals of the Anglo-Egyptian Sudan, by Henry Setzer. Quarterly Review of Biology, 33:82.
53. Technical names of the South American marsh deer and pampas deer. Proceedings of the Biological Society of Washington, 71: 13–16.
54. Type localities and nomenclature of some American Primates, with remarks on secondary homonyms. Proceedings of the Biological Society of Washington, 71:53–56.
55. Stabilization of zoological nomenclature by a "Law of prescription." Bulletin of Zoological Nomenclature, 15B(20/21):630–632.
56. A critique of Professor Chester Bradley's "Principle of conservation." Bulletin of Zoological Nomenclature, 15B(25/28):911–913.
57. The status of secondary homonyms and the concept of permanent rejection. Bulletin of Zoological Nomenclature, 15B(37/38):1242–1243.
58. A geographic classification of Neotropical mammals. Fieldiana: Zoology, 36(6):583–620.

59. The metatarsal glands in white-tailed deer and related forms of the Neotropical region. Mammalia, 22(4):537–546.

1959

60. The scientific names of the species of capuchin monkeys (*Cebus* Erxleben). Proceedings of the Biological Society of Washington, 72:1–4.
61. Two new genera of South American rodents (Cricetinae). Proceedings of the Biological Society of Washington, 72:5–10.
62. A new species of South American brocket, genus *Mazama* (Cervidae). Proceedings of the Biological Society of Washington, 72: 45–54.
63. A new race of red brocket deer (*Mazama americana*) from Colombia. Proceedings of the Biological Society of Washington, 72: 93–96.
64. The type locality of *Felix concolor concolor* Linnaeus. Proceedings of the Biological Society of Washington, 72:97–100.
65. Nomenclature and taxonomy of the Neotropical mammals described by Olfers, 1818. Journal of Mammalogy, 40(3):337–353.

1960

66. Supposed ape-man or "missing link" of South America. Chicago Natural History Museum Bulletin, 31(4):6–7.
67. [Review] The Mammals of North America. Chicago Natural History Museum Bulletin, 31(5):6–7.
68. Publication dates for names of the Anubis baboon. Journal of Mammalogy, 41(3):402–403.
69. Mammals of northern Colombia. Preliminary report no. 8: Arboreal rice rats, a systematic revision of the subgenus *Oecomys*, genus *Oryzomys*. Proceedings of the United States National Museum, 110:513–568.

1961

70. On the South American small-eared zorro *Atelocynus microtis* Sclater (Canidae). Fieldiana: Zoology, 39(44):505–523.
71. On the nomenclature of certain whales. Fieldiana: Zoology, 39(49):547–565.
72. "This is a mammal." Chicago Natural History Museum Bulletin, 32(6):3.

1962

73. Suriname zoological expedition. Chicago Natural History Museum Bulletin, 33(4):3, 7–8.
74. Bats and their menus. Chicago Natural History Museum Bulletin, 33(8):2–3, 5–8.
75. Evolution of Neotropical cricetine rodents (Muridae) with special reference to the phyllotine group. Fieldiana: Zoology, 46:1–524.

1963

76. A systematic and zoogeographic account of the monkeys of the genus *Callicebus* (Cebidae) of the Amazonas and Orinoco River basins. Mammalia, 27(1):3–79.
77. [Review] Primates. Comparative Anatomy and Taxonomy. Vol. V, Cebidae, part B., A Monograph; Edinburgh University Press. American Journal of Physical Anthropology, 21(1):92–98.
78. [Review] Primates. Comparative Anatomy and Taxonomy. Vol. V, Cebidae, part B., A Monograph; Edinburgh University Press. American Journal of Physical Anthropology, 21(3):391–398.
79. Notes on South American dolphins of the genera *Inia, Sotalia* and *Tursiops*. Journal of Mammalogy, 44(1):98–103.
80. The nomenclature of South American peccaries. Proceedings of the Biological Society of Washington, 76:85–88.
81. The Recent mammals of South America. Proceedings of the XVI International Congress of Zoology, Washington, D.C., Aug. 20–27, 1963.
82. Comments on the proposed suppression of *Zorilla* I. Geoffroy, 1826. Z.N.(S.) 758. Bulletin of Zoological Nomenclature, 20(4):242–244.

1965

83. Primate research and systematics. Science, 147(3662):1156–1157.
84. The importance of taxonomy in primate research and care. Illinois Society for Medical Research—Chicago Branch—Animal Care Panel Bulletin, 39:2 pp.

1966

85. Catalog of living whales. Bulletin of the United States National Museum, 246:1–259.

86. Taxonomic notes on tamarins, genus *Saguinus* (Callithricidae, Primates), with descriptions of four new forms. Folia Primatologica, 4:381–395.
87. On the identification of some marmosets, family Callithricidae (Primates). Mammalia, 30(2):327–332.
88. What ever happened to hairy man? Science, 153:362.
89. Comments on the proposal for conservation of *Pan* Oken, 1816, and *Panthera* Oken, 1816. Bulletin of Zoological Nomenclature, 23(2/3):67–69.
90. [Review] Evolutionary and Genetic Biology of Primates, vol. II; Academic Press. American Biology Teacher, 28(7):564–565.
91. Comments on the proposed suppression of *Meles montanus* Richardson, 1829, and *M. jeffersonii* Harlan, 1825. Z.N.(S.) 1639. Bulletin of Zoological Nomenclature, 22(5/6): 336–337.
92. On the status of *Procyon brachyurus* Wiegmann and *P. obscurus* Wiegmann. Z.N.(S.) 1640. Bulletin of Zoological Nomenclature, 22(5/6):338.
93. South American swamp and fossorial rats of the Scapteromyine group (Cricetinae, Muridae) with comments on the glans penis in murid taxonomy. Zeitschrift für Säugetierkunde, 31(2):81–149.
94. Status of the black-footed ferret in Wyoming. Journal of Mammalogy, 47(2):346–347.
95. Comments on the proposal on *Zorilla* by Dr. Van Gelder and the counter proposal by Dr. China. Z.N.(S.) 758. Bulletin of Zoological Nomenclature, 23(2/3):74–75.
96. Museum taxonomy serves medical research. Bulletin of the Field Museum of Natural History, 37(9):4–7.
97. Mice, land bridges and Latin American faunal interchange, pp. 725–751. *In* Wenzel, R. L., and V. J. Tipton, eds., Ectoparasites of Panama. Field Museum of Natural History, Chicago.

1967

98. [Review] Evolutionary and Genetic Biology of Primates, vol. I; Academic Press. American Biology Teacher, Nov. 1967:665.
99. Reply to Mayr's comment on the proposed preservation of *Pan* from Oken, 1816. Z.N.(S.) 482. Bulletin of Zoological Nomenclature 24(5):1 p.
100. Dynamics of rodent molar evolution: A study based on New World Cricetinae, family Muridae. Journal of Dental Research, Suppl. to 46(5):829–842.

1968

101. Metachromism or the principle of evolutionary change in mammalian tegumentary colors. Evolution, 22(3):556–575.
102. [Review] Dynamics of rodent molar evolution: A study based on New World Cricetinae, family Muridae. Oral Research Abstracts, May 1968.

1969

103. Comments on *Cynocephalus* Boddaert versus *Galeopithecus* Pallas. Z.N.(S.) 1792. Bulletin of Zoological Nomenclature, 25(6):202–203.
104. The evolution of mammals on southern continents. VI. The Recent mammals of the Neotropical Region: A zoogeographic and ecological review. Quarterly Review of Biology, 44(1):1–70.

1970

105. The decorative chin. Field Museum of Natural History Bulletin, 41(5):7–10.
106. Dental and periodontal diseases and abnormalities in wild-caught marmosets (Primates—Callithricidae). American Journal of Physical Anthropology, 32(3):377–392.
107. [Review] Taxonomy and Evolution of the Monkeys of Celebes (Primates: Cercopithecidae); Bibliotheca Primatologica, No. 10; Karger. Folia Primatologica, 13(1):75–76.
108. Metachromism like it is. Evolution, 24(3): 644–648.
109. Notes on Tertiary platyrrhine monkeys and description of a new genus for the Late Miocene of Colombia. Folia Primatologica, 12: 1–37.
110. Errata: Notes on Tertiary platyrrhine monkeys and description of a new genus for the Late Miocene of Colombia. Folia Primatologica, 12:1–37 (1970).
111. Cerebral fissural patterns in platyrrhine monkeys. Folia Primatologica, 13:213–240.
112. [Review] The Squirrel Monkey; Academic Press. Journal of Mammalogy, 51(4):836–839.
113. Supplementary notes on Neotropical *Oryzomys dimidiatus* and *Oryzomys hammondi* (Cricetinae). Journal of Mammalogy, 51(4): 789–794.

1971

114. Stapedial processes in tympanic cavities of capuchin monkeys (*Cebus*). Journal of Mammalogy, 52(3):607–609.
115. Basic crown patterns and cusp homologies of mammalian teeth, pp. 95–150. *In* Dahlberg, A. A., ed., Dental Morphology and Evolution. University of Chicago Press, Chicago.
116. A new rice rat of the *Oryzomys palustris* group (Cricetinae, Muridae) from northwestern Colombia, with remarks on distribution. Journal of Mammalogy, 52(4):700–709.

1972

117. Notes on New World monkeys. International Zoo Yearbook, 12:3–12.
118. The Recent mammals of the Neotropical Region: A zoogeographic and ecological review, pp. 311–431. *In* Keast, A., F. C. Erk, and B. Glass, eds., Evolution, Mammals and Southern Continents. State University of New York Press, Albany.

1974

119. The ectotympanic bone and origin of higher primates. Folia Primatologica, 22:237–242.
120. A new genus of Late Oligocene monkey (Cebidae, Platyrrhini) with notes on postorbital closure and platyrrhine evolution. Folia Primatologica, 21:1–35.

1975

121. [Review] Taxonomic Atlas of Living Primates; Academic Press. American Journal of Physical Anthropology, 41:155–156.
122. The scientific name of the lesser *Noctilio* (Chiroptera), with notes on the chauve-souris de la Vallee d'Ylo (Peru). Journal of Mammalogy, 56(1):242–247.
123. Comments on the taxonomy of Brazilian marmosets (*Callithrix,* Callitrichidae). Folia Primatologica, 24:137–172.

1976

124. The taxonomic status of "*Noctilio ruber* Rengger." Mammalia, 40(1):164–166.
125. Comments on generic names of four-eyed opossums (family Didelphidae). Proceedings of the Biological Society of Washington, 89(23):295–304.

1977

126. Comment: *Pan* and *Panthera* or Oken's Lehrbuch? Z.N.(S.) 482. Bulletin of Zoological Nomenclature, 33(3/4):135–136.
127. [Review] Catalogue of Primates in the British Museum (Natural History). I. Families Callitrichidae and Cebidae; British Museum (Natural History). Folia Primatologica, 28: 315.
128. Living New World Monkeys (Platyrrhini). With an Introduction to Primates. Volume I. University of Chicago Press, Chicago, xiv + 1117 pp.

1979

129. Races of the emperor tamarin, *Saguinus imperator* Goeldi (Callitrichidae, Primates). Primates, 20(2):277–287.
130. The species of sakis, genus *Pithecia* (Cebidae, Primates), with notes on sexual dichromatism. Folia Primatologica, 31:1–22.

1981

131. Comparative anatomy of platyrrhine mandibular cheek teeth dpm4, pm4, m1 with particular reference to those of *Homunculus* (Cebidae), and comments on platyrrhine origins. Folia Primatologica, 35:179–217.
132. *Philander* and four-eyed opossums once again. Proceedings of the Biological Society of Washington, 93(4):943–946.

1982

133. Supposed squirrel monkey affinities of the late Oligocene *Dolichocebus gaimanensis.* Nature, 298(5870):201–202.
134. Subspecies and geographic distribution of black-mantle tamarins *Saguinus nigricollis* Spix (Primates: Callitrichidae). Proceedings of the Biological Society of Washington, 95(4):647–656.
135. Neotropical deer (Cervidae). Part I. Pudus, genus *Pudu* Gray. Fieldiana: Zoology, n.s., 11:1–86.
136. The staggered marsupial lower third incisor (I_3), pp. 191–200. *In* Buffetaut, E., P. Janvier, J. C. Rage, and P. Tassy, eds., Phylogenie et Paleobiogeographie. Livre jubiliare en l'honneur de Robert Hoffstetter. Geobios, memoire spécial 6, Lyon.

1983

137. Two new species of night monkeys, genus *Aotus* (Cebidae, Platyrrhini): A preliminary report on *Aotus* taxonomy. American Journal of Primatology, 4:209–243.
138. On the validity of the family-group name Callitrichidae (Platyrrhini, Primates). Mammalia, 48:153.
139. Taxonomy of squirrel monkeys, genus *Saimiri* (Cebidae, Platyrrhini): A preliminary report with description of a hitherto unnamed form. American Journal of Primatology, 6:257–312.
140. [Review] Mammalian Biology in South America. M. A. Mares and H. H. Genoways, eds. Ecology, 65(6):1944–1945.

1985

141. A preliminary taxonomic review of the South American bearded saki monkeys, genus *Chiroptes* (Cebidae, Platyrrhini), with the description of a new subspecies. Fieldiana: Zoology, n.s., 27:1–46.

1986

142. [Review] Handbook of Squirrel Monkey Research. L. A. Rosenblum and C. L. Coe, eds. Quarterly Review of Biology, 61:286–287.
143. The piebald saki. Field Museum of Natural History Bulletin, 57(2):coverplate + 24–25.

1987

144. Uacaries, New World monkeys of the genus *Cacajao* (Cebidae, Platyrrhini): A preliminary taxonomic review with the description of a new subspecies. American Journal of Primatology, 12:1–53.
145. First South American record of Coue's marsh rice rat, *Oryzomys couesi*. Journal of Mammalogy, 68(1):152–154.
146. The titi. Field Museum of Natural History Bulletin, 58(6):11–15.
147. The taxonomy of South American sakis, genus *Pithecia* (Cebidae, Platyrrhini): A preliminary report and critical review with the description of a new species and a new subspecies. American Journal of Primatology, 12:387–468.

In Press

148. More on the *Homunculus* Dpm4 and m1 and comparisons with *Alouatta* and *Stirtonia* (Primates, Platyrrhini, Cebidae). American Journal of Primatology.

A History of the Recent Mammalogy of the Neotropical Region from 1492 to 1850

Philip Hershkovitz

ABSTRACTS

The history of Neotropical mammalogy began with the first voyage of Christopher Columbus in 1492. The earliest notices were purely anecdotal, recorded by Spanish chroniclers from the mouths of the sailors on their return from voyages of discovery during the 15th and 16th centuries. Colonization of the Guianan and Brazilian coasts during the 17th century provided opportunities for inventories and descriptions of the mammals by trained European naturalists and physicians. The systematization and scientific naming of the known Brazilian species by Carolus Linnaeus in 1758 were based primarily on the mammals described in the 17th century monograph of Brazilian biota by Georg Marcgraf. The actual collection and preservation of mammals for study, however, began in the 18th century with the Brazilian-born Alexandre Rodrigues Ferreira. The 18th and first half of the 19th century was an explosive and romantic period of independently or governmentally sponsored scientific expeditions for field observations, collections, preservations, and taxonomic studies of the specimens shipped to European museums and private collectors. Outstanding among the naturalists who made significant contributions to mammalogy during this period are Alexander von Humboldt, Johann Baptist Ritter von Spix (Brazil), Maximilian Prinz Wied-Neuwied (Brazil), Johann Natterer (Brazil), Sir Robert Herman Schomburgk and Richard Schomburgk (Guyana), Claudio Gay (Chile), Johann Jakob von Tschudi (Peru), Felix de Azara (Paraguay), Rudolph Rengger (Paraguay), Alcide Charles Victor d'Orbigny (Argentina, Bolivia), and Charles Robert Darwin (Patagonia and Galapagos). Their itineraries, collections of mammals, taxonomies, and some field notes are included in the accounts of these and other noteworthy naturalists. By the middle of the 19th century, the mammalian fauna of South America became the best known of any continent with exception of the western European part of Eurasia. The problems of origins and distribution of Neotropical mammals intrigued scholars from among the earliest chroniclers down to pre-evolutionary Darwin. Their concepts on these subjects are briefly discussed.

La historia de mastozoología neotropical empieza con el primer viaje transatlántico de Cristóbal Colón en 1492. Poco después de desembarcarse de sus viajes de regreso los descrubridores y conquistadores del Mundo Nuevo en los siglos quince y diez y seis contaron a los cronistas españoles de las cosas curiosas que encontraron. Colonización de las costas guyanas y brasileras durante el siglo diez y siete ofreció oportunidades a los naturalistas y médicos europeos residentes para levantar inventarios de los mamíferos y anotar y hacer informes sobre sus observaciones. La sistematización y el nombramiento científico de las especies brasileras conocidas por Carolus Linnaeus en 1758 fueron basadas primariamente sobre los mamíferos descritos y figurados por Jorge Marcgraf en su monografía del siglo diez y siete. La colección y preservación efectiva de mamíferos para el estudio empezó en comienzos del siglo diez y ocho con el "Viajem Filosófica" de Alejandro Rodriguez Ferreira, brasilero de nacimiento.

From the Division of Mammals, Department of Zoology, Field Museum of Natural History, Chicago, Illinois 60605-2496.

El siglo diez y ocho y las primeras décadas del siglo diez y nueve señalaron un período explosivo y romántico de expediciones científicas fomentadas por gobiernos europeos, o por naturalistas particulares con los objectos de hacer observaciones sobre la fauna, tomar notas de campo, y recoger y preservar ejemplares para estudios taxonómicos en los museos extranjeros. Entre los naturalistas europeos que hicieron contribuciones de consequencia a la mastozoología neotropical en este época, se cuentan Alejandro von Humboldt, Juan Baptista Ritter von Spix (Brasil), Maximiliano Príncipe de Wied-Neuwied (Brasil), Juan Natterer (Brasil), Lord Roberto Herman Schomburgk y Ricardo Schomburgk (Guyana), Claudio Gay (Chile), Juan Jacobo von Tschudi (Perú), Félix de Azara (Paraguay), Rodolpho Rengger (Paraguay), Alcidio Carlos Victor d'Orbigny (Argentina, Bolivia), y Carlos Roberto Darwin (Patagonia y Galápagos). Comprendido en este informe son los itinerarios, listas de mamíferos coleccionados y observados, taxonomías, y algunas experiencias de campo de los naturalistas mentados, y de otros digno de atención.

A mediados del siglo diez y nueve, la fauna mamífera de Sud America llegó a ser la mejor conocida de todos los continentes del mundo menos Europa. Problemas de origen y repartición geográfica de los mamíferos del Mundo Nuevo estimularon la imaginación de sabios desde los primeros cronistas del Descubrimiento hasta el joven Darwin pre-evolutionario. Los conceptos sobre estos temas son brevemente discutidos.

A história da mastozoología neotropical, começou com a primeira viagem transatlántica de Cristóvão Colombo, em 1492. Os primeiros relatórios, puramente anedotais, foram registrados pelos cronistas espanhóis, logo após o regresso das viagens de descobrimento durante os séculos XV e XVI. As colonizacões da costa Guianense e Brasileira durante o século XVII, ofereceram amplas oportunidades a naturalistas e médicos, de treinamento Europeu, para inventoriar e descrever os mamíferos encontrados. A sistemática e a nomenclatura científica das espécies Brasileiras conhecidas por Carolus Linnaeus em 1758 basearam-se primariamente nos mamíferos descritos por Georg Marcgraf, em sua monografia do século XVII. No entanto, as colecões e preservacões de mamíferos para estudos começaram, efetivamente, no século XVIII, com a "Viajem filosófica" do Brasileiro, Alexandre Rodrigues Ferreira.

Marcaram o século XVIII, e as primeiros décadas do século XIX, um período explosivo e romantico nas expedicões científicas. Estas foram patrocinadas tanto por naturalistas independentes, como por governos Europeus, a fim de realizarem observacões sobre a fauna e colecões para estudos taxonomicos nos museus Europeus. Entres os naturalistas Europeus que distinguiram-ce em suas contribuicões aos estudos de mamíferos neotropicais durante esta época, sobressaem: Alexandre von Humboldt, Juan Baptista Ritter von Spix (Brasil), Maximilian Príncipe de Wied-Neuwied (Brasil), Johan Natterer (Brasil), Sir Robert Herman Schomburgk e Richard Schomburgk (Guiana), Claudio Gay (Chile), Johan Jakob von Tschudi (Peru), Félix de Azara (Paraguai), Rudolph Rengger (Paraguai), Alcides Charles Victor d'Orbigny (Argentina, Bolívia) e Charles Robert Darwin (Patagonia e Galápagos). Os itinerários, as listas de mamíferos observados e colecionados, as taxonomias, e algumas notas de campo encontram-se incluídos nos relatórios aqui apresentados sobre estes e outros naturalistas importantes.

Nas meadas do século XIX, a fauna mamífera sul-americana tornouse a melhor conhecida de todos os continentes, exceto a da Europa. Os problemas de origem e da distribuicão geográfica dos mamíferos neotropicais estimularam a imaginacão de vários estudiosos, desde os primeiros cronistas até o pré-evolucionário Darwin. Seus conceitos sobre estes temas são brevemente discutidos.

Organization

I. Introduction

The gradual accumulation of knowledge of Neotropical mammals is recorded here from the time of the first voyage of discovery by Christopher Columbus in 1492 to the middle of the 19th century, or just before the Darwinian revolution in biological thought. The knowledge was mainly of species or kinds, the numbers of kinds, their behavior, habitat, geographic distribution, and relationship to man. Early voyagers to the New World followed by naturalist-travelers gathered the data used later by philosophers and scientists for the development of biological principles. Only the most important and better-known contributors are discussed here. At least as many more personages could be included in a more extended account.

The Neotropical Region Defined

The Neotropical Region, as defined by its mammalian fauna, includes all South America, Middle America except the dry and temperate zones of Mexico, continental islands of coastal Middle and South America, and the oceanic Bahamas, West Indies, Galápagos, and Falklands (Hershkovitz, 1972, p. 326).

With few exceptions, modern names for Neotropical countries and geographic features are used throughout the text. The map (fig. 1) shows the South America of the colonial period with colonial or precolonial names for political subdivisions and geographic features.

II. Voyages of Discovery: 15th and 16th Centuries

The inhabited islands found by Columbus on his first voyage across the Atlantic Ocean in 1492 were thought to be near the mainland of China or India. The islanders welcomed the ships' crews with food and drink, but the great stores of precious metals, stones, and artifacts the travelers expected to find were not seen. Nevertheless, the voyagers claimed the islands for the Spanish crown and returned with accounts told to awaiting reporters of their discoveries, including their descriptions of plants and animals of economic value or imputed medicinal virtues.

Zoological results of the four transatlantic voyages commanded by Christopher Columbus—the first (1492–1493) and second (1493–1496) to the Antilles, the third (1498) to the Antilles and Venezuela, and the last (1502–1504) to Middle America—included reports of a variety of mammals. The kinds seen were identified with such familiar Old World forms as lion, tiger, bear, fox, dog, ferret, rabbit, deer, boar, goat, sheep, rodent, monkey, and ape. Characterizations given were less descriptions of external morphology than of general mien, gross habitat, behavior in response to human confrontations or predation on human property, gastronomic qualities, and use, if any, in medical treatment, ceremonial rites or magic, or as household pets.

Those who followed Columbus in the discovery and exploration of the mainland returned with additional bits of information on mammals noted by the attendant Spanish chroniclers. Among the more important of these voyagers of discovery were Pinzón, who followed Columbus to the Venezuelan coast in 1500, and Amerigo Vespucci, who sailed first with Ojeda to Brazil in 1499 and independently again in 1502 and 1503 in the service of Portugal. Pedro Cabral, however, had already claimed Brazil for Portugal in 1500 on his way to India. In 1516 Juan Diaz de Solis discovered the estuary of the Río de la Plata, and Sebastian Cabot, in the service of Spain, sailed in 1526 up the Río Paraná. Vasco Nuñez de Balboa accompanied Enciso to Panama in 1510, and in 1513, with Francisco Pizarro, crossed the isthmus to behold the vast Pacific Ocean. Pizarro visited Panama again in 1531, recrossed the isthmus, and sailed south along the west coast of South America to the discovery and conquest of Peru. Cabeza Álvarez Nuñez de Vaca arrived in Buenos Aires in 1541 and continued overland into Paraguay. Pedro de Valdivia visited Venezuela in 1530, Peru in 1532, and Chile in 1540, 1541, and 1552. The explorations of Colombia by González Jiménez de Quesada from 1536 to 1539 and again in 1569 to 1571 signaled the end of the period of discovery and conquest.

III. Spanish Chroniclers of New World Discoveries

The recorders or chroniclers of New World discoveries, conquests, happenings, and natural phenomena were the clerics and scribes who accompanied the explorers or awaited their return to Spain for recording the news. Most of the accounts or records remained unpublished, but some of the manuscripts are reportedly preserved in the archives of Spain or the Vatican. The chroniclers whose published narratives contain interesting information on mammals include the following.

Peter Martyr of Anghiera (1455–1526), Italian by birth, and the first and most prestigious chronicler of the Discovery, was a member of the Royal Spanish Council of the Indies, Prothonotary of the Catholic Church, correspondent of Popes, confidant of Christopher Columbus, and friend of sea captains, clergymen, and other contemporary voyagers to the New World. News he received from his informants constitutes the first records of New World discoveries. His chronicles, known as the *Decades* and addressed to the Pope, began to arrive at the Vatican in 1494. The first *Decade de Orbe Novo,* with first notices of American mammals, was published in 1516, but pirated Italian

FIG. 1. Map, South America of the Colonial period from the Stevens (1726) translation of Herrera y Tordesillas.

editions appeared in 1504 and 1507. The fourth *Decade* was published in 1521, and the complete set of eight of the projected 10 appeared posthumously in 1587.

Gonzalo Fernández de Oviedo y Valdés (1478–1557) was appointed royal chronicler of news sent directly to him by provincial governors and other New World officials. Included were Oviedo's own observations and results of investigations during his residence as representative of the Spanish Crown in the Provinces of Darién, Panama, Guatemala, Cuba, and Santo Domingo. He published the first part of his *Historia de las Indias* in 1526, other parts in 1535 and 1547. The entire work was printed between 1851 and 1855 in Madrid.

The Spanish Jesuit José de Acosta (1539–1600) wrote his *Historia Natural y Moral de las Indias* during a residence in Peru from 1571 to 1587 and saw it published in 1590. Acosta's philosophical inquiries extended to all aspects of nature in the New World and greatly influenced the thinking of his contemporaries.

Antonio Vázquez de Espinosa (b. between 1560 and 1575, d. 1630), a Carmelite missionary, lived many years in Spanish America, most of them in Peru and Mexico. His natural history notes are compiled from many sources, including his personal observations and testimony of people he met in travels connected with his clerical duties. The forgotten manuscript of his *Compendium* was discovered in the Vatican library by Charles Upson Clark in the early part of the 20th century. Clark's English translation of the work was published in 1942 by the Smithsonian Institution, and his transcription of the original Spanish in 1948 by the same institution.

Antonio de Herrera y Tordesillas (1559–1625), historiographer to the King of Spain, compiled the *General History of the Vast Continent and Islands of America* from archived reports by the New World discoverers and conquistadores, governors, clergy, colonists, and travelers. He also borrowed heavily from published accounts, including those of other chroniclers. There is no indication that his notices on mammals were based on personal observations. The first edition of Herrera's *History* was published in 1601, another in 1601–1615. These and a 1728 Spanish edition in the Library of Congress are cited in the bibliography. I have not seen these works. The Stevens translation, published 1725–1726, was used here. Whatever the quality of the translation, I find no fault with the descriptions of mammals, and the stories about them are in line with similar accounts in other sources.

IV. First Mammals: Anecdotal Period

Island Mammals of the Discoverers

The first Columbian voyage, in 1492, resulted in the discovery of the Antillean islands of Cuba, Hispaniola, and part of the Bahaman Archipelago. According to Peter Martyr, who reported results of the voyage in his first *Decade* (1504, 1516), quadrupeds were not seen, but three kinds of "rabbits" were said to occur in Hispaniola (Haiti and Dominican Republic). The same animals, actually caviomorph rodents, were described later by Oviedo during his residence in Santo Domingo. The following accounts are freely translated or paraphrased from the Spanish of the Paraguayan (1944–1945) edition of Oviedo's work.

Hutia, the first "rabbit" (1944, libro XII, cap. I), is smaller than the ordinary rabbit, its ears smaller and tail ratlike. The *hutia* is said to be dark grayish in color and very good eating. It was hunted and killed by the barkless dogs of the natives, but is no longer found, except rarely.

Gerrit S. Miller (1929, p. 12), studied the remains of mammals in kitchen middens of the Samana Bay region, Dominican Republic, and concluded that the original description of the *hutia* "would apply as well to the species of *Plagiodontia,* and presumably also to the *Isolobodons* [*sic*] that there seems to be no reason to doubt that these were the animals Oviedo had in mind."

The *quemi,* second of the "rabbits" (1944, libro XII, cap. II), is said to be blackish like the *hutia* and similar in form, but larger like an ordinary hound. Natives of the island who saw and ate the animal found it savory. Oviedo believed them extinct.

All attributions to the *quemi,* according to Miller (1929, p. 13), agree with those of a "large rodent whose remains I found in the caves near St. Michel, Haiti, in 1925. Consequently, I proposed for it the generic name *Quemisia.* The presence of the same creature in the Boca del Infierno kitchen midden appears to confirm my guess."

The *mohuy* "rabbit" (1945, libro XII, cap. III), is smaller than the *hutia,* a paler brown or grayish in color, its flesh highly esteemed by the island's caciques and noblemen. The pelage, unlike that of the *hutia,* is stiff, sharply pointed, and erect. Oviedo saw no *mohuy,* but knew persons on the island who did and reportedly regarded its flesh as better than that of the other "rabbits."

"There be little if any doubt," says Miller (1929, p. 13), "that the animal Oviedo thus described was *Brotomys voratus* . . . its remains have been found in every kitchen midden that has been examined in the Dominican Republic. . . . The account of stiff, pointed, erect-standing hairs of the back seems especially applicable to a relative of the South American spiny-rats."

The *cori,* a fourth "rabbit," described by Oviedo

(1945, libro XII, cap. IV), is almost certainly the domestic guinea pig. Miller (1929, p. 14) questioned whether the guinea pig was pre-Columbian or a Spanish introduction. He inclined to the second alternative "chiefly because remains of the animal have been found in only one midden." It appears, however, that one Simone Verde, who accompanied Columbus on his first voyage, mentioned in a letter dated 20 March 1494 (cf. Martyr in Gaffarel, 1907 trans. p. 12, footnote 2; p. 14, footnotes 1, 2) the existence on the island of a black and white dormouse-like animal without tail. The guinea pig or *cui,* domesticated in Peru, was carried by pre-Columbian Indians for food and barter and introduced into islands and many parts of mainland South America where cavies do not naturally occur. Many of them, such as completely isolated colonies I saw in Colombia near Bogotá, had become feral, their coloration having reverted to the wild or agouti pattern.

Other Hispaniolan mammals mentioned by Oviedo are the barkless domestic dogs and house rats, the latter certainly brought by the Spaniards. Apart from the extinct insectivore *Nesophontes,* Miller found no remains of mammals the size of mice or rats in kitchen middens or owl pellets.

Two additional native West Indian mammals observed by Oviedo in 1523 or 1524 in Cuba differ from those of Hispaniola. My paraphrased translation of Oviedo's Spanish descriptions follows.

The *guabiniquinax* is somewhat larger than a rabbit, its feet similar, the tail long and ratlike, the pelage smoother than that of a badger, the skin white, the flesh savory. It lives and breeds in the mangroves along the coast. To capture it, the Indians position their canoes beneath the mangroves where it nests, then shake the tree to cause the animal to fall into the water where it is seized.

The animal as described above is certainly a form of *Capromys,* but Oviedo continues as follows: "The animal is the size of a hare, looks like a fox, its color is dark brown mixed with reddish, the tail hairy and the head shaped like that of a ferret. It abounds along the Cuban coast." The characterization and habitat are obviously out of place and probably were meant to be included with the description of the *ayre,* the second of the Cuban mammals reported by Oviedo. Herewith my paraphrased translation of his description of that animal.

The *ayre* is reddish brown, the size of a rabbit with pointed muzzle, its flesh exceedingly tough. Notwithstanding, the natives cook or roast as many of the animals as they can capture, for they are abundant. But no matter how long the meat may be cooked or roasted, it is no less tough to chew.

This characterization seems to fit the insectivore *Solenodon.* On the other hand, the flesh of *Capromys,* as of most if not all caviomorphs, is tender and, as a rule, delectable.

From his correspondents Oviedo received notice of still another mammal, the *guacabitinax,* an inhabitant of the islands near those of Las Perlas in the Golfo de San Miguel and the Isla de las Culebras or Gorgona, off the southwest coast of Colombia. The name, not to be confused with the preceding, and the description and details of the animal's habits, are unmistakably those of the paca (*Agouti paca* Linnaeus).

Manatees sighted at sea at various times by Columbus and his men were believed to be mermaids, albeit ugly ones. Martyr's narrative of a captive manatee as given in the available French translation of his third *Decade* (Gaffarel, 1907) is composite. The account by Herrera of the same manatee (in Stevens's translation, 1725, vol. 1, p. 278) appears to hew closer to the original source of information:

> The *Spaniards* at this Time found a new Sort of Fish, which was a considerable advantage to them; tho' in those Parts there is much Variety. It is call'd *Manati,* in shape like a Skin they use to carry Wine in, having only two Feet at the Shoulders, with which it swims, and it is found both in the Sea and in Rivers. From the Middle it sharpens off to the Tail, the Head of it is like that of an Ox, but shorter, and more fleshy at the Snout; the Eyes small, the Colour of it grey, the Skin very hard, and some scattering Hairs on it. Some of them are twenty Foot long, and ten in Thickness. The Feet are round, and have four Claws on each of them. The Females bring forth like the Cows, and have two Dugs to give suck. The Taste of it is beyond Fish; when fresh it is like Veal, and salted like Tunny-Fish, but better, and will keep longer; the Fat of it is sweet, and does not grow rusty. Leather for Shoes is dress'd with it. The Stones it has in the Head are good against the Pleurisy and the Stone. Sometimes they are taken ashore, grazing near the Sea, or Rivers, and when young they are taken with Nets. Thus the Cazique *Caramestex* took one, and fed it twenty-six Years in a Pond, and it grew sensible and tame, and would come when call'd by the Name of *Mato,*

which signifies Noble. It would eat whatsoever was given it by Hand, and went out of the Water to feed in the House, would play with the Boys, let them get upon him, was pleas'd with Musick, carry'd Men over the Pool, and took up ten at a Time, without any Difficulty.

Martyr's third *Decade* mentions many "rabbits" and deer encountered in 1516 by André Morales on the forested Isla Rica (now San José) of the Archipiélago de las Perlas in the Golfo de Panamá. The deer, most likely *Mazama gouazoubira permira* Kellogg, 1946, and rabbits (*Sylvilagus* sp.) were said to be so abundant that Spaniards could shoot them with arrows from horseback. The retiring tapeti, *Sylvilagus brasiliensis,* the only rabbit known from mainland Panama and the Pearl Islands, would have been an unlikely target for the equestrian Spaniards.

Mainland Mammals of the Discoverers

First knowledge of mainland American mammals was contained in reports of the Paria Peninsula, Venezuela, by Columbus on his third transatlantic voyage in 1498 and Vicente Yañez Pinzón, who followed in the tracks of Columbus. Martyr's first *Decade* carried the news of their encounters with the common opossum (*Didelphis marsupialis*), sloths, armadillos, anteaters, deer (*Odocoileus, Mazama*), peccaries, tapir, kinkajou (*Potos flavus*), barkless dog (*Canis familiaris*), jaguar, puma and their color varieties, vampire(?) bats, and red howler monkey (*Alouatta seniculus*).

On his fourth and last voyage (1502–1504), Columbus explored the Atlantic coast of Middle America from the Golfo de Urubá to Guatemala. Spanish emissaries charged with establishing settlements followed quickly. Mammals reported by them and noted in Martyr's second and third *Decades,* and by Oviedo, include the common opossum, bats, monkeys, three-toed sloths, anteaters, armadillos (*Dasypus novemcinctus*), whitetail deer, red brocket, collared and white-lipped peccaries, squirrels, a composite of carnivorous species identified as *raposas* (including *Didelphis*?), *zorros* (foxlike Carnivora), lobos (*Dusicyon* or *Lutra*), rabbits (*Sylvilagus brasiliensis*), "hares" (*Dasyprocta* sp., *Agouti paca,* and perhaps the newly introduced European hare or rabbit). The domestic dog, like that first seen in the Antilles, was barkless.

An encounter with vampire bats by Pinzón's men is reported by Martyr in the first *Decade.* Vampires are also mentioned in the second *Decade* in connection with Enciso's disastrous experiences in the Darién and in the third *Decade* in accounts of the animals of the Golfo de Urubá. The following characterization of a vampire bat by Herrera (in Stevens's translation, 1726, vol. 2, p. 71) is a translation from the original sources in Spanish.

"This venomous Creature has one quality that tho' it bites one man among an hundred one Night, the next Time it only bites in the very same Place, tho' the Person bit be among two hundred; which it does either on the Toes, the Fingers, or the Head, and much Blood runs from it."

That the same vampire bat should visit the same person sleeping in the same place on successive nights may not be unusual. An experience of mine in 1935 on the Río Napo in Ecuador is of interest in this regard. Two Indian families and I, altogether 10 persons including a five-year-old girl, traveled three days upstream in a large dugout canoe. The river was low and we could bivouac on sandbars at the end of each day's travel. On each of the three nights, a vampire bat visited the little girl, scraped the skin of her nose, and fed on the trickling blood. No other member of the party was attacked. It seems improbable that the same bat should have found the same victim at each of the three different bivouacs. Perhaps the child slept more soundly than the others of the party, or her blood was more attractive to the vampires which abounded in the region.

The last of Martyr's eight *Decades* includes descriptions of Spanish settlements in the Golfo de Paria, Venezuela. In addition to those mammals previously mentioned by Martyr (above) are the lesser anteater (*Tamandua tetradactyla*), capuchin monkey (*Cebus apella* or *C. nigrivittatus*), peccary, deer (*Odocoileus virginianus*), jaguar (*Felis onca*), spotted cats (*Felis pardalis* and/or *F. wiedii*), weasels (*Mustela frenata*), skunk (*Conepatus chinga*), porcupine (*Coendou prehensilis*), and manatee (*Trichechus manatus*). Oviedo described the same animals of the region in greater detail, but with no additions of species. Vázquez de Espinosa, who in 1628 presumably visited the northern Venezuelan coast and the town of Santo Tomas above the mouth of the Río Orinoco, reported the same mammals as well as squirrels (*Sciurus aestuans*) and many kinds of monkeys. He claimed that Isla Margarita, off the Venezuelan coast, was overrun with rabbits (*Sylvilagus floridanus*).

Many of the larger mammals of Colombia in the territories of the Muso and Colima Indians north of Bogotá were already known by 1544. With bats and other small mammals omitted, more kinds were reported by Herrera than could be recorded today from the same region on the basis of extant specimens preserved in museums. Herrera, in the English version by Stevens (1726, vol. 6, p. 191), states:

> There are a great number of grey Swine [*Tayassu pecari*] that have the Navel on the Back, and a smaller sort of several Colours [*Tayassu tajacu*] much like wild boars. Tigers (*Felis onca*) not numerous but very fierce; Lions (*Felis concolor*) that do no harm, except only among the Cattle and two other sorts of Tigers that were inoffensive besides another sort that are always in the water, like Greyhounds, and all their four feet are like those of a Goose [*Lutra annectens*]. The black wild cats [*Felis yagouaroundi*] seize the Hens, carry them away under one of their front legs and run away on the other three. The black Bears [*Tremarctos ornatus*] like those in Spain, do no hurt but only to the small Cattle. The Ant-Bears [*Myrmecophaga tridactyla*] when they go, lay their Tail, which is long, on their Heads, winding them about their Necks, and so walk from Ant-hill to Ant-hill, stretch out their Tongues near half a Yard which are soon cover'd with Pismires, then they draw them back and eat them. There are Dantas [*Tapirus pinchaque* or *T. terrestris*], Deer [*Odocoileus virginianus*] like ours in *Europe,* and others red like wild Goats [*Mazama rufina* or *Mazama americana*], and the Bezoar stones found in them are best. The Guadatinajas [*Agouti paca*] are like Hares; and the Zorillas [*Didelphis marsupialis*] or little foxes, that have a purse under their Belly, in which they carry their Cubs, the ever so many, are very mischievous to the Henroosts. The little Creatures call'd *Umazia* [*Marmosa*] have a dug growing out for every one of their young, and they stick to it till bred up. The Armadillo [*Dasypus novemcinctus*] which has been spoken of having five claws in each Forefoot, with which it throws up the Earth, is tame and eaten. The Perico Ligero [*Bradypus variegatus*] is three hours climbing a Tree, goes about in the Night, gives a cry every time it lifts a Foot, and is half an Hour, between every Step, is as big as a Barbary monkey, and fierce, yet does no harm. There are cats (?) that sleep all the Day, and all the Night catch Birds and Mice. The Pizma [*Nasua nasua*] about as big as a large Lap Dog, has a bad countenance, a long Snout, its voice like a Bird, defends itself against Dogs, and the Spaniards call them Badgers. The Hedgehogs [spiny echimyids] are like those in *Spain,* the largest like Porcupines [*Coendou* sp.] darting out their Prickles. There are many sort of Apes, squirrels.

Elsewhere, in the Province of Bogotá, Herrera (in Stevens's translation, 1726, vol. 6, p. 77) notes "innumerable apes, monkeys, ferrets [marsupials?], squirrels, weasels [*Mustela frenata*], deer [*Odocoileus virginianus*], roebuck [*Mazama rufina*] and Rabbits [*Sylvilagus brasiliensis*] . . . but not Hares." Manatees were reported from the Río Magdalena.

From coastal Colombia, at Zaragoza, 30 leagues from Cáceres in the lower Río Cauca Valley, Vázquez de Espinosa records jaguar, puma, danta (*Tapirus terrestris*), oso (*Myrmecophaga* or *Tamandua*), cuchumbi (*Nasua*), armadillo (*Dasypus*), raposa (*Dusicyon thous*), chucha (*Didelphis marsupialis*), "three" species of sahinos or peccary, perico ligero (*Bradypus variegatus*), nutria (*Lutra* or *Chironectes*), and guadatinaja (*Agouti paca*).

Acosta's long residence in Peru made him familiar with some of the mammals in the vicinities of Cuzco and Lima and others about which he may have learned from travelers or records. He described sahinos (peccaries), dantas (tapir), armadillos, perico ligero (three-toed sloth), osos (anteaters), otoronco (bear), chinchilla, vizcacha, cui (guinea pig). The "liebres verdaderas" or true hares are certainly the introduced European hare. He affirmed that conejos or rabbits (*Sylvilagus brasiliensis*) occur in the Reino de Quito (Ecuador).

Acosta declared there were monkeys of all kinds throughout America, but those he described were Middle American. At Capira near Nombre de Dios, Panama, he saw monkeys (presumably spider monkeys) swing by their tails from a tree on one side of a stream to another tree on the opposite side. Where the river was too wide for this maneuver, the monkeys of the troop, he related, crossed by forming a hanging chain holding on to each other's tail, then swinging so that the endmost could grab the branch of a tree on the other side of the river and let all the others clamber up.

The anecdote is less a fabrication than an exaggeration. Individual howlers and spider monkeys, usually the alpha male or an old suckling female, may bridge a narrow gap in the canopy pathway by holding with its prehensile tail the branch on one side of the gap and with it swinging the body to catch, with outstretched arms, the nearest branch of the far side. Monkeys too small or weak to hurdle the gap run or scramble over the bridging back of their elder. I have seen strong young adults take advantage of the same convenience.

Acosta also narrated the tale of a monkey that resided in the palace of a provincial governor. As related to him, the simian was trained to fetch wine from the town tavern. The animal would set off with the empty wine pitcher in one hand, the wine money gripped in the other. Not before the pitcher was filled to the brim did the sage monkey release his coins to the tavern keeper. There were times on these errands when taunting street urchins chased and hurled stones at the monkey. Annoyed by this sport, the simian halted, set down the pitcher, and returned the stones with sufficient force and accuracy to rout his tormentors. Retrieving his pitcher, he moved on serenely to deliver the wine at the palace.

Peruvian "sheep" or camelids were greatly admired by the Spaniards when first seen. Acosta's interesting account of them was suitably appreciated by Herrera, and the English translation by Stevens (1726, vol. 4, p. 36) is quoted herewith.

> There are no such *Vicuñas* and Sheep in *New Spain* [Mexico] As those of Peru, and those Sheep are Tame, and very serviceable; but the *Vicuñas* are wild, and have no Horns, the like of them not to be seen in the whole World, but only in *Peru* and *Chile,* bigger than Goats, but smaller than Calves, their Colour almost Murrey, breeding on the highest Mountains, in cold and desert Places, which they call *Punas.* They go in flocks, run swiftly, and when they see any Men, fly and drive their Young before them. Of their Wooll are made very valuable Mantles, which never lose their Colour, because it is natural; they are said to be good for Inflammations in the Kidneys, as are Quilts made of the Wooll, because they moderate the Heat, and the same in the Gout; and in them the Bezoar Stones are found.

The abundance and ubiquity of llamas may have inspired some Spaniards to attempt to raise Old World camels in Peru. According to Acosta, some brought from the Canary Islands were bred for a while.

Sebastian Cabot's journal of conquest and exploration of the Province of Río de La Plata, then consisting of modern northern Argentina, cisandean Bolivia, and southeastern Brazil, included data on natural history. As recorded by Herrera, the mammals seen were the hairy armadillo (*Chaetophractus* sp.) and several other kinds, cavies (*Cavia*), swamp deer (*Blastocerus dichotomus*), pampa deer (*Blastoceros bezoarticus*), brockets (*Mazama* sp.), tapirs (*Tapirus terrestris*), peccaries (any or all of the known species), howler monkeys (*Alouatta*), canids (*Dusicyon*), lesser anteater (*Tamandua tetradactyla*), jaguar (*Felis onca*), and puma (*Felis concolor*). Southern Brazilian mammals in particular included deer, peccary, tapir, "rabbits" with small, round ears (*Dolichotis*?), paca (*Agouti paca*), armadillo, sloth (*Bradypus torquatus*), opossum (*Didelphis albiventris*), monkeys, and coastal seals, most likely *Arctocephalus australis.*

Vázquez de Espinosa adds capybaras, armadillos (tatu and quirquincho specified), and guanacos. In the vicinity of Chuquisaca (La Plata), Bolivia, the missionary notes brockets (*Mazama*), vicuña, guanaco, dark gray wildcats known as oscollos, jaguar called "otorongo," puma locally called poma, a large beast called lilisto with a horselike head that lures cattle and humans, a ferret called siqui (*Mustela frenata*?), skunks or anatiria (*Conepatus*), bear (*Tremarctos ornatus*), antbears (probably *Tamandua*), vizcacha (*Lagidium*), and cuis (*Cavia porcellus*).

The occurrence of sea lions (*Otaria flavescens*) and fur seals (*Arctocephalus*) on both southern continental coasts was mentioned by Vázquez de Espinosa. The sea lions along the coast of Arequipa, Peru, he reported come out of the water onto the rocks and make low sounds at night. The animals were hunted by the Indians for their hides. In northern Chile, the natives of Arua and Atacama converted the hides into balloon-like floats for support of their seagoing fishing rafts.

The conquest of Chile by Pedro de Valdivia in 1541 provided the chroniclers with additional information on mammals. Vázquez de Espinosa reported huemul (*Hippocamelus bisulcus*), "fallow deer" (spotted fawns of huemul), guanaco, and vicuña in the vicinity of Osorno. According to the same authority, the Río Guasco valley (29°S) harbored "squirrels" (chinchillas) with very fine fur.

V. Brazil: Mammalogy Through 18th Century

André Thevet (1503-1592)

The French missionary André Thevet arrived in 1555 in Rio de Janeiro, the principal port of a French colony in the ephemeral France Antarctique. Thevet returned to France via the Antilles a year later, and the accounts of his travels were published in 1557 or 1558. Father Thevet's curiosity about all he saw in the New World knew no bounds, and he became an avid collector of Indian artifacts, local birds, and insects. Not all objects and events described in his book conformed to popular European prejudices or generally accepted misconceptions. The work stirred up considerable debate and was rejected by many not prepared to accept the realities that opossums had pouches or that the bodies of American Indians were not densely furred.

The Brazilian mammals described or mentioned by Thevet include the locally common opossum (*Didelphis albiventris*), tapeti (*Sylvilagus brasiliensis*), agouti (*Dasyprocta leporina*, declared good eating), peccaries, deer (probably *Mazama*), coati (*Nasua nasua*), tapir (*Tapirus terrestris*), capuchin monkey (*Cebus apella*), golden tamarin (*Leontopithecus rosalia*), armadillos, jaguar (*Felis onca*), and deer-hunting canids (*Speothos*?), but no lions or wolves. The three-toed sloth was abundant, but never observed eating or drinking. Thevet adds, however, that there are those who believe the beast sustains itself solely by the small, slender leaves of a very high tree called *amahut*.

Georg Marcgraf [or Marggrav or Marggraf] (1610-1644)

Most illustrious of the pre-Linnaean naturalist-explorers of Brazil was Georg Marcgraf. Born in Liebstad, Saxony, educated in Holland with emphasis on astronomy and botany, he sailed for Brazil in 1638 on a scientific expedition led by Johann Moritz, Count of Nassau-Siezen. The party, which included the young physician Piso (1611–1678), landed in Pernambuco. Explorations were restricted to northeastern Brazil in the present states of Pernambuco, Paraíba, and Rio Grande do Norte. Among Marcgraf's accomplishments were the construction of an astronomical observatory, the first of its class in the New World, and a monographic study of the plants and animals of the region. After turning over his notes and illustrations to Moritz, for preparation and publication, the naturalist sailed for Africa, where he died shortly after arrival. Marcgraf's monumental Historiæ Rerum Naturalia Brasiliæ, a part of Willen Piso's *Historia Naturalis Brasiliæ,* was published in 1648 in Amsterdam.

Of the mammals of the northeastern region of Brazil described by Marcgraf, 32 were native species, the others introduced. Their detailed descriptions and life history notes, together with crude but useful woodcuts (fig. 2), were among the primary references on which Linnaeus based binomials in the 10th (1758) and 12th (1766) editions of his *Systema Naturæ.*

The mammals are listed in Table 1 by the indigenous names used by Marcgraf and their current scientific names. Provenance of the forms which served as types for binomialists, mainly Linnaeus, was restricted for taxonomic purposes to Pernambuco by Thomas (1911).

Alexandre Rodrigues Ferreira (1756-1815)

The first Brazilian naturalist of European extraction, Alexandre Rodrigues Ferreira, was born in Salvador, Bahia. He pursued higher studies in Portugal, received his doctorate in 1779 from the University of Coimbra, and was then appointed Naturalist of the Museu Real d'Ajuda in Lisbon. He returned to Brazil in 1783 commissioned by the museum to collect samples of plants, animals, and minerals and to record all matters of scientific and political interest within his scope. The expedition, or "Viagem Filosófica," explored the provinces of Grão Pará, Rio Negro, Mato Grosso, and Cuiabá from 1783 to 1792 (fig. 3). Rodrigues Ferreira returned to Lisbon the following year.

The scientific materials collected in Brazil, with notes and illustrations, were deposited in the Museu d'Ajuda. Included were 417 species of animals represented by 592 specimens. Of these, 76 specimens represented 65 species of mammals. The whole collection was confiscated by the invading armies of Napoleon and taken to Paris for study

FIG. 2. Brazilian mammals illustrated and described by Marcgraf under Amerindian names; see Table 1 for bibliographic references and scientific names; from Marcgraf (1648).

TABLE 1. The Brazilian mammals of Marcgraf (1648) and their current Linnaean names.

Page	Vernacular name	Linnaean name	Figure
221	Ai sive Ignavus	*Bradypus variegatus* Schinz, 1825	2
222	Carigueya, female	*Didelphis albiventris* Lund, 1840	2
223	Tai-ibi (male)	*Didelphis albiventris* Lund, 1840	...
223	Aperea, type of the species	*Cavia aperea* Erxleben, 1777	2
223	Tapeti, type of the species	*Sylvilagus brasiliensis* Linnaeus, 1758*	2
224	Cavia Cobaya	*Cavia porcellus* Linnaeus, 1758	2
224	Paca, type of the species	*Agouti paca* Linnaeus, 1766	2
224	Aguti vel Acuti	*Dasyprocta leporina* Linnaeus, 1758	2
225	Tamandua guacu, type of the species	*Myrmecophaga tridactyla* Linnaeus, 1758	2
225	Tamandua-i, type of the species	*Tamandua tetradactyla* Linnaeus, 1758	2
226	Guariba (fig. misplaced on p. 228), type of the species	*Alouatta belzebul* Linnaeus, 1766	2
226	Lower figure only of caitaia misplaced with text of the guariba	*Cebus apella libidinosus* Spix, 1823 (fig. only)	2
227	Cagui minor, type of the species	*Callithrix jacchus* Linnaeus, 1758	2
227	Caitaia	*Cebus apella libidinosus* Spix, 1823	2
228	Coati, type of the species	*Nasua nasua* Linnaeus, 1766	2
228	Coatimondi	*Nasua nasua* Linnaeus, 1766	...
229	Tapiierete, type of the species	*Tapirus terrestris* Linnaeus, 1758	2
229	Mus araneus, type of the species	*Monodelphis americana* Müller, 1776	...
229	Tajacu Caaigoara	*Tayassu tajacu* Linnaeus, 1758	2
230	Capybara, Rio São Francisco, type of the species	*Hydrochaeris hydrochaeris* Linnaeus, 1766	2
230	Scyurus	*Sciurus aestuans* Linnaeus, 1766	...
231	Tatu	*Dasypus septemcinctus* Linnaeus, 1758	...
231	Tatu-ete	*Dasypus novemcinctus* Linnaeus, 1758	2
232	Tatu Apara	*Tolypeutes tricinctus* Linnaeus, 1758	2
233	Maraguo sive Maracaia	*Felis tigrina* Schreber, 1775	...
233	Cuandu, type of the species	*Coendou prehensilis prehensilis,* Linnaeus, 1758	2
234	Ibiya, type of the species	*Pteronura brasiliensis* Gmelin, 1788	2
235	Cuguacu-ete (female), type of the species	*Blastoceros bezoarticus* Linnaeus, 1758	...
235	Cuguacu-apara (male)	*Blastoceros bezoarticus* Linnaeus, 1758	...
235	Jaguara, type of the species	*Felis onca* Linnaeus, 1758	2
235	Jaguarete	*Felis onca* (melanistic)	2
235	Cuguacuarana	*Felis concolor* Linnaeus, 1771	...

* **Editors' Note:** Here and elsewhere in this paper, Article 51(c) of the International Code of Zoological Nomenclature, governing the use of parentheses in scientific names, is not followed.

by Étienne Geoffroy St.-Hilaire of the Muséum National de Histoire Naturelle in Paris.

Monkeys constituted a sizeable part of the loot, and the following were described as new by Étienne Geoffroy St.-Hilaire in 1812 and by others as noted in brackets; the current form of each name is used: *Callithrix jacchus penicillatus, Callithrix jacchus geoffroyi* [Humboldt], *Callithrix jacchus aurita, Callithrix humeralifer, Callithrix argentata melanura, Saguinus labiatus, Saimiri ustus* [I. Geoffroy], *Callicebus amictus, Callicebus personatus, Pithecia monachus, Alouatta fusca, Cebus apella cirrifer, Cebus flavus,* and *Lagothrix lagothricha canus.* Mounted specimens of previously named forms also brought to Paris from the Lisbon museum included *Callithrix jacchus* Linnaeus, *Leontopithecus rosalia* Linnaeus, *Chiropotes satanas* Hoffmannsegg, *Brachyteles arachnoides* É. Geoffroy, *Inia geoffrensis* Blainville, and probably others lost or discarded.

Except for the descriptions by the French zoologist, the specimens and manuscripts of Rodrigues Ferreira were largely neglected during the naturalist's lifetime. The several portions of the memoirs published posthumously were heavily edited. In 1972, however, the entire *Viagem Filosófica,* in two text volumes and two of colored plates, was published by the Conselho Federal de Cultura of the Brazilian Ministry of Education and Culture.

Treatment of mammals in the zoological memoir was a model of its kind for the times. Each species was described, with bibliographic references for the ones better known, external characters and what was learned of habitat, habits, reproduction, utilization by man, and gastronomic rating. With respect to the last, Rodriguez Ferreira grouped the Brazilian mammals according to those used most widely for food (peccary, deer, tapir, paca, agouti), those eaten only by Indians and some

ROTEIRO DA "VIAGEM FILOSÓFICA" REALIZADA POR ALEXANDRE RODRIGUES FERREIRA, NUMA DISTÂNCIA APROXIMADA DE 39.372 KM (1783-1792)

FIG. 3. Map of Brazil showing routes (bold lines) of Alexandre Rodrigues Ferreira, during the "Viagem Filosófica," 1783–1792; from Rodrigues Ferreira (1972).

TABLE 2. Mammals illustrated in the *Viagem Filosófica* by Rodrigues Ferreira (1971).

Plate no.	Brazilian name	Current scientific name	Figure
118	Gambá	*Didelphis marsupialis* Linnaeus	. . .
119	Macaco-da noite	*Aotus* sp.	. . .
120	Zogue-zogue; uapuca	*Callicebus moloch* Hoffmannsegg	. . .
121	Parauacu	*Pithecia monachus* É. Geoffroy	4
122	Cuxiu	*Chiropotes satanas chiropotes* Humboldt	. . .
123	Cuxiu-prêto	*Chiropotes satanas satanas* Hoffmannsegg	. . .
124	Guariba-vermelho	*Alouatta seniculus* Linnaeus	. . .
125	Guariba-da-mão-ruiva	*Alouatta belzebul* Linnaeus	. . .
126	Mico-de-cheiro	*Saimiri ustus* I. Geoffroy	4
127	Quatá-de-cara-vermelha	*Ateles paniscus* Linnaeus	. . .
128	Barrigudo-cinzento	*Lagothrix lagothricha* Humboldt	. . .
129	Sauitinga	*Callithrix argentata argentata* Linnaeus	. . .
130	Sauí dourado	*Callithrix humeralifer chrysoleuca* Wagner	. . .
131	Sauí	*Callithrix jacchus penicillata* É. Geoffroy	4
132	Sauí-de-mão-ruiva	*Saguinus midas midas* Linnaeus	. . .
133	Tamarin	*Saguinus midas tamarin* Link	. . .
134	Sauí-de-bigode-branco	*Saguinus labiatus labiatus* É. Geoffroy	4
135	Tamanduá-mirim	*Tamandua tetradactyla* Linnaeus	. . .
136	Tamanduaí	*Cyclopes didactylus* Linnaeus	. . .
137	Tamanduaí	*Cyclopes didactylus* Linnaeus	. . .
138	Preguiça-de-três-dedos	*Bradypus variegatus* Schinz	. . .
139	Tatu-galinha	*Dasypus novemcinctus* Linnaeus	. . .
140	Tatu peba	*Euphractus sexcinctus* Linnaeus	. . .
141	Guaraxaim	*Procyon cancrivorus* F. Cuvier	. . .
142	Janauíra	*Speothos venaticus* Lund	. . .
143	Guará	*Chrysocyon brachyurus* Illiger	. . .
144	Quati	*Nasua nasua* Linnaeus	. . .
145	Jupará	*Potos flavus* Schreber	. . .
146	Furão	*Galictis vittata* Schreber	. . .
147	Irara	*Eira barbara* Linnaeus	. . .
148	Ariranha	*Pteronura brasiliensis* Gmelin	. . .
149	Maracajá	*Felis geoffroyi* d'Orbigny and Gervais	. . .
150	Jaguartirica	*Felis pardalis* Linnaeus	. . .
151	Suçuarana	*Felis concolor* Linnaeus	. . .
152	Jaguar	*Felis onca* Linnaeus	. . .
153	Onça prêta	*Felis onca* Linnaeus	. . .
154	Peixe-boi, male & female	*Trichechus inunguis* Natterer	. . .
155	Caitetu	*Tayassu tajacu* Linnaeus	. . .
156	Veado vermelho	*Mazama americana* Erxleben	. . .
157	Cariacu	*Odocoileus virginianus cariacou* Boddaert	. . .
158	Quatipuru-vermelho	*Sciurus igniventris* Wagner	. . .
159	Quatipuru-preto	*Sciurus spadiceus* Olfers	. . .
160	Quatipuru-louro	*Sciurus igniventris* Wagner	. . .
161	Rato-d'agua	*Nectomys squamipes* Brants	. . .
162	Preá	*Cavia aperea* Erxleben	. . .
163	Cutia-vermelha	*Dasyprocta leporina* Linnaeus	. . .
164	Cutia-prêta	*Dasyprocta fuliginosa* Wagler	. . .
165	Acutiuaia	*Myoprocta exilis* Wagler	. . .
166	Paca	*Agouti paca* Linnaeus	. . .
167	Cuandu	*Coendou prehensilis* Linnaeus	. . .
168	Uiara	*Inia geoffrensis* Blainville	. . .
169	Tucuxí	*Sotalia fluviatilis* Gervais and Deville	. . .

white residents (anteaters, armadillos, sloths, porcupines, monkeys, jaguar), and animals not eaten by humans (marsupials, melanistic felids, squirrels, capybara). Bezoar stones and certain parts of the animal, usually tegumentary, were also cited for their medicinal merits, particularly as antivenins for headaches and female sterility, or as aphrodisiacs.

A memoir on the *peixe boi* or river manatee (*Trichecus inunguis* Natterer) provides detailed information on such topics as hunting, harpooning, reproduction, size, weight, blubber, butchery,

FIG. 4. Four monkeys of the "Viagem Filosófica" collections: **upper left,** parauaco (*Pithecia monachus* É. Geoffroy), possibly the holotype; **upper right,** saui-de-bigode-branco (*Saguinus labiatus labiatus* É. Geoffroy), possibly the holotype; **lower left,** mico-de-cheiro (*Saimiri ustus* I. Geoffroy), possibly the holotype; **lower right,** saui (*Callithrix jacchus penicillata* É. Geoffroy), possibly the holotype; from Rodrigues Ferreira (1972).

preservation, and market value of the flesh. The author decried the slaughter of the young and noted the disappearance of manatees in certain lakes.

Of all Brazilian mammals described or merely listed in the *Viagem Filosófica,* those depicted in color in the 50 plates (each 19 × 29 cm) are representative. They are listed in Table 2 by plate number with their Brazilian and current scientific names. The animals were postured as prepared by taxidermists (fig. 4). Many of the monkeys are those later described by É. Geoffroy.

VI. Brazil: Mammalogy to Middle of 19th Century

Introduction

Growth of science in South America during the first third of the 19th century shifted from the Spanish colonies, with their wars for independence and internal political turmoil, to the relatively stable Portuguese colony of Brazil. Following the invasion of Portugal by the Napoleonic armies, the royal family fled to Brazil and made Rio de Janeiro its capital and center of cultural activities. During previous years Brazil had been closed to foreigners to prevent the mines of precious metals and minerals from passing out of control of the ruling Portuguese. Dom João VI, however, opened the ports and changed the environment to one befitting an enlightened monarch in residence. Cultural institutions, including museums, libraries, and universities, were built, and scientific investigations were promoted. Betrothal of the Archduchess Leopoldina, daughter of the Emperor of Austria, with Dom Pedro, Crown Prince of Portugal and Brazil, became the most important single factor in the advancement of science in the New World during the first half of the 19th century. The entourage of the bride on her voyage to Brazil included some of the best and most adventurous of the younger scientists of Austria and Bavaria.

The Viennese naturalists of the party included the field collector Johann Natterer, and from the court of Munich, the zoologist Spix and the botanist Martius. Two years earlier, in 1815, the most accomplished of the naturalist-travelers, Maximilian Prinz Wied zu Neuwied of Prussia, arrived on the scene.

Modern Brazilian mammalogy begins with the scientific accounts of the collections and travels of these naturalists.

Johann Baptist Ritter von Spix (1781-1826) and Carl Friedrich von Martius (1794-1866)

The German naturalist Johann Baptist Ritter von Spix first studied for the priesthood, but after two years his attention turned to medicine and natural history. His doctorate was earned in 1806. That same year he was appointed assistant in the Museum of the Munich Academy of Science, with responsibility for the organization of the zoological collections. In 1816 he was ordered by the King of Bavaria to undertake a two-year scientific expedition to Brazil, together with the museum's assistant in botany, Carl Friedrich von Martius. The two departed on 10 April 1817 through the port of Trieste, and after considerable delay, they arrived in Rio de Janeiro on 15 July 1817.

The exuberance and variety of the native plant life in eastern Brazil at first awed and bewildered the two young naturalists. Everything they saw was new to them, and all they could possibly collect and preserve was easily reached along the trails they traveled from Rio de Janeiro to Minas Gerais and beyond. Real or fantasized dangers lurking in what they imagined as dark, brooding, impenetrable forests restrained their urges for stepping off the beaten path. The strange and wonderful wildlife encountered on the roads was enough to gratify their utmost expectations and inspired them to record their impressions in ecstatic prose. On the trip from Ipanema, São Paulo, to Vila Rica, Minas Gerais, they described, as translated into English by Lloyd in equally romanticized and tortured prose, the

> numerous flocks of little monkeys [that] run whistling and hissing to the recesses of the forest; the cavies, running about on the tops of the mountains, hastily secrete themselves under loose stones; the American ostriches (*Emas*), which herd in families, gallop at the slightest noise, like horses through the bushes, and over hills and valleys, accompanied by their young; the dicholopus (*Seriemas*), which pursues serpents, flies, sometimes sinking into the grass, sometimes rising into the trees, or rapidly climbing the summits of the hills, where it sends forth its loud deceitful cry, resembling that of the bustard; the terrified armadillo (*Tatú Canastra, Peba, Bola*) runs fearfully about to look for a hiding place, or, when the danger presses, sinks into its armour; the ant-eater (*Tamanduá, Bandeira mirim*) runs heavily through the

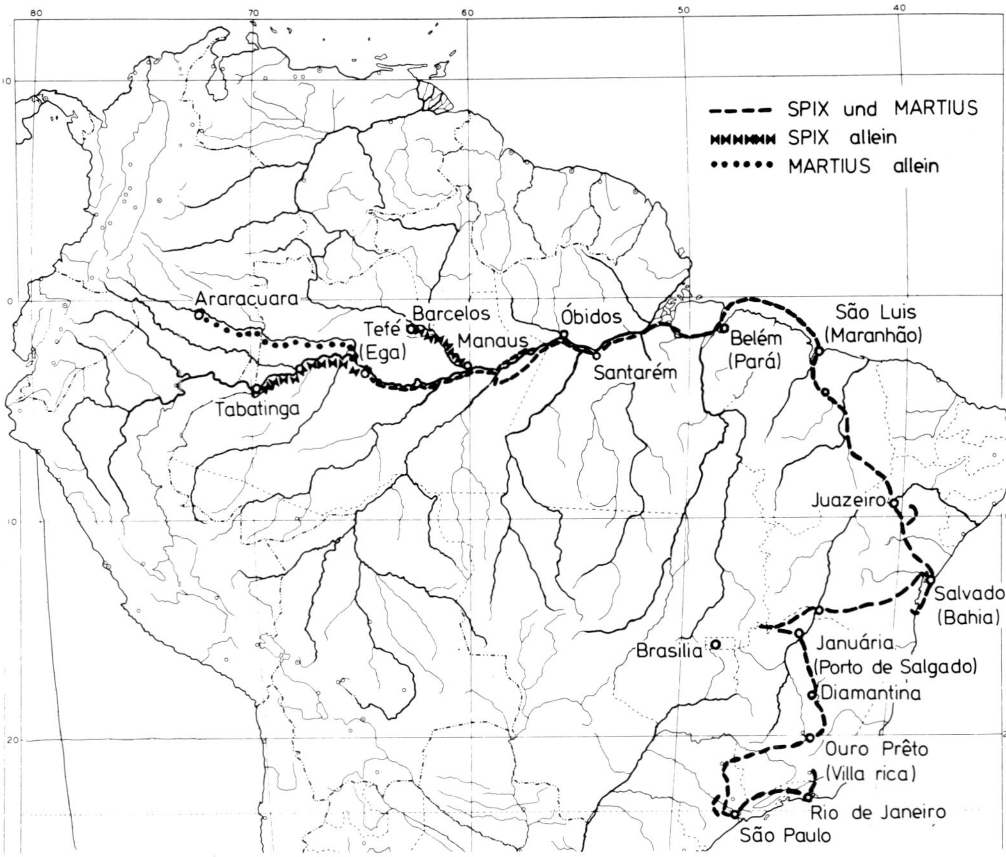

Karte von Brasilien mit dem eingezeichneten Reiseweg von Johann Baptist von Spix und Carl Friedrich Philipp von Martius anläßlich ihrer Expedition in den Jahren 1817–1820.

FIG. 5. Map of Brazil showing routes of the Spix and Martius expedition (1817–1820); only principal stations plotted; from Tiefenbacher (1983).

plain, and, in case of need, lying on its back, threatens its pursuers with its sharp claws. Far from all noise, the slender deer, the black tapir or a pecari, feed on the skirts of the forest. Elevated above all this, the red-headed vulture (*urubú*) soars in the higher regions; the dangerous rattle-snake (*Cascavel*), hidden in the grasses, excites terror by its rattle; the gigantic snake sports suspended from the tree with its head upon the ground; and the crocodile resembling the trunk of a tree, basks in the sun on the banks of the pools. After all this has passed during the day before the eyes of the traveler, the approach of night, with the chirping of the grasshoppers, the monotonous cry of the goat-sucker (*João corta pão*), the barking of the prowling wolf, and of the shy fox, or the roaring of the ounces, complete the singular picture of the animal kingdom in these peaceful plains.

For the next three years, the zoologist and botanist explored the eastern states of Brazil from São Paulo and Minas Gerais north to Pará. Most of July and August of 1819 was spent in Belém (Pará). On 21 August they shipped up the Rio Amazonas, making stopovers at the mouth of the Rio Tocantins, the Rio Xingu (10 September), Santarém on the Rio Tapajós (18 September), Óbidos (23 September), Parintins, and Vila Nova da Rainha (1 October). The mouth of the Rio Madeira was passed 15 October, and on 22 October they landed at Barra do Rio Negro (Manaus). Travel upstream continued in November with a stop at Tefé (formerly Ega) on 26 November. Spix then traveled alone up the Solimões to Tabatinga at the Peruvian border, arriving 9 January 1820. Martius, for his part, ascended the Rio Japurá to Araracuara in eastern Colombia.

Spix returned to Manaus on 3 February 1820. On 11 February he ascended the Rio Negro to Barcelos and was back again in Manaus 28 February to continue his travels downstream to Belém, where he arrived on 16 April. He embarked on 14 June 1820 for Europe from Rio de Janeiro (fig. 5).

In the *Reise,* Spix and Martius (1828, p. 541) made up an impressive list of the mammals of the sertão (scrub country) of Campos Gerais de São Felipe in the angle between the Rio São Francisco and its eastern tributary, the Rio Verde Grande, northern Minas Gerais. The data were evidently compiled uncritically from a number of sources, including local informers, personal observations, and publications based on the Wied-Neuwied (1826) collections. Their use and misuse of names are too involved to unravel here. Except for the missing bats (given elsewhere by Spix, 1823) and some small rodents, it is unlikely that a similar or larger number of mammalian species of the area, based on actual specimens, could be made today. The sertão mammals of the Spix and Martius expedition are listed in Table 3 by current scientific names of the species only, with the Spix and Martius equivalents omitted.

In his journey up the Amazon, Spix noted habits of the inia (*Inia geoffrensis*) (Spix & Martius, 1831, p. 1119) and of the manatees (*Trichechus inunguis*) (Spix & Martius, 1831, p. 1122).

The results of the expedition are recorded in several publications, including the *Simiarum et Vespertilionum Brasiliensium* by Spix (1823). The account of the nearly three-year journey or *Reise* in Brazil by Spix and Martius (1823–1831) is replete with observations on the biology, geography, geology, paleontology, mineralogy, meteorology, and the various human cultures and industries of the parts of the country they traveled. Many kinds of mammals are mentioned, but except for bats and monkeys, few of them were collected.

The zoological material actually collected consisted of thousands of invertebrates and 498 species of vertebrates, of which 34 were monkeys and 15 bats. Altogether, according to Avila Pires (1974, p. 139), 85 species of mammals were collected. Spix (1823) reported only on the monkeys and bats and illustrated in color the types of all species,

TABLE 3. Mammals of the sertão of Campos Gerais de São Felipe, Minas Gerais, recorded by Spix and Martius (1828, p. 541, footnote 3). Current scientific names to species only are used. The Spix and Martius usage of local, German, and scientific names is too confused for tabulation. The arrangement is phylogenetic.

MARSUPIALIA
Caluromys philander Linnaeus
Didelphis marsupialis Linnaeus
PRIMATES
Callithrix jacchus Linnaeus
Cebus apella Linnaeus
Alouatta fusca É. Geoffroy
Alouatta caraya Humboldt
EDENTATA
Tamandua tetradactyla Linnaeus
Myrmecophaga tridactyla Linnaeus
Bradypus torquatus Desmarest
Bradypus variegatus Schinz
Dasypus novemcinctus Linnaeus
Tolypeutes tricinctus Linnaeus
Priodontes maximus Kerr
Euphractus sexcinctus Linnaeus
CARNIVORA
Dusicyon thous Linnaeus
Chrysocyon brachyurus Illiger
Nasua nasua Linnaeus
Procyon cancrivorus G. Cuvier
Conepatus chinga Molina
Eira barbara Linnaeus
Pteronura brasiliensis Gmelin
Felis wiedii Schinz
Felis tigrina Schreber
Felis pardalis Linnaeus
Felis concolor Linnaeus
Felis onca Linnaeus
Felis yagouaroundi É. Geoffroy
PERISSODACTYLA
Tapirus terrestris Linnaeus
ARTIODACTYLA
Mazama gouazoubira Fischer
Mazama americana Erxleben
Blastoceros bezoarticus Linnaeus
LAGOMORPHA
Sylvilagus brasiliensis Linnaeus
RODENTIA
Sciurus aestuans Linnaeus
Wiedomys pyrrhorhinos Wied-Neuwied
Echimys and/or *Proechimys* species?
Myocastor coypus Molina
Kerodon rupestris Wied-Neuwied
Cavia aperea Linnaeus
Dasyprocta leporina Linnaeus
Agouti paca Linnaeus
Coendou insidiosus Kuhl
Chaetomys subspinosus Olfers

FIG. 6. Two mounted monkeys of the Spix and Martius expedition now in the Zoologische Staatssammlung München: **left,** bearded saki, *Brachyurus ouakary* Spix, the type (= *Cacajao melanocephalus ouakary*); **right,** *Callithrix personata* É. Geoffroy (= *Callicebus personatus* É. Geoffroy); from Spix (1823).

TABLE 4. Monkeys (Primates) of the Spix and Martius Expedition described by Spix (1823); the arrangement is phylogenetic.

Current name	Spix and Martius synonym	Figure
Cebuella pygmaea Spix, 1823	...	...
Callithrix jacchus jacchus Linnaeus, 1758	*Jacchus albicollis* Spix, 1823	...
Callithrix jacchus penicillatus É. Geoffroy, 1812	...	...
Saguinus bicolor bicolor Spix, 1823	...	...
Saguinus fuscicollis fuscicollis Spix, 1823	...	...
Saguinus mystax mystax Spix, 1823	...	...
Saguinus nigricollis nigricollis Spix, 1823	...	...
Saguinus oedipus geoffroyi Pucheran, 1845	*Midas oedipus* (*varietas*), Spix, 1823	...
Callicebus cupreus Spix, 1823	...	...
Callicebus personatus personatus É. Geoffroy, 1812	...	6
Callicebus personatus nigrifrons Spix, 1823	...	...
Callicebus personatus melanochir Kuhl, 1820	*Callithrix gigot* Spix, 1823	...
Callicebus torquatus torquatus Hoffmannsegg, 1807	*Callithrix amicta* É. Geoffroy, 1812	...
Callicebus cinerascens Spix, 1823	...	...
Aotus vociferans Spix, 1823	...	...
Aotus azarae infulatus Kuhl, 1820	*Nyctipithecus felinus* Spix, 1823	...
Pithecia monachus monachus É. Geoffroy, 1812	*Pithecia hirsuta* Spix, 1823; *Pithecia inusta* Spix, 1823	...
Pithecia pithecia pithecia Linnaeus, 1766	*Pithecia capillamentosa* Spix, 1823	...
Chiropotes satanas chiropotes Humboldt, 1812	*Brachyurus israelita* Spix, 1823	...
Cacajao melanocephalus ouakary Spix, 1823	...	...
Alouatta caraya Humboldt, 1812	*Mycetes barbatus* Spix, 1823	...
Alouatta belzebul discolor Spix, 1823	...	...
Alouatta fusca Spix, 1823	...	...
Alouatta seniculus stramineus Humboldt, 1812	...	...
Cebus albifrons unicolor Spix, 1823	*Cebus gracilis* Spix, 1823	...
Cebus apella libidinosus Spix, 1823	...	...
Cebus apella macrocephalus Spix, 1823	...	...
Cebus apella xanthosternos Wied-Neuwied, 1820	*Cebus cucullatus* Spix, 1823; *Cebus xanthocephalus* Spix, 1823	...
Lagothrix lagothricha lagothricha Humboldt, 1812	*Gastrimargus infumatus* Spix, 1823	...
Lagothrix lagothricha cana É. Geoffroy, 1812	*Gastrimargus olivaceus* Spix, 1823	...
Brachyteles arachnoides É. Geoffroy, 1806	*Brachyteles macrotarsus* Spix, 1823	6

most life-size. Separate reports on all groups of animals collected by Spix have been brought together in a Festschrift in his honor edited by Tiefenbacher (1983). The mammals are treated by Kraft (1983).

The 31 presently recognized species and subspecies of monkeys (15 new) and the 14 recognized species of bats (six new) are listed in Tables 4 and 5 by current names with synonyms in parentheses.

Maximilian Prinz von Wied-Neuwied (1782–1867)

Maximilian Prinz von Wied-Neuwied was born in Prussia and studied biological sciences at the University of Göttingen under the famous naturalist-anthropologist Blumenbach. His ambition to travel and study nature in South America was realized when he sailed for Rio de Janeiro from England the first week of May 1815, and arrived on 17 July.

After a few excursions in the surroundings of Rio de Janeiro, Wied-Neuwied left for Cabo Frio on 15 August 1815, stopping at many fazendas and villages along the way. He left Cabo Frio on 8 September for São Salvador dos Campos dos Goitacazes (now simply Campos) on the Rio Paraíba, and arrived on 25 September. After more excursions and more collections in the state of Rio de Janeiro, he crossed the Rio Itabapoana on 26 November into the state of Espírito Santo. A considerable amount of time was devoted there to explorations of the Rio Doce region. February 1816 saw Wied-Neuwied in Bahia, where he occupied himself until May 1817. The coastal town of Belmonte, where he arrived in August 1816, was the base for explorations of Botocudo Indian territory. In December 1816 Wied-Neuwied established Ilheus as center for travel westward to São Pedro de Alcantara, now Itabuna, and the border of Minas Gerais. On 10 May Wied-Neuwied embarked at Salvador for Lisbon, then transshipped to Germany through an English port.

Wied-Neuwied's itinerary is difficult to track be-

TABLE 5. Bats (Chiroptera) of the Spix and Martius Expedition described by Spix (1823); the arrangement is phylogenetic.

Current name	Spix and Martius synonym
Rhynchonycteris naso Wied-Neuwied, 1820	*Proboscidea rivalis* Spix, 1823; *Proboscidea saxatilis* Spix, 1823
Noctilio albiventris albiventris Desmarest, 1818	*Noctilio albiventer* Spix, 1823
Noctilio leporinus leporinus Linnaeus, 1758	*Noctilio rufus* Spix, 1823
Tonatia bidens Spix, 1823	...
Trachops cirrhosus Spix, 1823	...
Glossophaga soricina Pallas, 1766	*Glossophaga amplexicaudata* Spix, 1823
Carollia perspicillata Linnaeus, 1758	*Vampyrus soricinus* Spix, 1823
Artibeus planirostris Spix, 1823	...
Diphylla Spix, 1823	...
Diphylla ecaudata Spix, 1823	...
Thyroptera Spix, 1823	...
Thyroptera tricolor Spix, 1823	...
Eptesicus brasiliensis Desmarest, 1823	...
Promops nasutus Spix	*Molossus fumarius* Spix, 1823
Molossus ater É. Geoffroy, 1805	*Molossus ursinus* Spix, 1823

cause of his many roundabout journeys and short excursions with too few dates for fixing comings and goings. To add to the difficulty, the names of many localities he visited no longer exist or were never plotted on any official map; a few names have changed. Bokermann's (1957) gazetteer of nearly all localities of the *Reise,* with page references to their mention in Wied-Neuwied's works, is indispensable for study of the naturalist's operations in Brazil.

Wied-Neuwied was interested in all aspects of nature, but the fauna and Indians engaged most of his attention. His species accounts are models of precision, his descriptions detailed, and comparisons where needed are made with published descriptions by Humboldt, Azara, Buffon, and others. The bibliographic references to the species are complete. Observations of habitats and reproduction are carefully recorded, and geographic range is usually given with circumspection. Wied-Neuwied's account of Geoffroy's tufted-ear marmoset (his *Hapale leucocephalus*) is an example (my translation):

> I found it in the state of Espírito Santo. I am unable to determine if it extends north of the Rio Doce or beyond as I could not hunt often in the dark forests of this river because of the Botocudo Indians. I can therefore state that the habitat of this species lies between 20° and 21° south latitude. The animal is common in the forests of the Rio Espírito Santo, especially in the outlying bush and the mangue bush (*Conocarpus* and *Avicennis*) bordering the river, as well as in the low palm (*Allagoptera pumila* and others)-covered sandy coastal districts not far from the mouth of the Espírito Santo. . . .

The following excerpt of Wied-Neuwied's (1826, p. 161) observations on the golden lion tamarin (*Leontopithecus rosalia rosalia* Linnaeus) brings together his observations on distribution, habits, habitat, food, and reproduction:

> The sahuim vermelho is nowhere abundant; we saw only single individuals or family groups, particularly in the Serra da Inuá, the forests of São João, and in the hilly forest surrounding Ponta Negra and Gurupina. The animal lives just as well on bushy sandy plains as in the high mountain forests. It feeds on fruits and insects and hides from strangers by disappearing into the leafy treetops. One or two young are produced at a birth. The female carries the offspring on her back or at her breasts [when suckling] until they are strong enough to follow her on their own. . . . Any excitement causes them to erect the long hair surrounding their faces. In general, however, their habits are similar to those of other sahuis.

Wied-Neuwied also accurately delimited the distribution of the subspecies *Leontopithecus rosalia chrysomelas* and added information on habits and reproduction. Wied-Neuwied notes (1826, p. 159) that "sahuis born in Europe are carried by the father but I have never seen this here."

Although generally careful in interpreting his

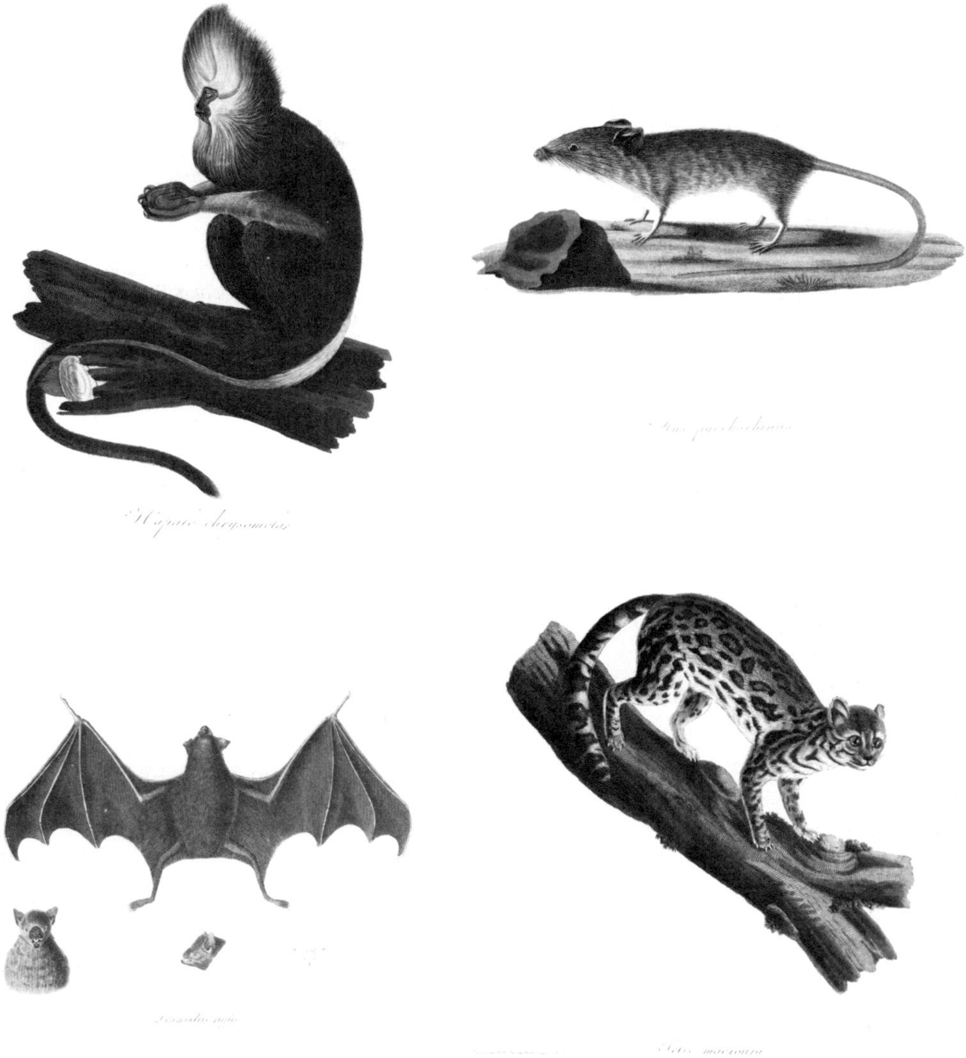

FIG. 7. Some animals of the Wied-Neuwied Brazilian expedition: **upper left,** *Hapale chrysomelas* Wied-Neuwied (= *Leontopithecus rosalia chrysomelas*), possibly the holotype; **upper right,** *Mus pyrrhorhinos* Wied-Neuwied (= *Wiedomys pyrrhorhinos*), possibly the holotype; **lower left,** *Desmodus rufus* Wied-Neuwied (= *Desmodus rotundus* É. Geoffroy); **lower right,** *Felis macroura* Wied-Neuwied (= *Felis wiedii* Schinz), possibly the holotype; from Wied-Neuwied (1822–1831).

data, Wied-Neuwied could arrive at unwarranted conclusions. Among the bats collected, the leaf-nosed *Phyllostomus hastatus* was largest and for this reason was regarded as a blood-sucking vampire, although Wied-Neuwied found only insects and no blood in the stomach of this or any other bat he had examined. After confessing he had never seen a bat feed on blood, he correctly blamed the large bats seen fluttering around the pack mules at night for causing them to appear next morning covered with blood. Convinced in his judgment, he described the wartlike excrescences around the mouth of innocent phyllostomine bats as adaptations for blood-sucking. Ironically, Wied-Neuwied (1824, 1826) later described and figured the external and dental characters of a bat he named *Desmodus rufus,* unaware it was a real blood-sucking vampire. Wied-Neuwied noted, however, that he had no opportunity to observe the live animal, because it had been captured and prepared as a specimen by assistants during his absence. The food and habits of this bat, he believed, were no different from those of other bats.

The mammals of Wied-Neuwied's Brazilian expedition are described or recorded in several publications. Those under Wied-Neuwied's own name are found in *Isis* (1820, 1821), the *Reise nach Brasilien* in two volumes (1820, 1821), the *Abbildungen zur Naturgeschichte Brasiliens* (1822–1831, see fig. 7 for some samples), and the four-volume *Beiträge zur Naturgeschichte von Brasilien.* The first volume of the last title is on reptiles, the second on mammals (1826), the third and fourth on birds.

Some diagnoses and binomials that Wied-Neuwied proposed for new forms received duly acknowledged advance publication by Kuhl (1820) and Schinz (1821). Authorship of such newly named forms continues to be attributed to Wied-Neuwied, according to Articles 11 and 50 of the International Code of Zoological Nomenclature. In the few cases where Kuhl or Schinz proposed names other than those used by Wied-Neuwied, authorship is determined by priority.

The actual number of mammals collected by Wied-Neuwied is unknown. According to him, they represented 82 species, but the number recognized today as valid is 71. The specimens were preserved in his private museum, but duplicates were distributed to the natural history museums of Berlin, Frankfurt, Leiden, and Paris. After Wied-Neuwied's death, the remainder of the collection was sold, and the American Museum of Natural History in New York acquired a part in 1869. Avila Pires (1965, p. 3) affirms that fewer than 600 specimens of the original collection are registered in the catalogue of mammals of the New York institution. Of these, only 38 skins and 16 skulls are of South American origin. Included are holotypes (or syntypes) of *Didelphis aurita* Wied-Neuwied, *Didelphis cinerea* Temminck, *Molossus plecotus* Wied-Neuwied, *Phyllostoma brevicaudum* Wied-Neuwied, *Vespertilio leucogaster* Wied-Neuwied, *Vespertilio naso* Wied-Neuwied, *Hypudeus dasytrichos* Wied-Neuwied, and *Mus pyrrhorhinos* Wied-Neuwied.

Table 6 lists all mammalian species recorded by Wied-Neuwied. Current names are used; synonyms used by Wied-Neuwied are included.

Johann Natterer (1787-1843)

Johann Natterer, born near Vienna, was well schooled in the sciences, especially biology, and in modern languages and illustration. Natterer's father, the imperial falconer and collector of birds and insects, taught him to hunt and preserve animals as museum specimens. In 1816 he was employed as assistant in the Imperial Natural History Museum of Vienna and in 1817 was appointed member of an expedition to investigate the Brazilian biota. He arrived in Rio de Janeiro on 5 November accompanied by Mikan and Pohl, both naturalists, and Schott, a botanist. Within a year Mikan, Sochor, a hunter, and two artists who were to accompany Natterer, returned to Europe. Pohl and Schott returned in 1821.

Natterer was primarily a bird collector, but his interest in collecting extended to mammals, other vertebrates, insects, and parasitic helminths. He traveled light and, as a rule, worked alone or with few native helpers (Ihering, 1902). He collected in most of the eastern coastal states and in Mato Grosso and the Amazonian region between the Rios Tapajós and Madeira and in the Rio Negro basin north of the Rio Amazonas (fig. 8). His main base for the first five years was Ipanema, São Paulo. His itinerary—with goings and comings, side trips, short stopovers in some sites, long delays in others—was arranged chronologically by Pelzeln (1871, 1883) into "Reisen" (or journeys), with dates given for all points visited, and is summarized below. Only general areas or terminal points and inclusive dates are given.

Johann Natterer's Brazilian Reisen, 1817–1835.

I. Rio de Janeiro, 5 November 1817 to 1 November 1818.
II. Eastern São Paulo, 2 November 1818 to March 1820.
III. Southern São Paulo to boundary between Rio Grande do Sul and Rio de Janeiro, July 1820 to 1 February 1821.
IV. Rio de Janeiro, São Paulo, 1 February to September 1822.
V. Northern São Paulo, Goiás, eastern Mato Grosso, Minas Gerais, October 1822 to 31 December 1824.
VI. Mato Grosso, January 1825 to July 1829.
VII. Mato Grosso, Rio Madeira, and upper tributaries to Borba in Amazonas (Capitania Rio Negro), 15 July 1829 to June 1830.
VIII. Borba to Rio Negro, Río Casiquiare, Venezuelan border, return to Barcelos and Borba, June 1830 to 31 August 1830.
IX. Rio Negro from Barcelos to Rio Branco, 5 September 1831 to 2 July 1832; Barra do Rio Negro, 29 August 1832 to 7 July 1834; Rio Tapajós, August 1834.
X. Pará, Maranhão, Rio Grande, Paraíba, Per-

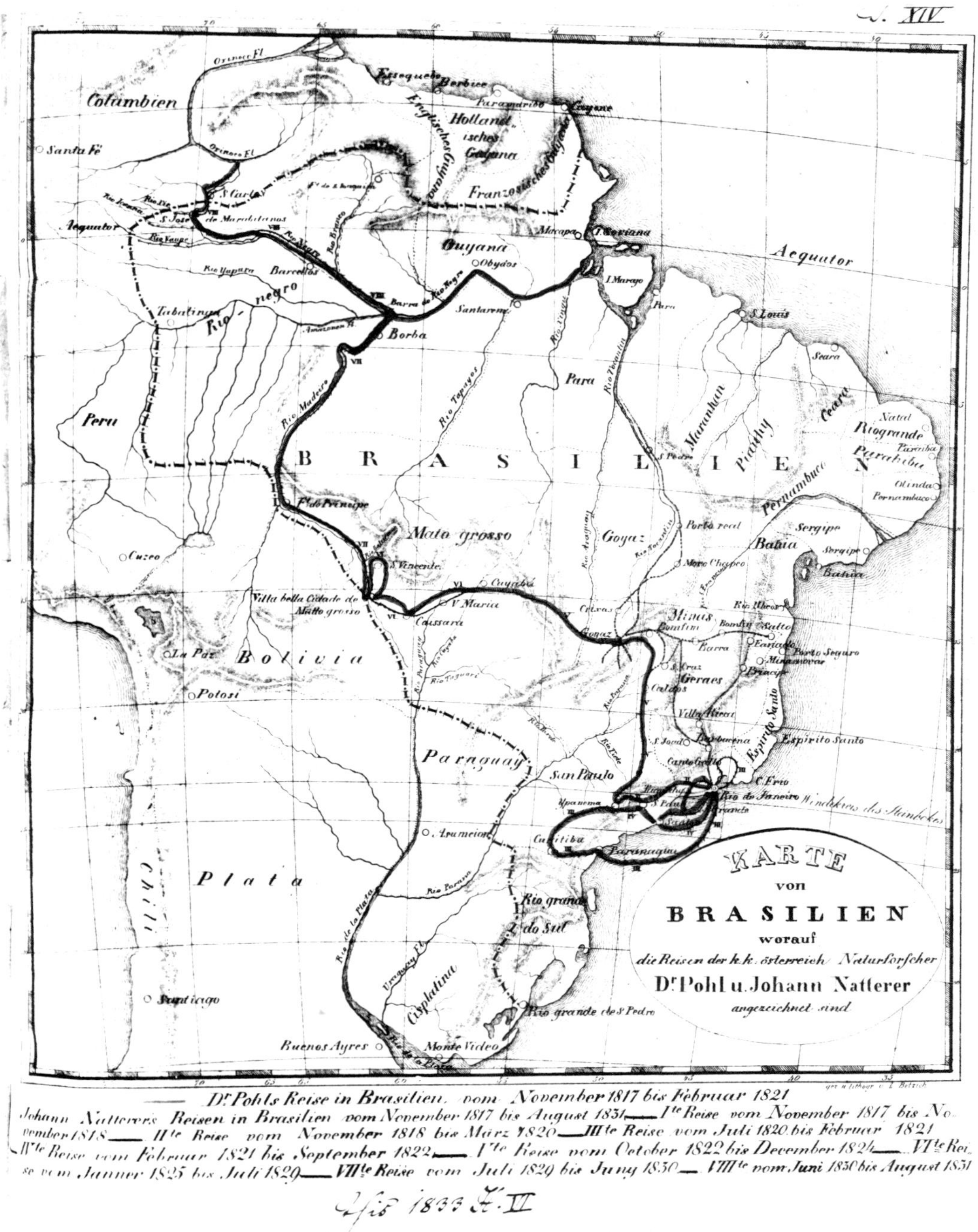

FIG. 8. Map of Brazil showing routes of Johann Natterer (bold line); from [brother of Johann] Natterer (1833, Oken's Isis, heft VI, pl. 14).

nambuco, Bahia, Rio de Janeiro, September 1834 to September 1835 (no mammal collections).

Sailed for Europe 15 September 1835.

Natterer's enormous collections were sent to the Vienna museum and, except for the birds and mammals, were never fully reported. His friend Andreas Wagner (1797–1861) described most of

TABLE 6. Brazilian mammals recorded by Wied-Neuwied (1826) with some figured in the Abbildungen (1822–1831); the arrangement is phylogenetic.

Current name	Wied-Neuwied synonym	Figure
MARSUPIALIA		
Marmosa murina Linnaeus, 1758	...	...
Marmosa cinerea Temminck, 1824	...	...
Philander opossum frenata Olfers, 1818	*Didelphys myosuros* Temminck, 1825	...
Didelphis marsupialis aurita Wied-Neuwied, 1826	*Didelphis marsupialis*, Wied-Neuwied, 1826, not Linnaeus	...
CHIROPTERA		
Rhynchonycteris naso Wied-Neuwied, 1820 (Reise)	...	...
Centronycteris maximiliani Fischer, 1829	*Vespertilio calcaratus* Wied-Neuwied, in Schinz, 1821, not Rafinesque, 1818	...
Peropteryx macrotis Wagner, 1843	*Vespertilio caninus* Wied-Neuwied, in Schinz, 1821, not Blumenbach, 1797	...
Diclidurus albus Wied-Neuwied, 1819	*Diclidurus freyreissii* Wied-Neuwied 1822, Abbild.	...
Noctilio leporinus Linnaeus, 1758	*Noctilio dorsatus* Desmarest, 1818; *Noctilio unicolor* Desmarest, 1818	...
Macrophyllum macrophyllum Wied-Neuwied, in Schinz, 1821	...	...
Phyllostomus hastatus Pallas, 1867	...	...
Phyllostomus obscurus Wied-Neuwied, in Schinz, 1821	*Artibeus planirostris* Spix, 1823	...
Glossophaga soricina Pallas, 1766	*Glossophaga amplexicaudata* É. Geoffroy, 1818	...
Anoura caudifera É. Geoffroy, 1818	...	...
Carollia brevicauda Wied-Neuwied, 1821	*Phyllostoma bernicaudum* (*sic*) Wied-Neuwied, in Schinz, 1821	...
Carollia perspicillata Linnaeus, 1758	*Phyllostoma brachyotos* (*sic*) Wied-Neuwied, in Schinz, 1821	...
Artibeus lituratus Olfers, 1818	*Phyllostoma superciliatum* Wied-Neuwied, in Schinz, 1821	...
Desmodus rotundus É. Geoffroy, 1810	*Rhinolophus ecaudatus* Wied-Neuwied, in Schinz, 1821; *Desmodus rufus* Wied-Neuwied, 1824	7
Myotis albescens É. Geoffroy, 1806	*Vespertilio leucogaster* Wied-Neuwied, in Schinz, 1821	...
Myotis nigricans Wied-Neuwied, in Schinz, 1821	...	...
Eumops perotis Wied-Neuwied, in Schinz, 1821	...	...
PRIMATES		
Callithrix jacchus penicillatus É. Geoffroy, 1812	*Hapale penicillatus kuhlii* Wied-Neuwied, 1826 p. 142)*	...
Callithrix jacchus geoffroyi Humboldt, 1812	*Hapale leucocephalus* Kuhlii (*sic*), Wied-Neuwied, 1826†	...
Leontopithecus rosalia chrysomelas Kuhl, 1820	...	7
Leontopithecus rosalia rosalia Linnaeus, 1758	...	...
Callicebus personatus personatus É. Geoffroy, 1812	...	...
Callicebus personatus melanochir Wied-Neuwied, 1820 (Reise)	...	...
Alouatta caraya Humboldt, 1812	*Mycetes niger* Kuhl, 1820	...
Alouatta fusca É. Geoffroy, 1812	*Mycetes ursinus* Humboldt, 1812, not Humboldt, 1805	...
Cebus apella nigritus Goldfuss, 1809	*Cebus cirrifer* É. Geoffroy, 1812, not *Cebus fatuellus* Linnaeus	...
Cebus apella robustus Kuhl, 1820	...	...
Cebus apella xanthosternos Wied-Neuwied, 1820 (Reise)	? *Cebus flavus* É. Geoffroy, 1812	...
Brachyteles arachnoides É. Geoffroy, 1806	*Ateles hypothanthus* Kuhl, 1820	...
EDENTATA		
Tamandua tetradactyla Linnaeus, 1758	...	...
Myrmecophaga tridactyla Linnaeus, 1758	*Myrmecophaga jubata* Linnaeus, 1766	...
EDENTATA		
Bradypus torquatus Desmarest, 1816	*Bradypus tridactylus* Wied-Neuwied, 1826, not Linnaeus, 1758	...

TABLE 6. *Continued.*

Current name	Wied-Neuwied synonym	Figure
Cabassous unicinctus Linnaeus, 1758	...	...
Euphractus sexcinctus Linnaeus, 1758	*Dasypus setosus* Wied-Neuwied, 1826; *Dasypus gilvipes* Illiger, 1815	...
Dasypus novemcinctus Linnaeus, 1758	*Dasypus longicaudus* Wied-Neuwied, 1826	...
Priodontes maximus Kerr, 1792	*Dasypus gigas* Cuvier, 1822	...
CARNIVORA		
Dusicyon thous brasiliensis Wied-Neuwied, in Schinz, 1821	*Canis azarae* Wied-Neuwied, 1823	...
Chrysocyon brachyurus Illiger, 1815	*Canis campestris* Wied-Neuwied, 1826	...
Nasua nasua solitaria Wied-Neuwied, in Schinz, 1821	*Nasua socialis* Wied-Neuwied, 1826	...
Procyon cancrivorus G. Cuvier, 1798	...	...
Potos flavus nocturnus Wied-Neuwied, 1826	...	...
Eira barbara Linnaeus, 1758	*Mustela gulina* Wied-Neuwied, 1821	...
Pteronura brasiliensis Gmelin, 1788	...	...
Felis wiedii Schinz, 1821	*Felix macroura* Wied-Neuwied, 1823	7
Felis pardalis mitis F. Cuvier, 1820	*Felis pardalis,* Wied-Neuwied, 1826	...
Felis yagouaroundi eyra Fischer, 1814	*Felis yaguarundi,* Wied-Neuwied, 1826	...
Felis concolor Linnaeus, 1771	...	...
Felis onca Linnaeus, 1758	*Felis brasiliensis* Wied-Neuwied, 1821	...
SIRENIA		
Trichechus manatus Linnaeus, 1758	*Manatus americanus* Link, 1795	...
PERISSODACTYLA		
Tapirus terrestris Linnaeus, 1758	*Tapirus americanus* Gmelin, 1788	...
ARTIODACTYLA		
Tayassu tajacu Linnaeus, 1758	*Dicotyles torquatus* Cuvier, 1817	...
Tayassu pecari Link, 1795	...	...
Mazama gouazoubira Fischer, 1814	*Cervus simplicicornis* Illiger, 1815	...
Mazama americana Erxleben, 1777	*Cervus rufus* Cuvier, 1817	...
Blastoceros bezoarticus Linnaeus, 1758	*Cervus campestris* Wied-Neuwied, 1826, not Cuvier, 1817	...
Blastocerus dichotomus Illiger, 1815	*Cervus paludosus* Desmarest, 1822	...
LAGOMORPHA		
Sylvilagus brasiliensis Linnaeus, 1758	...	...
RODENTIA		
Sciurus aestuans Linnaeus, 1766	...	...
Wiedomys pyrrhorhinos Wied-Neuwied, 1821 (Reise)	...	7
Oxymycterus rufus dasytrichos Wied-Neuwied, in Schinz, 1821	*Hypudeus dasytrichos* Wied-Neuwied, 1826	...
Proechimys myosuros Lichtenstein, 1818	...	...
Cavia aperea Erxleben, 1777	...	...
Kerodon rupestris Wied-Neuwied, 1820 (Isis)	...	...
Hydrochaeris hydrochaeris Linnaeus, 1766	...	...
Dasyprocta leporina aguti Linnaeus, 1766	...	...
Agouti paca Linnaeus, 1766	...	...
Coendou insidiosus Olfers, 1818	...	...
Chaetomys subspinosus Olfers, 1818	...	...

* The name is a correctly formed trinomial but this form was not in use at the time, and Wied-Neuwied used no trinomials elsewhere in his publications on Brazilian mammals.

† The name appears to be a trinomial although the patronymic, properly in the genitive, is not italicized. Most likely Wied-Neuwied meant to cite Kuhl for this and the preceding taxon as authority for his use of the names in question. It was common practice at the time to cite the author who replaced an earlier generic name with a different one.

the new mammalian species in a series of reports published in the *Archiv für Naturgeschichte* (1842, 1843), in the *Abhandlungen der Akademie München* (1847–1849), and his supplementary volumes of Schreber's *Säugethiere* (1840–1855). Finally, Pelzeln (1883) brought together most, if not all, available taxonomic, descriptive, and geographic data in a single report. Natterer intended to work up the entire collection himself, but died within a few years of his return to Europe. His journal, with notes on habits, reproduction, and anatomy of the Brazilian animals collected, was lost.

Natterer collected 781 specimens of mammals, representing more than half (58%) of the currently known Brazilian genera and nearly as many (44%) of the species (table 3). Most poorly represented are bats, mice, and mouse opossums. Had Natterer been equipped with suitable traps and trammel nets known at the time but not used in fieldwork, he might have collected nearly all the mammalian genera and species now known to occur in Brazil. Still, his collection represented more species and included types of more new species than had been collected in Brazil by anyone else in the century, or possibly at any time.

The numbers of genera and species of mammals collected by Natterer, as identified by Pelzeln (1883), are listed in Table 7. The totals are compared with the numbers currently recognized, some genera having been increased and some species eliminated by synonymy. The revised numbers of genera and species are shown, in turn, as percentages of the estimated total numbers of currently known Brazilian genera and species of mammals.

VII. Guianas: Mammalogy to End of 18th Century

Pierre Barrere (1690–1755)

The physician, botanist, and correspondent of the French Royal Academy of Sciences, Pierre Barrere, resided three years (1752–1755) in Cayenne, with instructions to prepare a detailed report on the natural history of French Guiana. The work he finally published in 1741, however, is no more than an abbreviated glossary of Guianan minerals, plants, mollusks, fishes, reptiles, birds, and mammals. The list of mammals was uncritically compiled from Marcgraf and others. Species previously recorded by early chroniclers from the lower Río Orinoco region which occur throughout the Guianas but were not mentioned by Barrere are the golden handed tamarin (*Saguinus midas*), red brocket (*Mazama americana*), red acuchi (*Myoprocta exilis*), tayra (*Eira barbara*), white-lipped peccary (*Tayassu pecari*), and silky anteater (*Cyclopes didactylus*).

José Gumilla (d. 1750)

A natural history and geography of the Río Orinoco region in Spanish, published by Father José Gumilla, provides interesting, but largely erratic, descriptions of the countryside and human inhabitants, but nothing of interest regarding native mammals. Gumilla's explorations of the interior led him to deny the reported existence of a connection between waters of the Orinoco and Negro rivers.

Jacques Nicolas Bellin (1703–1772)

The description of the Guianan possessions of France, Spain, Holland, and Portugal, from the Orinoco River to the Amazonas River, by Jacques Nicolas Bellin, published in 1763, contains information on natural history, but adds nothing noteworthy to the then-known mammalian fauna.

Edward Bancroft (1744–1821)

The English physician Edward Bancroft lived three years in Dutch Guiana, now Suriname, practicing medicine and gathering notes for his *Essay on the Natural History of Guiana*. The work, published in 1769, deals broadly with plants and animals, but the author's knowledge of mammals was mostly limited to hearsay, although he also made some observations on animals brought to him by natives or seen in captivity or during short walks into the countryside. Persistent reports of the existence of apes or ape-men in South America were recounted by Bancroft (p. 130) in these terms:

> The Orang-Outang of *Guiana* is much larger than either the *African* or the *Oriental,* if the accounts of the natives may be relied on; for I do not find that any of them have been seen by the White inhabitants of this coast, who never penetrate far into the woods. These animals, in all the different languages of the Natives, are called by names signi-

TABLE 7. Numbers of mammalian genera and species collected by Johann Natterer in Brazil, 1817–1835, based on Pelzeln (1883), and compared with currently known totals.

Taxon	Number reported by Pelzeln (1883)	Current equivalent number	Total currently known for Brazil (estimated, 1984)	Percentage of current total collected by Natterer
MARSUPIALIA				
Genera	2	6	8	75%
Species	18	15	30	50%
CHIROPTERA				
Genera	10	28	60	47%
Species	48	40	125	32%
PRIMATES				
Genera	12	14	16	87%
Species	45	28	50	56%
EDENTATA				
Genera	10	10	12	83%
Species	16	12	15	75%
CARNIVORA				
Genera	11	10	14	71%
Species	17	14	25	56%
PINNIPEDIA				
Genera	0	0	2	0%
Species	0	0	2	0%
SIRENIA*				
Genera	1	1	1	100%
Species	1	1	2	50%
PERISSODACTYLA				
Genera	1	1	1	100%
Species	1	1	1	100%
ARTIODACTYLA				
Genera	4	4	5	80%
Species	7	6	7	86%
LAGOMORPHA				
Genera	1	1	1	100%
Species	1	1	1	100%
RODENTIA				
Sciuropmorpha				
Genera	1	1	3	33%
Species	5	3	6	50%
Myomorpha (Murinae excluded)				
Genera	3	5	20	25%
Species	17	17	45	24%
Caviomorpha				
Genera	11	15	23	65%
Species	24	22	47	47%
CETACEA*				
Genera	2	2	2	100%
Species	2	2	2	100%
TOTALS				
Genera	69	99	170	58%
Species	202	156	358	44%

* Fresh water only.

fying a Wild Man. They are represented by the *Indians* as being near five feet in height, maintaining an erect position, and having a human form, thinly covered with short black hair; but I suspect that their height has been augmented by the fears of the *Indians,* who greatly dread them, and instantly flee as soon as one is discovered, so that none of them have ever been taken alive, much less attempts made for taming them. The *Indians* relate many fabulous stories of these animals; and, like the inhabitants of *Africa* and the *East,* assert, that they will attack the males, and ravish the females of the human species.

Philippe Fermin (1720–1790)

Philippe Fermin, the author of an account published in 1769 of the history, geography, and natural objects of colonial Suriname, was one of those European men who "never penetrate far into the woods." Indeed, Fermin believed that all Europeans and Creoles were physically incapable of coping with the difficulties of surveying the natural fauna of the countryside, let alone the wilderness, or resisting the diseases generated by the "foul" air of forests and swamps. Notwithstanding this, Fermin compiled a fair list of the mammals. The didelphids included *Didelphis marsupialis, Philander opossum,* and *Marmosa* spp. All three kinds of anteaters and the two- and three-toed sloths are mentioned. The two native squirrels, *Sciurus aestuans* and *Sciurillus pusillus,* are distinguished. Other rodents are the capybara, paca, a porcupine, cavy, spiny rats or echimyids (most likely of the genera *Proechimys* and *Echimys*), and a water rat (probably *Nectomys squamipes*). The carnivores include tayra, otter, jaguar, puma, margay, ocelot, and two kinds of bush dogs (*Dusicyon thous* and possibly *Speothos venaticus*). Monkeys are *Saguinus midas, Saimiri sciureus, Pithecia pithecia, Chiropotes satanas, Cebus apella, Alouatta seniculus, Ateles paniscus,* and variants of some of them regarded as distinct species. African simians introduced with the slave trade and mentioned by Marcgraf are included. The ungulates are tapir, brocket (*Mazama americana*), and the collared and white-lipped peccaries.

Regarding white-lipped peccaries, Fermin affirms they form herds of as many as 300 individuals. Hunters, he states, tremble when they hear the sound of their clicking tusks. When attacked, only two avenues of escape are open: The first is a tree, if it can be climbed; the second and surest is standing ground and urinating, the odor of the urine, he affirms, being a powerful peccary repellant.

Monsieur Bajon (1763?)

The French physician, surgeon, and anatomist Bajon, with 12 years' residence in French Guiana, investigated climate, agriculture, natural history, and human diseases. The knowledge he gained was acquired firsthand, much of it new or supplementary to what was already contained in the encyclopedic volumes on natural history by Buffon and Daubenton.

In the second of his two-volume work, Bajon (1778, p. 178) declared that, contrary to popular belief, the jaguar feared man and did not attack without provocation. His accounts of habits and detailed descriptions of intestinal morphology and female genitalia of peccaries supplement Daubenton's (in Buffon) gross anatomy of a male collared peccary. Bajon clarified the differences between the agouti (*Dasyprocta leporina*) and acouchi (*Myoprocta exilis*). He described the male agouti penis, with its peculiar complement of spines, erectile spears, and sharp blades. Descriptions with life history notes are given for the chien sauvage (*Dusicyon thous*), eira (*Eira barbara*), and chien crabier (*Procyon cancrivorus*). Marsupials fascinated him, particularly the role of the pouch in females of the pean (*Didelphis marsupialis*), quatre-ouel (*Philander opossum*), and also the pouchless rat de bois (*Marmosa* sp.). The commonly held belief that each didelphid young is born and develops at the end of a teat was rejected by Bajon, but despite numerous observations and dissections, he failed to solve the mystery of marsupial birth.

Bajon's monographic account of the tapir (*Tapirus terrestris*) includes detailed, but not always accurate, descriptions of anatomy, reproduction, development, behavior, food, vocalization, hunt, and human utilization.

John Gabriel Stedman (1744–1797)

A soldier of the Scots Brigade of the Netherlands, John Gabriel Stedman arrived in Suriname in 1773 to help subdue the uprising of the African slaves. Most of the fighting was already over when he landed, so Stedman devoted much of his time to recording his observations of life in the country and wilderness. His *Narrative,* published in 1796 in two volumes, contains much on the natural history of Suriname, with illustrations by his own hand (fig. 9). The mammals, some only listed, others described, often with anecdotes, are the following. Stedman used local names, current binomials are in parentheses.

Volume I, p. 14. Narwhal (*Monodon monoceros*).
Sighted from shipboard at Devil's Island off Cayenne. ". . . appeared but six or eight feet in length, and its horn about four. . . . The narwhal . . . is more frequently found in cold than warm climates. The female is said to be unprovided with that protuberance so remarkable in the male. It appears that some authors have confounded this animal with the *sword-fish,* to which however it does not prove to have the very smallest resemblance."
The locality record for the circumpolar narwhal is unexpected, and no doubt erroneous. Nevertheless, Stedman's description is accurate albeit the dimensions given seem small. At the same time, Stedman provided a detailed description and good figure of a swordfish or sawfish to prove it was not a sawfish he saw!

Volume I, p. 153, pl. 16. Sicapo (*Bradypus tridactylus*).

Volume I, p. 153, pl. 16. Dago luyaree (*Choloepus didactylus*).

Volume I, p. 153. Ourang-outang.
"I have never seen, nor heard described, while I was in this country. . . ."

Volume I, p. 166, pl. 18. Micoo or mecoo (*Cebus apella*) (fig. 9).

FIG. 9. Animals from Suriname depicted by Stedman: **left**, mecoo (*Cebus apella* Linnaeus) and the smaller kishee kishee (*Saimiri sciureus* Linnaeus); **upper right**, tapir (*Tapirus terrestris* Linnaeus) with juvenal spotted coat; **lower right**, manatee (*Trichechus manatus* Linnaeus); from Stedman (1796).

Volume I, p. 166, pl. 18. Kessee-keesee or kishee kishee (*Saimiri sciureus*) (fig. 9).

Volume I, p. 167. Monkee-monkee (*Saguinus midas*).

"One morning I saw from my barge a monkey of this kind come down to the water's edge, rinsing his mouth, and appearing to clean his teeth with one of his fingers."

Volume I, p. 168. Tavous (*Lutra enudris*).

Volume I, p. 221. Sea-cow (*Trichechus manatus*).

About three in the morning while asleep in a boat, Stedman and a companion were suddenly thrown from their bunks.

"By the account of the negroes [a manati had] slept under the boat, which, by the creature's awakening, had been lifted up and thrown upon one side, and again replaced when the manati made its escape from underneath. I did not so much as see the creature, nor indeed hardly had the negro, owing to the darkness of the night."

Volume I, p. 222, pl. 24. Capasce (*Dasypus novemcinctus*); largest armadillo (*Priodontes maximus*).

Volume I, p. 223, pl. 24. Adjora (*Coendou prehensilis*).

Volume I, p. 224. Hedge-hog (spiny rat, family Echimyidae).

Volume I, p. 308, pl. 33. Bajew (*Odocoileus virginianus,* adult male and spotted fawn); boosee-cabritta (Negro), wirrebocerra (Indian) (*Mazama americana,* large spotted female, smaller spotted fawn).

Volume I, p. 347. Coney coney (Negro), puccarara (*Dasyprocta leporina*).

Volume I, p. 355, pl. 37. Pingo (*Tayassu pecari*).

"They live in herds of sometimes above three hundred and run always in a line, the one closely following the other; when the foremost or leader is shot, the line is instantly broken, and the whole herd is in confusion, for which reason the Indians take care (if possible) to knock their captain on the head before the rest; after this the others even often stand still, stupidly looking at one another, and allowing themselves to be killed one by one, of which I have been a witness. They do not attack the human species, or make any resistance at all, like the European wild-boar, when wounded, as has been by some authors erroneously asserted."

Volume I, p. 355. Peccary (*Tayassu tajacu*).

Volume I, p. 356. Cras pingo (*Sus scrofa*).

Volume II, p. 10, pl. 42. Quata or Quato (*Ateles paniscus*).

"Their throwing short sticks and excrements seems to be no more than a mimicking of the human actions without any purpose, as they neither have strength to throw far, nor dexterity to hit their objects, and if they befoul them it is by accident only. But what appears to be peculiarly remarkable is, that when one is hurt by a musket or arrow, the poor animal instantly claps its hand on the wound, looks at the blood, and with the most piteous lamentations ascends to the very top of the tree, in which he is assisted by his companions; where, hanging by the tail, he continues to bewail his fate, till by the loss of blood he grows totally faint, and drops down dead at the feet of his adversaries.

"It is not so extraordinary that one of this species, when wounded, should be assisted by one of his companions in climbing; but that they should have so much knowledge in botany, as to procure vulnerary herbs, and chew and apply them to the wound, is what I cannot credit, though it is so confidently asserted by a late traveller; and as to the assistance they give in passing a river, by holding each others tails, and swinging till the lowermost is thrown up to the branch of a high tree . . . I must take the liberty to doubt this fact. . . .

Volume II, p. 12. Wanacoe (*Pithecia pithecia,* male).

"This monkey is the only one of the species [monkeys] that is not sociable, being constantly found alone, and so despicable is this solitary animal, that he is continually beaten and robbed of his food by all the others, from whom he is too lazy to escape, though too cowardly to fight."

Volume II, p. 12. Saccawinkee (*Callithrix jacchus jacchus*).

[Common marmosets were brought from Brazil for the pet market. They are not native to the Guianan region.]

Volume II, pp. 16–17. Brown squirrel (*Sciurus aestuans*); white squirrel (*Sciurus aestuans,* albinotic); flying squirrel (probably mistaken impression of a leaping pygmy squirrel, *Sciurillus pusillus*).

Volume II, p. 40, pl. 46. Taibo, woodrat (*Myoprocta exilis*).

The description is better than the figure which suggests a doglike marsupial.

Volume II, p. 41, pl. 46. Crabbo-dago (*Galictis vittata*).
Volume II, p. 49, pl. 48. Tyger (*Felis onca*).
"It has even happened that the *jaguar* has carried off young negro women at work in the field, and too frequently their children."
Volume II, p. 50. Red tyger (*Felis concolor*).
Volume II, p. 50, pl. 48. Tyger-cat (*Felis pardalis*).
Volume II, p. 51. Jaguaretta (melanistic *Felis onca*).
"I have never seen one."
Volume II, p. 135. Cabiai (*Hydrochaeris hydrochaeris*).
Volume II, p. 142, pl. 57. Vampire or spectre (*Vampyrum spectrum*).
Figured are a flying bat and a side view of a truncated head that had been preserved in spirits. Stedman, while asleep, had been bitten on his toe by a true vampire bat, likely *Desmodus rotundus*. He had not seen his attacker, but like others believed most bats were vampires, particularly the larger species, most certainly the largest, *Vampyrum spectrum*.
Volume II, p. 144, pl. 47. Murine or mouse opossum (*Philander opossum*).
Volume II, p. 152, pl. 58. Paca (*Agouti paca*).
"Nothing can be better eating than the Paca or spotted Cavey."
Volume II, p. 153, pl. 58. *Agouti pacarara*, Indian coney (*Dasyprocta leporina*).
Volume II, p. 153. Indian rat-coney (*Myoprocta exilis*).
"This I never saw, unless it is the same animal . . . that I have described under the name of bush-rat."
Volume II, p. 175, pl. 59. Sea-cow or manatee (*Trichechus manatus*) (fig. 9).
Volume II, p. 176, pl. 59. Tapir (*Tapirus terrestris*) (fig. 9).
Volume II, p. 176. Mermaid.
"Major Abercromby . . . declared that a *mermaid* was lately seen in the River Surinam. Lord *Monboddo* also positively affirms the existence of sea-women and sea-men, while he asserts that they were seen so late as 1720. But, however respectable his lordship's judgement and authority may be on other subjects, I can no more agree with him, as to men and women, having fins and scales, than to their having *tails*."
Volume II, p. 235. Howling baboon (*Alouatta seniculus*).
Volume II, pp. 325–326. Awaree (*Didelphis marsupialis*).
Volume II, p. 327, pl. 74. Quacy-quacy (*Nasua nasua*).
Volume II, p. 328, pl. 74. Great ant-eater (*Myrmecophaga tridactyla*).
Volume II, p. 329. Tamandua (*Tamandua tetradactyla*).
Volume II, p. 329. Fourmillier (*Cyclopes didactylus*).

VIII. Guianas: Mammalogy of First Half of 19th Century

Sir Robert Herman Schomburgk (1804-1865) and Richard Schomburgk (1811-1891)

Robert Herman Schomburgk was born in Freiburg, Germany, son of a Protestant minister. In 1829 he went to the United States and in 1830 to Anegada of the Virgin Islands. His survey of the island, submitted to the Royal Geographical Society of London, won him the command of an exploring expedition to British Guiana (Guyana) in 1835.

Robert Schomburgk's accounts of his travels in the colony and bordering parts of Brazil and Venezuela during 1835–1839 were published by the Royal Geographical Society in its *Journal* for volumes 6 (1836), 7 (1837), and 10 (1840). The reports were translated into German by O. A. Schomburgk and published in 1841 as a single volume. This, in turn, was translated back into English by Roth (1931). A brief description of the colony by Robert Schomburgk was published in 1840. Some notes on natural history by Schomburgk were included in his reports to the Geographical Society; others appeared in several numbers of the *Annals of Natural History* (London) for 1840.

Upon the successful conclusion of his explorations in 1839 and return to England, Robert Schomburgk was commissioned in 1840 by the government to survey the colony and fix its boundary with Venezuela. He was knighted in 1845 after his return to England.

Richard Schomburgk, with the patronage of the King of Prussia, accompanied his older brother on the second journey to British Guiana. Plants and animals collected by the expedition were sent to the Berlin museum for scientific study where they were examined by Richard Schomburgk and other specialists; the mammals were studied by Schomburgk and Cabanis. Richard Schomburgk's three-volume account in German of the travels from 1840 to 1844 was published 1847–1848. Roth's English translation of the first two volumes appeared in 1922–1923.

As a result of the Schomburgk expeditions, British Guiana advanced from a practically unknown South American country to one of the then best known for its geography, biota, and ethnology. Virtually all major physical features of the Guianan region, from the Corentyne River between British and Dutch Guiana (Suriname), west across the colony and headwaters of the Rios Branco and Negro in Brazil to headwaters of the Río Orinoco in Venezuela, were traversed, described, and mapped in detail (fig. 10). The reported observations on mammals are as good and as welcome today as when first published.

The following excerpts of observations on mammals taken from the three volumes of the *Reise* originated with Richard, Robert, or both. Those from the earlier published *Annals* are Robert's. The mammals were first identified by Cabanis. Current scientific names, however, are used here with the Cabanis synonyms in parentheses. None of the animals were figured in the original works cited.

MARSUPIALIA

Didelphis marsupialis marsupialis Linnaeus (*Didelphys cancrivora*)
Reise I:192 (human breast-fed young).
Annals 5:343. "... if we could reconcile the geographical distribution of *D. virginiana* over a space so different in temperature, I should consider the specimen [of *D. marsupialis*] I am now describing a variety of that species; the circumstance that the ears are of uniform black would scarcely constitute a specific difference."
Reise III:777 (behavior in captivity).

Philander opossum opossum Linnaeus (*Didelphys quica*)
Reise III:777 (distribution).
Annals 5:344 (description; habits).

Caluromys philander philander Linnaeus (*Didelphys philander*)
Reise III:777 (distribution).
Annals 5:344 (description; habits).

Marmosa murina murina Linnaeus (*Didelphys dorsiguera, D. musculus* Cabanis)
Reise III:777 (distribution; characters; habitat).
Annals 5:345 (description; habits).

Lutreolina crassicaudata turneri Gunther
Reise III:777 (predation).

Chironectes minimus minimus Zimmermann (*Chironectes variegatus*)
Reise III:777 (distribution).

CHIROPTERA

Molossus molossus Pallas (*Molossus obscurus*)

Tonatia bidens Spix (*Phyllostoma bidens*)
Reise III:772 (habitat; colony size; characters).

EDENTATA

Myrmecophaga tridactyla tridactyla Linnaeus (*Myrmecophaga jubata*)
Reise II:44, 214, 223, 374 (characters; habits; defense; chase; flesh).
Reise III:782 (distribution).
Annals 4:203–207 (characters; habits; capture):

"The young Ant-bear was quite wild at first, and sought for some dark corner in the room in which it was confined, in order to hide itself. When we approached it, it put itself immediately in defense like the adult ones, and struck out with its right paw, uttering at the same time a growl like that of an incensed puppy. After a few days, however, it became accustomed to its situation, and an Indian woman took upon her to feed it with milk and Cassada [cassava] and sometimes White Ants. It soon showed great attachment to her and followed her like a dog.

"It appeared to be of a very cold nature; not only the extremities, but the whole body felt cold to the touch, although we kept it wrapped up in a blanket. It preferred, however, to be nestled, and to be taken up, and on putting it down it uttered a whining but not unpleasant sound; when it did not succeed in attracting attention, and was not taken up again, the whining sound was raised to a harsh and grating noise. In following a person, it directed its course more by the smell than by sight, and carried its snout close to the ground. If it found itself at fault, it wheeled round at right angles upon the hind legs, and snuffed the air in all directions, until it found the right scent again. Of the dimness of its sight we had various proofs; it hurt itself frequently against objects that stood in its way, not observing them until it came in contact with them. Its power of smelling was exquisite, and it could discover its nurse, or any person to whom it had taken a liking, at a considerable distance. Upon these occasions it would immediately commence the whining sound so peculiar to

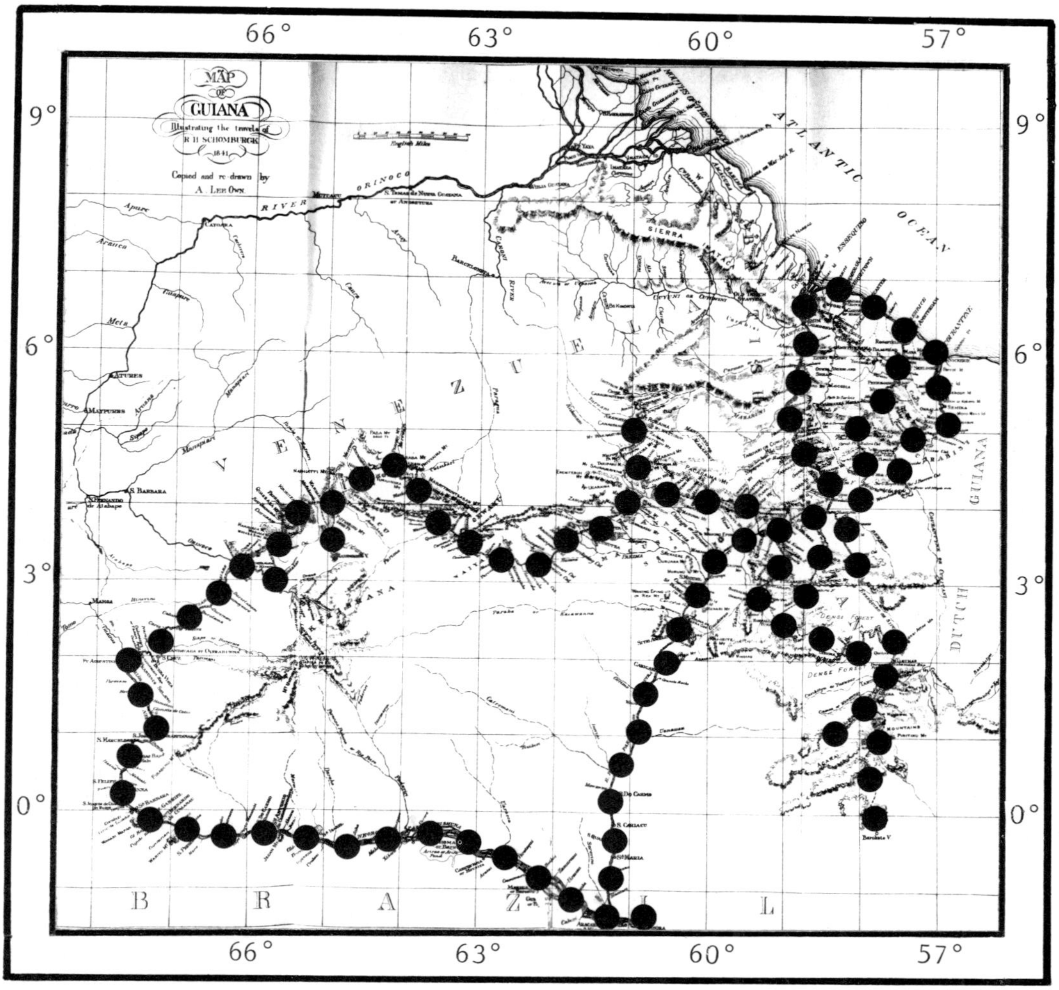

FIG. 10. Map of British Guiana (Guyana) and bordering parts of Venezuela, Dutch Guiana (Suriname), and Brazil. Robert Herman Schomburgk's routes and surveying areas (1835–1839) shown by large dots. The map (without the dots) was copied and redrawn by A. Lee Owen for the Roth translation of R. H. Schomburgk (1841).

this animal. It was an expert climber; it happened that I was one of its favourites, and whilst writing on my table it used to come softly behind me, and as soon as it was sure it had found me out, it climbed up my legs with great dexterity. It showed its attachment by licking, and was very gentle and even sportive; we all prized it highly. . . .

"It secretes a liquid substance, transparent like water, which drops down almost constantly out of its nostrils and mouth; this is the more remarkable, as it used very little water. . . ."

Tamandua tetradactyla tetradactyla Linnaeus (*Myrmecophaga tamandua*)
Reise III:782 (distribution).

Cyclopes didactylus didactylus Linnaeus
Reise III:782 (distribution).

Bradypus tridactylus Linnaeus (*Bradypus gularis*)
Reise I:142, 258, 455 (capture; swimming; climbing; mother-infant).
Reise III:781 (distribution).

Bradypus variegatus Schinz (*Bradypus torquatus*)

Priodontes giganteus É. Geoffroy
Reise II:97 (characters; flesh).
Annals 5:32–33 (habits; description).

Dasypus novemcinctus novemcinctus Linnaeus (*Dasypus peba*)
Reise II:24, 29 (excavation).
Annals 5:34 (description; habits; reproduction).

Cabassous unicinctus unicinctus Linnaeus (*Dasypus tatouay*)

Reise III:782 (distribution).
Annals 5:34 (description).

Euphractus sexcinctus sexcinctus Linnaeus (*Dasypus encoubert*)
Reise III:782 (distribution).

[?]*Dasypus sabanicola* Mondolfi (*Dasypus minutus* not Desmarest [= *Zaedyus pichiy* Desmarest])
Reise II:491. "Of the fauna of the Sandhills [formation across Guyana 2 to 40 miles from coast], the genus *Dasypus* seemed to be the most numerous among the mammals and of the species present in Guiana three are found on the sandhills alone: *Dasypus Peba* Desm. ([= *D. novemcinctus*] lessy of the Arawaks); *D. minutus* Desm. (lessy Barakatta of the Arawaks); and *D. tatouay* Desm. [= *Cabassous unicinctus*]."

[?]"*Dasypus villosus*" (*Chaetophractus villosus* Desmarest [not Guianan]
Reise II:24. "One of the boys brought me an armadillo (*Dasypus villosus* Desm.) which he had surprised on his way across the savannah [south of the Kanuku Mountains]."
Reise II:97. "The sharp eyes of a Wapisiana again noticed something alive moving about in the savannah below; he quickly ran to the spot and soon returned carrying another but smaller [than *Priodontes giganteus*] armadillo by the tail. It was *Dasypus villosus* Desm. According to the statements of the Indians this species is particularly distinguished by a peculiar growth of hair that covers not only the body but also the plates on the back, is solely present in the savannahs, and for the most part lives on carrion . . . a characteristic that is ascribed only to this one species amongst the seven met with in Guiana."
Reise III:782 (distribution).
Annals 5:34. "The savanna armadillo is Desmarest's *Dasypus villosus*; and, as we were assured by the Indians, it inhabits only the plains, and is never to be met in the forest, the Indians accuse it of feeding occasionally on carrion."

PRIMATES

Ateles paniscus paniscus Linnaeus
Reise II:93. "One finds them mostly in companies of 16 to 20; often also in lesser number. I never noticed them on the ground but always on the highest trees. When exposed to the full rays of the sun, they lie at full length stretched out on the branches, to bathe themselves in it."
Reise III:767 (troop size; reproduction; mother-infant).

Alouatta seniculus seniculus Linnaeus
Reise I:278, 352 (vocalization; habits; flesh).
Reise III:768 (characters; distribution; vocalization; social organization; mother-infant).

Cebus apella apella Linnaeus (and other monkey species)
Reise I:354 (sociability; troop size, 400–500 individuals).
Reise II:247. "It is only in the Canuku Ranges that I can call to mind having met troops of monkeys that consisted solely of *Cebus apella*: their haunts seem generally limited to particular localities because except in the Ranges just mentioned, I have only seen them on the coast and then always among *C. capucinus* [*C. nigrivittatus*] with which the neat little *Callithrix* [= *Saimiri*] *sciurea* had also often associated itself. I invariably found *Mycetes* [= *Alouatta*], *Ateles, Pithecia* and *Hapale* [= *Saguinus midas*] absolutely separate from one another and even among *Pithecia leucocephala* [= *P. pithecia*] never a specimen of *Pithecia* [= *Chiropotes satanas*] *chiropotes*."
Reise III:768–770 (characters; troop size; behavior; urine washing; tool use): "I placed some fruit near the chained monkey out of arms reach so he tried to sweep it nearer with his tail. This failing, he searched around as far as he could and found a stick and with it managed to roll the fruit to himself."

Cebus nigrivittatus olivaceus Schomburgk (*Cebus capucinus* not Linnaeus)
Reise I:247, 437 (variation; mother-infant; domestication).
Reise III:770 (most common and widely distributed Guianan monkey).

Saimiri sciureus sciureus Linnaeus
Reise I:333 (social relations; mother-infant).
Reise II:247, 366 (associations).
Reise III:770 (distribution; not viable in captivity).

Chiropotes satanas chiropotes Humboldt (*Pithecia chiropotes, P. satanas*)
Reise I:351, 352 (description; social relations; flesh).
Reise III:771 (distribution).

Pithecia pithecia pithecia Linnaeus (*Pithecia leucocephalus*)
Reise I:352 (social relations).
Reise III:771 (troop size; distribution).
Royal Geographical Society of London, 6:265 (1836):

". . . numerous monkeys jumped from branch to branch, and, astonished at the uncommon visit, accompanied us for a considerable distance [along the banks of the upper Essequibo River]. Our Caribbees called this species arieghi, or yahriae; the male has straight long hair of a shining black, the head rather round, the forehead and part of the face and neck covered with short, yellowish hair, part of the front, the nose, and mouth black, the latter slightly bearded, hands black, nails claw-like, except the thumb. The female is different in colour, and her fur resembles that of the European hare; her hands are likewise black, and covered with short yellowish hair, from under the eyes to the chin extends hair of a similar colour, but somewhat longer than those of the front and cheeks, the breast is nearly naked, and the oshyoides [(*sic*) oschyoides or scrotal pad] visible. They jumped with great agility from tree to tree, the female and sometimes the male carrying the young ones upon the back. . . ."

The strongly marked sexual dichromatism described in 1836 by Robert Schomburgk was not discerned by taxonomists until late in the century. Twelve different names had been bestowed on *Pithecia pithecia*, five of them based on males, the others on females.

Aotus sp. (*Nyctipithecus trivirgatus*)
 Reise II:460 (house pet seen at Asacota, Barimani River, NW Br. Guiana).

Saguinus midas midas Linnaeus (*Midas rufimanus*)
 Reise II:366, 367, 505 (distribution; behavior).
 Reise III:772 (distribution; vocalization; captivity).

CARNIVORA

Nasua nasua vittata Tschudi (*Nasua socialis*; *Nasua solitaria*)
 Reise II:247–248. "The new *Nasua* I discovered here . . . suffered a strange fate in its identification . . . we took it for a new species, but unfortunately possessing too few natural-history books to confirm our subspecies, forwarded it to Berlin with the next assignment undescribed. I was accordingly all the more surprised to find that very same *Nasua* determined as *N. vittata* by von Tschudi in his *Untersuchung über die Fauna Peruana*. The specimen was shown him on its arrival and he, recognizing it as new, took the required notes, and before it was yet described in Berlin, published it in his Fauna Peruana, although it does not occur there."
 Annals 5:431–432:
 "They live in large societies, and know how to defend themselves bravely if attacked by dogs; indeed they fall often *en masse* upon them and kill the assailants. They are excellent climbers, and in descending a tree they always come down head foremost. Their food consists of insects, fruits, roots and such small prey as they are able to secure. They are destructive to young birds, and expert in digging after large beetles, for which their claws, which are very strong, are admirably adapted. They do not burrow in the ground for a residence."

Procyon cancrivorus cancrivorus Cuvier
 Reise II:443 (behavior).
 Annals 4:433–434:
 "Although the Racoon [*sic*] is not an animal which inhabits the savannahs, its relation to the preceding genus induces me to give now the few particulars which I know about its habits. It frequents the sea coast, and is generally found in the neighbourhood of inhabited spots, where it is destructive to poultry.
 "Among the favourite haunts of these animals are the thickets of Curida bushes (*Avicennia tomentosa*), which extend along the sea coast, where they feed upon crabs which they are expert in killing, first tearing off their claws or nippers; and being thus disabled from doing harm, the crab dog or racoon uses its sharp teeth to break the shell. In their native state they sleep by day, and issue at dusk in search of food; birds, insects, roots, and vegetables, nothing comes amiss; and as they possess a particular fondness for sweets, I have been told by practical planters that the injury which they do to sugar plantations is very considerable.
 "They take their food with both paws like the squirrel, and are fond of dipping it in water. I have noted with astonishment that they drink as well by lapping like the dog as by sucking. I have had several in a domesticated state, all of which possessed this peculiarity."

Potos flavus flavus Schreber (*Cercoleptes caudivolvulus*)
 Reise II:435 (habits; food; predation).
 Annals 5:29 (habits; distribution).

Eira barbara poliocephala Traill (*Galictis barbara*)
 Reise II:99 (chase; characters).
 Annals 5:30 (habits; distribution): ". . . like the coati or *Nasua*, are able to run down a tree

. . . head first. They are sometimes tamed and are thus gentle and playful; but they are easily excited, and when preparing for defense or war they erect the hair of their tail."

Galictis vittata vittata Schreber (*Galictis allamandi*)
Reise II:447 (characters).
Annals 5:31–32 (habits).

[?]*Lutreolina crassicaudata* Desmarest, or perhaps *Mustela africana* Desmarest (*Mustela brasiliensis*?)
Reise III:775: "I have only a few incomplete stuffed specimens found among the Arekuna Indians who wore them as ornaments."

Lutra enudris enudris F. Cuvier (*Lutra enydris* [*sic*])
Reise I:340 (encounter).
Reise III:775 (distribution).

Pteronura brasiliensis brasiliensis Zimmermann (*Pterura* [*sic*] *sambachii*)

Otters (*Lutra* and/or *Pteronura*)
Reise I:340; II:35 (habits).
Annals 5:284–285 (habits):

"We watched a pack of Otters at the Great Cataracts of the Corentyn, where, at the basin which one of the cataracts formed, they appeared to carry on their pursuits with great success. One had secured a Haimura at least from ten to twelve pounds weight, and carried it in its mouth to a rock which was partly over water. Here it began devouring its prey without taking much notice of us, although we were not twenty yards from it on the opposite shore. It did not care for our shouting; its success was however disputed by the Indians, who got into the canoe and paddled so rapidly towards the rock, that the Otter saw itself obliged to retreat and to leave the better half of the fish to the Indians. Although the Otters were numerous round the rock, none of them showed any intention to share the prey with the successful hunter or to dispute its possession.

"I have already alluded to their having their holes on the edge of rivers, sheltered by the impending bank. Every rock in the vicinity of their residence bears the mark of their excrements; and their feeding-places are so devoid of vegetation, if we except the larger bushes and trees, that they cannot be mistaken, even if the number of scales and fish-bones did not point out the frequency of their visits. A complete path leads up to these places, which, in consequences of their ascending and descending in single file, is hollowed out.

"The young remain for a considerable time under the protection of their parents, the mother teaching them to plunge and dive at approaching danger.

"We had entered the upper Essequibo by its tributary the Cuyuwini, and passed at the foot of a ridge of mountains, when we observed on a large ledge of rocks a family of Otters, consisting of about fifteen, including old and young. At our approach they broke out into their peculiar noisy cry, and the parents seizing the young with their mouth they plunged into the water and disappeared,—and having placed their young in security, we saw them shortly after appearing at the head of our canoe. They raised themselves with half their body out of the water, snoring for rage and showing their formidable teeth. At approaching danger or when apprehensive of it, they collected in a body, deputing the most courageous in advance; as our canoe came nearer, they sank under as if by a preconcerted sign, and appeared the next moment within a few yards of it. We saw nothing again of the young; but the adults and larger-sized young ones accompanied us, threatening and snoring, until no doubt we were so far out of reach of their stronghold that they considered their progeny now safe. In other instances, when we attempted to find out their holes, they became so outrageous that they bit our paddles and left the print of their teeth. The Indians know nevertheless how to surprise the young ones, who are then taken home alive, and become in a short time so tractable that they follow their masters like dogs."

Dusicyon thous thous Linnaeus (*Canis cancrivorus*; *Canis azarae*)
Reise II:196, 338 (habits).
Annals 4:430–431 (characters; habits):

"They vie in cunning and art with the European fox, and the depredations which they commit on the hen-roosts are considerable. Their favourite haunts are thickets near open savannahs, and if a pack succeed in entering the village and in surprising the Indians' poultry, few escape, as they completely surround the roosting-place, and generally carry off their spoil before the inhabitants have any idea of their presence. I have been assured by the Indians that they soon run down deer, and pursue their game under full cry. They destroy in other ways large quantities of game. . . . They seldom lose, even when domesticated, their depredatory habits, and those Indians who raise them for the sake of procuring a cross breed with the dog, are obliged to keep them tied, as otherwise, they would kill all the fowls and parrots. It is called by the Macusis *Maikang,* in Warrau *Warityou.*

"The variety which has sprung from the breed between the Indian domestic dog and the *Carasissi* more resembles the dog, its body is however longer in proportion to its size, and its ears are pricked up. Their progeny become prolific. They are hardy, and many of them prove excellent hunters; they are therefore very much prized by the Indians, who pay great attention to their training."

Reise III:775 (distribution).

Felis concolor discolor Schreber

Reise II:86 (characters; predation).

Annals 4:325–326 (characters; habits; predation):

"It is very destructive to the cattle farms, and it is so powerful an animal, that I have been told by an eye witness, that it killed a mule and dragged it across a trench to the opposite side, although the trench was not quite full of water, and the Puma had to drag it a few feet up hill, after it landed with its prey on the other side. My informant, who had watched its proceedings, had meanwhile sent for his gun, and shot him while attempting to pull the mule into the wood. They seem to be particularly partial to dogs, and a great number of those which are kept by the settlers for the purpose of hunting, are killed and eaten by them. They follow in the woods the herds of Peccaries, and watch their motion in order to seize upon the stragglers, being well aware that if they attacked the flock, they would be overpowered and torn to pieces. They hunt as well by day as in the night, and feed also on deer and the smaller domestic animals. They give birth to two young ones, seldom three, which have spots of a darker hue, more or less visible, according as the lights fall upon them, and which I have been told they lose after the first year. . . ."

Felis onca onca Linnaeus (*Felis nigra*)

Reise I:436 (encounter).

Reise II:34, 85–90, 504 (encounter; characters; predation; distribution; vocalization; artifacts of teeth and hides): "Except during the period when the female has her young, the jaguar does not seem to possess any particular lair. . . . It swims over the widest rivers. . . . When circling round a camp or cattle-pen, it is always with a continual purring; not until hunting at night for its prey does it set up a frightful roar, that booms through the whole forest."

Annals 4:262–263:

"I consider the number of wild cattle scattered over the savannahs at about 4000, but I doubt whether they are on the increase, as man and jaguars commit fearful ravages among them. . . . Their most deadly enemy is the greater jaguar, *Felis onca,* Linn., which hovers in such quantities about Fort San Joaquim, that during the month of June 1838, twelve individuals were killed by the cattle-drovers. They are very daring, and sometimes kill cattle within a few yards of houses that are inhabited. They care very little for the fires which are made to prevent their encroachments. If one or a pair of these animals would take up their quarters in the vicinity of a cattle farm, scarcely a night passes in which they do not commit ravages. They do not eat much of any they kill, perhaps ten or twelve pounds, and principally of the breast; but they prefer killing fresh every time they are hungry. When out of the reach of cattle farms or the wild herds of the savannahs, they subsist on Peccaris, Capybaras, Tapirs, and Deer. . . ."

Felis pardalis melanurus Ball [or ♀ *Felis onca onca* Linnaeus]

Reise II:83 (characters; predation).

Annals 4:263:

"Not less destructive is the Turtle-tiger, a species or variety of the former [*Felis onca*]. They are of the same strong build as the greater jaguar, and very much resemble it both in form, colour, and disposition of its spots, but they are about a third less in size. In the vicinity of human habitations they commit great ravages among domestic animals; Hogs, Sheep, Goats, &c. are alike exposed to their attacks, but I never heard of an authenticated instance of their attacking man, although they will come boldly to his habitation, and even enter the houses and carry away the dogs from the fireside."

Felis tigrina tigrina Schreber

Felis wiedii vigens Thomas (*Felis macroura*)

Reise I:85 (characters).

Felis yagouaroundi yagouaroundi É. Geoffroy (*Felis jaguarundi*; *F. unicolor*)

Reise II:227 (encounter).

Annals 4:327 (description; predation).

RODENTIA

Sciurus aestuans aestuans Linnaeus
Reise II:491 (characters).
Reise III:778 (distribution).

Echimys armatus armatus I. Geoffroy (*Echinomys hispidus* not Desmarest)
Reise II:498–499: "The strange hedgehog *Echinomys hispidus* Geoffr. seems most plentiful, especially in the neighborhood of the small forest streams. It appears to reside upon the trees: at least I have never come across it on the ground. In climbing and springing from branch to branch it can vie with the smartest squirrel. The female drops 4 young in the hollow limb of a tree, and these soon follow at their mother's heels: they constitute a special dainty for the Indians. It seems to be spread all over British Guiana, because I at least found it everywhere."

Echimys chrysurus chrysurus Zimmermann
Reise III:779 (distribution).

Coendou prehensilis prehensilis Linnaeus (*Cercolabes insidiosus* not Kuhl)
Reise III:779 (habits).

Dasyprocta leporina cayana Lacépède (*Dasyprocta agouti*)
Reise II:80 (food).
Reise III:779 (distribution; predation; chase).

Myoprocta acouchy acouchy Erxleben (*Dasyprocta acuchy*)
Reise III:779 (distribution).

Agouti paca paca Linnaeus
Reise II:491, 492 (chase).
Reise III:780 (distribution; habitat; food).

Hydrochaeris hydrochaeris hydrochaeris Linnaeus (*Hydrochoerus capybara*)
Reise I:418: "Among the many domesticated animals met with at the settlement [was a] full grown water-haas. The creature was so tame that it regularly stuck to the heels of the women. Although the river Nappi flowed past the houses not fifty paces away, it never visited its favorite element otherwise than in company with the women when they went to draw water and even then only to drink."
Reise II:29:
"I often found 6 to 8 of them together [along the Essequibo River] forming a line in the middle of which the young were to be seen. But unless we killed it outright the wounded animal every time escaped us by immediately rushing into the water the neighborhood of which it seldom left."
Reise III:780 (distribution).

Cavia porcellus guianae (*Cavia leucopyga* Cabanis not Brandt)
Reise II:249:
"Six to eight living specimens would often be brought to us but without our being able to keep them alive. The Indians' statement that they could never by any manner or means be tamed, was confirmed. Had we ten or twelve together, none would be alive after the third day. They live in holes out of which they are driven by pouring water in, and then easily caught. . . . Its silky fur is attached so delicately to the skin that even the slightest touch of the hand knocks it off and leaves a bare space."

PERISSODACTYLA

Tapirus terrestris terrestris Linnaeus
Reise II:167 (chase), 169: ". . . since I could not override the definite instructions given me not to forward any skins of the larger mammals to Berlin, I handed the hide over to the Indians to make sandals of. I prepared the skeleton for the Anatomical museum."
Reise III:783 (distribution; habitat; forage; jaguar; flesh).

ARTIODACTYLA

Tayassu tajacu patira Sonnini (*Dicotyles torquatus*)
Reise II: 100, 164 (description; habits; chase).
Reise III:783 (distribution; characters).
Annals 5:401 (description; habits; chase).

Tayassu pecari pecari Link (*Dicotyles labiatus*)
Reise II:98, 164 (habits; chase): "June and July would seem to be the time when they drop [give birth]."
Reise III:784 (distribution; herd size).
Annals 5:402 (description; habits; chase).

Mazama americana americana Erxleben (*Cervus rufus*)
Reise II:57 (ectoparasites; habits).
Reise III:784 (distribution).

Mazama gouazoubira nemorivaga F. Cuvier (*Cervus simplicicornis*)
Reise III:785 (distribution; species not seen).

Mazama sp.? [= *M. gouazoubira*?] (*Cervus humilis* not Bennett [= *Pudu puda* Molina])
Reise II:58. "The fourth and smallest species is known under the name of Wilibisiri (*Cervus humilis*?): its home is also in the dense forest."
Reise II:363. "In the evening the hunters brought us [in camp at mouth of Aripai, upper Ru-

pununi, Kanaku Mts.] one of those pretty deer which the Indians call Walibisiri. It is the smallest species met with in Guiana, hardly 1⅓ ft. high."

Odocoileus virginianus gymnotis Wiegmann (*Cervus savannarum* Cabanis and Schomburgk; *Cervus mangivorus*)

Reise II:57. "The female must throw her young in March or April because we found amongst our lot four specimens very advanced in pregnancy; but as I have killed deer in a similar condition during September or October, they must either throw twice a year, or else they are not usually limited to any fixed breeding season. The deer is never present in the forests."

Reise II:157. "In cutting up the venison [secured in savannas of Rio Cotinga, upper Rio Branco] we found does well advanced in pregnancy, which helped to strengthen my previously expressed opinion that they either throw twice, or else have no particular pairing season."

Reise III:785 (description; distribution).

SIRENIA

Trichechus inunguis Natterer (*Manatus australis* not Tilesius)

Reise II:141, 156:

"The *Peixe Boys,* as the vaqueiros [cowhands] call the Sea cow (*Manatus*) had already left the neighborhood of the Fazenda [Rio Branco above Fort São Joaquim] several days before, the water having commenced falling; that during high water they usually travel up as far as the mouth of the Maku which so many had visited this rainy season, and that ten had been harpooned. . . . As soon as the Takutu begins to fall a few feet, the Manatis disappear and make their way back to below the rapids of the Rio Branco. The search for more abundant food probably brings them to the Takutu where their favorite grasses, species of *Panicum* and *Paspelum,* grow in abundance."

[I have observed that when the river drops a few feet and manatees cannot reach forage growing on the edge of the embankments, they move elsewhere, usually downstream.]

CETACEA

Inia geoffrensis Blainville (*Delphinus amazonicus*)

Reise II:18: "They would not only raise their pointed snouts out of the water but mostly a large portion of their seven to eight foot long body."

Reise III:786 (Rio Tacutu, upper Rio Branco, Brazil, near Guianan border).

IX. Alexander von Humboldt (1769–1859) and Aimé Bonpland (1773–1858)

Alexander von Humboldt and Aimé Bonpland were rigorously trained scientists highly qualified to survey the natural resources and native peoples of a major part of tropical America. Their investigations and discoveries in the New World from 1799 to 1803 resulted in numerous publications of primary importance.

Alexander von Humboldt was born into a wealthy and distinguished family and could pursue his cultural interests without stint. His studies in the arts and sciences prepared him to develop into one of the most innovative and versatile scientific investigators of his time, if not all time. He was at once botanist, zoologist, anthropologist, ecologist, geologist, cartographer, biogeographer, physicist, chemist, astronomer, demographer, historian, mountain climber, poet, artist, and linguist. He excelled in every field and gained recognition and prominence in all. Humboldt raised geography to a science. Knowledge of the fundamental principles of climatology is due to him. Last but not least of his many talents appears in Humboldt's writings, which inspired a generation of naturalist-travelers, including Charles Darwin.

The young Humboldt's greatest desire was for an opportunity to apply his skills, knowledge, and the scientific instrumentation accumulated at his own expense to the exploration of little-known lands. After disappointing starts on a number of prospective expeditions, he visited Spain in June 1799 accompanied by the young French botanist Aimé Bonpland. While in Madrid he had the good fortune to meet an influential friend who helped him secure royal orders for travel throughout the Spanish colonies in America to study natural resources and collect samples of scientific interest.

Humboldt and Bonpland sailed for South America on 5 June 1799 and landed 16 July 1799 at Cumaná, capital of Nueva Andalucía (Venezuela). The remainder of that year and part of the following were spent in exploration of the coastal region. Of prime interest, however, was the planned

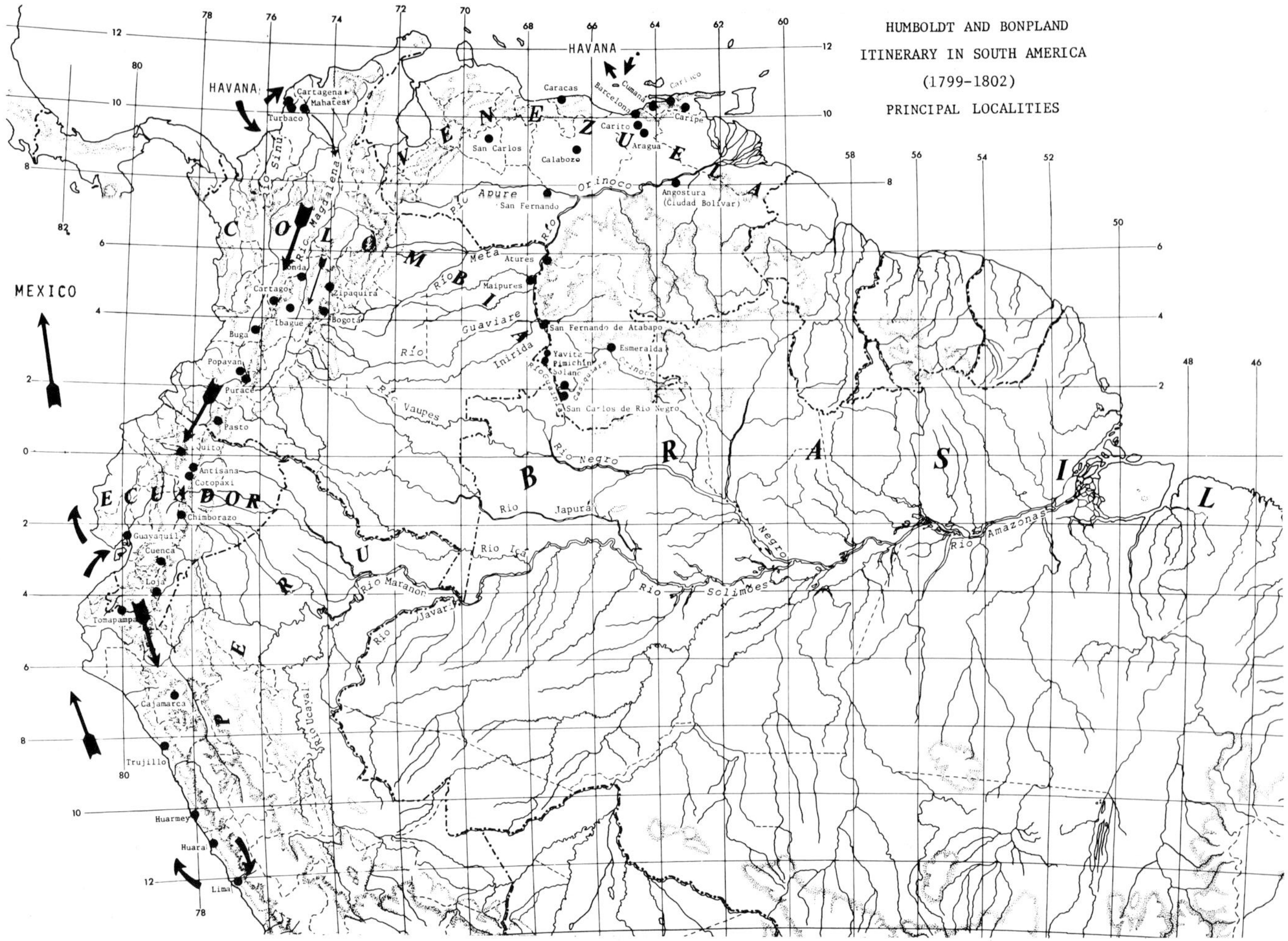

FIG. 11. Map of northwestern South America showing itinerary of Humboldt and Bonpland in Venezuela, Colombia, Ecuador, and Peru (1799–1802); principal stations plotted, arrows indicate direction of travels.

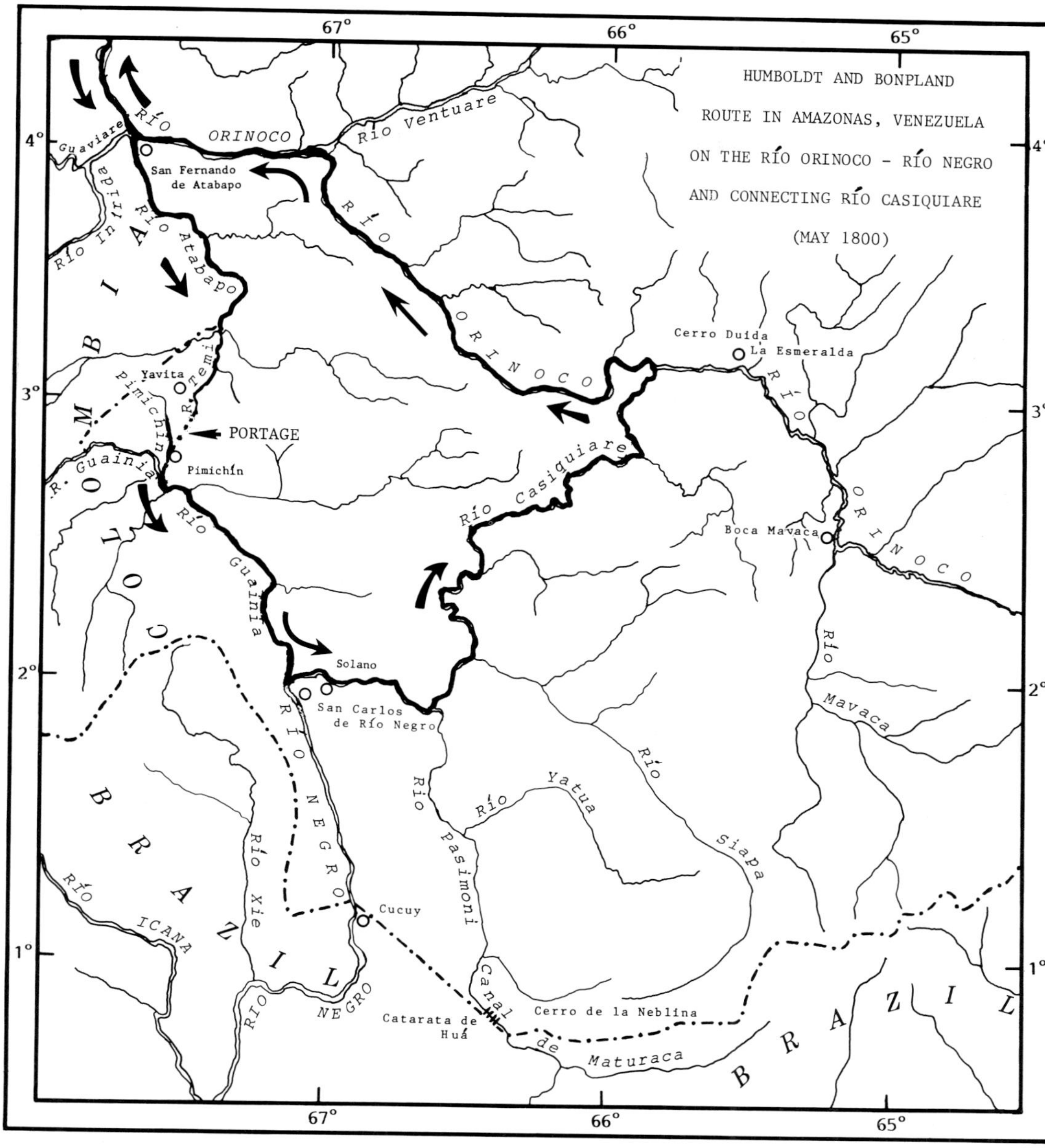

FIG. 12. Route of Humboldt and Bonpland in Amazonas, Venezuela, from the Río Orinoco-Atabapo to the Río Guainia-Negro via portage between the Río Temi and Río Pimichín and the Río Casiquiare connecting the Negro and Orinoco.

expedition up the Río Orinoco for verification of its reputed connection with the Amazonian Rio Negro. The exploration began on 27 March 1800 with a three-day inspection of a western tributary, the Río Apure. The journey then continued up the mainstream to the Spanish mission of San Fernando de Atabapo near the confluence of the Ríos Atabapo and Guaviare with the Orinoco. At this point, the travelers left the Orinoco and continued up the Atabapo to the tributary Temi, which they followed to the tiny mission of Yavita, arriving on 1 May. On 10 May, after portage to the Río Pimichín, a tributary of the Guainia, they attained San Carlos de Rio Negro at the mouth of the Río Casiquiare. The next day they headed up the Casiquiare and, after 10 days' travel by water, reentered the Orinoco on 21 May (figs. 11–12).

Having confirmed the connection between the

waters of the Orinoco and Amazon rivers, the explorers shipped 750 miles downstream to arrive at Angostura (Ciudad Bolívar) in mid-June 1800. After more work on the coast, Humboldt and Bonpland departed Venezuela on 24 November 1800 for Havana, Cuba. They remained there until 17 March 1801, then sailed for Colombia with landings along the Río Sinú on 25 March and Cartagena on 30 March. The journey thereafter was devoted mainly to explorations of the Cordilleras of Colombia and Ecuador, then through mountains, deserts, and the upper Amazonia of Peru south to Lima. The few mammals observed or described during this part of the journey are mentioned in Humboldt's (1805–1811) *Recueil.*

From Lima, Humboldt and Bonpland embarked on 24 December 1802 for Guayaquil and left 15 February 1803 for Mexico.

Humboldt's lively *Personal Narrative* evokes visions of Venezuelan life and landscapes from coastal plains to the headwaters of the Río Orinoco. The narrative is replete with descriptions of geography, ecology, astronomical orientations, widths, depths, and volumes of rivers, histories, languages and customs of Indians, Catholic missions, missionaries, and the human interest trials and tribulations of the travelers. Information on mammals, however, is comparatively meager, but some interesting bits can be quoted or paraphrased from the Ross translation of the original French (Humboldt, 1884).

Humboldt and Bonpland found manatees abundant in the Río Orinoco and tributaries Meta and Apure, but absent above the cataracts of Maipures. Some of the animals they caught were 10 to 12 feet long and weighed 500 to 800 pounds. Humboldt's dissection of one (fig. 13) revealed "no vestige of nails on the external surfaces of the fins which were quite smooth, but little rudiments of nails appear at the third phalanx when the skin of the fins is taken off." The lungs, they observed, consisted of "large cells resembling immense swimming bladders; they [the lungs] are 3 feet long. Filled with air they have a bulk of more than a thousand cubic inches [Humboldt, Ross translation, 1884, vol. II, p. 169]." Its distinction from *T. manatus* was not appreciated, however, until 1883 when described by Natterer (in Pelzeln, 1883). There is also considerable doubt that a clawless manatee does occur in the Río Orinoco basin or anywhere outside the Amazonian watershed.

Dolphins (*Sotalia*) were seen above and below the great cataracts of the Orinoco and often swam alongside the canoe. In the inundated forest of the divide between the waters of the Orinoco and Negro, the travelers "were astonished by an extraordinary noise. On beating the bushes a shoal of toninas (fresh-water dolphins) four feet long surrounded our boat. They fled across the forest, throwing out those spouts of compressed air and water. . . ."

Other Venezuelan mammals mentioned in the narrative include the expected jaguar, otter, deer, peccaries, capybara, and vampire bats.

Monkeys, however, absorbed more of Humboldt's attention than other animals. He carried with him a number of live simians captured in the upper Río Orinoco region for shipment to the Jardin des Plantes in Paris via the Antillean island of Guadeloupe. The newly discovered bearded saki (*Chiropotes satanas chiropotes* Humboldt; fig. 14) died before transshipment, but its skin was saved and arrived in Paris. The type specimen of red howler, *Simia ursina* Humboldt (= *Alouatta seniculus arctoides* Cabrera) survived the journey, whereas the first-known douroucouli or night monkey (*Aotus trivirgatus* Humboldt; fig. 14) succumbed in Guadeloupe.

Humboldt often mentioned the ubiquitous, highly visible howler or araguato (*Alouatta seniculus*). At one time he saw from the road below troops of 30 to 40 individuals crossing through the trees. In a carefully deployed experiment in Aragua, he calculated the distance the howler's vocalization could be heard as 800 toises (6 ft 4.73 inches × 800 = 5,115 ft) or nearly 1 mile (5,280 ft).

Humboldt (Ross translation, 1884, vol. II, p. 453) recounts the Indian tale of bearded sakis (*Chiropotes*) and uacaries (*Cacajao*) of the Orinoco "placing themselves in a circle and, by striking the shell [of the Brazil nut pericarp] with a stone, succeed in opening it so as to take out the triangular nuts." Although Humboldt dismissed the story as fabulous, he did believe that the monkeys cracked the shell of the *Bertholletia* nut with their teeth to obtain the meat which they devoured with gusto.

Belief in the existence of a hairy man of the woods was practically universal. The missionary Father Gili gravely related to Humboldt the tale of a woman "in the town of San Carlos in the Llanos of Venezuela who much praised the gentle character and attentions of the man of the woods. She is stated to have lived several years with one in great domestic harmony, and only requested some hunters to take her back because she and the children (a little hairy also) were weary of living so far from the church and the sacraments." Hum-

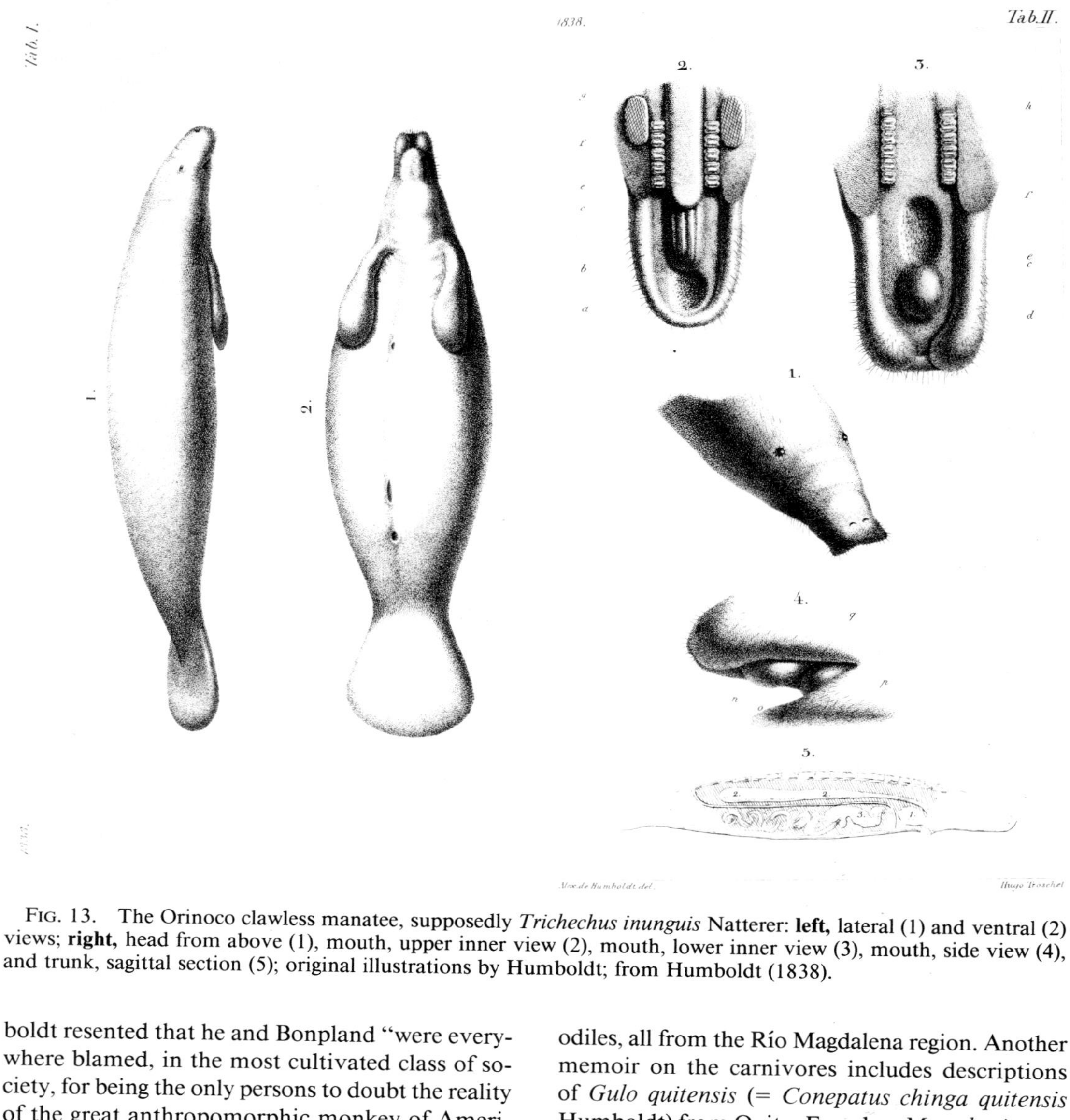

FIG. 13. The Orinoco clawless manatee, supposedly *Trichechus inunguis* Natterer: **left,** lateral (1) and ventral (2) views; **right,** head from above (1), mouth, upper inner view (2), mouth, lower inner view (3), mouth, side view (4), and trunk, sagittal section (5); original illustrations by Humboldt; from Humboldt (1838).

boldt resented that he and Bonpland "were everywhere blamed, in the most cultivated class of society, for being the only persons to doubt the reality of the great anthropomorphic monkey of America."

Humboldt's *Recueil d'Observations de Zoologie et d'Anatomie Comparée,* a collection of memoires published as a volume in 1811–1812, deals with many species of invertebrates and vertebrates, but a large share of the text is about monkeys. One memoir with excellent illustrations by Humboldt is on the comparative anatomy of the hyoid bone and larynx of the cotton-top tamarin (*Saguinus oedipus oedipus* Linnaeus; fig. 14), and that of the red howler (*Alouatta seniculus seniculus* Linnaeus), the Colombian squirrel (*Sciurus granatensis granatensis* Humboldt; fig. 14), birds, and crocodiles, all from the Río Magdalena region. Another memoir on the carnivores includes descriptions of *Gulo quitensis* (= *Conepatus chinga quitensis* Humboldt) from Quito, Ecuador, *Mustela sinuensis* (= *Eira barbara sinuensis* Humboldt), from the Río Sinú, Colombia, and a discourse on other mustelids and the kinkajou (*Potos flavus* Schreber). The memoir on monkeys of the upper Río Orinoco and connecting Ríos Casiquiare and Negro includes the original descriptions of *Aotus trivirgatus, Chiropotes satanas chiropotes, Cacajao melanocephalus, Callicebus torquatus lugens, Lagothrix lagothricha,* and *Cebus albifrons.* A chapter on the monkeys of Colombia and the upper Amazonian region includes the description of a representative each of *Cebus capucinus* Linnaeus from the Río Sinú, *Ateles belzebuth marginatus* É.

FIG. 14. Monkeys and anatomical dissections from Humboldt (1811): **upper left,** *Simia melanocephala* Humboldt (= *Cacajao melanocephalus*), holotype; **lower left,** two views of *Simia trivirgata* Humboldt (= *Aotus trivirgatus*), holotype; **upper right,** *Simia satanas* Hoffmannsegg (= *Chiropotes satanas satanas*), lectotype; **lower right,** throat cartilages of *Psittacus araurana* Linnaeus (= *Ara araurana*), *Sciurus granatensis* Humboldt, and *Simia oedipus* Linnaeus (= *Saguinus oedipus oedipus*).

TABLE 8. New World monkeys (Platyrrhini) recorded by Humboldt (1812); the arrangement is phylogenetic.

Current name	Humboldt synonym	Figure
CALLITRICHIDAE		
Callithrix jacchus jacchus Linnaeus, 1758	...	...
Callithrix jacchus penicillata É. Geoffroy, 1812	...	...
Callithrix jacchus geoffroyi Humboldt, 1812	*Jacchus leucocephalus* Geoffroy, 1812	...
Callithrix jacchus aurita É. Geoffroy, 1812	...	...
Callithrix humeralifer humeralifer É. Geoffroy, 1812	...	...
Callithrix argentata melanura É. Geoffroy, 1812	...	...
Callithrix argentata argentata Linnaeus, 1771	...	...
Saguinus fuscicollis fuscus Lesson, 1840	*Simia leonina* Humboldt, 1805, not Shaw, 1800	13
Saguinus labiatus labiatus É. Geoffroy, 1812	...	...
Saguinus midas niger É. Geoffroy, 1803	*Simia ursula* Hoffmannsegg, 1808	...
Saguinus midas midas Linnaeus, 1758	...	...
Saguinus oedipus oedipus Linnaeus, 1758	...	14
Leontopithecus rosalia rosalia Linnaeus, 1766	...	...
CEBIDAE		
Saimiri sciureus cassiquiarensis Lesson, 1840	Not *Simia sciurea* Linnaeus	...
Callicebus moloch moloch Hoffmannsegg, 1808	...	...
Callicebus torquatus lugens Humboldt, 1811	...	...
Callicebus torquatus torquatus Hoffmannsegg, 1808	*Simia amicta* Humboldt, 1811	...
Callicebus personatus personatus É. Geoffroy, 1812	...	...
Aotus trivirgatus Humboldt, 1811	...	13
Aotus azarae azarae Humboldt, 1811	...	...
Pithecia monachus monachus É. Geoffroy, 1812	...	...
Pithecia pithecia pithecia Linnaeus, 1766	*Pithecia rufiventer* É. Geoffroy, 1812; *Simia leucocephala* É. Geoffroy, 1812	...
Chiropotes satanas satanas Hoffmannsegg, 1808	...	13
Cacajao melanocephalus Humboldt, 1811	...	...
Alouatta caraya Humboldt, 1812	...	...
Alouatta seniculus arctoidea Cabrera, 1940	*Simia ursina* Humboldt, 1805, not Bechstein, 1800	...
Alouatta seniculus straminea Humboldt, 1812	...	...
Cebus capucinus capucinus Linnaeus, 1758	*Simia hypoleuca* Humboldt, 1811	...
Cebus nigrivittatus nigrivittatus Wagner, 1848	*Simia capucina* Humboldt, 1812, not Linnaeus, 1758	...
Cebus apella apella Linnaeus, 1758	*Cebus barbatus* Humboldt, 1812, attributed to É. Geoffroy	...
Cebus apella xanthosternos Wied-Neuwied, 1820	*Simia variegata* Humboldt, 1812, not Kerr, 1792	...
Cebus apella nigritus Goldfuss, 1810	*Simia cirrifera* Humboldt, 1812; *Cebus niger* É. Geoffroy, 1812	...
Lagothrix lagothricha lagothricha Humboldt, 1812	...	...
Lagothrix lagothricha cana É. Geoffroy, 1812	...	...
Lagothrix flavicauda Humboldt, 1811	...	...
Ateles paniscus chamek Humboldt, 1812	...	...
Ateles paniscus paniscus Linnaeus, 1766	...	...
Ateles belzebuth belzebuth É. Geoffroy, 1806	...	...
Ateles belzebuth marginatus É. Geoffroy, 1809	*Simia chuva* Humboldt, 1811, p. 340; 1812, p. 362, footnote 2	...
Brachyteles arachnoides É. Geoffroy, 1806	...	...

Geoffroy from lower Amazonia, *Alouatta seniculus* Linnaeus from the Río Magdalena, and *Lagothrix flavicauda* Humboldt from northern Peru. In an addendum, Humboldt listed all platyrrhine monkeys known to 1812. They are arranged in Table 8 by current scientific names with Humboldt's synonyms.

X. PARAGUAY

The Paraguayan province, claimed by Spain, was first visited in 1526 by Sebastian Cabot and then explored by Cabeza Álvarez Nuñez de Vaca in 1541. For the next two centuries, waves of mis-

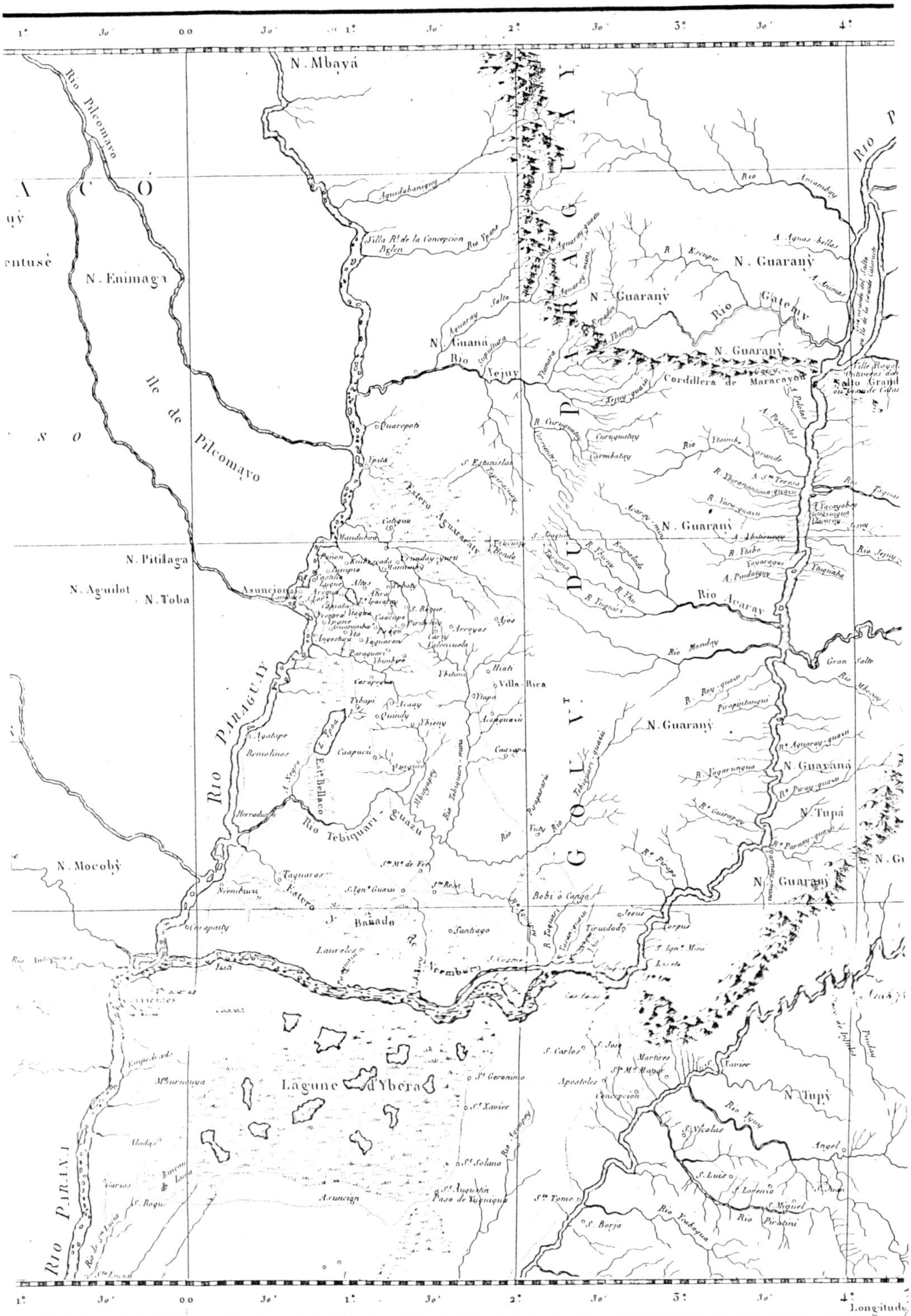

FIG. 15. Map of Azara's Paraguay and adjacent parts of Brazil and Argentina; from Azara (1809).

sionaries and colonists penetrated to the remotest corners of the province in quest of climates or environments that resembled or could be transformed into the familiar ones of Spain.

The monumental *Histoire du Paraguay* by the Jesuit missionary Pierre Francois Xavier de Charlevoix (1682–1761), published in 1757, describes the land that extended from the Atlantic to the eastern base of the Andes between latitudes 15° and 35° in the drainage basin of the Río Paraná-Paraguay. It relates the history of the province from the time of the conquest, describes native customs, conversions to Christianity, and establishment of missions. The little of natural history in the text adds nothing about wild mammals not already recorded by others. Two decades later Felix de Azara wrote the most complete natural history account of the mammalian fauna of Paraguay for its time and ever since.

Felix de Azara (1746–1811)

The Spaniard Don Felix de Azara (1746–1811), an army engineer, was commissioned in 1781 to assist in defining the boundaries between Spanish and Portuguese territories. Unmapped territories between Brazil and Paraguay were assigned to Azara, but the Portuguese showed no interest in their share of the work. With time on his hands and a disposition toward the natural sciences, Azara devoted nearly the full 20 years, from 1781 to 1800, of his American residence to the study of geography, Guaraní Indians, and the birds and mammals of Paraguay and northeastern Argentina between 24° and 36°S and about 54°31′ to 56°W (or 60°W of Greenwich) (fig. 15).

With no schooling in the natural sciences and no books for reference or guidance, Azara depended on his own resources. They proved adequate. Azara recorded his observations with care, precision, meticulous attention to detail, and rigorous exclusion of speculation and fantasy. His anatomical descriptions, measurements, and accounts of behavior were based on animals observed in the wild or in captivity, usually in his own home or garden. Useful information received from others was credited to the informants. Popular beliefs and hearsay were labeled as such. Without other sources of information, Azara used the Guaraní names for most of the amimals he described and Spanish epithets for the remainder.

The manuscript of the mammals or *quadrúpedos* of Paraguay contained accounts of 66 species. Shortly after its completion, the author received a shipment of several volumes of a Spanish translation of Buffon's *Histoire Naturelle.* Not surprisingly, Azara found in them much with which to disagree, but some of his adverse criticism was unfair. Azara knew Paraguayan mammals better than anyone else, but only a minority of the species were the same as the Neotropical species described in the *Histoire Naturelle,* and those that were the same did not always behave in the same way at different times or in different places.

Azara sent a copy of the manuscript of the *quadrúpedos* to his brother, José Nicolás, then Spanish ambassador to Paris, who arranged for publication in that city after translation into French by M.-L.-E. Moreau de Saint-Mery. A year after his return to Spain in 1801, Azara secured publication in Madrid of the original Spanish manuscript with emendations and addition of 11 species, for a total of 77.

Azara may not have been aware that as many as 62 of the 77 species he described were still unknown to science. His clear and precise characterization of each of the species or subspecies, however, provided contemporary cataloguers and systematists with the bases for the descriptions of 50 new species, many with their vernacular appellations in the binomial. Actual specimens served as types for the remaining 12 species.

The mammals described by Azara are listed below, with the scientific name of each given first followed by its local name(s). The page references are to Azara's works in French (*Essais,* 1801), Spanish (*Apuntamientos,* 1802), and the *Voyage* (1809). The last is a French translation in four volumes of Azara's travels in Paraguay with separate atlas, but only the first volume and atlas contain information on mammals.

Tapirus terrestris Linnaeus, 1758
Mborébi, *Essais* I, p. 1; Mborebi, *Apunt.,* I, p. 1; Mborebi ou tapir, *Voyage,* p. 246.

Tayassu G. Fischer, 1814
Couré ou Tayazou, *Essais,* I, p. 18; Curés o Tayazús, *Apunt.,* I, p. 14; Curé ou tayazú, *Voyage,* p. 248.

Tayassu pecari albirostris Illiger, 1815
Tagnicati, *Essais,* I, pp. 21, 25; Tañicati, *Apunt.,* p. 19; Tañicati, *Voyage,* p. 249.
Bibliographic type of the subspecies.

Tayassu tajacu Linnaeus, 1758
Taytétou, *Essais,* I, pp. 21, 31; Taytetú, *Apunt.,* I, p. 23; Taytetú, *Voyage,* p. 249.

CERVIDAE

Gazou, *Essais,* I, p. 43; Venados, *Apunt.,* I, p. 29; Guazú, *Voyage,* p. 250.

Blastocerus dichotomus Illiger, 1815

Gouazoupoucou, *Essais,* I, p. 70; Güazú-pucú, *Apunt.,* I, p. 33; guazú-pucú, *Voyage,* p. 250.

Bibliographic type of the species.

Blastoceros bezoarticus leucogaster Goldfuss, 1817

Gouazouti, *Essais,* I, p. 77; Güazú-ti, *Apunt.,* I, p. 41; Guazú-ti, *Voyage,* p. 251.

Bibliographic type of the subspecies.

Mazama americana gouazoupita Fischer, 1814

Gouazoupita, *Essais,* I, p. 82; Güazú-pitá, *Apunt.,* I, p. 51; Guazú-pitá, *Voyage,* p. 252.

Bibliographic type of the subspecies.

Mazama gouazoubira gouazoubira Fischer, 1814

Gouazoubira, *Essais,* I, p. 86; Güazú-birá, *Apunt.,* I, p. 57; Guazú-birá, Voyage, p. 252.

Bibliographic type of the species.

DIDELPHIDAE

Micourés, *Essais,* I, p. 240; Fecundos, *Apunt.,* I, p. 204; Féconds, *Voyage,* p. 281.

Didelphis albiventris Lund, 1840

Micouré premier, ou micouré propement dit, *Essais,* I, p. 244; Micuré, *Apunt.,* I, p. 209; Micure, *Voyage,* p. 283.

Caluromys lanatus Olfers, 1818

Micouré second, ou Micouré laineux, *Essais,* I, p. 175; Lanoso, *Apunt.,* I, p. 221; Lanoso, *Voyage,* p. 287.

Holotype in alcohol, no. 528, Museo de Ciencias Naturales, Madrid, captured 22 July 1789, by Felix d'Azara (Cabrera, 1916, Bol. Real Soc. española Hist. Nat., 16, p. 1).

Lutreolina crassicaudata Desmarest, 1804

Micouré troisième, ou micouré a queue grosse, *Essais,* I, p. 284; Coligrueso, *Apunt.,* I, p. 229; Coligrueso, *Voyage,* p. 290.

Bibliographic type of the species.

Marmosa pusilla Desmarest, 1804

Micouré quatrième, ou micouré a queue longue, *Essais,* I, p. 290; Colilargo, *Apunt.,* I, p. 251; Colilargo, *Voyage,* p. 291.

Bibliographic type of *Marmosa macrura* Olfers, 1818 (= *M. pusilla* Desmarest).

Micouré sixième, ou micouré nain, *Essais,* I, p. 304; Enano, *Apunt.,* I, p. 262; Enano, *Voyage,* p. 284.

Bibliographic type of *Marmosa pusilla* Desmarest, 1804.

Monodelphis brevicaudis Olfers, 1818

Micouré cinquième, ou micouré a queue courte, *Essais,* I, p. 295; Colicorto, *Apunt.,* I, p. 258; Colicorto, *Voyage,* p. 293.

Bibliographic type of the species.

MYRMECOPHAGIDAE

Hormigueros, *Apunt.,* I, p. 61.

Myrmecophaga tridactyla Linnaeus, 1758

Gnouroumi, ou Yoquoui, *Essais,* I, p. 89; Ñurumi o Yoqui, *Apunt.,* I, p. 66; Ñurumi ou tamanduá, *Voyage,* pp. 253, 255.

Tamandua tetradactyla Linnaeus, 1758 (fig. 16)

Cagouare, *Essais,* I, p. 103; Cagüaré, *Apunt.,* I, p. 74; Cagüaré, *Voyage,* pp. 253, 256; Atlas, pl. VII (tamandua noir), pl. VIII (Cagouaré).

FELIDAE

Gatos, *Apunt.,* I, p. 85.

Felis onca Linnaeus, 1758

Yagouarété, *Essais,* I, p. 114; Yagüareté, *Apunt.,* I, p. 91; Yaguareté, *Voyage,* p. 258; Atlas, pl. IX.

Yagüareté negro, *Apunt.,* I, p. 114; Yaguareté noir, *Voyage,* p. 267.

Felis concolor Linnaeus, 1771

Gouazouara, *Essais,* I, p. 133; Güazuará, *Apunt.,* I, p. 120; Guazúará, *Voyage,* p. 268.

Felis geoffroyi D'Orbigny and Gervais, 1844

Mbaracayá, *Apunt.,* I, p. 147; Baracayá, *Voyage,* p. 271.

Note: Said not to exist in Paraguay.

Felis species?

Negro, *Apunt.,* I, p. 154; Chat noir, *Voyage,* p. 273.

Felis pardalis Linnaeus, 1758

Chibigouazou, *Essais,* I, p. 152; Chibi-güazú, *Apunt.,* I, p. 132; Chibi-guazu, *Voyage,* p. 269.

Herpailurus yagouaroundi eyra Fischer, 1814 (fig. 16)

Yagouaroundi, *Essais,* I, p. 171; Yaguarundi, *Apunt.,* I, p. 156; Yaguarundi, *Voyage,* p. 273, Atlas, pl. X (Yagouarondi, black phase); Eyra, *Essais,* I, p. 177; Eyra, *Apunt.,* I, p. 159; Eyra, *Voyage,* p. 274 (red phase).

Bibliographic type of the subspecies.

Felis colocolo pajeros Desmarest, 1816

Chat pampa, *Essais,* I, p. 179; Pajero, *Apunt.,* I, p. 160; Pajero, *Voyage,* p. 274.

Bibliographic type of the species.

Note: Said not to exist in Paraguay.

FIG. 16. Two of Azara's Paraguayan animals: **top,** le yagouarondi (= *Eira yagouarondi eyra* Fischer); **bottom,** le cagouare or cagüaré (= *Tamandua tetradactyla* Linnaeus); from Azara (1809).

MUSTELIDAE

Furets, *Essais,* I, p. 185; Hurones, *Apunt.,* I, p. 167.

Galictis cuja furax Thomas, 1907
Petit furet, *Essais,* I, p. 190; Hurón menor, *Apunt.,* I, p. 182; Hurón minor, *Voyage,* p. 276.

Eira barbara Linnaeus, 1758
Grand furet, *Essais,* I, p. 197; Hurón mayor, *Apunt.,* I, p. 172; Hurón major, *Voyage,* p. 275, Atlas, pl. XI.

Conepatus chinga suffocans Illiger, 1815
Yagouré, *Essais,* I, p. 211; Yagüaré, *Apunt.,* I, p. 187; Yaguaré, *Voyage,* p. 277.
Bibliographic type of the subspecies.

Pteronura brasiliensis paranensis Rengger, 1830
Loutre, *Essais,* I, p. 348; Nutria, *Apunt.,* I, p. 304; Loutre, *Voyage,* p. 304.

CANIDAE

Zorro, *Apunt.,* I, p. 264; Renard, *Voyage,* p. 295.

Chrysocyon brachyurus Illiger, 1815
Agouara-gouazou, *Essais,* I, p. 307; Agüara-guazú, *Apunt.,* I, p. 266; Aguará-guazú, *Voyage,* p. 296.
Bibliographic type of the species.

Dusicyon gymnocercus Fischer, 1814
Agouarachay, *Essais,* I, p. 317; Agüarachai, *Apunt.,* I, p. 271; Aguarachay, *Voyage,* p. 298, Atlas, pl. XII.
Bibliographic type of the species.

PROCYONIDAE

Procyon cancrivorus nigripes Mivart, 1886
Agouarapope, *Essais,* I, p. 324; Popé, *Apunt.,* I, p. 278; Popé, *Voyage,* p. 299.

Nasua nasua spadicea Olfers, 1818
Couati, *Essais,* I, p. 334; Cuati, *Apunt.,* I, p. 293; Cuati, *Voyage,* p. 301.
Bibliographic type of the subspecies.

LEPORIDAE

Sylvilagus brasiliensis paraguensis Thomas, 1901
Tapiti, *Essais,* II, p. 57; tapiti, *Apunt.,* II, p. 32; Tapity, *Voyage,* p. 313.

RODENTIA

Roedores, *Apunt.,* II, p. 68.

Myocastor coypus bonariensis Commerson, 1805
Quouiya, *Essais,* II, p. 5; Quiyá, *Apunt.,* II, p. 1; Quiyá, *Voyage,* p. 308.

Hydrochaeris hydrochaeris dabbenei Rovereto, 1913
Capiygoua, *Essais,* II, p. 12; Capibara, *Apunt.,* II, p. 8; Capibara, *Voyage,* p. 309.

Agouti paca Linnaeus, 1758
Pay, *Essais,* II, p. 20; Pai, *Apunt.,* II, p. 14; Pay, *Voyage,* p. 310.

Dasyprocta azarae paraguayensis Liais, 1872
Acouti, *Essais,* II, p. 26; Acuti, *Apunt.,* II, p. 21; Acuty, *Voyage,* p. 312.

Vizcacia maximus Desmarest, 1817
Vizcache, *Essais,* II, p. 41; Vizcacha, *Apunt.,* II, p. 45; Vizcacha, *Voyage,* p. 316.
Bibliographic type of the species.
Note: *Vizcacia* Schinz, 1824 (Naturg. Abbild. Säugeth., p. 243) antedates *Viscaccia* Schinz, 1825 and *Lagostomus* Brookes, 1825.

Dolichotis patagonum Zimmermann, 1780
Lièvre pampa, *Essais,* II, p. 51; Liebre patagona, *Apunt.,* I, p. 51; Lièvre patagon, *Voyage,* p. 318.
Note: Said not to exist in Paraguay.

Cavia aperea hypoleuca Cabrera, 1953
Apéreá, *Essais,* II, p. 65; Apereá, *Apunt.,* II, p. 37; Apereá, *Voyage,* p. 314.

Euryzygomatomys spinosus Fischer, 1814
Rat premier, ou rat épineus, *Essais,* II, p. 73; Espinoso, *Apunt.,* II, p. 76; Épineuse, *Voyage,* p. 326, Atlas, pl. XIII.
Bibliographic type of the species.

Oryzomys megacephalus Fischer, 1814
Rat second, ou rat á grosse tête, *Essais,* II, p. 82; Cola igual al cuerpo, *Apunt.,* II, p. 87; Cola igual al cuerpo, *Voyage,* p. 330.
Bibliographic type of the species.
Note: Antedates *Oryzomys capito* Olfers.

Oryzomys angouya Fischer, 1814
Rat troisième, ou rat angouya, *Essais,* II, p. 86; Anguyá, *Apunt.,* II, p. 89; Anguya, *Voyage,* p. 331.
Bibliographic type of the species.
Note: Antedates *Oryzomys buccinatus* Olfers.

Reithrodon auritus Fischer, 1814
Rat quatrième, ou rat oreillard, *Essais,* II, p. 91; Orejón, *Apunt.,* II, p. 83; Orejon, *Voyage,* p. 329.
Bibliographic type of the species.

Oxymycterus rufus Fischer, 1814
Rat cinquième, ou rat roux, *Essais,* II, p. 94; Hocicudo, *Apunt.,* II, p. 80; Hocicudo, *Voyage,* p. 328.
Bibliographic type of the species.

Oryzomys nigripes Olfers, 1818
Rat sixième, ou rat a tarse noir, *Essais,* II, p. 98; Colilargo, *Apunt.,* II, p. 91; Colilargo, *Voyage,* p. 331.
Bibliographic type of the species.
Calomys laucha Olfers, 1818
Rat septième, ou rat laucha, *Essais,* II, p. 102; Laucha, *Apunt.,* II, p. 96; Laucha, *Voyage,* p. 333.
Bibliographic type of the species.
Coendou insidiosus Olfers, 1818
Couiy, *Essais,* II, p. 105; Cuiy, *Apunt.,* II, p. 55; Cuiy, *Voyage,* p. 320.
Bibliographic type of the species.
Akodon colibreve Brants, 1827
Colibreve, *Apunt.,* II, p. 86; Colibreve, *Voyage,* p. 329.
Bibliographic type of the species.
Note: *Akodon obscurus* Waterhouse, 1837, is probably a junior synonym, but see Langguth (1978).
Ctenomys tucotuco Brants, 1827
Tucutuco, *Apunt.,* II, p. 89; Tucutuco, *Voyage,* p. 324.
Bibliographic type of the species.
Akodon agreste Brants, 1827
Agreste, *Apunt.,* II, p. 94; Agreste, *Voyage,* p. 332.
Bibliographic type of the species.
Note: Antedates *Akodon azarae* Fischer.
"Mus dubius", Fischer, 1829 [= ?]
Blanco debaxo, *Apunt.,* II, p. 97; Blanco-debaxo, *Voyage,* p. 333.
Bibliographic type of the species.

DASYPODIDAE

Tatous, *Essais,* II, p. 122; Tatús, *Apunt.,* II, p. 99; Tatoús, *Voyage,* p. 334.
Priodontes maximus giganteus, É. Geoffroy, 1803
Tatou premier, ou grand tatou, *Essais,* II, p. 132; Máximo, *Apunt.,* II, p. 110; Grand tatou ou geant, *Voyage,* p. 336.
Bibliographic type of the subspecies.
Euphractus sexcinctus flavimanus Desmarest, 1804
Tatou second, tatou poyou, ou tatou a main jaune, *Essais,* II, p. 142; Poyú, *Apunt.,* II, p. 118; Tatú-poyú, *Voyage,* p. 338.
Bibliographic type of the subspecies.
Cabassous tatouay Desmarest, 1804
Tatou troisième, ou tatou tatouay, *Essais,* II, p. 155; Tatuai, *Apunt.,* II, p. 131; Tatuai, *Voyage,* p. 341.
Bibliographic type of the species.
Chaetophractus villosus Desmarest, 1804
Tatou quatrième, ou tatou velu, *Essais,* II, p. 164; Peludo, *Apunt.,* II, p. 140; Tatou velu, *Voyage,* p. 343.
Bibliographic type of the species.
Dasypus novemcinctus niger Desmarest, 1804
Tatou cinquième, ou tatou noir, *Essais,* II, p. 175; Negro, *Apunt.,* II, p. 144; Tatou noir, *Voyage,* p. 346.
Bibliographic type of the subspecies.
Dasypus hybridus Desmarest, 1804
Tatou sixième, ou tatou mulet, *Essais,* II, p. 186; Mulita, *Apunt.,* II, p. 156; Tatou mulita, *Voyage,* p. 348.
Bibliographic type of the species.
Zaedyus pichiy Desmarest, 1804
Tatou septième, ou tatou pichiy, *Essais,* II, p. 192; Pichiy, *Apunt.,* II, p. 158; Tatú-pichy, *Voyage,* p. 345.
Bibliographic type of the species.
Tolypeutes matacus Desmarest, 1804
Tatou huitième, ou tatou mataco, *Essais,* II, p. 197; Matáco, *Apunt.,* II, p. 161; tatou-mataco, *Voyage,* p. 350.
Bibliographic type of the species.

PLATYRRHINI

Singes, *Essais,* II, p. 206; Micos, *Apunt.,* II, p. 167; Singes, *Voyage,* p. 351.
Alouatta caraya Humboldt, 1812
Caraya, *Essais,* II, p. 208; Carayá, *Apunt.,* II, p. 169; Carayá, *Voyage,* p. 351.
Bibliographic type of the species.
Cebus apella cay, Illiger, 1815
Cay, *Essais,* II, p. 230; Cay, *Apunt.,* II, p. 182; Cay, *Voyage,* p. 354.
Bibliographic type of the subspecies.
Aotus azarae azarae Humboldt, 1812
Miriquouina, *Essais,* II, p. 243; Miriquiná, *Apunt.,* II, p. 195; Miriquiná, *Voyage,* p. 356.
Bibliographic type of the species.
Callithrix jacchus penicillatus É. Geoffroy, 1812
Titi, *Essais,* II, p. 254; Titi, *Apunt.,* II, p. 200; Titi, *Voyage,* p. 359. "N'est pas du Paraguay, mais du Bresil." [Description is of a captive pair seen in the province of Buenos Aires.]

MICROCHIROPTERA

Chauve-souris, *Essais,* II, p. 264; Murcielagos, *Apunt.,* II, p. 288; chauve-souris, *Voyage,* p. 382.

Artibeus lituratus Olfers, 1818
Chauve-souris premiere, ou chauve-souris obscure et rayée, *Essais,* II, p. 269; Obscuro listado, *Apunt.,* II, p. 291.
Bibliographic type of the species.
Vampyrops lineatus É. Geoffroy, 1810
Chauve-souris seconde, ou chauve-souris brune et rayée, *Essais,* II, p. 271; Pardo listado, *Apunt.,* p. 292
Bibliographic type of the species.
Desmodus rotundus É. Geoffroy, 1810
Chauve-souris troisième, ou chauve-souris brune, *Essais,* II, p. 273; Mordedor, *Apunt.,* II, p. 293.
Bibliographic type of the species.
[Azara was first to distinguish true vampires from other bats, particularly *Vampyrum spectrum.*]
Sturnira lilium É. Geoffroy, 1810
Chauve-souris quatrième, ou chauve-souris brun-rougeâtre, *Essais,* II, p. 277; Pardo roxizo, *Apunt.,* II, p. 299.
Bibliographic type of the species.
Noctilio leporinus rufescens Olfers, 1818
Chauve-souris cinquième, ou chauve-souris rougeâtre, *Essais,* II, p. 280; Roxizo, *Apunt.,* II, p. 301.
Bibliographic type of the subspecies.
Molossus ater castaneus É. Geoffroy, 1805
Chauve-souris sixième, ou chauve-souris chataine, *Essais,* II, p. 282; Castaño, *Apunt.,* II, p. 302.
Bibliographic type of the subspecies.
Lasiurus cinereus villosissimus É. Geoffroy, 1806
Chauve-souris septième, ou chauve-souris brun-blanchâtre, *Essais,* II, p. 284; Pardo blanquizco, *Apunt.,* II, p. 303.
Bibliographic type of the subspecies.
Histiotus velatus I. Geoffroy, 1824
Orejón, *Apunt.,* II, p. 304.
Tadarida laticaudata É. Geoffroy, 1805
Chauve-souris huitième, ou chauve-souris obscure, *Essais,* II, p. 286; Obscuro, *Apunt.,* II, p. 305.
Bibliographic type of the species.
Molossus molossus Pallas, 1766
Chauve-souris neuvième, ou petit chauve-souris obscure, *Essais,* II, p. 288; Obscuro menor, *Apunt.,* II, p. 306.
Molossus crassicaudatus É. Geoffroy, 1805
Chauve-souris dixième, ou chauve-souris brun-cannelle, *Essais,* II, p. 290; Pardo acanelado, *Apunt.,* II, p. 307.
Bibliographic type of the species.
Myotis ruber É. Geoffroy, 1806
Chauve-souris onzième, ou chauve-souris cannelle, *Essais,* II, p. 292; Acanelado, *Apunt.,* II, p. 308.
Bibliographic type of the species.
Myotis albescens É. Geoffroy, 1806
Chauve-souris douzième, ou chauve-souris brun-obscur, *Essais,* II, p. 294; Pardo obscuro, *Apunt.,* II, p. 309.
Bibliographic type of the species.

Johann Rudolph Rengger (1795–1832)

Azara was followed by Johann Rudolph Rengger, a Swiss pharmacist and naturalist, who arrived in Paraguay in 1819 and devoted himself to the study of its mammals. His six-year study culminated in the *Naturgeschichte der Saeugethiere von Paraguay,* published 1830. A total of 59 species was described, including four as new of which only *Calomys callosus* and *Proechimys longicaudatus* survived revisions. Azara distinguished 77 species, or 18 more, but several are not strictly Paraguayan. Among the Paraguayan forms missed by Rengger but recognized by Azara are the murine opossum (*Marmosa*), hairy armadillo (*Chaetophractus*), three-banded armadillo (*Tolypeutes*), skunk (*Conepatus*), tucotuco (*Ctenomys*), four cricetine rodents, and two bats. Well over 100 species are presently known from Paraguay.

No doubt Azara set standards for the high quality and accuracy of Rengger's descriptions and behavioral accounts. The wealth of information in the *Naturgeschichte* has hardly been tapped by modern mammalogists.

XI. Chile

Giovanni Ignazio Molina (1737–1829)

Knowledge of Chilean land mammals as a regional fauna begins with publication of the *Saggio* in 1782 by the Jesuit priest Don Giovanni (Juan) Ignazio Molina, who lived in Chile the first 30 years of his life. Expulsion of the Jesuits from the country obliged Molina to emigrate in 1768 and settle in his ancestral Italy. What Molina knew about Chilean mammals he learned before 1768; much of what he wrote about them thereafter suffered from a decayed memory.

Molina was a naturalist in the broadest sense

and was familiar with the *Systemæ* of Linnaeus. He was not, however, particularly dedicated to any one branch of science, and his descriptions of the Chilean mammals are, for the most part, vague, inaccurate, and sometimes composite. A few of his subjects were fanciful, and none of the remainder were closely examined. Nevertheless, by dint of elimination and stretches of the imagination, modern mammalogists have come to agreement on the application of most of the Linnaean names proposed by Molina for the likeliest species he may have had in mind.

Thirty kinds of mammals were described in the *Saggio*. According to Osgood (1943, p. 15), five of them are unidentifiable, four (armadillos) are extraterritorial, two are but one and the same, and one is duplicated. The 14 still valid, with names dating from Molina, 1782, are *Lutra felina, Myocastor coypus, Conepatus chinga, Galictis cuja, Dusicyon culpaeus, Felis guigna, Felis colocolo, Felis concolor puma, Spalacopus cyanus, Octodon degus, Vizcacia vizcacia, Pudu puda, Vicugna vicugna,* and *Hippocamelus bisulcus.* Remaining species, notably the larger mammals, recorded by Molina were well known to early voyagers, chroniclers, and naturalists and had already received Linnaean names.

Eduard Friedrich Poeppig (1798–1868)

The German naturalist Eduard Poeppig is known for his *Reise* in Chile, Peru, and on the Rio Amazonas during the years 1827–1832. The account of his travels, in two volumes, was published 1835–1836. The Chilean mammals recorded include seals, sea lions, and elephant seals, the degu, *Spalacopus cyanus* Molina (*Psammomys noctivagus* Poeppig, a synonym), the coypu, and a small canid, probably *Dusicyon griseus* Gray. In Antuco, Province of Bío Bío, he encountered the pudu, huemul, and two species of bats, one described as *Nyticyus varius* (= *Lasiurus borealis bonariensis* Lesson & Garnot, 1827), the other as *Nycticyus macrotus* (currently *Histiotus macrotis* Poeppig, 1835).

Claudio Gay (1800–1873)

Between the years 1844 and 1871, Claudio Gay, French naturalist and longtime resident of Chile, produced 25 volumes, including two of plates, on the history, geography, and biota of Chile. The first volume (1847) of eight on zoology contains virtually all Chilean mammals known at the time. Fifty-four species are described, with accounts of habits, habitat, and geographic distribution of each. For the most part, Gay worked from actual specimens brought to him by natives or observed by him on his travels throughout the country. On his return to France, Gay included in his studies the Chilean material preserved in the Paris Natural History Museum.

The species recorded by Gay include Marsupialia, 2 (4% of the total); Chiroptera, 7 (13%); Carnivora, 12 (22%); Pinnipedia, 6 (11%); Rodentia, 23 (43%; myomorphs, 24%, caviomorphs, 18%); Artiodactyla, 3 (5%). Among the 30 species recorded by Molina, only 3 or 10% are rodents. Of the 20 Chilean species collected by Darwin, 12 or 60% are rodents. In this volume Patterson and Feigl recognize 93 living Chilean species, of which 53 or 57% are rodents (33% myomorphs, 24% caviomorphs), and 10 or 11% are bats.

XII. Peru

Johann Jacob von Tschudi (1818–1889)

The Swiss biologist Johann von Tschudi was born in the town of Glarus and studied the sciences at Swiss, French, and German universities. Inspired by the accounts of the travels of Humboldt and Darwin in South America, Tschudi sailed on 27 February 1838 from Le Havre for Peru. The first landing on the continent was made 5 June 1838 on the Chilean island of Chiloé. After a delay of about three weeks and many observations of the natural history of the island, von Tschudi reembarked for Callao, Peru, with short stopovers in Valdivia and Juan Fernández.

From August 1838 through most of 1843, von Tschudi traveled over much of Peru. Of particular interest to him were the higher vertebrates and the physical factors controlling their geographic distribution. He distinguished faunal zones based on ruling ecological features. The major zones were Pacific coast, Andean altitudinal zones of western and eastern slopes, and the tropical Amazonian selva. Apparently, no one had preceded von Tschudi in the recognition of definable biogeographic areas in the New World.

The narrative of von Tschudi's travels in Peru was published in 1846 in German, followed in 1847 by Thomasina Ross's English translation.

FIG. 17. Mammals of Peru: **upper left,** vicuña, *Auchenia vicuna* (= *Vicugna vicugna* Molina); **lower left,** agouti, *Dasyprocta variegata* Tschudi; **upper right,** wooly opossum, *Didelphys ornata* Tschudi (= *Caluromys lanatus ornatus*); **lower right,** coatimundi, *Nasua montana* Tschudi (= *Nasua nasua montana*); from Tschudi (1844b).

The scientific accounts of the mammals are found in a preliminary report (1844a) and first part of the *Untersuchungen über die Fauna Peruana,* published later the same year (1844b).

Although von Tschudi attempted to provide the fullest account possible of Peruvian mammals, it appears he had little or no contact with the majority of them. Most of his characterizations and life history accounts are taken from Humboldt, Spix, Wied-Neuwied, other European travelers and natives. Camelids, the dominant animals of the Peruvian landscape fascinated von Tschudi, and he wrote more about them than of other animals. His description of a vicuña hunt is quoted below from the Ross translation (Tschudi, 1847, pp. 219–220).

> The Indians seldom employ fire-arms in hunting the vicuñas. They catch them by what they term the *chacu.* In this curious hunt, one man at least belonging to each family in the Puna villages takes a part, and women accompany the train, to officiate as cooks to the hunters. The whole company, frequently amounting to seventy or eighty individuals, proceeds to the Altos (the most secluded parts of the Puna), which are the haunts of the vicuñas. They take with them stakes, and a great quantity of rope and cord. A spacious open plain is selected, and the stakes are driven into the ground in a circle, at intervals of from twelve to fifteen feet apart, and are connected together by ropes fastened to them at the height of two or two and a half feet from the ground. The circular space within the stakes is about half a league in circumference, and an opening of about two hundred paces in width is left for entrance. On the ropes by which the stakes are fastened together the women hang pieces of colored rags, which flutter about in the wind. The chacu being fully prepared, the men, some of whom are mounted on horseback, range about within a circuit of several miles, driving before them all the herds of vicuñas they meet with, and forcing them into the chacu. When a sufficient number of vicuñas is collected, the entrance is closed. The timid animals do not attempt to leap over the ropes, being frightened by the fluttering rags suspended from them, and, when thus secured, the Indians easily kill them by the *bolas.* These bolas consist of three balls, composed either of lead or stone; two of them heavy, and the third rather lighter. They are fastened to long, elastic strings, made of twisted sinews of the vicuña, and the opposite ends of the strings are all tied together. The Indian holds the lightest of the three balls in his hand, and swings the two others in a wide circle above his head; then taking his aim at the distance of about fifteen or twenty paces, he lets go the hand-ball, upon which all the three balls whirl in a circle, and twine round the object aimed at. The aim is usually taken at the hind legs of the animals, and the cords twisting round them they become firmly bound. It requires great skill and long practice to throw the bolas dexterously, especially when on horseback: a novice in the art incurs the risk of dangerously hurting either himself or his horse, by not giving the balls the proper swing, or by letting go the hand-ball too soon.
>
> The vicuñas, after being secured by the bolas, are killed, and the flesh is distributed in equal portions among the hunters. The skins belong to the Church. The price of a vicuña skin is four reals. When all the animals are killed, the stakes, ropes, &c., are packed up carefully, and conveyed to another spot, some miles distant, where the chacu is again fixed up. The hunting is continued in this manner for the space of a week. The number of animals killed during that interval varies according to circumstances, being sometimes fifty or sixty, and at other times several hundred. During five days I took part in a chacu hunt in the Altos of Huayhuay, and in that space of time 122 vicuñas were caught. With the money obtained by the sale of the skins a new altar was erected in the church of the district. The flesh of the vicuña is more tender and better flavored than that of the llama. Fine cloth and hats are made of the wool. When taken young, the vicuñas are easily tamed, and become very docile; but when old, they are intractable and malicious. At Tarma I possessed a large and very fine vicuña. It used to follow me like a dog whenever I went out, whether on foot or on horseback.
>
> The frequent hunting seems not to have the effect of diminishing the numbers of these animals. If in the vicinity of the villages where chacus are frequently established, they are less numerous than in other parts, it is because, to elude the pursuit of the hunters,

TABLE 9. Peruvian mammals according to Tschudi (1844a,b); current scientific names are used followed by Tschudi's synonym or misidentification, local names, and figure in this text; extralimital species are bracketed; arrangement of taxa follows Tschudi.

Current name	Tschudi synonym or misidentification	Local name	Figure
Ateles paniscus chamek Humboldt	*Ateles marginatus*; *Ateles ater*; *Ateles pentadactylus*	Chuva; maquisapa; chamek; marimonda; machucusillo; supaya	...
Lagothrix lagothricha poeppigi Schinz	*Lagothrix humboldti*; *Lagothrix canus*	Mono oki; choko	...
Lagothrix flavicauda Humboldt	*Mycetes flavicaudatus* (*sic*)	...	...
Alouatta seniculus Linnaeus	*Mycetes stramineus*	Coro [= coto?]	...
[*Mycetes rufimanus* Kuhl]	*Alouatta belzebul*	...	...
Cebus apella Linnaeus	*Cebus robustus*	Macaquito	...
Cebus albifrons Humboldt	...	...	...
[*Cebus capucinus* Linnaeus]	...	...	...
Saimiri boliviensis peruviensis Hershkovitz	*Chrysothrix sciureus*	...	...
Callicebus torquatus Hoffmannsegg (subsp.?)	*Callithrix amictus*	...	...
[*Callicebus personatus* É. Geoffroy]	*Callithrix personatus*	...	...
Aotus nigriceps Dollman	*Nyctipithecus trivirgatus*	Tocon	...
[*Chiropotes*]	?*Pithecia**	...	...
Saguinus mystax mystax Spix	[*Midas labiatus*]	...	...
Saguinus nigricollis Spix	[*Midas labiatus*]	...	...
Saguinus fuscicollis Spix	[*Midas labiatus*]	...	...
[*Saguinus midas midas* Linnaeus]	[*Midas rufimanus*]	...	...
[*Leontopithecus rosalia chrysomelas* Kuhl]	[*Midas chrysomelas*]	...	...
CHIROPTERA			
Phyllostomus elongatus É. Geoffroy	...	...	...
Phyllostomus hastatus Pallas	...	Hatummasu	...
Phyllostomus discolor Wagner, 1843	*Phyllostomus innominatum* Tschudi	...	...
Artibeus cinereus Gervais	*Phyllostomus* (*Artibeus*) *pusillum*	...	...
Sturnira erythromos Tschudi	...	...	...
Sturnira oporophilum Tschudi	*Phyllostomus* (*Sturnira*) *oporophilum* Tschudi	...	...
Glossophaga soricina Pallas	*Glossophaga amplexicauda*	...	...
Anoura geoffroyi peruana Tschudi	*Glossophaga* (*Choeronycteris*) *peruana* Tschudi	...	...
Eptesicus innoxius Gervais	*Vespertilio innoxius*	...	...
Histiotus macrotus Poeppig	*Vespertilio* (*Vesperugo*) *velatus*	...	...
Noctilio leporinus Linnaeus	*Noctilio unicolor*	...	...
Noctilio albiventris Desmarest	*Noctilio affinis*	...	...
Tadarida brasiliensis I. Geoffroy	*Molossus* (*Dysopes*) *naso*	...	...
Molossus molossus Pallas?	*Molossus* (*Dysopes*) *velox*	...	...
Eumops auripendulus Shaw	*Molossus* (*Dysopes*) *ferox*; *Dysopes longimanus*	...	...
Molossus ater É. Geoffroy	*Molossus* (*Dysopes*) *myosuros* Tschudi; *Molossus anonymus* Tschudi	...	...
[*Promops nasutus* Spix]	*Dysopes fumarius*	...	...
CARNIVORA			
Tremarctos ornatus F. Cuvier	*Ursus frugilegus* Tschudi	Hucamari	...
Nasua nasua montana Tschudi	*Nasua socialis*; *Nasua solitaria*; *Nasua leucorhynchos* Tschudi	Achuna, mishash	17
Potos flavus Schreber	*Cercoleptes caudivolvulus*	Cushumbi	...
Eira barbara Linnaeus	*Galictis barbara*	Omeyro	...

TABLE 9. *Continued.*

Current name	Tschudi synonym or misidentification	Local name	Figure
CARNIVORA (*continued*)			
Mustela frenata agilis Tschudi	...	...	...
Conepatus chinga Molina	*Molina* (*Thiosmus*) *mapurita*; *Mephitis furcata*; *Mephitis amazonica*	...	...
Lutra felina Molina	*Lutra chilensis*	...	...
Lutra montana Tschudi†	...	...	...
Dusicyon thous Linnaeus	*Canis azarae*	...	...
Felis concolor Linnaeus	...	Poma, león	...
Felis onca Linnaeus	*Felis onza*	Choque china, yana choque, tigre	...
Felis pardalis Linnaeus	...	Uturunco	...
Felis wiedii Schinz	*Felis macrura* (*sic* = *Felis macroura*); *Felis celidogaster*	...	...
Felis yagouaroundi É. Geoffroy	*Felis yaguarundi*	...	...
PINNIPEDIA			
Otaria flavescens Shaw	*Otaria jubata*; *Otaria ulloae* Tschudi; *Otaria aurita* Humboldt (in Tschudi)	...	...
MARSUPIALIA			
Didelphis marsupialis Linnaeus	*Didelphys azarae*	Mucamuca, jarachupa	...
Metachirus nudicaudatus É. Geoffroy	*Didelphys myosuros*	...	...
Philander opossum Linnaeus	...	...	...
Marmosa noctivaga Tschudi	...	...	...
Marmosa impavida Tschudi	...	...	...
Marmosa murina Linnaeus	...	...	...
Caluromys lanatus ornatus Tschudi	...	...	17
RODENTIA			
Sciurus aestuans Linnaeus	...	...	...
Sciurus pyrrhinus Thomas	[*Sciurus variabilis*]	...	...
Sciurus stramineus Eydoux and Souleyet	...	...	...
Sciurus spadiceus tricolor Tschudi	...	...	...
Proechimys sp.?	[*Echinomys leptosoma*]	...	...
Chinchilla brevicaudata Waterhouse	*Eriomys chinchilla*	...	...
Lagidium peruanum Meyen	*Lagidium peruvianum* (*sic*)	...	...
Lagidium viscacia Molina‡	*Lagidium pallipes*	...	...
[*Octodon degus* Molina]	[*Octodon cummingii*]	...	...
[*Myocastor coypus* Molina]	[*Myopotomus coypus*]	...	...
Coendou bicolor Tschudi	*Sphingurus* (*sic*) *bicolor*	...	...
Dasyprocta leporina Linnaeus	*Dasyprocta aguti* Linnaeus	...	...
Dasyprocta variegata Tschudi	...	Cutspi or cushpi	17
Akodon boliviensis Meyen	*Acodon boliviense*	...	...
Phyllotis darwini Waterhouse	*Hesperomys darwini*	...	...
Oryzomys longicaudatus destructor Tschudi	*Hesperomys destructor*	...	...
Oryzomys melanostoma Tschudi	*Hesperomys melanostoma*	...	...
Rhipidomys leucodactylus Tschudi	*Hesperomys* (*Rhipidomys*) *leucodactylus*	...	...
Agouti paca Linnaeus	*Coelogenys fulvus*	...	...
Hydrochaeris hydrochaeris Linnaeus	*Hydrochoerus capybara*	...	...
Cavia porcellus Linnaeus	*Cavia cutleri*	Cuy del monte	...

TABLE 9. *Continued.*

Current name	Tschudi synonym or misidentification	Local name	Figure
LAGOMORPHA			
Sylvilagus brasiliensis Linnaeus	*Lepus brasiliensis*	...	...
EDENTATA			
Bradypus variegatus Schinz	*Bradypus infuscatus*	...	...
[*Bradypus torquatus* Illiger]	...	...	...
Dasypus novemcinctus Linnaeus	*Dasypus 9-cinctus* (*sic*)	...	...
Cabassous unicinctus Linnaeus	*Dasypus tatuay* (*sic* = tatouay)	Quirquincho	...
Tamandua tetradactyla Linnaeus	*Myrmecophaga tamandua*	...	...
Cyclopes didactylus Linnaeus	*Myrmecophaga didactyla*	...	...
PERISSODACTYLA			
Tapirus terrestris Linnaeus	*Tapirus americanus*	...	...
Tapirus pinchaque Roulin	*Tapirus villosus*	...	...
ARTIODACTYLA			
Tayassu tajacu Linnaeus	*Dicotyles torquatus*	...	...
Tayassu pecari Link	*Dicotyles labiatus*	...	...
Lama glama Linnaeus	*Auchenia lama*	Llama	...
Lama pacos Linnaeus	*Auchenia paco*	Alpaca	...
Lama guanicoe Müller	*Auchenia huanaco*	...	...
Vicugna vicugna Molina	*Auchenia vicuna*	Vicuña	17
Mazama americana Erxleben	*Cervus rufus*	...	...
Mazama gouazoubira peruana Tschudi	*Cervus nemorivagus* var. *peruana*	Liucho, venado	...
Hippocamelus antisensis d'Orbigny	*Cervus antisiensis*	Tarush, taruga	...

* Sakis (*Pithecia*) evidently not seen by von Tschudi. His descriptions are of bearded sakis (*Chiropotes*) after Humboldt (1811), which do not occur in Peru.
† May not be an otter, according to Thomas (1908, p. 393).
‡ The species was known to occur in parts of formerly southwestern Peru now in Chile.

they seek refuge in the Altos, where they are found in vast numbers. Several modern travelers have lamented the diminution of the vicuñas, but without reason. In former times those animals were hunted more actively than at present.

Von Tschudi's journeys in the puna inspired him to poetic descriptions of the habits, particularly the visual propensities, of its denizens.

> Herds of vicuñas approached me with curious gaze, and then on a sudden fled with the swiftness of the wind. In the distance I observed stately groups of huanacos turning cautiously to look at me, and then passing on. The Puna stag (tarush) slowly advanced from his lair in the mountain recesses, and fixed on me his large, black, wondering eyes, whilst the nimble rock rabbits (viscachas) playfully disported and nibbled the scanty herbage growing in the mountain crevices. (Tschudi, Ross translation, 1847, p. 249)

On descending the eastern slope of the cordilleras to the subtropical zone inhabited by a greater variety of different kinds of mammals, von Tschudi (Ross translation, 1847, p. 275) romanticized that:

> . . . the swift-footed roe [*Mazama* sp.] of the Cordillera roams here and dwells in the thickets, avoiding the warm forest. The dark brown coati (*Nasua montana,* Tsch.) howls and digs at the root of trees in search of food, the shy opossum crawls fearfully under the foliage; the lazy armadillo creeps into his hole, but the ounce [*Felis onca*] and the lion [*Felis concolor*] seldom stray hither to contest with the black bear (*Ursus frugilegus* Tsch.) the possession of his territory. The

little hairy tapir (*Tapirus villosus,* Wagn.) ventures only at twilight out of his close ambush to forage in the long grass.

The systematic arrangement in the *Untersuchungen* is said to include all mammals known at the time to occur in Peru. By von Tschudi's count, the fauna consists of 119 species in 48 genera. These totals include domestic animals, the introduced house mouse, some duplicated names of native species, and a number of others not known to occur in Peru. In terms of currently recognized species found in Peru, von Tschudi's combined lists (1844a, pp. 244–255; 1844b, pp. 6–20; 21–264) consist of 87 species in 58 genera. The species are listed in Table 9 with von Tschudi's synonyms or misidentifications. Author attributions of the synonyms are omitted unless they are to von Tschudi himself. Vernacular names, if given, are included. Extralimital species are shown in brackets. In the case of unrevised groups or where two or more subspecies occur in Peru without possibility of determining which were described by von Tschudi, only the specific names are given.

XIII. Patagonia

Alcide Charles Victor d'Orbigny (1802–1857)

The French-born Alcide d'Orbigny was educated by his country's leading naturalists. His aptitudes were recognized by authorities of the Muséum National d'Histoire Naturelle, and with that institution's financial and material assistance, he sailed for South America charged with making a scientific survey of the southern half of the continent. Circumstances restricted his studies and collections of mammals almost entirely to Argentina and Bolivia.

D'Orbigny left France 31 July 1826 and arrived in Rio de Janeiro 24 September 1826 on his way to Montevideo where he landed on 29 September. The natural history of the region between Maldonado east of Montevideo and Buenos Aires engaged his attention for several months.

On 14 February 1827, d'Orbigny ascended the Río Paraná and arrived 15 March at the important fluvial port of Corrientes, capital of the province of the same name. With the town as base, d'Orbigny explored the province throughout much of one year.

On his return to Buenos Aires in April 1828, he made stops in Entre Ríos and Santa Fé. Beginning June 1828 and continuing through 1829, his attentions were devoted to faunal studies in the provinces of Buenos Aires and Río Negro. The chronology of the early part of 1829, as given by d'Orbigny (1835–1847) in the *Voyage,* confuses time spent in the two provinces with that spent in Corrientes. In any event, d'Orbigny was clearly in Buenos Aires and Río Negro during the last half of 1829. He returned to Montevideo in December 1829 and on 29 December sailed on to Patagonia and Chile.

Cape Horn was rounded on 19 January 1830 and Valparaiso, Chile, was reached 16 February. Because of the political unrest in the country, d'Orbigny sailed to the then Bolivian port of Cobija, where he landed on 8 April; 20 April found him in Arica and Tacna, both ports then in Peru's possession. After some investigation of the coast, d'Orbigny left Tacna on 19 May for La Paz, the mountain capital of Bolivia, arriving there 28 or 29 May.

For the next three years, d'Orbigny explored, mapped, and sampled the natural resources of the country. He crisscrossed Bolivia from La Paz east to the Paraguayan border and from Potosí in the south to the lower Río Mamoré in the north. D'Orbigny's actual itinerary is almost impossible to track because of the inaccuracies of the then available maps. Modern maps aided Pilleri and Arvy (1977) in their reconstruction of the itinerary in chronological sequence (fig. 18).

A complete account of d'Orbigny's South American journey with observations on and descriptions of the geology, paleontology, living plants, animals, and Indians is contained in seven huge volumes published serially from 1835 through 1847 in Paris under the title *Voyage dans l'Amérique Méridionale.* A full report on the mammals was reserved for the last, or perhaps a separate publication, but a turn in d'Orbigny's fortunes interrupted the work. A number of colored plates of mammals believed new to science and a few short articles on others had already been published. So that all would not be lost, a synoptic systematic report on the mammals collected was published in 1847 jointly with the distinguished mammalogist Paul Gervais, as number 2 of volume 4 of the *Voyage.* Brief notes on distribution and behavior accompany the abbreviated descriptions of each species. The species are listed in Table 10 with abstracted locality data. Scientific names used are current with synonyms and misidentifications added. The specimens are deposited in the Muséum National d'Histoire Naturelle in Paris.

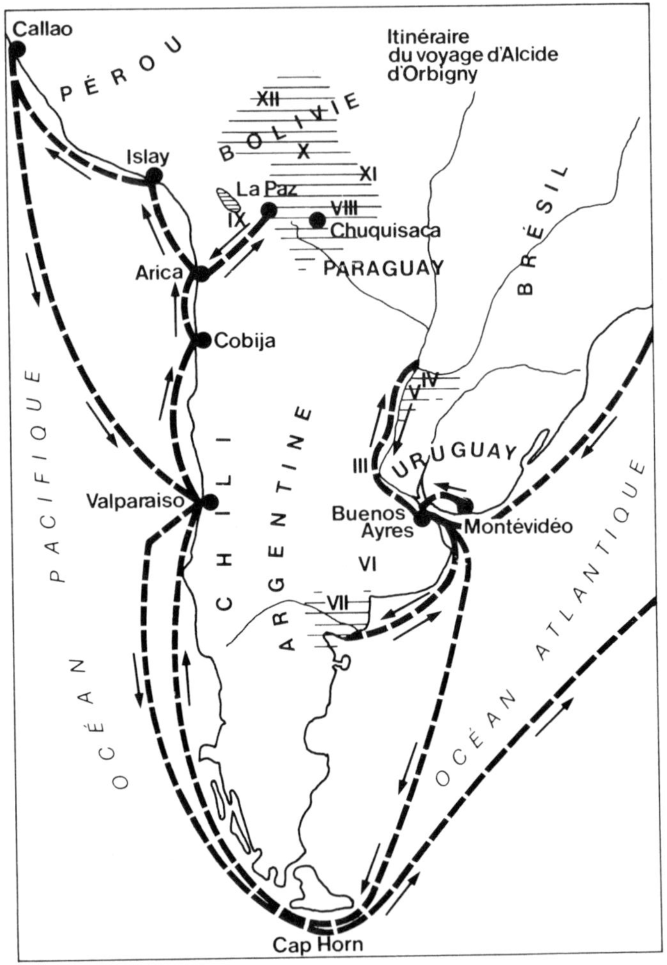

Fig. 2. Itinerario del viaje de Alcide d'Orbigny en Sudamérica.

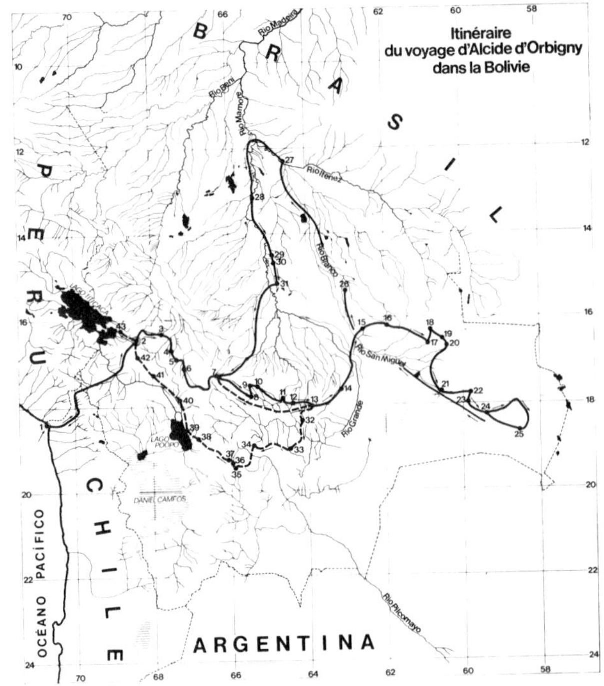

Fig. 3. Itinerario de la expedición de Alcide d'Orbigny en Bolivia.

1 Arica (IV. 1830)
2 La Paz
3 Chulumani
4 Cajuata
5 Inquisivi
6 Cavari
7 Cochabamba
8 Mizque
9 Pocona
10 Totora
11 Chilon
12 Pulquina
13 Samaipata
14 Santa Cruz (XI. 1830)
15 San Javier
16 Concepción (VII. 1831)
17 San Miguel
18 San Ignazio
19 Santa Ana
20 San Rafael (VIII. 1831)
21 San José
22, 23 San Juan
24 Santiago
25 Santa Ana (X. 1831)
26 Río Blanco
27 Fuerte Príncipe da Beira (1832)
28 Exaltación
29 San Xavier
30 Trinidad
31 Loreto
32 Vallegrande
33 Toncina
34 Sucre
35 Potosí
36 Tarapaya
37 Yocalla
38 Ancacato
39 Lago Poopo
40 Oruro
41 Sica Sica
42 Calamarca
43 Lago Titicaca (1832)

Fig. 18. **Left,** map of southern South America, and **right,** map of Bolivia, showing routes of d'Orbigny Expedition and localities visited; from Pilleri and Arvy (1977).

FIG. 19. Animals of the d'Orbigny Bolivian Expedition: **upper left,** *Callithrix entomophagus* d'Orbigny (= *Saimiri boliviensis boliviensis* I. Geoffroy and Blainville); **upper right,** *Callithrix donacophilus* d'Orbigny (= *Callicebus donacophilus donacophilus*); **lower left,** *Felis geoffroyi* d'Orbigny and Gervais (= *Felis colocolo geoffroyi*); **lower right,** *Mephitis humboldtii* (= *Conepatus chinga suffocans* Illiger); from d'Orbigny and Gervais (1847).

TABLE 10. Mammals of the southern half of South America, mostly Bolivia and Argentina, recorded by d'Orbigny and Gervais (1847); the arrangement is phylogenetic.

Current name	d'Orbigny and Gervais synonym	Locality	Figure
CHIROPTERA			
Noctilio albiventris Desmarest, 1818	*Noctilio affinis* d'Orbigny, 1835	BOLIVIA: Moxos Province	...
Noctilio leporinus rufipes d'Orbigny, 1835	...	BOLIVIA: Chiquitos and Moxos provinces	...
Tonatia sylvicola d'Orbigny, 1835	...	BOLIVIA: Yuracaré territory, base of eastern Cordillera	...
Artibeus planirostris Spix, 1823	Not *Vespertilio perspicillatus* Linnaeus, 1758	BOLIVIA: Chiquitos Province	...
Desmodus rotundus É. Geoffroy, 1810	*Desmodus rufus* Wied-Neuwied, 1824; *Edostoma cinerea* d'Orbigny, 1835	BOLIVIA: Chiquitos	...
Myotis nigricans Wied-Neuwied, 1821	*Vespertilio hypothrix* d'Orbigny and Gervais, 1847	BOLIVIA: Moxos	...
Eptesicus furinalis d'Orbigny and Gervais, 1847	...	ARGENTINA: Corrientes	...
Myotis albescens É. Geoffroy, 1806	*Vespertilio isidori* d'Orbigny and Gervais, 1847	ARGENTINA: Corrientes	...
Myotis ruber É. Geoffroy, 1806	...	ARGENTINA: Corrientes	...
Histiotus velatus I. Geoffroy, 1824	...	BOLIVIA: Chuquisaca	...
Tadarida brasiliensis I. Geoffroy, 1824	*Molossus rugosus* d'Orbigny, 1835, not *Molossus nasutus* Spix, 1823	ARGENTINA: Corrientes	...
Molossus crassicaudatus É. Geoffroy, 1805	*Molossus moxensis* d'Orbigny, 1835; *Molossus velox* Temminck, 1827	BOLIVIA: Moxos and Chiquitos provinces	...
PRIMATES			
Saimiri boliviensis boliviensis I. Geoffroy and Blainville, 1834	*Calithrix* (*sic*) *entomophagus* d'Orbigny, 1835	BOLIVIA: Chiquitos; Moxos; Santa Cruz	19
Callicebus donacophilus donacophilus d'Orbigny, 1835	...	BOLIVIA: Moxos Province	19
Alouatta seniculus sara Elliot, 1910	Not *Stentor stramineus* É. Geoffroy	BOLIVIA: Santa Cruz; Chiquitos; Moxos	...
Cebus apella paraguayanus Fischer, 1829	*Cebus fulvus* var.	BOLIVIA: near Santa Cruz de la Sierra	...
CARNIVORA			
Dusicyon gymnocercus Fischer, 1814	Not *Canis cancrivorus* Desmarest, 1820	BOLIVIA: Chiquitos	...
Chrysocyon brachyurus Illiger, 1815	*Canis jubatus* Desmarest, 1820	Tropical South America to 41°S	...
Tremarctos ornatus F. Cuvier, 1825	...	BOLIVIA: Cochabamba; Chuquisaca	...
Procyon cancrivorus nigripes Mivart, 1886	Not *Procyon cancrivorus* Cuvier, 1798	BOLIVIA: Chiquitos; ARGENTINA: Corrientes	...
Nasua nasua solitaria Wied-Neuwied, 1821	*Nasua fusca* Desmarest, part	BOLIVIA: tropics to 30°S	...
Potos flavus Schreber, 1774	*Cercoleptes caudivolvulus* Schreber, 1774	BOLIVIA: foot of eastern cordillera	...
Lyncodon patagonicus Blainville, 1842	...	ARGENTINA: Río Negro	...
Galictis cuja furax Thomas, 1907	Not *Mustela brasiliensis* Gmelin, 1788	...	...
Conepatus chinga suffocans Illiger, 1815	*Mephitis castaneus* d'Orbigny and Gervais, 1847, not *Mephitis humboldtii* Gray, 1837	...	19
Lutra platensis Waterhouse, 1838	...	ARGENTINA: Río Paraná in Provinces Buenos Aires and Corrientes	...

TABLE 10. *Continued.*

Current name	d'Orbigny and Gervais synonym	Locality	Figure
CARNIVORA			
Felis colocolo pajeros Desmarest, 1816	...	ARGENTINA: from 35°–45°S	...
Felis geoffroyi d'Orbigny and Gervais, 1847	...	ARGENTINA: Pampas to 44°S	19
Felis concolor Linnaeus, 1771	...	BOLIVIA; ARGENTINA: to Straits of Magellan	...
Felis onca Linnaeus, 1758	...	Tropical South America not beyond 40°S; ARGENTINA: Pampas; Serranía de Tandil	...
PINNIPEDIA			
Otaria flavescens Shaw, 1800	*Otaria jubata* Schreber, 1776	ARGENTINA: S mouth Río Negro	...
Arctocephalus australis Zimmermann, 1782	*Otaria porcina* Molina, 1782	ARGENTINA: coast; PERU: coast	...
Mirounga leonina Linnaeus, 1758	*Phoca proboscidea* Peron, 1817	ARGENTINA: Río Negro, near mouth	...
ARTIODACTYLA			
Mazama gouazoubira Fischer, 1814	*Cervus simplicicornis* Illiger, 1815	Tropical South America to 28°S	...
Blastoceros bezoarticus Linnaeus, 1758	Not *Cervus campestris* F. Cuvier, 1817	Lowland savannas to northern Patagonia	...
Hippocamelus antisensis d'Orbigny, 1834	...	BOLIVIA: La Paz; Cochabamba; Chuquisaca; rarely below 3500 m	20
Blastocerus dichotomus Illiger, 1815	*Cervus paludosus* Desmarest, 1822	ARGENTINA: Corrientes; BOLIVIA: Chiquitos	...
RODENTIA			
Sciurus spadiceus Olfers, 1818	Not *Sciurus igniventris* Wagner, 1842	BOLIVIA: Chiquitos	...
Eligmodontia typus F. Cuvier, 1837	...	ARGENTINA: Corrientes	...
Octodon degus Molina, 1782	...	CHILE: Santiago de Chile	...
Octodontomys gliroides, Gervais and d'Orbigny, 1844	...	BOLIVIA: La Paz	...
Ctenomys boliviensis Waterhouse, 1848	Not *Ctenomys brasiliensis* Blainville, 1826	ARGENTINA: Corrientes; BOLIVIA: Santa Cruz de la Sierra	...
Ctenomys magellanicus Bennett, 1835	...	ARGENTINA: northern Patagonia	...
Microcavia australis Gervais and d'Orbigny, 1833	...	ARGENTINA: Río Negro	...
Galea flavidens Brandt, 1835	...	BOLIVIA: Cochabamba; Chuquisaca; La Paz	...
Dolichotis patagonum Zimmerman, 1780	*Dasyprocta patachonica* Desmarest, 1820	ARGENTINA: northern Patagonia; Corrientes	...
Dasyprocta azarae Lichtenstein, 1827	Not *Dasyprocta nigricans* Wagner, 1842	Tropical South America	...
CETACEA			
Inia boliviensis d'Orbigny, 1834	...	BOLIVIA: rivers of Moxos and Chiquitos	20
[*Pontoporia blainvillei* Gervais and d'Orbigny, 1844; not part of d'Orbigny collection]	...	URUGUAY: Montevideo	...
Lagenorhynchus cruciger Quoy and Gaimard, 1824	...	Atlantic Ocean (57°–76°S, E and S of Cape Horn)	...
Lissodelphis peroni Lacépède, 1804	...	Atlantic Ocean (48°–64°S); Atlantic-Pacific Oceans around Cape Horn	...

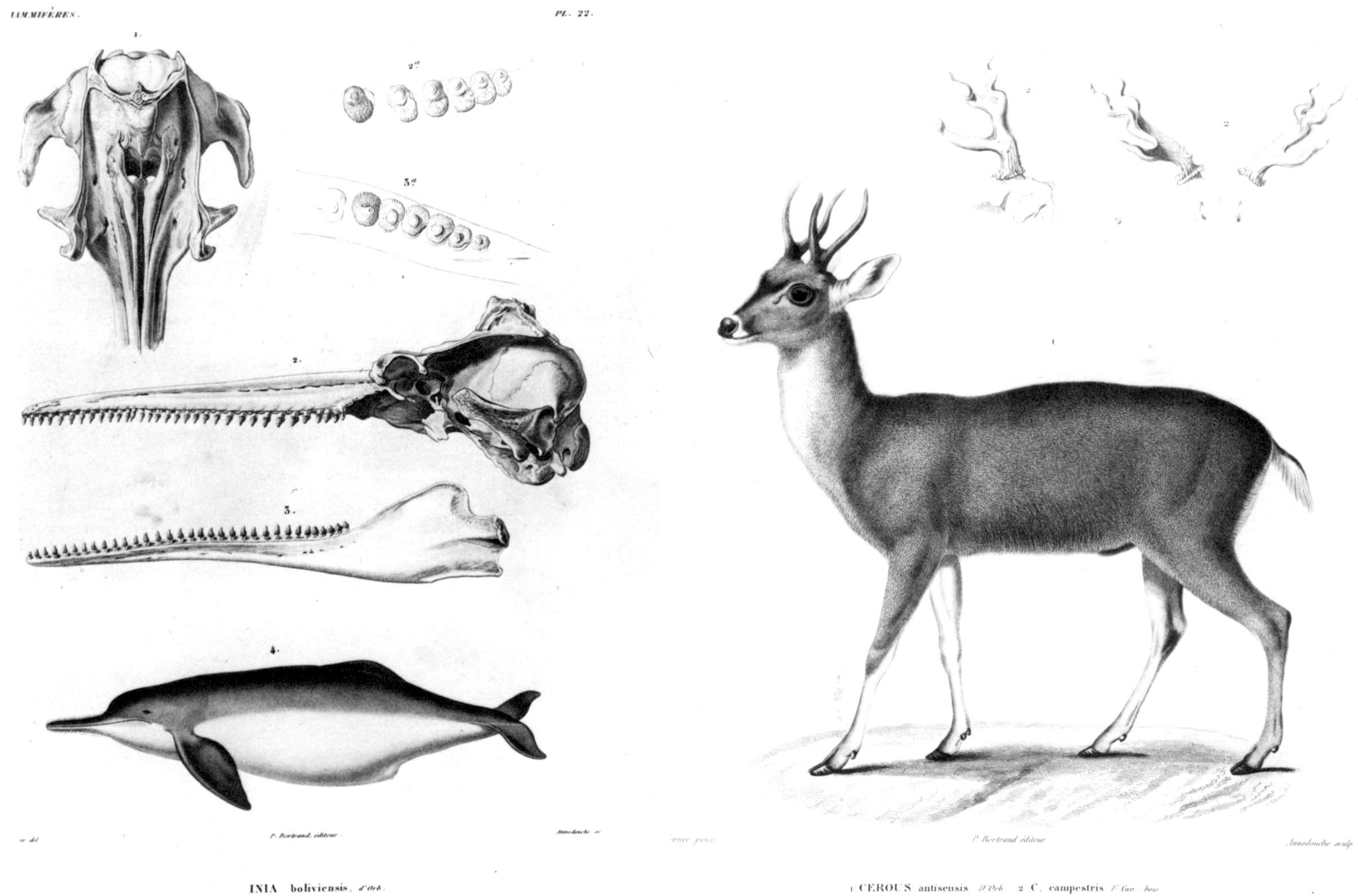

FIG. 20. Animals of the d'Orbigny Bolivian Expedition: **left,** *Inia boliviensis* d'Orbigny, skull, teeth, and left side of animal; **right,** *Cervus antisensis* d'Orbigny (= *Hippocamelus antisensis*) and variable antlers; from d'Orbigny and Gervais (1847).

Charles Robert Darwin (1809–1882)

Charles Robert Darwin was born in Shrewsbury, England, to a wealthy and distinguished family. Although his early schooling emphasized the classics, Darwin's interests since boyhood were in natural history, particularly of the insects he collected, and in hunting as a sport. As a university student, he dropped out of medical school after two years, then took up theology, and abandoned that after three years. Nevertheless, through the influence of his teachers, he developed and sharpened his interests in biology and geology, and his reading of Humboldt's *Personal Narrative of Travels to the Equinoctial Regions of America* fired him with a zeal for travel and discoveries in distant and unexplored lands.

The opportunity for travel in exotic parts soon came. At age 22, with his mostly self-acquired knowledge of geology and systematic biology and experience as a collector and hunter, Charles Darwin accepted the unsalaried post of Naturalist on H.M.S. *Beagle* for a five-year cruise of chronometrical explorations round the world. The experiences on the voyage, which began 27 December 1831 (fig. 21), transformed Darwin into the leading naturalist of his time and were the prime source of inspiration for Darwin's theory of organic evolution by natural selection.

The *Beagle* touched the South American mainland at Bahia (now Salvador), Brazil, on 29 February 1832 for a short stay. Before the ship left for Rio de Janeiro in March, Darwin captured and prepared for study a specimen of the very common phyllostomid bat, *Carollia perspicillata* Linnaeus, his first mammalian specimen of the expedition.

In Rio de Janeiro, Darwin was taken on a hunt by an old Portuguese priest. Two howler monkeys (*Alouatta fusca* É. Geoffroy), described by Darwin (1839, p. 32) as "two large bearded monkeys," had been shot the day before by his companion. Darwin wrote:

> These animals have prehensile tails, the extremity of which, even after death, can support the whole weight of the body. One of them thus remained fast to a branch, and it was necessary to cut down a large tree to procure it. This was soon effected and down came tree and monkey with an awful crash.

The priest later presented Darwin with an eyra cat (*Herpailurus yagouaroundi eyra* Fischer) that had just been killed in the Gavea mountain.

In July 1832 the *Beagle* left Brazil for the Patagonian subregion. Up to this time, Darwin's zoological collections consisted mainly of insects and mollusks. Because only negligible contact with the rich mammalian fauna of Brazil had been made, Darwin was deprived of a basis for direct comparisons with the comparatively poor but largely unique mammalian fauna of the Patagonian subregion, which he studied zealously. As a result, his attention focused on morphological and ecological differences between the individual species (or subspecies) he collected or observed in La Plata, Bahía Blanca, Patagonia, the Falklands, Chile, and the Galápagos and the same or nearly related species of Paraguay and Chile described by Azara and Molina. How much would Darwin's concept of the origin of life been affected if his thoughts had been directed primarily to faunas and faunal regions rather than to species and their geographic variation?

The *Beagle* remained in the area of La Plata from July 1832 to July 1833, affording Darwin opportunities to collect near Maldonado, a short distance up the coast from Montevideo. The *Beagle* then sailed south to the mouth of the Río Negro. While the vessel's crew mapped and took soundings up and down the coast between the Río Negro and Río Plata, Darwin made a number of excursions into the Pampas, Bahía Blanca, Sierra de la Ventana, Río Colorado, Río Paraná, and Río Uruguay. Many observations were made on the behavior and habitat of mammals characteristic of the region, but few animals were actually collected. Among the species mentioned are armadillos (known as pichiy, peludo, apar, and mulita), the Patagonian hare or mara (misnamed "agouti"), the capybara, cavia, skunk, puma, jaguar, guanaco, and pampas deer. Darwin (1839, p. 144) was fascinated by the viscacha's packrat-like habits such as:

> dragging every hard object to the mouth of its burrow; around each group of holes many bones of cattle, stones, thistle stalks, hard lumps of earth, dry dung, etc., are collected into an irregular heap, which frequently amounts to as much as a wheelbarrow would contain. I was credibly informed that a gentleman, when riding on a dark night, dropped his watch; he returned in the morning, and by searching the neighborhood of every bizcacha hole on the line of the road, as he expected, soon found it.

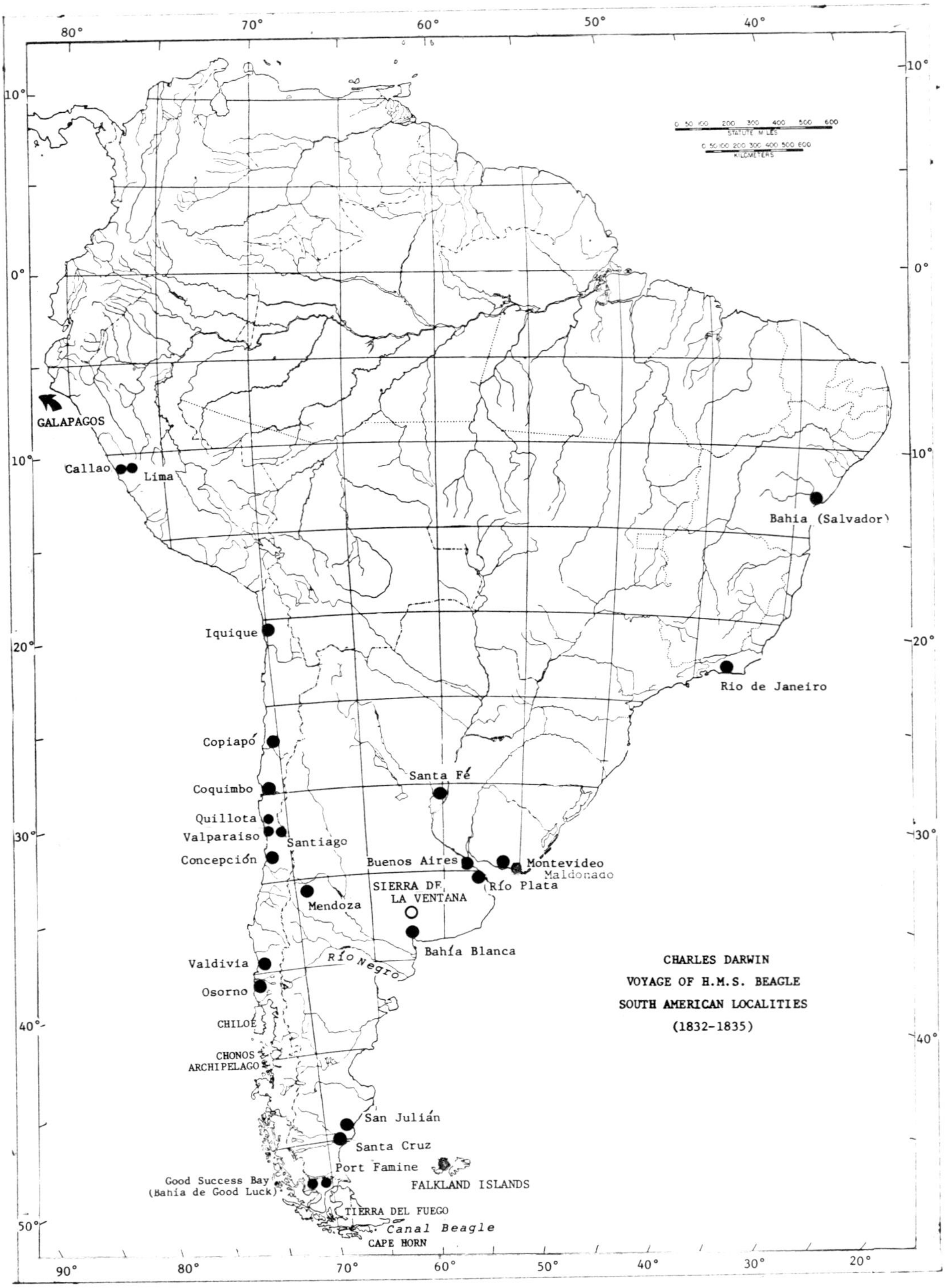

FIG. 21. Map showing principal South American stations visited by Charles Darwin (1832–1835) on world cruise of H.M.S. *Beagle* (1832–1836).

FIG. 22. H.M.S. *Beagle* laid ashore at Santa Cruz, Argentina; from Darwin (1839).

TABLE 11. Mammals collected or observed by Darwin in the Maldonado Region, Uruguay and parts of Argentina, and those recorded by Waterhouse (1838–1839); the arrangement is phylogenetic.

Current name	Waterhouse synonym or misidentification	Locality	Figure
MARSUPIALIA			
Didelphis albiventris Lund	*Didelphis azarae* Auct.	URUGUAY: Maldonado	...
Lutreolina crassicaudata Desmarest	...	URUGUAY: Maldonado	...
Monodelphis dimidiata Wagner	*Didelphis brachyura* Auct.	URUGUAY: Maldonado	22
CHIROPTERA			
Tadarida brasiliensis I. Geoffroy	Not *Dysopes nasutus* Spix	URUGUAY: Maldonado	...
EDENTATA			
Dasypus hybridus Desmarest	...	URUGUAY: Maldonado	...
Zaedyus pichiy Desmarest	*Dasypus minutus* Auct.	ARGENTINA: Banda Oriental, Entre Ríos	...
Chaetophractus villosus Desmarest	...	ARGENTINA: Bahía Blanca (observed)	...
Tolypeutes matacus Desmarest	...	ARGENTINA: Bahía Blanca (observed)	...
CARNIVORA			
Dusicyon gymnocercus Fischer	Not *Canis azarae* Wied-Neuwied	ARGENTINA: La Plata (observed)	...
Felis colocolo pajeros Desmarest	...	ARGENTINA: Bahía Blanca	...
Galictus cuja furax Thomas	Not *Galictis vittata* Schreber	URUGUAY: Maldonado	...
Lutra platensis Waterhouse	...	URUGUAY: Maldonado	...
Conepatus chinga gibsoni Thomas	...	ARGENTINA: Bahía Blanca (observed)	...
Felis concolor acrocodia Goldman	...	ARGENTINA: the pampas (observed)	...
Felis onca palustris Ameghino	...	ARGENTINA: in the Río Paraná (observed)	...
ARTIODACTYLA			
Blastoceros bezoarticus Linnaeus	Not *Cervus campestris* Cuvier	URUGUAY: Maldonado; ARGENTINA: Bahía Blanca; Río Negro	...
Lama guanicoe Müller	...	ARGENTINA: Río Negro (observed)	...
RODENTIA			
Myomorpha			
Oryzomys flavescens Waterhouse	...	URUGUAY: Maldonado	...
Calomys laucha Olfers	*Mus bimaculatus* Waterhouse; *Mus gracilipes* Waterhouse	ARGENTINA: Bahía Blanca	...
Eligmodontia typus Cuvier	*Mus elegans* Waterhouse	ARGENTINA: Bahía Blanca	...
Holochilus brasiliensis darwini Thomas	...	ARGENTINA: Bahía Blanca	...
Reithrodon physodes typicus Waterhouse	...	URUGUAY: Maldonado	...
Akodon azarae Fischer	*Mus arenicola* Waterhouse	URUGUAY: Maldonado	...
Akodon colibreve Brants	*Mus obscurus* Waterhouse	URUGUAY: Maldonado	...
Scapteromys tumidus Waterhouse	...	URUGUAY: Maldonado	...
Oxymycterus rufus nasutus Waterhouse	...	URUGUAY: Maldonado	...
Caviomorpha			
Cavia porcellus Linnaeus	*Cavia cobaia* Auct.	URUGUAY: Maldonado	...
Hydrochaeris hydrochaeris Linnaeus	*Hydrochoerus capybara* Auct.	URUGUAY: Maldonado	...

TABLE 11. *Continued.*

Current name	Waterhouse synonym or misidentification	Locality	Figure
Caviomorpha (*continued*)			
Dolichotis patagonum Zimmerman	. . .	ARGENTINA: Río Negro (observed)	. . .
Vizcacia maximus Desmarest	*Lagostomus trichodactylus* Brookes	URUGUAY: Maldonado	. . .
Ctenomys brasiliensis Blainville	. . .	URUGUAY: Maldonado	. . .

Darwin could not explain the viscacha's behavior.

Pampas deer (*Blastoceros bezoarticus*) were abundant throughout the La Plata region. Darwin (1839, p. 55) saw

> very many small herds, containing from five to seven animals each, near the Sierra Ventana, and among the hills north of Maldonado. If a person crawling close along the ground, slowly advances toward a herd, the deer frequently, out of curiosity, approach to reconnoitre him. I have by this means killed, from one spot, three out of the same herd. Although so tame and inquisitive, yet when approached on horseback, they are exceedingly wary. In this country nobody goes on foot, and the deer knows man as its enemy only when he is mounted and armed with the bolas.

The jaguar by some accounts is a man-killer, by others, fears man. Darwin (1839, p. 159) records several instances reported to him of man-killing jaguars of the Río Paraná region.

The *Beagle* left the Río Plata on December 1833 for Puerto Deseado, or Port Desire, on the Patagonian coast. The mammals collected by Darwin and reported by Waterhouse (1838–1839), with descriptions and supplementary notes by Darwin, are listed in Table 11, with the Waterhouse synonyms (misidentifications included). Added are the few species Darwin mentioned in his *Journal* but did not collect. Unless otherwise indicated, all species are from the neighborhood of Maldonado, Uruguay.

The geology and natural history of Patagonia investigated by Darwin included those of the Straits of Magellan and Tierra del Fuego (December 1832–January 1833; May–June 1834), Puerto Deseado (Port Desire) (December 1833–January 1834), Santa Cruz (April–May 1834), and the Falkland Islands (March 1834). The *Beagle* itself (fig. 22) sailed up the Río Santa Cruz to a point 140 miles from its mouth in the Atlantic Ocean to about 60 miles from the nearest arm of the Pacific Ocean on the opposite side of the cordillera.

Darwin was greatly impressed by the number, variety, and great size of fossil mammals, mostly Pleistocene, exposed on the Patagonian plains. These, he (1839, p. 209) believed, were confirmation of the "law" that existing animals in an area have a close relation in form with extinct species in the same area. The natural causes for extinction, however, eluded Darwin. After proposing and rejecting a number of explanations, the nonevolutionist Darwin (1839, p. 212) concluded that

> the whole series of animals, which have been created with peculiar kinds of organization, are confined to certain areas; and we can hardly suppose these structures are only adaptations to peculiarities of climate or country; for otherwise, animals belonging to a distinct type, and introduced by man, would not succeed so admirably even to the extermination of the aborigines. On such grounds it does not seem a necessary conclusion that the extinction of species, more than their creation, should exclusively depend on the nature (altered by physical change) of their country. All that at present can be said with certainty, is that, as with the individual, so with the species, the hour of life has run its course, and is spent.

The small number of extant large mammals and great number and variety of small mammals, also impressed Darwin (1839, p. 215).

> Patagonia, poor as she is in some respects, can, however, boast of a greater stock of small rodents than perhaps, any other country in the world. Several species of mice are externally characterized by large thin ears and a very fine fur. These little animals

swarm amongst the thickets in the valleys, where they cannot for months together taste a drop of water. They all seem to be cannibals, for no sooner was a mouse caught in one of my traps than it was devoured by others. A small and delicately-shaped fox which is likewise very abundant, probably derives its entire support from these small animals.

The guanaco was regarded as the characteristic quadruped of the Patagonian plains (Darwin, 1839, p. 215).

> Herds of fifty or a hundred were common, and, as I have said, we saw one which must have contained at least five hundred. The puma with the condor in its train, follows and preys upon these animals. The footsteps of the former were to be seen almost everywhere on the banks of the river [Santa Cruz]; and the remains of several guanaco, with their necks dislocated, and bones broken, showed how they had met their death.

In March 1834, Darwin visited the Falkland Islands. On a tour he encountered large numbers of horses, cattle, swine, and rabbits (*Oryctolagus cuniculus* Linnaeus [= *Lepus magellanicus* Lesson and Garnot] in domestic and feral states. The animals had been brought by French colonists in 1764. Darwin wrote (p. 249),

> The only quadruped native to the island, is a large wolf-like fox [*Dusicyon* (?*culpaeus*) *australis* Kerr] which is common to both East and West Falkland. I have no doubt it is a peculiar species and confined to this archipelago. . . . These wolves are well known . . . [for] their tameness and curiosity. . . . To this day their manners remain the same. . . . As far as I am aware, there is no other instance in any part of the world, of so small a mass of broken land, distant from a continent, possessing so large a quadruped peculiar to itself. Their numbers have rapidly decreased; they are already banished from that half of the island which lies to the eastward of the neck of land between St. Salvador Bay and Berkeley Sound. Within a very few years after these islands shall have become regularly settled, in all probability this fox will be classed with the dodo, as an animal which has perished from the face of the earth.

The mammals collected by Darwin in the Argentine Patagonia (including Falkland Islands) and bordering parts of the Chilean Straits of Magellan are listed in Table 12.

The Chilean leg of the cruise began in May 1834 with the passage of the *Beagle* into the eastern mouth of the Straits of Magellan and ended July 1835 with departure from Copiapó in northern Chile. While the *Beagle* sailed up and down the Chilean coast, Darwin explored the coast, islands, archipelagos, and cordillera. He crossed the Andes on 21 March 1835 through the Portillo Pass south of Santiago, and proceeded to the town of Mendoza in Argentina. Differences observed between the biota of eastern and western versants of the cordillera impressed Darwin. The mountains, he (1839, p. 399) reasoned,

> have existed as a great barrier, since a period so remote that whole races of animals must subsequently have perished from the face of the earth. Therefore, unless we suppose the same species to have been created in two different countries, we ought not to expect any closer similarity between the organic beings on opposite sides of the Andes, than on shores separated by a broad strait of the sea.

The correlation between geographic isolation and faunal peculiarity was noted in other circumstances. Darwin (1839, p. 439) observed that

> next to lizards, mice appear to be able to support existence on the smallest and driest portions of the earth—even on islets in the midst of great oceans. I believe it will be found, that several islands, which possess no other warm-blooded quadruped, have small rodents peculiar to themselves.

Ratadas or rat plagues in Chile also caught Darwin's attention. One of the earliest recorded for *Oryzomys longicaudatus longicaudatus,* viewed through the eyes of Darwin (in Waterhouse, 1838, p. 40), "overran the wooded country south of Concepción, in swarms of infinite numbers."

The mammals of Tierra del Fuego tallied by Darwin (1839, p. 300) included, besides cetaceans and phocids,

> one bat [not named but likely *Histiotus montanus magellanicus* Philippi], a mouse with grooved front teeth (*Reithrodon* of Waterhouse) and two other species, the tu-

TABLE 12. Patagonian mammals collected by Darwin and recorded by Waterhouse (1838–1839); arrangement is phylogenetic.

Current name	Waterhouse synonym or misidentification	Locality	Figure
CARNIVORA			
Dusicyon australis Kerr	*Canis antarcticus* Shaw	Falkland Islands	...
Dusicyon griseus Gray	Not *Canis azarae* Wied-Neuwied	ARGENTINA: Patagonia	...
Felis colocolo pajeros Desmarest	...	ARGENTINA: Santa Cruz	22
ARTIODACTYLA			
Lama guanicoe Müller	*Auchenia llama* Desmarest	ARGENTINA: Patagonia	...
RODENTIA			
Oryzomys longicaudatus magellanicus Bennett	...	ARGENTINA: Puerto de Hambre (Port Famine); CHILE: Straits of Magellan	...
Akodon xanthorhinus Waterhouse	...	CHILE: Península de Hardy	...
Akodon canescens Waterhouse	...	ARGENTINA: Puerto Deseado (Port Desire); Santa Cruz	...
Auliscomys micropus Waterhouse	...	ARGENTINA: Río Santa Cruz, Santa Cruz	...
Graomys griseoflavus Waterhouse	...	ARGENTINA: Río Negro	...
Phyllotis xanthopygus Waterhouse	...	ARGENTINA: Puerto Deseado (Port Desire); Santa Cruz	...
Reithrodon physodes cuniculoides Waterhouse	...	ARGENTINA: Puerto Deseado (Port Desire); San Julián; Río Santa Cruz, Santa Cruz	...
Euneomys chinchilloides Waterhouse	...	ARGENTINA: Eastern entrance to Straits of Magellan	...
Myocastor coypus Molina	...	ARGENTINA: Río Chubut	...
Microcavia australis I. Geoffroy and d'Orbigny	...	ARGENTINA: 41°S to Straits of Magellan	...
Dolichotis patagonum Zimmerman	*Cavia patachonica* Shaw	ARGENTINA: Patagonia	...
CETACEA			
Lagenorhynchus cruciger Quoy and Gaimard	*Delphinus fitzroyi* Waterhouse	ARGENTINA: Golfo San José, 42°30′S, Chubut	...

> cotuco (the greater number of these rodents are confined to the eastern and dry part), a fox, sea-otter, guanaco, and one deer [unnamed but likely *Hippocamelus bisulcus*]. The latter animal is rare, and is not, I believe, to be found south of the Straits of Magellan, as happens with the others.

With respect to geographic distribution, Darwin (1839, p. 300),

> observing the general correspondence of the cliffs of soft sandstone, mud, and shingle, on the opposite side of the Strait, together with those on some intervening islands [was] strongly tempted to believe that the land was once joined and thus allowed animals so delicate and helpless as the tucotuco, and Reithrodon to pass over.

The tucotuco in question is *Ctenomys magellanicus fueginus* Philippi (Osgood, 1943, p. 119). Darwin (1839, p. 327) also mentioned the occurrence of the puma (*Felis concolor*) in Tierra del Fuego, and related something of its habits in other parts of Chile and Argentina.

The type specimen of Darwin's zorro (*Dusicyon fulvipes* Martin), peculiar to the island of Chiloé, was discovered by Darwin (p. 341) on 6 December 1834 sitting on the rocks and so intently absorbed in watching the maneuvers of two ship's officers engaged in surveying,

> that I was able, by quietly walking up behind, to knock him on the head with my geological hammer. This fox, more curious or more scientific, but less wise, than the generality of his brethren, is now mounted in the museum of the Zoological Society.

Sea otters (*Lutra felina* Molina) were described by Darwin (in Waterhouse, 1838, p. 24) as exceedingly common amongst the innumerable channels and bays which form the Chonos Archipelago.

> . . . they may generally be seen quietly swimming with their heads just out of water amidst the great entangled beds of kelp, which abounds on this coast. They burrow in the ground, within the forest, just above the rocky shore, and I was told, that they sometimes roam about through the woods. This otter does not, by any means, live exclusively on fish. One was shot whilst running to its hole with a large volute-shell in its mouth; another (I believe the same species) was seen in Tierra del Fuego devouring a cuttle fish. But in the Chonos Archipelago perhaps the chief food of this animal, as well as that of the immense herds of great seals, and flocks of terns and cormorants, is a red-coloured crab (belonging to the family Marrouri) of the size of a prawn, which swims near the surface in such dense bodies, that the water appears of a red colour. This specimen weighed nine pounds and a half.

The vampire bat which Darwin (1839, p. 25) recognized as a species of d'Orbigny's genus *Edostoma* (= *Desmodus*) was singled out as

> often the cause of much trouble, by biting the horses on their withers. The injury is generally not so much owing to the loss of blood, as to the inflammation which the pressure of the saddle afterwards produces. The whole circumstance has lately been doubted in England; I was therefore fortunate in being present when one was actually caught on a horse's back. We were bivouacking late one evening near Coquimbo, in Chile, when my servant, noticing that one of the horses was very restive, went to see what was the matter, and fancying he could distinguish something, suddenly put his hand on the beast's withers, and secured the vampire. In the morning, the spot, where the bite had been inflicted, was easily distinguished from being slightly swollen and bloody. The third day afterwards we rode the horse, without any ill effects.

The Chilean mammals collected and others, only observed by Darwin, are listed in Table 13.

Departing Chile on 12 July 1835, the *Beagle* sailed north along the Peruvian coast before turning west to the Galápagos Islands. Darwin's accounts of the stopovers in Iquique, Callao, and Lima make no mention of indigenous mammals.

The Galápagos Archipelago, it seemed to Darwin (1839, p. 454), was "a little world within itself; the greater number of its inhabitants both vegetable and animal being found nowhere else." Darwin (p. 464) "endeavoured to make as nearly a perfect collection in every branch as time permitted" but the only land mammals he found were the Chatham Island rice rats described by Waterhouse as *Oryzomys galapagoensis* (fig. 23), and the introduced *Rattus* on James Island.

Darwin's investigations of the Galápagos fauna, particularly the birds, lizards, tortoises, and certain plants stirred old beliefs and generated new and conflicting thoughts. However, at the time he wrote his journal in October 1835, Darwin (1839, p. 474) made no

> attempt to come to any definite conclusions, as the species have not [as yet] been accurately examined; but we may infer, that, with the exception of a few wanderers, the organic beings found on this archipelago are peculiar to it; and yet their general form strongly partakes of an American character. . . . This similarity in type between distant islands and continents, while the species are distinct, has scarcely been noticed. The circumstances would be explained according to the views of some authors, by saying that the creative power had acted according to the same law over a wide area.

Writers on Darwin, quoting from his revised (1845) edition of the *Journal,* attribute to Darwin more foresight on the origin of species than is apparent in the first (1839) edition quoted here. At the time of its publication, two years delayed, Darwin, still a creationist and believer in the immutability of species, had yet to know the identity or specific affinities of the vast majority of the plants and animals he had collected. This knowledge served him later for definition and elaboration of thoughts expressed in the second and other revised editions of the *Journal,* but not in the first.

The following impressions of the biota of the Galápagos Islands in the second edition (p. 372 of an 1899 "authorized edition") and oft quoted in whole or in part by various authors, are absent in the first.

> The natural history of the islands is eminently curious and well deserved attention. Most of the organic productions are aboriginal creations, found nowhere else; there is

TABLE 13. Chilean mammals collected or only observed by Darwin, and those identified by Waterhouse (1838–1839); the arrangement is phylogenetic.

Current name	Waterhouse synonym or misidentification	Locality	Figure
MARSUPIALIA			
Marmosa elegans Waterhouse	...	Valparaiso	...
CHIROPTERA			
Histiotus montanus magellanicus Philippi	...	Tierra del Fuego (observed)	...
Myotis chiloensis Waterhouse	...	Chiloé	...
Tadarida brasiliensis I. Geoffroy	Not *Dysopes nasutus* Spix	Valparaiso	...
Desmodus rotundus dorbignyi Waterhouse	...	Coquimbo	...
CARNIVORA			
Dusicyon culpaeus magellanicus Gray	...	Copiapó; Straits of Magellan	...
Dusicyon fulvipes Martin	...	Chiloé	22
Dusicyon griseus Gray	Not *Canis azarae* Wied-Neuwied	Copiapó; Straits of Magellan	...
Lutra felina Molina	*Lutra chilensis* Bennett	Chonos Archipelago	...
Felis concolor Linnaeus	...	Tierra del Fuego and central Chile to 10,000 ft elevation (observed)	...
ARTIODACTYLA			
Hippocamelus bisulcus Molina	...	Tierra del Fuego (observed)	...
RODENTIA			
Oryzomys longicaudatus longicaudatus Bennett	...	Concepción	...
Oryzomys longicaudatus magellanicus Bennett	...	Puerto de Hambre, Straits of Magellan	...
Akodon olivaceus olivaceus Waterhouse	*Mus renggeri* Waterhouse	Valparaiso; Coquimbo	...
Akodon olivaceus brachiotus Waterhouse	...	Chonos; Chiloé	...
Akodon xanthorhinus xanthorhinus Waterhouse	...	Hardy Peninsula, Tierra del Fuego	...
Abrothrix longipilis longipilis Waterhouse	...	Coquimbo	...
Phyllotis darwini darwini Waterhouse	...	Coquimbo	22
Reithrodon chinchilloides Waterhouse	...	Straits of Magellan	...
Abrocoma bennetti Waterhouse	*Abrocoma cuvieri* Waterhouse	Valparaiso; Aconcagua	...
Spalacopus cyanus Molina	*Poephagomys ater* Cuvier	Valparaiso	...
Myocastor coypus Molina	...	Chonos Archipelago	...
Octodon degus Molina	*Octodon cummingii* Bennett	Valparaiso	...
Ctenomys magellanicus fueginus Philippi	...	Tierra del Fuego (observed)	...

even a difference between the inhabitants of the different islands; yet all show a marked relationship with those of America, though separated from that continent by an open space of ocean, between 500 and 600 miles in width. The archipelago is a little world within itself, or rather a satellite attached to America, whence it has derived a few stray colonists, and has received the general character of its indigenous productions. Considering the small size of these islands, we feel the more astonished at the number of their aboriginal beings, and at their confined range. Seeing every height crowned with its crater, and the boundaries of most of the lava-streams still distinct, we are led to believe that within a period, geologically recent, the unbroken ocean was here spread out. Hence, both in space and time, we seem to be brought somewhat near to that great fact—that mystery of mysteries—the first appearance of new beings on this earth.

FIG. 23. Some mammals collected by Darwin on the voyage of H.M.S. *Beagle*: **upper left,** *Canis fulvipes* Martin (= *Dusicyon fulvipes*); **upper right,** *Felis pajeros* Desmarest (= *Felis colocolo pajeros*; **lower left,** *Mus darwini* Waterhouse (= *Phyllotis darwini darwini*); **lower center,** *Mus galapagoensis* Waterhouse (= *Oryzomys galapagoensis*); **lower right,** *Didelphis brachyura* (= *Monodelphis dimidiata* Wagner); from Waterhouse (1838–1839).

XIV. Georges Louis Leclerc de Buffon (1707-1788)

Georges Louis Leclerc de Buffon was born into wealth and devoted his life to scientific labors; he won recognition as the leading naturalist of his time. In 1739 he was appointed keeper of the Jardin du Roi in Paris (now the Jardin des Plantes), which he turned into one of the most important centers of biological research during the 18th century. Buffon's lifetime work was a general natural history in 36 volumes. The first volume dealt with science in general, the second with man, the next 13 with nonhuman mammals (1750–1767). These were followed by nine volumes on birds, seven volumes (1789) supplementary to the preceding, and the last five on minerals, including fossils.

Treatment of most species in the *Histoire Naturelle* is usually monographic. Gross descriptions, including measurements and weights, are based on individuals received in the Jardin du Roi. Geographic distribution of the species is included with the description. Habits observed in captivity and mentioned in the literature are recorded. Anatomical descriptions by Daubenton, Buffon's collaborator, are of the skeleton, with soft parts and tegumentary structures of particular interest. Complete bibliographic references and synonymies, including those to the 10th edition of the Linnaean *Systema Naturæ,* accompany each species account.

Buffon drew together much if not everything known of a species, often an indiscriminate composite of species. Most of the information was compiled, some of it original. Many life history notes were received from correspondents, particularly M. de la Borde, the royal physician resident in Cayenne, French Guiana. Another correspondent, M. Saint Lurrent of Trinidad, believed he had solved the mystery of marsupial birth (cf. p. 40). At a certain stage of development, he informed Buffon, the embryonic opossum crawled from the uterus through a tube at the end of which it found a long teat to which it remained attached until fully developed. An easily verifiable discovery by Daubenton (in Buffon) was that tapirs have simple stomachs, not the complex ruminant type claimed by Bajon (above). Buffon reported that domestic cats kill but do not eat shrew- or short-tail opossums of the genus *Monodelphis.* House cats do indeed kill these animals and usually deposit them whole in the middle of the path leading from the house to the garden.

Most of the illustrations of mammals and all anatomical drawings of the *Histoire Naturelle* are original. A small sampling is reproduced here (figs. 24–25).

Buffon was the first naturalist to recognize regional faunas as such and to discriminate between Old World species and different but similar appearing or like-named species of the New World. He perceived the platyrrhine-catarrhine dichotomy of primates, and the phylogenetic distance between the groups. He further distinguished prehensile-tailed monkeys from non-prehensile-tailed species, and cebids from callitrichids by their ungues and teeth.

Buffon's sense of rivalry with the contemporary Linnaeus led him to find fault with and cast scorn on the binomial system used in the *Systema Naturæ.* Buffon argued for retention of vernacular names for species as well as a makeshift vernacular terminology for generic or supergeneric groups.

Lack of a scientific system of nomenclature in Buffon's work, and the almost universal adoption of the Linnaean binomial system by contemporary and later authors caused the *Histoire Naturelle* to be regarded as no better than a layman's encyclopedia of science. It has been republished with many revisions in many editions and languages. It is unfortunate that Buffon's important contributions to life histories, morphology, and evolutionary biology were largely ignored by Darwin and are little appreciated today. It seems that the greater luster credited to Darwin owes much to the dimming of Buffon's because of his lack of organization and consistency in his writings.

XV. Faunal Origins and Distribution

Early attempts to explain observed similarities and differences between Old and New World mammals all supposedly descended from occupants of Noah's ark, began with the 16th century philosopher and chronicler Acosta and in some quarters continues to this day.

José de Acosta (1539-1600)

José de Acosta argued that the animals of the New World had not been carried there by man. His evidence indicated that New World man brought nothing but himself over a land route. The possibility that animals migrating from the ark might have crossed the Atlantic Ocean by swim-

FIG. 24. Mammals figured in the *Histoire Naturelle* of Buffon: **upper left,** le saki (= *Pithecia pithecia* Linnaeus, male; from Buffon, 1767); **upper right,** le sagouin singe de nuit (= *Pithecia pithecia* Linnaeus, female; from Buffon, 1789); **lower left,** la grande chauve-souris fer-de-lance de la Guyanne (= *Phyllostomus hastatus* Pallas; from Buffon, 1789); **lower right,** le cabiai (= *Hydrochaeris hydrochaeris* Linnaeus; from Buffon, 1764).

FIG. 25. Mammals figured in the *Histoire Naturelle* of Buffon: **upper left,** la mouffette du Chili (= *Conepatus chinga* Molina); **upper right,** la grande marte de la Guyanne (= *Eira barbara* Linnaeus); **lower left,** le tamarin nègre (= *Saguinus midas* Linnaeus); **lower right,** hyoid apparatus and thyroid cartilage of the throat of *Alouatta seniculus* Linnaeus); from Buffon (1789).

ming or island hopping was also dismissed because, as Acosta pointed out, none of the animals was known to occur on oceanic islands. The legendary island of Atlantis, which might have been part of a former transatlantic archipelago, was treated as fable. Other conjectures discarded, Acosta resolved that New and Old World northern continents were or had been connected, or very nearly approximated, at their polar extremities. Differences between New and Old World species, he affirmed, could be explained by the disappearance of connecting Old World populations, mutations among the New World species engendered by their isolation in different environments, or by degeneration. No accounting or explanations were needed for "imperfect" organisms such as rats, frogs, insects, or vermin in general. These, it was commonly held, arose spontaneously from decaying matter.

Antonio Vázquez de Espinosa (1560/1575–1630)

Antonio Vázquez de Espinosa agreed with the explanation of a northern migratory or connecting route but added (1948, chap. 36), in the awkward phrasing of Clark's translation, that

> near the Straits of Magellan in what is called Tierra del Fuego, which is still not well known or explored, and there are numerous other quarters where the mainland of the New World could have communicated with that of the Old, or at least have lain so close as to afford passage not merely for the peoples who settled the New World, but the various kinds of animals which live in them—many of species well known in Europe and elsewhere, and others peculiar and unique in the New World, like the Peruvian sheep [llamas], the guanacos [regarded as the wild form of llamas], vicuñas and tarugas [*Hippocamelus bisulcus*].

Carolus Linnaeus (1707–1778)

Linnaeus played no direct role in the development of Neotropical mammalogy apart from providing scientific names for some species discovered or described by others. His impact on the scientific world, however, was enormous. Like the philosophers before him, he believed all species were created as they are now in one place from which they spread in search of the habitats for which they were specially created. This idea of a staging area as the center of origin and dispersal, still dominant today in the minds of some students of Neotropical mammalogy, was critically reexamined by Buffon.

Georges Louis Leclerc de Buffon (1707–1788)

Buffon, on comparing faunas of New and Old Worlds, was impressed by similarities between some of the species and differences between others. His explanation for similar-appearing species, like that of the chroniclers, was that a land connection permitted passage of animals from Old to New World. Differences between New and Old World species, he suggested, in agreement with the chronicler José Acosta, could have resulted from degeneration, environmental pressures, or isolation of the New World derivatives. On the other hand, Buffon argued, species peculiar to the New World or without Old World analogs must have arisen *in situ,* an opinion already intimated by Vázquez de Espinosa.

Buffon was the first naturalist to envision the mammals of the region as a community or fauna that might well have originated independently of other regional faunas. Noah's ark had no place in his concept of faunal origins, and he rejected as too short the scripturally based 6,000-year estimate of the earth's age. Buffon's ideas of organic evolution and multiple centers of origin were novel and prepared the minds of his and succeeding generations for the acceptance of Darwinian evolution.

Linnaeus, the arch exponent of the fixity of species and their origin and dispersal from a single center, conceived the elements of his binomial system as symbols for nailing down his credo. The system was so good it proved to be the best yet devised for the expression of genetic relationships between species and the surest base for the construction of evolutionary sequences in nominate terms. On the other hand, Buffon, independent of the religious constraints of his time and evolutionist in thought if not always in words, never attained the stature of his contemporary for lack of a competing system, key, code, or standard that would bring cohesion to his rambling philosophies.

Johann Andreas Wagner (1797-1861)

Johann Andreas Wagner, the foremost mastozoologist of his generation and author of a monograph on the geographic distribution of mammals, summarized (1844, p. 13) the three current but disparate opinions on mammalian origin and dispersal. First, all species were created in one and the same region and spread from there to all corners of the earth. Second, the species could have been created in separate localities in the same or different regions. Finally, each species could have arisen spontaneously anywhere and developed according to its peculiar constraints.

Zoogeographers of the early half of the 19th century divided the world into major faunal regions correlated primarily with climate. Wagner (1844) separated the earth into four provinces: the *Nordliche* north of 30°N, the *Mittlere* between 30°N and 30°S, excluding the *Australische* roughly between 0°S–55°S and 130°W–200°W, and the South American *Magellanische,* south of 30°S. The South American portion of the pantropical Mittlere Province extended from Mexico southward to southern Brazil and central Chile. Wagner's descriptions of the provincial faunas included tabulations of their respective genera and included species.

Maximilian Prinz von Wied-Neuwied (1782–1867)

Scriptural constraints were not evident in the thinking of the field naturalists. Maximilian Prinz von Wied-Neuwied recognized the limitations of geographic range as a property of a species.

Johann Jacob von Tschudi (1818-1889)

Tschudi attempted to follow Wied-Neuwied in defining specific ranges, but nearly all were based on the presumption that the geographic range of the species coincided with ecological life zones. The ecological life zones of Peru described by von Tschudi on the basis of fauna, flora, and climate, are the first of their kind for any Neotropical region.

Charles Robert Darwin (1809-1882)

The young Darwin also recognized geographic limitations of distribution in the light of physical barriers such as mountains and large bodies of water.

XVI. Inventories to Middle of 19th Century

Systema Naturæ of Linnaeus, 1758, 1766

The 10th edition of the *Systema Naturæ* published in 1758 by the Swedish naturalist Carolus Linnaeus (1707–1778), marks the beginning of the consistent application of his binomial system of zoological nomenclature. According to the universally accepted International Code of Zoological Nomenclature, names for animals published before 1758 are not available, no matter how clearly defined the species. Likewise, zoological names for species published after 1757 that are not binomial or do not satisfy all provisions of the Code are not available. The effect of the Code in practice is that species without Linnaean names are treated as unknown to science.

The 10th edition of the *Systema Naturæ* lists a total of 172 species of mammals, exclusive of marine cetaceans, each with its binomen consisting of a defined generic and defined specific name. Subsequent revisions of the bases for the names revealed that some represented more than a single species, others were duplicates or synonyms, and a few were equivocal or belonged to unidentifiable animals. The revisions, however, made no significant change in the total number of real mammalian species known to Linnaeus in 1758.

The 12th and last revised edition of the *Systema Naturæ* by Linnaeus himself, published in 1766, lists a world total of 208 mammalian species. Table 14 compares the relative numbers of world, Neotropical, and Nearctic genera and species in the Linnaean 10th and 12th editions of the *Systema Naturæ* with the totals in Buffon's *Histoire Naturelle.* Cetaceans are omitted because they are oceanic species known before the discovery of America.

Primary sources for the definition and naming of the Linnaean New World species were specimens preserved in the Swedish museums, particularly the Adolphi Friderici Regis Museum, and primary bibliographic references. Such references for the Neotropical mammals were the works of Marcgraf (1648), Anson (1748), Browne (1756), and Seba (1734–1765). For both Neotropical and

TABLE 14. Number of world and New World mammals known to Linnaeus (1758, 1766) and to Buffon (1750–1789) and their percentage of world total.

Author	Date	World total	Neotropica	Nearctica
Linnaeus	1758	35 genera	18 (51%)	15 (43%)
		172 species	45 (26%)	23 (13%)
Linnaeus	1766	36 genera	20 (55%)	15 (42%)
		208 species	55 (26%)	28 (13%)
Buffon	1750–1789	251 species	78 (31%)	47 (19%)

Nearctic species, Linnaeus cited Hernández (1651) and Edwards (1743–1751), and for Nearctic species alone, Catesby (1731) and Kalm (1753).

Histoire Naturelle of Buffon, 1750–1789

In volume 9 of his *Histoire Naturelle,* published in 1761, Buffon estimated a world total of approximately 200 known mammalian species of which 130 were Old World, 70 New World. When the species of all 13 volumes on mammals and the supplementary volumes are counted, the total is 251, of which 78 are Neotropical, 47 Nearctic.

The greater number of species recognized by Buffon as compared with those of Linnaeus reflects his better knowledge of mammals and wider use of the available literature. Neither authority searched the works of the chroniclers for descriptions or figures of New World mammals.

Synopsis Mammalium of Schinz, 1844

The Schinz catalogue of Recent mammals of the world, published 1844, brings the inventory of mammalian species to near the cutoff date of this part of the history of Neotropical mammalogy. The totals indicate that about 50% of all New World mammalian species now known had been described. The vast majority of the remaining 50% described since the middle of the 19th century are as small as or smaller than common tree squirrels. With respect to Neotropical mammals, by mid-19th century about 90% of known species larger than common tree squirrels had been described. In contrast, no more than about 10% of the smaller forms, mainly marsupials, bats, and rodents, had been named.

The Schinz catalogue is summarized in Table 15. A first glance at the figures of the first order, the Marsupialia, may suggest the list is skewed. Only one marsupial species, *Didelphis virginiana,* is known to occur in Nearctica. Schinz recorded three because the species had been overnamed at the time. It is believed, however, that "undernamed" or composite species, as well as overnamed identical species, are more or less evenly distributed in all three columns. As a result, the bottom line totals, particularly the percentages, are fair estimates of the real number of species known to science at the time of Schinz's compilation. The percentages have not changed significantly since.

XVII. Summary

Knowledge of Neotropical mammals from 1492 to the mid-17th century was mainly an aggregation of anecdotes often riddled with myth, folklore, and untested generalizations. Europeans identified New World species with similar-appearing or similar-behaving Old World species and used the same names for them. Descriptions of mammals were usually comparisons with familiar European animals; measurements, rarely given, were rough estimates. Habitat when mentioned was usually on the order of "forest," "plain," or "river."

Descriptions of animals often included use as food or pets, medicinal merits, or value of rawhide or bones in the manufacture of artifacts. Habits were usually described in terms of reactions to man when hunted or in captivity, or as harmful or benign to his person or property. Mammals—the term had not yet been concocted—were the hairy beasts of the earth. Whales and manatees were fish and could be eaten on Fridays. Bats were something else, mostly vampires; mice and other small mammals were vermin, in a class with frogs and cockroaches.

Mammal collecting during this period was generally limited to capture of live individuals for domestication, as pets, or for exhibition in zoos, circuses, or fairs. Dead animals were sometimes skinned and stuffed or bottled in brandy. The crudely prepared or pickled specimens, if not live animals, often served as models for the woodcut drawings of early treatises on natural history. Some specimens were purchased for museums or cabinets of collectors, including those of Linnaeus, King Frederick Adolph of Sweden, Reaumur of Paris, the King of France (Jardin du Roi), or the shelves of the Dutch pharmacist, Albert Seba. Most of the Neotropical specimens probably originated in the South American possessions of Holland and France.

The few crude attempts at classification of mammals during the 16th and 17th centuries were hardly more than random arrangements equivalent to shopping lists. Species, being individually created kinds, were unrelated to other created kinds, or simply arose spontaneously from putrefying matter.

The scientific study of mammals, or mammalogy, of the Neotropical Region began with the explorations of northeastern Brazil by Georg Marcgraf and culminated with the publication in 1648 of his *Historiæ Rerum Naturalium Brasiliæ*. His accounts of the included 32 species of mammals reveal the glimmer of an attempt at natural groupings of kinds or the beginnings of a classification of Neotropical mammals. Insofar as is known, none of Marcgraf's animals were preserved. Linnaean names for the species of the *Historiæ* were based on bibliographic references to their descriptions and figures (cf. fig. 2, table 1).

The first expedition to the Neotropical Region actually committed to the collection and permanent preservation of mammals (and other objects) for scientific study was the Brazilian *Viagem Filosófica,* 1783 to 1792, conceived by the Portuguese government and conducted by the Brazilian-born naturalist Alexandre Rodrigues Ferreira. The large number of specimens gathered by the expedition was deposited in Lisbon's Museu d'Ajuda. The specimens of monkeys that had been carried away to the Paris Natural History Museum were studied by the French scientist Étienne Geoffroy St.-Hilaire. His descriptions were published without reference to source of material.

Alexander von Humboldt followed on the heels of the *Viagem Filosófica* with his explorations of northwestern South America from 1799 through 1802. His expedition was highly successful and in scope has rarely been equaled by other "one-man" surveys of a large portion of the South American continent. The personal narrative of his travels inspired successive naturalist-travelers, most notably the explorers of Brazil, Spix and Martius, Maximilian Wied-Neuwied, and Johann Natterer.

TABLE 15. Number of world, Neotropic, and Nearctic species (subspecies) of mammals known to Schinz (1844); species common to both regions are included in both.

Order	World	Neotropica	Nearctica
Marsupialia	138	31 (22%)	4 (3%)
Insectivora	114	2 (2%)	21 (18%)
Chiroptera	326	110 (34%)	21 (6%)
Primates	281	73 (26%)	0 (0%)
Edentata	31	24 (77%)	0 (0%)
Carnivora	303	58 (19%)	41 (13%)
Pinnipedia	39	11 (28%)	2 (5%)
Sirenia	5	2 (40%)	1 (20%)
Perissodactyla	23	2 (9%)	0 (0%)
Artiodactyla	186	11 (6%)	12 (6%)
Lagomorpha	52	4 (8%)	14 (27%)
Rodentia	563	152 (27%)	104 (18%)
Cetacea	4	2 (50%)	0 (0%)
Proboscidea	2	0 (0%)	0 (0%)
Hyracoidea	5	0 (0%)	0 (0%)
TOTALS	2,072*	482 (23%)	220 (11%)

* The estimated number of species in 1972 (Hershkovitz, p. 332, table 3) is Neotropica 810, Nearctica 442 or an approximate doubling since 1844 in both regions, with a slightly greater increase (less than 2%) in Nearctica relative to Neotropica. Increase since then has been almost exclusively Neotropical.

Later there was von Tschudi, who traveled in Peru; d'Orbigny, who journeyed in Patagonia but did his finest and most lasting work in Bolivia on a scale almost equal to that of Humboldt's; and Darwin, who voyaged around the southern half of South America and the Galápagos Islands in H.M.S. *Beagle.*

The brothers Schomburgk, motivated by Humboldt's trip up the Río Orinoco to its connection via the Casiquiare Canal with the Rio Negro tributary of the Rio Amazonas, completed the trajectory by plotting the course of the upper Rio Negro to its connection with the Casiquiare. Their explorations of the British Guianas and bordering parts of Brazil and Venezuela yielded the first large collection of Guianan mammals, all deposited in the Berlin Natural History Museum.

Chilean mammals became fairly well known through the reports of Molina (1782), Poeppig (1836), and Gay (1847). The mammals of Paraguay, their distribution, habits, or biology in general, became better known through the efforts of Felix de Azara than those of other Neotropical countries.

The 200-year period from Marcgraf's (1648) report to the last of those of Schomburgk (1848) was one of survey and inventory of South American mammals. The published reports and personal narratives of travel provided much reliable data on geographic distribution, habitat, life histories, ecological backgrounds, itineraries and maps of the expeditionary routes, and stopping and collecting localities. Descriptions of the collected mammals, most of them by the naturalist-travelers themselves, were often based on skeletal, dental, and soft parts in addition to purely tegumentary characters. Their classifications were putatively natural groupings on the ordinal, family, and, as a rule, the generic levels. The prevailing belief in the biblical version of creation and fixity of species, not confessed in writing, did not blind systematists of the period to evident relationships between species and their clusterings into supraspecific groups. Descriptions of species were, nevertheless, typological. Subspecies or geographic races were, at best, vaguely conceived but described as species. The infrequent or rare use of trinomials was accidental or equivocal and not certainly intended for a clearly defined geographic race. The term usually used for deviates from "types" was "variety."

Controversies regarding origin of species or faunas centered on where, not how. Philosopher-chroniclers of the first era accepted Noah's ark literally as the one place of origin and dispersal of the Recent fauna. Acosta may have been the first to suggest the former existence of intercontinental connections for passage of Old World animals into the New World.

More and better knowledge of the world's fauna during the second era revealed the weaknesses or fallacy of the ark dogma. Staunch creationists such as Linnaeus pointed instead to a vaguely located region as the place from which all species dispersed to occupy predestined habitats for which they had been created. Other authorities like Buffon argued for multiple centers of origin, with species originating in the habitats for which they were adapted. Darwin also believed in multiple places of origin, or faunal regions separated by geographic barriers but with some trepidation. The belief in multiple creations was heretical.

Inconsistencies between religious dogma and realities did not prevent Wied-Neuwied from recognizing the geographic range of a species (or subspecies) as a property of that species. Another advance beyond scriptures was the concept of ecological life zones contributed by von Tschudi, who plotted them for Peru on the basis of plants, animals, and climate.

The total of named Neotropical species of mammals counted from 1758, the year of publication of the 10th edition of the Linnaean *Systema Naturæ* and starting date of zoological nomenclature, to mid-19th century, exceeded by far that of the Nearctic region and any other equivalent area of the world. Neotropical mammals were also better known than those of other continents except western Europe.

By mid-19th century, about 90% of currently known Neotropical mammalian species larger than common tree squirrels had already been described, but no more than about 10% of the smaller forms.

The great number and variety of Neotropical mammals (and animals generally) known to science by mid-19th century and the accumulated knowledge gained from study of living and preserved specimens in field and laboratory, much of it contributed by Charles Darwin, helped pave the way to the Darwinian revolution of the next half century.

XVIII. Acknowledgments

I am indebted to Benjamin W. Williams, Associate Librarian and Librarian of Rare Books, Field Museum of Natural History, for permission to consult at pleasure in the Museum's Mary W. Runnells Rare Book Room the books needed for writing this article; and to Bruce D. Patterson, Robert M. Timm, Ronald H. Pine, Debra Moskovitz, and J. A. Gagliano for reviewing the manuscript. Maps shown in Figures 11, 12, and 21 were prepared by the author with assistance of Mary Anne Rogers from accounts of the travelers cited and other sources. Photographic reproductions of the figures are by Field Museum of Natural History Staff Photographer Ron Testa. Technical Assistants Barbara Brown and Mary Anne Rogers typed the manuscript and contributed in other ways toward its completion.

XIX. Literature Cited

ACOSTA, JOSÉ DE. 1590. Historia natural y moral de las Indias en que se tratan las cosas notables del cielo, elementos, metales, plantas y animales dellas y los ritos y ceremonías, leyes y gobierno y guerras de los Indios. Juan de León, Seville, 535 + (36) pp. [not seen].

———. 1894. Historia natural y moral de las Indias. Escrita por el P. Joseph de Acosta de la Compañia de Jesus. Publicada en Sevilla en 1590 y ahora fielmente reimpresa de la primera edición. Ramón Angles, Madrid, 2 vols.

ANGHIERA, see Martyr.

ANSON, GEORGE. 1748. A Voyage Round the World in the Years MDCCXL, I, II, III, IV by George Anson . . . Esq. Compiled and published by Richard Walter. J. & P. Knapton, London.

AVILA PIRES, FERNANDO DIAS DE. 1965. The type specimens of Brazilian mammals collected by Prince Maximilian zu Wied. American Museum Novitates, no. **2209:** 1–21.

———. 1974. Caracterização zoogeográfica da Provincia Amazônica. I. Expedições científicas na Amazônia Brasiliera. Anales Academia Brasiliera de Ciencias, **46**(1): 133–158.

AZARA, FELIX DE. 1801. Essais sur l'histoire naturelle des quadrupèdes de la Province du Paraguay. Translated from the original Spanish by M. L. E. Moreau-Saint-Mery. Charles Pougens, Paris, vol. 1, xxvii + 366 pp.; vol. 2, 499 pp.

———. 1802. Apuntamientos para la historia natural de los quadrúpedos del Paraguay y Río de la Plata. La Viuda de Ibarra, Madrid, vol. 1, xiii + 318 pp.; vol. 2, (3) + 328 pp.

———. 1809. Voyages dans l'Amérique Méridionale. Contenant la description géographique politique et civile du Paraguay et de la rivière de La Plata; l'histoire de la découverte et de la conquête de ces contrées; des details nombreaux sur leur histoire naturelle, et sur les peuples sauvages qui les habitent, le recit des moyens employes par les Jesuits pour assujétir et civiliser les indigenes, etc. Publiés d'apres les manuscrits de l'auteur, avec une notice sur sa vie et ses ecrits. Translated from the original Spanish and annotated by M. Sonnini, with notes by G. Cuvier; edited by C. A. Walckenaer. Dentu, Paris, 4 vols. with atlas.

BAJON, M. 1777–1778. Mémoires pour servir a l'histoire de Cayenne et de la Guiane Française. Grange, Paris, vol. 1, 1777; vol. 2, 1778.

BANCROFT, EDWARD. 1769. An Essay on the Natural History of Guiana in South America. T. Beckett and P. A. DeHondt, London, (8) + 402 + (2) pp.

BARRERE, PIERRE. 1741. Essai sur l'histoire naturelle de la France Équinoxiale ou de'nombrement des plantes, des animaux, y des mineraux, que se trouvent dans l'Isle de Cayenne, les Isles de Remire, sur les côtes de la mer, & dans le continent de la Guyane. Piget, Paris.

BELLIN, JAQUES NICOLAS. 1763. Description géographique de la Guiane. Didot, Paris.

BOKERMANN, WERNER C. A. 1957. Atualização do itinerario de viajem do Principe de Wied do Brasil (1815–1817). Arquivos de Zoología (Rio de Janeiro), **10**(3): 209–251.

BROWNE, PATRICK. 1756. The Civil and Natural History of Jamaica. Privately printed, London.

BUFFON, GEORGE LOUIS LECLERC, COMTE DE. 1753–1789. Histoire naturelle, générale et particuliere, avec la description du cabinet du roy. L'Imprimerie Royale, Paris, 44 vols., 1750–1804; vols. 4–15, Quadrupedes par M. de Buffon et M. Daubenton, 1753–1767; vol. 9, 1761; supplement, vol. 107 par M. de Buffon, 1774–1789.

CATESBY, MARK. 1731–1743. The Natural History of Carolina, Florida and the Bahama Islands. Privately printed, London, vol. 1, (2) + vii + (1) + xliv + 100 + 2 pp., 100 pls., fold. map; vol. 2, 2 + 120 + (6) + 2 pp., 120 pls.

CHARLEVOIX, PIERRE FRANCOIS XAVIER DE. 1757. Histoire du Paraguay. Desaint, David, Durand, Paris, 6 vols.

DARWIN, CHARLES ROBERT. 1839. Journal of Researches into the Geology and Natural History of the Various Countries Visited by H.M.S. *Beagle,* Under the Command of Captain Fitzroy, R.N., from 1832 to 1836. Henry Colburn, London, xiv + 615 pp., 16 pls.

EDWARDS, GEORGE. 1743–1751. A Natural History of Uncommon Birds, and of Some Other Rare and Undescribed Animals. Printed for the author at the College of Physicians, London, 4 vols.

FERMIN, PHILIPPE. 1769. Déscription générale, historique, géographique et physique de la colonie de Surinam. E. van Harrevelt, Amsterdam, 2 vols.

FERREIRA RODRIGUEZ, ALEXANDRE, see Rodriguez Ferreira, Alexandre.

GAFFAREL, PAUL. 1907. See Martyr, Anghiera Pierre.

GAY, CLAUDIO. 1847. Historia física y política de Chile. Vol. 1, Zoología. Museo de Historia Natural de Santiago [de Chile], Malde y Renow, Paris.

GUMILLA, JOSÉ. 1741. El Orinoco ilustrado, historia natural, civil y geográfica, de este gran río, y de sus caudalosas vertientes; gobierno, usos y costumbres de los Indios, sus habitadores con nuevas, y útiles noticias de animales, árboles, frutos, aceytes, resinas, yervas, y singulares a nuestra santa fé, y casos de mucha edificación. M. Fernández, Madrid.

———. 1758. Histoire naturelle, civile et géographique de l'Orenoque. Translated from the Spanish. 2nd edition by M. Eidous. Veuve de F. Firard, Avignon, 3 vols.

HERNÁNDEZ, FRANCISCO. 1651. Nova plantarum, animalium et mineralium mexicanorum historia. B. Deuersini & Z. Masotti, typis V. Masacardi, Rome, (16) + 950 + 90 pp.

———. 1959. Historia natural de Nueva España, vol. 2. Translated by José Rojo Navarro. Universidad Nacional de México, D.F.

HERRERA Y TORDESILLAS, ANTONIO DE. 1601. Historia general de las Indias occidentalis: o, de los hechos de los castellanos en las islas y tierra firme de mar océano, escrita por Antonio de Herrera, cronista mayor de Su Magestad de las Indias y de Castilla. En la Emplenterea, 4 vols. [work not seen, cited from reviewer comments].

———. 1601–1615. Historia general de las Indias occidentalis: o, de los hechos de los castellanos en las islas y tierra firme del mar océano, escrita por Antonio de Herrera, cronista mayor de Su Magestad de las Indias y de Castilla. Madrid, 9 parts in 4 vols. [work not seen, cited from reviewer comments].

———. 1725–1726. The General History of the Vast Continent and Islands of America, Commonly Called, the West Indies, from the First Discovery Thereof: With the Best Accounts the People Could Give of the Antiquities. Collected from the original relations sent to the Kings of Spain. Translated into English by Capt. John Stevens. Batley, London, vol. 1, 1725; vols. 2–6, 1726.

———. 1728. Historia general de las Indias occidentalis: o, de los hechos de los castellanos en las islas y tierra firme del mar océano, escrita por Antonio de Herrera, cronista mayor de Su Magestad de las Indias y de Castilla. En ocho decadas. Sigue a la ultima decada, la Description de las Indias, por el mismo autor. J. B. Verdussen, Amberes [work not seen, cited from Library of Congress catalogue].

Hershkovitz, Philip. 1972. The Recent mammals of the Neotropical Region: A zoogeographic and ecological review, pp. 311–431. *In* Keast, A., F. C. Erk, and B. Glass, eds., Evolution, Mammals and Southern Continents. State University of New York Press, Albany.

Humboldt, Alexander von. 1811–1812. Voyage de Humboldt et Bonpland. Observations de zoologie et d'anatomie comparée. Pt. 2, vol. 1, Recueil d'observations de zoologie et d'anatomie comparée, faites dans l'ocean Atlantique dans l'intérieur du nouveau continent et dans la mer du sud pendant les années 1799, 1800, 1801, 1802 et 1803. Levrault Schoell, Paris, viii + 368 pp., 40 pls.

———. 1838. Ueber den Manati des Orinoko. Translated from the original manuscript in French by the editor. Wiegmann's Archiv für Naturgeschichte, **4**(1): 1–10.

———. 1884. Personal Narrative of Travels to the Equinoctial Regions of America During the Years 1799–1804 by Alexander von Humboldt and Aimé Bonpland. Translated from the original French and edited by Thomasina Ross. George Bell and Sons, London, 3 vols.

Ihering, Herman von. 1902. Natterer e Landsdorff. Exploradores antigos do estado de São Paulo. Revista do Museu Paulista (São Paulo), **5**: 13–34.

Kalm, Pehr. 1753–1761. En resa til Norra America. Pa Kong. Svenska Vetenskapsacademiens befallning, och publici Kostnad. Stockholm, 3 vols.

Kraft, Richard. 1983. Die von J. B. von Spix beschriebenen neotropischen Primaten und Chiropteren. Verzeichnis der in der Zoologischen Staatssammlung München aufbewahrten Typusexemplare. Spixiana Suppl., **9**: 429–441.

Kuhl, Heinrich. 1820. Beiträge zur Zoologie und vergleichenden Anatomie, Abt. 1. Frankfurt am Main, 152 + unnumbered pp.

Langguth, A. 1978. Revived application in the case of the names for South American rodents published by Brants (1827). Z.N.(S.) 1775. Bulletin of Zoological Nomenclature, **35**(2): 115–120.

Linnaeus, Carolus. 1758. Systema Naturæ. Regnum animale, ed. 10, vol. 1. Holmiae, 4 + 824 + 3 pp.

———. 1766. Systema Naturæ per regna tria naturæ, secundum classis, genera, species, cum characteribus, differentius, synonymis locis . . . , ed. 12 reformata, vol. 1. L. Salvii, Holmiae, 1 + 327 + 36 pp.

Marcgraf [or Marcgrav, or Marggraf] de Liebstad, Georg. 1648. Historiæ rerum naturalium Brasiliæ, libro octo cum appendice de Tapuyis, et Chilensibus. Ioannes de Laet, Antvverpianus, In ordinem digessit & annotationes addidit multas, & varia ab auctore omissa supplevit & illustravit. *In* Piso, W., Historia Naturalis Brasilliæ, asspicio et beneficio Illustras. I. Maritii Com. Nassau. Issius provinciæ et Maris summi praefecti adornata in qua non tantum plantæ et animalia sed et indiginarum morbi, ingenia et mores describuntur et iconibus scipta quingentas illustrantus. Fransiscum Hackium, et Amstelodami, Lud, Elizevirium, Leiden.

Martyr, Anghiera Pierre. 1907. De orbe novo. Les huit décades. Translated from Latin with notes and comments by Paul Gaffarel. Recueil de voyages et de documents pour servir a l'histoire de la géographie depuis le XIIIe jusqu'a la fin du XVIe siecle. No. XXI. Imprimerie Nouvelle Noel Texier, La Rochelle, France.

———. 1912. De orbe novo. The Eight Decades of Peter Martyr d'Anghiera. Translated from the Latin with notes and introduction by Francis Augustus MacNutt. G. P. Putnam's Sons, New York & London, 2 vols.

Miller, Gerrit S., Jr. 1929. Mammals eaten by Indians, owls, and Spaniards in the coast region of the Dominican Republic. Smithsonian Miscellaneous Collections, **82**(5): 1–16.

Molina, Giovanni Ignazio. 1782. Saggio sulla storia naturale del Chili. S. Tommaso d'Aquino, Bologna, 367 + 1 pp.

Orbigny, Alcide Dessalines d'. 1835–1847. Voyage dans l'Amérique Méridionale (le Brésil, la République orientale de Uruguay, la République Argentine, la Patagonie, la République du Chile, la République de Bolivia, la République du Perou) executé pendant las années 1826, 1827, 1828, 1829, 1830, 1831, 1832, et 1833. Pitois-Levrault et cie, Paris, Strasbourg, 9 vols.

Orbigny, Alcide d', and Paul Gervais. 1847. Mammifères, vol. 4, pt. 2. *In* Orbigny, A. d', Voyage dans l'Amérique Méridionale (le Brésil, la République orientale de l'Uruguay, la République Argentine, la Patagonie, la République du Chili, la République de Bolivia, la République du Perou) executé pendant les années 1826, 1827, 1828, 1829, 1830, 1831, 1832, 1833. Pitois-Levrault et cie, Paris, Strasbourg.

Osgood, Wilfred H. 1943. The mammals of Chile. Publications of Field Museum of Natural History, Zoological Series, **30**: 1–268.

Oviedo y Valdés, Gonzalo Fernández de. 1944–1945. Historia general y natural de las Indias, Islas y Tierra-Firme del Mar Océano. Prólogo de J. Natalicio González. Notas de José Amador de los Ríos. Editorial Guaranía, Asunción (Paraguay), 14 vols. [text and illustrations said to be reproduced from the original manuscript].

Pelzeln, August von. 1871. Ueber Ornithologie Brasiliens. Resultate von Johann Natterer's Reisen in den Jahren 1817 bis 1835. A. Pichler's Witwe & Sohn, Vienna.

———. 1883. Brasilische Säugethiere. Resultate von Johann Natterer's Reisen in den Jahren 1817 bis 1835. Verhandlungen der kaiserlich-königlichen zoologisch-botanischen Gesellschaft in Wien, **33**: 1–140.

Pilleri, G., and Lucie Arvy. 1977. La expedición de Alcide d'Orbigny en Bolivia y el descubrimiento del delphín boliviano de agua dulce *Inia boliviensis* (d'Orbigny, 1834). Hirnanatomisches Institut, Bern, Switzerland, 15 pp., 4 figs., 16 pls., fold. map.

Poeppig, Eduard. 1835–1836. Reise in Chile, Peru und auf dem Amazonenstrome während der Jahre 1827–1832. Friedrich Fleischer, Leipzig, vol. 1, xii + 466 pp.; vol. 2, v + 464 pp.

Rengger, J. R. 1830. Naturgeschichte der Saeugethiere von Paraguay. Schweighauserschen Buchhandlung, Basel, Switzerland, 16 + 394 pp.

Rodrigues Ferreira, Alexandre. 1971. Viagem filosófica pelas capitanias do Grão-Pará, Rio Negro, Mato Grosso e Cuiabá. Iconografia. Vol. 2, Zoología. Conselho Federal de Cultura, Rio de Janeiro. Editora Monumento S. A., São Paulo. Artes Graficas Gomes de Souza S. A., Rio de Janeiro, 168 color plates.

———. 1972. Viagem filosófica pelas capitanias do Grão-Pará, Rio Negro, Mato Grosso e Cuiabá. Memorias. Zoología, Botánica [nomenclatorial notes by Luiz Carlos Souto; introduction by José Candido de Melo Carvalho]. Conselho Federal de Cultura. Departamento de Imprenta Nacional, Rio de Janeiro, 246 pp.

Roth, Walter E. 1931. See Schomburgk, Richard, and Schomburgk, Robert Hermann.

Schinz, Heinrich R. 1821. Das Thierreich eingetheilt nach dem Bau der Thiere als Grundlage ihrer Naturgeschichte und der vergleichenden Anatomie von dem Herrn Ritter von Cuvier. Vol. 1, Säugethiere und Vogel. J. G. Cotta'schen Buchhandlung, Stuttgart and Tübingen.

———. 1844–1845. Systematisches Verzeichniss aller bis jetzt bekannten Säugethiere oder Synopsis Mammalium nach dem Cuvier'schen System. J. Gassmann, Sohn, Solothurn, Switzerland, 2 vols.

Schomburgk, Richard. 1847–1848. Reisen in British-Guiana in den Jahren 1840–1844. Im auftrag Sr. Majestät des Königs von Preussen. Nebst einer Fauna und Flora Guiana's nach Vorlagen von Johannes Müller, Ehrenberg, Erichson, Klotzsch, Troschel, Cabanis und Andern. Mit Abbildungen und einer Karte von British-Guiana afgenommen von Sir Robert Schomburgk. Versuch einer Fauna und Flora von British-Guiana (1848). J. J. Weber, Leipzig, vol. 1, 1847; vols. 2–3, 1848.

———. 1922–1923. Travels in British Guiana During the Years 1840–1844 Carried out Under the Commission of His Majesty the King of Prussia. Together with a Fauna and Flora of Guiana According to the Works of Johannes Müller, Ehrenberg, Erichson, Klotzsch, Troschel, Cabanis and Others. Including illustrations and a map of British Guiana drawn by Sir Robert Schomburgk. Translated and edited, with geographical and general indices and route maps, by Walter E. Roth. "Daily Chronicle," Georgetown, British Guiana, vol. 1 (1922), 38 + 402 pp., 8 pls., 4 maps, 2 figs.; vol. 2 (1923), (12) + 443 pp., 8 pls., 7 maps.

Schomburgk, Robert Herman. 1840. A Description of British Guiana, Geographical and Statistical: Exhibiting Its Resources and Capabilities, Together with the Present and Future Condition and Prospects of the Colony. Simpkin, Marshall, and Co., London.

———. 1840. Remarks on the greater ant-bear (*Myrmecophagus jubata*). Proceedings of the Zoological Society. Annals of Natural History, ser. 1, **4**(23): 202.

———. 1840. Information respecting botanical travellers. Recent expedition in Guiana. Annals of Natural History, ser. 1, **4**(24): 262–266; (25): 322–328; (26): 429–434; **5**(28): 29–35; (31): 282–288; (32): 343–348; (33): 399–403.

———. 1841. Reisen in Guiana und am Orinoko während der Jahre 1835–1839. Nach seinen Berichten und Mitteilungen an die geographische Gesellschaft in London. Edited by O. A. Schomburgk, with a foreword by Alexander von Humboldt. Georg Wigand, Leipzig.

———. 1931. Travels in Guiana and on the Orinoco During the Years 1835–1839. According to his Reports and Communications to the Geographical Society of London. Edited by O. A. Schomburgk, with a preface by Alexander von Humboldt. English translation by Walter E. Roth. "The Argosy" Company, Limited, Georgetown, British Guiana.

Seba, Albertius. 1734–1765. Locuplestissimi rerum naturralium thesauri accurata descripti et iconibus artificiosissimis expressio, per universam historiam. . . . Jansson-Waesberg, Amsterdam, 4 vols., 449 color pls.

Spix, Jean de. 1823. Simiarum et vespertilionum brasiliensium species novæ: ou histoire naturelle des especies nouvelles de singes et de chauve-souris observées et recueillies pendant le voyage dans l'intérieur du Bresil. Typis Francisci Serephici Hübschmanni, Monaco, 16 + 72 pp., 38 color pls.

Spix, Johann Baptist von, and Carl Friedrich von Martius. 1823–1831. Reise in Brasilien auf Befehl Sr. Majestat Maximilian Joseph I., Königs von Baiern, in den Jahren 1817 bis 1820. M. Lindauer, Munich, vol. 1, 1823; vol. 2, 1828; vol. 3, 1831.

———. 1824. Travels in Brazil in the Years 1817–1820, Undertaken by Command of His Majesty the King of Bavaria. Translated from the original German by H. E. Lloyd. Longman, Hurst, Rees, Orme, Brown, and Green, London, vol. 1, 24 + 327 pp., 4 pls.; vol. 2, 8 + 298 pp., 5 pls.

Stedman, John Gabriel. 1796. Narrative of a Five Years' Expedition Against the Revolted Negroes of Surinam in Guiana on the Wild Coast of South America; from the Year 1772 to 1777: Elucidating the History of that Country, and Describing Its Productions, *viz.* Quadrupeds, Birds, Fishes, Reptiles, Trees, Shrubs, Fruits, & Roots; with an Account of the Indians of Guiana, & Negroes of Guinea. J. Johnson and J. Edwards, London, vol. 1, 18 + 423 pp., 18 pls., 3 maps; vol. 2, 9 + 419 + (5) pp., 38 pls., 2 maps.

Stevens, John. 1726. See Herrera y Tordisillas, Antonio de.

Tamayo H., Manuel, and C. Daniel Frassinetti. 1980. Catálogo de los mamíferos fósiles y vivientes de Chile. Boletín Museo Nacional de Historia Natural de Chile, **37:** 323–399.

Thomas, Oldfield. 1908. On certain African and S.-American otters. Annals and Magazine of Natural History, ser. 8, **1:** 387–395.

———. 1911. The mammals of the tenth edition of Linnaeus; an attempt to fix the type of the genera and the exact bases and localities of the species. Proceedings of the Zoological Society of London, **1911:** 120–158.

TIEFENBACHER, L., ED. 1983. Festschrift zu Ehren von Dr. Johann Baptist Ritter von Spix. Spixiana, Suppl. 9.

TSCHUDI, JOHANN JACOB VON. 1844a. Mammalium conspectus quæ in Republica Peruana reperiuntur et pleraque observata vel collecta sunt in itinerere. Archiv für Naturgeschichte, **10**(1): 244–255.

———. 1844b. Untersuchungen über die Fauna Peruana. Physiognomik von Peru (1844); Therologie (1844b); Ornithologie [with J. Cabanis] (1845–1846); Herpetologie (1845 [1846]). Scheitlin und Zollikofer, St. Gallen, Switzerland.

———. 1846. Peru. Reiseskizzen aus den Jahren 1838–1842. Scheitlin und Zollikofer, St. Gallen, Switzerland.

———. 1847. Travels in Peru During the Years 1838–1842. On the Coast, in the Sierra, Across the Cordilleras and the Andes into the Primeval Forests. English translation from the original German by Thomasina Ross. Wiley and Putnam, New York.

VÁZQUEZ DE ESPINOSA, ANTONIO. 1942. Compendium and description of the West Indies. Translated by Charles Upson Clark. Smithsonian Miscellaneous Collections, **102:** i–xii; 1–862.

———. 1948. Compendio y descripción de las Indias Occidentales. Transcrito del manuscrito original por Charles Upson Clark. Smithsonian Miscellaneous Collections, **108:** i–xii; 1–801.

WAGNER, JOHANN ANDREAS. 1840–1855. Die Säugethiere in Abbildungen nach der Natur mit Beschreibungen von Dr. Johann Christian Daniel von Schreber, Supplementenbanden 1–5. Erlangen, Germany.

———. 1842. Diagnosen neuer Arten brasilischer Säugethiere. Archiv für Naturgeschichte, **8**(1): 356–362.

———. 1843. Diagnosen neuer Arten brasilischer Handflugler. Archiv für Naturgeschichte, **9**(1): 365–368.

———. 1844–1846. Die geographische Verbreitung der Säugethiere. Abhandlungen der Mathem.-physikalischen Classe der Königlich Bayerischen Akademie der Wissenschaften, **4**(1): 1–116 (1844); (2): 39–108 (1845); (3): 1–114 (1845); (3): 241–244, explanation of maps, plates I–IX (1846).

———. 1850. Beiträge zur Kenntniss der Säugethiere America's. Abhandlungen der Mathem.-physikalischen Classe der Königlich Bayerischen Akademie der Wissenschaften (München), **5:** Erste Abt. [Marsupialia, Chiroptera], 119–208 (1847); Zweiten Abt. [Rodentla], 269–332 (1847); Dritte Abt. Vierte Ordnung Affen, 405–480 (1848).

WATERHOUSE, GEORGE R. 1838–1839. Mammalia. *In* Darwin, C., The Zoology of the Voyage of H.M.S. *Beagle,* Under the Command of Captain Fitzroy, During the Years 1832–1836. With Notes by Charles Darwin, part II. Smith, Elder and Co., London, no. 1, 1838; no. 2, 1838; no. 3, 1838; no. 4, 1839.

WIED-NEUWIED, MAXIMILIAN ALEXANDER PHILIPP, PRINZ VON. 1820. Ueber ein noch unbeschriebenes Säugethier aus der Familie der Nager. Isis (Oken's), **1:** 43.

———. 1820–1821. Reise nach Brasilien in den Jahren 1815 bis 1817. Heinrich Ludwig Brönner, Frankfurt A.M., vol. 1, 1820; vol. 2, 1821.

———. 1821. Ueber die Brasilianischen Hirschart. Isis (Oken's), **2:** 649–651.

———. 1822–1831. Abbildungen zur Naturgeschichte Brasiliens. (Recueil de Planches Coloriées d'Animaux du Brésil.) Weimar, Germany, 90 color pls. with descriptive text issued in 15 parts with 6 pls. each (30 pls. of mammals in parts 1–5, 1822–1824).

———. 1826. Beiträge zur Naturgeschichte von Brasilien, vol. 2. Gr. H. S. priv. Landes-Industrie-Comtpoirs, Weimar, Germany.

A New Superfamily in the Extensive Radiation of South American Paleogene Marsupials

Rosendo Pascual and Alfredo A. Carlini

ABSTRACTS

Significant new mammals have been recovered from the Colhuehuapian mammal-bearing beds (latest Oligocene) exposed in the Gaiman region of Patagonia (Chubut Province, Argentina). Some fragmentary mandibles and isolated teeth belong to a new genus and species, **Patagonia peregrina.** The mandibular and dental specializations of this form are so distinctly convergent on those of some fossorial rodents that it is regarded as a distinct clade of South American marsupials. It represents the type of a new family, **Patagoniidae,** which is assigned to a new superfamily, **Patagonioidea,** which represents a natural evolutionary group in the same sense as other superfamilies of marsupials recognized by Simpson. Its systematic position within the superorder Marsupialia awaits comprehensive analysis of those enigmatic marsupials (Groeberioidea and Argyrolagoidea) apparently most closely related to it.

Varios nuevos y significativos mamíferos han sido recogidos de capas mamalíferas del Colhuehuapense (Oligoceno tardío) expuestas en la región de Gaiman, Patagonia (Chubut, Argentina). Algunos fragmentos mandibulares y dientes aislados pertenecen a un nuevo género y especie, **Patagonia peregrina.** Esta forma presenta especializaciones mandibulares y dentarias tan distintamente convergentes hacia las de albunos roedores cavadores que es considerada como un distinto clado de marsupiales sudamericanos. Representa el tipo de una nueva Familia, **Patagoniidae,** que es asignada a la nueva Superfamilia **Patagonioidea,** porque representa un grupo evolutivo natural como los de otras Superfamilias de marsupiales reconocidas por Simpson. Su posición sistemática dentro del Superorden Marsupialia depende del análisis integrado de aquellos marsupiales enigmáticos (Groeberioidea y Argyrolagoidea) aparentemente más estrechamente relacionados a él.

Novos mamíferos foram recuperados dos leitos de Colhuehuapian (do alto Oligoceno), expostos na região de Gaiman, Patagonia (Província de Chubut, Argentina). Fragmentos mandibulares e dentes isolados pertencem a um novo genero e espécie, **Patagonia peregrina.** As especializações mandibulares e dentais encontradas são tão claramente convergentes às de alguns roedores fossórios, que esta forma é considerada uma classe distinta de marsupiais sulamericanos. A espécie representa o tipo de uma nova família, **Patagoniidae,** a qual é designada a uma nova superfamília, **Patagonioidea,** por formar um grupo evolutivo bem definido, como o formam as outras famílias de marsupiais, reconhecidas por Simpson. A posição sistemática dos Patagonioidea, dentro da superordem Marsupialia, aguarda uma análise mais compreensiva dos marsupiais ainda enigmáticos (como Groeberioidea e Argyrolagoidea) aparentemente e seus relativos mais próximos.

From the División Paleontología Vertebrados, Museo de La Plata, Paseo del Bosque, 1900 La Plata, Argentina; and CONICET, Argentina.

Introduction

The taxon described in this paper is yet another example of the great adaptive radiation and dispersal of marsupials in South America. It represents a second line of marsupials that is convergent on the rodent adaptive zone (cf. Groeberioidea—Patterson, 1952; Simpson, 1970c; Clemens & Marshall, 1976). However, it is distinct from previously named forms, not only phylogenetically but also ecologically.

The new form does not suggest that marsupials attained the breadth and diversity of rodent adaptations, but it does show that marsupials occupied the rodent adaptive zone in previously unimagined ways. This new marsupial indicates that marsupial radiations in South America were almost as broad and reached as great extremes as those in Australia. The find is consistent with the view that "A complete record of South American marsupials would certainly include a large number of taxa, probably some of high categorical rank, now unknown" (Simpson, 1970a, p. 59). This and other forms recently found in northwestern Argentina (Pascual, 1980a, b, 1981, 1983) validate Simpson's prophetic suggestion that ". . . major parts of marsupial evolution were occurring in areas and facies inadequately sampled, if at all, by the known fossil deposits and the collections so far made" (Simpson, 1970a, p. 58). These deposits indicate the value of applying new sample-collecting techniques at mammal-bearing localities that are supposedly well known; it is only necessary to find new, appropriate facies.

The new ecological type from the Paleogene provides evidence to support Gould's (1983) view of "early experimentation, later standardization," with a consequent reduction in diversity. As in therians (Pascual et al., 1985) the diversification of South American marsupials took place principally in the Paleogene.

Measurements reported in Table 1 are depicted in Figure 3 and are given in millimeters. The abbreviation MACN CH is used for the Museo Argentino de Ciencias Naturales "Bernardino Rivadavia" (Buenos Aires), Colección Chubut.

Classification

Superfamily PATAGONIOIDEA nov.

The only known family of this taxon is the Patagoniidae. The superfamily is sufficiently characterized by the diagnoses of that family and its only known species. Justification for its superfamilial rank is given in a later section on affinities.

Family Patagoniidae nov.

TYPE—*Patagonia* gen. nov. The only known genus.

KNOWN DISTRIBUTION—Late Oligocene. Colhuehuapian from Central Patagonia (Chubut Province, Argentina).

DIAGNOSIS.—Small marsupials with the same reduced number of lower teeth as the Groeberiidae, but with a different dental formula: 1.1.0.3. Open-rooted and rodent-like lower incisor, oval in cross section, strongly curved, although not as much as in the Groeberiidae, and with the intra-alveolar portion differently arranged. The incisor extends lingually along the ventral border of the mandible to the root of the inflected crest beneath the last molar, where it forms a prominence similar to that of hystricognathous rodents, but ventrally. Lower canine smaller, procumbent, apparently incisor-like and closed-rooted, separated from the cheekteeth by a short, crested diastema at alveolar level. Lower cheekteeth rectangular in cross section, decreasing in size posteriorly, hypselodont, rootless, wholly surrounded by enamel, and slightly curved, with the concavity forward. Horizontal ramus of the mandible short and deep, with the highest part posterior, beneath the masseteric fossa, where the body of the mandible becomes strongly convex and inflected; deep pterygoid fossa, limited ventrally by a flange like that found in Argyrolagidae and in some Australian marsupials (e.g., Macropodidae); strong, salient coronoid process; masseteric fossa relatively deep but reduced, dorsally situated with a prominent masseteric crest; subvertical symphysis unfused, with nearly smooth symphyseal surfaces.

PATAGONIA gen. nov.

ETYMOLOGY—From Patagonia, its geographical record.

TYPE—*Patagonia peregrina* sp. nov.

KNOWN RANGE AND DIAGNOSIS—Same as that of the family.

Patagonia peregrina sp. nov. Figures 1–3

ETYMOLOGY—From Latin *peregrinus,* strange or rare.

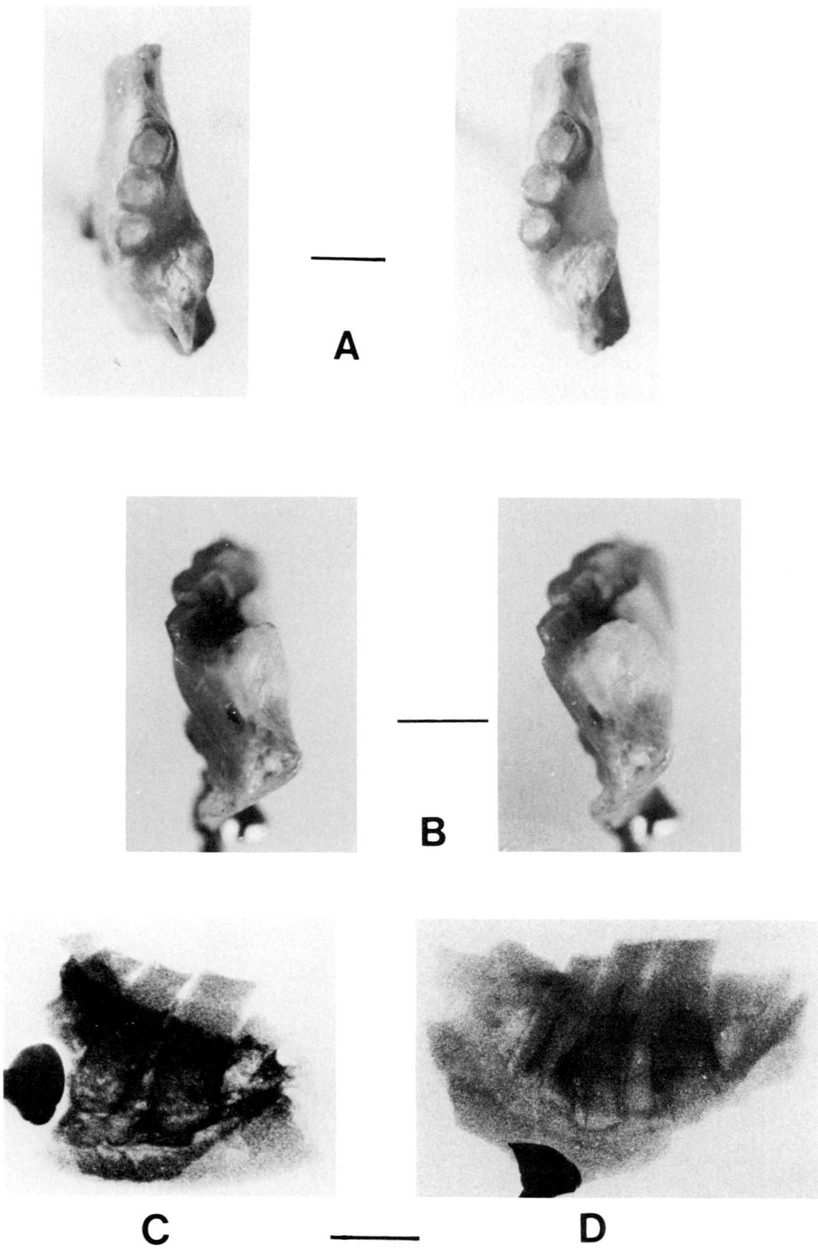

FIG. 1. *Patagonia peregrina* gen. et sp. nov. **A–B,** Stereopairs of MACN CH-865, a fragment of a right mandibular ramus with m_{1-3}: **A,** occlusal view; **B,** posterior view; **C–D,** X-ray of fragments of two right mandibular rami with i_1, alveolus of c_1, and m_{1-3} complete (**C,** holotype; MACN CH-869) and with alveoli of i_1, and c_1, and m_{1-3} complete (**D,** MACN CH-865). Graphic scale = 2 mm.

HOLOTYPE—MACN CH-869 (fig. 2A–B). Fragment of right mandibular ramus with three cheekteeth, intra-alveolar portion of the incisor, and alveolus of the canine.

HYPODIGM—Holotype and the following: MACN CH-864, part of right mandibular ramus with first and second cheekteeth, part of alveolus of the third, and part of alveoli of incisor and canine; MACN CH-865, part of right ramus with three cheekteeth and alveoli of the incisor and

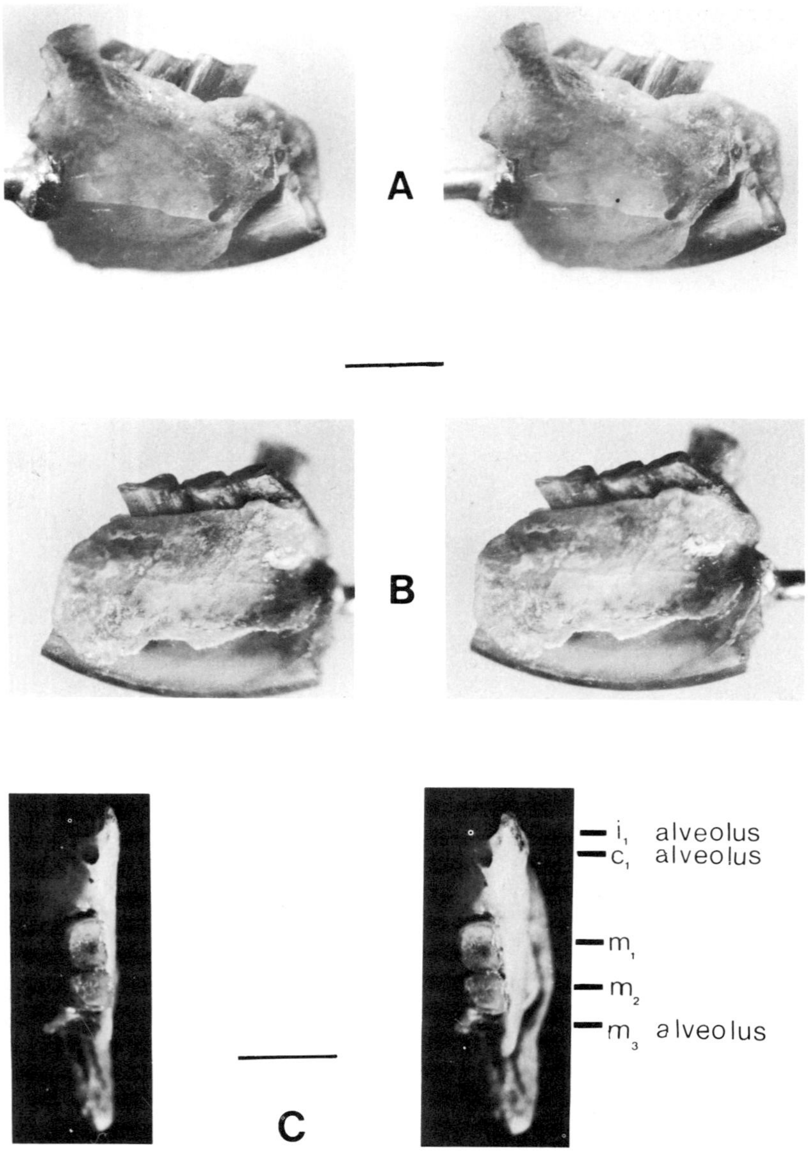

FIG. 2. *Patagonia peregrina* gen. et sp. nov. **A-B,** Stereopairs of MACN CH-869 (holotype), a fragment of a right mandibular ramus, with i_1, alveolus of c_1, and m_{1-3}: **A,** labial view; **B,** lingual view. Graphic scale = 2 mm. **C,** Stereopairs of MACN CH-867, a fragment of a left mandibular ramus, with alveoli of i_1, c_1 and m_3, and m_{1-2}; occlusal view.

canine; MACN CH-866, part of left ramus with the second and third cheekteeth, and part of alveoli of the first cheektooth and the incisor; MACN CH-867, part of the left ramus with the first and second cheekteeth, and part of alveoli of the third cheektooth, incisor, and canine; MACN CH-868, part of the right ramus with three cheekteeth and alveolus of the incisor; MACN CH-870, part of left ramus with first and second cheekteeth and part of the alveolus of the third; MACN CH-874, part of right ramus with the second cheektooth, alveoli of the first and third cheekteeth, and part of the alveolus of the incisor; MACN CH-875, part of right ramus with the second and third cheekteeth and part of the alveolus of the incisor; and MACN CH-876, three isolated upper(?) cheekteeth.

HORIZON AND LOCALITY—Both the holotype and the hypodigm come from the Trelew Member of the Sarmiento Formation (see Mendía & Bayarsky, 1981) and are Colhuehuapian (Late Oligocene) in age. Apparently they were found in the upper unit, exposed on the south side of the Chubut River valley, Chubut Province, Argentina (Central Patagonia; see Fleagle & Bown, 1983, pp. 242–244). Quite probably this corresponds to Simpson's "stratum *F* of Fig. 1," which is part of his "Trelew beds" (= "Trelewense"). The material was recovered by O. E. Donadío, M. Soria, J. G. Fleagle, and T. M. Bown (see Fleagle & Bown, 1983) through dry-screening local deflation lag deposits.

DIAGNOSIS—The only known species of the family.

DESCRIPTION—**Dentition**—See Figures 1A,C–D; 2–3. Each side of the lower jaw has one fully rodent-like gnawing incisor, only incompletely preserved in the holotype; it is posteriorly bordered by a relatively shallow and conical alveolus (the tooth being absent in all specimens at hand) separated from the medial one by bone and set at a relatively oblique angle (figs. 1C–D; 2C). Homologies of these teeth are uncertain, but the rodent-like medial tooth is surely an incisor, designated for description as i_1. The shape and disposition of the second alveolus agrees with the procumbent canine of Polydolopidae (Epidolopinae; cf. Paula Couto, 1952, 1961; Pascual & Bond, 1981) and Prepidolopidae (Pascual, 1980b, fig. 2D–E); it thus appears that this tooth is c_1. This alveolus is followed by a short diastema at alveolar level, then three cheekteeth, all rectangular in cross section (with some differences among them) and in close approximation, forming a molariform series. They are surrounded by enamel on all sides and are not strictly lobate, nor are the trigonid, talonid, or original cusps clearly indicated, as occlusion with the uppers was mediated through practically flat areas. The dentine forms a shallow basin surrounded by the highest enamel layer, which is slightly higher on the lingual side. There is a slightly deeper anteroposterior wear groove, extending from the anterolabial corner to the posterolingual one (fig. 1A). Grinding involved a longer propalinal movement and a shorter ectental stroke. The homologies of these teeth with the more numerous ancestral series cannot be determined. Plausibly

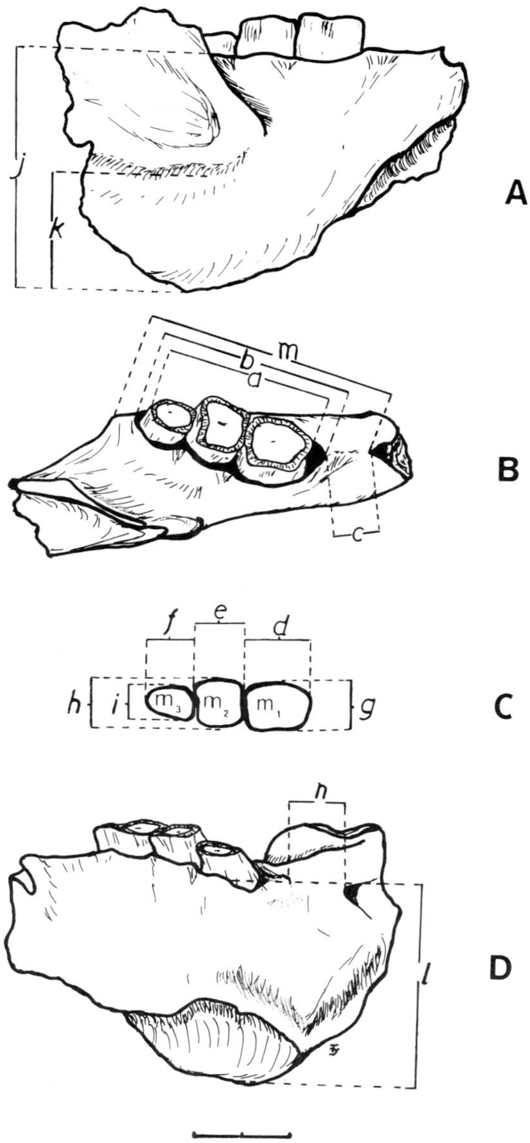

FIG. 3. *Patagonia peregrina* gen. et sp. nov. Outline of a right mandibular ramus fragment, with alveoli of i_1 and c_1, and m_{1-3} complete (MACN CH-865), showing the measurements of Table 1. **A,** Labial view; **B,** occlusal view; **C,** cheekteeth series (m_{1-3}); **D,** lingual view. Graphic scale = 2 mm.

they are homologous with those typically designated m_{1-3} in marsupials and are so designated here, yielding the lower dental formula 1.1.0.3. which is provisionally homologized as i_1, c_1, m_{1-3}. However, many specialized marsupials from the South American fossil record show tendencies (1)

TABLE 1. Dimensions of specimens of *Patagonia peregrina* gen. et sp. nov. (see fig. 3 for measurement references).

Specimen	Dimensions													
	a	b	c	d	e	f	g	h	i	j	k	l	m	n
MACN CH-864	...	...	...	...	1.00	...	1.00	0.96	...	...	...	...	...	...
MACN CH-865	3.68	4.08	1.40	1.40	1.20	1.08	1.08	1.12	0.96	5.40	2.80	...	6.20	1.80
MACN CH-866	...	...	...	...	1.00	1.08	...	1.00	0.92	...	...	...	...	1.60
MACN CH-867	...	...	1.12	1.20	1.00	...	0.96	0.96	...	...	...	...	...	...
MACN CH-868	...	4.08	...	...	1.00	1.00	1.04	1.00	0.92	...	...	...	...	1.80
MACN CH-869	3.60	...	...	1.36	1.08	1.00	0.96	0.96	0.92	5.12	3.00	4.60	...	...
MACN CH-870	...	...	...	1.40	1.00	...	1.00	1.00	...	...	...	...	...	...
MACN CH-874	...	...	...	...	1.08	...	...	0.96	...	4.80	2.20	4.40	...	...
MACN CH-875	...	...	...	...	1.28	1.16	...	1.24	1.08	6.76	...	...	...	2.40

to elongate or modify either the p_3 (Polydolopidae and Parabderitini caenolestids; see Marshall, 1980) or the m_1 (Abderitinae and Palaeothentinae caenolestids), and (2) to reduce (e.g., Caenolestidae) or lose (Polydolopidae) the m_4. All teeth except the c_1 are completely hypselodont and rootless. The cheekteeth are slightly curved, with the concavity forward (fig. 1C–D).

Incisor (i_1)—Incompletely preserved in the holotype (fig. 2A–B). The anterior end is broken, but the posterior end is unaltered, showing an open pulp cavity with no sign of root formation. It is as elongate and curved as in the most specialized caviomorphs (e.g., Ctenomyidae), although not upcurved posteriorly. The extra-alveolar part would have been nearly vertical, although not so much as in the Groeberiidae (cf. fig. 4B,D). The tooth extends along the ventral border lingually, first beneath the m_1 and then lingually to far below the m_3, terminating where the ventral border becomes an inflected flange (MACN CH-865, fig. 1B) reminiscent of kangaroos; the base of the alveolus shapes a superficial prominence similar to that of hystricognathous rodents, although enveloped by the inflected ventral border. It is approximately oval in cross section, with the long axis oriented dorsoventrally and with a flatter medial surface. Apparently enamel covers the entire tooth, but a noticeably heavier layer extends as a ventral band.

Canine (c_1)—None of the specimens in the hypodigm include this tooth, but its alveolus (figs. 1C–D; 2C) is oblique, tapered, and relatively shallow, indicating a closed-rooted tooth; its oblique orientation suggests that the occlusal apex was appressed against the occlusal tip of i_1.

First Molar (m_1)—The first molar is separated from the c_1 by a crested diastema (fig. 1A) as long as m_1. It is the largest cheektooth, almost rectangular in cross section, with the longer lateral faces slightly concave; the lingual face sometimes bears a very shallow groove along the intra-alveolar portion. The anterior face is convex, occasionally somewhat pointed; the posterior face is almost flat, forming angles with the lateral faces slightly greater than 90°.

Second Molar (m_2)—The second molar is irregularly quadrate, with the lateral faces slightly convex and the anterior and posterior ones flatter. The posterolingual angle is less than 90°, whereas the others are almost 90°. Its width is similar to that of m_1 (fig. 1A).

Third Molar (m_3)—The third molar is the smallest cheektooth, being subtriangular in cross section rather than square. The anterior face is slightly convex, lingually flatter, and labially more strongly curved; the labial face converges posterolingually with the lingual face, forming a rounded pillar rather than a well-defined posterior face (fig. 1A).

Mandible—No nearly complete mandible is known, but parts of the horizontal ramus and base of the coronoid process are known. These parts indicate the mandible is extremely short and deep, like that of *Groeberia minoprioi*, although very different in other respects (cf. fig. 4B,D). The symphysis is subvertical and unfused, with a nearly smooth symphyseal surface (i.e., normal in structure instead of fused and forming the odd medial posterior projection peculiar to Groeberiidae). The depth of the mandible increases abruptly toward the m_3. The deep masseteric fossa appears to be peculiarly confined to a dorsal position, as the masseteric crest is situated at a level between the alveolar rim and the lowest level of the rounded and inflected ventral border (figs. 1B; 2A). A similar condition is found in some Abderitinae caenolestids (e.g., *Parabderites bicrispatus*; Marshall, 1976, fig. 8a), although in *P. bicrispatus* the mandibular body is not as deep and the alveolar border not as extensively inflected. The coronoid process has its root beneath the m_2, forming a strong, salient lamina (known only by its root), so that a conspicuous diagonal valley is formed between the coronoid and the alveolar border behind m_3 (fig. 1A); a similar structure is present in *Groeberia minoprioi* (see Patterson, 1952, p. 41); the valley is open labially and lingually limited by a prominence similar to that present in Australasian Potoroinae.

In many respects this strong, salient, ascending ramus and correlated features are reminiscent of highly fossorial caviomorphs, such as burrow-inhabiting Ctenomyidae. Although the mandibular angle is not preserved in any of the specimens, it probably was inflected, as suggested by the inflection of the ventral border, beginning at the level of m_3, which defines a lingual flanged crest (figs. 1B; 2B) similar to that producing the extremely inflected angle in the Macropodidae. This lingual ventral flanged crest seems to be the lingual border of an expanded and relatively deep pterygoid fossa, resembling that of argyrolagids (see Simpson, 1970a). There is a relatively large alveolar foramen within the pterygoid fossa, level with the alveolar border and within a pit (fig. 3D), and a mental foramen beneath the anterior face of m_1 at the level of the alveolus of i_1 (fig. 2A).

Affinities

As in the case of *Groeberia* (see Simpson, 1970c), the conclusion that *Patagonia* is a marsupial rests on a combination of definite, negative, and indirect evidence. The most definite evidence for its being a marsupial is the inflected ventral border of the mandible and probably the related inflected angle. This evidence alone is inconclusive, as a few marsupials lack an inflected angle and a few placentals have one. However, no known placental has such an extended and upturned flange-shaped inflection, and even in marsupials it is rarely so well developed (e.g., Groeberiidae [Patterson, 1952]. Argyrolagidae [Simpson, 1970a,b], and the Australasian Macropodidae). Unlike *Groeberia, Patagonia* has other characters supporting its marsupial affinities, namely the lower procumbent incisor-like canine. In the Epidolopinae (Pascual & Bond, 1981) there are three procumbent lower teeth, the third being unquestionably the canine. Within the more advanced Polydolopidae (Polydolopinae), there are one or three procumbent lower teeth; in the latter case, evidence suggests they consist of two incisors and a canine, the medial incisor being quite reduced and the canine well developed, single, and closed-rooted.

As in *Groeberia* the negative evidence is that *Patagonia* has no features precluding its reference to the Marsupialia. It does exhibit characters making reference to any Eutheria highly improbable. Its habitus is rodent-like, but its two differentially procumbent lower teeth rule out reference to the Rodentia. While the incisor is rodent-like in shape, it is oriented differently than that in rodents, extending along the ventral border of the horizontal ramus, first below the m_1, then lingually to other molars, without curving upward. It apparently shapes the ventral border of the mandible. In addition, the short diastema extends at the level of alveoli. Among known rodents, only Paramyidae and Ischyromyidae developed diastemas at the alveolar level, but even in these groups, the incisor extends as in other rodents, not as in *Patagonia.* A more-or-less rodent-like habitus was also characteristic of some notoungulates, especially among Typotheria and Hegetotheria, but insofar as known not so extreme in development as in *Patagonia.* Neither the enlargement of the incisor nor the reduction of the cheekteeth is known in any primitive Paleocene notoungulates or in other South American "ungulates." Even later rodent-like notoungulates were much less specialized than the Oligocene *Patagonia.* South American marsupials diverged very early into unique evolutionary lineages (see Simpson, 1970a–c, 1971, 1980; Pascual, 1980a,b, 1981; Paula Couto, 1979; Reig, 1981).

Patagonia peregrina is unquestionably a marsupial because its unique and diagnostic combination of characters are unknown in any eutherian. Nevertheless, it could be regarded as another of the extinct South American mammals considered by some as *incertae sedis* and by others as a *tertium quid* with regard to the eutherian-marsupial dichotomy (McKenna, 1980; Reig, 1981). However, the marsupial affinities of other peculiar fossil mammals from South America remain unquestioned, despite weaker support than that offered here for *Patagonia.* For example, the basis for considering the Polydolopidae as marsupials is the combination of an inflected mandibular ramus, palatal vacuities, and a cheektooth formula of $P\frac{1\text{-}3}{1\text{-}3}$ and $M\frac{1\text{-}4}{1\text{-}4}$. These characters were formerly used to exclude the polydolopids from the Allotheria. But, as these characters are present in primitive therians outside South America, their diagnosis of marsupials can be considered an "act of

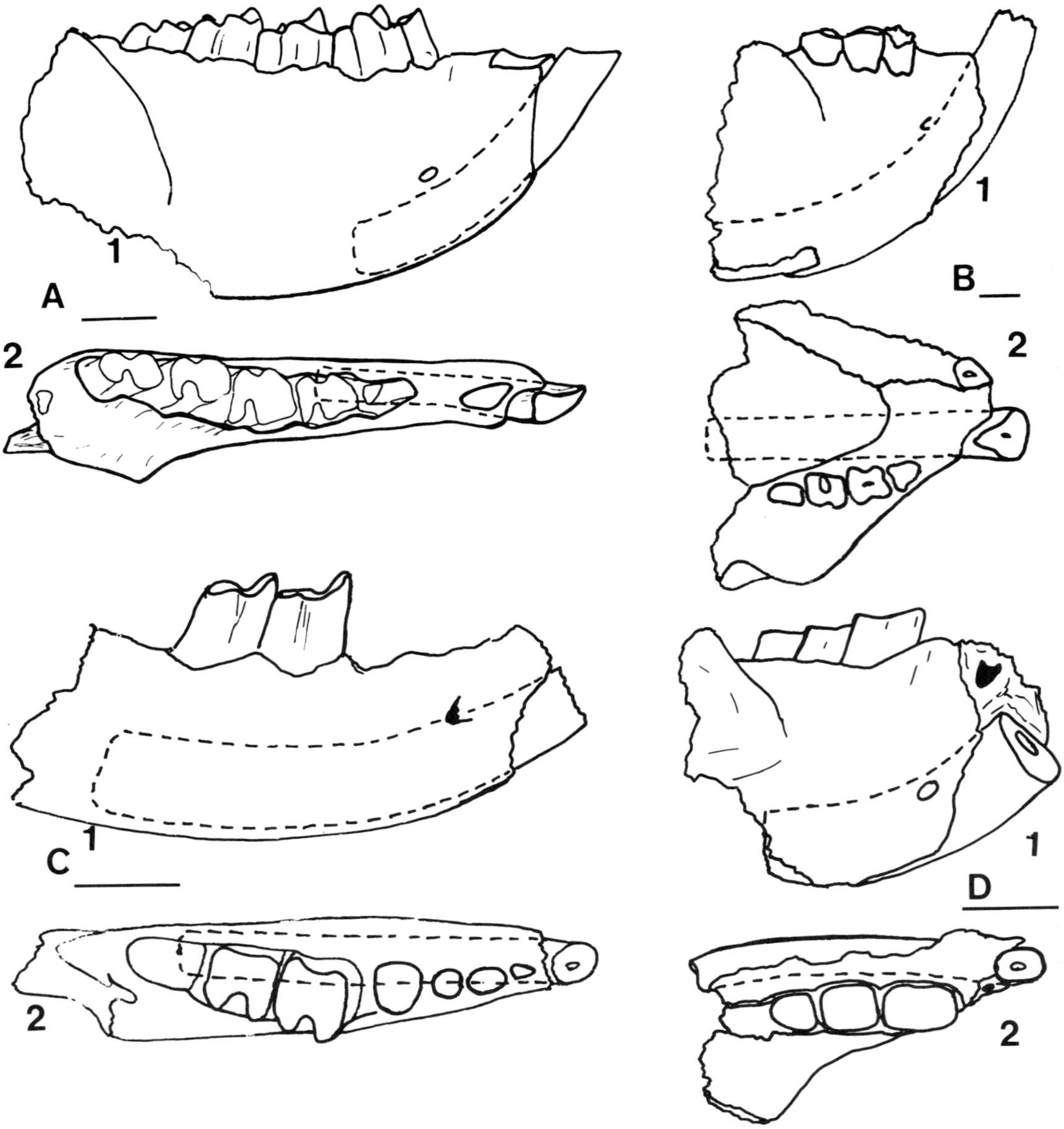

FIG. 4. Labial (1) and occlusal (2) outlines of mandibles, showing the different development of incisor. **A,** *Argyrolagus parodii* Rusconi; **B,** *Groeberia minoprioi* Patterson; **C,** *Proargyrolagus bolivianus* Wolff; **D,** *Patagonia peregrina* gen. et sp. nov. Graphic scale = 2 mm.

faith based on . . . geography and stratigraphic position rather than on . . . biology" (McKenna, 1980, pp. 58–59). We believe that assignment of polydolopids to marsupials represents the most parsimonious conclusion.

Like the newly described *Proargyrolagus bolivianus* (Wolff, 1984), *Patagonia peregrina* is another peculiar marsupial that appears in the fossil record without known ancestors (see Simpson, 1970c, p. 16) only to vanish again soon afterward: Groeberiidae (Divisaderan Age, Late Eocene); Patagoniidae (Colhuehuapian Age, Late Oligocene); Necrolestidae (Santacrucian Age, Early Miocene); Argyrolagidae (Huayquerian to Uquian Ages, Late Miocene to Early Pleistocene). We believe there are cogent reasons to think of *Proargyrolagus bo-*

livianus Wolff, 1984, described as a Deseadan argyrolagid, as possibly representing a distinct family of Argyrolagoidea.

This raises the question of the position of *Patagonia* among the varied ranks of South American marsupials. The previous descriptions and illustrations demonstrate that *Patagonia peregrina* has many peculiarities that are rare, differently developed, or completely absent in other marsupials (cf. fig. 4). The most striking of these are:

1. Mandible extremely short and deep, with unfused subvertical symphysis, dorsally positioned masseteric fossa, ventral border inflected at level of the m_3, enveloping there the alveolus of the incisor.
2. Presence in each ramus of mandible of one rodent-like rootless incisor that extends lingually along ventral border of mandible to below the m_3.
3. Presence in each ramus of one procumbent canine, single- and closed-rooted, scarcely separated from the incisor and with the occlusal apex probably appressed to the incisive apex.
4. Three rectangular and continuously growing cheekteeth arranged in close sequence.

These and other less striking characters underscore the unique specializations of *Patagonia peregrina,* leading to its assignment to a new family, Patagoniidae. But the distinctive combination of characters in the Patagoniidae identify it as a distinct evolutionary group, that is, a different cladogenetic unit. Simpson (1945, 1970a, 1980) designated natural evolutionary groups of marsupials as superfamilies. Following this line of reasoning, Patagoniidae should be allocated to a new superfamily, the Patagonioidea.

What are the affinities of this new superfamily to other superfamilies within the superorder Marsupialia? Any discussion of its affinities depends on the systematics of other taxa, many of which are problematic. The systematics of fossil and extant South American marsupials, including the merits of recognizing Marsupialia as a superorder, are discussed by Simpson (1970a, 1971) and Pascual (1980b).

The majority of South American marsupials represent the order Polyprotodonta; this is roughly equivalent to Ride's (1964) Marsupicarnivora, but also includes Ameghino's Paucituberculata (see Pascual, 1980b; *contra* Kirsch, 1977a,b; Reig, 1981). There is as yet no compelling argument to include any South American families within the Australasian order Diprotodonta (Reig, 1981), despite some suggestions to the contrary (e.g., Pascual & Herrera, 1973, 1975). While the allocations of these groups seems unambiguous, the positions of most remaining groups (e.g., Argyrolagidae, Necrolestidae, and Groeberiidae) remain uncertain. With some reservation, Kirsch (1977b) included the Necrolestidae in the polyprotodont Borhyaenoidea (as did Patterson, 1958), and the Groeberioidea and Argyrolagoidea within the Paucituberculata. Independently, Clemens and Marshall (1976) also treated these animals as marsupials, recognizing each as superfamilies: Argyrolagoidea, Necrolestoidea, and Groeberioidea. Like Simpson, they made these assignments with disclaimers that the interrelationships of these groups were far from clear.

Reig (1981, p. 60) not only questioned whether the Argyrolagidae (his Microtragulidae) were marsupials, as none of its known characters are typologically diagnostic, but conjectured probable affinities to the Anagalida. Further, without rigorous analysis, he suggested that the Argyrolagidae could be treated as an independent order, proposing the name Argyrolagida. He concluded that only more intensive study or additional records could substantiate allocation of this order to the Metatheria or the Eutheria.

Remains of Patagoniidae exhibit a unique mosaic of characters, some of which are absent or differently developed in Groeberiidae and Argyrolagidae. Despite their similarities, each of these taxa appears *prima facie* to represent independent evolutionary trends. To assess their interrelationships, common and distinctive characters of each must be carefully weighed. Remains of Argyrolagoidea obtained in the same horizon and locality as the hypodigm of *Patagonia peregrina* should be particularly useful in this regard and are now under study. Ordinal and subordinal allocation of the Patagonioidea await this more comprehensive analysis. Known representatives of this taxon are so highly derived, as is the case with other peculiar marsupials, that their relationships to other marsupial groups are obscure and can only be clarified by an expanded record of earlier forms.

Ecology and Historical Biogeography

Biological inferences of *Patagonia* are necessarily limited to the mandibular fragments thus far known. These demonstrate unique characters among marsupials, living or extinct, which are obviously related to a particular mode of life. No known eutherian possesses such mandibular features. Superficially it is similar to *Groeberia,* both being rodent-like marsupials: each has a short and deep mandible with a single enlarged, open-rooted incisor, deeply extended along the mandible, with the extra-alveolar part apparently nearly vertical. These represent functional not phylogenetic similarities, as similar states were attained by different routes: in *Groeberia* this tooth extended within an odd medial posterior projection of the symphysis, whereas in *Patagonia* the intra-alveolar portion is truly rodent-like, in being extended along the horizontal ramus (cf. fig. 4B,D). No doubt both were powerful gnawers as the lower incisor worked almost vertically, much more so than in most rodents. The unknown face and snout of *Patagonia* was probably short and deep; whether it had two pairs of lagomorph-like upper incisors like *Groeberia* remains unknown. Related to this gnawing specialization, both *Groeberia* and *Patagonia* show a short diastema near the alveolar level and a reduced number of postincisive teeth, four in both; however, *Patagonia* has three cheekteeth, whereas *Groeberia* has four. The rodent-like habitus of *Patagonia* is especially advanced, because the three cheekteeth are truly hypselodont, rectangular-shaped in cross section, with at most only shallow lateral grooves representing the remnants of ancestral bilobate cheekteeth.

This combination of features suggests food was obtained by gnawing and prepared for swallowing by grinding. It represents extraordinary convergence on some desert-adapted and fossorial forms, such as the Octodontidae. The evolution of cheekteeth toward a rectangular shape and numerically reduced sequence has been recognized as occurring within the Octodontoidea (from the Octodontidae to the Ctenomyidae; Pascual et al., 1965). The dental features of *Patagonia* are also convergent on those of the desert-dwelling African Bathyergidae, particularly to the sand rat *Heterocephalus glaber,* and to the North American Geomyidae. These convergent anatomical features suggest that Patagoniidae were probably fossorial marsupials.

Anatomical convergence of *Patagonia* on desert-dwelling fossorial rodents is curious, because prevailing conditions in central Patagonia during the Colhuehuapian Age were not highly favorable to desert dwellers. The first record of platyrrhine monkeys in Patagonia occurs at the same locality and level (Fleagle & Bown, 1983) as *Patagonia.* Many other vertebrate remains recovered at this site (see Bordas, 1939; Donadío, 1983) suggest an environment of well-watered tropical woodlands. Conversely, however, both Argyrolagoidea and very advanced Cephalomyidae rodents from this site (currently under study) show dental features reminiscent of desert or at least drier environments. Generally, the Colhuehuapian vertebrate fauna from central Patagonia (see Pascual, 1970; Pascual & Odreman Rivas, 1971; Marshall et al., 1983) is composed of both forest and open-country types, presumably brought together in a subtropical savanna. Thus, the Patagoniidae, Cephalomyidae, and Argyrolagoidea occurred in apparently inappropriate environments, probably restricted to xeric patches in the subtropical savanna mosaic. Because the Colhuehuapian Patagonioidea were already highly specialized for xeric habitats, they probably evolved earlier in the Paleogene. It therefore seems likely that ancestral forms existed in the Deseadan (Early Oligocene). Another highly specialized group of marsupials, the Argyrolagoidea, suggests this hypothesis. Formerly believed present in the record from the Huayquerian (Late Miocene) to the Uquian (Early Pleistocene; see Marshall et al., 1983), argyrolagoids have now been reported from the Deseadan of Bolivia (Wolff, 1984), and here from the Colhuehuapian beds of central Patagonia.

The pre-Deseadan record contains no potential ancestor for either Argyrolagoidea or Patagonioidea. Simpson (1970c, p. 17) proposed that "these groups (including Groeberioidea) evolved in what are now (and quite likely were then) the tropics and are picked up in our record only when they spread rather briefly to what was for them a marginal area." It seems quite probable that the environments responsible for their initial divergence were poorly or not represented in the known fossil record.

Global diastrophic movements in the Late Eocene, and apparently related climatic and environmental changes, are thought to be responsible for the cosmopolitan turnover in Early Oligocene mammal communities (Kurtén, 1971). This turnover also occurred in South America (Pascual, 1984). Mammal communities in the Deseadan (Early Oligocene) are substantially different from Eocene communities in composition (see Pascual et al., 1985), apparently reflecting Stehlin's "*grande coupure.*" The apparently sudden occurrence of the Argyrolagoidea, and probably the Patagonioidea, in the Deseadan Age is probably another example of this global turnover.

It is remarkable that, to the numerous successive parallel trends ("successive trends" or "iteration"; Simpson, 1953, pp. 248–259; 1961, p. 127) in the evolutionary history of South American mammals, especially from the Deseadan on, can be added the convergence of Oligocene patagoniid marsupials and Pliocene to Recent ctenomyid rodents on a common morphology. These convergences are products of similar responses to repeated environmental conditions. The anatomical and functional similarities of *Patagonia peregrina* with the extant *Ctenomys* are so striking that we are tempted to call the former the "marsupial tucotuco."

Acknowledgments

All of the material studied here was discovered by 1983 and 1984 expeditions of the Museo Argentino de Ciencias Naturales "Bernardino Rivadavia" (MACN), in which Lic. Oscar E. Donadío and Lic. Miguel Soria (both of MACN) and the American paleontologists John G. Fleagle and Thomas M. Bown participated. Dr. José F. Bonaparte, Chief of the Sección Paleontología Vertebrados, MACN, and responsible for these expeditions, generously put this and other marsupial material at our disposal. The x-ray plates were made by Dr. Roberto Guevara, Profesor de Odontología, Universidad Nacional de La Plata, by the authority of the Dean, Dr. Oscar Barletta. We thank all of them very much.

Literature Cited

BORDAS, A. F. 1939. Diagnosis sobre algunos mamíferos de las capas con *Colpodon* del valle del Río Chubut. Physis, **14:** 413–433.

CLEMENS, W. A., AND L. G. MARSHALL. 1976. American and European Marsupialia. Pars 123. Marsupialia. *In* Westphal, F., ed., Fossilium Catalogus. I: Animalia. W. Junk, The Hague, 114 pp.

DONADÍO, O. E. 1983. Los lacertilios del Colhuehuapense de la provincia del Chubut, Argentina. Circular Informativa de la Asociación Paleontológica Argentina, **11:** 5–6.

FLEAGLE, J. G., AND T. M. BOWN. 1983. New Primate fossil from Late Oligocene (Colhuehuapian) localities of Chubut Province, Argentina. Folia Primatologica, **41:** 240–266.

GOULD, S. J. 1983. Nature's Great Era of experiments. Natural History, **7:** 12–21.

KIRSCH, J. A. W. 1977a. The classification of marsupials, pp. 1–48. *In* Hunsaker II, D., The Biology of Marsupials. Academic Press, New York, San Francisco, London, 537 pp.

———. 1977b. The comparative serology of Marsupialia, and a classification of marsupials. Australian Journal of Zoology, Supplementary Series, **52:** 1–152.

KURTÉN, B. 1971. The Age of Mammals. Columbia University Press, New York, 250 pp.

MARSHALL, L. G. 1976. Revision of the South American fossil marsupial subfamily Abderitinae (Mammalia, Caenolestidae). Publicaciones Museo Municipál de Ciencias Naturales de Mar del Plata "Lorenzo Scaglia," **2**(3): 57–90.

———. 1980. Systematics of the South American marsupial family Caenolestidae. Fieldiana: Geology, n.s., **5:** i–vii, 1–145.

MARSHALL, L. G., R. HOFFSTETTER, AND R. PASCUAL. 1983. Mammals and stratigraphy: Geochronology of the continental mammal-bearing Tertiary of South America. Palaeovertebrata, Mémoire Extraordinaire, Montpellier, **1983:** 1–93.

MCKENNA, M. C. 1980. Early history and biogeography of South America's extinct land mammals. *In* Ciochon, R. L., and A. B. Chiarelli, eds., Evolutionary Biology of the New World Monkeys and Continental Drift. Plenum Publishing Corp., New York, xvi + 528 pp.

MENDÍA, J. E., AND A. BAYARSKY. 1981. Estratigrafía del Terciario en el valle inferior del río Chubut. VII Congreso Geológico Argentino, San Luis (20–26 Septiembre 1981), Actas **3:** 593–606.

PASCUAL, R. 1970. Evolución de comunidades, cambios faunísticos e integraciones biocenóticas de los vertebrados cenozoicos de Argentina. Actas IV Congresso Latinoamericano de Zoología (Caracas, 10–16 de Noviembre de 1968), **2** (Paleontología Sudamericana): 991–1088.

———. 1980a. Nuevos singulares tipos ecológicos de marsupiales extinguidos de América del Sur (Paleoceno tardío o Eocene temprano) del Noroeste Argentino. Actas II Congreso Argentino de Paleontología y Bioestratigrafia y I Congreso Latinoamericano de Paleontología (Buenos Aires, 2–6 de Abril 1978), **2:** 151–173.

———. 1980b. Prepidolopidae, nueva familia de Marsupialia Didelphoidea del Eoceno sudamericano. Ameghiniana, **17**(3): 216–242.

———. 1981. Adiciones al conocimiento de *Bonapartherium hinakusijum* (Marsupialia, Bonapartheriidae) del Eoceno temprano del Noroeste Argentino. Anais II Congresso Latino-Americano de Paleontología (26 a 30 de Abril, 1981, Porto Alegre-Brasil), **2:** 507–520.

———. 1983. Novedosos marsupiales paleógenos de la Formación Pozuelos (Grupo Pastos Grandes) de la Puna, Salta, Argentina. Ameghiniana, **20**(3–4): 265–280.

———. 1984. La sucesión de las Edades-mamífero, de los climas y del diastrofismo sudamericanos durante el Cenozoico: fenómenos concurrentes. Anales de la Academia Nacional de Ciencias Exactas, Físicas y Naturales, **36:** 15–37.

Pascual, R., and M. Bond. 1981. Epidolopinae subfam. nov. de los Polydolopidae (Marsupialia, Polydolopoidea). Anais II Congresso Latino-Americano de Paleontologia (26 a 30 de Abril, 1981, Porto Alegre-Brasil), **2:** 479–488.

Pascual, R., and H. E. Herrera. 1973. Adiciones al conocimiento de *Pliolestes tripotamicus* Reig, 1955 (Mammalia, Marsupialia, Caenolestidae) del Plioceno superior de la Argentina. Ameghiniana, **10**(1): 36–50.

———. 1975. *Stilotherium* Ameghino, 1887, el más primitivo Caenolestidae conocido. Consideraciones sobre la transición Didelphidae-Caenolestinae (Marsupialia). Actas I Congreso Argentino de Paleontología y Bioestratigrafía (12–16 Agosto, 1974, Tucumán, Argentina), **2:** 417–430.

Pascual, R., and O. E. Odreman Rivas. 1971. Evolución de las comunidades de los vertebrados del Terciario argentino. Los aspectos paleozoogeográficos y paleoclimáticos relacionados. Ameghiniana, **8**(3–4): 372–412.

Pascual, R., J. Pisano, and E. J. Ortega. 1965. Un nuevo Octodontidae (Rodentia, Caviomorpha) de la Formación Epecuén (Plioceno medio) de Hidalgo (Provincia de La Pampa). Ameghiniana, **4**(1): 19–30.

Pascual, R., M. G. Vucetich, G. J. Scillato-Yané, and M. Bond. 1985. Main pathways of mammalian diversification in South America, pp. 219–247. *In* Stehli, F., and S. D. Webb, eds., The Great American Biotic Interchange. Plenum Publishing Corp., New York.

Patterson, B. 1952. Un nuevo y extraordinario marsupial deseadiano. Revista del Museo Municipal de Ciencias Naturales, Mar del Plata, **1:** 39–44.

———. 1958. Affinities of the Patagonian fossil mammal *Necrolestes*. Breviora (Museum of Comparative Zoology), **49:** 1–14.

Paula Couto, C. de. 1952. Fossil mammals from the beginning of the Cenozoic in Brasil. Marsupialia: Polydolopidae and Borhyaenidae. American Museum Novitates, **1559:** 1–27.

———. 1961. Marsupiais fósseis do Paleoceno do Brasil. Anais Academia Brasileira Ciências, Rio de Janeiro, **33**(3/4): 321–333.

———. 1979. Tratado de Paleomastozoologia. Academia Brasileira de Ciências, Rio de Janeiro, 590 pp.

Reig, O. A. 1981. Teoría del origen y desarrollo de la fauna de mamíferos de América del Sur. Monographiae Naturae (Publicadas por el Museo Municipal de Ciencias Naturales "Lorenzo Scaglia"), Mar del Plata, **1:** 1–162.

Ride, W. D. L. 1964. A review of Australian fossil marsupials. Journal of the Royal Society of Western Australia, **47**(4): 91–131.

Simpson, G. G. 1945. The principles of classification and a classification of mammals. Bulletin of the American Museum of Natural History, **85:** i–xxvi, 1–350.

———. 1953. The Major Features of Evolution. Columbia University Press, New York, 248 pp.

———. 1961. Life of the Past. Yale Paper Bound, Yale University Press, New Haven, Conn.

———. 1970a. The Argyrolagidae, extinct South American marsupials. Bulletin of the Museum of Comparative Zoology, **139:** 1–86.

———. 1970b. Additions to knowledge of the Argyrolagidae (Mammalia, Marsupialia) from the Late Cenozoic of Argentina. Breviora (Museum of Comparative Zoology), **361:** 1–9.

———. 1970c. Addition to knowledge of *Groeberia* (Mammalia, Marsupialia) from the Mid-Cenozoic of Argentina. Breviora (Museum of Comparative Zoology), **362:** 1–17.

———. 1971. The evolution of marsupials in South America. Anais Academia Brasileira de Ciências, Suplemento, **43:** 103–118.

———. 1980. Splendid Isolation. The Curious History of South American Mammals. Yale University Press, New Haven and London, IX + 266 pp.

Wolff, R. G. 1984. A new Early Oligocene Argyrolagid (Mammalia: Marsupialia) from Salla, Bolivia. Journal of Vertebrate Paleontology, **4**(1): 108–113.

An Additional 14-Chromosome Karyotype and Sex-Chromosome Mosaicism in South American Marsupials

Milton H. Gallardo and Bruce D. Patterson

ABSTRACTS

The karyotype of *Rhyncholestes* Osgood is described for the first time. The karyotype has 2n = 14 and is similar in most respects to karyotypes of similar number found in other American and Australasian genera in several families. The karyotype of somatic (bone marrow) tissues from male *Dromiciops* Thomas is presented for the first time; surprisingly, it differs from the 2n = 14 complement previously reported from female bone marrow and male gonads. The 2n = 13 karyotype found in bone marrow of male *Dromiciops* lacks a minute element thought to be the Y chromosome. This instance of somatic chromosome elimination represents the first case reported for American marsupials and presents an interesting parallel to sex-chromosome mosaicism among Australasian Peramelidae and Petauridae.

El cariotipo de *Rhyncholestes* Osgood es descrito por primera vez. El cariotipo consta de 2n = 14 y es muy similar a cariotipos de igual número encontrados en otros géneros americanos y australoasiáticos de varias familias. El cariotipo de tejidos somáticos (médula ósea) de un *Dromiciops* Thomas macho es presentado por primera vez; sorprendentemente, difiere del complemento 2n = 14 reportado previamente de médula ósea femenina y gónadas masculinas. El cariotipo 2n = 13 encontrado en médula ósea del *Dromiciops* macho carece de un diminuto elemento que supuestamente corresponde al cromosoma Y. Este ejemplo de eliminación somática de cromosomas representa el primer caso reportado en marsupiales americanos y presenta un interesante paralelo con el extenso mosaicismo de los cromosomas sexuales descrito entre las formas australoasiáticas.

Descreve-se pela primeira vez, o cariótipo de *Rhyncholestes* Osgood. O cariótipo é de 2n = 14, e, na maioria de seus aspectos, assemelha-se aos cariótipos de números similares encontrados em outros generos americanos e australásios. O cariótipo de tecidos somáticos (da medula óssea) de *Dromiciops* Thomas machos é descrito pela primeira vez. Supreendentemente, este cariótipo difere do complemento de 2n = 14, prèviamente descrito para a medula óssea das femeas e para as gonadas dos machos. No cariótipo de 2n = 13, encontrado na medula óssea de *Dromiciops* machos, falta um elemento miúdo, possìvelmente o cromossomo Y. Este é o primeiro exemplo documentado da eliminação somática de um cromossomo em marsupiais americanos, e apresenta um paralelo interessante ao mosaico frequentemente encontrado nos cromossomos sexuais de outras formas australásias.

From the Instituto de Ecología y Evolución, Universidad Austral de Chile, Casilla 567, Valdivia, Chile; and Division of Mammals, Field Museum of Natural History, Chicago, IL 60605-2496.

Introduction

Several unusual cytological features, including low diploid number (Hayman & Martin, 1969; Reig et al., 1977), paternally derived X inactivation (Lyon, 1974a,b), multiple sex-chromosome systems (Hayman & Martin, 1969; Schneider, 1977), somatic elimination of sex-chromosomes (Schneider, 1977; Close, 1984), and sperm conjugation (Biggers & Creed, 1962; Biggers & DeLamater, 1965), have made marsupials interesting subjects of cytological research. These studies have clarified fundamental cytological mechanisms. Additionally, results of the research have shed light upon directions of chromosomal evolution and upon interrelationships of lineages (Hayman & Martin, 1969; Reig et al., 1977; Sharman, 1982).

A 14-chromosome karyotype occurs in several distinct lineages in all living American families: Didelphidae (Reig et al., 1977), Microbiotheriidae (Spotorno & Fernández, 1971; Reig et al., 1972), and Caenolestidae (Hayman et al., 1971). This karyotype also occurs in several Australian marsupial lineages (Hayman & Martin, 1969) and is therefore considered the primitive chromosome number for Metatheria (Reig et al., 1977). Direction of chromosome evolution in Metatheria has proceeded via centromeric dissociations—with pericentric inversions superimposed on the basic Robertsonian mechanism—to give rise to the remaining 2n = 18 and 2n = 22 karyotypes known for American forms (Hayman & Martin, 1969; Reig et al., 1977). Extremes of karyotypic variation in Australasian marsupials are 2n = 10 to 32 (Schneider, 1977).

Two autochthonous and endemic South American genera, *Rhyncholestes* and *Dromiciops,* are especially interesting from an evolutionary viewpoint. Both are represented by a single species and occur only in the temperate Valdivian rainforests of southern Chile and Argentina. *Rhyncholestes,* one of three extant genera of Caenolestidae, is widely isolated from its relatives in the northern Andes and presents some striking morphological specializations. *Dromiciops,* thought by some to have special affinities with Australasian lineages (Sharman, 1982; Szalay, 1982), is the only living genus of the otherwise extinct Microbiotheriidae (Marshall, 1982). Its affinities with other marsupial genera are currently uncertain. In this note we present the first somatic karyotypes of male *Rhyncholestes raphanurus* and *Dromiciops australis.* Additionally, we document the first instance of somatic sex-chromosome mosaicism in South American marsupials.

Materials and Methods

Seven specimens of *D. australis* (five males and two females) from Valdivia (39°32′S, 72°52′W), Osorno (41°06′S, 72°30′W), and Concepción (37°26′S, 73°19′W) provinces, Chile, were analyzed by the in vivo colchicine-hypotonic citrate technique using bone marrow as a source of mitoses (Patton, 1967). Modifications of the same procedure were used for the one *R. raphanurus* collected at La Picada, Volcán Osorno (41°06′S, 72°30′W); incubation with colchicine lasted 2.5 hours and a slightly more hypotonic solution of sodium citrate was used. A total of 419 mitotic plates was examined: 291 from male and 120 from female *D. australis* and 10 from *R. raphanurus.* Museum specimens were deposited in the Collection of Mammals, Instituto de Ecología y Evolución, Universidad Austral de Chile, and Field Museum of Natural History.

Results and Discussion

Rhyncholestes raphanurus presents a 2n = 14 complement, consisting of three pairs of large metacentric, one pair of medium-sized metacentric, and two pairs of small metacentric autosomes. The sex-chromosomes are an acrocentric X and a minute Y (fig. 1). This karyotype differs morphologically from the didelphid 2n = 14 in not showing a clear break between chromosome groups A and B. It also differs in arm ratios (table 1) from the other living caenolestids, *Lestoros* and *Caenolestes* (see Hayman et al., 1971). Moreover, the interstitial region of the short arm of pair two shows an achromatic area, resembling a secondary constriction, not described in other caenolestids (but see discussion in Sharman, 1982). Nevertheless, a 2n = 14 karyotype characterizes all three genera of Caenolestidae, which supports previous claims that this karyotype is primitive for Metatheria (Hayman & Martin, 1969; Hayman et al., 1971; Reig et al., 1977) and reinforces the pattern of low karyotypic variation within marsupial families.

Secondary constrictions can serve as chromosome markers and are thus useful, in the absence of banding data, for phylogenetic reconstruction. However, the secondary constriction evident in the karyotype of *Rhyncholestes* is unreported in other South American marsupials, although secondary constrictions are widespread among Australasian marsupials (Hayman & Martin, 1969). Considering commonality and in-group and out-group comparisons, we regard the secondary constriction of *Rhyncholestes* as apomorphic. Thus, the similar structures of Australasian marsupials were apparently independently derived and cannot be traced back to some marsupicarnivorous or other common ancestor.

Chromosome counts from all four male *D. australis* consistently indicated 2n = 13 chromosomes. The diploid number for females was 2n = 14 as was previously reported (Spotorno & Fernández, 1971; Reig et al., 1972). No differences among our specimens from geographically isolated localities were detected, nor were secondary constrictions evident.

Electron microscope studies of sex-chromosomes in spermatocytes of *D. australis* and the didelphid *Marmosa elegans* demonstrate striking similarities (Fernández et al., 1979). These similarities suggest that a 2n = 14 karyotype should be present in *D. australis,* its Y chromosome should resemble that of *M. elegans,* and both genera should exhibit an XX/XY sex-chromosome system.

We have consistently found 2n = 13 chromosomes in somatic tissues of male *Dromiciops* and 2n = 14 in female somatic tissue. The missing chromosome in males is dotlike and probably the Y chromosome (fig. 1). Translocation of the Y to an autosome is an unlikely mechanism for the differences between sexes because males have 2n = 14 in germinal cells and because the sex vesicle appears normal (Fernández et al., 1979). While it is possible that such a small chromosome might be overlooked in one or a few chromosomal spreads, its universal absence in all counted plates makes this alternative highly unlikely. Available data favor a somatic elimination of the Y chromosome.

Previous studies have shown that both constitutive and facultative heterochromatin can be deleted from marsupial cells in vivo without apparent deleterious effects on cell replication and survival (Hayman & Martin, 1969). Most examples of somatic elimination of sex-chromosomes in marsupials involve the X chromosome in dosage compensation (e.g., peramelids and petaurids; Close, 1984). Mitotic figures from the testes of

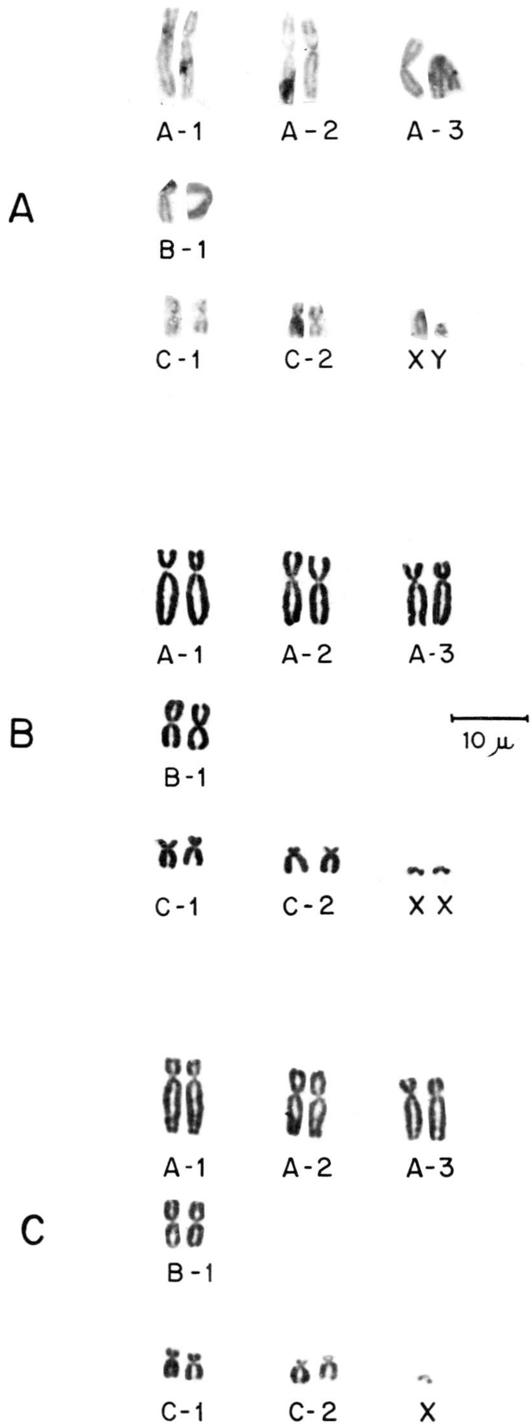

FIG. 1. Karyotypes from bone marrow cells of **A,** *Rhyncholestes raphanurus,* male; **B,** *Dromiciops australis* (2n = 14), female; **C,** *Dromiciops australis* (2n = 13), male.

TABLE 1. Arm ratios (long arm/short arm) of *Rhyncholestes* autosomes (ratios are based on 10 counted plates; sex chromosomes are acrocentric).

Pair 1:	1.43
Pair 2:	1.53
Pair 3:	1.39
Pair 4:	1.44
Pair 5:	1.42
Pair 6:	1.46

Dromiciops do show the XY constitution. Therefore, male zygotes begin development as XY, and the Y is retained in the germinal cell line, but is lost in at least some somatic tissues. More studies will be needed to determine the extent of this mosaicism in other tissues.

We believe this instance of sex-chromosome mosaicism probably represents a parallel, independently derived case from that in Australasian forms. However, it could be used to support Szalay's (1982) assertion that *Dromiciops* is more closely related to Australasian lineages than any other American form, belonging in the Australasian cohort Australidelphia. In this regard it is noteworthy that Sharman's (1982) analysis of gross chromosomal morphology suggested that the 2n = 14 karyotype of *Dromiciops* (virtually identical to those of some burramyids, peramelids, and *Vombatus ursinus*) might be highly similar to that of the common ancestor of Australian marsupials. This instance of sex-chromosome mosaicism also bears on Archer's (1976) contention that peramelids, which also exhibit sex-chromosome mosaicism, appear to be derivatives of didelphids in basicranial anatomy. Banding studies of chromosomal morphology in these groups are needed to help resolve these various suggestions.

A "ratchet" model for the evolution of the Y chromosome and dosage compensation has been suggested (Charlesworth, 1978). Initially an active chromosome, the Y is homologous to the X, but chiasmata formation (and thus recombination between the two) is suppressed (e.g., Ohno, 1967). A gradual accumulation of deleterious mutations could account for its erosion over time, leading to minute size. In the didelphid *Monodelphis dimidiata,* the synaptonemal complex is absent in the X-Y pairing region. Structural elements of the complex are present, but their assembly seems inhibited by the shortness of the Y chromosome. It could be argued that, in *Monodelphis* and other metatherians with dotlike Y chromosomes, Y function is apparently reduced to sex determination, unnecessary in at least some somatic tissues. We favor the evolutionary erosion of the Y chromosome and its lack of function in the bone marrow tissue of *Dromiciops australis* as ultimate causes for this sex-chromosome mosaicism. Late replication of highly heterochromatinized DNA, a proximate mechanism for sex-chromosome mosaicism suggested in dosage compensation (Hayman & Martin, 1974), may account for the accidental loss of the minute Y chromosome during mitotic divisions of somatic cells of *Dromiciops.*

Acknowledgments

We thank Brian K. Lang and Peter L. Meserve for assistance in obtaining specimens at La Picada. We received financial support from the Dirección de Investigacion, Universidad Austral de Chile (S-83-03), Field Museum of Natural History, American Philosophical Society (Johnson Fund #1646), and National Geographic Society (#2582-82). Dr. R. Fernández kindly facilitated study of *Dromiciops* material. The constructive criticisms of J. A. W. Kirsch, P. Myers, and J. L. Patton on an earlier draft are thankfully acknowledged.

Literature Cited

ARCHER, M. 1976. The basicranial region of marsupicarnivores (Marsupialia), inter-relationships of carnivorous marsupials, and affinities of the insectivorous marsupial peramelids. Zoological Journal of the Linnean Society, **59:** 217–322.

BIGGERS, J. D., AND R. F. S. CREED. 1962. Conjugate spermatozoa of the North American opossum. Nature (London), **196:** 1112–1113.

BIGGERS, J. D., AND E. D. DELAMATER. 1965. Marsupial spermatozoa pairing in the epididymis of American forms. Nature (London), **208:** 402–404.

CHARLESWORTH, B. 1978. Model for evolution of Y chromosomes and dosage compensation. Proceedings of the National Academy of Science, USA, **75:** 5618–5622.

CLOSE, R. L. 1984. Rates of sex chromosome loss during development in different tissues of the bandicoots *Perameles nasuta* and *Isoodon macrourus* (Marsupialia: Peramelidae). Australian Journal of Biological Science, **37:** 53–61.

FERNÁNDEZ-D., R., S. BERRIOS, AND J. PINCHEIRA. 1979. Position of the nucleolus within the nuclei of pachytene spermatocytes of *Dromiciops australis* and *Marmosa elegans* (Didelphoidea-Marsupialia). Experientia (Basel), **35:** 1021–1023.

HAYMAN, D. L., AND P. G. MARTIN. 1969. Cytogenetics

of marsupials, pp. 191–217. *In* Benirschke, K., ed., Comparative Mammalian Cytogenetics. Springer, New York.

———. 1974. Mammalia. I: Monotremata and Marsupialia, pp. 1–110. *In* John, B., ed., Animal Cytogenetics. Vol. 4: Chordata 4. Gebrüder Borntraeger, Berlin.

Hayman, D. L., J. A. W. Kirsch, P. G. Martin, and P. F. Waller. 1971. Chromosomal and serological studies of the Caenolestidae and their implications for marsupial evolution. Nature (London), **231**: 194–195.

Lyon, M. F. 1974a. Evolution of X-chromosome inactivation in mammals. Nature (London), **250**: 651–653.

———. 1974b. Mechanisms and evolutionary origins of variable X-chromosome activity in mammals. Proceedings of the Royal Society of London, Series B, **187**: 243–268.

Marshall, L. G. 1982. Systematics of the South American marsupial family Microbiotheriidae. Fieldiana: Geology, n.s., **10**: 1–75.

Ohno, S. 1967. Sex Chromosomes and Sex-Linked Cells. Springer-Verlag, Berlin, 192 pp.

Patton, J. L. 1967. Chromosome studies of certain pocket mice, genus *Perognathus* (Rodentia: Heteromyidae). Journal of Mammalogy, **48**: 27–37.

Reig, O. A., R. Fernández, and O. A. Spotorno. 1972. Further occurrence of a karyotype of 2n = 14 chromosomes in two species of Chilean didelphoid marsupials. Zeitschrift für Säugetierkunde, **37**: 37–42.

Reig, O. A., A. L. Gardner, N. O. Bianchi, and J. L. Patton. 1977. The chromosomes of the Didelphidae (Marsupialia) and their evolutionary significance. Biological Journal of the Linnean Society, **9**: 191–216.

Schneider, L. K. 1977. Marsupial chromosomes, cell cycles, and cytogenetics, pp. 51–93. *In* Hunsaker II, D. D., ed., The Biology of Marsupials. Academic Press, New York.

Sharman, G. B. 1982. Karyotypic similarities between *Dromiciops australis* (Microbiotheriidae, Marsupialia) and some Australian marsupials, pp. 711–714. *In* Archer, M., ed., Carnivorous Marsupials, Vol. II. Royal Zoological Society of New South Wales, Sydney, 804 pp.

Spotorno, O. A., and D. R. Fernández. 1971. The chromosomes of the "monito del monte" *Dromiciops australis* Philippi. Mammalian Chromosomes Newsletter **12**(2): 40–41.

Szalay, F. S. 1982. A new appraisal of marsupial phylogeny and classification, pp. 621–640. *In* Archer, M., ed., Carnivorous Marsupials, Vol. II. Royal Zoological Society of New South Wales, Sydney, 804 pp.

Notes on the Black-Shouldered Opossum, *Caluromysiops irrupta*

Robert J. Izor and Ronald H. Pine

ABSTRACTS

Caluromysiops is distinct from the three species of *Caluromys* in external, cranial, dental, skeletal, and phallic characters, although the two genera are certainly more closely related to each other than to any other extant genus. Much uncertainty remains regarding the ecology and distribution of this rare opossum.

Caluromysiops es distinto de las tres especies de *Caluromys* en caracteres externos, craneales, dentales, esqueletales y fálicos, aunque los dos géneros son por cierto mas cercanamente relacionados entre sí que lo es ningun otro género existente. Todavía hay mucha incertidumbre en relación a la ecología y distribución de este rara raposa.

Caluromysiops difere das tres espécies de *Caluromys* en caráteres externos, craniais, dentais, esqueletais e fálicos, embora sejam os dois generos claramente mais próximos entre si do que entre qualquer outro genero atualmente existente. A ecologia e a distribuição desta rara espécie continuam muito pouco conhecidas.

Introduction

The black-shouldered opossum, *Caluromysiops irrupta* Sanborn, is the rarest of the larger didelphids. Its history as a subject of scientific study is peculiar, beginning with a very late discovery (1951); also, many more specimens have been displayed in zoos (15) than have been collected for museums directly from the wild (2). Despite the paucity of associated data and other shortcomings, zoo animals have been the source of some valuable information during this study.

Materials and Methods

All zoos known or suspected to have kept *Caluromysiops* were contacted for information on the acquisition, history in captivity, and eventual disposition of animals. All known specimens preserved in collections were examined by one or both of us, and all tag data were recorded.

Results

Small sample sizes have hampered previous work on this species, and have affected this study to some extent. Most published information is based on single specimens. Some of the characters described by Sanborn (1951) as diagnostic are individually or ontogenetically variable and not reliable for identification in all cases.

For example, the extent of the hair on the dorsum of the tail is distinctive in *Caluromysiops,* although not as extreme as originally described. On immature animals such as the holotype, the furred area reaches nearly to the end of the tail. Adults, however, lack fur on the distal 15–20 mm.

From the Division of Mammals, Field Museum of Natural History, Chicago, Illinois 60605-2496. Dr. Pine's present address is Illinois Mathematics and Science Academy, Aurora, Illinois 60506-1039.

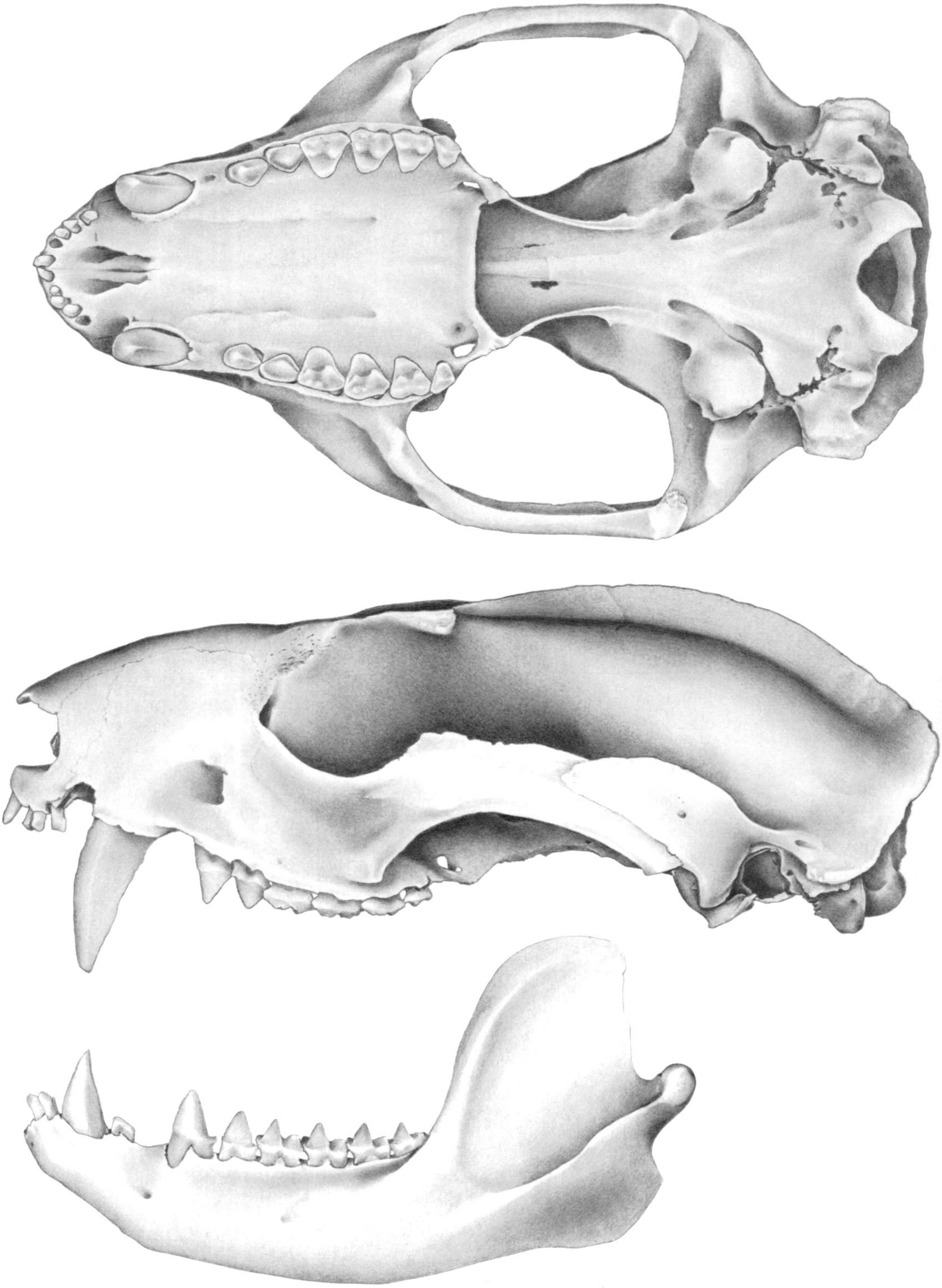

FIG. 1. Ventral and lateral views of the cranium and lateral view of the mandible of adult male *Caluromysiops irrupta*, FMNH 60698. Certain incisors have fallen out and have been lost.

FIG. 2. Adult female *Caluromysiops irrupta* with two young (one is specimen no. CVG M-17 BE 173). Photo courtesy of Edward T. Maruska and the Cincinnati Zoo.

The portion of the tail covered dorsally with fur is still much more extensive than even in *Caluromys lanatus,* in which only the proximal 50%–70% is covered. Except perhaps for some *Glironia, Caluromysiops* is unique among didelphids in that the fur extends onto a distal unpigmented portion of the tail. The distal one-quarter to one-third of the tail fur is also white. In other genera of didelphids, individuals with some distal portion of the tail skin unpigmented have fur of the tail confined to the proximal pigmented area of the tail skin.

The most striking external feature of *Caluromysiops* is probably the pair of dark lateral and dorsal stripes. These typically arise on the back of the hand and run up the inner side of the forelimb onto the shoulder, where they reach their greatest width of 15–30 mm. They approach each other middorsally but usually do not merge, and run in narrowing parallel bands to the rump. In one old individual, CVG M-30 BE 95, which had been displayed for six years and eight months at the Cincinnati Zoo, the pattern is obscured by a general grizzling. A common variant of the pattern has the back of the hand white, with the dark stripe beginning as a sharply delineated black band around the wrist. This feature may occur on one or both forefeet.

As Sanborn and others have noted, some individuals of the woolly opossums *Caluromys derbianus* and *C. lanatus* have coloration suggesting the characteristic dorsal markings of *Caluromysiops.* In the species of *Caluromys,* there is typically a darker brown or reddish dorsal area which grades into the paler, grayer sides of the body. In some individuals, this darker region is bisected on the back of the head, neck, and shoulders by a middorsal gray streak. The supposed similarity to *Caluromysiops,* however, is not at all close. The darker dorsal areas in *Caluromys* are most separated in the place where in *Caluromysiops* they are closest to merging. Moreover, the individuals of *Caluromys* having the gray middorsal stripe

TABLE 1. *Caluromysiops irrupta* formerly exhibited in zoos.

Sex	Date arrived	Date died	Disposition of remains	Acquisition data
BRONX ZOO (New York Zoological Society)				
F	10 Sept. 1962	28 July 1969	Incinerated?	"Iquitos, Ecuador" (presumably Peru)
F	20 Nov. 1963	26 Dec. 1964	AMNH 208101	...
NATIONAL ZOO (Smithsonian Institution, Washington, D.C.)				
M	4 Nov. 1969	...	(Sent to Lincoln Park Zoo, 11 Oct. 1972)	C. Chase, Miami
M	31 Mar. 1971	12 Apr. 1971	USNM 396160	From Oklahoma City Zoo
OKLAHOMA CITY ZOO				
F	23 Nov. 1965	20 Oct. 1970	Discarded?	C. Chase, Miami
M	28 Jan. 1967	...	(Sent to National Zoo, 31 Mar. 1971)	...
F	19 Dec. 1965	5 Aug. 1967	Discarded?	...
LINCOLN PARK ZOO, CHICAGO				
F	18 Aug. 1972	?	Discarded?	...
M	18 Aug. 1972 (pouch young)	4 Apr. 1973	FMNH 121522	...
F	18 Aug. 1972 (pouch young)	25 Mar. 1973	FMNH 60154	...
M	11 Oct. 1972	16 Sept. 1974	FMNH 60398	...
BROOKFIELD ZOO (Chicago Zoological Society)				
M	21 Apr. 1970	19 Aug. 1971	FMNH 60698	R. Baudy, Center Hill, Fla.
TARPON SPRINGS ZOO				
M	(1 Aug. 1972 "rec'd in lab")	...	USNM 397626	...
CINCINNATI ZOO				
F	25 July 1965	8 Mar. 1967	Discarded?	Peru, via Animated Shippers, Miami
F	25 July 1965 (young w/F)	6 Nov. 1965	CVG M-17 BE 173	Peru, via Animated Shippers, Miami
F	25 July 1965 (young w/F)	11 Dec. 1967	Discarded?	Peru, via Animated Shippers, Miami
M	1 July 1967	27 Feb. 1974	CVG M-30 BE 95	Cuxio (= Cuzco?), Peru, via C. Chase, Miami

Data from E. Maruska, M. Jones, J. Eisenberg, A. Dittmar, A. Hamer, C. Chase (all in litt.), and Collins (1973). AMNH = American Museum of Natural History, New York; USNM = National Museum of Natural History, Washington, D.C.; FMNH = Field Museum of Natural History, Chicago; CVG = personal collection of E. Maruska, Director Cincinnati Zoo.

TABLE 2. Measurements of *Caluromysiops irrupta*.

Sex	No.	Total length	Tail length	Hind foot	Ear	Greatest skull length (incl. incisors)	Condylo-incisive length	Basal length (incl. incisors)	Medial palatal length (incl. incisors)
♂	CVG M-30 BE 95	617	333.5	...	29	63.7	61.0	57.3	30.0
♂	USNM 397626	630+	330+	52	34	64.5	63.4	59.1	32.3
♂	USNM 396160	590	340	51	32	63.6	62.5	57.3	30.5
♂	FMNH 60698	...	...	...	...	63.4	60.6	56.5	30.5
♀	AMNH 208101	570	310	47	37	62.6	60.7	56.4	29.7

also strongly tend to have the palest extremities, whereas *Caluromysiops* has extremities with broad blackish bands (on the inner side of the forelimbs and outer side of the hind limbs).

Other differences in pelage include the *Marmosa*-like eye rings and the median facial stripe of all *Caluromys,* which are completely lacking in most *Caluromysiops* and only faintly suggested in a few. There is no feature of the color pattern indicating that *Caluromys* and *Caluromysiops* represent simple variants of a single evolutionary trend.

Cranially, the extant didelphids present a rather restricted array of morphologies. All have the same dental formula. The skulls differ primarily in size, in the presence and arrangement of palatal vacuities, and in details of the masticatory apparatus such as sagittal crests, shape of the zygomata, and the postorbital processes. To our knowledge, a key to the skulls of the genera has never been constructed. It is not surprising, therefore, that it is difficult to find trenchant cranial characters supporting the distinctiveness of *Caluromysiops* as a genus. In the context of the family's relative uniformity, this does not necessarily argue against generic distinction. Pine, however, indicated in Honacki et al. (1982) that he prefers to regard *Caluromysiops* as a subgenus of *Caluromys,* mostly because of similarity in skull shape.

The dentition of *Caluromysiops irrupta* was described by Sanborn (1951) as having larger M^{1-2} and m_{1-3} than *Caluromys.* He noted the absence in the holotype of M^3, M^4, and m_4 and attributed the lack of an M^3 to its probable loss in the cleaning of the skull, but did not discuss the absence of the other molars. The holotype is a juvenile and the developing alveolus of the m_4 is quite evident, so the tooth is probably unerupted. The larger size of the molars is generally a valid character distinguishing *Caluromysiops* from *Caluromys.* Some individuals of *Caluromysiops* may never have had the minute P^1, which is frequently lost in adults, but otherwise the dental formula conforms to that of the other didelphids. The single root of the usually spicule-like P^1 differs from the condition in *Caluromys,* in which the tooth is double rooted, or at least very broad with an incipient division. There is a strong tendency in *Caluromys* for the small cusps on the labial stylar shelf to be subdivided into as many as nine small, low cusps. *Caluromysiops* typically has five such cusps, each being higher and more distinct than in *Caluromys.*

Caluromys and *Caluromysiops* are united by the apparently derived character (Archer, 1982) of closure of the maxillary palatal fenestrae. This feature alone is sufficient to distinguish them from all other living New World marsupials, with the possible exception of some *Marmosa.* Archer apparently erred in attributing such closure to *Glironia. Caluromysiops* is slightly farther along in the process than *Caluromys,* with only small, round, paired foramina remaining at the maxillopalatal suture. Species of *Caluromys* have more or less elongate foramina.

Several cranial features of *Caluromysiops* suggest adaptations for strong biting forces. The sagittal crest in adults is very pronounced, and the zygomatic arches are robust and widely bowed outward. Rostral length is relatively shorter than in *Caluromys,* and the mandible is deeper, with the ascending ramus broader and more upright. This seems incongruous in view of the description by Janson et al. (1981) of nectarivorous behavior. Zoo animals, however, have readily accepted a varied diet including animal products (Collins, 1973), and the species probably only exploits nectar and pollen opportunistically.

Cranial asymmetry is prevalent in our sample. About half of the skulls examined had some sort of deviation of the rostral axis relative to that of the braincase, or deflection of the sagittal crest from the midline.

TABLE 2. *Continued.*

Interorbital constriction	Breadth postorbital processes	Postorbital constriction	Zygomatic breadth	Postzygomatic braincase width	Length longer nasal	Depth braincase (incl. bullae)	Length of mandible	Length mand. ramus	Maxillary toothrow	M^1-M^3	Mandibular toothrow
12.7	22.1	8.2	38.2	23.4	25.1	22.2	45.8	47.2	22.8	9.5	28.3
13.9	21.1	···	36.9	23.2	24.4	20.6	46.8	48.6	23.2	8.7	28.8
12.4	21.4	7.8	39.2	22.9	25.1	22.1	47.5	···	23.1	···	29.4
14.0	20.1	9.5	37.0	23.6	23.9	···	···	···	···	9.3	···
11.3	18.3	9.2	38.0	23.3	25.8	···	···	···	22.3	9.1	···

Postcranial anatomy of the black-shouldered opossum displays some interesting but as yet inexplicable differences from that of woolly opossums. The hind limbs of *Caluromysiops* are relatively much shorter than the forelimbs. The forearm is especially long. In addition, all of the skeletal elements are more heavily built than in *Caluromys,* with larger articular surfaces. Both genera exhibit a slightly offset articulation of the second metacarpal, which allows the animals to spread the second and third digits and grasp small branches between them. This schizodactylous grip, also found in phalangeroids, is useful for slow, deliberate climbers which may back up along a branch rather than turn around to proceed headfirst. The tail has 30–31 vertebrae, compared to 36–38 in *Caluromys,* and has well-developed chevron bones throughout its length.

Rosenthal (1972, 1975) noted that a female *Caluromysiops* was received at the Lincoln Park Zoo with pouch young, which 40 days later still lacked markings and body hair. Details of pouch anatomy were not provided.

All of the didelphids examined to date have a more or less cleft glans penis. Biggers (1966) noted that *Caluromys derbianus* differed from other species he examined in the greater extent of the cleft (half the length of the penis), in the continuation of medial urethral grooves to the apices, and in the rounded, slightly bulbous ends of the glans. The single available dissected-out specimen of a *Caluromysiops* penis (FMNH 60698) suffered some postmortem deterioration and may not be completely representative, but still shows clearly a very deeply split glans (ca. 4 cm) with distinctly enlarged, rounded tips. The urethral grooves also seem to extend nearly to the ends.

These characters of the genitalia would seem to ally *Caluromys* and *Caluromysiops.* However, *Caenolestes* also has a deeply cleft glans penis (Osgood, 1921), and many Australian marsupials exhibit some version of the same phenomenon, so it may represent a shared primitive character. Moreover, a large majority of didelphid species have not been evaluated in this regard, and the significance cannot be properly assessed. Genitalia of mammals lacking bacula generally have been less studied, even though soft tissue structure can be equally informative (Woolley, 1982), and our cursory survey of preserved material indicates considerable undocumented variety.

A remarkable feature, poorly preserved on FMNH 60698, but manifest on the protruding penial apex of FMNH 60398, is a dense covering of small (ca. 1 mm), cornified, recurved spines. These are distributed primarily on the rather rugose tip and medial sides of the glans, along the urethral groove. Osgood (1921) described the glans of *Caenolestes* as rugose proximally and covered distally by small circular papillae, but Biggers (1966) noted no such structures on *Caluromys* or other didelphids examined.

The taxonomic affinities of *Caluromysiops irrupta* have been controversial at both the generic and suprageneric levels. Cabrera (1958), Hershkovitz (in Marshall, 1982), and Pine (in Honacki et al., 1982) have suggested that its evident relationship to *Caluromys* might be better expressed by including it in the latter genus. The present authors are divided on the question of whether this change would improve the current arrangement.

Reig's (1955) assertion that this species belongs in the Microbiotheriidae has received adequate refutation (Segall, 1969; Szalay, 1982). Kirsch's (1977) attempt to subdivide the Didelphidae is undermined by the fact that his subfamily names Caluromyinae and Dactylopsilinae, as proposed, are *nomina nuda.* Given his uncertainty about the contents of the supposed subfamilies of didelphids, this fact could spare future workers considerable confusion, although the names may have since become available inadvertantly in subsequent publications.

As most zoo animals have changed hands several times before reaching their final destinations, there is little likelihood of accurate field data accompanying them. Among dubious origins reported for zoo-held *Caluromysiops* are São Paulo, Brazil, and Iquitos, Ecuador (sic). According to J. A. Davis, Jr. (in litt.), the latter animal "was said by the dealer to have been captured in a backyard on the outskirts of Iquitos, Peru"; see also Bridges (1968) and Davis (1965). Another purported locality, Cuxio, Peru, has not been located and may represent a transcription error for Cuzco.

There are only three unquestioned locality records, all from southern Amazonian Peru, as follows:

Peru: Depto. Cuzco; Prov. Quispicanchis, Quince Mil (13°16′S, 70°38′W), 680 m, FMNH 68336 (the holotype).

Peru: Depto. Madre de Dios; Itahuania (12°47′S, 71°13′W), skull is FMNH 84426, skin is in the Museo Nacional de Historia Natural "Javier Prado", Lima.

Peru: Depto. Madre de Dios; Manu National

Park, Cocha Cashu Biological Station (11°55′S, 71°18′W) (Janson et al., 1981; Terborgh et al., 1984; Emmons, 1984).

These three localities are within 150 km of each other, along the western margin of the Amazon basin, between 400–700 m elevation. The only sympatric species of *Caluromys* recorded is *C. lanatus.*

Simonetta's (1979) report of a *Caluromysiops* near Leticia, Colombia, is a problem. Although we are unable to locate the original account, it is our opinion that this record is best discounted. The photograph appears to have been staged with a captive specimen, since the species is nocturnal (Collins, 1973; Janson et al., 1981; Terborgh et al., 1984). Leticia is at least 900 km from the three well-documented localities, and one of the museum specimens we examined (USNM 397626) is known to have passed through Leticia from an unknown source en route to a zoo in Florida. Leticia is the location of a major animal dealership, and the point of exportation of many Amazonian species to the U.S. The dusky brown color on the crown of the head, which Simonetta suggests may differentiate his Colombian specimen subspecifically, is variable in the material we examined, and is probably of no taxonomic importance.

Conclusions

Caluromysiops irrupta is a species which has often been erroneously or incompletely characterized in the scientific literature. There are now enough specimens in collections to allow reasonably complete treatments of its morphology, although it remains an almost complete ecological and behavioral enigma.

Acknowledgments

The authors thank those individuals and institutions listed in Table 1 for their invaluable assistance in compiling these data, and for loans of specimens in their care. Anita McQuaig, Linda E. Pine, Nobuko Etoh Pine, Joyce Shaw, and Mary Reed helped with the manuscript. Joseph A. Davis and the editors and reviewers made many helpful suggestions.

Literature Cited

ARCHER, M. 1982. A review of Miocene thylacinids (Thylacinidae, Marsupialia), the phylogenetic position of the Thylacinidae and the problem of apriorisms in character analysis, pp. 445–476. *In* Archer, M., ed., Carnivorous Marsupials. Royal Society of New South Wales.

BIGGERS, J. D. 1966. Reproduction in male marsupials, pp. 251–280. *In* Rowlands, I. W., ed., Symposia of the Zoological Society of London, **15:** 1–559.

BRIDGES, W. 1968. The Bronx Zoo Book of Wild Animals. New York Zoological Society and Golden Press, New York, 8 unnumbered pp. + 304 pp.

CABRERA, A. 1958. Catálogo de los mamíferos de América del Sur. I. (Metatheria-Unguiculata-Carnivora). Revista del Museo Argentino de Ciencias Naturales "Bernardino Rivadavia": Zoología (1957), **4:** 1–307.

COLLINS, L. R. 1973. Monotremes and Marsupials. Smithsonian Publication 4888, Smithsonian Institution, Washington, D.C., 323 pp.

DAVIS, J. A., JR. 1965. A great year for rarities. Animal Kingdom, **68**(5): 130–133.

EMMONS, L. H. 1984. Geographic variation in densities and diversities of non-flying mammals in Amazonia. Biotropica, **16:** 210–222.

HONACKI, J. H., K. E. KINMAN, AND J. W. KOEPPL, EDS. 1982. Mammal Species of the World. Allen Press and Association of Systematics Collections, Lawrence, Kansas, 694 pp.

JANSON, C., J. TERBORGH, AND L. H. EMMONS. 1981. Non-flying mammals as pollinating agents in the Amazonian forest. Biotropica, **12**(Suppl.): 1–6.

KIRSCH, J. A. W. 1977. The comparative serology of Marsupialia, and a classification of marsupials. Australian Journal of Zoology, supp. ser., **52:** 1–152.

MARSHALL, L. G. 1982. Evolution of South American Marsupialia, pp. 251–272. *In* Mares, M. A., and H. H. Genoways, eds., Mammalian Biology in South America. Special Publication Series, Pymatuning Laboratory of Ecology, University of Pittsburgh, **6:** 1–539.

OSGOOD, W. H. 1921. A monographic study of the American marsupial, *Caenolestes.* Field Museum of Natural History, Zoological Series, **14:** 1–156.

REIG, O. A. 1955. Noticia preliminar sobre la presencia de microbiotherinos vivientes en la fauna sudamericana. Investigaciones Zoológicas Chilenas, **2:** 121–130.

ROSENTHAL, M. A. 1972. Observations on the water opossum or yapok (*Chironectes minimus*). Proceedings 48th Annual Conference American Association of Zoological Parks and Aquariums held in Portland, Oregon, Oct. 1–5, **1972:** 95–98.

———. 1975. Observations on the water opossum or yapok *Chironectes minimus* in captivity. International Zoo Yearbook, **15:** 4–6.

SANBORN, C. C. 1951. Two new mammals from southern Peru. Fieldiana: Zoology, **31:** 473–477.

SEGALL, W. 1969. The middle ear region of *Dromiciops.* Acta Anatomica, **72:** 489–501.

SIMONETTA, A. M. 1979. First record of *Caluromysiops* from Colombia. Mammalia, **43:** 247–248.

SZALAY, F. S. 1982. A new appraisal of marsupial phylogeny and classification, pp. 621–640. *In* Archer, M., ed., Carnivorous Marsupials. Royal Society of New South Wales.

TERBORGH, J. W., J. W. FITZPATRICK, AND L. EMMONS. 1984. Annotated checklist of bird and mammal species of Cocha Cashu Biological Station, Manu National Park, Peru. Fieldiana: Zoology, n.s., **21:** 1–29.

WOOLLEY, P. A. 1982. Phallic morphology of the Australian species of *Antechinus* (Dasyuridae: Marsupialia): A new taxonomic tool?, pp. 767–781. *In* Archer, M., ed., Carnivorous Marsupials. Royal Society of New South Wales.

Feeding Habits of the Opossum (*Didelphis marsupialis*) in Northern Venezuela

Gerardo A. Cordero R. and Ruben A. Nicolas B.

ABSTRACTS

The food items in the annual diet of the opossum (*Didelphis marsupialis*) in northern Venezuela are reported by season, sex, and dental age. One hundred eight opossums were sampled in 21 different sites on a monthly basis from March 1983 to March 1984. The number of food items recorded varies seasonally. By volume, animal foods (63.5%) are more important than plant foods (22.9%) throughout the year. Birds (21.5%), mammals (15.3%), insects (14.8%), and fruits (12.8%) are the most prominent foods, by volume. Feeding habits of males and females do not differ significantly. However, diets of young and old animals are different.

Se señalan los componentes de la dieta anual del rabopelado (*Didelphis marsupialis*) en el norte de Venezuela por época del año, sexo y edad. El muestreo se hizó mensualmente colectándose 108 animales desde Marzo 1983 a Marzo 1984 de 21 localidades diferentes. El número de componentes de la dieta varía estacionalmente. En términos de volumen, los alimentos de origen animal (63.5%) son más importantes que los de origen vegetal (22.9%) a través del año. Las aves (21.5%), los mamíferos (15.3%), los insectos (14.8%) y las frutas (12.8%) son las alimentos más sobresalientes, en términos de volumen. Los hábitos alimentarios de los machos y las hembras no difieren significativamente. Sin embargo, las dietas de los animales jóvenes y viejos son diferentes.

Relata-se os componentes da dieta anual do gambá (*Didelphis marsupialis*) no norte da Venezuela, por época, sexo e idade. Amostras foram coletadas mensalmente de março de 1983 a março de 1984, e de 21 locais diferentes, para um total de 108 animais examinados. O número dos componentes da dieta varia sasonalmente. Em termos de volume, os alimentos de origem animal (63,5%) sao mais importantes do que os de origem vegetal (22,9%) através do ano. Aves (21,5%), mamíferos (15,3%), insetos (14,8%), e frutos (12,8%) foram os alimentos mais abundantes por volume. Apesar dos hábitos alimentares não diferirem entre machos e femeas, a dieta dos animais jovens difere da dieta dos adultos.

Introduction

Feeding habits of neotropical didelphid marsupials are poorly known, in spite of their high diversity and broad geographical distribution. Diets of at least seven of 70 species are known (Fleming, 1972; Hunsaker, 1977; Atramentowicz, 1982; Streilein, 1982; Charles-Dominique, 1983; cf. Kirsch & Calaby, 1977). However, the information reported for most species is based on qualitative data. This paper reports the food items ingested by opossums (*Didelphis marsupialis*) in northern Venezuela throughout the year by season, sex, and dental age.

From the Facultad Ciencias Instituto de Zoología Tropical, Apartado 47058, Caracas 1041-A, Venezuela.

Study Area

Fieldwork was conducted mainly in the Barlovento region of the State of Miranda and within the city of Caracas and its surroundings in northern Venezuela (10°00′–10°30′N, 66°00′–67°00′E). The climate is highly seasonal, with a humid period of nine months (May–January) and a dry period of three months (February–April) in Barlovento and seven months of rainfall (May–November) and five months of drought (December–April) in Caracas. Annual mean temperature of Barlovento is 26° C versus 20.6° C in Caracas and its surroundings. Rainfall is 2,053 mm at Barlovento and 1,011 mm at Caracas. Elevations sampled range from 40 m to more than 1,000 m above sea level. According to the Holdridge Life Zones (Ewel et al., 1976), the vegetation of Barlovento is primarily a humid tropical forest, whereas that of Caracas is mostly in premontane humid forest.

Materials and Methods

The sample of 108 opossums was assembled from March 1983 to March 1984, either from road-kills or hunting. Fifty-two animals were taken from nine localities at Barlovento, whereas 56 specimens were taken from 12 localities in or near Caracas. Body measurements, sex, and dental age of each animal were recorded. Age determination was based on tooth eruption and wear (Petrides, 1949; Tyndale-Biscoe & MacKenzie, 1976), permitting their grouping into seven age classes (Cordero, unpubl. data, see Appendix 1). Stomach contents were analyzed according to Korschgen's (1980) recommendations. Each stomach and its contents were placed in a fine sieve (1-mm diameter mesh screen) and thoroughly washed under running water in order to separate fine from coarse material. After measuring the entire volume of the contents, each item was separated under a dissecting microscope and its volume recorded. A reference collection was used for the identification of insects.

Results

Opossum Foods and Seasonal Variation

Six (5.6%) of the 108 stomachs we examined were empty. Numbers of stomachs with items were: dry season, 16 (15.7%) and wet season, 86 (84.3%). Data for these 102 stomachs appear in Table 1 and Figure 1. Percentage of volume and frequency of occurrence are shown for each class of items in Table 1.

Considerable seasonal variation exists in the number of food items recorded. During the dry season, the most important food items are mammals, birds, and insects. In the wet season, birds are more important by volume than mammals or insects, and fruits seem to be of greater importance. Gastropods are ingested in a higher proportion during dry season than wet season. Snakes, toads, and earthworms are consumed only in the latter period.

Food of animal origin is more important (63.5% by volume) than plant food (22.9%) in the diet of opossums throughout the year. By volume, birds (21.5%), mammals (15.3%), insects (14.8%), and fruits (12.8%) are the principal foods ingested by opossums. In terms of frequency, insects (49.1%), fruits (18.6%), birds (12.7%), and mammals (8.8%) contribute to the annual diet.

Domestic cats (*Felis catus*) and rats (*Rattus rattus*) were considered as prey items of opossums because no dipteran carrion larvae were observed in stomach contents. However, unidentified mammalian remains are more important than those of cat and rat by both volume and frequency. Birds ingested by opossums were either chickens (*Gallus* sp.) or young birds which were more numerous in the wet season; during the dry season, chickens were recorded as carrion. Avian material accounted for 12.7% of the stomach contents and 21.4% by volume. Snakes and toads are consumed at low levels in relation to their abundance in study sites, suggesting that these food items are of little importance for opossums in northern Venezuela.

Insects of at least nine families occurred in 49.0% of the stomachs, with an annual volume of 14.8%. Beetles and grasshoppers accounted for the majority of insects consumed.

Slugs (Veronicellidae) were recorded in the rainy season (1.7% by volume), whereas Vulimulidae are important in the dry season (6.3% by volume). Centipedes and earthworms were poorly represented in the stomachs.

Fruits such as *Psidium guajava* and *Guazuma ulmifolia* are very important in the diet of opossums. By both volume and frequency of occurrence, fruits are more important in the rainy season.

Miscellaneous foods such as garbage (paper, plastic bags, felt, thread filaments), particulate ma-

TABLE 1. Percentages of volume (V) and frequency (F) of food items of opossums in northern Venezuela in 1983 and 1984, by season and for the year.

Food items	Wet season % V	Wet season % F	Dry season % V	Dry season % F	Annual % V	Annual % F
ANIMALS	65.95	90.65	51.22	114.40	63.48	97.03
Mammalia	15.42	5.81	14.75	20.19	15.31	8.82
Felis catus	4.14	2.32	...	...	3.44	1.96
Rattus rattus	...	...	0.41	7.69	0.07	0.98
Mammal remains	11.28	3.49	14.34	12.50	11.80	5.88
Aves	23.52	13.95	11.27	7.69	21.46	12.74
Gallus sp.	6.42	2.32	...	...	5.34	1.96
Young birds	10.04	3.49	...	...	8.35	4.90
Bird remains	7.06	8.14	11.27	7.69	7.77	5.88
Reptilia	0.41	1.16	...	...	0.34	0.98
Snake remains	0.41	1.16	...	...	0.34	0.98
Amphibia	1.97	1.16	...	...	1.64	0.98
Bufo sp.	1.97	1.16	...	...	1.64	0.98
Insecta	14.82	47.65	14.70	63.45	14.81	49.01
Coleoptera	7.36	23.24	3.69	23.07	6.76	21.56
Passalidae	0.21	1.16	1.20	7.69	0.28	1.96
Scarabaeidae	5.07	11.63	1.54	7.69	4.48	9.80
Coccinelidae	...	...	1.54	7.69	0.26	0.98
Curculionidae	0.31	2.32	...	...	0.26	1.96
Meloidae	0.31	1.16	...	...	0.26	0.98
Carabidae	0.12	1.16	...	...	0.10	0.98
Remains	1.34	5.81	...	...	1.12	4.90
Orthoptera	5.12	16.28	0.87	7.69	4.41	14.71
Acrididae	5.12	16.28	0.87	7.69	4.41	14.71
Cursores	0.64	2.32	8.09	25.00	1.89	5.88
Blattaria	0.43	1.16	8.09	25.00	1.72	4.90
Phasmida	0.21	1.16	...	...	0.17	0.98
Lepidoptera	1.66	4.65	2.05	7.69	1.72	5.88
Larvae	1.66	4.65	2.05	7.69	1.72	5.88
Homoptera	0.04	1.16	...	...	0.03	0.98
Cicadidae	0.04	1.16	...	...	0.03	0.98
Mollusca	1.71	8.14	6.25	7.69	2.47	5.88
Veronicellidae	1.71	8.14	...	...	1.42	4.90
Vulimulidae	...	...	6.25	7.69	1.05	0.98
Chilopoda	1.73	5.81	0.15	7.69	1.46	4.90
Annelida	1.20	5.81	...	...	1.00	4.90
Lumbricidae	1.20	5.81	...	...	1.00	4.90
Carrion	5.17	1.16	4.10	7.69	4.99	3.92
Dendrophidion parcarinatum	5.17	1.16	...	...	4.30	0.98
Gallus sp.	...	...	4.10	7.69	0.69	2.94
PLANTS	22.05	43.01	27.26	32.69	22.92	39.22
Fruits	14.19	20.92	6.15	7.69	12.84	18.63
Psidium guajava	5.70	13.95	6.15	7.69	5.77	13.73
Guazuma ulmifolia	6.21	2.32	...	...	5.17	1.96
Passiflora sp.	1.91	3.49	...	...	1.59	1.96
Mangifera sp.	0.37	1.16	...	...	0.31	0.98
Grass remains	0.83	2.32	...	...	0.69	1.96
Plant remains	7.03	19.77	21.11	25.00	9.39	18.63
MISCELLANEOUS	2.38	19.76	8.71	12.40	1.98	18.62
Paper	trace	11.63	...	...	trace	9.80
Plastic bags	trace	4.65	trace	6.20	trace	4.90
Felt	trace	1.16	...	...	trace	0.98
Thread filaments	trace	2.32	trace	6.20	trace	2.94
PARTICULATE MATERIAL	9.62	12.79	21.52	18.75	11.62	11.76

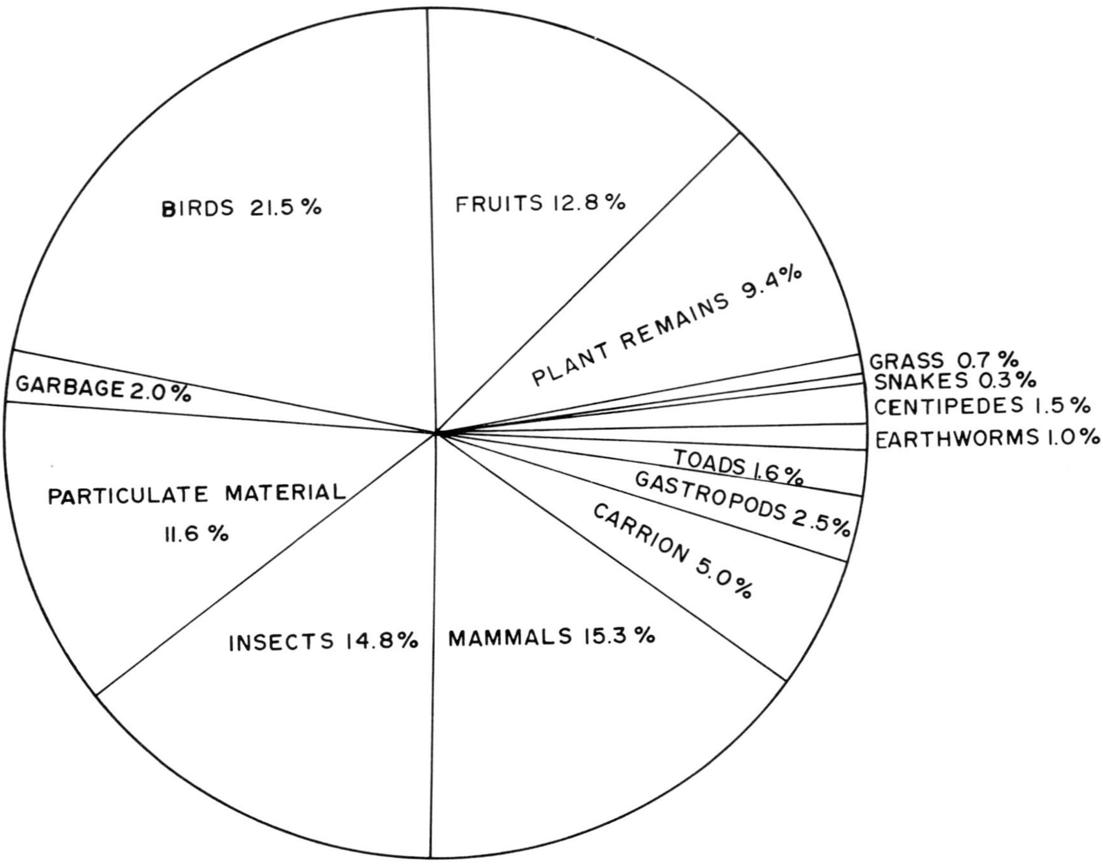

FIG. 1. Proportionate annual volumes of major groups of items from 102 stomach contents of opposums from northern Venezuela between March 1983 and March 1984.

terial, and plant remains comprised 2.0%, 11.6%, and 9.4% by volume, respectively. Garbage items were only recorded for those animals collected in or near Caracas.

Variation of Food Items by Sex

Feeding habits of male and female opossums are compared in Table 2. By volume, males consume mainly fruits (22.8%), birds (17.1%), plant remains (15.4%), and insects (14.5%), whereas females consume mammals (31.4%), birds (14.5%), insects (11.6%), and fruits (8.8%). However, by frequency of occurrence, males consume primarily insects (30%), fruits (19.2%), and plant remains (15.6%); females consume insects (28.6%), plant remains (12.2%), mammals (10.2%), and fruits (10.2%). Both comparisons by means of a Mann-Whitney *U* test indicate no significant differences between the sexes.

TABLE 2. Food items, by sex, in terms of volume (V) and frequency (F).

	Males (N = 53)		Females (N = 31)	
Food items	% V	% F	% V	% F
Mammalia	9.6	3.6	31.4	10.2
Aves	17.1	6.0	14.5	8.1
Reptilia	0.7	1.2	...	...
Amphibia	3.3	1.2	...	...
Insecta	14.5	30.0	11.6	28.6
Mollusca	2.3	6.0	4.0	6.1
Chilopoda	1.2	3.6	0.08	2.0
Annelida	0.6	2.4	1.8	4.1
Carrion	...	...	2.0	2.0
Fruits	22.8	19.2	8.8	10.2
Plant remains	15.4	15.6	7.6	12.2
Miscellaneous	0.7	3.6	5.2	6.1
Particulate material	12.4	7.2	13.0	10.2

N = Sample size.

A Mann-Whitney *U* test was also used to compare volumes of principal food groups (mammals, birds, amphibians, fruits, insects, and plant remains) in the diets of male and female opossums; no significant differences were detected. Despite this, the composition of the diet suggests that males are more arboreal than females. However, a *t* test comparing the capture frequencies of both sexes on the ground in a 26-hectare grid indicates no significant differences ($P > 0.05$; Cordero, unpubl. data). The grid contained 18 × 18 National live-traps, with a distance of 30 m between stations and rows, and was run from December 1981 to May 1984.

Variation of Food Items by Age

Food of opossums by age classes appears in Table 3. Note that the number of food items increases as animals become older. By volume and frequency of occurrence, animals of younger ages (I, II, III, and IV) consume mainly invertebrates, fruits, and plant remains, while older animals (ages V, VI, and VII) take those items plus mammals and birds, which become more important as the animal ages.

Diets of opossums were compared by successive ages, that is, II with III, III with IV, and so on, by Mann-Whitney *U* tests. No significant differences were detected. However, when diets of young and old animals were compared, significant differences were demonstrated. Nothing has been published on age-related diet variation for *D. virginiana, D. albiventris,* or any other marsupial.

Discussion

These results provide a preliminary view of the annual diet of *Didelphis marsupialis* in northwestern Venezuela. This study shows that opossums, while omnivorous, are more carnivorous and insectivorous than herbivorous or frugivorous. However, we accept these patterns guardedly because they may represent methodological artifacts: (1) most of our specimens (84.3%) were taken in the wet season, so that trophic habits during the dry season are imprecisely known; and (2) the rinsing step in processing stomach contents may have inadvertently washed away traces of fruit pulp that might have been studied using other methods.

Our results indicate that insects, fruits, birds, and mammals figure prominently in the annual diet. These figures contrast with those reported by

TABLE 3. Food items, by age, in terms of volume (V) and frequency (F).

Food items	Dental age													
	I (N = 1)		II (N = 10)		III (N = 3)		IV (N = 11)		V (N = 15)		VI (N = 13)		VII (N = 10)	
	% V	% F	% V	% F	% V	% F	% V	% F	% V	% F	% V	% F	% V	% F
Mammalia	. . .	. . .	0.3	14.3	. . .	. . .	. . .	. . .	. . .	. . .	27.8	11.7	41.3	11.2
Aves	. . .	. . .	. . .	. . .	. . .	. . .	0.91	6.2	23.2	3.8	37.8	23.5	. . .	. . .
Reptilia	. . .	. . .	. . .	. . .	. . .	. . .	. . .	. . .	2.3	3.8	. . .	. . .	. . .	. . .
Amphibia	. . .	. . .	. . .	. . .	. . .	. . .	. . .	. . .	. . .	. . .	. . .	. . .	15.7	5.8
Insecta	100	100	28.9	28.5	47.1	66.6	20.0	43.7	17.6	26.9	7.9	29.4	7.4	23.5
Mollusca	. . .	. . .	7.2	7.1	5.8	16.6	1.8	12.5	1.2	3.8	. . .	. . .	0.8	5.8
Chilopoda	. . .	. . .	. . .	. . .	. . .	. . .	. . .	. . .	1.3	3.8	. . .	. . .	1.6	5.8
Annelida	. . .	. . .	5.4	7.1	. . .	. . .	. . .	. . .	. . .	. . .	. . .	. . .	3.3	5.8
Fruits	. . .	. . .	9.4	21.4	. . .	. . .	. . .	. . .	26.4	26.9	6.8	23.5	12.4	11.7
Plant remains	. . .	. . .	10.8	14.3	47.1	16.6	68.2	25.0	17.4	23.1	1.7	5.9	12.4	11.7
Miscellaneous	. . .	. . .	. . .	. . .	. . .	. . .	. . .	. . .	10.4	7.7	. . .	. . .	3.3	11.7
Particulate material	. . .	. . .	3.6	7.1	. . .	. . .	9.1	12.5	. . .	. . .	6.8	5.9	1.6	5.8

N = Sample size.

Molins de la Serna and Lorenzo (1982) in a study of stomach contents of 47 *Didelphis marsupialis* sampled from February 1981 to May 1982 in the lowlands of Sierra de Perija in the State of Zulia, northwestern Venezuela. In their study, the order of importance of food items, by frequency, is as follows: plant leaves (68.3%), fruits (56.2%), reptiles (42.6%), insects (29.2%), amphibians (28.8%), birds (14.3%), mammals (15.1%), mollusks (12.2%), and seeds (11.4%). The effects of seasonal and habitat differences in the two studies may explain these differences, since the main vegetation types of the lowlands of Perija are dry and humid tropical forests, with eight months of rainfall (April–November) and four months of drought (December–March).

Other studies have also shown that opossums feed on vertebrates. The volume we recorded for mammalian prey (15.3%) is low in comparison with diets determined for the Virginia opossum (*Didelphis virginiana*), except for Lay's (1942) 7% value. Hopkins and Forbes (1980) also recorded cats and rats in low frequencies and volumes in the diets of opossums in Oregon. Similarly, domestic chickens figured prominently in the diet of our specimens and have been reported as prey or carrion of *D. virginiana* in New York (Hamilton, 1951, 1958), Missouri (Reynolds, 1945), Iowa (Wiseman & Hendrickson, 1950), Michigan (Taube, 1947), and Kansas (Sandidge, 1953). In contrast, snakes and toads were taken infrequently, paralleling the results of Blumenthal and Kirkland (1976), who reported traces of amphibians in the diets of Pennsylvania *Didelphis,* and of Wiseman and Hendrickson (1950), who showed reptiles have a frequency of 1% in the diet of Iowa opossums. The importance of insects in the diet of our animals is somewhat lower than that previously reported for opossums in Michigan (30.4%; Gardner, 1982, citing Dearborn, 1932), Missouri (34.2%; Reynolds, 1945), and Kansas (42.2%; Sandidge, 1953). However, the volumes we report are higher than those in literature records for New York (Hamilton, 1951, 1958), Oregon (Hopkins & Forbes, 1980), and Pennsylvania (Blumenthal & Kirkland, 1976). Records for other invertebrates are also similar to those in existing literature reports (e.g., Taube, 1947; Hamilton, 1951, 1958; Reynolds, 1945; Sandidge, 1953).

Our data and literature records indicate that *Didelphis* species have similar diets, embracing a wide range of food items. More detailed studies, especially of food-use in relation to availability, will be needed to establish the degree of euryphagy.

Acknowledgments

This study was partly granted by CONICET Project S1-1158. We thank J. Ojasti for suggestions and review of the manuscript. We greatly appreciate the editorial assistance of B. Patterson. The staff members of the Estación Experimental Río Negro, Universidad Simón Rodriguez provided logistical support during fieldwork. L. Duque and R. Martinez helped us in the identification of snakes and slugs, and E. Pannier provided some stomach contents. To all of them, our thanks.

Literature Cited

ATRAMENTOWICZ, M. 1982. Influence du milieu sur l'activite locomotrice et la reproduction de *Caluromys philander* (L). Revue d'Ecologie Appliquée (Terre Vie), **36:** 373–395.

BLUMENTHAL, E. M., AND G. L. KIRKLAND. 1976. The biology of the opossum, *Didelphis virginiana* in south-central Pennsylvania. Proceedings of the Pennsylvania Academy of Science, **50:** 81–85.

CHARLES-DOMINIQUE, P. 1983. Ecology and social adaptations in didelphid marsupials: Comparison with eutherians of similar ecology, pp. 395–422. *In* Eisenberg, J. F., and D. G. Kleiman, eds., Advances in the Study of Mammalian Behavior. Special Publication of the American Society of Mammalogy, no. 7.

EWEL, J. J., A. MADRIZ, AND J. A. TOSI. 1976. Zonas de vida de Venezuela. Fondo Nacional de Investigaciones Agropecuarias, Caracas, 265 pp.

FLEMING, T. H. 1972. Aspects of the population dynamics of three species of opossums in the Panama Canal Zone. Journal of Mammalogy, **53:** 619–623.

GARDNER, A. L. 1982. Virginia opossum (*Didelphis virginiana*), pp. 3–36. *In* Chapman, J. A., and G. A. Feldhamer, eds., Wild Mammals of North America. Johns Hopkins University Press, Baltimore.

HAMILTON, W. J. 1951. The food of the opossum in New York State. Journal of Wildlife Management, **15:** 258–264.

———. 1958. Life history and economic relations of the opossum (*Didelphis marsupialis virginiana*) in New York State. Cornell University Agricultural Station Memoir, **354:** 1–48.

HOPKINS, D. D., AND R. B. FORBES. 1980. Dietary patterns of the Virginia opossum in an urban environment. The Murrelet, **61:** 20–30.

HUNSAKER, D. 1977. Ecology of New World marsupials, pp. 95–156. *In* Hunsaker II, D., ed., Academic Press, New York.

KIRSCH, J. A. W., AND J. H. CALABY. 1977. The species of living marsupials: An annotated list, pp. 9–26. *In* Stonehouse, B., and D. Gilmore, eds., The Biology of Marsupials. The Macmillan Press Ltd., London and Basingstoke.

KORSCHGEN, L. J. 1980. Procedures for food-habits analyses, pp. 113–127. *In* Schemnitz, S. D., ed., Wildlife Management Techniques. The Wildlife Society, Washington, D.C.

LAY, D. W. 1942. Ecology of the opossum in eastern Texas. Journal of Mammalogy, **23:** 147–159.

MOLINS DE LA SERNA, M., AND J. LORENZO PRIETO. 1982. Alimentación del rabipelado (*Didelphis marsupialis*) de la Sierra de Perija. Acta Científica Venezolana, **33:** 410.

PETRIDES, G. A. 1949. Sex and age determination in the opossum. Journal of Mammalogy, **30:** 364–378.

REYNOLDS, H. C. 1945. Some aspects of the life history and ecology of the opossum in central Missouri. Journal of Mammalogy, **26:** 361–379.

SANDIDGE, L. L. 1953. Food and dens of the opossum (*Didelphis virginiana*) in northeastern Kansas. Kansas Academy of Science, **56:** 97–106.

STREILEIN, K. E. 1982. Ecology of small mammals in the semiarid Brazilian Caatinga. I. Climate and faunal composition. Annals of Carnegie Museum, **51:** 79–107.

TAUBE, C. M. 1947. Food habits of Michigan opossums. Journal of Wildlife Management, **11:** 97–103.

TYNDALE-BISCOE, C. H., AND R. B. MACKENZIE. 1976. Reproduction in *Didelphis marsupialis* and *D. albiventris* in Colombia. Journal of Mammalogy, **57:** 249–265.

WISEMAN, G. L., AND G. D. HENDRICKSON. 1950. Notes on the life history and ecology of the opossum in southeast Iowa. Journal of Mammalogy, **31:** 331–337.

APPENDIX 1. Dental age classes for *Didelphis marsupialis.*

Tooth eruption		Wear	Age class	Age (months)
dP^3	M^1	0	I	3.0–3.5
dP^3	M^2	0	II	4.5–5.0
dP^3	M^3	0	III	6.2–6.7
P^3	M^3	0	IV	7.9–8.7
P^3	M^4	0	V	10.9–11.7
P^3	M^4	P^3 M^{1-2}	VI	12.8–14.1
P^3	M^4	P^3 M^{3-4}	VII	> 16.1

Source: G. A. Cordero (unpublished data).

Notes on Distribution of Some Bats from Southwestern Colombia

Michael S. Alberico

ABSTRACTS

Noteworthy range extensions are presented for *Noctilio albiventris, Rhinophylla alethina, Sturnira aratathomasi,* and *Lonchophylla handleyi,* including the second Colombian report for the last. A previous report of *Molossops brachymeles* is clarified as representing *M. abrasus.*

Se presentan algunas notables extensiones del rango de distribución para las especies *Noctilio albiventris, Rhinophylla alethina, Sturnira aratathomasi* y *Lonchophylla handleyi,* este último siendo el segundo reporte para Colombia. Un reporte anterior de *Molossops brachymeles* se clarifica como representativo de *M. abrasus.*

Apresentam-se notáveis extenções mas distribuições das espécies *Noctilio albiventris, Rhinophylla alethina, Sturnira aratathomasi,* e *Lonchophylla handleyi,* esta última sendo apenas o segundo registro para a Colombia. Clarifica-se o registro anterior de *Molossops brachymeles* como representativo de *M. abrasus.*

Introduction

Despite considerable interest in Neotropical mammals, southwestern Colombia remains poorly understood in this respect. This is mainly a result of a lack of adequate collections caused by the inaccessible nature of much of the zone. Early collecting expeditions to which we owe much of our knowledge were undertaken around the turn of the century by personnel of the American Museum of Natural History and summarized by Allen (1916). Bats were typically underrepresented in these early collections because of inadequate collecting techniques in use at the time. Now, with the aid of Japanese mist nets, we are able to obtain more complete samples of bat communities. In this report I present results of a continuing collecting effort during the past five years in this poorly known region, extending the known distribution of *Noctilio albiventris, Lonchophylla handleyi, Rhinophylla alethina, Sturnira aratathomasi,* and *Molossops abrasus.*

All specimens mentioned were collected in mist nets, prepared as standard study skins with skulls, and deposited in the mammal collection of the Departamento de Biología, Universidad del Valle, Cali, Colombia (UV).

From the Departamento de Biología, Universidad del Valle, Cali, Colombia.

Distribution

Noctilio albiventris

The lesser bulldog bat was recently reviewed by Davis (1976) and by Hood and Pitocchelli (1983). Both mapped the distribution as including eastern Colombia across the Llanos and Amazonas and the northern Caribbean coast. Davis (1976) reported the altitudinal range of the species as extending up to 1,100 m. We have found this species to be common in the upper Cauca valley, between the Cordillera Central and the Cordillera Occidental of the Andes, where the elevation reaches this approximate limit. Fifteen specimens from the Departamento (= state) del Valle del Cauca and adjacent Departamento del Cauca were compared with the descriptions and measurements of all subspecies recognized by Davis (1976). This population is indistinguishable from *N. a. minor* in all characters examined and undoubtedly follows the Río Cauca south from the Caribbean lowlands. A similar southern extension is most probable in the valley of the Río Magdalena to the Departamento de Huila, but has yet to be confirmed by collections.

SPECIMENS EXAMINED—**Cauca:** Río Palo, 18 km S, 5 km E Puerto Tejada, 3°04′N, 76°22′W, 1,050 m (3 ♀♀, UV313, 324, 325); **Valle del Cauca:** 2 km S, 4 km W Candelaria, 3°23′N, 76°23′W, 1,000 m (1 ♂, UV676); Universidad del Valle (Melendez Campus), 8 km S Cali, 3°22′N, 76°32′W, 1,000 m (5 ♂♂, UV2602, 2603, 2604, 2608, 2609; 2 ♀♀, UV2605, 2607); 13 km S, 1 km E Cali, 3°22′N, 76°32′W, 1,000 m (2 ♂♂, UV2620, 2611; 1 ♀, UV2612).

Lonchophylla handleyi

This species was described on the basis of specimens from Peru and southern Ecuador by Hill (1980), who suggested that some individuals in existing collections might be misidentified as *L. robusta. Lonchophylla handleyi* was first reported for Colombia by Alberico and Orejuela (1982), who collected a single individual from near the Ecuadorian border at 850 m. A specimen recently collected from the Departamento del Valle del Cauca at 480 m provides the second record for Colombia. Both specimens are larger (greatest length of skull, 28.4 and 28.6 mm, respectively) than the largest *L. robusta* reported by Hill (1980) for Peru and Ecuador and are larger than any *L. robusta* in our collections from western Colombia. Both Colombian specimens of *L. handleyi* are from the lower slope Andean forests, probably one of the last habitats to be intensively sampled for mammals in this country. The presence of this species in a relatively narrow elevational band within this habitat type attests to the importance of continued collecting in the Pacific slope of the Andes in southwestern Colombia.

SPECIMENS EXAMINED—**Nariño:** 5 km E Junín, 1°20′N, 78°08′W, 850 m (1 ♂, UV3007); **Valle del Cauca:** Río Cajambre, approx. 60 km S Buenaventura, 3°20′N, 77°00′W, 480 m (1 ♀, UV3694).

Rhinophylla alethina

This species was described based on specimens from western Colombia in the Departamento del Valle del Cauca (Handley, 1966) and until recently was known only from the type locality. Alberico and Orejuela (1982) reported it from Nariño near the Ecuadorian border and suggested that it might have a broader geographic range than previously thought, which was confirmed by Baud (1982) who reported the species for Ecuador. Our collections show *R. alethina* to be relatively common in the Pacific lowlands and the adjacent lower slopes of the western Andes up to 850 m. That this species was only recently described and remains poorly known is undoubtedly due to insufficient collecting in the forests of this zone.

SPECIMENS EXAMINED—**Nariño:** 5 km E Junín, 1°20′N, 78°08′W, 850 m (3 ♂♂, UV3029, 3033, 3036; 5 ♀♀, UV3030, 3031, 3032, 3034, 3035). **Valle del Cauca:** Alto Anchicayá, 35 km S, 20 km E Buenaventura, 3°34′N, 76°54′W, 400 m (2 ♂♂, UV3166, 3167); Río Azul, 5 km N, 25 km W Darién, 3°59′N, 76°44′W, 560 m (1 ♀, UV3391); Río Cajambre, approx. 60 km S Buenaventura, 3°20′N, 77°00′W, 480–520 m (1 ♂, UV3702; 1 ♀, UV3703); Río Calima, 13 km N, 14 km E Buenaventura, 4°00′N, 76°59′W, 40 m (1 ♀, UV2809).

Sturnira aratathomasi

In their description of this species, Peterson and Tamsitt (1968) reported three specimens, the holotype from the Departamento del Valle del Cauca in western Colombia and two from an unknown locality in Ecuador. They stated that it might be restricted to the Pacific side of the Andes. Thomas and McMurray (1974) provided measurements for the holotype and six individuals collected near the type locality and suggested that this species may be common at high elevations in the western Andes of Colombia. Our recent collections extend the known range some 150 km to the north in the Cordillera Occidental and, more importantly, record the presence of *S. aratathomasi* in the Cordillera Central, where it was previously unknown. This species appears to inhabit medium to high elevation forests which are relatively continuous in Colombia, and its occurrence both farther to the north and in the Cordillera Oriental is likely.

SPECIMENS EXAMINED—**Valle del Cauca:** Cordillera Central: Hacienda "Los Alpes," 6 km S, 11 km E Florida, 3°16′N, 76°09′W, 2,400 m (1 ♀, UV3482); Cordillera Occidental: Betania, 10 km N, 15 km W Bolívar, 4°26′N, 76°19′W, 1,800 m (1 ♀, UV3876); Parque Nacional "Los Farallones de Cali," 10 km S, 16 km W Cali, 3°22′N, 76°41′W, 2,600 m (1 ♀, UV3373); Paso de Galápagos, 8 km N, 4 km E El Cairo, 4°50′N, 76°12′W, 1,800 m (2 ♂♂, UV4131, 4133).

Molossops abrasus

This species was reported for Colombia by Alberico and Naranjo-H. (1982) as *M. brachymeles,* based on specimens from the Cauca valley in northern Valle del Cauca. Although often referred to by this latter specific epithet (see Cabrera, 1958; Freeman, 1981), the holotype of *Dysopes abrasus* from Brazil has been shown to represent this species (Husson, 1962; Carter & Dolan, 1978). The Colombian record extends the known distribution of *M. abrasus* in western South America from Andean Peru some 1,600 km to the north.

SPECIMENS EXAMINED—**Valle del Cauca:** 11 km S, 2 km W Cartago, 4°39′N, 75°56′W, 930 m (2 ♂♂, UV2451, 2452; 1 ♀, 2453).

Acknowledgments

This report is the result of the combined efforts of many friends and students, too numerous to mention by name, who have collaborated either by accompanying the author in the field, by sharing specimens collected during other activities, or both. However, a few individuals have contributed more than could be expected in the normal turn of events, and their support in the field and out has been especially important in the present study: Eduardo Velasco, Gloria Giral, Alonso González, Guillermo Cantillo, and Luz Marina Alberico. To these, the author is most appreciative.

Literature Cited

ALBERICO, M., AND L. G. NARANJO-H. 1982. Primer registro de *Molossops brachymeles* (Chiroptera: Molossidae) para Colombia. Cespedesia, **11:** 141–143.

ALBERICO, M., AND J. E. OREJUELA. 1982. Diversidad específica de dos comunidades de murciélagos en Nariño, Colombia. Cespedesia, Suplemento no. **3**(41–42): 31–40.

ALLEN, J. A. 1916. List of mammals collected in Colombia by the American Museum of Natural History expeditions, 1910–1915. Bulletin of the American Museum of Natural History, **35:** 191–238.

BAUD, F. J. 1982. Présence de *Rhinophylla alethina* (Mammalia, Chiroptera) en Equateur et répartition actuelle du genre en Amérique du Sud. Revue Suisse de Zoologie, **89:** 815–821.

CABRERA, A. 1958. Catálogo de los mamíferos de América del Sur. Revista del Museo Argentino de Ciencias Naturales "Bernardino Rivadavia", Ciencias Zoológicas, **4:** 1–307.

CARTER, D. C., AND P. G. DOLAN. 1978. Catalogue of type specimens of neotropical bats in selected European museums. Special Publications, The Museum, Texas Tech University, **15:** 1–136.

DAVIS, W. B. 1976. Geographic variation in the lesser noctilio, *Noctilio albiventris* (Chiroptera). Journal of Mammalogy, **57:** 687–707.

FREEMAN, P. W. 1981. A multivariate study of the family Molossidae (Mammalia, Chiroptera): Morphology, ecology, evolution. Fieldiana: Zoology, n.s., **7:** 1–173.

HANDLEY, C. O., JR. 1966. Descriptions of new bats (*Choeroniscus* and *Rhinophylla*) from Colombia. Proceedings of the Biological Society of Washington, **79:** 83–88.

HILL, J. E. 1980. A note on *Lonchophylla* (Chiroptera: Phyllostomatidae) from Ecuador and Peru, with the description of a new species. Bulletin of the British Museum (Natural History), Zoology Series, **38:** 233–236.

HOOD, C. S., AND J. PITOCCHELLI. 1983. Noctilio albiventris. Mammalian Species, **197:** 1–5.

HUSSON, A. M. 1962. The bats of Suriname. Rijksmuseum van Natuurlijke Historie, Leiden, **58:** 1–282.

PETERSON, R. L., AND J. R. TAMSITT. 1968. A new species of bat of the genus *Sturnira* (family Phyllostomatidae) from northwestern South America. Life Sciences Occasional Papers, Royal Ontario Museum, **12:** 1–8.

THOMAS, M. E., AND D. N. MCMURRAY. 1974. Observations on *Sturnira aratathomasi* from Colombia. Journal of Mammalogy, **55:** 834–836.

Distributional Records of Bats from the Caribbean Lowlands of Belize and Adjacent Guatemala and Mexico

Timothy J. McCarthy

ABSTRACTS

Thirty new species records are presented for the bat fauna of Belize, along with secondary records for eight bats that had been recorded previously from that country. Contiguous lowland localities in Guatemala provided new department records: nine for El Petén, five for Izabal, and two for Alta Verapaz. The El Petén records include the first confirmation of *Vampyrum spectrum* in Guatemala. One state record for Quintana Roo, Mexico, is reported. These species represent the genera *Saccopteryx, Balantiopteryx, Diclidurus, Noctilio, Pteronotus, Mormoops, Micronycteris, Lonchorhina, Macrophyllum, Tonatia, Mimon, Phyllostomus, Phylloderma, Trachops, Chrotopterus, Vampyrum, Glossophaga, Uroderma, Vampyrops, Vampyrodes, Vampyressa, Chiroderma, Artibeus, Centurio, Diphylla, Natalus, Myotis, Eptesicus, Lasiurus, Bauerus, Eumops,* and *Molossus.* Range extensions are acknowledged for *Saccopteryx leptura, Diclidurus virgo, Noctilio leporinus, Micronycteris nicefori, Macrophyllum macrophyllum, Phyllostomus discolor, Vampyrum spectrum, Glossophaga commissarisi, Uroderma bilobatum, Vampyrodes caraccioli, Artibeus toltecus,* and *Bauerus dubiaquercus.* A checklist of the bat fauna of Belize, which stands at 66 species, is presented.

Se registran 30 especies que no habían sido citadas antes para la fauna de murciélago de Belice, con registros secundarios para ocho murciélagos ya conocidos de ese país. En ciertas localidades contiguas de las tierras bajas de Guatemala, se obtuvieron nuevos registros departamentales: nueve de El Petén, cinco de Izabal, y dos de Alta Verapaz. Los registros de El Petén incluyen la primera confirmación de *Vampyrum spectrum* en Guatemala. Además, se presenta un nuevo registro estatal para Quintana Roo, México. Las especies obtenidas están segregadas en los géneros *Saccopteryx, Balantiopteryx, Diclidurus, Noctilio, Pteronotus, Mormoops, Micronycteris, Lonchorhina, Macrophyllum, Tonatia, Mimon, Phyllostomus, Phylloderma, Trachops, Chrotopterus, Vampyrum, Glossophaga, Uroderma, Vampyrops, Vampyrodes, Vampyressa, Chiroderma, Artibeus, Centurio, Diphylla, Natalus, Myotis, Eptesicus, Lasiurus, Bauerus, Eumops,* y *Molossus.* Para cada una de las siguientes especies de murciélagos se anota el alcance geográfico de su distribución conocida: *Saccopteryx leptura, Diclidurus virgo, Noctilio leporinus, Micronycteris nicefori, Macrophyllum macrophyllum, Phyllostomus discolor, Vampyrum spectrum, Glossophaga commissarisi, Uroderma bilobatum, Vampyrodes caraccioli, Artibeus toltecus,* y *Bauerus dubiaquercus.* Se incluye una lista de 66 especies que representan la fauna de murciélagos de Belice.

Apresenta-se records de 30 novas espécies de morcêgos para Belice, e de oito espécies pouco conhecidas no país. Áreas adjacentes, na Guatemala, providenciaram novos records para: El

From the Department of Mammalogy, American Museum of Natural History, Central Park West at 79th Street, New York, NY 10024.

Petén (nove espécies), Izabal (cinco espécies), e Alta Verapaz (dois espécies). Os records de El Petén incluem as primeiras confirmações de *Vampyrum spectrum* na Guatemala. Um novo record para Quintana Roo, México, é incluido. Estas espécies representam os generos *Saccopteryx, Balantiopteryx, Diclidurus, Noctilio, Pteronotus, Mormoops, Micronycteris, Lonchorhina, Macrophyllum, Tonatia, Mimon, Phyllostomus, Phylloderma, Trachops, Chrotopterus, Vampyrum, Glossophaga, Uroderma, Vampyrops, Vampyrodes, Vampyressa, Chiroderma, Artibeus, Centurio, Diphylla, Natalus, Myotis, Eptesicus, Lasiurus, Bauerus, Eumops,* e *Molossus.* Reconhece-se extensões nas áreas onde são encontrados *Saccopteryx leptura, Diclidurus virgo, Noctilio leporinus, Micronycteris nicefori, Macrophyllum macrophyllum, Phyllostomus discolor, Vampyrum spectrum, Glossophaga commissarisi, Uroderma bilobatum, Vampyrodes caraccioli, Artibeus toltecus,* e *Bauerus dubiaquercus.* Apresenta-se uma lista da fauna de morcêgos em Belice, que agora conta com 66 espécies.

Introduction

Inventories of bat communities in Mexico and Central America have increased significantly during the last twenty-five years (Jones et al., 1977). Although the resulting data have enhanced our knowledge of the distributions and the zoogeographical relationships of species, incomplete surveys exist for certain regions. The northern lowlands along the Caribbean coast of Honduras, Guatemala, Belize, and Quintana Roo, in Mexico, is one such region. Travel within this coastal versant has improved with agricultural and settlement expansion. The isolation of Belize from its neighbors has been reduced with the construction of roads in Guatemala's frontier of El Petén and Mexico's former territory of Quintana Roo. A paved road from Izabal now connects El Petén and Belize with the Pan-American Highway in western Guatemala. Road development continues within Belize for all-weather travel.

Belize is situated within the Caribbean lowland drainage of northern Central America. Contiguous with Belize on this eastern slope is the eastern portion of the department of El Petén to the west and, to the south, the department of Izabal, both of Guatemala. Southern Quintana Roo of peninsular Mexico borders to the north (see fig. 1 and Gazetteer). The topography of these Caribbean lowlands extends from the lower ranges (600 m and below) of the eastern Sierra de Chamá, the Sierras de las Minas, the Sierra de Santa Cruz, the Sierra del Meredón, and the Montañas del Mico in Alta Verapaz and Izabal, and the Maya Mountains of southern Belize and southeastern El Petén to the low undulating relief of southern Quintana Roo. The Maya Mountains represent a heavily eroded Paleozoic formation that now ranges at the top from 671 to 853 m in elevation, with the highest peak at 1113 m (Wright et al., 1959). Annual rainfall in portions of Izabal averages from 3,000 to nearly 5,000 mm (Portig, 1976). Over 4,500 mm of rainfall was reported from the most southern coastal area of Belize. North and northwestward of the Maya Mountains, rainfall decreases appreciably to less than 1,500 mm in north-central El Petén and northern Belize, where less than 1,400 mm was recorded near the Quintana Roo border (Walker, 1973). The severity of this northward reduction of rainfall is intensified by the increased lack of surface drainage into the Yucatán Peninsula of Mexico. Because the limestone shelf of northern Belize has geological affinities with the Yucatán Peninsula (Wright et al., 1959), the southern limit of this peninsula can be considered the fault line extending from north of the Maya Mountains westward through the northern shore of Lake Petén-Itzá, El Petén (Wadell, 1938; West, 1964). Effectively, the northern plain of Belize and northern El Petén are portions of the Yucatán Peninsula. The northward shift from alluvial soils to shallow calcareous soils, along with the mentioned climatic changes, create edaphic conditions that affect the composition and the structure of the vegetation that can be supported (Lundell, 1934, 1937; Standley & Record, 1936; Wright et al., 1959; Pennington & Sarukhan, 1968). The potential effect of this transitional physiography on the distribution and relative abundance of bats in this Caribbean lowland region will require further inventory studies.

This paper documents 30 new records for Belize. A checklist of the known bat fauna for this country is annotated in the Appendix. Sixty-six species are cited. Included here are also records from nearby localities for El Petén, Izabal, and Alta Verapaz, Guatemala, and Quintana Roo, Mexico. Nine species records from El Petén, five records from

Izabal, and two records from Alta Verapaz increase the number of reported species for these departments to 35, 31, and 40, respectively (see Jones, 1966; Carter et al., 1966; Rick, 1968; Smith, 1972; LaVal, 1973a; Martínez R., 1980; McCarthy, 1982). Jones et al. (1973) and Birney et al. (1974) summarized the records for 31 species from Quintana Roo, and this paper provides one additional record.

Materials and Methods

The bats that I collected during the years 1974–1984 in Belize and El Petén (Parque Nacional Tikal), Guatemala, were obtained principally with mist nets set at ground level; aerial netting and the use of a bat trap were limited. Unless otherwise stated, mist netting was carried out during the first half of the night. A limited number of specimens were obtained with hand nets or plastic funnel traps at roost sites. Specimens were prepared as standard museum skins with skulls and/or skeletons, or as fluid-preserved specimens. These vouchers are housed in Field Museum of Natural History, Chicago (FMNH); The Museum, Michigan State University, East Lansing (MSU); Carnegie Museum of Natural History, Pittsburgh (CM); and American Museum of Natural History, New York (AMNH).

A survey of 45 museum collections in the United States, Canada, Mexico, Guatemala, and England resulted in additional specimens from Belize, El Petén, Izabal, Alta Verapaz, and Quintana Roo. Pertinent specimens (147) have been included in this report from the following institutions [collectors in brackets]: American Museum of Natural History, New York [N. Sullivan]; British Museum (Natural History), London, England (BM) [R. H. L. Disney; P. Williams; A. M. Hutson; R. E. Stebbings]; Carnegie Museum of Natural History [N. A. Bitar]; Field Museum of Natural History [L. de la Torre]; Florida State Museum, University of Florida, Gainesville (FSM) [F. J. Bonaccorso]; Museum of Zoology, Louisiana State University, Baton Rouge (LSUMZ) [D. M. Lay]; Royal Ontario Museum, Toronto, Canada (ROM) [R. L. Peterson; J. Kamstra; J. Fragoso]; Texas Cooperative Wildlife Collections, Texas A&M University, College Station (TCWC) [D. C. Carter; M. D. Engstrom]; Texas Tech University, Lubbock (TTU) [P. Diamond]; and United States National Museum of Natural History, Washington, D.C. (USNM) [E. L. Tyson].

Systematic arrangement of species accounts and nomenclature, unless otherwise indicated, follow Jones et al. (1977) and Handley (1980). Disney (1968) did not provide data for the first records of *Pteronotus davyi, Tonatia minuta,* and *Eptesicus furinalis* from Belize. Those data are presented in the respective accounts of this report, with additional records. Further secondary records from Belize of *Mimon crenulatum, Trachops cirrhosus, Glossophaga commissarisi, Vampyressa pusilla,* and *Eumops auripendulus* are also included. All of the species accounts are discussed in the context of their range and elevational distributions in Mexico and Central America. Hall (1981) was the primary reference for this unless cited otherwise.

Forest types in Belize follow Wright et al. (1959), whose classification was partially based on the seasonal formation series (Beard, 1944), which refers to structural appearance. The correct political alignments between the states of the Yucatán Peninsula are inconsistent among a number of published maps. The state boundary between Quintana Roo and Campeche on the map in Figure 1 (see also Gazetteer) is based on a number of Government of Mexico (Secretaría de Programación y Presupuesto) maps, including "Carta Topográfica, Mérida" (1:1,000,000; 1979 and 1983) and "Mapa Geográfica" (1:5,000,000; 1980).

Species Accounts

Family EMBALLONURIDAE

Subfamily EMBALLONURINAE

Saccopteryx leptura (Schreber, 1774)

SPECIMEN EXAMINED—**BELIZE. Toledo:** 2.1 km NNE Salamanca Camp, Columbia Forest, 1 ♀ (CM).

The known distribution of this small sac-winged bat north of Panama extends through Costa Rica and Nicaragua to Chiapas along the Pacific versant. The presence of predominantly lowland *Saccopteryx leptura* in southern Belize represents a country record and an extension of its distribution along the Caribbean side from southeastern Nicaragua.

Small bats were observed foraging up to heights of 13–13.5 m during the twilight period of the evening. Flight appeared to be concentrated within a small, open area below the lower canopy of evergreen seasonal forest. A short mist net was hand-hoisted to capture (24 March) this adult specimen. *Saccopteryx bilineata* was collected shortly after the capture of *S. leptura.*

Balantiopteryx io Thomas, 1904

SPECIMENS EXAMINED—**GUATEMALA. El Petén:** Poptún, Finca Ixobel, 18 ♂♂, 18 ♀♀ (CM).

The restricted distribution of *Balantiopteryx io* ranges from the Gulf lowlands of Veracruz, Oaxaca, and Tabasco to the lowlands of Belize and eastern Guatemala. Kirkpatrick et al. (1975), Cartwright and Kirkpatrick (1977), and Sanborn (1936) represent the previous records for Belize and Izabal. The Poptún locality represents the first record for El Petén.

The specimens reported here were collected (12 June) by N. A. Bitar as they exited from a cave surrounded by secondary forest. The distribution of this colonial species may be restricted in part by the availability of adequate cave habitats as roosting sites.

Subfamily DICLIDURINAE

Diclidurus virgo Thomas, 1903

SPECIMEN EXAMINED—**BELIZE. Cayo:** 1.5 km SSW Roaring Creek, 1 ♂ (FMNH).

The white bat is represented by relatively few localities in Middle America, which extend from western (Nayarit) and eastern (southern Veracruz) Mexico through Central America. Specimens from southwestern El Petén were reported by Jones (1966). The single specimen from Belize represents a northward range extension in the Caribbean lowlands from northwest Honduras (Carter & Dolan, 1978) and a record for the country.

The single bat apparently was roosting on the trunk of a fig tree (*Ficus insipida*) overhanging a pool along the Roaring Creek River. It was captured (May) by C. Tzul after being observed on a number of occasions roosting near, but not among, a group of *Rhynchonycteris naso.* Jones (1966), Starrett and Casebeer (1968), and Handley (1976) commented on the high foraging habits of *Diclidurus.* Similar to the molossid bats, *Diclidurus* probably concentrates its foraging efforts at levels well above the tree canopy and beyond the reach of conventional collecting techniques, except firearms. This may explain why there are few specimens available in collections.

Goodwin (1969) considered *Diclidurus virgo* at best not more than subspecifically different from *D. albus.* Both species were recognized by Ojasti and Linares (1971), who questioned Goodwin (1969) because they believed that his South American comparative material represented *D. virgo* and not *D. albus.*

Family NOCTILIONIDAE

Noctilio leporinus mastivus (Vahl, 1797)

SPECIMENS EXAMINED—**BELIZE. Cayo:** Banana Bank, 1 ♂, 1 ♀ (FMNH); Barton Creek, at Western Hwy., 1 ♀ (FMNH). **Stann Creek:** Melinda, Stann Creek River, 1 ♂ (FMNH). **Toledo:** 1.2 km E Aguacate, Aguacate River, 2 ♂♂ (CM), 1 ♀ (BM); Big Fall, vicinity Río Grande Bridge, 1 ♀ (FMNH); Salamanca Camp, 1 ♀ (BM).

The fishing bat occurs along the riparian habitats of river systems, inland lakes, and coastlines in primarily lowland regions from northwestern (northern Sinaloa), eastern (southern Veracruz), and peninsular (Yucatán) Mexico throughout Central America (Davis, 1973; Hellebuyck et al., 1985). Dickerman et al. (1981) reported a locality for *Noctilio leporinus* from Alta Verapaz as in the Caribbean drainage when it was clearly in the Río Usamacinta drainage of the Gulf lowlands. The Belizean localities extend northward the recorded occurrence of *N. leporinus* from Izabal and northwestern Honduras (Carter et al., 1966).

All of the specimens were obtained (March, April, May, July, August) over rivers and a pond except for one individual, which was mist netted (28 August) low over a pasture adjacent to a flooded river. This bat was foraging primarily for insects since its feces contained the chitinous remains of these prey. Additional fishing bats from the localities in Cayo and Stann Creek districts were captured, banded, and released. This bat was common along the South Stann Creek drainage, Cockscomb Basin. A specimen belonging to M. Craig, Belize Audubon Society, was collected at Indian Church (Lamanai), New River Lagoon, Orange Walk District. An old specimen of *N. leporinus* in the collections of British Museum (Natural History) was registered in 1909 without pertinent field data. The two peninsular records from Campeche (Jones et al., 1973) and Yucatán (Birney et al., 1974) were obtained in coastline habitat along the

Gulf of Mexico. Although subsurface drainage predominates north of Belize into Quintana Roo, shallow inland "lagunas" are fairly common and probably support *Noctilio* populations.

Family MORMOOPIDAE

Pteronotus davyi fulvus (Thomas, 1892)

SPECIMENS EXAMINED—**BELIZE. Cayo:** Central Farm, 1 ♂ (CM), 1 ♂ (FMNH); Ontario, 5.5 km W Teakettle, 1 ♂ (FMNH); Unitedville, 9 km WSW Teakettle, 1 ♂ (FMNH). **Orange Walk:** Tower Hill, B.S.I. compound, 3 ♀♀ (FMNH). **Toledo:** Aguacate, 1 ♂ (CM); 1.2 km E Aguacate, 1 ♂ (BM), 1 ♂ (CM); Rice Station, 2 ♂♂ (FMNH); 0.4 km W Rice Station, 1 ♂ (FMNH); San Antonio, 1 ♂ (FMNH). **GUATEMALA. El Petén:** Parque Nacional Tikal, 1 ♂ (MSU).

Smith (1972) summarized the majority of the capture localities for this subspecies of naked-backed bat, which ranges from northwestern (Sonora), northeastern (Tamaulipas), and peninsular (Yucatán) Mexico southeastward to Honduras and El Salvador, but omitted the only record for Belize (Disney, 1968). Parque Nacional Tikal is the first record for El Petén, and the Belizean specimens provide additional records for Belize.

Disney (1968) did not present data for his single *Pteronotus davyi* specimen. This male was obtained (1 November) in Cayo District, at Listowel along the Belize River, and is housed in British Museum (Natural History). The subsequent specimens reported here were collected (October–December, May, July, August) in open areas, bordering on vegetation and buildings, and over water. The specimen from El Petén was captured (25 March) along a trail in upland deciduous forest. An additional *P. davyi* from Tikal was captured, banded, and released.

Pteronotus personatus psilotis (Dobson, 1878)

SPECIMENS EXAMINED—**BELIZE. Toledo:** 1.2 km E Aguacate, Aguacate River, 1 ♀ (BM), 4 ♂♂, 1 ♀ (CM); Big Fall, 1.5 km WSW Río Grande Bridge, 1 ♂ (FMNH); 0.8 km NW Blue Creek, 1 ♀ (FSM); Crique Jute, 1 ♂ (FMNH); Crique Lagarto, 1 km NW San Antonio, 1 ♀ (FMNH); Jacinto Creek, at Punta Gorda Road, 1 ♂ (MSU); 0.4 km W Rice Station, 1 ♀ (FMNH); Salamanca Camp, 1 ♀ (USNM); San Antonio, 1 ♀ (FMNH); 0.9 km WNW San Pedro Columbia, 1 ♀ (FMNH).

The distribution of *Pteronotus personatus psilotis* extends from western (southern Sinaloa) and eastern (Tamaulipas) Mexico southeastward to Honduras and El Salvador (Smith, 1972), with Caribbean lowland localities in Campeche (Jones et al., 1973), El Petén (Jones, 1966), and Alta Verapaz (Jones, 1966). Elevations range from 123 to 984 m. These localities from southern Belize are the first records for the country.

Fifty-three percent of the small moustache bats were collected (March, May, July) over open water; the remainder were foraging (January, April, August, December) in open areas adjacent to buildings or corralled cattle.

Mormoops megalophylla megalophylla Peters, 1864

SPECIMENS EXAMINED—**BELIZE. Belize:** 6.6 km N Churchyard, 1 ♀ (CM). **Cayo:** 1.6 km NW Augustine, Río Frío, 1 ♂ (TTU). **Stann Creek:** Melinda, 1 ♂ (FMNH). **Toledo:** Forest Home, 1 ♂ (MSU); Pueblo Viejo, 1 ♂ (FMNH). **GUATEMALA. Izabal:** 25 km SSW Puerto Barrios, 1 ♂ (TCWC).

The leaf-chinned bat has been reported throughout Mexico, Guatemala, El Salvador, and Honduras (Smith, 1972). Davis and Carter (1962), Jones (1966), and Taibel (1977) provided lowland records for El Petén and Alta Verapaz. Elevations range from near sea level to 2270 m. These localities are the first records for Belize and Izabal.

Except for one Belizean specimen, which was captured (9 June) in a cave, these leaf-chinned bats were associated (March, April, December) with open areas bordering on forest or orchard edges, including pine savanna. One *Mormoops* specimen, which was registered into the British Museum (Natural History) collections in 1892, may have been obtained in the vicinity of Belize City. The Guatemalan specimen was collected by D. C. Carter.

Family PHYLLOSTOMIDAE

Subfamily PHYLLOSTOMINAE

Micronycteris brachyotis (Dobson, 1878)

SPECIMENS EXAMINED—**BELIZE. Cayo:** 1 km NW Augustine, 2 ♂♂ (FMNH). **Toledo:** Crique Negro, Columbia Forest, 1 ♂ (BM).

The first Middle American specimen of *Micro-*

nycteris brachyotis was initially reported from Nicaragua as *M. sylvestris* by Goodwin (1946), but was correctly identified by Sanborn (1949). Subsequent records are from the Gulf-Caribbean lowlands of southern Veracruz (Medellín L. et al., 1983), Oaxaca (Schaldach, 1964), Chiapas (Davis et al., 1964), and El Petén (Jones, 1966; Rick, 1968; McCarthy, 1982); and the Pacific-Caribbean versants of Costa Rica (Howell & Burch, 1974; Starrett, 1976; LaVal & Fitch, 1977) and Panama (Handley, 1966; Fleming et al., 1972; Bonaccorso, 1979). Reported elevations range from 40 to 594 m. The present specimens are the first records of the yellow-throated bat for Belize.

Two specimens were captured (29 July) as they exited from a cave into a low deciduous seasonal forest, and a third bat was taken (28 May) along a path in an evergreen seasonal forest. The two specimens from Cayo, which I obtained while assisting a histoplasmosis survey, were listed as "*M. bardyotis*" in a preliminary report (Quinones et al., 1978, p. 559) and no specific locality data were provided.

Micronycteris megalotis mexicana Miller, 1898

SPECIMENS EXAMINED—**BELIZE. Corozal:** San Antonio, 2 km NW Corozal, 1 ♂ (FMNH). **Orange Walk:** San Antonio, Río Hondo, 2 ♂♂, 1 ♀ (FMNH). **Toledo:** Aguacate, 1 ♂ (CM); Big Fall, 2 km E Río Grande Bridge, 1 ♀ (BM); Cuevas Creek Bridge, 10 km NW Punta Gorda, 1 ♂, 1 ♀ (BM), 1 ♂ (AMNH), 1 ♀ (MSU); Nimli Punit, 1 ♀ (CM); Rocky Run Ranch, 4.8 km NW Punta Gorda, 1 ♂, 1 ♀ (BM); Unión Camp, 2 ♀♀ (BM); Vista Hermosa Ranch, 3.7 km WNW Punta Gorda, 1 ♀ (CM). **GUATEMALA. El Petén:** Parque Nacional Tikal, 1 ♂ (FMNH).

The distribution of this subspecies of big-eared bat extends from western (Jalisco), eastern (southern Tamaulipas), and peninsular (Yucatán) Mexico, along the Pacific coastal and highland regions, to Costa Rica. Gardner et al. (1970) suggested that the southern extent of *Micronycteris megalotis mexicana* is in the Cordillera Talamanca of Costa Rica. This species has been recorded most often at lowland–moderate elevations, up to 2870 m. Specimens from Isla Cozumel, Quintana Roo, represent the only record for Quintana Roo (Jones et al., 1973). The records of *M. m. mexicana* which are reported here are the first for Belize and El Petén.

Belizean specimens were obtained (May, July, August, November) in diurnal roost sites (shallow caves and limestone chambers, bridge approachments, abandoned rum factory boiler) and collected in forest habitats (riparian marsh, evergreen and semi-evergreen, deciduous semi-evergreen, and deciduous seasonal). The Tikal specimen was captured (6 June) roosting in a passageway of an excavation tunnel within a ruin complex. A second juvenile male was captured, banded, and released (29 July) in escobal palm (*Cryosophila argentea*) forest, 1.9 km SE Tikal Reservoir.

Micronycteris nicefori Sanborn, 1949

SPECIMEN EXAMINED—**BELIZE. Toledo:** 0.4 km NE Aguacate, 1 ♀ (FMNH).

Handley (1966) documented the first specimens of *Micronycteris nicefori* north of South America, from Panama. Subsequently, it has been reported from southeastern Nicaragua (Baker & Jones, 1975) and both the dry Pacific (Starrett, 1976) and wet Caribbean (LaVal, 1977) lowlands of Costa Rica. These Central American localities range from near sea level to over 100 m. This first record from Belize also represents a significant Central American range extension along the Caribbean versant.

The *M. nicefori* specimen reported here was mist netted on 15 December along a track in hilltop, evergreen seasonal forest.

Micronycteris schmidtorum Sanborn, 1935

SPECIMENS EXAMINED—**BELIZE. Corozal:** Patchakan, 2 ♀♀ (FMNH). **Orange Walk:** 1.3 km W San Antonio, Río Hondo, 1 ♂ (FMNH). **Toledo:** Big Fall, 1 km E Río Grande Bridge, 1 ♂ (CM).

Micronycteris schmidtorum was described (Sanborn, 1935) from specimens collected in the Caribbean lowlands of Izabal. An additional Guatemalan specimen was recorded in the Pacific piedmont (Dickerman et al., 1981). The remaining Central American records represent both the Pacific and Caribbean lowland slopes from Honduras (Sanborn, 1941), Nicaragua (Davis et al., 1964; Baker & Jones, 1975), Costa Rica (Starrett & Casebeer, 1968; Fleming et al., 1972; Howell & Burch, 1974; LaVal & Fitch, 1977), and Panama (Handley, 1966). Specimens from Yucatán assigned to *M. schmidtorum* by Villa-R. (1966) were reidentified as *M. megalotis* by Jones et al. (1973). An identification of *M. schmidtorum* (Jones et al., 1973) for a specimen from Isla Cozumel, Quintana Roo, was questioned by Hall (1981) because this specimen previously was identified as *M. megalotis* (Jones & Lawlor, 1965). I examined this specimen (University of Kansas 91539) and agree that it is *M. schmidtorum*. The northern distribution of this big-eared bat extends to the Caribbean coast of the Yucatán Peninsula. The specimens reported here are the first records for Belize.

At Parque Nacional Tikal, one juvenile and two adult females, which were captured (30 July) in a hollow tree (*Bursera semirouba*) of an upland deciduous seasonal forest, were photographed, banded, and released. This site was revisited during the following March, but no *Micronycteris* were found. These individuals of *M. schmidtorum* were the first seen in El Petén. Similarly, Sanborn (1935) and Starrett and Casebeer (1968) reported individuals from tree hollows. The Belizean specimens were captured (February, September, November) in the orchard vegetation of a village, along a secondary forest edge, and in riparian secondary vegetation.

Lonchorhina aurita aurita Tomes, 1863

SPECIMENS EXAMINED—**BELIZE. Stann Creek:** 5.3 km WNW Quam Bank, Cockscomb Basin, 1 ♀ (CM). **Toledo:** 0.8 km NW Blue Creek, 1 ♂, 1 ♀ (AMNH); Crique Jute Village, 1 ♀ (CM); Crique Negro, Columbia Forest, 1 ♂ (BM), 1 ♂ (USNM); 2.1 km NNE Salamanca Camp, Columbia Forest, 3 ♂♂ (CM). **GUATEMALA. El Petén:** Poptún, Finca Ixobel, 2 ♂♂ (CM).

Lonchorhina aurita was first recorded in Middle America from Panama (Miller, 1912). Subsequent collecting has found this cave-dwelling bat northward through Central America to southeastern (southern Veracruz, Oaxaca, Tabasco) and peninsular (Quintana Roo) Mexico. Predominately lowland, this distinctive leaf-nosed bat extends up to more than 1500 m in representative habitats. Jones et al. (1973) reported the only record from Quintana Roo, while specimens from Izabal (Sanborn, 1936) are apparently the next Caribbean versant record north of eastern Costa Rica (Nelson, 1965); records from Nicaragua and Honduras are lacking. The specimens examined for this account are the first records from Belize and El Petén.

All specimens from Belize were captured (March, April, May, August) in deciduous seasonal and evergreen seasonal forests. The Guatemalan bats were captured by N. A. Bitar as they exited from the cave discussed in the *Balantiopteryx io* account.

Macrophyllum macrophyllum (Shinz, 1821)

SPECIMENS EXAMINED—**BELIZE. Cayo:** Sibun River at Indian Creek, 1 ♂ (FMNH). **Toledo:** Big Fall, 1.7 km NE Río Grande Bridge, 1 ♀ (CM).

Tabasco, Mexico, represents the northernmost occurrence for the long-legged bat, which is known from both the Caribbean and Pacific regions of Central America. Primarily a lowland inhabitant, *Macrophyllum macrophyllum* ranges from 40 to almost 600 m. These specimens represent a Caribbean lowlands range extension from northwestern Honduras (Valdez & LaVal, 1971) and the first records for Belize.

Harrison and Pendleton (1974), Gardner (1977), and Dickerman et al. (1981) indicated that long-legged bats may be closely associated with aquatic habitats. Similarly, the Belizean specimens were obtained (17 March, 1 April) from along the Sibun River, although not directly above water, and over the surface of the Río Grande. The first bat was taken at approximately 0340 in a stand of shade trees, dominated by cohune palms (*Orbignya cohune*), at the edge of an open pasture.

Tonatia bidens bidens (Spix, 1823)

SPECIMENS EXAMINED—**BELIZE. Cayo:** Río Frío, 1.6 km W Augustine, 1 ♀ (CM). **Toledo:** Nimli Punit, 1 ♂ (CM); Orange Creek, 1.5 km SW Punta Gorda, 1 ♂ (MSU); 2.1 km NNE Salamanca Camp, Columbia Forest, 1 ♂ (CM); 2.2 km NNE Salamanca Camp, Columbia Forest, 1 ♀ (CM).

Goodwin (1946) first recorded *Tonatia bidens* in Central America from the Pacific lowlands of Costa Rica. Other humid lowland records include both the Caribbean and Pacific versants of Panama, continuing along the Caribbean corridor of Nicaragua, Honduras, and Guatemala. The northernmost record is from eastern Chiapas (Medellín L., 1983). The Guatemalan records are from the Caribbean lowlands of El Petén (McCarthy, 1982) and Izabal (Carter et al., 1966). Elevations range from near sea level to around 660 m. The present specimens constitute the first records from Belize.

Four adult males were taken (March, April) over a creek in a low transitional forest, in a high evergreen seasonal forest, and in a deciduous seasonal forest. A subadult male was captured (24 September) in the courtyard of a Mayan archaeological site located in a high deciduous seasonal forest.

Tonatia evotis Davis and Carter, 1978

SPECIMEN EXAMINED—**GUATEMALA. El Petén:** Parque Nacional Tikal, 1 ♂ (FMNH).

Davis and Carter (1978) described *Tonatia evotis* on the basis of its smaller size in comparison to *T. sylvicola*; a female from Izabal was designated as the holotype. El Petén is part of a Gulf-Caribbean distribution which extends from southern Veracruz, Tabasco, Chiapas, and Campeche to Belize, and continues along northern Honduras (Davis & Carter, 1978). Martínez R. (1980) recorded an additional eastern Guatemalan locality in Alta Verapaz. All recorded elevations are less than 100 m. The *T. evotis* from Tikal represents the first record for El Petén.

Two adult males and one pregnant female were mist netted (20 February, 29 and 25 March) in Tikal along the Uaxactún Road, at a permanent water pool in escobal palm forest, and in an upland deciduous seasonal forest. One male and the female were banded and released.

Tonatia minuta Goodwin, 1942

SPECIMENS EXAMINED—**BELIZE. Cayo:** 1.1 km W Augustine, 1 ♀ (FMNH); Central Farm, at Belize River, 1 ♀ (FMNH); 1.2 km E Macaw Bank, 1 ♀ (FMNH). **Toledo:** Big Fall, 1.7 km NE Río Grande Bridge, 1 ♀ (MSU); San Lucas, 1 ♀ (MSU).

This small *Tonatia* was originally described from the Caribbean coast of Nicaragua as *T. nicaraguae* (Goodwin, 1942a). Its Middle American distribution is lowland (15 to 610 m) along Caribbean and Pacific versants, from southern Veracruz (Lackey, 1970) to El Petén, Guatemala (McCarthy, 1982) and Belize (Disney, 1968), continuing through Honduras (LaVal, 1969; Valdez & LaVal, 1971; Greenbaum & Jones, 1978), Nicaragua (Jones et al., 1971; Greenbaum & Jones, 1978), and Costa Rica (Gardner et al., 1970; LaVal, 1977), to Panama (Davis et al., 1964; Handley, 1966). This account represents additional records for the small round-eared bat in Belize.

Disney (1968) reported no data for the first *Tonatia minuta* specimen from Belize, which was a female collected (25 November) in Cayo District, at Listowel, along the Belize River. This specimen was deposited in British Museum (Natural History). The additional specimens reported here were captured (November, January, February, April, May) over rivers or in a deciduous seasonal forest.

The name *minuta* is applied in accordance with the discussion by McCarthy (1982). Gardner (1976) referred to a personal communication with C. O. Handley, Jr., who suggested that all small *Tonatia* (including *minuta*) represent a single species, *T. brasiliense.* Because the taxonomy is poorly understood, a systematic review of this group would be useful.

Mimon cozumelae Goldman, 1914

SPECIMENS EXAMINED—**BELIZE. Belize:** Churchyard, Sibun River, 1 ♀ (FMNH). **Cayo:** "Mountain Pine Ridge", 2 ♂♂, 1 ♀ (BM); 0.8 km W Augustine, 1 ♂ (CM); 1 km NW Augustine, 2 ♂♂ (FMNH); Barton Creek, at Western Hwy., 2 ♂♂, 3 ♀♀ (FMNH). **Toledo:** vicinity Aguacate, 2 ♂♂, 2 ♀♀ (CM), 1 ♂ (FMNH); Crique Negro, Columbia Forest, 1 ♀ (BM); Pueblo Viejo, 1 ♂, 1 ♀ (FMNH); 2.1 km NNE Salamanca Camp, Columbia Forest, 2 ♂♂ (CM); 2.2 km NNE Salamanca Camp, Columbia Forest, 1 ♂ (CM); vicinity Unión Camp, 2 ♂♂, 1 ♀ (BM), 2 ♀♀ (CM).

This spear-nosed bat ranges from southeastern (northern Oaxaca, southern Veracruz) and peninsular (Yucatán, Quintana Roo) Mexico southeastward along the humid Caribbean side of Central America. Specimens from Isla Cozumel, Quintana Roo, provided the original description for this species (Goldman, 1914). Recorded elevations extend to 495 m. The Belizean localities reported here are the first records for the country.

Mimon cozumelae were collected (January, March, May, July, August, September, December) along the edge of deciduous and semi-evergreen seasonal forests bordered with pasture, on riparian flood plains, over rivers, along paths in high deciduous, semi-evergreen seasonal forests, and in caves.

Schaldach (1964), Villa-R. (1966), and Hall (1981) considered *cozumelae* a subspecies of *bennettii.* I tentatively accept *cozumelae* at the specific level.

Mimon crenulatum keenani Handley, 1960

SPECIMENS EXAMINED—**BELIZE. Cayo:** Listowel, Baking Pot, 1 ♂ (FMNH). **Toledo:** Crique Negro, Columbia Forest, 1 ♂ (USNM).

There are few records for *Mimon crenulatum keenani* from Middle America. The distribution of this distinctive spear-nosed bat extends along the Caribbean versant, from Panama (Handley,

1966; Bonaccorso, 1979), Costa Rica (Gardner et al., 1970; LaVal, 1977), Nicaragua (Greenbaum & Jones, 1978), Belize (Ruiz, 1983), El Petén (McCarthy, 1982), and Campeche (Jones, 1964) to the Gulf lowlands of eastern Chiapas (Medellín L., 1983). All recorded elevations range below 265 m. These specimens are the second and third records from Belize. The first record (Ruiz, 1983) was obtained near Blue Hole, 14 km SE Belmopan, Cayo District.

One *Mimon crenulatum* was captured (8 October) in a house after it flew through an open window. The house was situated along the Belize River in an agricultural area. The second specimen was netted (29 March) along a path in evergreen seasonal forest. E. L. Tyson collected the specimen from Toledo District.

Phyllostomus discolor verrucosus Elliot, 1905

SPECIMENS EXAMINED—**BELIZE. Toledo:** Crique Lagarto, 1 km NW San Antonio, 1 ♂ (FMNH); 1 km NNE Salamanca Camp, Columbia Forest, 1 ♂ (CM). **GUATEMALA. Alta Verapaz:** Lanquin, Lanquin Cave, approx. 149 km WSW Puerto Barrios, 1 ♂, 1 ♀ (FMNH).

Records of *Phyllostomus discolor* extend from southern (Oaxaca, Veracruz) Mexico along both the Pacific and Caribbean corridors of Central America. Records are more common at lower elevations, less than 600 m. The new records from southern Belize provide a limited range extension northward from eastern Izabal (Sanborn, 1936).

An adult from Crique Lagarto was captured (1 January) along the edge of low secondary forest bordering this settlement. The head of the bat was covered with yellow pollen. The second specimen was netted (21 March) in secondary vegetation, which resulted from slash-burn agriculture. Whitish pollen dusted the face, chest, and ventral wing surfaces. A male subadult *Phyllostomus discolor* that was taken (13 July) along a fenceline of secondary vegetation between two pastures, 1.9 km ENE Río Grande Bridge, Big Fall, Toledo District, was photographed, banded, and released. L. de la Torre apparently captured (3 June) the two *Phyllostomus* from Alta Verapaz inside the entrance of Lanquin Cave.

I tentatively follow Jones et al. (1977) in assigning the specimens of *Phyllostomus discolor* from the Caribbean lowlands to the subspecies *verrucosus*. Sanborn (1936, p. 98) recognized *verrucosus* subspecifically, stating the "available measurements of *discolor* would place them much closer to *verrucosus*." He suggested the Panamanian *P. d. discolor* are assignable to *verrucosus* based on larger size. Felten (1956) and Burt and Stirton (1961) concurred with his statement by referring a large series from El Salvador to *verrucosus*; with the availability of greater series of specimens, Davis and Carter (1962) indicated they could not recognize two subspecies of *P. discolor* in Central America and northern South America, acknowledging only *P. d. verrucosus*. Handley (1966) apparently disagreed as he recognized the subspecies *discolor* in Panama. Multivariate analysis of morphological data (Power & Tamsitt, 1973) suggested this species might be monotypic.

Phylloderma stenops septentrionalis Goodwin, 1940

SPECIMENS EXAMINED—**BELIZE. Toledo:** Crique Negro, Columbia Forest, 1 ♀ (USNM); 2.1 km NNE Salamanca Camp, Columbia Forest, 2 ♂♂ (CM).

This rarely encountered species has been recorded north of Panama from the Caribbean coast of Costa Rica (LaVal, 1977), the highlands of Honduras (Goodwin, 1940), the Caribbean lowlands of Guatemala (McCarthy, 1982), and the Gulf lowlands of Chiapas (Carter et al., 1966). Limited elevational data are from lowland to approximately 1320 m. The specimens of *Phylloderma stenops* from Belize represent the eighth, ninth, and tenth specimens north of Panama and the first records from Belize.

All specimens were mist netted (March, December) in similar evergreen seasonal forest habitats. E. L. Tyson collected the specimen from Crique Negro.

Handley (1966) regarded the Panamanian specimens to be *Phylloderma stenops stenops*, and those from northward into Middle America were thought to be subspecifically different from the nominal species. LaVal (1977) did not designate a subspecies for his Costa Rican specimen.

Trachops cirrhosus coffini Goldman, 1925

SPECIMENS EXAMINED—**BELIZE. Orange Walk:** Richmond Hill (Goat Hill), 8.9 km SSW Orange Walk Town, 1 ♂, 1 ♀ (CM). **Toledo:** 2.2 km NNE Salamanca Camp, Columbia Forest, 1 ♂ (CM).

GUATEMALA. Izabal: 25 km SSW Puerto Barrios, 1 ♂ (TCWC).

This lowland subspecies of the fringe-lipped bat is recognized from eastern (southern Veracruz) and southeastern (eastern Oaxaca) Mexico southeastward to Nicaragua. Recorded elevations are from near sea level to approximately 330 m. Jones (1966), Rick (1968), and McCarthy (1982) provided records for El Petén. The description of this subspecies was based on specimens from eastern El Petén (Goldman, 1925). The first Belizean records were reported from Belize District in the vicinity of Belize City (Sanborn, 1941) and Rockstone Pond (Pendergast, 1979). The specimen from Izabal is the first record for that Guatemalan department.

D. C. Carter obtained the single specimen from Izabal on 19 February. The additional Belizean specimens were mist netted (March, April) in deciduous marsh and evergreen forests.

Chrotopterus auritus (Peters, 1856)

SPECIMENS EXAMINED—**BELIZE. Toledo:** vicinity Crique Negro, Columbia Forest, 1 ♀ (FMNH); 1.6 km NNE Salamanca Camp, Columbia Forest, 1 ♀ (FMNH).

Chrotopterus was first reported in Central America from El Salvador (Burt & Stirton, 1961). Subsequently, this carnivorous bat has been recorded from southern (southern Veracruz, northern Oaxaca, Chiapas) and peninsular (Yucatán, Quintana Roo) Mexico southeastward throughout Central America at lowland and upland elevations (40 to over 1880 m). *Chrotopterus auritus* has been reported from Quintana Roo (Jones et al., 1973) and El Petén (Rick, 1968; McCarthy, 1982). These specimens from southern Belize provide the first records for the country.

The Belizean specimens were netted (10 April, 28 July) in an evergreen seasonal forest at ground level along a path and at a height of about 13.7 m over an intermittent stream bed. Both were active during the morning hours, 0418 and 0330, respectively.

The subspecific name *Chrotopterus auritus auritus* has been applied to Middle American populations (Jones et al., 1971). Carter and Dolan (1978) stated the type specimen for *Vampyrus auritus* Peters, 1856, actually was collected in Santa Catarina, Brazil, not in Mexico. The discussion by Carter and Dolan (1978, p. 37) suggested that Peters based his description on one or more specimens from Brazil and compared these with a specimen from an unrecorded locality in Mexico as the "verwandten Art aus Mexico." Handley (1966) doubted that subspecies were recognizable.

Vampyrum spectrum (Linnaeus, 1758)

SPECIMEN EXAMINED—**BELIZE. Toledo:** Santa Elena, 1 ♂ (FMNH).

Two localities in southern Veracruz, Mexico (Goldman, 1917; Navarro L., 1979) are the northwesternmost records of the false vampire bat's Middle American distribution, which continues in Nicaragua (Dobson, 1878; Allen, 1910), Costa Rica (Casebeer et al., 1963; Armstrong, 1969; Gardner et al., 1970; Howell & Burch, 1974; Vehrencamp et al., 1977; LaVal & Fitch, 1977), and Panama (Handley, 1966; Peterson & Kirmse, 1969; Bonaccorso, 1979). Although primarily lowland in distribution, its highest recorded elevation was about 1815 m. The occurrence of *Vampyrum spectrum* in the Caribbean lowlands of Belize is documented by this specimen.

There appears to be no definite record of this carnivorous bat from Guatemala (Jones, 1966). Dobson (1878, p. 471) recorded "Guatemala" as part of the Central American range for *Vampyrum,* but did not list any examined specimens. Alston (1879–1882, p. 39) stated Dobson (pers. comm.) saw specimens from Guatemala, although Alston realized the collector, O. Salvin, had not obtained specimens of *Vampyrum*; hence, the identification of this species is doubtful. Five false vampire bats were mist netted on three separate dates in Parque Nacional Tikal, El Petén. Two females were captured during the dry season (22 and 24 March) in an upland deciduous seasonal forest, in the vicinity of Central Plaza of the archaeological site, and at a permanent water pool in escobal palm forest, 2.6 km SE Central Plaza. Two females and one male were netted during the wet season (22 July) at a location along an archaeological transect in escobal palm forest, 1 km SE Tikal Reservoir. All of these bats were released after being observed, measured, and/or photographed. These individuals provide the first record for Guatemala and, along with the specimen from Belize, bridge an intermittent distribution that now extends northward toward peninsular Mexico.

The *Vampyrum spectrum* from Belize was captured (8 April) during the early morning (0300) in an open field. We were "trapping" *Desmodus rotundus* during a vampire bat control effort in the village. This large bat was captured after it made a number of low passes over horses and mules, which were encircled by mist nets. The bat died while enroute to captivity via an assistant.

The Central American population of *Vampyrum* was described as a distinct subspecies, *V. s. nelsoni* (Goldman, 1914), but Handley (1966) argued that the species was monotypic.

Subfamily GLOSSOPHAGINAE

Glossophaga commissarisi commissarisi Gardner, 1962

SPECIMENS EXAMINED—**BELIZE. Belize:** Rockstone Pond, 2 ♂♂, 3 ♀♀ (ROM). **Toledo:** Aguacate, 1 ♀ (FMNH), 1 ♀ (CM); Big Fall, 1 km SE Río Grande Bridge, 2 ♂♂ (CM); Forest Home, 1 ♀ (FMNH); 2.8 km NNW Punta Gorda, 1 ♀ (FMNH). **GUATEMALA. Izabal:** 25 km SSW Puerto Barrios, 7 ♂♂, 6 ♀♀ (TCWC).

Webster and Jones (1982) summarized the distribution for this subspecies of nectivorous bat, which was documented from eastern (Veracruz) and southern (Oaxaca, Chiapas) Mexico and southern Belize southeastward throughout Central America. Hellebuyck et al. (1985) recently reported records from El Salvador. The specimens from Izabal are the first records from this Guatemalan department. The specimens from Belize District extend northward the distribution of *Glossophaga commissarisi* along the Caribbean lowlands.

According to D. C. Carter's field notes, the majority of the Guatemalan *Glossophaga commissarisi* were mist netted (February, March) over a stream and in the adjacent undisturbed forest. Many of these nectivorous bats were captured in association with night-blooming "bat flowers" bordering on a stream. The Belizean specimens reported (Webster & Jones, 1982) from Lubaantun, Toledo District, were collected (18 April) in a disturbed semi-evergreen seasonal forest. Additional specimens were secured (January, July, September, December) in secondary and orchard vegetation of villages, in riparian secondary vegetation, and from the hollow of a mamey tree (*Pouteria mammosa*).

Subfamily STENODERMATINAE

Uroderma bilobatum molaris Davis, 1968

SPECIMEN EXAMINED—**MEXICO. Quintana Roo:** 2 km N, 8 km W Bacalar, 1 ♂ (TCWC).

Davis (1968) recognized this subspecies of the tent-making bat from the Gulf-Caribbean versant of southern Veracruz, Tabasco, northeast Oaxaca, northern Chiapas, Belize, Honduras, Nicaragua, Costa Rica, and northwest Panama. Disney (1968) and Pendergast (1979) also reported the occurrence of *Uroderma bilobatum* from Belize. The specimen reported here represents the first record for Quintana Roo and a marginal range extension into the Mexican peninsula of Yucatán.

The above specimen was taken in a net on 6 August by M. D. Engstrom along a path leading to an inland lagoon.

Vampyrops helleri helleri Peters, 1866

SPECIMENS EXAMINED—**BELIZE. Cayo:** Banana Bank, 5 ♀♀ (FMNH); 0.8 km W Macaw Bank, 1 ♂ (FMNH). **Toledo:** Big Fall, 1.9 km ENE Río Grande Bridge, 1 ♀ (AMNH), 1 ♀ (CM), 1 ♂ (MSU); Crique Negro, Columbia Forest, 1 ♂ (BM); Forest Home, 1 ♂ (FMNH), 1 ♂ (MSU); Salamanca Camp, 1 ♂ (BM), 1 ♂ (FMNH), 1 ♀ (USNM); 1.8 km NNE Salamanca Camp, Columbia Forest, 1 ♀ (FMNH); vicinity Unión Camp, 1 ♀ (BM), 2 ♀♀ (CM).

The Middle American records of this fruit bat indicate a distribution from sea level to elevations of over 1300 m and a range from southeastern Mexico (southern Veracruz, Oaxaca, Tabasco) throughout Central America. Lowland records have been reported from El Petén (Rick, 1968) and Izabal (Carter et al., 1966). This account constitutes the first records from Belize.

Eighty-seven percent of the *Vampyrops helleri* specimens were captured along or in proximity to waterways. Eleven additional individuals were released at Banana Bank, where a concentration of stenodermatines (*Sturnira, Uroderma, Vampyressa, Chiroderma, Artibeus,* and *Vampyrops*) was observed. The remaining localities were in upland evergreen seasonal forest and in disturbed village vegetation. A specimen in the collection of St. John's College, Belize City, was collected by E. L. Tyson in Columbia Forest.

I follow Dickerman et al. (1981) for the taxonomic assignment of the subspecific epithet.

Vampyrodes caraccioli major G. M. Allen, 1908

SPECIMENS EXAMINED—**BELIZE. Toledo:** Aguacate, 1 ♂ (CM); Big Fall, 1.9 km ENE Río Grande Bridge, 1 ♂ (CM), 1 ♂ (FMNH); Big Fall, 2.1 km E Río Grande Bridge, 1 ♂ (BM); Crique Negro, Columbia Forest, 1 ♂ (BM), 1 ♂, 1 ♀ (MSU); Salamanca Camp, 1 ♂ (USNM); 1.6 km N Salamanca Camp, Columbia Forest, 1 ♂ (FMNH); 2.1 km NNE Salamanca Camp, Columbia Forest, 4 ♂♂, 1 ♀ (CM); San Antonio, 1 ♀ (FMNH).

The published distribution of *Vampyrodes caraccioli major* northwestward of Costa Rica and Panama is confined to the Gulf-Caribbean lowlands as far as southern Mexico (Oaxaca, southern Veracruz, Chiapas); elevational data are less than 300 m. The records from Belize extend the range of this stenodermatine north of Izabal (Sanborn, 1936).

The Belizean localities represent habitats of riparian lowland and upland evergreen seasonal forests and village secondary vegetation. The capture dates cover both the dry and wet seasons (March, April, May, July–September, December).

I follow Carter and Dolan (1978) for the correct spelling of *Vampyrodes caraccioli.*

Vampyressa pusilla thyone Thomas, 1909

SPECIMENS EXAMINED—**BELIZE. Cayo:** 1.6 km NW Augustine, 3 ♂♂, 1 ♀ (CM); Banana Bank, 1 ♀ (FMNH); Blancaneaux, 8.3 km NNE Augustine, 1 ♀ (FSM). **Toledo:** vicinity Aguacate, 1 ♀ (BM), 3 ♀♀ (CM); 1.2 km E Aguacate, 1 ♂, 1 ♀ (CM); Big Fall, 1 km E Río Grande Bridge, 1 ♀ (CM); Big Fall, 2.1 km E Río Grande Bridge, 1 ♂ (CM); Big Fall, 1.9 km ENE Río Grande Bridge, 1 ♂, 1 ♀ (CM), 1 ♀ (FMNH); Crique Negro, Columbia Forest, 1 ♂ (MSU), 1 ♂ (USNM); Forest Home, 1 ♂ (MSU); Pueblo Viejo, 1 ♀ (FMNH); 1.6 km NNE Salamanca Camp, Columbia Forest, 1 ♂, 2 ♀♀ (FMNH).

The general distribution of the little yellow-eared bat extends from southern (Oaxaca, southern Veracruz, Chiapas) and peninsular (Campeche) Mexico and continues southeastward along the Caribbean slope to both the Pacific and Caribbean corridors of southern Nicaragua, Costa Rica, and Panama, into South America. Elevational data are primarily lowland, from sea level up to a recorded 2200 m. Peterson (1966) reported the only record of *Vampyressa pusilla* in Belize, from Rockstone Pond, Belize District. There are also previous records from El Petén (Rick, 1968) and southeastern Campeche (Jones et al., 1973). This account provides additional records of this species.

These specimens of *Vampyressa pusilla* were collected (February–May, July–September, December) in moist habitats, the majority of which were associated directly with riparian vegetation or in village and pasture-edge vegetation situated near rivers. Evergreen seasonal forest provided an upland habitat.

Chiroderma villosum jesupi J. A. Allen, 1900

SPECIMENS EXAMINED—**BELIZE. Cayo:** Banana Bank, 1 ♂, 5 ♀♀ (FMNH). **Corozal:** Chan Chen, 1 ♂ (FMNH). **Toledo:** Big Fall, vicinity Río Grande Bridge, 1 ♂ (FMNH); Big Fall, 1.7 km NE Río Grande Bridge, 1 ♀ (MSU); Big Fall, 1.9 km ENE Río Grande Bridge, 1 ♂, 1 ♀ (CM); San Antonio, 1 ♂ (FMNH); 1 km WNW San Pedro Columbia, 1 ♀ (FMNH). **GUATEMALA. El Petén:** Parque Nacional Tikal, 1 ♂ (FMNH).

The Middle American occurrence of *Chiroderma villosum* has been documented in southern (Oaxaca, southern Veracruz, Chiapas) and peninsular (Campeche, Quintana Roo) Mexico, Guatemala, Nicaragua, Costa Rica, and Panama. Hellebuyck et al. (1985) recently reported this fruit bat from El Salvador. Locality records reach from the coastal lowlands to upland habitats at 1300 m. Southeastern Campeche (Jones et al., 1973) and northern Quintana Roo (Birney et al., 1974) are previous Caribbean lowland localities, in addition to these first records from Belize and El Petén.

All but one of the Belizean *Chiroderma* were associated either directly with or in the vicinity of riparian evergreen or semi-evergreen seasonal forests (April, May, August, September, December). One individual was captured (15 November) in village orchard vegetation. Five additional individuals were released at Banana Bank. The Tikal specimen was captured (24 March) along the permanent water pool mentioned in the *Tonatia evotis* account.

Artibeus toltecus toltecus (Saussure, 1860)

SPECIMENS EXAMINED—**BELIZE. Cayo:** vicinity Augustine, 2 ♂♂, 4 ♀♀ (FSM); 1.6 km NW Augustine, Río Frío, 1 ♂, 1 ♀ (FMNH), 5 ♂♂ (TTU), 4 ♂♂ (CM); "Río On," ? km N Augustine, 1 ♀ (TTU); 1.1 km S Baldy Beacon, Bald Hills, 3 ♀♀ (CM); vicinity San Luis, 7.1 km SSW Augustine, 1 ♀ (TTU). **Toledo:**

Orange Point, 1 ♀ (FMNH); Pueblo Viejo, 3 ♀♀ (FMNH); Unión Camp, 5 ♂♂, 4 ♀♀ (CM).

In his revision of the small *Artibeus* of Middle America, Davis (1969) recognized the range of *Artibeus toltecus toltecus* from southern Tamaulipas, Mexico, southeastward along the mountainous region of the Gulf versant, upland of southern Mexico, Guatemala, Honduras, Nicaragua, and Costa Rica. He did not examine Panamanian specimens. Handley (1966) summarized the Panamanian localities for *A. toltecus*. This bat primarily occurs at elevations between 328 and 1640 m, although elevations near sea level were recorded (Davis, 1969). Consequently, the occurrence of *A. toltecus* in the Maya Mountain range of southern Belize and southeastern El Petén was not unexpected. These Belizean localities represent the first northern Caribbean lowland records.

The Belizean localities range in elevation from near sea level to approximately 720 m. *Artibeus toltecus* is more common at the higher elevations. These dark-colored *Artibeus* were captured (December–February, April, June, September) in habitats of deciduous seasonal forest, semi-evergreen seasonal forest, transitional forest, and pine forest-savanna.

The subspecies *toltecus* is applied, based on the proximity of Belize to its distribution as defined by Davis (1969).

Centurio senex senex Gray, 1842

SPECIMENS EXAMINED—**BELIZE. Belize:** 1.4 km S San Pedro, Ambergris Caye, 1 ♂, 1 ♀ (FMNH). **Cayo:** 1.6 km NW Augustine, Río Frío, 1 ♂ (TTU); vicinity Augustine, Río On, 1 ♀ (TTU); Blancaneaux, 8.3 km NNE Augustine, 1 ♀ (FSM); Central Farm, 1 ♂, 1 ♀ (FMNH); Teakettle, Young Gal Road at Belize River, 1 ♂, 1 ♀ (FMNH); Xunantunich, 1 ♂ (FMNH). **Corozal:** 1.2 km E, 1.6 km N Corozal, 1 ♂ (LSUMZ). **Orange Walk:** 1.6 km NW San Antonio, Río Hondo, 1 ♀ (FMNH). **Toledo:** Big Fall, 1.9 km ENE Río Grande Bridge, 1 ♂ (CM); Crique Negro, Columbia Forest, 1 ♂, 1 ♀ (USNM); Forest Home, 1 ♀ (AMNH); vicinity Unión Camp, 2 ♀♀ (BM), 1 ♀ (CM). **GUATEMALA. Alta Verapaz:** Lanquin, vicinity Lanquin Cave, approx. 149 km WSW Puerto Barrios, 1 ♂ (AMNH). **Izabal:** 25 km SSW Puerto Barrios, 1 ♂, 5 ♀♀ (TCWC).

The recorded distribution of the wrinkle-faced bat extends from western (southern Sinaloa), northeastern (southern Tamaulipas), and peninsular (Campeche and Quintana Roo) Mexico and continues southeastward through Central America at principally lower to upland elevations (sea level to 1882 m). The records given here are the first for Belize, Alta Verapaz, and Izabal.

The distribution of this unusual bat in Belize reflects apparent ecological flexibility. *Centurio senex* has been captured in low littoral forest and mangrove swamp edge on the coastal sand strip of Ambergris Caye, to about 720 m in evergreen and semi-evergreen seasonal forest on the southern slope of the Maya Mountains. Evergreen seasonal and transitional forests, secondary forest, and agriculturally disturbed areas provide additional habitats. This bat was captured throughout the year. Two males and one female were mist netted and released at Orange Point, Toledo District. Brother N. Sullivan collected (15–17 January) the specimen from Alta Verapaz, but I assume the bat was captured outside of Lanquin Cave. The specimens from Izabal were obtained (February, March) by D. C. Carter and field party. Field data are limited, but four *Centurio* were captured over a stream.

Diphylla ecaudata Spix, 1823

SPECIMENS EXAMINED—**BELIZE. Cayo:** vicinity Augustine, 1 ♂ (ROM); San Antonio, 1 ♂ (FMNH). **Toledo:** Crique Jute, 1 ♂ (AMNH); San Antonio, 1 ♀ (FMNH); Santa Elena, 1 ♀ (FMNH).

The distribution of *Diphylla ecaudata* appears primarily restricted along the Gulf side and in the Yucatán Peninsula of Mexico southeastward throughout Central America, where this bat occurs from the coastal lowlands up into the mountainous highlands (1880 m). The hairy-legged vampire bat has been recorded from El Petén (McCarthy, 1982) and Quintana Roo (Jones et al., 1973). The specimens reported here are the first records from Belize.

Four of the localities represent village environments where *Diphylla* was captured (April, July, August, December) along with *Desmodus rotundus* during vampire bat control activities. Mist netting was carried out in direct immediacy to domestic livestock and homes. The feeding activities of *Diphylla* in these villages were not documented, although one blood meal was obtained for analysis. P. Boreham, Imperial College Field Station, England, reported (in litt.) a weak precipitin reaction for a mammal host from the blood meal sample without a response for bird or reptile. It is not known if this blood meal was obtained in the village (Santa Elena). Gardner (1977) summarized the sanguivorous preference of *Diphylla* as for primarily avian hosts. The hairy-legged vampire from Augustine was apparently taken (22 February) in a deciduous seasonal forest.

Family NATALIDAE

Natalus stramineus saturatus
Dalquest and Hall, 1949

SPECIMENS EXAMINED—**BELIZE. Cayo:** 1.6 km NW Augustine, Río Frío, 2 ♂♂, 2 ♀♀ (FSM); 0.8 km W Augustine, 2 ♂♂, 3 ♀♀ (CM); 1.5 km N Augustine, 5 ♀♀ (CM); Sibun Camp, Hummingbird Hwy. at Silver Creek, 1 ♀ (FMNH). **Orange Walk:** Richmond Hill (Goat Hill), 8.9 km SSW Orange Walk Town, 1 ♀ (CM). **Stann Creek:** Kendal, 1 ♂ (FMNH). **Toledo:** vicinity Aguacate, 1 ♂, 3 ♀♀ (CM); 1.2 km E Aguacate, 1 ♀ (CM); Vista Hermosa, 3.7 km WNW Punta Gorda, 8 ♂♂, 6 ♀♀ (FMNH).

The northern range of *Natalus stramineus saturatus* extends from both northwestern (Sinaloa) and northeastern (Nuevo León) Mexico, including the Yucatán Peninsula, southeastward through Central America where the number of records for this species is noticeably reduced beyond Guatemala to Panama. Although predominately a lowland species, elevations were recorded as high as 2400 m. The presence of the funnel-eared bat in Belize was anticipated, as it appears to be well reported throughout the Gulf-Caribbean versant.

Those specimens obtained (April, August, September) at roost sites in Belize were from caves. Other capture localities include low riparian forest and open areas bordering on forest, in orchard habitats, and alongside a building.

Family VESPERTILIONIDAE

Subfamily VESPERTILIONINAE

Myotis elegans Hall, 1962

SPECIMENS EXAMINED—**BELIZE. Belize:** Belize City, Landivar, 1 ♀ (AMNH), 1 ♂, 1 ♀ (FMNH), 1 ♂ (MSU); Mussel Creek, 7.5 km W Burrell Boom, 1 ♂, 1 ♀ (FMNH).

LaVal (1973a) summarized the lowland distribution of *Myotis elegans,* ranging from the Gulf (eastern San Luis Potosí, Veracruz), Pacific coastal (Chiapas), and peninsular (southeastern Campeche) regions of Mexico to Honduras, Nicaragua, and northeastern Costa Rica. Subsequent records were reported from the Pacific side of Costa Rica and the Caribbean lowlands of El Petén (LaVal, 1977; McCarthy, 1982). The majority of elevations are less than 120 m, ranging to 750 m. These additional Caribbean lowland localities are the first records from Belize.

Two elegant *Myotis* were netted (1 July) along a tractor track, in low riparian vegetation dominated by bamboo and thistle palms. Four individuals were obtained (January, February, May, December) at a coastal locality in low vegetation bordering on disturbed mangrove (*Rhizophora mangle, Avicennia germinans*) habitat.

Eptesicus furinalis gaumeri
(J. A. Allen, 1897)

SPECIMENS EXAMINED—**BELIZE. Belize:** Belize City, Landivar, 1 ♀ (CM). **Cayo:** Central Farm, 2 ♀♀ (CM), 5 ♂♂, 16 ♀♀ (FMNH), 2 ♂♂, 1 ♀ (TTU); Little Vaquero Creek, 9.3 km NNW Augustine, 1 ♂, 1 ♀ (FSM); Ontario, 5.5 km W Teakettle, 1 ♀ (FMNH); Teakettle, 1 ♂ (FMNH). **Corozal:** Estero Lagoon, 4 km W Patchakan, 1 ♂, 1 ♀ (FMNH); Santa Clara, 1 ♀ (FMNH). **Orange Walk:** Honey Camp Lagoon, 1 ♂, 2 ♀♀ (FMNH); Tower Hill, B.S.I. compound, 3 ♂♂ (CM), 1 ♂, 4 ♀♀ (FMNH); 2 km SSW Tower Hill Bridge, 1 ♀ (CM). **Stann Creek:** Melinda, 3 ♀♀ (FMNH); Dangriga (Stann Creek), 1 ♂ (USNM). **Toledo:** Orange Creek, 1.5 km S Punta Gorda, 1 ♀ (MSU); Punta Gorda, 1 ♀ (MSU).

The Mexican distribution of *Eptesicus furinalis gaumeri* ranges from the western (Jalisco) and the eastern (San Luis Potosí) versants southeastward to South America. Davis (1965), Disney (1968), and Starrett and Casebeer (1968) reported records from all of the Central American countries except El Salvador. Lowland elevations range from near sea level to 1800 m, the majority being below 500 m. This tropical brown bat has been reported from El Petén (Rick, 1968; McCarthy, 1982) and Quintana Roo (Jones et al., 1973). The localities here are additional records for Belize.

Disney (1968) did not present locality data for his two specimens of *Eptesicus furinalis.* Both were males, captured (16 November, 29 December) in Cayo District, near Central Farm and Esperanza (4.5 km W Central Farm). These are located in British Museum (Natural History). An additional 196 individuals were captured from three of the localities reported here; the majority of these were banded and released during a behavioral study. The majority was found in direct association with buildings, utilizing the infrastructure of the walls or floors and the space behind window shutters as roost sites. Individuals have been taken over water (creeks and a swimming pool) at three localities and in riparian vegetation along two lagoons.

Lasiurus borealis (Müller, 1776)

SPECIMENS EXAMINED—**BELIZE. Orange Walk:** Tower Hill, B.S.I. compound, 1 ♀ (FMNH). **Stann Creek:** 5.3 km NNW Quam Bank, Cockscomb Basin, 1 ♀ (CM). **GUATEMALA. El Petén:** Parque Nacional Tikal, 1 ♀ (FMNH).

The subspecies *teliotis* ranges southward from both the western and eastern regions of Mexico to Oaxaca and the northern Yucatán Peninsula. Specimens of *Lasiurus borealis* from the Guatemalan central highlands were assigned by Jones (1966) to the Central American subspecies *frantzii*, based on Handley (1960). Carter et al. (1966) assigned specimens from both lowland and highland localities in Chiapas to *frantzii*, suggesting that the region of the Isthmus of Tehuantepec represents the break between *frantzii* and *teliotis*. Hall (1981) concurred with this arrangement. Similarly, Jones et al. (1973) suggested that southern Mexico, including the Yucatán Peninsula, may represent a zone of intergradation between *frantzii* and *teliotis*. Few specimens of *L. borealis* are available from El Salvador (Burt & Stirton, 1961), Honduras (Goodwin, 1942b), Nicaragua (Davis & Carter, 1962—as *L. b. teliotis*), Costa Rica (Goodwin, 1946; Gardner et al., 1970) and Panama (Handley, 1966). Recorded elevations (near sea level to about 2540 m) are primarily low or moderate (< 1155 m). Koopman (1959) reported the only record from Quintana Roo. This account represents the first records for Belize and eastern Guatemala from El Petén.

The red bats captured in Belize (April, May) were netted over a stream and a swimming pool. The Tikal specimen was taken (30 July) while it was flying in an open area near a large man-made reservoir.

I hesitate to assign a subspecific designation because I see no practical purpose in doing so until adequate series of specimens from throughout the range of *Lasiurus borealis* become available. Handley (1960) had fewer specimens of *L. borealis* at hand for a proper evaluation of subspecific variation. Consequently, the limits of the distributions for the recognized subspecies remain unresolved.

Lasiurus ega (Gervais, 1855)

SPECIMENS EXAMINED—**BELIZE. Belize:** Tropical Park, Mi. 14.5 Western Hwy., 1 ♂ (FMNH). **Orange Walk:** Tower Hill, B.S.I. compound, 2 ♂♂ (FMNH), 1 ♂ (CM). **Stann Creek:** 5.3 km WNW Quam Bank, Cockscomb Basin, 2 ♂♂, 1 ♀ (CM). **Toledo:** Big Fall, 1.7 km NE Río Grande Bridge, 1 ♀ (CM); Orange Creek, 1.5 km SW Punta Gorda, 1 ♂ (MSU).

Similar to *Lasiurus borealis*, the distribution for the two recognized subspecies of the yellowish bat is not well understood. While *L. e. panamensis* was recognized along the Pacific versant of Chiapas (Baker & Patton, 1967) and Guatemala (Dolan & Carter, 1979; Dickerman et al., 1981), Goodwin (1969) identified *panamensis* from the moderate elevations of the Gulf drainage in northern Oaxaca and suspected *L. e. xanthinus* may occur in the drier Pacific portion of that state. Baker et al. (1971) determined the variation in karyotypes and pelage color of *L. ega* from near Brownsville, Texas, resembled those from eastern coastal and southern Mexico and referred the Texas specimens to *L. e. panamensis*. Meanwhile, *L. e. xanthinus* was recognized in the Yucatán Peninsula (Jones et al., 1973; Birney et al., 1974). The yellow bat is poorly represented from the remainder of Central America, which includes Honduras (Goodwin, 1942b; LaVal, 1969; Greenbaum & Jones, 1978), Costa Rica (Goodwin, 1946; Starrett & Casebeer, 1968; Gardner et al., 1970; LaVal & Fitch, 1977), and Panama (Handley, 1966). Where designated, the subspecies *panamensis* has been applied to these preceding Central American localities, although Hall (1981) did not acknowledge *panamensis* north of Costa Rica. Elevational data are similar to those for *L. borealis*. Ingles (1958) reported two *L. ega* from Quintana Roo. Jones et al. (1973, p. 23) translated Ingles's locality from Spanish as "Puerto Morelos" when it was actually a collection site only 16 km east of the state border with Yucatán, along the highway from Valladolid (Yucatán) to Puerto Morelos (Quintana Roo). Alvarez and Ramírez-P. (1972) cited an additional Caribbean lowland record from southeastern Campeche. This account provides the first *L. ega* records from Belize.

Eight yellowish bats were captured (April, May) over streams, a river, and a swimming pool. Another was netted (18 August) at about 5 m above the ground while circling a building located in grass-sedge savanna.

Lasiurus intermedius intermedius
(H. Allen, 1862)

Specimen Examined—**BELIZE. Toledo:** Crique Jute, 1 ♀ (CM).

The range of this subspecies of the large yellow bat extends southeastward from Mexico to Honduras (Handley, 1960; Carter et al., 1966), El Salvador (Hellebuyck et al., 1985), and Guatemala (Carter et al., 1966). *Lasiurus intermedius* has been recorded in Mexico from the northern Yucatán Peninsula and Chiapas northwestward to Texas along the eastern coast and to Sinaloa on the Pacific side. Recorded elevations range from lowland to highland (1620 m) habitats. A single specimen of *L. intermedius* from northern Quintana Roo (Birney et al., 1974) provided the only record for that Mexican state. This Belizean specimen represents the first record for the country.

The above specimen was obtained on 30 March over the stream Crique Jute surrounded by secondary vegetation.

Bauerus dubiaquercus (Van Gelder, 1959)

Specimens Examined—**BELIZE. Cayo:** 1.6 km NW Augustine, Río Frío, 1 ♀ (ROM). **Toledo:** 2.1 km NNE Salamanca Camp, Columbia Forest, 1 ♂ (CM).

The published localities of the rarely encountered *Bauerus dubiaquercus* are scattered from the Islas Tres Marías (Nayarit), Jalisco, and southern Veracruz in Mexico to eastern Honduras and Costa Rica (Engstrom & Wilson, 1981; Dinerstein, 1985). Mainland elevations range from approximately 460 to 1450 m and appear to represent mid-elevation and montane forest habitats (Pine, 1966; Pine et al., 1971; Engstrom & Wilson, 1981; Dinerstein, 1985). These first occurrences of *Bauerus* in Belize extend northward a scattered distribution along the northern Caribbean lowlands in Central America.

J. Kamstra and J. Fragoso collected (8 July) one specimen inside the main Río Frío cave, located in a deciduous seasonal forest at approximately 410 m. The second *Bauerus* was netted (26 March) along an open forestry track in an evergreen forest at about 180 m.

Engstrom and Wilson (1981) and Martin and Schmidly (1982) evaluated the taxonomic status of *Antrozous* (*Bauerus*) *dubiaquercus* and concluded the chromosomal, cranial, postcranial, and phallic differences between this bat and *Antrozous* (*Antrozous*) *pallidus* were sufficient to recognize *Bauerus* as a distinct genus. I follow their conclusions and agree that the species is monotypic since the mainland sample size that previously was assigned to *A. d. meyeri* Pine, 1971, was limited to a total of five specimens representing both sexes.

Family MOLOSSIDAE

Eumops auripendulus auripendulus (Shaw, 1800)

Specimens Examined—**BELIZE. Orange Walk:** Orange Walk Town, 1 ? (CM); Tower Hill, B.S.I. compound, 1 ♀ (FMNH).

The recorded distribution of *Eumops auripendulus auripendulus* includes both moist uplands and drier lowland coastal and plateau areas, ranging from eastern Oaxaca, Tabasco, Quintana Roo, and Belize, through Guatemala, western Honduras, El Salvador, western Nicaragua, Costa Rica, and Panama, into South America (Eger, 1974; Greenbaum & Jones, 1978). Villa-R. (1956) and Eger (1974), respectively, reported this free-tailed bat from Quintana Roo and Belize (Belize District: Rockstone Pond). This account provides the second and third records for Belize.

The Orange Walk specimen consists of a mandible and partial skull, which were recovered from an owl (*Tyto alba*) roost in a church tower. The second specimen was discovered (July) alive by L. G. Hoevers, after it apparently was attacked by a bird.

Eumops bonariensis nanus (Miller, 1900)

Specimens Examined—**BELIZE. Orange Walk:** Orange Walk Town, 2 ?? (CM).

Eger (1977) summarized the few available Middle American localities for this small mastiff bat which are limited to southeastern Mexico (southern Veracruz, Tabasco, Yucatán), eastern Honduras, and Panama. These and additional localities in Panama (Dolan & Carter, 1979) and Nicaragua (Hall, 1981) are restricted to coastal lowland environments. This is the first recording of *Eumops bonariensis* for Belize.

Entire specimens of *Eumops bonariensis* as yet are unavailable from Belize. Documentation is based on two fragmented sets of maxillary toothrows, which were sifted from regurgitated rubble beneath an owl (*Tyto alba*) roost in a church tower.

Molossus ater nigricans (Miller, 1902)

SPECIMENS EXAMINED—**GUATEMALA. Izabal:** 25 km SSW Puerto Barrios, 2 ♂♂, 4 ♀♀ (TCWC).

The black mastiff bat is a common inhabitant of roof spaces throughout its lowland Middle American range, from western (Sinaloa) and eastern (Tamaulipas) Mexico southeastward into South America. This species has been reported from the Caribbean lowlands of Quintana Roo (Jones et al., 1973) and Belize (Murie, 1935; Pendergast, 1979). These *Molossus ater* from Izabal are the first record for that department. Apparently, all of these specimens were collected on 15 February over a stream by D. C. Carter and his field party.

Addendum

While this paper was in press, other papers and additional information concerning bats in Belize came to my attention. Two recent papers provide new country records of the glossophagines *Lichonycteris obscura* (Hill, 1985) and *Hylonycteris underwoodi* (McCarthy & Blake, 1987), which increase the known bat fauna to 68 species. Both records are from Toledo District. McCarthy and Blake (1987) also reported the occurrence of the following bats: *Rhynchonycteris naso, Saccopteryx bilineata, Balantiopteryx io, Noctilio leporinus, Pteronotus parnellii, Micronycteris megalotis, M. nicefori, M. schmidtorum, Tonatia evotis, Mimon cozumelae, Phyllostomus discolor, Trachops cirrhosus, Chrotopterus auritus, Vampyrum spectrum, Glossophaga soricina, Carollia brevicauda, C. perspicillata, Sturnira lilium, Uroderma bilobatum, Vampyressa pusilla, Artibeus jamaicensis, A. lituratus, A. phaeotis, A. watsoni, Centurio senex, Natalus stramineus, Rhogeessa tumida, Bauerus dubiaquercus,* and *Eumops underwoodi.*

HILL, J. E. 1985. The status of *Lichonycteris degener* Miller, 1931 (Chiroptera: Phyllostomidae). Mammalia, **49**(4): 579–582.

MCCARTHY, T. J., AND M. BLAKE. 1987. Noteworthy bat records from the Maya Mountains Forest Reserve, Belize. Mammalia, **51**(1): 161–164.

A bat specimen from Stann Creek was reported (Miller & Allen, 1928, p. 180) as *Myotis nigricans nigricans.* I examined this specimen at USNM and found it to be an immature *Eptesicus furinalis.* It is included in this paper under the species account for the latter species.

MILLER, G. S., JR., AND G. M. ALLEN. 1928. The American bats of the genera *Myotis* and *Pizonyx.* Bulletin of the United States National Museum, **144:** 1–218.

Silva-Taboada and Koopman (1964, p. 3) reported specimens of *Tadarida laticaudata* (= *Nyctinomops laticaudatus*) from Corozal District. Most of the bat species discussed in an unpublished dissertation by A. M. Cartwright were also reported by Kirkpatrick et al. (1975) and Cartwright and Kirkpatrick (1977). The remaining identifications (Cartwright, 1977, pp. 240, 242–246, 250, 251), which were from Belize District, included *Rhynchonycteris naso, Saccopteryx bilineata, Carollia brevicauda, C. perspicillata, Sturnira lilium, Artibeus lituratus, A. phaeotis, Desmodus rotundus, Eptesicus furinalis, Rhogeessa tumida,* and *Molossus molossus.* Baker et al. (1985, p. 236) reported cytogenetic data from specimens of *Rhogeessa tumida* that I collected in Belize District.

BAKER, R. J., J. W. BICKHAM, AND M. L. ARNOLD. 1985. Chromosomal evolution in *Rhogeessa* (Chiroptera: Vespertilionidae): Possible speciation by centric fusions. Evolution, **39**(2): 233–243.

CARTWRIGHT, A. M. F. 1977. Patterns of Neotropical chiropteran reproduction including histological and ecological aspects of bats collected in Belize. Ed.D. diss., Ball State University, Muncie, Ind., 278 pp.

SILVA-TABOADA, G., AND K. F. KOOPMAN. 1964. Notes on the occurrence and ecology of *Tadarida laticaudata yucatanica* in eastern Cuba. American Museum Novitates, **2175:** 1–6.

A mammal checklist was included in a resource paper on Belize (Hartshorn et al., 1984). The list of bats supposedly was a compilation of known and expected species. The result is inaccurate and undocumented. The reader is referred to the checklist in the Appendix as correct.

HARTSHORN, G., ET AL. 1984. Belize. Country Environmental Profile. A Field Study. United States

Agency for International Development, San José, Costa Rica, 151 pp.

Additional specimens of *Micronycteris megalotis* (2: Belize District, Cayo District), *Artibeus toltecus* (11: Cayo District), *Centurio senex* (2: Cayo District), and *Diphylla ecaudata* (3: Cayo District) were found in the mammal collection of Royal Ontario Museum. D. J. Tallman collected specimens of *M. megalotis* (1) and *Mimon cozumelae* (2) from Orange Walk District, which were deposited in Bell Museum of Natural History, University of Minnesota. I secured further voucher specimens (AMNH) of *Mimon cozumelae* (1), *Vampyrodes caraccioli* (1), *Vampyressa pusilla* (1), and *Bauerus dubiaquercus* (1) from Toledo District.

Certain specimens (*Sturnira lilium* and *Rhogeessa tumida*) that were catalogued by Dobson (1878, pp. 540, 246) were listed as collected in "Honduras." These were obtained by D. Dyson and H. Cuming between November 1844 and late 1845. During that time, "Honduras" corresponded to the present region that extends from southern Quintana Roo, Mexico, southeastward to northern Honduras. Many early collectors did not differentiate between the area of Belize ("British settlement in Honduras") and that of the Republic of Honduras ("Spanish Honduras"), but recorded only "Honduras" or "Bay of Honduras" without further locality data. The above specimens did not originate from present day Honduras, but were collected in Belize. Additional specimens of *Micronycteris megalotis* (Dobson, 1878, p. 479) from the "Bay of Honduras" and *Rhynchonycteris naso* (Dobson, 1878, p. 368) from "Honduras" remain orphaned records of the historical literature.

Uroderma bilobatum was reported (Sanchez-H. et al., 1986) from southern Quintana Roo while this volume was delayed. Ten specimens were collected at Ruinas de Kohunlich (18°23′N; 88°42′W), about 16 km W Estevez on the Belizean border. Four other species (*Pteronotus davyi, Mormoops megalophylla, Tonatia evotis, T. minuta*) were documented for the first time from Quintana Roo, from localities within 35 km of the northern border of Belize. The known bat fauna of Quintana Roo is now represented by 36 species.

SANCHEZ-H., O., G. TELLEZ-G., R. A. MEDELLÍN, AND G. URBANO-V. 1986. New records of mammals from Quintana Roo, México. Mammalia, **50**(2): 275–278.

Acknowledgments

Grateful acknowledgments are due H. Flowers, O. Rosado, R. Belisle, and E. O. Bradley, Department of Forestry, Belize, and H. Topsey, W. Branche, E. A. Graham, and J. Palacio, Department of Archaeology, Belize, for the necessary permission to conduct fieldwork. The Department of Agriculture, Belize, provided opportunities and support while I headed the Vampire Bat Education and Control Program (1975–1978). Division of Mammals, Field Museum of Natural History, provided partial equipment needs (1974) and partial funding (1979). Mellon North American Mammal Research Institute, O'Neil Fund, and Section of Mammals, Carnegie Museum of Natural History, supported further fieldwork (1982, 1984). Numerous colleagues and friends have been of direct assistance to my needs in Belize, and the following are but a few: E. Awe, P. Boreham, H. Burns, W. D. Burton, J. Cab, G. Cal, G. C. Canto, J. Chun, M. Craig, L. de la Torre, L. Dieckman, R. Foster, P. W. Freeman, P. Gamble, E. García, J. P. and M. García, H. H. Genoways, E. Gomez, L. Henderson, R. W. Henderson, L. G. and E. Hoevers, A. M. Hutson, R. J. Izor, T. Juring, E. King, K. Leacock, D. Owen-Lewis, N. MacKenzie, G. T. McCarthy, H. Pastor, R. H. Pine, A. Rabinowitz, M. L. Reed, C. J. Rushin, K. and T. Salisbury, R. Schmidt, B. M. Silva, M. F. Valentine, L. Waight, L. Wilkins, and A. C. S. Wright. Various regiments and the Royal Air Force of Her Majesty's British Forces provided certain logistical support over the years.

My fieldwork in Parque Nacional Tikal, El Petén, Guatemala, was made possible while assisting D. J. Howell and J. G. Cant during their respective visits to the park. M. Dary-R., Universidad de San Carlos, and F. Polo-Sifontes, Instituto de Antropología y Historia, graciously assisted and permitted my work in 1979. J. R. Martínez R. kindly provided a copy of his thesis.

The courtesies rendered to me by the personnel of the following institutions while providing either information, the loan of specimens, or assistance during visits are greatly appreciated: American Museum of Natural History; British Museum (Natural History); Carnegie Museum of Natural History; Field Museum of Natural History; Florida State Museum, University of Florida; Museum of Natural History, University of Kansas; Museum of Zoology, Louisiana State University; The Museum, Michigan State University; Royal Ontario Museum; Texas Cooperative Wildlife Collection, Texas A & M University; The Museum, Texas Tech University; United States National Museum of Natural History. I am grateful to N. A. Bitar, F. J. Bonaccorso, D. C. Carter, M. D. Engstrom, A. M. Hutson, J. Kamstra, R. L. Peterson, and E. L. Tyson for sharing information concerning their field efforts in either Belize, El Petén, Izabal, or Quintana Roo.

Besides J. R. Choate, J. K. Jones, Jr., K. F. Koopman, and R. M. Timm, two anonymous reviewers enhanced earlier manuscripts with their comments and editorial skills. E. Méndez kindly reviewed my Spanish abstract. Verification of the state boundaries in the Yucatán Peninsula was provided by T. Dachtera, National Geographic Society, and G. de la Torre M., Secretaría de Programación y Presupuesto, México. M. A. Schmidt, R. A. Rollin, V. Risoli, and S. Cramer typed various versions of the manuscript. S. M. Vélez guided the manuscript during my absence. This report is a contribution of the Mammals of Belize Program.

Literature Cited

ALLEN, J. A. 1910. Additional mammals from Nicaragua. Bulletin of the American Museum of Natural History, **28:** 87–115.

ALSTON, E. R. 1879–1882. Biologia Centrali-Americana, Mammalia. Taylor and Francis, London, xx + 220 pp., 22 pls.

ALVAREZ, T., AND J. RAMÍREZ-P. 1972. Notas acerca de murciélagos mexicanos. Anales de la Escuela de Ciencias Biológicas, México, **19:** 167–178.

ANDERSEN, K. 1908. A monograph of the chiropteran genera *Uroderma, Enchisthenes,* and *Artibeus.* Proceedings of the Zoological Society of London, **1908:** 204–319.

ARMSTRONG, D. M. 1969. Noteworthy records of bats from Costa Rica. Journal of Mammalogy, **50**(4): 808–810.

BAKER, R. J., AND J. K. JONES, JR. 1975. Additional records of bats from Nicaragua, with a revised checklist of Chiroptera. Occasional Papers of The Museum, Texas Tech University, **32:** 1–13.

BAKER, R. J., T. MOLLHAGEN, AND G. LOPEZ. 1971. Notes on *Lasiurus ega.* Journal of Mammalogy, **52**(4): 849–852.

BAKER, R. J., AND J. L. PATTON. 1967. Karyotypes and karyotypic variation of North American vespertilionid bats. Journal of Mammalogy, **48**(2): 270–286.

BEARD, J. S. 1944. Climax vegetation in tropical America. Ecology, **25**(2): 127–158.

BIRNEY, E. C., J. B. BOWLES, R. M. TIMM, AND S. L. WILLIAMS. 1974. Mammalian distributional records in Yucatán and Quintana Roo, with comments on reproduction, structure, and status of peninsular populations. Occasional Papers, Bell Museum of Natural History, University of Minnesota, **13:** 1–25.

BONACCORSO, F. J. 1979. Foraging and reproductive ecology in a Panamanian bat community. Bulletin of the Florida State Museum, Biological Sciences, **24:** 359–408.

BURT, W. H., AND R. A. STIRTON. 1961. The mammals of El Salvador. Miscellaneous Publications of the Museum of Zoology, University of Michigan, **117:** 1–69.

CARTER, D. C., AND P. G. DOLAN. 1978. Catalogue of type specimens of Neotropical bats in selected European museums. Special Publications of The Museum, Texas Tech University, **15:** 1–136.

CARTER, D. C., R. H. PINE, AND W. B. DAVIS. 1966. Notes on Middle American bats. Southwestern Naturalist, **11**(4): 488–499.

CARTWRIGHT, A. M., AND R. D. KIRKPATRICK. 1977. A range extension of *Peropteryx kappleri* (family Emballonuridae) in Central America. Proceedings of the Indiana Academy of Sciences, **86:** 466.

CASEBEER, R. S., R. B. LINSKY, AND C. E. NELSON. 1963. The phyllostomid bats, *Ectophylla alba* and *Vampyrum spectrum,* in Costa Rica. Journal of Mammalogy, **44**(2): 186–189.

DAVIS, W. B. 1965. Review of the *Eptesicus brasiliensis* complex in Middle America with the description of a new subspecies from Costa Rica. Journal of Mammalogy, **46**(2): 229–240.

———. 1968. Review of the genus *Uroderma* (Chiroptera). Journal of Mammalogy, **49**(4): 676–698.

———. 1969. A review of the small fruit bats (genus *Artibeus*) of Middle America. Southwestern Naturalist, **14**(1): 15–29.

———. 1970a. A review of the small fruit bats (genus *Artibeus*) of Middle America. Part II. Southwestern Naturalist, **14**(4): 389–402.

———. 1970b. The large fruit bats (genus *Artibeus*) of Middle America, with a review of the *Artibeus jamaicensis* complex. Journal of Mammalogy, **51**(1): 105–122.

———. 1973. Geographic variation in the fishing bat, *Noctilio leporinus.* Journal of Mammalogy, **54**(4): 862–874.

———. 1984. Review of the large fruit-eating bats of the *Artibeus* "*lituratus*" complex (Chiroptera: Phyllostomidae) in Middle America. Occasional Papers of The Museum, Texas Tech University, **93:** 1–16.

DAVIS, W. B., AND D. C. CARTER. 1962. Notes on Central American bats with description of a new subspecies of *Mormoops.* Southwestern Naturalist, **7**(1): 64–74.

———. 1978. A review of the round-eared bats of the *Tonatia silvicola* complex, with descriptions of three new taxa. Occasional Papers of The Museum, Texas Tech University, **53:** 1–12.

DAVIS, W. B., D. C. CARTER, AND R. H. PINE. 1964. Noteworthy records of Mexican and Central American bats. Journal of Mammalogy, **45**(3): 375–387.

DICKERMAN, R. W., K. F. KOOPMAN, AND C. SEYMOUR. 1981. Notes on bats from the Pacific lowlands of Guatemala. Journal of Mammalogy, **62**(2): 406–411.

Dinerstein, E. 1985. First records of *Lasiurus castaneus* and *Antrozous dubiaquercus* from Costa Rica. Journal of Mammalogy, **66**(2): 411–412.

Disney, R. H. L. 1968. Observations on a zoonosis: Leishmaniasis in British Honduras. Journal of Applied Ecology, **5**(1): 1–59.

Dobson, G. E. 1878. Catalogue of the Chiroptera in the Collection of the British Museum. British Museum, London, xlii + 567 pp., 30 pls.

Dolan, P. G., and D. C. Carter. 1979. Distributional notes and records for Middle American Chiroptera. Journal of Mammalogy, **60**(3): 644–649.

Eger, J. L. 1974. A new subspecies of the bat *Eumops auripendulus* (Chiroptera: Molossidae), from Argentina and eastern Brazil. Life Sciences Occasional Papers, Royal Ontario Museum, **25:** 1–8.

———. 1977. Systematics of the genus *Eumops* (Chiroptera: Molossidae). Life Sciences Contributions, Royal Ontario Museum, **110:** 1–69.

Engstrom, M. D., and D. E. Wilson. 1981. Systematics of *Antrozous dubiaquercus* (Chiroptera: Vespertilionidae), with comments on the status of *Bauerus* Van Gelder. Annals of Carnegie Museum, **50**(15): 371–383.

Felten, H. 1956. Fledermäuse (Mammalia, Chiroptera) aus El Salvador. Senckenbergiana Biologica, **37**(3–4): 179–212.

Fleming, T. H., E. T. Hooper, and D. E. Wilson. 1972. Three Central American bat communities: Structure, reproductive cycles, and movement patterns. Ecology, **53**(4): 555–569.

Gardner, A. L. 1976. The distributional status of some Peruvian mammals. Occasional Papers of the Museum of Zoology, Louisiana State University, **48:** 1–18.

———. 1977. Feeding habits, pp. 293–350. *In* Baker, R. J., J. K. Jones, Jr., and D. C. Carter, eds., Biology of bats of the New World family Phyllostomatidae. Part II. Special Publications of The Museum, Texas Tech University, **13:** 1–364.

Gardner, A. L., R. K. LaVal, and D. E. Wilson. 1970. The distributional status of some Costa Rican bats. Journal of Mammalogy, **51**(4): 712–729.

Gaumer, G. F. 1917. Monografía de los mamíferos de Yucatán. Departamento de Talleres Gráficos de la Secretaría de Fomento, México, xli + 331 pp., 1 map.

Goldman, E. A. 1914. A new bat of the genus *Mimon* from Mexico. Proceedings of the Biological Society of Washington, **27:** 75–76.

———. 1917. New mammals from North and Middle America. Proceedings of the Biological Society of Washington, **30:** 107–116.

———. 1925. A new bat of the genus *Trachops* from Guatemala. Proceedings of the Biological Society of Washington, **38:** 23–24.

Goodwin, G. G. 1940. Three new bats from Honduras and the first record of *Enchisthenes harti* (Thomas) for North America. American Museum Novitates, **1075:** 1–3.

———. 1942a. A summary of recognizable species of *Tonatia,* with descriptions of two new species. Journal of Mammalogy, **23**(2): 204–209.

———. 1942b. Mammals of Honduras. Bulletin of the American Museum of Natural History, **79:** 107–195.

———. 1946. Mammals of Costa Rica. Bulletin of the American Museum of Natural History, **87:** 275–473.

———. 1969. Mammals from the state of Oaxaca, Mexico, in the American Museum of Natural History. Bulletin of the American Museum of Natural History, **141:** 1–269.

Greenbaum, I. F., and J. K. Jones, Jr. 1978. Noteworthy records of bats from El Salvador, Honduras, and Nicaragua. Occasional Papers of The Museum, Texas Tech University, **55:** 1–7.

Hall, E. R. 1981. The Mammals of North America, vol. 1. John Wiley and Sons, New York, xvii + 600 + 90 pp.

Handley, C. O., Jr. 1960. Descriptions of new bats from Panama. Proceedings of the United States National Museum, **112**(3442): 459–479.

———. 1966. Checklist of the mammals of Panama, pp. 753–795. *In* Wenzel, R. L., and V. J. Tipton, eds., Ectoparasites of Panama. Field Museum of Natural History, Chicago, 861 pp.

———. 1976. Mammals of the Smithsonian Venezuelan Project. Brigham Young University Science Bulletin, Biological Series, **20**(5): 1–91.

———. 1980. Inconsistencies in formation of family-group and subfamily-group names in Chiroptera, pp. 9–13. *In* Wilson, D. E., and A. L. Gardner, eds., Proceedings Fifth International Bat Research Conference. Texas Tech Press, Lubbock, 434 pp.

Harrison, D. L., and N. Pendleton. 1974. A second record of Wied's long-legged bat (*Macrophyllum macrophyllum* Schinze, 1821, Chiroptera: Phyllostomatidae) in El Salvador, with notes on the palate, reproduction, and diet of the species. Mammalia, **38**(4): 689–693.

Hellebuyck, V., J. R. Tamsitt, and J. G. Hartman. 1985. Records of bats new to El Salvador. Journal of Mammalogy, **66**(4): 783–788.

Hershkovitz, P. 1951. Mammals from British Honduras, Mexico, Jamaica, and Haiti. Fieldiana: Zoology, **31**(47): 547–569.

Howell, D. J. 1977. Time sharing and body partitioning in bat-plant pollination systems. Nature, **270:** 509–510.

Howell, D. J., and D. Burch. 1974. Food habits of some Costa Rican bats. Revista de Biología Tropical, **21**(2): 281–294.

Ingles, L. G. 1958. Notas acerca de los mamíferos mexicanos. Anales del Instituto de Biología, México, **29:** 379–408.

Jones, J. K., Jr. 1964. Bats from western and southern Mexico. Transactions of the Kansas Academy of Science, **67**(3): 509–516.

———. 1966. Bats from Guatemala. University of Kansas Publications, Museum of Natural History, **16**(5): 439–472.

Jones, J. K., Jr., and T. E. Lawlor. 1965. Mammals from Isla Cozumel, Mexico, with description of a new species of harvest mouse. University of Kansas Publications, Museum of Natural History, **16**(3): 409–419.

JONES, J. K., JR., J. D. SMITH, AND H. H. GENOWAYS. 1973. Annotated checklist of mammals of the Yucatan Peninsula, Mexico. I. Chiroptera. Occasional Papers of The Museum, Texas Tech University, **13:** 1–31.

JONES, J. K., JR., J. D. SMITH, AND R. W. TURNER. 1971. Noteworthy records of bats from Nicaragua, with a checklist of the chiropteran fauna of the country. Occasional Papers of the Museum of Natural History, University of Kansas, **2:** 1–35.

JONES, J. K., JR., P. SWANEPOEL, AND D. C. CARTER. 1977. Annotated checklist of the bats of Mexico and Central America. Occasional Papers of The Museum, Texas Tech University, **47:** 1–35.

KIRKPATRICK, R. D., A. M. CARTWRIGHT, J. C. BRIER, AND E. J. SPICKA. 1975. Additional mammal records for Belize. Mammalia, **39**(2): 330–331.

KOOPMAN, K. F. 1959. The zoogeographical limits of the West Indies. Journal of Mammalogy, **40**(2): 236–240.

LACKEY, J. A. 1970. Distributional records of bats from Veracruz. Journal of Mammalogy, **51**(2): 384–385.

LAVAL, R. K. 1969. Records of bats from Honduras and El Salvador. Journal of Mammalogy, **50**(4): 819–822.

———. 1973a. A revision of the Neotropical bats of the genus *Myotis.* Natural History Museum of Los Angeles County, Science Bulletin, **15:** 1–54.

———. 1973b. Systematics of the genus *Rhogeessa* (Chiroptera: Vespertilionidae). Occasional Papers of the Museum of Natural History, University of Kansas, **19:** 1–47.

———. 1977. Notes on some Costa Rican bats. Brenesia, **10/11:** 77–83.

LAVAL, R. K., AND H. S. FITCH. 1977. Structure, movements and reproduction in three Costa Rican bat communities. Occasional Papers of the Museum of Natural History, University of Kansas, **69:** 1–27.

LUNDELL, C. L. 1934. Preliminary sketch of the phytogeography of the Yucatan Peninsula. Carnegie Institute of Washington, **436:** 257–321.

———. 1937. The vegetation of Peten. Carnegie Institute of Washington, **478:** 1–207.

MARTIN, C. O., AND D. J. SCHMIDLY. 1982. Taxonomic review of the pallid bat, *Antrozous pallidus* (Le Conte). Special Publications of The Museum, Texas Tech University, **18:** 1–48.

MARTINEZ R., J. R. 1980. Detección de virus rabico en cerebros de murciélagos de la cuenca del Río Polochic, Alta Verapaz e Izabal. Tésis de grado, Facultad de Medicina Veterinaria y Zootecnia, Universidad de San Carlos, Guatemala, 54 pp.

MCCARTHY, T. J. 1982. Bat records from the Caribbean lowlands of El Petén, Guatemala. Journal of Mammalogy, **63**(4): 683–685.

MEDELLÍN L., R. A. 1983. *Tonatia bidens* and *Mimon crenulatum* in Chiapas, Mexico. Journal of Mammalogy, **64**(1): 150.

MEDELLÍN L., R. A., D. NAVARRO L., W. B. DAVIS, AND V. J. ROMERO. 1983. Notes on the biology of *Micronycteris brachyotis* (Dobson) (Chiroptera) in southern Veracruz, Mexico. Brenesia, **21:** 7–11.

MILLER, G. S., JR. 1912. A small collection of bats from Panama. Proceedings of the United States National Museum, **42**(1882): 21–26.

MURIE, A. 1935. Mammals from Guatemala and British Honduras. Miscellaneous Publications of the Museum of Zoology, University of Michigan, **26:** 1–30.

NAVARRO L., D. 1979. *Vampyrum spectrum* (Chiroptera, Phyllostomatidae) in Mexico. Journal of Mammalogy, **60**(2): 435.

NELSON, C. E. 1965. *Lonchorhina aurita* and other bats from Costa Rica. Texas Journal of Science, **17:** 303–306.

OJASTI, J., AND O. J. LINARES. 1971. Adiciones a la fauna de murciélagos de Venezuela con notas sobre los especies del género *Diclidurus* (Chiroptera). Acta Biológica Venezolana, **7**(4): 421–441.

PENDERGAST, D. M. 1971. Excavations at Eduardo Quiroz Cave, British Honduras (Belize). Occasional Paper, Art and Archaeology, Royal Ontario Museum, **21:** 1–123.

———. 1979. Excavations at Altun Ha, Belize, 1964–1970, vol. 1. Royal Ontario Museum, Toronto, 226 pp.

PENNINGTON, T. D., AND J. SARUKHAN. 1968. Manual para la identificación de campo de los principales arboles tropicales de México. Instituto Nacional de Investigaciones Forestales y Organización de las NN.UU. para la Agricultura y la Alimentación, México, D.F., 413 pp.

PETERSON, R. L. 1966. The genus *Vampyressa* recorded from British Honduras. Journal of Mammalogy, **46**(4): 676.

PETERSON, R. L., AND P. KIRMSE. 1969. Notes on *Vampyrum spectrum,* the false vampire bat, in Panama. Canadian Journal of Zoology, **47**(1): 140–142.

PINE, R. H. 1966. *Baeodon meyeri* (Chiroptera: Vespertilionidae) a new species of bat from Veracruz. Southwestern Naturalist, **11**(2): 308–310.

———. 1972. The bats of the genus *Carollia.* Technical Monograph of the Texas Agricultural Experimental Station, Texas A & M University, **8:** 1–125.

PINE, R. H., D. C. CARTER, AND R. K. LAVAL. 1971. Status of *Bauerus* Van Gelder and its relationships to other nyctophiline bats. Journal of Mammalogy, **52**(4): 663–669.

PORTIG, W. H. 1976. The climate of Central America, pp. 405–478. *In* Schwerdtfeger, W., ed., Climates of Central and South America. Elsevier Scientific Publ. Co., New York, 532 pp.

POWER, D. M., AND J. R. TAMSITT. 1973. Variation in *Phyllostomus discolor* (Chiroptera: Phyllostomatidae). Canadian Journal of Zoology, **51**(4): 461–468.

QUINONES, E., J. P. KOPLAN, L. PIKE, F. STAINE, AND L. AJELLO. 1978. Histoplasmosis in Belize, Central America. American Journal of Tropical Medicine and Hygiene, **27**(3): 558–561.

RICK, A. M. 1968. Notes on bats from Tikal, Guatemala. Journal of Mammalogy, **49**(3): 516–520.

RUIZ, R. G. 1983. *Mimon crenulatum keenani* (Chiroptera: Phyllostomidae) from Belize. Southwestern Naturalist, **28**(3): 374.

SALVIN, O. 1864. A fortnight amongst the sea birds of British Honduras. Ibis, **1864:** 372–387.

SANBORN, C. C. 1935. New mammals from Guatemala and Honduras. Field Museum of Natural History, Zoology Series, **20**(11): 81–85.

———. 1936. Records and measurements of Neotropical bats. Field Museum of Natural History, Zoology Series, **20**(13): 93–106.

———. 1937. American bats of the subfamily Emballonurinae. Field Museum of Natural History, Zoology Series, **20**(24): 321–354.

———. 1941. Descriptions and records of Neotropical bats, pp. 371–387. *In* Papers on mammalogy. Published in honor of Wilfred Hudson Osgood. Field Museum of Natural History, Zoological Series, **27:** 1–395.

———. 1949. Bats of the genus *Micronycteris* and its subgenera. Fieldiana: Zoology, **31**(27): 215–233.

SANDERSON, I. T. 1941. Living Treasure. The Viking Press, New York, 290 pp.

SCHALDACH, W. J., JR. 1964. Notas breves sobre algunos mamíferos del sur de México. Anales del Instituto de Biología, México, **35:** 129–137.

SMITH, J. D. 1972. Systematics of the chiropteran family Mormoopidae. Miscellaneous Publications of the Museum of Natural History, University of Kansas, **56:** 1–132.

STANDLEY, P. C., AND S. J. RECORD. 1936. The forests and flora of British Honduras. Field Museum of Natural History, Botanical Series, **12:** 1–432.

STARRETT, A. 1976. Comments on bats newly recorded from Costa Rica. Natural History Museum of Los Angeles, Contributions in Science, **277:** 1–5.

STARRETT, A., AND R. S. CASEBEER. 1968. Records of bats from Costa Rica. Los Angeles County Museum, Contributions in Science, **148:** 1–21.

TAIBEL, A. M. 1977. Mammiferi del Guatemala, con speciale riguardo alla regione del Petèn, raccolti dal Maggio al Settembre 1932. Atti Societa Italiana di Scienze Naturali, Museo Civico di Storia Naturale di Milano, **118**(3–4): 379–401.

VALDEZ, R., AND R. K. LAVAL. 1971. Records of bats from Honduras and Nicaragua. Journal of Mammalogy, **52**(1): 247–250.

VEHRENCAMP, S. L., F. G. STILES, AND J. W. BRADBURY. 1977. Observations on the foraging behavior and avian prey of the Neotropical carnivorous bat, *Vampyrum spectrum.* Journal of Mammalogy, **58**(4): 469–478.

VILLA-R., B. 1956. Otros murciélagos nuevos para la fauna de México. Anales del Instituto de Biología, México, **26:** 543–545.

———. 1966. Los murciélagos de México. Universidad Nacional Autónoma de México, Instituto de Biología, México, D.F., xvi + 491 pp.

WADELL, H. A. 1938. Physical-geological features of Peten, Guatemala. Carnegie Institute of Washington, **437:** 336–348.

WALKER, S. H. 1973. Summary of climatic records for Belize. Land Resources Division, Surrey, England, Supplementary Report, **3:** 1–145.

WEBSTER, W. D., AND J. K. JONES, JR. 1982. A new subspecies of *Glossophaga commissarisi* (Chiroptera: Phyllostomidae) from western Mexico. Occasional Papers of The Museum, Texas Tech University, **76:** 1–6.

WEST, R. C. 1964. Surface configuration and associated geology of Middle America, pp. 33–83. *In* West, R. C., ed., Handbook of Middle American Indians, vol. 1. University of Texas Press, Austin, 570 pp.

WRIGHT, A. C. S., D. H. ROMNEY, R. H. ARBUCKLE, AND V. E. VIAL. 1959. Land in British Honduras. Report of the British Honduras Land Use Survey Team. Colonial Research Publication, London, **24:** 1–327.

Gazetteer

The numbers for localities are plotted in Figure 1.

MEXICO

QUINTANA ROO

1. Bacalar 18°43′N; 88°22′W

GUATEMALA

EL PETÉN

2. Parque Nacional Tikal 17°20′N; 89°39′W
3. Poptún 16°21′N; 89°26′W

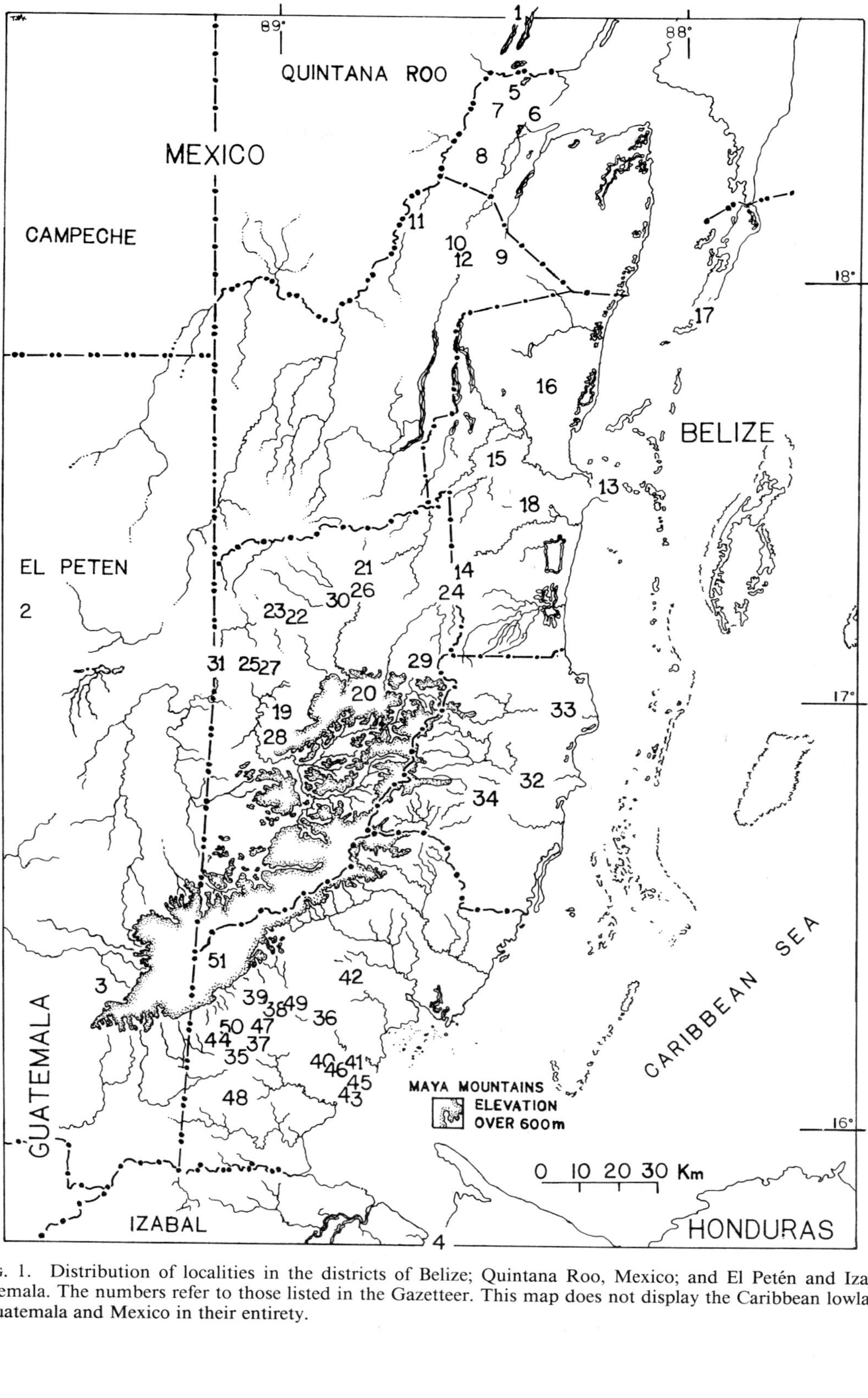

FIG. 1. Distribution of localities in the districts of Belize; Quintana Roo, Mexico; and El Petén and Izabal, Guatemala. The numbers refer to those listed in the Gazetteer. This map does not display the Caribbean lowlands of Guatemala and Mexico in their entirety.

Izabal

4. Puerto Barrios 15°43′N; 88°36′W

BELIZE

Corozal District

5. Chan Chen 18°26′N; 88°27′W
6. Corozal 18°24′N; 88°24′W
7. Patchakan 18°24′N; 88°29′W
8. Santa Clara 18°18′N; 88°30′W

Orange Walk District

9. Honey Camp Lagoon 18°03′N; 88°27′W
10. Orange Walk Town 18°05′N; 88°34′W
11. San Antonio, Río Hondo 18°11′N; 88°39′W
12. Tower Hill, Belize Sugar Industries (B.S.I.) 18°02′N; 88°34′W

Belize District

13. Belize City 17°30′N; 88°12′W
14. Churchyard 17°18′N; 88°33′W
15. Mussel Creek 17°39′N; 88°24′W
16. Rockstone Pond (Altun Ha) 17°46′N; 88°22′W
17. San Pedro, Ambergris Caye 17°55′N; 87°58′W
18. Tropical Park 17°28′N; 88°23′W

Cayo District

19. Augustine 16°58′N; 88°59′W
20. Baldy Beacon, Bald Hills 17°01′N; 88°47′W
21. Banana Bank 17°15′N; 88°48′W
22. Barton Creek at Western Hwy. 17°13′N; 89°57′W
23. Central Farm and Listowel 17°11′N; 89°00′W
24. C.I.T.A., Sibun River at Indian Creek 17°16′N; 88°34′W
25. Macaw Bank 17°05′N; 89°04′W
26. Roaring Creek 17°15′N; 88°47′W
27. San Antonio 17°05′N; 89°01′W
28. San Luis 16°54′N; 89°00′W
29. Sibun Camp 17°05′N; 88°39′W
30. Teakettle 17°13′N; 88°51′W
31. Xunantunich 17°05′N; 89°08′W

Stann Creek District

32. Kendal 16°49′N; 88°22′W
33. Melinda 17°00′N; 88°18′W
34. Quam Bank, Cockscomb Basin 16°47′N; 88°28′W

Toledo District

35. Aguacate 16°10′N; 89°06′W
36. Big Fall 16°15′N; 88°53′W
37. Blue Creek 16°12′N; 89°03′W
38. Crique Jute and Salamanca Camp (Forestry Camp) 16°16′N; 89°01′W
39. Crique Negro, Columbia Forest 16°17′N; 89°02′W
40. Cuevas Creek and Jacinto Creek Bridges, at Punta Gorda Road 16°09′N; 88°53′W
41. Forest Home 16°08′N; 88°50′W
42. Nimli Punit 16°20′N; 88°48′W
43. Orange Point 16°04′N; 88°49′W
44. Pueblo Viejo 16°13′N; 89°09′W
45. Punta Gorda 16°07′N; 88°48′W
46. Rice Station (Agricultural Station) 16°08′N; 88°51′W
47. San Antonio 16°15′N; 88°02′W
48. San Lucas (deserted) 16°05′N; 89°06′W
49. San Pedro Columbia 16°17′N; 88°58′W
50. Santa Elena 16°14′N; 89°06′W
51. Unión Camp 16°24′N; 89°08′W

Appendix

This district checklist of the bat fauna of Belize is based on published accounts. The citations refer to the initial taxonomic treatments of specimens. Districts are arranged from north (left) to south (right). Abbreviations are as follows: Cz = Corozal; OW = Orange Walk; Bz = Belize; Cy = Cayo; SC = Stann Creek; Td = Toledo.

In order to give an accurate list of bats, it is necessary to present certain discrepancies that have appeared in the literature. Dobson (1878) referred to certain early specimens that may have originated from Belize. One reference (Dobson, 1878,

p. 520) to "Half-Moon Key, Honduras" for a specimen of *Artibeus perspicillatus* (= *jamaicensis*) belongs to Belize, since this specimen was collected by O. Salvin (see Salvin, 1864). Sanderson (1941, p. 228) recalled "*Anoura* sp." in his descriptive narrative of a visit to then British Honduras. This species was not identified (Hershkovitz, 1951) in the Sanderson bat collection. *Diaemus youngi* was cited from Belize (Villa-R., 1966, p. 340), but R. L. Peterson (pers. comm.) stated that the specimen in question was actually from Guyana. Specimens of 17 species of bats were listed in Disney (1968) without locality data. These specimens, which are housed in British Museum (Natural History) and Royal Ontario Museum, are all from Cayo District. J. L. Eger (pers. comm.) identified the questionable specimen of *Molossus bondae* in Disney (1968, p. 7) as *M. molossus*. Quinones et al. (1978, p. 559) reported six species, which I collected and identified, without the exact locality information other than "the Maya Mountains region." This locality is 1 km NW Augustine, Cayo District. Sixty-six bat species are recognized in Belize.

Species	Cz	OW	Bz	Cy	SC	Td	References
Rhynchonycteris naso	...	...	...	X	...	...	Murie, 1935, pp. 17–18; Disney, 1968, p. 7
Saccopteryx bilineata	...	X	X	X	X	...	Sanborn, 1937, p. 331; Hershkovitz, 1951, p. 553; Disney, 1968, p. 7; Pendergast, 1979, p. 10
Saccopteryx leptura	...	...	...	...	...	X	This paper
Peropteryx kappleri	...	...	...	X	...	...	Cartwright & Kirkpatrick, 1977, p. 466
Peropteryx macrotis	...	...	...	...	X	...	Hershkovitz, 1951, p. 553
Centronycteris maximilliani	...	...	...	...	X	...	Sanborn, 1941, p. 372
Balantiopteryx io	...	...	X	X	...	...	Kirkpatrick et al., 1975, p. 330; Cartwright & Kirkpatrick, 1977, p. 466
Diclidurus virgo	...	...	...	X	...	...	This paper
Noctilio leporinus	...	X	...	X	X	X	This paper
Pteronotus davyi	...	X	...	X	...	X	Disney, 1968, p. 7; this paper
Pteronotus parnellii	...	...	X	X	X	...	Hershkovitz, 1951, pp. 553–554; Smith, 1972, p. 74; Cartwright & Kirkpatrick, 1977, p. 466; Quinones et al., 1978, p. 559; Pendergast, 1979, p. 10
Pteronotus personatus	...	...	...	...	...	X	This paper
Mormoops megalophylla	...	...	X	X	X	X	This paper
Micronycteris brachyotis	...	...	...	X	...	X	This paper
Micronycteris megalotis	X	X	...	...	...	X	This paper
Micronycteris nicefori	...	...	...	...	...	X	This paper
Micronycteris schmidtorum	X	X	...	...	...	X	This paper
Lonchorhina aurita	...	...	...	...	X	X	This paper
Macrophyllum macrophyllum	...	...	...	X	...	X	This paper
Tonatia bidens	...	...	...	X	...	X	This paper
Tonatia evotis	...	...	...	X	X	...	Sanborn, 1941, pp. 372–373; Disney, 1968, p. 7
Tonatia minuta	...	...	...	X	...	X	Disney, 1968, p. 7; this paper
Mimon cozumelae	...	...	X	X	...	X	This paper
Mimon crenulatum	...	...	...	X	...	X	Ruiz, 1983, p. 374; this paper
Phyllostomus discolor	...	...	...	...	...	X	This paper
Phylloderma stenops	...	...	...	...	...	X	This paper
Trachops cirrhosus	...	X	X	...	...	X	Sanborn, 1941, p. 374; Pendergast, 1979, p. 10; this paper
Chrotopterus auritus	...	...	...	...	...	X	This paper
Vampyrum spectrum	...	...	...	...	...	X	This paper
Glossophaga commissarisi	...	...	X	...	...	X	Webster & Jones, 1982, p. 5; this paper
Glossophaga soricina	...	...	X	X	...	...	Villa-R., 1966, p. 231; Disney, 1968, p. 7; Cartwright & Kirkpatrick, 1977, p. 466; Howell, 1977, p. 510; Quinones et al., 1978, p. 559; Pendergast, 1979, p. 10

Continued on next page

Species	Cz	OW	Bz	Cy	SC	Td	References
Carollia brevicauda	X	X	X	X	X	...	Hershkovitz, 1951, pp. 555–556; Villa-R., 1966, p. 270; Disney, 1968, p. 7; Pine, 1972, p. 42; Quinones et al., 1978, p. 559; Pendergast, 1979, p. 10
Carollia perspicillata	...	...	X	X	X	...	Hershkovitz, 1951, p. 555; Disney, 1968, p. 7; Pine, 1972, p. 42; Quinones et al., 1978, p. 559; Pendergast, 1979, p. 10
Sturnira lilium	...	...	X	X	...	...	Disney, 1968, p. 7; Pendergast, 1979, p. 10
Uroderma bilobatum	...	...	X	X	X	...	Disney, 1968, p. 7; Davis, 1968, p. 697; Pendergast, 1979, p. 10
Vampyrops helleri	...	...	...	X	...	X	This paper
Vampyrodes caraccioli	...	...	...	...	...	X	This paper
Vampyressa pusilla	...	...	X	X	...	X	Peterson, 1966, p. 676; this paper
Chiroderma villosum	X	...	...	X	...	X	This paper
*Artibeus intermedius**	X	...	...	...	...	...	Davis, 1984, p. 14
*Artibeus jamaicensis**	X	X	X	X	...	...	Dobson, 1878, p. 520; Andersen, 1908, p. 263; Gaumer, 1917, p. 298; Disney, 1968, p. 7; Davis, 1970b, p. 118; Pendergast, 1971, p. 102
*Artibeus lituratus**	...	...	X	X	X	...	Murie, 1935, p. 19; Hershkovitz, 1951, pp. 556–557; Disney, 1968, p. 7; Pendergast, 1979, p. 10
Artibeus phaeotis	X	X	X	X	...	...	Disney 1968, p. 7; Davis, 1970a, p. 398; Pendergast, 1979, p. 10
Artibeus toltecus	...	...	...	X	...	X	This paper
Artibeus watsoni	...	X	X	...	...	...	Davis, 1970a, p. 393; Pendergast, 1979, p. 10
Centurio senex	X	X	X	X	...	X	This paper
Desmodus rotundus	...	...	X	X	X	...	Hershkovitz, 1951, p. 557; Disney, 1968, p. 7; Cartwright & Kirkpatrick, 1977, p. 466; Quinones et al., 1978, p. 559; Pendergast, 1979, p. 10
Diphylla ecaudata	...	...	...	X	...	X	This paper
Natalus stramineus	...	X	...	X	X	X	This paper
Thyroptera tricolor	...	...	...	...	X	...	Sanborn, 1941, p. 382
Myotis elegans	...	...	X	...	...	...	This paper
Myotis keaysi	...	...	...	X	X	...	Hershkovitz, 1951, pp. 557–558; LaVal, 1973a, p. 25; Quinones et al., 1978, p. 559
Eptesicus furinalis	X	X	X	X	X	X	Disney, 1968, p. 7; this paper
Lasiurus borealis	...	X	...	...	X	...	This paper
Lasiurus ega	...	X	X	...	X	X	This paper
Lasiurus intermedius	...	...	...	...	...	X	This paper
Rhogeessa tumida	...	...	X	X	...	...	LaVal, 1973b, p. 29; Kirkpatrick et al., 1975, p. 331
Bauerus dubiaquercus	...	...	...	X	...	X	This paper
Nyctinomops laticaudatus	...	...	...	X	...	...	Murie, 1935, p. 19
Eumops auripendulus	...	X	X	...	...	...	Eger, 1974, p. 5; this paper
Eumops bonariensis	...	X	...	...	...	...	This paper
Eumops glaucinus	...	...	...	X	...	...	Eger, 1977, p. 42
Eumops underwoodi	X	...	...	...	...	...	Eger, 1977, p. 55
Molossus ater	...	...	X	X	...	...	Murie, 1935, p. 19; Pendergast, 1979, p. 10
Molossus molossus	...	...	...	X	...	...	Murie, 1935, p. 19
Molossus sinaloae	...	...	X	X	X	...	Hershkovitz, 1951, p. 559; Disney, 1968, p. 7

* Davis (1984) examined the *Artibeus* "*lituratus*" complex in Middle America and restored *Artibeus intermedius* J. A. Allen to specific status. Specimens cited in the publications listed for *A. jamaicensis* and *A. lituratus* should be reevaluated.

New Species of Mammals from Northern South America: Fruit-Eating Bats, Genus *Artibeus* Leach

Charles O. Handley, Jr.

ABSTRACTS

The larger species of *Artibeus* of the Amazon Basin are defined, and a new giant species is named and described from Venezuela and Colombia. *Artibeus fallax, A. hercules,* and *A. planirostris* are regarded as subspecies of *Artibeus jamaicensis,* by far the most variable of the larger *Artibeus* of the region.

The smaller *Artibeus* are keyed and arranged in six species groups. A new dwarf species is described from Brazil, Ecuador, Guyana, Peru, and Venezuela. Distribution and diversity of the smaller species are discussed. *Artibeus cinereus,* once thought to range throughout Central America and much of South America and to include all of the smaller taxa except *A. concolor* and *A. hartii,* is restricted to include only the nominate form and *A. quadrivittatus* of the lower Amazon Basin and adjacent coastal areas.

With these additions and changes in status, at least nine species of *Artibeus* now are known to occur in northeastern South America.

Las especies de gran tarnaño de *Artibeus* de la Cuenca del Río Amazonas son definidas y una nueva especie gigante de Venezuela y Colombia es nombrada y descrita. *Artibeus fallax, A. hercules,* y *A. planirostris* son consideradas como subespecies de *Artibeus jamaicensis,* que es el más variable de los grandes *Artibeus* de la región.

Una clave es preparada para las especies de *Artibeus* menores, y las especies son arregladas en seis grupos. Una nueva especie enana de Brasil, Ecuador, Guyana, Perú, y Venezuela es descrita. La distribución y la diversidad de las especies menores son discutidas. *Artibeus cinereus,* que antes se pensó estaba distribuida en Centro América y una gran parte de Sudamérica, y que incluyera todas las taxa más pequeñas (a excepción de *A. concolor* y *A. hartii*), es ahora restringuida para incluir solamente la especie nominal y *A. quadrivittatus* a la Cuenca baja del Río Amazonas y a las areas costeras adyacentes.

Con estas adiciones y cambios de "status," por lo menos nueve especies de *Artibeus* ya son conocidas y se encuentran en el nordeste de Sudamérica.

São definidas as espécies maiores de *Artibeus* que ocorrem na Bacia Amazônica, e uma espécie nova, gigante, é descrita. *Artibeus fallax, A. hercules,* e *A. planirostris* são consideradas subespécies de *Artibeus jamaicensis,* certamente a espécie mais variável dos *Artibeus* maiores da região.

Uma chave para os *Artibeus* menores, os quais foram designados a seis grupos de espécies, é fornecida. Uma espécie nova aña é descrita do Brasil, Equador, Guiana, Peru, e Venezuela. A diversidade, e as distribuições geográficas destas espécies, são discutidas. *Artibeus cinereus,* o qual acreditava-se abranger toda América Central e grande parte da América do Sul, além de incluir todos taxa menores com excessão de *A. concolor* e *A. hartii,* é reduzido a um único taxon, restrito ao sul da Bacia Amazônica e às suas áreas adjacentes.

Incluindo as adições e mudanças de status propostas neste trabalho, são reconhecidas, atualmente, ao menos nove espécies de *Artibeus* na região nordeste da América do Sul.

From the National Museum of Natural History, Smithsonian Institution, Washington, D.C. 20560.

Introduction

Mammals and their ectoparasites were collected in Venezuela between 1965 and 1968 by the Smithsonian Venezuelan Project (SVP), supported in part by a contract (DA-49-MD-2788) of the Medical Research and Development Command, Office of the Surgeon General, U.S. Army. Numerous papers have described the ectoparasites and mammals of the Project. Throughout these papers undescribed species of mammals have been referred to by alphabetical designations. Some of these have been named subsequently by Handley and Ferris (1972), Handley and Gordon (1980), and Handley (1984). This paper deals with fruit-eating bats of the genus *Artibeus* Leach.

The cranial measurements reported here were taken as outlined by Handley (1959, p. 98). Hind foot, tibia, calcar, and forearm were measured on dry museum specimens or on specimens preserved in alcohol; all other external dimensions were measured on fresh specimens in the field. All measurements are in millimeters. Coloration was determined under Examolites (Macbeth Corp., Newburg, NY 12533) with natural light excluded.

A New Giant *Artibeus*

It is now generally agreed that in and around the Amazon Basin there are three large species of *Artibeus*. Handley (1976) recognized them as: (1) *A. fuliginosus* Gray—smaller, molars 3/3, rostrum arched, postorbital process poorly developed, fur long, coloration blackish, facial stripes faint or absent, interfemoral membrane (IM) naked; (2) *A. jamaicensis* Leach—larger, molars 3/3, rostrum arched, postorbital process poorly developed, fur short, coloration gray-brown, facial stripes present but not sharply defined, IM naked; and (3) *A. lituratus* Olfers—larger, molars 2/3, rostrum flattish, postorbital process well developed, fur short, coloration chocolate brown, facial stripes prominent and well defined, IM hairy. However, as shown by Koopman (1978) and Honacki et al. (1982), there is no consensus on the delimitation of these species.

The difficulty in defining the species arises primarily from the fact that *Artibeus jamaicensis* is unusually variable geographically in morphology. The other species show very little variation in this region. *Artibeus jamaicensis* is large in the Amazon Basin, so large in fact that the subspecies there, *A. j. fallax* Peters and *A. j. hercules* Rehn, until recently have been aligned by most authors with the universally large *A. lituratus,* although they differ from it in many characteristics other than size. To the southeast of the Amazon Basin (*A. j. planirostris* Spix) and to the north of it (*A. j. trinitatis* Andersen), *A. jamaicensis* is dramatically smaller, in fact similar in size to *A. fuliginosus.* Everywhere east of the Andes *A. jamaicensis* has 3/3 molars; west of the Andes and in Central America it has 2/3 molars.

Specimens in the SVP collection show that the large *Artibeus jamaicensis fallax* and small *A. j. trinitatis* apparently intergrade in the Llanos of Venezuela where the habitat is marginal for *A. jamaicensis* and where it is an uncommon bat. Furthermore, intergradation between the small, 12-molar *A. j. trinitatis* and the slightly larger, 10-molar *A. j. aequatorialis* Andersen of the northwest coast of South America can be seen in specimens from northern Colombia.

These two zones of intergradation are of crucial importance in the nomenclature of *Artibeus,* for they serve to link "*A. jamaicensis*" of the West Indies and Central America and "*A. planirostris*" of eastern South America. They are especially important in the present context because of the discovery of a fourth large *Artibeus,* superficially similar to but larger than *A. j. fallax,* occurring together with it in southern Venezuela and with the small *A. j. trinitatis* in western Venezuela and northern Colombia. It can be recognized as follows:

Artibeus amplus new species

HOLOTYPE—USNM 440932, adult female with suckling young, skin and skull, collected 15 April 1968 by Norman E. Peterson, F. P. Brown, Jr., and J. O. Matson at Kasmera, 21 km SW Machiques, Estado Zulia, Venezuela, 270 m, in a damp cave in a cliff across the Río Yasa from the Kasmera Biological Station, eastern foothills of the Sierra de Perijá. Original number, SVP 22086.

ETYMOLOGY—Latin *amplus,* large, referring to the large size of this bat, one of the largest *Artibeus.*

DISTRIBUTION—Northern foothills of the Central Andes in Colombia; lower eastern slopes of the Sierra de Perijá and the Venezuelan Andes in western Venezuela; and the vicinity of Cerro Duida and the low mountains of southeastern Bolívar in southern Venezuela. It probably occurs in adjacent parts of Guyana and Brazil as well. The SVP collectors found *A. amplus* near streams and in other moist areas (98%); in evergreen forest (90%) and in forest openings such as yards and orchards (10%). Most specimens were mist netted (86%), but some (14%) were found roosting in the twilight zone of caves. Elevational range, 24–1200 m. Holdridge life zones (Ewel & Madriz, 1968): Tropical humid forest (10%), Tropical very humid forest (22%), Premontane humid forest (12%), Premontane very humid forest (2%), Premontane rain forest (4%), Lower montane very humid forest (10%), and Lower montane rain forest (40%). Ridge slopes and valley floor in the area where the holotype was collected were clothed with second growth evergreen forest, while lawns, shrubbery, banana and papaya plants, and scattered grapefruit trees characterized the grounds of the biological station.

DESCRIPTION—Size large (forearm 70.0, greatest length of skull 31.3, maxillary toothrow 11.2—averages of Venezuelan specimens). Coloration of fur as in sympatric *Artibeus jamaicensis* (dorsum blackish brown to brown; facial stripes present but obscure; underparts blackish brown, usually frosted with white; underarms with abundant long, usually whitish hairs); ears dark fuscous to black, paler basally; lips and noseleaf blackish; membranes blackish; wing tips undifferentiated or grayish, never white. Horseshoe of noseleaf bound down mediobasally; legs and interfemoral membrane slightly hairy, the latter particularly medioventrally, where hairs extend as a short fringe beyond edge of membrane; forearm long.

Skull superficially like that of *Artibeus jamaicensis,* but relatively longer and narrower; rostrum long and flattish; supraorbital ledges subparallel and together with postorbital processes often poorly developed or even ill-defined; zygomata not very flared from skull, usually subparallel to one another, and in side view, thin and fragile; posterolateral angle of skull not particularly flared; palate relatively narrow and toothrows ovoid in outline; postpalatal extension usually long, narrow, and parallel sided; dentition as in *A. jamaicensis,* except I^1 only weakly bilobed; dental formula 2/2–1/1–2/2–3/3 × 2 = 32. This bat is the only known host of *Strebla paramirabilis* Wenzel and *Trichobius assimilis* Wenzel (Diptera: Streblidae), so it can be distinguished from other *Artibeus* by its parasites as well as its morphology.

Measurements of the holotype, an adult female: total length 101, tail vertebrae 0, hind foot (dry) 17, ear from notch 25, forearm 69.2, tibia 24.1, calcar 6.2, weight 70.4 g. Greatest length of skull 31.9, zygomatic breadth 18.3, postorbital breadth 7.7, breadth of braincase 13.3, depth of braincase 11.6, length of maxillary toothrow 11.2, postpalatal length 9.8, palatal breadth outside of M^3 12.9, rostral breadth at base of canines 8.3. See Table 1 for additional measurements.

COMPARISONS—Four large species of *Artibeus* occur in Venezuela, all of them together in the southern part of the country. Among these, *Artibeus amplus* and *A. jamaicensis* are most alike; but despite the superficial resemblance, the two can be distinguished by many characters, both external and cranial. All *A. amplus* examined have the lower edge of the noseleaf horseshoe bound down, while about 95% of *A. jamaicensis* from the same localities have it free; all *A. amplus* have the interfemoral membrane slightly hairy and fringed medially, but *A. jamaicensis* never does; and while *A. jamaicensis* often has the wings white-tipped, *A. amplus* never does. Cranially, *A. amplus* differs from *A. jamaicensis* in having a longer, narrower skull; longer, somewhat more flattened rostrum (most easily seen in dimensions of rostral shield); less arched nasals; margins of supraorbital nearly parallel, rather than converging posteriorly, and usually not as well developed; zygomata thinner and more fragile and usually subparallel rather than diverging markedly posteriorly; posterolateral angle of skull not so flaring; palate narrower and toothrows usually less nearly circular in outline; and postpalatal extension usually longer and narrower, parallel sided (not flaring posteriorward). The two species are hosts of different species of parasitic streblid flies.

SPECIMENS EXAMINED—Total 55. **COLOMBIA. Antioquia:** La Tirana, 33 km SW Zaragoza, 520 m (2 USNM). **VENEZUELA. Apure:** Nulita, Selvas de San Camilo, 29 km SSW Santo Domingo, 24 m (2 USNM). **Bolívar:** 21 to 33 km NE Icabarú, 775–851 m (6 USNM); Km 125, 85 km SSE El Dorado, 826–1165 m (5 USNM). **T.F. Amazonas:** Belén, Río Cunucunuma, 56 km NNW Esmeralda, 150 m (9 USNM); Cabecera del Caño Culebra, Cerro Duida, 40 km NNW Esmeralda, 1140–1200 m (21 USNM); Caño Culebra, Cerro Duida, 50 km NNW Esmeralda, 800 m (2 USNM); Tamatama, Río Orinoco, 2 km above Boca del Casiquiare, 135 m (2 USNM). **Zulia:** Kasmera, 21 km SW Machiques, 270 m (3 USNM, 1 UCV); 15 km W Machiques (1 AMNH); Novito, 19 km WSW Machiques, 1135 m (1 USNM).

REMARKS—In previous publications of SVP, *Artibeus amplus* has been known as "*Artibeus* sp. D".

Table 1. Measurements of adult *Artibeus amplus* and *A. jamaicensis*. For each measurement, line 1 includes the mean plus or minus two standard errors, line 2 the extremes, and line 3, in parentheses, the number of specimens measured. All specimens are from Venezuela unless otherwise stated.

Total length	Hind foot (dry)	Ear	Forearm	Greatest length	Zygomatic breadth	Postorbital breadth
Artibeus amplus, males and females, Zulia and Colombia						
100.4 ± 3.88	18.4 ± 0.60	23.7 ± 1.28	70.8 ± 1.78	31.4 ± 0.26	18.6 ± 0.20	7.9 ± 0.20
93–104	17–19	22–26	68.6–75.3	31.0–31.9	18.1–18.8	7.6–8.3
(5)	(7)	(7)	(7)	(7)	(7)	(7)
A. amplus, males and females, T.F. Amazonas and Bolívar						
89.9 ± 2.22	18.3 ± 0.30	23.0 ± 0.98	69.1 ± 0.90	31.2 ± 0.24	18.4 ± 0.14	7.8 ± 0.10
80–100	17–20	18–26	65.0–73.2	30.3–32.8	17.4–19.1	7.3–8.4
(21)	(22)	(21)	(22)	(29)	(30)	(31)
A. jamaicensis, females, Zulia						
86.4 ± 2.90	15.9 ± 0.40	22.6 ± 0.96	61.1 ± 0.64	27.7 ± 0.24	16.9 ± 0.30	6.8 ± 0.08
77–95	14–18	17–25	58.9–64.2	26.7–28.5	16.0–17.7	6.5–7.1
(19)	(19)	(19)	(19)	(19)	(12)	(19)
A. jamaicensis, males, Zulia						
83.8 ± 4.14	15.4 ± 0.32	22.2 ± 0.98	59.3 ± 0.92	27.4 ± 0.14	17.0 ± 0.28	6.8 ± 0.14
73–91	15–16	20–25	56.2–61.4	27.1–27.7	16.6–17.6	6.3–7.0
(10)	(10)	(10)	(10)	(10)	(6)	(10)
A. jamaicensis, females, T.F. Amazonas						
88.1 ± 1.62	17.8 ± 0.32	24.6 ± 0.46	66.8 ± 1.12	30.7 ± 0.28	19.1 ± 0.14	7.5 ± 0.10
80–93	16–19	23–26	62.1–70.1	29.4–31.3	18.7–19.6	7.1–7.7
(17)	(17)	(17)	(17)	(14)	(13)	(14)
A. jamaicensis, males, T.F. Amazonas						
86.3 ± 1.60	18.3 ± 0.32	24.6 ± 0.52	65.4 ± 1.42	30.7 ± 0.26	19.3 ± 0.32	7.6 ± 0.16
82–90	18–19	24–26	62.4–68.6	30.2–31.4	18.6–20.2	7.2–7.9
(8)	(8)	(8)	(8)	(8)	(8)	(8)

A New Dwarf *Artibeus*

The taxonomy of the smaller *Artibeus* is in a state of flux. As recently as 35 years ago all of the smaller species except *A. concolor* Peters and *A. hartii* Thomas were believed to be variants of *A. cinereus* Gervais. Since then, first one and then another of the supposed subspecies of *A. cinereus* has been shown to be independent species. Today only *A. bogotensis* Andersen, *A. glaucus* Thomas, *A. pumilio* Thomas, *A. quadrivittatus* Peters, *A. rosenbergi* Thomas, and *A. watsoni* Thomas remain associated with *A. cinereus* (Honacki et al., 1982). However, except for *A. quadrivittatus,* these do not properly belong with *A. cinereus* either.

Artibeus glaucus and *A. bogotensis* intergrade in Ecuador and form an Andean-northern South American species sympatric with *A. cinereus* in southern Venezuela. *Artibeus glaucus* thus has two subspecies, the nominate form and *A. g. bogotensis. Artibeus watsoni* Thomas of northwestern South America and Central America is closely related, but intergradation with *A. g. glaucus* or *A. g. bogotensis* has not been observed.

Artibeus pumilio is an enigmatic taxon. Many museum specimens bear the name *A. pumilio,* but perhaps the only specimen properly associated with the name is the holotype. This specimen may be only an odd variant of one of the other species, but not of the species described here. For the present, *A. pumilio* must be regarded as unplaceable. The same can be said for *A. rosenbergi,* characterized by a curiously long, narrow skull such as can be found occasionally in large samples of most species of *Artibeus*. Because of their equivocal status, neither *A. pumilio* nor *A. rosenbergi* is included in the appended list of species and key. The characteristics and status of these taxa will be the subject of another paper.

Thus, *A. cinereus* now has been shorn of all of its supposed subspecies except *A. c. quadrivittatus.* Its supposed range has been reduced from encompassing most of Central America and tropical South America to occupying only the Amazon Basin (possibly only the lower basin) and adjacent coastal areas. Sympatric with *A. cinereus* in much of its range is a distinctive dwarf species which can be known as:

TABLE 1. *Continued.*

Braincase breadth	Braincase depth	Maxillary toothrow	Postpalatal length	Width at molars	Width at canines	Tibia
			Artibeus amplus, males and females, Zulia and Colombia			
13.5 ± 0.20 13.2–14.0 (7)	11.0 ± 0.26 10.6–11.6 (7)	11.2 ± 0.16 11.1–11.5 (7)	9.7 ± 0.20 9.3–10.0 (7)	13.2 ± 0.26 12.7–13.5 (7)	8.4 ± 0.16 8.2–8.8 (7)	25.9 ± 1.28 24.1–28.2 (6)
			A. amplus, males and females, T.F. Amazonas and Bolívar			
13.3 ± 0.12 12.9–14.0 (30)	11.2 ± 0.06 10.7–11.5 (30)	11.2 ± 0.10 10.7–11.8 (31)	9.8 ± 0.14 9.1–10.6 (28)	13.3 ± 0.10 12.8–13.9 (30)	8.6 ± 0.08 8.3–8.9 (30)	24.8 ± 0.38 23.1–26.2 (22)
			A. jamaicensis, females, Zulia			
12.3 ± 0.10 12.0–12.8 (19)	10.2 ± 0.16 9.5–10.9 (19)	10.0 ± 0.12 9.4–10.5 (17)	8.6 ± 0.14 8.2–9.3 (18)	12.1 ± 0.16 11.2–12.7 (19)	7.6 ± 0.10 7.2–8.0 (19)	22.4 ± 0.50 20.1–23.8 (19)
			A. jamaicensis, males, Zulia			
12.1 ± 0.14 11.8–12.4 (10)	10.3 ± 0.18 10.0–10.9 (10)	10.1 ± 0.16 9.7–10.4 (8)	8.4 ± 0.12 8.2–8.8 (10)	12.2 ± 0.18 11.8–12.6 (10)	7.7 ± 0.12 7.4–7.9 (9)	21.7 ± 0.52 20.1–23.0 (10)
			A. jamaicensis, females, T.F. Amazonas			
13.2 ± 0.12 12.9–13.5 (14)	10.8 ± 0.18 10.2–11.5 (14)	11.4 ± 0.20 11.0–12.0 (13)	9.4 ± 0.18 8.9–10.0 (14)	13.7 ± 0.22 13.2–14.4 (13)	8.6 ± 0.12 8.3–8.9 (13)	24.1 ± 0.46 22.3–25.4 (17)
			A. jamaicensis, males, T.F. Amazonas			
13.4 ± 0.22 12.8–13.7 (8)	11.0 ± 0.18 10.7–11.4 (8)	11.4 ± 0.18 11.1–11.8 (8)	9.3 ± 0.24 8.7–9.8 (8)	13.8 ± 0.30 13.0–14.3 (8)	8.8 ± 0.12 8.6–9.1 (8)	23.3 ± 0.52 22.0–24.4 (8)

Artibeus gnomus new species

HOLOTYPE—USNM 387534, adult female, skin and skull, collected 14 June 1966 by A. L. and M. D. Tuttle at El Manaco (= Km 74), 59 km SE El Dorado, Bolívar, Venezuela, 150 m, in a mist net in an orchard. Original number, SVP 9298.

ETYMOLOGY—Latin *gnomus,* diminutive fabled being, dwarf, alluding to the small size of this species, one of the smallest *Artibeus.*

DISTRIBUTION—The Amazon Basin and bordering regions; from northern Amazonas Territory (14 km SSE Pto. Ayacucho) and northern Bolívar State (28 km SE El Manteco) in Venezuela and northern Guyana, to Pará (Belém) and Mato Grosso (Serra do Roncador), Brazil, and Loreto (Santa Rosa), Peru. SVP collectors netted *A. gnomus* mostly in moist sites (92%) in evergreen forest (52%) or openings such as savannas (25%) and yards and orchards (23%). Elevations range 119–161 m in Venezuela, sea level to 530 m in Brazil. Holdridge life zones: Tropical dry forest (22%), Tropical humid forest (67%), Tropical very humid forest (2%), and Premontane humid forest (9%).

DESCRIPTION—Body size small (forearm averages 36–38, greatest length of skull 18.5–18.7, and maxillary toothrow 5.7–6.0). Dorsal coloration gray-brown to brown; underparts paler; facial stripes very white and sharply defined. Soft parts coloration in life (USNM 361742, male, Belém, Brazil): ear narrowly edged with yellow, brightest toward base; antitragus entirely yellow; tragus yellow, brightest distally and on posterior basal lobe; noseleaf and horseshoe gray-brown medially, cream color laterally; lips and chin gray-brown; iris brown; forearm and fingers brownish flesh color; wings blackish, except membrane between fingers II and III transparent, grayish; interfemoral membrane sooty brown; legs and feet dark brown; claws horn color. Face short; shape and proportions of ears, noseleaf, horseshoe, lips, chin, and interfemoral membrane as in *Artibeus cinereus*; noseleaf minutely hirsute; lower edge of horseshoe free; basal part of forearm hairy; hind extremities (except for short hairs on feet) appear naked.

Skull small, short, and broad; zygomata subparallel; rostrum narrow, very short, moderately

TABLE 2. Measurements of adult male and female (combined) *Artibeus gnomus* and *A. glaucus bogotensis*. For each measurement, line 1 includes the mean plus or minus two standard errors, line 2 the extremes, and line 3, in parentheses, the number of specimens measured. All specimens are from Venezuela.

Total length	Hind foot (dry)	Ear	Forearm	Greatest length	Zygomatic breadth	Postorbital breadth
Artibeus gnomus, Río Supamo, Los Patos, and El Manaco						
47.5 ± 1.40	9.5 ± 0.28	16.9 ± 0.62	36.7 ± 0.54	18.5 ± 0.18	11.0 ± 0.18	4.9 ± 0.10
44–54	9–10	14–19	34.0–38.3	17.9–19.1	10.4–11.2	4.5–5.2
(13)	(13)	(13)	(13)	(14)	(8)	(14)
A. glaucus bogotensis, Km 125, 85 km SSE El Dorado						
52.2 ± 0.86	10.6 ± 0.22	17.4 ± 0.38	39.6 ± 0.60	20.3 ± 0.10	11.6 ± 0.10	5.0 ± 0.04
49–56	10–11	16–19	36.8–41.9	19.4–21.2	10.8–12.1	4.6–5.3
(19)	(19)	(19)	(19)	(50)	(44)	(51)

deep and arched, and much swollen posterolaterally (part on rostral shield, part within orbit, above eye); excavation for orbital nerve large and deep; braincase short and deep, with swelling at posterodorsal apex interrupting junction of sagittal and lambdoidal crests; postpalatal extension relatively short; internal edge of pterygoid fossa strongly ridged, narrowing mesopterygoid fossa and cupping pterygoid fossa which opens straight back; vomerine ridge visible in mesopterygoid fossa; vacuities in roof of posterior nares much anterior to mesopterygoid fossa and not easily seen; outline of maxillary toothrows nearly circular; upper canine small (especially in basal diameter); M^1 with accessory internal ridge on lateral cusps, and with relatively wide talon; m_3 present (75 of 79 specimens examined).

Measurements of the holotype, an adult female: total length 47, tail vertebrae 0, hind foot (dry) 9, ear from notch 18, forearm 36.5, tibia 12.6, calcar 4.9, weight 10.5 g. Greatest length of skull 18.2, zygomatic breadth 10.8, postorbital breadth 4.8, breadth of braincase 8.5, depth of braincase 7.2, length of maxillary toothrow 5.5, postpalatal length 6.5, palatal breadth outside of M^3 7.1, rostral breadth at base of canines 4.6. See Table 2 for additional measurements.

COMPARISONS—*Artibeus gnomus* differs from *A. concolor* and *A. hartii* in many ways, but most significantly in lack of M^3. From all other small *Artibeus* (*A. anderseni, A. cinereus,* and *A. glaucus bogotensis*) that occur within its range, *A. gnomus* can be distinguished by its possession of m_3. Among the specimens examined, m_3 is consistently absent in these other taxa while it is consistently present in *A. gnomus* (except in southern Venezuela, where it is absent from both mandibles in four of 53 specimens and from one mandible only in two others). In addition, *A. gnomus* differs from all of the sympatric taxa in its more prominent white facial stripes; more colorful ears, noseleaf, and lips; average browner, less grayish coloration of pelage; shorter face and rostrum (except when compared with *A. concolor*); more swollen supraorbital region; average larger and deeper orbital nerve excavation (sometimes equally large and deep in *A. g. bogotensis*); and more cupped pterygoid fossa, with internal ridge so enlarged as to significantly narrow the mesopterygoid fossa.

Artibeus gnomus differs from the sympatric taxa individually in several other ways. It is much smaller than *A. concolor* (forearm averages 36–38 vs. 46–48). In contrast to *A. hartii* it has notched inner upper incisors, brownish rather than dark chocolate coloration, and a wide, unfringed interfemoral membrane. Compared with *A. anderseni* (including the holotype, FMNH 21331), *A. gnomus* is similar in size (slightly larger than Rio Madeira *A. anderseni*); has rostrum much deeper, more arched, narrower, and shorter; face not dished; orbit larger; zygomata more nearly parallel; and vacuities in roof of posterior nares far forward of mesopterygoid fossa, rather than opening in it or close to it.

At Belém, Brazil, both *Artibeus gnomus* and *A. cinereus* were numerous and were often taken in the same nets. There, fresh specimens of the two species were compared. *Artibeus gnomus* is smaller in size, and has a smaller head and shorter face; facial stripes much brighter, more sharply defined, and more prominent; ears, noseleaf, and lips more brownish, less grayish; ear edgings, antitragus, and tragus bright yellow, rather than cream; and noseleaf edged with cream, rather than plain gray-brown. Furthermore, it has zygomata more nearly parallel; rostrum deeper and shorter; supraorbital area much swollen and its edges nearly parallel; and smaller teeth.

TABLE 2. *Continued.*

Braincase breadth	Braincase depth	Maxillary toothrow	Postpalatal length	Width at molars	Width at canines	Tibia
Artibeus gnomus, Río Supamo, Los Patos, and El Manaco						
8.5 ± 0.14 8.1–9.0 (14)	7.4 ± 0.14 7.1–8.0 (14)	5.7 ± 0.06 5.5–5.9 (14)	6.3 ± 0.10 6.0–6.7 (14)	7.5 ± 0.12 7.1–7.9 (14)	4.9 ± 0.08 4.6–5.1 (14)	13.2 ± 0.46 11.2–14.4 (13)
A. glaucus bogotensis, Km 125, 85 km SSE El Dorado						
9.0 ± 0.06 8.5–9.5 (48)	7.9 ± 0.06 7.2–8.3 (48)	6.5 ± 0.04 6.0–6.8 (51)	7.0 ± 0.08 6.5–7.5 (47)	8.0 ± 0.06 7.5–8.7 (50)	5.1 ± 0.06 4.9–5.6 (50)	13.6 ± 0.36 12.3–15.8 (19)

In southern Venezuela *Artibeus gnomus* is sympatric with *A. glaucus bogotensis.* Compared with Venezuelan specimens and with the holotype (BM 99.11.4.35) of this taxon, *A. gnomus* is much smaller and shorter faced; has a deeper, shorter rostrum; disproportionately wider zygomatic spread; and smaller teeth.

In addition to comparisons of *A. gnomus* with sympatric species, two other small *Artibeus* need to be considered:

1. ***Artibeus g. glaucus***—This species occurs nearby in the Andes. It (including the holotype, BM 94.8.6.13) possesses m_3, and its skull has the basic shape of *A. gnomus.* However, it is much larger and darker in color, has the hind extremities much hairier, the supraorbital region usually less swollen, and the pterygoid fossa much less cupped and opening to the mesopterygoid fossa.

2. ***Artibeus watsoni***—West of the Andes and extending into Central America is another small species, *A. watsoni* Thomas, which like *A. gnomus* possesses m_3. It (including its holotype, BM 0.7.11.19) is larger than *A. gnomus*; has larger teeth; longer rostrum, with reduced supraorbital swelling; shallower and less well-defined orbital nerve excavation; and like *A. glaucus* has the pterygoid fossa not cupped and opening into the mesopterygoid fossa (which consequently is not narrowed by the inner pterygoid ridge).

DISCUSSION—The ten small species of *Artibeus* recognized here can be associated in six species groups:

1. ***Artibeus concolor* Group**—Amazon and upper Orinoco basins and Guianas. Includes only *Artibeus concolor.*

2. ***Artibeus hartii* Group**—Mexico and Central America, across northern South America to Trinidad, and south to Peru east of the Andes and to Ecuador west of the Andes. Includes only *Artibeus hartii.*

3. ***Artibeus glaucus* Group**—Mexico, Central America, and South America to Mato Grosso and Peru. Includes *Artibeus glaucus* (with two subspecies, *A. g. bogotensis* and *A. g. glaucus*), *A. gnomus,* and *A. watsoni.*

4. ***Artibeus toltecus* Group**—Mexico and Central America. Includes *Artibeus aztecus* Andersen and *Artibeus toltecus* Saussure, each with several subspecies.

5. ***Artibeus cinereus* Group**—Guiana region, coastal Brazil, and lower Amazon Basin (dubiously also upper Amazon Basin). Includes only *Artibeus cinereus,* with *A. c. quadrivittatus* as a subspecies.

6. ***Artibeus phaeotis* Group**—Mexico, Central America, and South America to upper Amazon Basin and western Ecuador. Includes *Artibeus anderseni* Osgood and *Artibeus phaeotis* Miller,[1] with several subspecies.

Diversity in the small *Artibeus* is greatest in eastern South America, where representatives of five of the six groups occur and where three of the groups are endemic. Altogether six species occur in and around the Amazon Basin, while only one is known with certainty in the central portion of the Basin; there are three in the lower Amazon

[1] Until recently (Koopman, p. 152, in Honacki et al., 1982) it has not been generally recognized that *Artibeus phaeotis* and *A. ravus* are conspecific. They intergrade in eastern Panama and western Colombia. Both names date from Miller (1902). Although *A. ravus* was named first, on an earlier page, *A. phaeotis* became embedded in the literature as the name of this species.

and on the southern fringes in Brazil and Bolivia, four or five in southern Venezuela, and five in eastern Peru, Ecuador, and Colombia. In contrast, only three of the species groups occur in Central America, and only one of them is endemic there.

Several distributional patterns are represented in the complex of Amazonian species. *Artibeus concolor* is found throughout the Basin but scarcely beyond it; *A. cinereus* occurs in the lower Amazon and along the coast for some distance north and south of the river; *A. anderseni* is known from the upper Amazon and an isolated area in northern Colombia; *A. glaucus* is higher up, in the Andes, and eastward around the northern edge of the Basin in Venezuela; the range of *A. hartii* resembles that of *A. glaucus,* but extends on into Central America; and the dwarf *A. gnomus* has a peculiar circular range, completely ringing the Amazon Basin but apparently not extending into its interior.

Key to the Smaller Species of *Artibeus*

1. Molars 3/3 (m_3 large) 2
1′. Molars 2/3 (m_3 minute) or 2/2 3
2. I[1] notched; facial stripes absent; coloration pale brown; interfemoral membrane broad and naked; forearm 43–52 mm .... *Artibeus concolor*
2′. I[1] not notched; facial stripes present; coloration dark chocolate brown; interfemoral membrane narrow and fringed; forearm 36–42 mm .... *Artibeus hartii*
3. Supraorbital region much swollen; molars 2/3 (2/2 in *A. g. bogotensis* and occasionally in the others) ... *Artibeus glaucus* Group, 4
3′. Supraorbital region little, or not at all, swollen; molars 2/2 7
4. Rostrum short and moderately arched; pterygoid fossa cupped and opening back, causing mesopterygoid fossa to be narrowed; forearm 34–38 mm .... *Artibeus gnomus*
4′. Rostrum long and much or only moderately arched; pterygoid fossa not cupped, opening into and not narrowing mesopterygoid fossa 5
5. Rostrum much arched; orbital nerve excavation shallow and often ill-defined; dorsum pale brownish; ears pale; forearm 35–41 mm .... *Artibeus watsoni*
5′. Rostrum moderately arched; orbital nerve excavation deep and well defined 6
6. Molars usually 2/3; dorsum dark grayish or blackish; ears dark; forearm 38–42 mm ... . *Artibeus glaucus glaucus*
6′. Molars 2/2; dorsum pale brownish or grayish; ears pale; forearm 37–41 mm .... *Artibeus glaucus bogotensis*
7. Interfemoral membrane narrow and fringed; coloration blackish .... *Artibeus toltecus* Group, 8
7′. Interfemoral membrane broad, naked; coloration brownish 9
8. Larger, forearm 42–48 mm .... *Artibeus aztecus*
8′. Smaller, forearm 37–41 mm .... *Artibeus toltecus*
9. Rostrum deep and arched; palate long and moderately wide .... *Artibeus cinereus*
9′. Rostrum shallow and flattened; palate short and very wide .... *Artibeus phaeotis* Group, 10
10. Maxillary toothrow 5.2–6.2 mm; rostrum often tilted up anteriorly .... *Artibeus anderseni*
10′. Maxillary toothrow 6.7–7.1 mm; rostrum usually not tilted up anteriorly[2] .... *Artibeus phaeotis*

SPECIMENS EXAMINED—***Artibeus anderseni***—**BRAZIL. Amazonas:** Borba, Rio Madeira (1 AMNH). **Rondônia:** Pôrto Velho (2 AMNH, 2 FMNH, including holotype of *A. anderseni*); Sto. Antonio do Hauayara (4 AMNH). **COLOMBIA. Bolívar:** Catival, Upper Río San Jorge, 120 m (16 FMNH). **Antióquia:** Aljibos, 26 km S and 22 km W Zaragoza, 630 m (2 USNM); nr. La Tirana, 24 km S and 22 km W Zaragoza, 520 m (2 USNM). **ECUADOR. Napo:** Río Suno (Abajo) (4 AMNH). **Pastaza:** Montalvo, Río Bobonaza (1 FMNH); Río Pindo Yacu (1 FMNH); Río Yana Rumi (1 FMNH). **PERU. Huánuco:** Monte Alegre (1 AMNH). **Loreto:** Boca Río Curaray (1 AMNH); Boca Río Peruate, Río Amazonas, 90 m (1 FMNH); Lagarto, Alto Ucayali (1 AMNH); Mazán (1 AMNH); 59 km W Pucallpa (1 USNM); Puerto Indiana, Río Amazonas (2 AMNH); Río Morona (Quebr. Pushaga), Alto Amazonas, 220 m (2 FMNH); Río Yavari Mirim (Quebr. Esperanza), 200 m (2 FMNH); Santa Cecilia, Río Maniti, Iquitos, 110 m (3 FMNH); Santa Luisa, Río Nanay, Iquitos, 160 m (1 FMNH); Sarayacu, Río Ucayali (1 AMNH). **Pasco:** San Juan, Oxapampa,

[2] Couplet 10 will separate *Artibeus anderseni* and *A. phaeotis* in South America and in southern Central America, but it will not distinguish *A. anderseni* from Mexican *A. phaeotis nanus.* In such a comparison, *A. anderseni* can be recognized by its relatively broader skull.

274 m (3 USNM). **Departamento (?):** Yuhucumayo, 1200 ft [= Puno: Yahuaramayo, 366 m?] (1 MCZ).

***Artibeus cinereus cinereus*—BRAZIL. Amazonas:** Sta. Clara, Vila Bela Imperatriz [nr. Parintins] (1 AMNH). **Pará:** Fordlândia, Rio Tapajós (2 AMNH); Maracano, Rio Jamundá [= Nhamundá?], Faro (5 AMNH); Rio Yumundá, Faro (1 BM).

***Artibeus cinereus quadrivittatus*—BRAZIL. Maranhão:** Juryassú [= Turiaçu?] (1 BM). **Pará:** Belém (10 USNM); Benevides (1 BM); Pará [= Belém] (1 BM); Ilha do Taiuna, Rio Tocantins (3 AMNH). Pernambuco: Pernambuco [= Recife] (2 BM). Rio Grande do Norte: Natal (1 USNM). **SURINAME. Surinam** (1 BM). **VENEZUELA. Bolívar:** Hato San Felipe, Serranía de Nuria (1 UCV); Hato San José, 20 km W La Paragua, 300–324 m (2 USNM).

***Artibeus glaucus bogotensis*—COLOMBIA. Cundinamarca:** Bogotá (2 BM); nr. Bogotá (1 BM); Curiche, nr. Bogotá (2 BM, including holotype of *A. bogotensis*); Fómeque (1 AMNH); Fusagasuga (2 MCZ); Río Negro, nr. Bogotá (2 BM). **GUYANA.** Kanuku Mts. (3 BM). **VENEZUELA. Bolívar:** El Manaco, 59 km SE El Dorado, 150 m (3 USNM); Hato San José, 20 km W La Paragua, 300–324 m (3 USNM); 23 to 45 km NE Icabarú, 824–851 m (3 USNM); Km 125, 85 km SSE El Dorado, 826–1165 m (120 USNM); Río Supamo, 50 km SE El Manteco, 150 m (2 USNM). **T.F. Amazonas:** Belén, Río Cunucunuma, 56 km NNW Esmeralda, 150 m (1 USNM); Caño Culebra, Cerro Duida, 50 km NNW Esmeralda, 800 m (3 USNM).

***Artibeus glaucus glaucus*—BOLIVIA. Santa Cruz:** Buenavista, 400 m (1 FMNH). **ECUADOR. Napo:** Baeza (1 BM). **PERU. Cuzco:** Collpa de San Lorenzo, Quincemil, 700 m (11 FMNH); Hda. Cadena, Quincemil, 1000 m (9 FMNH). **Junin:** Chanchamayo, 1000 m (2 BM, including holotype of *A. glaucus*); Huacapistana (1 FMNH). **Puno:** Río Inambari, 670 m (3 AMNH); Santo Domingo (1 AMNH); Yahuaramayo, 366 m (1 BM, 1 USNM).

Artibeus gnomus—Total 104. **BRAZIL. Mato Grosso:** Serra do Roncador, 264 km N (by road) Xavantina, 533 m (17 USNM). **Pará:** Belém, Sta. A, IPEAN (7 USNM); Belém, Utinga (5 USNM); Belém, Benevides (2 USNM). **ECUADOR. Pastaza:** Canelos, upper Río Bobonaza (1 AMNH). **GUYANA. E. Berbice District:** Wikki River (3 USNM). **Mazaruni-Potaro District:** Issano Road, 12 mi W of Bartica-Potaro Road (1 USNM). **PERU. Loreto:** 59 km SW Pucallpa (1 USNM); Santa Rosa, Alto Ucayali (10°42′S, 73°50′W) (2 AMNH). **VENEZUELA. Bolívar:** El Manaco, 59 km SE El Dorado, 150 m (12 USNM); Km 38, SE El Dorado, 100 m (1 UCV); Los Patos, 28 km SE El Manteco, 150 m (4 USNM); Río Supamo, 50 km SE El Manteco, 150 m (1 USNM); Salto Chalimaha, Río Paramichí, Río Paragua (1 UCV); Salto Ichun, Río Paragua (2 UCV). **T.F. Amazonas:** Belén, Río Cunucunuma, 56 km NNW Esmeralda, 150 m (2 USNM); Boca Mavaca, 84 km SSE Esmeralda, 138 m (1 USNM); Caño León, Cerro Duida, 325 m (1 AMNH); Capibara, Brazo Casiquiare, 106 km SW Esmeralda (1 USNM); Esmeralda, Cerro Duida, 325 m (3 AMNH); 14 to 65 km S, SSE, and SSW Pto. Ayacucho, 119–161 m (16 USNM); Río Mavaca, 108 km SSE Esmeralda, 140 m (7 USNM); San Juan, Río Manapiare, 163 km ESE Pto. Ayacucho, 155 m (6 USNM); Tamatama, Río Orinoco, 135 m (7 USNM).

Artibeus phaeotis—Holotypes of *A. phaeotis* and *A. ravus*, plus hundreds of other specimens from Mexico, Central America, and NW South America.

***Artibeus pumilio*—PERU. Loreto:** Masisea, Tushemo, Río Ucayali, 328 m (1 BM, holotype of *A. pumilio*).

***Artibeus watsoni*—PANAMA. Chiriquí:** Bogava, 250 m (5 BM, including holotype of *A. watsoni*); Progreso (34 USNM); Puerto Armuelles (2 USNM).

REMARKS—In previous publications of SVP, *Artibeus gnomus* has been known as "*Artibeus* sp. A".

Acknowledgments

Among the persons who helped me put together this paper I am especially grateful to Sally DeMott, who measured the SVP skulls; Linda Gordon, who compiled the tables and worked with me on the comparisons; and Jane Ailes Small, who read and criticized the manuscript and did the word processing. Curators of several collections kindly permitted me to study specimens under their care in the preparation of these descriptions: American Museum of Natural History (AMNH), British Museum (Natural History) (BM), Field Museum of Natural History (FMNH), Museum of Comparative Zoology, Harvard University (MCZ), and Universidad Central de Venezuela (UCV). The SVP collection is in United States National Museum of Natural History (USNM); a portion of its specimens have been returned to Venezuela.

Literature Cited

EWEL, J. J., AND A. MADRIZ. 1968. Zonas de Vida de Venezuela. Ministerio de Agricultura y Cría, Caracas, 265 pp., map.

HANDLEY, C. O., JR. 1959. A revision of American bats of the genera Euderma and Plecotus. Proceedings of the United States National Museum, **110:** 95–246.

———. 1976. Mammals of the Smithsonian Venezuelan Project. Brigham Young University Science Bulletin, Biological Series, **20**(5): 1–91.

———. 1984. New species of mammals from northern South America: A long-tongued bat, genus *Anoura* Gray. Proceedings of the Biological Society of Washington, **97:** 513–521.

HANDLEY, C. O., JR., AND K. C. FERRIS. 1972. Descriptions of new bats of the genus *Vampyrops*. Proceedings of the Biological Society of Washington, **84:** 519–523.

HANDLEY, C. O., JR., AND L. K. GORDON. 1980. New species of mammals from northern South America. Mouse possums, genus *Marmosa* Gray, pp. 65–72. *In* Eisenberg, J. F., ed., Vertebrate Ecology in the Northern Neotropics. Smithsonian Institution Press, Washington, D.C.

HONACKI, J. H., K. E. KINMAN, AND J. W. KOEPPL, EDS. 1982. Mammal Species of the World. Allen Press, Inc., and the Association of Systematic Collections, Lawrence, Kansas, 694 pp.

KOOPMAN, K. F. 1978. Zoogeography of Peruvian bats with special emphasis on the role of the Andes. American Museum Novitates, **2651:** 1–33.

MILLER, G. S., JR. 1902. Twenty new American bats. Proceedings of the Academy of Natural Sciences of Philadelphia, **54:** 389–412.

Seasonality of Reproduction in Peruvian Bats

Gary L. Graham

ABSTRACTS

The reproductive conditions of 3,489 specimens were used to determine seasonal patterns of pregnancy and parturition in Peruvian bats. More species that are trophic generalists yield birth records for the dry season than do trophic specialists. Relatively more highland than lowland species have births recorded for both seasons (dry and wet). Presumably, trophic generalists and highland species experience less seasonal variation in food supplies compared to the other groups. A larger percentage of nectarivorous species than frugivores have birth records for the dry season; the reverse is true for the wet season. These patterns are associated with greater floral resource abundance during the dry season and a greater abundance of fruit resources during the wet season.

Las condiciones reproductivas de 3489 especimenes fueron usadas para describir patrones estacionales de embarazo y alumbramiento en murciélagos peruanos. Mas especies de generalistas alimenticios tienen registros de nacimiento durante los dos estaciones (seca y mojada) que tienen especialistas. Relativamente mas especies desde tierras altas que desde tierras bajas tienen registros de nacimiento durante la estación seca. Possiblamente, generalistas alimenticios y especies de tierras altas sufrir menos variacion estacional de provisiones alimentos que los otros grupos. Un mayor porcentaje de especies nectivoras que frugivoras tienen registros de nacimiento durante la estación seca pero, el opuesto exista para la estación mojada. Estos patrones son asociados con un mayor abundancia de recursos de flores durante la estación seca y con un mayor abundancia de recursos de frutas durante la estación mojada.

As condições reprodutivas de 3489 especimes de morcêgos peruanos foram usadas para descrever seus padrões de parturição. Os morcêgos de hábitos alimentares generalizados possuem uma proporção maior de espécies que possuem registros de partos durante a época da sêca, do que os morcêgos com hábitos alimentares especializados. Proporcionalmente mais espécis de morcêgos de areas elevadas, do que das planícies, possuem registros de partos durante as duas épocas (sêca e chuvosa). Morcêgos generalizados e os de areas elevadas provavelmente experimentam menos variação epocal em a quantidade da comida, do que os morcêgos das planícies ou morcêgos com hábitos alimentares especializados. Uma porcentagem maior das espécies nectarívoras, do que das espécies frugívoras, possuem registros de partos durante a época da sêca, são a regra ao invés da exceção entre os morcêgos que consumen fruta. Estes padrões podem ser interpretados como adpatações de indivíduos a recursos alimentares que são regularmente, ou ocasionalmente, obteníveis durante a época da sêca.

From the Department of Biology, University of New Mexico, Albuquerque, New Mexico 87131.

Introduction

Although the timing of reproductive events is an important life history adaptation, there are few studies of the reproductive phenology of South American bats (Racey, 1982). Most of these studies are of individual species and cover only part of the year. Gestation, parturition, lactation, and weaning should be timed to correspond with variations in the abundance and diversity of food supplies (Fleming et al., 1972; Bradbury & Vehrencamp, 1977; Bonaccorso, 1979; Wilson, 1979; August & Baker, 1982). These variations are known to be seasonal in most of the Neotropics (Janzen, 1967, 1973; Foster, 1982; Wolda, 1982; Smythe, 1982; Terborgh, 1983). Which reproductive event is actually synchronized with peak food supplies is likely to be determined by the relative cost of each event. Lactation is the most costly period for most females, but weaning and dispersal pose the greatest survival problem for young bats (Wilson, 1979; Racey, 1982). These authors, and Tuttle and Stevenson (1982), agree that the weaning of young is the most critical period for individuals of most bat populations.

Most species of Neotropical bats that have been studied are polyestrous (Fleming et al., 1972; Thomas, 1972; Bradbury & Vehrencamp, 1977; Myers, 1977; Wilson, 1979; Bonaccorso, 1979; Humphrey & Bonaccorso, 1979). Each year, female bats confront the problem of timing two (rarely three in a few vespertilionids) periods of lactation and weaning with variations in food availability. Individuals of most species handle this problem by producing their young so that the first is weaned at the beginning of the wet season as fruit supplies are reaching their peak, and the second later in the wet season when fruit availability reaches a second peak or remains relatively high (Wilson, 1979; Tuttle & Stevenson, 1982). Exceptions to this general pattern of seasonal polyestry have been observed in most studies (Fleming et al., 1972; Thomas, 1972; Myers, 1977; Myers & Wetzel, 1983; Bonaccorso, 1979; see Mares & Wilson, 1971; Bradbury & Vehrencamp, 1977; and August & Baker, 1982 for good discussions of dry season birth periods).

The purpose of this study is to examine some of the details of dry season reproduction in Peruvian bats. I demonstrate that the different feeding assemblages and the faunas of different zoogeographical regions differ in the proportion of species with birth records during the dry season. I also offer suggestions as to how dry season births may be adaptive.

Methods

The bat specimens from Peru housed in the collections of the American Museum of Natural History, Field Museum of Natural History, Louisiana State University Museum of Zoology, National Museum of Natural History, and the Texas Cooperative Wildlife Collection were examined in late 1977 and early 1978 for information on reproduction. Data were taken directly from Tuttle (1970) and Bowles et al. (1979) for specimens not in the above collections but included in their reports. Information on reproductive condition of bats was also obtained from the notes of A. L. Gardner and J. L. Patton. Fieldwork at several localities in Peru, conducted from June through August 1977 and from June through November 1978, enabled me to collect and record the reproductive data for many of the bats now in the Louisiana State University collection. Diet information was taken from the literature (Heithaus et al., 1975; Paradiso, 1975; Gardner, 1977) and is given in Graham (1983). Nomenclature follows Graham (1983) except for *Artibeus glaucus,* which I now consider a distinct species.

Recorded data consisted of species identification, locality, elevation, sex, age, and macroscopic reproductive condition. Individuals with incompletely ossified phalangeal epiphyses were classified as juveniles. Females were designated as pregnant or lactating if this information was included on specimen tags or indicated by dissection of fluid preserved females. The crown-rump lengths of embryos (including extra-embryonic membranes) were either measured (in mm) or taken from specimen tags. Length of testes was also noted when it was recorded on specimen tags, but was not used to establish reproductive patterns because of uncertainties in the relationship between testes size and sexual activity (Taddei, 1976; Thomas, 1972). Monthly pregnancy frequencies are based only on those female specimens for which I am confident that the presence (or absence) of embryos had been properly recorded. This limitation excludes most of the specimens collected prior to 1960.

I classified each species with adequate data (those with evidence of at least one birth period) into those that can give birth in the dry (May–September) and/or wet (October–April) seasons. I assumed that a birth between 1 May and 31 August also indicates weaning of the young during the dry season. Species with births recorded only for the month of September (usually a dry month) were not included in either breeding season, since these young could be weaned either as the dry season ended or as the wet season began. The above classifications were accomplished by determining the distribution of births across all months, as indicated by the distribution of juveniles, lactating females, and embryo sizes. If a single birth was indicated for a given season, then that species was identified as having the potential to produce young during that season. The presence of juveniles or lactating females was assumed to indicate parturition during the month prior to capture. If a female was carrying a large, near-term embryo, parturition was assumed to take place in that month. On the other hand, if the embryo was very small relative to the adult body length, birth was assumed to take place four months later for most species, three months later for small insectivorous species (Findley & Wilson, 1974), and five months later for emballonurids (Bradbury & Vehrencamp, 1977) and *Desmodus rotundus* (Wilson, 1979).

Classification of a species as one that can give birth during the dry season does not necessarily mean that it does so each year. I was unable with this method to determine the actual frequency of births during either season. A dry season classification simply means that individuals can at least occasionally produce young during the dry season.

The lack of birth records for a given species for either season can be the result of an actual absence of birthing during that season, or it may be due to an inadequate sample from that season. If I assume that the lack of a breeding record for a given season is not due to inadequate sampling, then I can use the *G* test (Sokal & Rohlf, 1981) to assess the significance of the differences between the groups of species compared. This is a safe assumption, since most of the specimens were collected during the Peruvian dry season (the season with which this study is principally concerned) and because the groups of bats predicted to have the greatest proportion of dry season births are the groups most poorly sampled (see Appendix and below).

Regions and Climate

I follow Koopman's (1978) division of Peru into three zoogeographical regions. He lists species as lowland if they were collected below 1000 m east of the Andes and highland if collected above this elevation. Pacific coastal species are those collected either along the arid coast or adjacent Andean foothills (including the mesic areas of northwestern Peru).

The Pacific coastal plain is characterized by low (< 50 mm) annual rainfall, most of which falls between December and April. The mesic areas on the northwestern Andean slopes experience greater rainfall over an extended period. Rainfall east of the Andes (fig. 1) is seasonal and abundant, with the greatest amounts deposited between the elevations of 1000 and 3000 m (Bowman, 1916). The wet season begins in October and continues through April and the dry season begins in May and extends through September.

Most investigators agree that food resources are primarily modulated by rainfall seasonality (Janzen, 1967, 1973; Frankie et al., 1974; Ricklefs, 1975; Buskirk & Buskirk, 1976; Wolda, 1978a,b, 1982; Foster, 1982; Smythe, 1982; Terborgh, 1983). Many of the following generalizations on seasonal changes in food supplies for bats of middle and higher elevations are based on the assumption that Peruvian plants and insects respond to environmental conditions in a manner similar to closely related organisms in Costa Rica, Panama, or Puerto Rico (as described by the above authors). Terborgh's (1983) study of the changes in fruit, flower, and insect supplies in Manu National Park indicates that in the lowlands of southeastern Peru, fruit abundance and diversity usually increase in October with the onset of the wet season, peak in November and December and again in March, and decrease to the lowest level in May and June. Not only is the first peak greater but it also includes a greater proportion of plants in the family Moraceae (Terborgh, pers. comm.), which have fruit favored by bats (Gardner, 1977). Flower abundance peaks in the dry season, but some flowers are present throughout the year (Terborgh, 1983). At middle to higher elevations, the phenology of fruit and flower species is probably similar to lowland species, but less seasonal (Terborgh, 1977; Nevling, 1971). Changes in insect resources are more difficult to generalize because the species in different size classes (Smythe, 1982; Wolda, 1982) and in different habitats (Bradbury & Vehrencamp, 1976; Terborgh, 1983; Janzen, 1983) have different periods of peak abundance and diversity. If the supply across size classes and habitats is considered, then it is probably rather high throughout the year. I will not discuss the

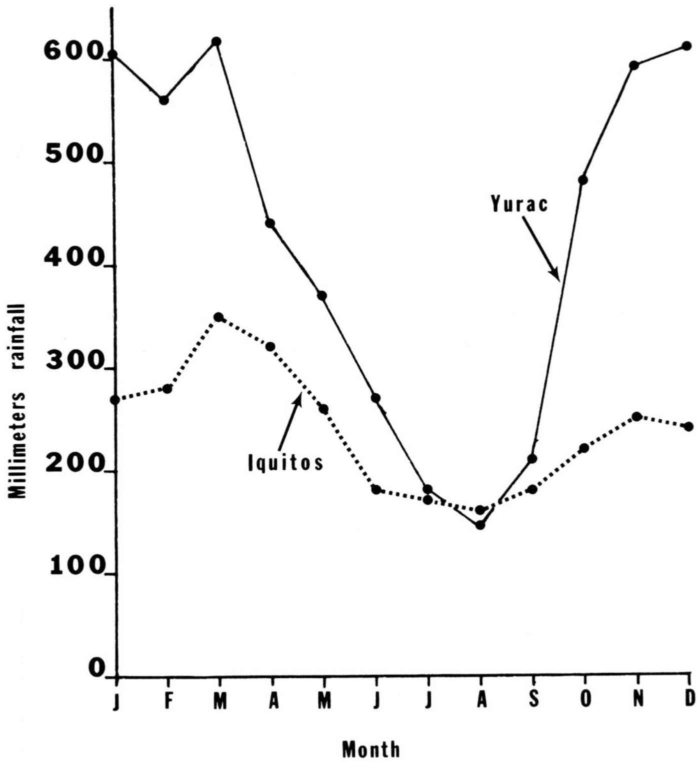

FIG. 1. Average rainfall amounts at Iquitos (lowland) and Yurac (middle elevation), Department of Loreto, Peru. Data are taken from a map published in 1971 by the Servicio Nacional de Meteorología e Hidrología in Peru and are averaged from 10 years of records.

seasonal changes in food levels west of the Andes in Peru because less is known of this region.

Results

More than 3,400 female specimens divided among 109 species were included in this study. Twins were recorded for only one female, a *Carollia perspicillata,* collected in November bearing two well-developed embryos (24 and 29 mm). Specimens were collected almost exclusively between May and December, with June through August having the largest samples (see sample sizes in Appendix). The reproductive records are also unevenly distributed among the species (Appendix). These sampling problems made it difficult to identify the reproductive patterns for most species. If, however, the percentages of all the females that were pregnant are determined for each species for each month, many species have data that fit the pattern of seasonal polyestry. This pattern is well illustrated by the pregnancy curve of *Carollia perspicillata* (fig. 2). Pregnancy levels peak in September and December and are followed by periods of parturition, as suggested by the October drop in the frequency of pregnant females and by the juveniles collected in January (Appendix). The beginning of the second reproductive cycle of the season is also indicated by the females of *Carollia castanea* (1), *Vampyrops dorsalis* (7), and *Artibeus planirostris* (1) that were simultaneously lactating and pregnant in November and December (Appendix). On the other hand, the pregnancy curve of *Myotis* (fig. 2) and the records of juveniles and lactating females of *Carollia* (fig. 2) and *Artibeus* (Appendix) for May–September indicate that births can occur outside of the wet season. Individuals of some species (i.e., *Desmodus rotundus, Myotis nigricans,* and perhaps *Glossophaga soricina* and *Artibeus planirostris*; see Appendix) may be able to produce young during any month of the year.

Table 1 lists the number of species in each family or subfamily with births recorded during the seasons. I was able to identify birth periods for 79 (72.5%) of the 109 species listed in the Appendix. Most species (63 of the 79, 79.7%) have birth records for the wet season but a surprisingly large proportion of all of the species (46 out of 79, 58.2%) have records of dry season parturition. The absence of birth records during the wet season

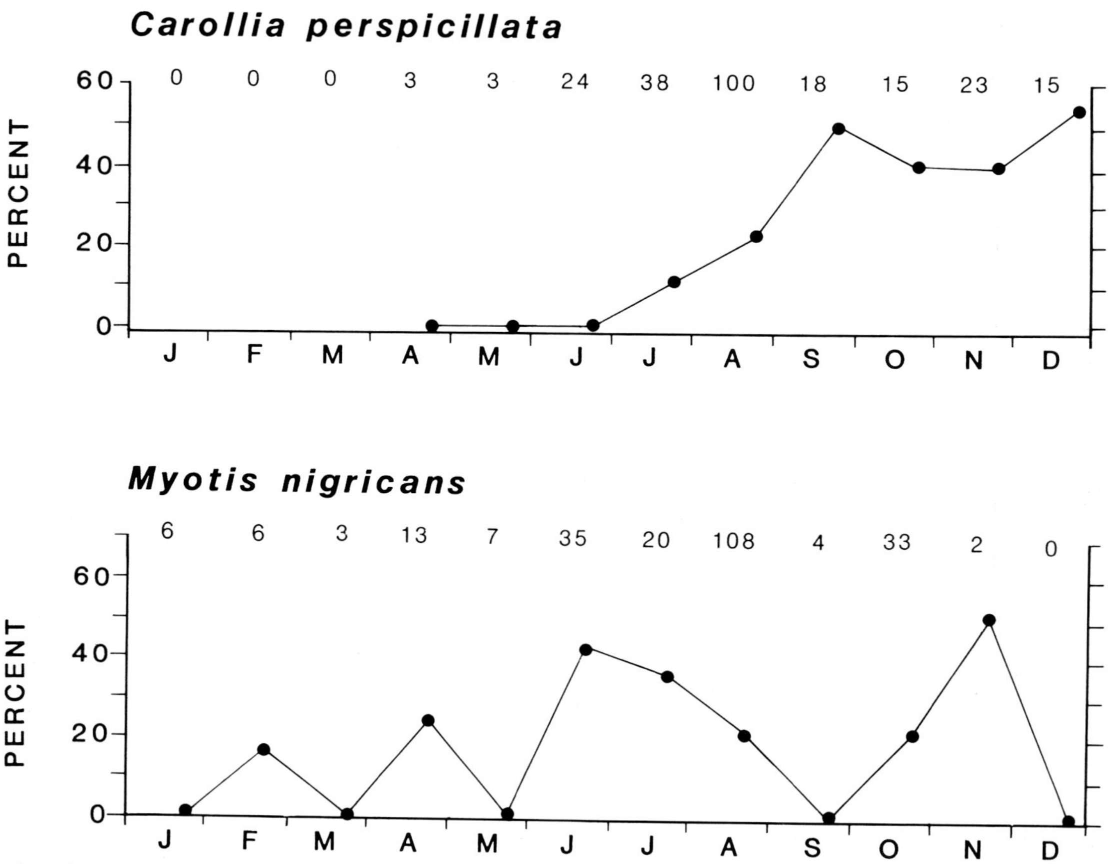

FIG. 2. Changes in pregnancy levels of *Carollia perspicillata* and *Myotis nigricans*. The numbers across the top refer to the sample sizes for each month. The ordinate represents the percentage of all females for each month that were recorded as pregnant.

(Mormoopidae) and the dry season (Furipteridae and Thyropteridae) may represent real periods of no births but may also be the result of inadequate sampling.

Are there groups of bats that have a greater tendency than other groups toward parturition during the dry season? If seasonal fluctuations in food supplies are less in the highlands than in the lowlands, then proportionately more highland than lowland species could be expected to produce and wean young during the dry season and, if most species are polyestrous, during both seasons. Thirty-one of the 58 lowland species (53.4%) and 17 of the 27 highland species (63.0%) have records of dry season births (table 2). Although the 9.6% difference is in the predicted direction, it is not significant ($P > 0.25$). The highland region has a greater proportion of species with birth records from both seasons (51.9% vs. 34.5%, $P < 0.05$). The highland *Sturnira bogotensis* that were lactating and pregnant in June and August (Appendix) provide evidence that some species with dry season birth records are polyestrous, becoming pregnant again during the dry season. Bats of the Pacific coastal and slope region also show a strong tendency toward dry season births.

If trophic generalists (those that consume more than one type of food, such as fruit and insects) can switch to another resource when one type becomes scarce, they should be less vulnerable to seasonal fluctuations in their food supply than are trophic specialists (those species that use only one major type of food). This reduced seasonality of food resources should be reflected by a tendency for births to occur during the dry season and, if polyestry is common, for both seasons. Nine of the 11 generalist species (81.8%) and 23 of the 68 specialists (33.8%, table 2) have records of individuals that have given birth during the dry season, and relatively more generalists (63.6%) than

TABLE 1. Number of species of Peruvian bats with births recorded for each season.

Family	Season* Dry	Wet	Both	Total species†
Emballonuridae	1	3	...	4
Noctilionidae	2	1	1	2
Mormoopidae	2	...	...	2
Phyllostomidae				
Phyllostominae	6	6	3	9
Glossophaginae	6	4	3	7
Carolliinae	2	5	2	5
Sturnirinae	4	6	4	6
Stenodermatinae	10	22	10	22
Desmodontinae	1	2	1	2
Furipteridae	...	1	...	1
Thyropteridae	...	1	...	1
Vespertilionidae	7	7	4	10
Molossidae	5	5	2	8
TOTALS	46	63	30	79

* The number of species with records of birth in either the dry or the wet season or in both. A species is listed in all three columns if it has birth records for both seasons.

† The total number of species that have data indicating birth periods. The sum of the dry and wet season records does not equal the total, because some species have records for both seasons.

specialists (29.7%) have records for births during both seasons (table 2). These large proportional differences are significant ($P < 0.01$ and $P < 0.05$, respectively).

Since floral resources are greatest in the dry season when fruit resources are lowest, nectarivorous species should show a greater tendency for individuals to produce and wean young during the dry season compared to frugivores. Six of the seven (85.8%) bats that are at least partially nectarivorous have birth periods in the dry season compared with less than half (17 of 40, 42.5%) of those that are partially frugivorous (table 2). This large relative difference is significant ($P < 0.05$). Dry season production and weaning of young is probably the common pattern for most bats that eat nectar. This is indicated by the large number of juveniles (13) and lactating females (35) recorded for this group from May through August. Since the resource abundances are reversed during the wet season, the pattern of frugivorous species having relatively more birth records (39 of the 40 species that are partially frugivorous, 97.5%) than nectarivores (3 of the 7 species that are partially nectarivorous, 42.9%) was expected ($P < 0.001$). Percentages of the species in these trophic groups that have birth records for both seasons are similar (40.0% and 42.9% for frugivores and nectarivores, respectively, $P > 0.50$).

The number of insectivorous species that have birth records for each season is almost equal (17 for the dry, 18 for the wet). This is the expected pattern if insect abundance and diversity remain high throughout the year.

If a species is seasonally polyestrous, the probability of recording a birth for the wet season for that species is greater, since both young are produced during that season. If most species are seasonally polyestrous, then the majority of species should have wet season birth records. All of the ecological groups mentioned above (except the nectarivores) show considerable evidence of parturition during the wet season (table 2).

Discussion

Seasonal polyestry is the dominant pattern in other studies of Neotropical bat reproduction (Fleming et al., 1972; Thomas, 1972; Taddei, 1976; LaVal & Fitch, 1977; Myers, 1977; Bonaccorso, 1979; Wilson, 1979; August & Baker, 1982). The majority of bats in Peru, especially frugivores, give birth during the wet season. Although this pattern is expected if most species are seasonally polyestrous most of the time, this conclusion is compromised by the possibility that some species may be monestrous, producing and weaning their young during the wet season. I was unable, for most species, to distinguish between these two reproductive patterns. However, the females collected from November and December that were simultaneously lactating and pregnant provide evidence for seasonal polyestry. Females that follow this pattern wean their first young concurrently with the onset of the heavy rains, as fruit supplies are reaching their peak, and wean their second young near the middle of the wet season when food resources remain plentiful.

In Peru, as in many of the other Neotropical areas discussed in the above studies, young bats are at least occasionally born and weaned in the dry season. These bats are members of zoogeographical and ecological groups whose tendencies toward dry season births can be predicted. If food availability for a particular group of bats does not fluctuate greatly throughout the year relative to a second group, then the first group should include a larger percentage of species with birth records for the dry season and, if the species are polyes-

TABLE 2. Number of species of Peruvian bats with births recorded for each season, by geographical region and ecological group.

Region and group	Season* Dry % (N)	Wet % (N)	Both % (N)	Total species*
GEOGRAPHICAL REGIONS				
Lowlands	53.4 (31)	79.3 (46)	34.5 (19)	58
Highlands	63.0 (17)	88.8 (24)	51.9 (14)	27
Pacific coastal	80.0 (8)	60.0 (6)	30.0 (3)	10
TROPHIC GROUPS				
Insectivores	58.6 (17)	62.1 (18)	22.2 (6)	29
Frugivores	38.7 (12)	100 (33)	38.7 (12)	33
Nectarivores	66.7 (2)	33.3 (1)	...	3
Piscivores	100 (1)	...	...	1
Vampires	50.0 (1)	100 (2)	50.0 (1)	2
Frugivore/insectivores	60.0 (3)	80.0 (4)	40.0 (2)	5
Nectarivore/insectivores	100 (3)	66.7 (2)	66.7 (2)	3
Piscivores/insectivores	100 (1)	100 (1)	100 (1)	1
Omnivores†	100 (2)	100 (2)	100 (2)	2
TROPHIC SPECIALISTS‡	33.8 (23)	79.4 (54)	29.7 (19)	68
TROPHIC GENERALISTS§	81.8 (9)	90.9 (10)	63.6 (7)	11

* See notes to Table 1. N = Number of species; % = % of total species.
† Includes species that eat more than two major types of items.
‡ Includes insectivores through vampires (see text, p. 177).
§ Includes frugivores/insectivores through omnivores.

trous, for both seasons. Two groups of Peruvian bats, highland species and trophic generalists, possibly experience less seasonal fluctuation in food supplies relative to lowland species and trophic specialists. However, the reasons for the dampened fluctuations differ with each group.

At high elevations in Peru, even during the dry season, clouds cover the forests most of the time (Terborgh, 1977; pers. obs.), providing moisture that is largely unavailable to lowland plants. This yearlong availability of moisture presumably keeps plant productivity and ultimately food availability from fluctuating as greatly as in the lowlands. This reduced fluctuation in food supplies is reflected by a weak trend toward relatively more highland birth records for the dry season and by a stronger highland trend for birth records for both seasons.

Species with generalized feeding habits are thought to be able to switch resources when one becomes reduced. This dampens the effect of seasonal changes in food availability and explains their larger percentages of birth records for the dry season and for both seasons. Fleming et al. (1972), Heithaus et al. (1975), and Bonaccorso (1979) provide data from Costa Rica and Panama that illustrate seasonal switches made by several of the same species included in my study as feeding generalists (i.e., *Phyllostomus discolor* and *Glossophaga soricina*).

Relatively more nectarivorous than frugivorous species have individuals with dry season birth and weaning records, but the records for both seasons are not different. These were the expected patterns, since both groups presumably experience seasonal variations in their food supplies; floral resources are greatest and fruit resources lowest during the dry season (Terborgh, 1983). Hence, dry season births may actually be the rule for bats that consume nectar but the exception for frugivores. The reverse is true for the wet season. These seasonality differences explain the almost equal proportions of species with birth records for both seasons.

Thus, for trophic generalists, highland, and nectarivorous bat species, many females produce their first young in the dry season and become pregnant again, producing their second young during the wet season. Obviously, many more data are needed to clarify these patterns and to address the possibility of differential survival of young between the two seasons.

In Peru, the onset and termination of the rainy season can vary annually (Terborgh, 1983). Food supplies track these variations (Terborgh, 1983; Foster, 1982), and so may bat reproduction. I do not believe that climatic variability is the principal factor responsible for the births recorded in Peru between May and August because the ecological groups that I compared (frugivores and nectarivores, trophic specialists and generalists) showed different seasonal birth patterns. If unusual dry season rains were the principal factor, then equal proportions of the different ecological groups should have responded by producing young during the normally dry season.

My data support the hypothesis that dry season births are adaptations for producing and perhaps weaning young when food resources are available predictably or occasionally during this time. These patterns need to be confirmed by long term studies of actual bat communities, and of the resources used by individuals at both highland and lowland localities and during both seasons.

Acknowledgments

I am grateful to the many people who assisted me in Peru and in the United States. Antonio Brack E., Eric Cardich Briceno, Richard Bustamante M., and Susana Moller-H. of the Dirección General Forestal y de Fauna, Ministerio de Agricultura, Lima, provided the necessary collecting permits. Hernando de Macedo R. and Ramon Ferreyra of the Museo de Historia Natural "Javier Prado" were also helpful. I thank Arturo Koenig R., Manuel A. Plenge, and Gustavo del Solar for their hospitality and help in making this study possible. I appreciate the assistance provided by Faucet and Aero Peru. Without the reliable assistance of my Peruvian associates, Manuel Sanchez, Klaus Wehr, and Reyes Rivera, the fieldwork would have been much more difficult.

Karl F. Koopman (American Museum of Natural History), Don E. Wilson (National Museum of Natural History), David J. Schmidly and William B. Davis (Texas Cooperative Wildlife Collection), and Patricia W. Freeman (formerly of Field Museum of Natural History) were very helpful during trips to their museums. Koopman and James L. Patton (Museum of Vertebrate Zoology) loaned fluid-preserved bats. James B. Cope (Joseph Moore Museum), Alfred L. Gardner (National Museum of Natural History), and Patton were kind enough to send me unpublished manuscripts and field notes. For valuable assistance in the field I thank Linda J. Barkley, J. William Eley, Gary R. Graves, John P. O'Neill, Theodore A. Parker, III, Thomas S. Schulenberg, and Morris D. Williams. The excellent suggestions on the manuscript by Philip Myers were appreciated. I thank Susan T. Graham for her support during the several revisions of this work.

I gratefully acknowledge the financial support of the LSUMZ Peruvian fieldwork by John S. McIlhenny, Irving and Laura R. Schweppe, E. W. Mudge, and the late Babette M. Odom. Travel to other museums was made possible in part by an LSU Foundation-Graduate Student Travel Award.

Literature Cited

AUGUST, P. V., AND R. J. BAKER. 1982. Observations on the reproductive ecology of some Neotropical bats. Mammalia, **46:** 177–181.

BONACCORSO, F. J. 1979. Foraging and reproductive ecology in a Panamanian bat community. Bulletin of the Florida State Museum, Biological Sciences, **24:** 359–408.

BOWLES, J. B., J. B. COPE, AND E. A. COPE. 1979. Biological studies of selected Peruvian bats of Tingo María, Departamento de Huánuco. Transactions of the Kansas Academy of Sciences, **82:** 1–10.

BOWMAN, I. 1916. The Andes of Southern Perú: Geographical Reconnaissance Along the Seventy-Third Meridian. H. Holt and Co., New York, 336 pp.

BRADBURY, J. W., AND S. L. VEHRENCAMP. 1976. Social organization and foraging in emballonurid bats. I. Field studies. Behavioral Ecology and Sociobiology, **1:** 337–381.

———. 1977. Social organization and foraging in emballonurid bats. IV. Parental investment patterns. Behavioral Ecology and Sociobiology, **2:** 19–29.

BUSKIRK, R. E., AND W. H. BUSKIRK. 1976. Changes in arthropod abundance in a highland Costa Rican forest. American Midland Naturalist, **95:** 288–298.

FINDLEY, J. S., AND D. E. WILSON. 1974. Observations on the Neotropical disk-winged bat, *Thyroptera tricolor* Spix. Journal of Mammalogy, **55:** 562–571.

FLEMING, T. H., E. T. HOOPER, AND D. E. WILSON. 1972. Three Central American bat communities: Structure, reproductive cycles, and movement patterns. Ecology, **53:** 555–569.

FOSTER, R. B. 1982. The seasonal rhythm of fruitfall of Barro Colorado Island, pp. 151–172. *In* Leigh, E. G., Jr., A. S. Rand, and D. M. Windsor, eds., The Ecology of a Tropical Forest. Smithsonian Institution Press, Washington, D.C., 468 pp.

FRANKIE, G. W., H. G. BAKER, AND P. A. OPLER. 1974. Comparative phenological studies of trees in tropical wet and dry forests in the lowlands of Costa Rica. Journal of Ecology, **62:** 881–919.

GARDNER, A. L. 1977. Feeding habits, pp. 293–350. *In* Baker, R. J., J. K. Jones, Jr., and D. C. Carter, eds., Biology of bats of the New World family Phyllostomatidae, Part II. Special Publications, The Museum, Texas Tech University, **13:** 1–364.

GRAHAM, G. L. 1983. Changes in bat species diversity along an elevational gradient up the Peruvian Andes. Journal of Mammalogy, **64:** 559–571.

HEITHAUS, E. R., T. H. FLEMING, AND P. A. OPLER. 1975. Foraging patterns and resource utilization in seven species of bats in a seasonal tropical forest. Ecology, **56:** 841–854.

HUMPHREY, S. R., AND F. J. BONACCORSO. 1979. Population and community ecology, pp. 409–441. *In* Baker, R. J., J. K. Jones, Jr., and D. C. Carter, eds., Biology of bats of the New World family Phyllostomatidae, Part III. Special Publications, The Museum, Texas Tech University, **16:** 1–441.

JANZEN, D. H. 1967. Synchronization of sexual reproduction of trees within the dry season in Central America. Evolution, **21:** 620–637.

———. 1973. Sweep samples of tropical foliage insects: Effects of seasons, vegetation types, elevation, time of day, and insularity. Ecology, **54:** 687–708.

———. 1983. Insects: Introduction, pp. 619–645. *In* Janzen, D. H., ed., Costa Rican Natural History. The University of Chicago Press, Chicago, 816 pp.

KOOPMAN, K. F. 1978. Zoogeography of Peruvian bats with special emphasis on the role of the Andes. American Museum Novitates, **2651:** 1–33.

LAVAL, R. K., AND H. S. FITCH. 1977. Structure, movements and reproduction in three Costa Rican bat communities. Occasional Papers, Museum Natural History, University of Kansas, **69:** 1–27.

MARES, M. A., AND D. E. WILSON. 1971. Bat reproduction during the Costa Rican dry season. BioScience, **21:** 471–477.

MYERS, P. 1977. Patterns of reproduction of four species of vespertilionid bats in Paraguay. University of California Publications in Zoology, **107:** 1–41.

MYERS, P., AND R. M. WETZEL. 1983. Systematics and zoogeography of the bats of the Chaco Boreal. Miscellaneous Publications, Museum of Zoology, University of Michigan, **165:** 1–59.

NEVLING, L. I., JR. 1971. The ecology of an elfin forest in Puerto Rico, 16. The flowering cycle and an interpretation of its seasonality. Journal of the Arnold Arboretum, **52:** 586–613.

PARADISO, J. L., ED. 1975. Walker's Mammals of the World, 3rd ed., vols. 1 and 2. Johns Hopkins University Press, Baltimore, 1,500 pp.

RACEY, P. A. 1982. Ecology of bat reproduction, pp. 57–104. *In* Kunz, T. H., ed., Ecology of Bats. Plenum Press, New York, 425 pp.

RICKLEFS, R. E. 1975. Seasonal occurrence of night-flying insects on Barro Colorado Island, Panama Canal Zone. Journal of the New York Entomological Society, **83:** 19–32.

SMYTHE, N. 1982. The seasonal abundance of night-flying insects in a Neotropical forest, pp. 309–318. *In* Leigh, E. G., Jr., A. S. Rand, and D. M. Windsor, eds., The Ecology of a Tropical Forest. Smithsonian Institution Press, Washington, D.C., 468 pp.

SOKAL, R. R., AND F. J. ROHLF. 1981. Biometry, 2nd ed. W. H. Freeman, San Francisco, Calif., 859 pp.

TADDEI, V. A. 1976. The reproduction of some Phyllostomidae (Chiroptera) from the northwestern region of the State of São Paulo. Boletim de Zoologia, Universidade de São Paulo, **1:** 313–330.

TERBORGH, J. 1977. Bird species diversity on an Andean elevational gradient. Ecology, **58:** 1007–1019.

———. 1983. Five New World Primates. Princeton University Press, Princeton, N.J., 260 pp.

THOMAS, M. E. 1972. Preliminary study of the annual breeding patterns and population fluctuations of bats in three ecologically distinct habitats in southwestern Colombia. Ph.D. diss., Tulane University, New Orleans, La., 161 pp.

TUTTLE, M. D. 1970. Distribution and zoogeography of Peruvian bats, with comments on natural history. University of Kansas Science Bulletin, **49:** 45–86.

TUTTLE, M. D., AND D. STEVENSON. 1982. Growth and survival of bats, pp. 105–150. *In* Kunz, T. H., ed., Ecology of Bats. Plenum Press, New York, 425 pp.

WILSON, D. E. 1979. Reproductive patterns, pp. 317–378. *In* Baker, R. J., J. K. Jones, Jr., and D. C. Carter, eds., Biology of bats of the New World family Phyllostomatidae, Part III. Special Publications, The Museum, Texas Tech University, **16:** 1–441.

WOLDA, H. 1978a. Fluctuations in rainfall, food and abundance of tropical insects. Journal of Animal Ecology, **47:** 369–381.

———. 1978b. Fluctuations in abundance of tropical insects. American Naturalist, **112:** 1017–1045.

———. 1982. Seasonality of Homoptera on Barro Colorado Island, pp. 319–330. *In* Leigh, E. G., Jr., A. S. Rand, and D. M. Windsor, eds., The Ecology of a Tropical Forest. Smithsonian Institution Press, Washington, D.C., 468 pp.

Appendix

Summary of the reproductive data for Peruvian bats (numbers are percentages of females pregnant; numbers in parentheses are numbers of females examined).

Species	Region*	Jan.	Feb.	Mar.	Apr.	May	Jun.	Jul.	Aug.	Sep.	Oct.	Nov.	Dec.	Season†
Emballonuridae														
Rhynchonycteris naso	L	...	...	...	0(1)	25(4)[l]	...	71(7)	[j]33(3)	100(2)	...	...	...	D
Saccopteryx bilineata	L	0(1)	...	[j]...	0(1)	...	40(5)	8(13)	65(51)	...	100(2)	...	...	W
Saccopteryx leptura	L	...	[j]...	0(2)	...	0(1)	...	0(5)	0(2)	1(1)	...	...	...	W
Peropteryx macrotis	L	...	...	...	[j]40(5)	...	...	...	53(15)	67(15)	...	...	...	W
Noctilionidae														
Noctilio albiventris	L	...	...	0(1)	0(3)	...	11(28)	[j]50(10)	4(24)[l]	...	0(2)	...	...	D
Noctilio leporinus	L	...	...	...	...	...	...	25(4)	[j]20(10)[l]	...	...	...	...	D/W
Mormoopidae														
Pteronotus parnellii	L	...	...	...	...	...	67(3)	...	[j]19(16)	...	...	...	...	D
Pteronotus personatus	L	...	...	...	...	...	...	...	0(1)[l]	...	...	...	...	D
Pteronotus davyi	P	...	...	...	...	0(3)	...	...	...	...	...	...	...	?
Pteronotus gymnonotus	L	...	...	...	...	...	...	0(1)	38(8)	...	...	...	...	?
Mormoops megalophylla	P	...	...	...	...	...	0(1)	...	...	...	0(4)	...	...	?
Phyllostominae														
Micronycteris megalotis	L	...	...	...	...	...	[j]...	0(3)	[j]21(24)[l]	0(1)	...	0(1)[l]	...	D/W
Micronycteris megalotis	H	...	...	...	...	0(2)	[j]100(3)	...	[j]0(1)[l]	...	...	...[l]	...	D/W
Micronycteris minuta	L	...	...	...	...	...	...	0(3)	...	...	...	...	...	?
Micronycteris hirsuta	L	...	...	...	...	...	...	50(2)	...	...	...	...	...	?
Micronycteris nicefori	L	...	...	...	...	...	...	0(1)	...	...	...	...	...	?
Micronycteris daviesi	L	...	...	...	...	...	...	...	100(1)[n]	...	...	...	...	D
Lonchorhina aurita	L	...	...	...	...	...	0(1)	...[l]	[j]33(6)	...	...	...	...	D
Macrophyllum macrophyllum	L	...	...	...	...	[j]...	...	...	0(1)	...	...	...	...	D
Tonatia bidens	L	...	...	...	0(2)	...	...	67(3)	...	...	...	...	...	?
Tonatia carrikeri	L	...	...	...	...	...	0(2)	...	...	...	...	...	...	?
Tonatia sylvicola	L	...	...	[j]...	...	...	...	63(8)[n]	50(6)	100(1)	[j]100(1)	[j]...	...	D/W
Tonatia sylvicola	P	...	...	...	0(1)	...	...	[j]0(2)	...	...	...	...	...	D
Mimon crenulatum	L	...	...	...	...	...	...	[s]75(8)	...	...	0(1)	...	...	W
Mimon koepckeae	H	...	...	...	...	0(1)	100(1)	...	...	...	...	...	...	?
Phyllostomus discolor	L	...	...	...	...	...	0(1)	...	...	...	...	67(3)	...	W
Phyllostomus hastatus	L	...	...	0(2)	...	...	50(4)	11(9)	[j]40(40)	80(10)	68(6)[l]	0(2)	...	D/W
Phyllostomus elongatus	L	...	...	0(3)	0(1)	...	0(1)	57(14)	88(8)	...	...	0(3)[l]	[j]...	W
Phylloderma stenops	L	...	...	0(1)	...	...	...	...	...	...	...	...	...	?
Trachops cirrhosus	L	...	...	...	...	...	0(3)	33(3)	0(3)	...	...	...	...	?
Chrotopterus auritus	L	...	...	...	...	...	0(2)	...	...	...	...	...	...	?
Vampyrum spectrum	L	...	...	0(1)	...	...	0(1)	0(2)	...	...	...	...	...	?

APPENDIX. *Continued.*

Species	Region*	Jan.	Feb.	Mar.	Apr.	May	Jun.	Jul.	Aug.	Sep.	Oct.	Nov.	Dec.	Season†
Glossophaginae														
Glossophaga soricina	L	...	...	0(1)	...	...	17(12)	29(14)	35(17)[l]	...	...	75(4)	...	D/W
Glossophaga soricina	H	...	...	...	0(1)	0(1)	20(5)[n]	0(1)	...	...	73(15)	...	...	D/W
Glossophaga soricina	P	...	...	0(4)[l]	...	0(6)	0(5)[l]	...	0(6)[l]	29(21)[l]	14(28)	14(7)[l]	0(2)	D/W
Lonchophylla hesperia	P	...	...	...	...	...	0(2)	...	...	...	...	...	...	?
Lonchophylla robusta	L	...	...	...	...	...	...	...	...	...	...	100(1)	...	W
Lonchophylla handleyi	L	...	...	...	...	...	...	0(8)	...	...	...	...	...	?
Lonchophylla handleyi	H	...	...	...	...	...	...	0(2)[l]	0(2)	...	...	...	...	D
Lonchophylla thomasi	L	...	...	0(1)	...	...	0(3)	33(6)	...	0(1)	...	...	...	?
Lionycteris spurrelli	L	...	...	...	...	...	...	...	29(7)	...	...	0(1)	...	?
Anoura geoffroyi	L	...	...	...	...	...	50(2)	43(7)	...	...	0(2)	...	...	?
Anoura geoffroyi	H	...	0(2)	...	...	50(2)	[j]11(35)[l]	[j]7(27)[l]	5(22)[l]	0(5)	[j]0(4)	0(3)[l]	0(3)	D/W
Anoura caudifer	L	...	...	0(1)	...	100(1)	...	40(5)[l]	...	40(5)	...	50(2)[l]	...	D/W
Anoura caudifer	H	...	...	...	...	...	[j]...	100(1)	[s]33(3)	0(1)	...	0(2)	...	D/W
Anoura cultrata	L	...	...	...	...	...	100(1)	...	...	...	...	...	...	?
Anoura cultrata	H	...	...	...	100(1)	100(1)	0(1)	[j]0(8)[l]	[j]0(6)[l]	...	...	...	...	D
Lichonycteris obscura	L	...	...	...	...	...	0(1)	...	...	...	...	...	...	?
Platalina genovensium	P	...	...	...	...	...	...	...	...	50(2)	...	...	...	?
Choeroniscus minor	L	...	...	0(1)	...	...	100(2)	...	[j]...	...	...	...	...	D
Choeroniscus intermedius	L	...	...	...	...	...	...	0(2)	...	...	...	...	...	?
Carolliinae														
Carollia castanea	L	[j]...	...	...	0(3)	...	[s]21(19)	[s]14(14)	[s]8(13)	77(13)	...	56(9)[n]	100(2)	W
Carollia brevicauda	L	...	0(1)	...	0(2)	...	0(19)	0(12)	41(17)	0(1)	80(5)	71(14)	...	W
Carollia brevicauda	H	[j]...	...	...	...	[j]...	0(7)	0(31)[l]	9(11)	100(2)	100(2)	50(4)[l]	...	D/W
Carollia perspicillata	L	[j]...	...	...	0(3)	0(2)[l]	0(24)	[j]13(32)	21(99)	50(18)	29(7)	[j]37(19)	57(14)[l]	D/W
Carollia perspicillata	H	...	...	...	...	0(1)	...	0(6)	100(1)	...	50(8)	25(4)	0(1)	W
Rhinophylla pumilio	L	...	...	...	0(2)	...	...	14(22)	[s]33(6)	50(2)	0(1)	50(6)	...	W
Rhinophylla fischeri	L	...	...	...	...	...	50(2)	[s]9(23)	0(12)	...	0(4)	...	...	W
Sturnirinae														
Sturnira lilium	L	...	0(1)	0(1)	0(1)	0(7)	3(29)	31(26)	62(29)[l]	100(1)	100(1)	[j]3(38)	14(14)[l]	D/W
Sturnira tildae	L	...	...	0(1)	0(1)	...	[s]100(1)	0(7)	33(3)	...	...	67(3)[l]	...	W
Sturnira magna	L	...	...	...	...	...	0(1)	0(7)[l]	...	...	75(4)	...	100(1)[n]	D/W
Sturnira magna	H	...	...	...	...	0(2)	0(2)	[j]33(6)	...	...	...	...	...	D
Sturnira bidens	H	...	...	...	...	0(5)	[s]50(2)	[s]13(8)	29(14)	...	[s]75(12)	...	...	W
Sturnira nana	H	...	...	...	...	0(3)	...	0(1)	100(1)	...	...	...	...	?
Sturnira bogotensis	L	...	...	...	...	...	0(7)[l]	0(1)	0(6)	100(1)	40(5)[n]	[s]50(2)	...	D/W
Sturnira bogotensis	H	...	...	...	0(3)	...	[j]40(15)[l]	0(5)	39(26)[l]	100(1)	...	100(1)	...	D/W
Sturnira erythromos	H	0(1)	0(1)	...	...	0(1)	4(23)[l]	47(17)	53(45)	[j]10(20)	50(4)	0(1)	[j]40(25)[l]	D/W

Appendix. *Continued.*

Species	Region*	Jan.	Feb.	Mar.	Apr.	May	Jun.	Jul.	Aug.	Sep.	Oct.	Nov.	Dec.	Season†
Stenodermatinae														
Uroderma bilobatum	L	50(2)	...	0(1)	0(1)	0(4)[l]	14(7)	54(11)[l]	60(15)	[j]83(6)	100(1)	[j]83(2)	57(7)	D/W
Uroderma biobatum	H	...	...	...	...	...	...	...	100(1)	...	...	...	...	?
Uroderma magnirostrum	L	...	0(2)	0(2)[l]	0(1)	0(1)	0(5)	...	0(3)	...	...	...	...	W
Vampyrops infuscus	L	...	...	0(1)	...	...	0(5)	11(9)	10(10)	...	...	100(4)	...	W
Vampyrops infuscus	H	...	...	...	...	...	...	0(5)	0(1)	...	...	...	...	?
Vampyrops vittatus	H	...	...	...	...	...	0(4)	0(1)	50(8)	...	...	...	...	?
Vampyrops dorsalis	H	...	...	...	...	0(2)	[s]100(2)	35(17)[l]	63(16)	100(1)	100(1)	62(21)[l]	...	D/W
Vampyrops dorsalis	L	...	...	...	...	0(2)	...	...	...	...	...	...	...	?
Vampyrops lineatus	L	...	...	...	...	...	0(1)	...	...	...	...	...	...	?
Vampyrops lineatus	H	...	...	...	...	0(5)	0(3)	53(15)	[s]57(30)	...	...	...	...	W
Vampyrops brachycephalus	L	...	0(2)	0(1)[l]	...	...	0(4)	17(12)	33(12)	29(7)	50(2)[n]	60(5)	...	W
Vampyrops helleri	L	...	...	0(2)	...	0(1)	0(2)	67(3)	50(4)	67(3)	83(6)	38(8)[l]	50(2)	W
Vampyrodes caraccioli	L	...	...	...	...	...	0(2)	100(1)[l]	...	...	...	100(3)[l]	...	D/W
Vampyressa pusilla	L	...	...	...	...	...	0(1)	0(5)	0(1)	...	...	92(13)	20(15)[l]	W
Vampyressa pusilla	H	...	...	...	...	...	0(1)	...	0(1)	...	...	...	...	?
Vampyressa melissa	H	...	...	...	...	0(3)	...	...	50(6)	...	...	100(1)	...	W
Vampyressa bidens	L	...	...	...	...	...	0(2)	...	100(1)	...	50(2)	94(17)	69(13)[l]	W
Chiroderma villosum	L	...	...	0(1)	0(2)	0(1)	20(5)	25(4)	33(3)	100(1)[l]	50(2)	50(2)	100(1)	W
Chiroderma trinitatum	L	...	...	...	0(1)	0(6)	...	0(1)	...	...	...	50(8)[l]	0(4)[l]	W
Ectophylla macconnelli	L	...	...	...	0(2)	...	25(4)	50(4)	33(3)	...	33(3)[l]	0(1)	...	W
Ectophylla macconnelli	H	...	...	...	...	0(2)[l]	...	...	50(2)	...	...	...	...	D/W
Artibeus cinereus	L	...	...	...	...	...	...	...	...	...	...	0(1)[l]	0(1)	W
Artibeus glaucus	L	...	...	...	...	...	...	100(2)	...	...	...	...	...	?
Artibeus glaucus	H	...	...	0(2)	0(8)	0(3)	38(8)[l]	30(20)[l]	62(13)	100(1)	...	[s]100(1)	...	D/W
Artibeus anderseni	L	...	...	...	...	...	7(15)	0(3)	[j]22(9)	...	...	[j]50(2)	0(1)	D/W
Artibeus fraterculus	P	...	...	...	...	...	33(6)	...	75(4)[l]	67(9)	[j]42(24)[l]	75(4)	73(30)[l]	D/W
Artibeus fuliginosus	L	...	...	0(1)	...	...	27(11)[l]	20(10)	50(8)	67(6)	88(8)	[j]25(8)	50(2)	D/W
Artibeus fuliginosus	H	...	...	...	...	...	10(10)	...	...	...	...	...	...	?
Artibeus planirostris	L	100(1)	...	0(1)[l]	100(1)	...	6(35)	4(28)	32(22)	83(6)	[j]0(2)	[j]37(46)[l]	100(1)	D
Artibeus planirostris	H	...	...	...	...	0(1)	...	50(8)	[s]67(3)	...	...	[j]67(12)[l]	...	W
Artibeus planirostris	P	...	...	...	...	...	...	...	...	...	0(1)	...	...	?
Artibeus lituratus	L	...	...	...	100(1)	...	0(1)	33(3)	33(3)	0(2)	50(6)[l]	[j]29(24)[l]	100(1)	D/W
Artibeus lituratus	H	...	...	...	...	...	...	100(2)	...	...	...	...	...	?
Enchisthenes hartii	L	...	...	...	...	...	...	...	...	...	...	100(1)	...	W
Enchisthenes hartii	H	...	...	...	...	...	0(1)	0(1)	[s]20(5)	100(1)	...	100(2)	100(1)	W

APPENDIX. *Continued.*

Species	Region*	Jan.	Feb.	Mar.	Apr.	May	Jun.	Jul.	Aug.	Sep.	Oct.	Nov.	Dec.	Season†
Desmodontidae														
Desmodus rotundus	L	···	···	0(2)	···	···	···	0(9)	29(7)[l]	100(1)[n]	···	···	···	D
Desmodus rotundus	H	···	···	···	···	[j]···	[j]67(3)	[j]23(13)	50(2)	[j]50(8)	[j]45(33)	[j]14(7)	0(3)	D/W
Desmodus rotundus	P	···	···	···	[j]0(5)	0(1)	0(2)	50(2)	0(1)	···	···	100(1)	0(3)	W
Desmodus youngi	L	···	···	···	···	···	100(1)	···	···	···	···	···	···	?
Diphylla ecaudata	L	···	···	···	···	0(1)	···	···	···	···	···	···	···	?
Diphylla ecaudata	H	···	···	···	···	0(3)	0(1)	···	···	···	100(1)	···	···	W
Furipteridae														
Amorphochilus schnablii	P	···	···	···	[j]0(2)	0(23)	20(5)	···	[s]91(11)	···	71(7)	···	···	W
Thyropteridae														
Thyroptera tricolor	L	···	···	[j]0(3)	···	···	···	100(3)	0(1)	···	[j]0(5)	···	0(9)[l]	W
Vespertilionidae														
Myotis nigricans	L	0(2)	···	[j]0(3)	[j]23(13)	[j]0(7)	[j]43(37)[l]	35(20)	[j]17(36)[l]	[j]0(2)	[j]0(2)	[j]50(2)	···	D/W
Myotis nigricans	H	[j]0(4)	[j]17(6)[l]	···	···	···	···	50(2)	[j]23(69)[l]	[j]0(2)	[j]23(31)[l]	···	···	D/W
Myotis nigricans	P	···	···	···	···	···	···	0(2)	···	···	···	···	···	?
Myotis keaysi	L	···	0(2)	···	···	···	0(2)	78(9)	0(1)	···	···	···	0(1)	?
Myotis albescens	L	···	···	···	[j]0(2)[l]	0(1)	[j]10(10)	[j]0(5)	[sj]35(29)	[j]···	···	···	···	D/W
Myotis albescens	H	···	···	0(1)	···	···	···	···	···	···	···	···	···	?
Myotis albescens	P	···	···	···	···	···	···	[j]0(3)	···	···	···	···	···	D
Myotis simus	L	···	···	0(3)	0(1)	···	···	100(1)	67(3)	···	···	···	···	?
Myotis riparius	L	···	···	···	···	···	···	0(3)	[j]33(3)	···	···	···	···	D
Myotis oxyotus	H	···	0(1)	···	···	···	50(2)	0(3)	[j]0(1)	···	···[l]	0(1)	···	D/W
Eptesicus andinus	H	···	···	···	···	···	0(1)	···	[s]73(15)	0(1)	···	···	···	W
Eptesicus brasiliensis	L	0(1)	···	···	···	0(1)[l]	33(3)[l]	0(8)	0(5)[l]	···	···	···	···	D
Eptesicus innoxius	P	···	···	···	···	···	···	0(3)	···	···	0(2)	···	···	?
Eptesicus furinalis	L	···	···	···	···	···	···	0(1)	···	···	···	···	···	?
Histiotus macrotus	H	···	···	0(1)	···	···	···	···	···	···	60(5)[l]	···	···	W
Histiotus montanus	H	···	···	···	[j]0(1)	[j]0(4)	[j]0(10)[l]	···	···	0(1)	···	···	···	D/W
Lasiurus borealis	L	···	···	···	···	···	0(1)	0(1)	0(1)	···	···	0(1)[l]	···	W
Lasiurus borealis	P	···	···	···	0(1)	···	···	···	···	···	···	···	···	?
Lasiurus cinereus	H	···	···	···	···	0(1)	···	···	···	···	···	···	···	?
Tomopeas ravus	P	···	···	···	···	···	···	···	[j]0(6)	···	···	0(1)	···	D

Appendix. *Continued.*

Species	Region*	Jan.	Feb.	Mar.	Apr.	May	Jun.	Jul.	Aug.	Sep.	Oct.	Nov.	Dec.	Season†
Molossidae														
Molossops abrasus	L	...	...	...	...	...	...	[s]100(1)	...	...	...	0(2)[l]	...	W
Molossops temminckii	L	...	...	...	...	...	...	...	100(2)	...	...	...	...	?
Tadarida brasiliensis	H	...	...	...	...	...	...	...	100(3)	...	43(7)	...	...	W
Tadarida brasiliensis	P	0(1)	...	...	...	0(1)	...	...	...	...	...	...	...	?
Nyctinomops macrotis	P	...	...	...	...	...	...	...	0(1)	...	...	...	...	?
Nyctinomops laticaudatus	P	...	...	...	...	...	...	0(1)	...	...	0(5)	...	...	?
Mormopterus kalinowskii	P	...	0(1)	...	0(1)	0(1)[l]	...	...	...	...	...	...	...	D
Molossus ater	L	...	...	...	...	...	...	...	...	...	...	100(1)	0(5)[l]	W
Molossus molossus	L	...	...	0(2)	[s]25(4)	...	0(1)	8(13)	[j]58(38)[l]	...	...	18(11)[l]	42(12)[l]	D/W
Molossus molossus	P	...	...	...	...	0(6)	0(3)	0(11)	...	[s]10(20)	9(11)	...	...	W
Promops centralis	P	...	...	...	100(1)	0(1)	...	0(6)	...	0(1)	[s]17(6)[l]	...	...	D/W
Eumops auripendulus	L	...	...	...	...	...	...	...	...	[j]50(2)	...	...	...	D
Eumops auripendulus	P	...	...	...	...	...	...	0(2)	0(1)	...	...	...	...	?
Eumops glaucinus	P	...	...	...	...	0(1)	0(2)	0(4)	[j]0(4)	...	0(1)	...	...	D
Eumops perotis	L	...	...	...	...	...	0(2)	...	...	...	0(2)	...	...	?

* L = Eastern lowlands; H = Andean highlands; and P = Pacific slope.
† D = Parturition possible during dry season; W = parturition possible during wet season; D/W = parturition possible during both seasons; and ? = insufficient data for pattern.
[j] Juveniles. [l] Lactating females.
[s] Small embryos. [n] Large, near-term embryos.

Tent Construction by Bats of the Genera *Artibeus* and *Uroderma*

Robert M. Timm

ABSTRACTS

Herein, I describe new styles of tents cut and utilized by *Artibeus anderseni, A. glaucus, A. gnomus, A. phaeotis, A. toltecus, A. watsoni, Uroderma bilobatum,* and *U. magnirostrum*; review and summarize the literature on tent use by *Artibeus* and *Uroderma*; and discuss the effectiveness of tents as diurnal roosts. *Artibeus anderseni* alters the shape of *Heliconia* leaves by cutting the lateral nerves and interconnected tissue extending out from the midrib. *Artibeus glaucus* cuts the basal lateral nerves in *Xanthosoma,* causing the two sides of the leaf to collapse downward around the midrib. *Artibeus phaeotis* cuts the lateral nerves and interconnected tissues in both banana and *Heliconia imbricata*; the basal cuts veer out from the midrib such that a distinctive V-shaped enclosure is formed by the hanging leaf. *Artibeus toltecus* cuts the basal nerves on *Anthurium,* causing the sides of the leaf to fold down around the midrib to form a pyramid-shaped tent. *Artibeus watsoni* was found to make four distinctive styles of tents, including simple V-shaped cuts on bifurcated palms, cuts of a few side veins on aroids to produce a rounded pyramid, elongate J-shaped cuts on banana and *Heliconia,* and polygonal cuts on *Carludovica palmata. Artibeus watsoni* has the greatest repertoire in tent styles, and uses the most diverse array of plant species and leaf shapes. Two styles of tents constructed by *Uroderma bilobatum* are reported for the first time, one on the large pinnately leafed palm *Scheelea rostrata* and the second on banana. The common denominator between the *Uroderma bilobatum* tents reported herein and those previously described is that all are on large, broad leaves and all have a distinctive V-shaped pattern cut by the bats. *Uroderma magnirostrum* also creates an inverted elongate V-shaped tent on pinnately leafed palms.

All New World tent-makers described to date are tropical members of the phyllostomid subfamily Stenoderminae. Each species of tent-making bat has one or more distinctive style of tent. Bats select leaves of specific shapes, sizes, and angles for tent construction. Most species appear to be obligate tent-roosters. Tents provide bats with a cryptic diurnal roost site, in addition to providing shelter from both the sun and rain and an early warning to the approach of predators.

Aqui yo describo nuevos estilos de carpas cortadas y utilizadas por *Artibeus anderseni, A. glaucus, A. gnomus, A. phaeotis, A. toltecus, A. watsoni, Uroderma bilobatum,* y *U. magnirostrum* reviso y hago un sumario de la literatura acerca del uso de carpas por filostomidos; y discuto la efectividad de las carpas como perchas diurnas. *Artibeus anderseni* altera la forma de las hojas de *Heliconia* cortando las nervaduras centrales y tejido interconectado que se extiende

---

From the Division of Mammals, Field Museum of Natural History, Chicago, Illinois 60605-2496. The author's present address is Museum of Natural History and Department of Systematics and Ecology, University of Kansas, Lawrence, Kansas 66045.

exteriormente desde la nervadura central. *Artibeus glaucus* corta la base de las nervaduras laterales en *Xanthosoma,* causando asi que los dos lados de la hoja colapsen hacia abajo al rededor de la nervadura central. *Artibeus phaeotis* corta las nervaduras laterales y tejidos interconectados en banana y *Heliconia imbricata*; los cortes basales viran hacia afuera desde la nervadura central de tal modo que una distintiva cavidad en forma de V es formada por la hoja colgante. *Artibeus toltecus* corta las nervaduras basales de *Anthurium,* adi que los lados de la hoja doblen en redor de la nervadura central de tal modo que una cavidad es formada en forma de una piramide. *Artibeus watsoni* fue encontrada haciendo cuatro distintos estilos de carpas, incluyendo simples cortes en forma de V en palmas bifurcadas, cortes en unas pocas nervaduras laterales en araceas para producir una piramide redondeada, cortes alargados en forma de J en banana y *Heliconia,* y cortes poligonales en *Carludovica palmata. Artibeus watsoni* tiene el mas grande repertorio en estilos de carpas y usa la mas diversa serie de plantas y formas de hojas. Dos estilos de carpas construidas por *Uroderma bilobatum* son reportadas por primera vez; una en la larga y pinnada hoja de palma *Scheelea rostrata* y la segunda en banana. El comun denominador entre carpas de *Uroderma bilobatum* reportadas aqui y aquellas previamente descritas es que todas usan hojas grandes y anchas y todas tienen un distintivo patron en forma de V cortado por los murciélagos. *Uroderma magnirostrum* también corta una carpa en forma de V-invertido en hojas pinnadas de palmas.

Todos los filostomidos cortadores y utilizadores de carpas descritos del Nuevo Mundo son membros tropicales de la subfamilia Stenoderminae. Cada especie tiene uno o mas estilos distintos de carpas. Los murciélagos escogen hojas de formas especificas, y constróen sus carpas en angulos peculiares. La mayor parte de las especies parecen utilizar las carpas obligatoriamente. Las carpas oferecen una percha oculta durante el dia, asi como un abrigo del sol, de la lluvia, y de predadores.

Neste trabalho, (1) descrevo novos estilos de tendas cortadas e utilizadas por *Artibeus anderseni, A. glaucus, A. gnomus, A. phaeotis, A. toltecus, A. watsoni, Uroderma bilobatum,* e *U. magnirostrum*; (2) reviso e resumo a literatura sobre o uso de tendas pelos morcegos da família Phyllostomidae, e (3) discuto a eficiência de tendas como alojamentos diurnos. *Artibeus anderseni* altera a forma das fôlhas de *Heliconia,* cortando as veias laterais e os tecidos interligados que estendem da veia central. *Artibeus glaucus* corta as veias basilares laterais em *Xanthosoma,* causando com que os dois lados da fôlha caiam contra o centro. *Artibeus phaeotis* corta as veias laterais e os tecidos interligados nas fôlhas de bananas e de *Heliconia imbricata.* Os cortes basilares partem da veia central, formando um abrigo distinto em forma de "V". *Artibeus toltecus* corta as veias basilares de *Anthurium,* causando com que os lados da fôlha dobrem em volta da veia central, criando um abrigo em forma de pirâmide. *Artibeus watsoni* constrói quatro tipos diferentes de tendas, incluindo simples cortes em forma de "V" em fôlhas de palmeiras bifurcadas; cortes em algumas das veias laterais em fôlhas de trepadeiras (resultando em pirâmides redondas); cortes alongados, em forma de "J", em fôlhas de bananas e de *Heliconia*; e cortes polígonos em fôlhas de *Carludovica palmata. Artibeus watsoni* possue o maior repertório de estilos de tendas, e usa o conjunto mais diverso de espécies de plantas e de configurações de fôlhas. Dois estilos de tendas construídas por *Uroderma bilobatum* são descritos pela primeira vez; um nas fôlhas grandes da palmeira *Scheelea rostrata,* e outro nas fôlhas de bananas. Fatôres comuns entre as tendas construídas por *Uroderma bilobatum* aqui descritas, e as descritas prèviamente, são a forma distinta em "V" cortada pelos morcêgos, e o uso de fôlhas grandes e largas para a construção das tendas. *Uroderma magnirostrum* também constróe tendas em forma de "V" invertido nas fôlhas de palmeiras.

Todos morcêgos construidores de tendas no Novo Mundo pertencem à subfamília Stenoderminae (família Phyllostomidae), e cada espécie exibe um ou mais estilos característicos de construção. A maioria destas espécies de morcêgos parecem alojar-se obrigatòriamente em tendas, as quais oferecem não só um abrigo diurno camuflado, mas também proteção contra sol, chuva, e predadores.

Introduction

The use of cut leaves for diurnal roosting sites by bats was first described by Thomas Barbour (1932), who reported on diurnal roosts of *Uroderma bilobatum* near the Panama Canal. He found these bats roosting under the leaves of two cultivated palms identified as *Livistona chinensis* and *Prichardia pacifica*. Not only were these bats found roosting under palm fronds, but they had also altered the leaf to produce a diurnal roosting structure. Barbour (1932, p. 307) stated that "by nipping the ridges of the plications on the under side the leaf is weakened and as the bitten spots are skillfully and serially distributed the leaf finally is sufficiently weakened so that the distal portion droops sharply downward." Chapman (1932, p. 555) discovered *Artibeus watsoni* roosting under the cut veins of a bifurcated palm, *Geonoma cuneata* (reported as *G. decurrens*), on Barro Colorado Island, Panama, and first called these modified leaves "tents." Ingles (1953) also reported *A. watsoni* constructing tents in the palm *Geonoma oxycarpa* (reported as *G. binervia*) on Barro Colorado Island. Goodwin and Greenhall (1961, p. 262) found *Artibeus cinereus* roosting "under cut leaves of palm trees and on the under side of banana leaves" and *Uroderma bilobatum* roosting under cut leaves of the carat palm, *Sabal glaucescens*, and coconut palm, *Cocos nucifera*, on Trinidad. *Ectophylla alba* was reported by Timm and Mortimer (1976) to alter the leaves of five species of *Heliconia* in Costa Rica; the bats selected specific leaves for both size and angle to the ground. *Artibeus jamaicensis* was found by Foster and Timm (1976) roosting under the cut leaflets on a pinnately leafed palm, *Scheelea rostrata*, in a tropical dry forest in Costa Rica. Recently, Timm (1984) reported tent construction by another phyllostomid, *Vampyressa pusilla*, in Costa Rica, and Koepcke (1984) found *Mesophylla macconnelli* utilizing similar tents in Peru. Only one Old World bat, *Cynopterus sphinx* (Pteropodidae), has been reported to alter the shape of palm leaves to produce a diurnal roosting structure (Goodwin, 1979). Reviews of roosting site selection by bats were provided by Tuttle (1976) and Kunz (1982).

To date, seven species of phyllostomids (*Artibeus cinereus, A. jamaicensis, A. watsoni, Ectophylla alba, Mesophylla macconnelli, Uroderma bilobatum*, and *Vampyressa pusilla*) have been reported to modify leaves of plants to produce diurnal roosting sites herein referred to as tents. All are tropical members of the phyllostomid subfamily Stenodermatinae.

The phyllostomid genus *Artibeus*, which includes some 15 species, is widespread in the Neotropics from northern Mexico southward to Argentina and Chile. These bats range in size from 10 g (*A. anderseni* and *A. watsoni*) to 70 g (*A. lituratus*). *Uroderma*, a closely related genus of medium-sized stenodermatines that includes only two species, is found from southern Mexico through the Amazon Basin of South America. The better known of the two species, *U. bilobatum*, weighs from 13 to 21 g.

Herein I describe tent construction and utilization by *Artibeus anderseni, A. glaucus, A. gnomus, A. phaeotis, A. toltecus, A. watsoni, Uroderma bilobatum*, and *U. magnirostrum*; describe several new styles of tents; review and summarize the literature on tent use by *Artibeus* and *Uroderma*; discuss the effectiveness of tents as diurnal roosts; and suggest directions for future research.

Methods

Descriptions of Study Areas

Costa Rica—Bosque Brrancia, also known locally as Bosque Blanco, is located 0.8 km west of Cuarto Cruces on the south side of the Pan American Highway (Route 1) in Guanacaste Province, in the Pacific lowlands of western Costa Rica. Bosque Brrancia is a restricted area of undisturbed lowland forest classified as Tropical Moist Forest; the dominant vegetation includes *Anacardium excelsum* and *Scheelea rostrata*. This stand of forest probably has not been logged previously, and represents a close approximation to the original (pre-Columbian) forests of this part of Guanacaste. Further descriptions of this forest can be found in Janzen (1971) and Wilson and Janzen (1972).

Parque Nacional de Corcovado is located on the Osa Peninsula of southwestern Costa Rica, Puntarenas Province (between 08°27′N and 08°39′N, and 83°25′W and 83°45′W); the elevation ranges from sea level to 400 m. Corcovado lies within the Tropical Wet Forest Life Zone (Holdridge, 1967), with lowland evergreen forest being the dominant forest type. Mean annual rainfall is 3,800+ mm and the wettest months are from August through November; mean monthly temperatures range from 25.0°C to 26.5°C. Vegetation and habitat types at Corcovado have been described by Herwitz (1981) and Hartshorn (1983). Areas surveyed included both primary forest and secondary scrub along the coast.

The La Selva Biological Station is the field station of the Organization for Tropical Studies located 1 km SW of Puerto Viejo de Sarapiquí, Heredia Province in the Caribbean lowlands of northeastern Costa Rica (10°27′N,84°00′W); elevation ranges from 29 to 100 m. Mean annual rainfall is 3,990 mm, with the wettest months being November, December, and February; mean monthly temperatures range from 24.5°C (December) to 26.1°C (April). La Selva lies within the Tropical Wet Forest Life Zone, with lowland evergreen forest being the dominant forest type. Vegetation and habitat types of La Selva have been described by Slud (1960), Holdridge et al. (1971), Sawyer and Lindsey (1971), and Hartshorn (1983). One unusual feature of the subcanopy of the La Selva forest is the diversity and abundance of dwarf palms (Hartshorn, 1983), especially the broad-leaved species, *Geonoma congesta* and *Asterogyne martiana*. These species are regularly utilized by *Artibeus watsoni* for tent construction. The transect surveys, which included all *Artibeus* tents observed, were restricted to primary forest. The *Uroderma* tents described from this site were restricted to an open banana patch.

Palo Verde (Refugio Nacional de Fauna Silvestre Dr. Rafael Lucas Rodríguez Caballero) is a wildlife refuge located 2 km S and 12 km E of Bolsón, in the Pacific lowlands of Guanacaste Province, northwestern Costa Rica (10°30′N, 85°20′W); the elevational range is from 3 to 183 m. Palo Verde lies within the Tropical Dry Forest Life Zone, with lowland deciduous forest and riverine swamp forest being the dominant forest types. The vegetation and habitat types of Palo Verde have been described by Slud (1980) and Hartshorn (1983). Mean annual rainfall is 1,700+ mm, with the wettest months being April, May, September, and October; mean monthly temperatures range from 26.0°C (November) to 29.7°C (April). The immediate vicinity of the survey at this site was in a mosaic of primary forest and secondary scrub that included considerable bananas scattered throughout.

The newly expanded Parque Nacional Braulio Carrillo is located in northeastern Costa Rica (between 10°05′N and 10°25′N, and 83°54′W and 84°05′W); the elevation of the park ranges from 100 to 2600 m. Braulio Carrillo is located on the eastern, Caribbean slope of Volcan Barba in Heredia Province. The elevational range at which *Artibeus toltecus* and associated tents were observed was from 700 to 1400 m, within the Premontane Rain Forest Zone with midelevational evergreen forest and tall palms being the dominant forest types. Mean annual rainfall at this elevational range is perhaps as much as 5,000 mm, although no exact measurements are yet available. The vegetation and habitat types at the midelevational ranges also have yet to be described. On 14–15 April 1986, 3 km of trail ranging from 700 to 1100 m were surveyed for bat tents. Additionally, an intensive netting effort with Richard K. LaVal and Don E. Wilson was conducted in this area over a 7-day period to determine what species of bats were present and their relative abundance.

At Finca Las Cruces (2 km S of San Vito, Puntarenas Province, 08°45′N, 82°58′W, 1200 m) in the Premontane Wet Forest–Rain Forest transition area, approximately 2 km of trail leading down to the Río Jaba was surveyed for tents on 13 August 1982.

ECUADOR—Cascada San Rafael lies 17 km (by road) west of the village of Reventador, Napo Province, in northeastern Ecuador (00°5.8′S, 77°34.4′W). Rainfall averages 1,500 to 2,000 mm; average temperatures range from 18°C to 22°C. The elevation is 1200 m. Reventador lies within the Humid Subtropical Life Zone.

Lagarto Cocha and Zancudo Cocha are military encampments along the Río Aguarico of Amazonian Ecuador named for prominent lagoons. Both areas are undisturbed primary lowland rain forests classified as Moist Forest, with an annual rainfall averaging from 2,000 to 3,000 mm; the elevation is approximately 200 m.

PERU—Hacienda Amazonia lies just north of the confluence of the Alto Río Madre de Dios and the Río Pinipini in the department of Madre de Dios, southeastern Peru (12°58′S, 71°09′W), just north of Atalaya. The Hacienda is located just east of Parque Nacional del Manu in the Upper Tropical Zone on the eastern foothills of the Andes. On 25 July 1985 Barbara L. Clauson searched the ridge above the Hacienda for bat tents at an elevation of 825 m in primary rain forest that had received some selective timber harvest; she returned to this site again on 3 November 1985.

Cerro de Pantiacolla lies above the Río Palotoa, 10–15 km NNW of Shintuya, in the department of Madre de Dios, southeastern Peru (12°35′N, 71°18′W). On 15 November 1985 Clauson searched a steep sloping ridge at 600 m. The sur-

TABLE 1. Individual measurements (in cm or degrees) collected from four tents constructed by *Artibeus anderseni* on *Heliconia* in Ecuador.

Blade			Petiole		Basal	
Length	Width	Angle	Length	Angle	height	Remarks
176	34	40	80	80	200	1 adult male *A. anderseni*
158	38	15	75	55	220	3 male *A. anderseni*
154	36	50	88	80	260	. . .
140	27	. . .	60	70	140	. . .

rounding forest was primary rain forest in the Upper Tropical Zone on the eastern Andean foothills.

Reference specimens of the bats have been deposited at Field Museum of Natural History, Chicago; Escuela Politécnica Nacional, Quito, Ecuador; and Universidad Nacional de Costa Rica and Servicio de Parques Nacionales, San José, Costa Rica. Voucher specimens of plants have been deposited in the herbaria at Field Museum, Duke University, Missouri Botanical Garden, and/or University of Wisconsin.

Accounts of Species

Artibeus

Artibeus anderseni Osgood, 1916

Artibeus anderseni occupies an extensive range in western Amazonia; however, little is known of its biology. This species has long been considered a junior synonym of *A. cinereus*. In resurrecting *A. anderseni* as a distinct species, Koopman (1978, p. 14) stated, "Besides its shorter face and more abrupt forehead, *A. anderseni* apparently always lacks the last lower molars, which *A. cinereus* in western Amazonia almost always has." I concur with Koopman in recognizing *A. anderseni* as a distinct species.

In late October and early November of 1983, the trails and forest surrounding the military encampments were searched at Lagarto Cocha and Zancudo Cocha in eastern Ecuador. *Artibeus anderseni* was found to alter the shape of leaves of several small, forest *Heliconia* species to produce diurnal roosting structures. To create a tent from a *Heliconia* leaf, the bat severs the lateral nerves and interconnecting veins that extend along both sides of the midrib. The cuts ran along the central 90% of the leaf from near (0 to 14 cm) the base to near (10 to 20 cm) the tip. Nerves and interconnected tissues were severed, but not completely, so that they did afford some support for the sides of the leaf. Cuts ran parallel to the midrib for most of its length, but did flare outward slightly toward the base. The lateral nerves were cut from 3 to 8 mm from the midrib; the midrib was not cut. Claw marks where the bats roosted started 50 cm from the base in one tent and ran for 16 cm distally; in another they started at 70 cm and ran for 10 cm distally. Measurements of the blade length, blade width, blade angle, petiole length, petiole angle, and basal height of four tents are provided in Table 1.

At Zancudo Cocha one *Heliconia* tent was unoccupied for two days in succession, then on the third day was occupied by an adult male with enlarged testes, an adult lactating female, and a juvenile male *Artibeus anderseni* (fig. 1). Another *Heliconia* tent was unoccupied.

At Lagarto Cocha 13 tents were found in *Heliconia*. One was occupied by three subadult males not in breeding condition; a second tent contained a single adult male with enlarged testes.

Artibeus cinereus (Gervais, 1856)

Artibeus cinereus, Gervais's fruit-eating bat, is found on the islands of Grenada, Trinidad, and Tobago, and throughout the Amazon Basin and adjacent coastal areas. Surprisingly little has been published on roosting behavior or ecology of this widely distributed species. On Trinidad, Goodwin and Greenhall (1961, p. 262) stated, "It roosts in small colonies of a few individuals under the cut leaves of palm trees and on the under side of banana leaves." On Tobago, Husson (1954, p. 64) reported a single male *Artibeus cinereus* "hanging in a banana tree in cultivated country near the shore."

FIG. 1. Photograph of an adult male, adult female, and juvenile *Artibeus anderseni* roosting in a *Heliconia* leaf tent. Details of the cut side nerves can be seen along the midrib of the leaf.

Artibeus glaucus Thomas, 1893

Artibeus glaucus is found at midelevations along the eastern slopes of the Andes from Venezuela to Peru. The status of the name *glaucus* has long been in a state of flux; it often has been considered a subspecies of *A. cinereus.* I concur with Handley (1987) in regarding *A. glaucus* as a distinct species.

A single old adult male *Artibeus glaucus* (FMNH 124844) was observed roosting under a cut *Xanthosoma* leaf at Cascada San Rafael, Ecuador, on 21 September 1983. Four cut leaves on separate plants were observed in close proximity to each other; a fifth cut leaf was observed approximately 10 to 15 m to the south. Only the one cut leaf was occupied by the single bat. An adult female *A. glaucus* was netted in the vicinity that evening. All tent leaves were cut down at the time and four were measured. On 26 November, two more cut leaves were found in this *Xanthosoma* population; one contained three *A. glaucus.* The *Xanthosoma* in which the bats were roosting were part of a population of *Xanthosoma* that occupied approximately 1 ha on a steep east-facing hillside.

The basal lateral nerves from 2 to 5 (usually 3) of the *Xanthosoma* leaves were severed near the midrib, causing the two sides of the leaf to collapse downward around the midrib. The midrib was not cut. Four of the five cut leaves were measured in September; one found in November was measured (table 2). The leaves that were selected by *Artibeus glaucus* for tents all had the midrib running approximately parallel to the ground, whereas the majority of unaltered leaves in the population stood at more vertical angles.

Artibeus gnomus (Handley, 1987)

Artibeus gnomus, the dwarf fruit-eating bat, is found in a peculiar circular range ringing the western edge of the Amazon Basin. Although this species has a wide distribution from Venezuela and Guyana to Peru, it was only recently recog-

TABLE 2. Individual measurements (in cm or degrees) collected from five tents constructed by *Artibeus glaucus* on *Xanthosoma* in Ecuador.

Blade			Petiole	Basal	
Length	Width	Angle	length	height	Remarks
57	68	60	140	230	3 basal cut veins on each side, 2–3 cm from midrib
53	...	40	...	230	3 basal cut veins on each side, 1.5–3.0 cm from midrib
55	62	40	...	230	2 veins cut left, 4 cut right side, 2–3 cm from midrib
64	36	45	137	300	3 basal veins cut on each side, 1.5–3.0 cm from midrib; adult male *A. glaucus* hanging 16 cm from base
61	66	20	145	275	5 basal veins cut on each side, 3.5–4.0 cm from midrib; 3 *A. glaucus*

nized as a distinct species, and little biological information is available (Handley, 1987).

On 15 November 1985 Barbara L. Clauson found a single adult male *Artibeus gnomus* roosting under a cut *Monstera lechleriana* leaf. The *Monstera* was growing as an epiphyte approximately 10 m off the ground on a tree on a sloping hillside at 600 m elevation at Cerro de Pantiacolla, southeastern Peru. The single cut leaf was green and healthy and hung horizontally. No other cut leaves were observed in the immediate vicinity.

The altered leaf was 70 cm long and 38.5 cm at its widest point. All lateral nerves along the basal nearly two-thirds of the leaf were severed immediately adjacent to the thick midrib; this included the basal 12 nerves on one side and 14 on the other. The midrib was severed at 44.5 cm from the base which caused the apical third of the leaf to droop downward perpendicular to the midrib. All nerves proximal to the midrib cut were severed. The lateral nerves along the apical, drooping 25.5 cm were unaltered.

The tent resulting from these cuts was quite enclosed, being formed by the sides of the leaf collapsing downward around the midrib and the distal third of the leaf folding down, perpendicularly to the midrib. The lone *Artibeus gnomus* was roosting 9 cm toward the base from the severed midrib of the leaf.

Artibeus jamaicensis Leach, 1821

Artibeus jamaicensis, the Jamaican fruit-eating bat, is found throughout much of tropical Central America, the northern half of South America, and the Greater and Lesser Antilles. At many localities this is one of the most common species of bats encountered and consequently has received more study than any other phyllostomid.

Artibeus jamaicensis was reported roosting under the cut leaflets of *Scheelea rostrata* in Costa Rica by Foster and Timm (1976). *Scheelea rostrata* is a large, pinnately leafed palm with the leaflets extending out at right angles from the horizontal rachis. Leaflets within the middle 1.3 m region of the frond were cut at varying distances that increased going up to the center of the cut area, then decreased. "As a result . . . the distal parts of the leaflets folded perpendicularly, hung vertically below the frond, and formed a broadly lanceolate tent" (Foster & Timm, 1976, p. 266).

Although several *Artibeus jamaicensis* occupied the roost, only two males were captured, one an adult with enlarged testes, and the second a smaller male not in breeding condition. *Artibeus jamaicensis* apparently has a harem mating system, in which a single breeding male defends a roost used by several females and their offspring; nonbreeding males may be found either singly or in small groups (Morrison, 1979; Kunz et al., 1983). *Artibeus jamaicensis* has been found roosting in a wide variety of situations, including caves, hollow trees, buildings, and under unaltered leaves (Tuttle, 1976), and thus is certainly not an obligate tent-roosting species, as apparently are the smaller species of *Artibeus.*

Artibeus phaeotis (Miller, 1902)

Artibeus phaeotis, the pygmy fruit-eating bat, is found from central Mexico to northern South America (Timm, 1985). Most accounts of habitat for pygmy fruit-eating bats mention their being netted in close proximity to stands of bananas, *Musa* × *paradisiaca* (Ramírez-Pulido et al., 1977; Watkins et al., 1972). Davis (1970) suggested that they might roost under the leaves of bananas.

During the summer of 1982, *Artibeus phaeotis*

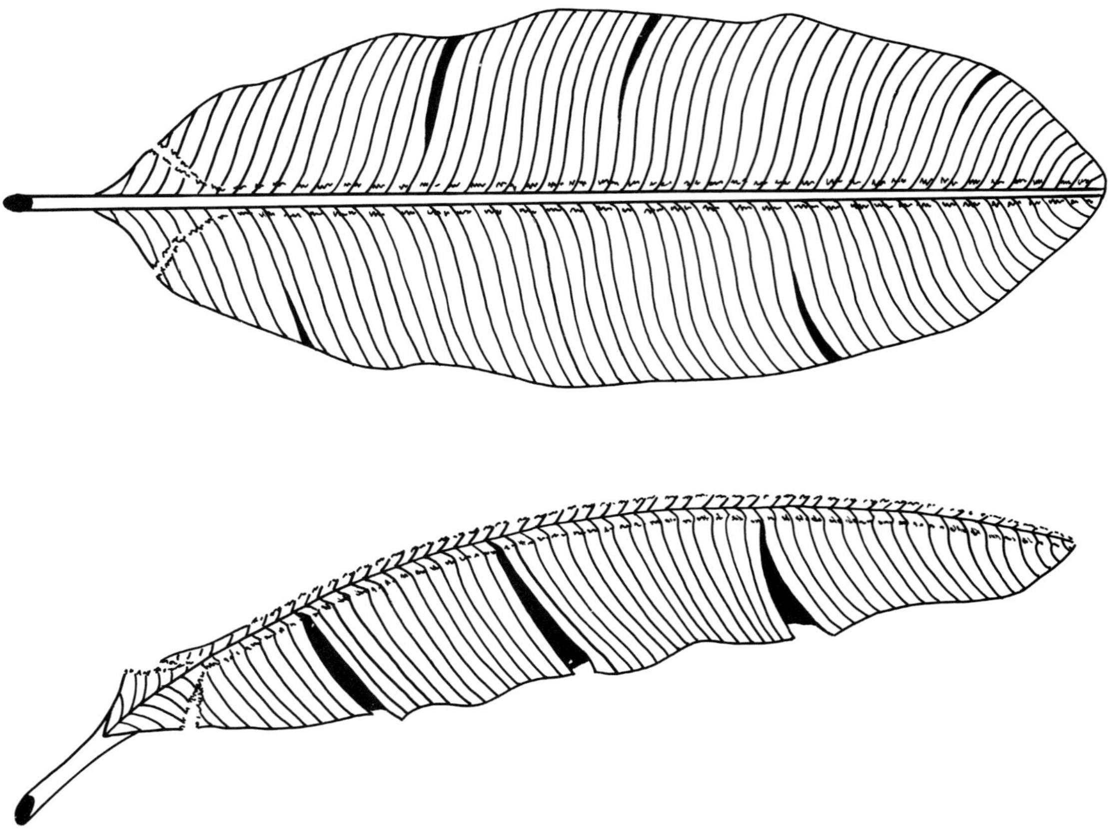

FIG. 2. **Top,** Dorsal view of a banana leaf (*Musa* × *paradisiaca*) showing the cut nerves running along the midrib and flaring out toward the base; **bottom,** tent of *Artibeus phaeotis* made from a banana leaf.

was observed roosting under the leaves of banana and *Heliconia imbricata* at Palo Verde and La Selva, respectively, in Costa Rica. In all cases the leaves had been altered to produce tents.

Artibeus phaeotis constructs roosts in both bananas and *Heliconia imbricata* by biting the lateral nerves and interconnecting tissue that extend at right angles from the midrib, causing the blade to fold over in a V-shaped enclosure. The two sides of the leaf collapse downward, hanging beneath the midrib (fig. 2). Nerves and interconnected tissues are not completely severed, thus the sides of the leaf provide some support for the entire length of the cut. The cuts ran from the base of the leaf to near the tip (table 3). Near the base, the cuts flared out from the midrib to the sides to form an elongate J-shaped pattern. The uncut tip and basal portion of the leaf provide additional strength. The undersides of roost leaves are obscured from view from almost all angles except from directly beneath the tent.

To characterize *Musa* clumps, each of which presumably represented an individual plant, the number of stalks (ramets) per clump was counted; the height of each ramet was estimated to the nearest half meter. This was done for all clumps of *Musa* in the patches that could be located, both those with tents and without bat tents. When a bat tent was located, the following measurements were taken in centimeters: the angle of the petiole (midway from stalk to blade), angle of blade (midway on blade), petiole length, blade length, blade width, height of base of blade, height of tip of blade, height of roost, isolation distance (distance from nearest solid object on the same vertical level), length of uncut basal portion of blade, and length of uncut distal portion. I also noted whether the roosts were in direct sunlight or in shade (table 3). Also for all tents the number of stalks on that column was counted, and the age of the leaf relative to other expanded leaves was recorded. The oldest (lowest leaf) on a plant was assigned the number 1, then the rest counted up from there.

I scored 232 individual stalks of *Musa,* which

TABLE 3. Individual measurements (in cm or degrees) collected from 19 tents constructed by *Artibeus phaeotis* on leaves of banana, *Musa* × *paradisiaca,* in Costa Rica.

Petiole		Blade			Basal height	Height		Isolation	Uncut	
Length	Angle	Length	Angle	Width		Tip	Roost		Base	Tip
43	40	110	−27	20	196	156	182	81	0	15
53	38	136	−14	19	230	146	225	120	10	45
43	42	146	−40	22	213	170	228	125	1.5	48
49	12	120	−55	23	250	180	240	53	5	8
45	52	100	−16	18	205	. . .	200	70	0	30
50	55	120	−3	21	200	165	210	25	3	8
54	50	142	−36	20	235	160	230	115	1	44
40	75	80	+15	19	190	195	206	80	0, 25	20
36	68	175	−12	26	320	260	320	> 300	0	1.5
60	70	195	+25	30	360	400	400	> 300	0, 35	0
44	66	130	+23	25	250	260	290	120	0	40
41	50	94	+5	20	140	155	170	40	0, 40	25
34	45	97	−22	20	205	175	200	100	0	0
50	68	130	0	23	180	220	230	60	0	2
56	60	147	−37	27	260	180	240	35	0	28
33	55	85	−12	21	205	215	220	18	0	5
16	52	60	+28	13	130	. . .	140	35	0	12
46	50	129	+15	21	250	220	260	120	0	65
40	54	88	+15	14	170	. . .	180	220	0	3

were in 41 clumps scattered in 100 m of second growth forest. The mean number of ramets per clump was 5.4; the range from 1 to 23. I located 26 leaves that were cut by bats, a ratio of roughly one tent per 8.1 ramets. Fourteen clumps had cut leaves; if a clump had cut leaves it had a mean of 1.9 cut leaves (range 1 to 2). Of the 26 altered leaves, complete data were taken on 19. The remaining tents were decomposing; in a few cases the bats had not completed the tents. On the partially completed tents, only one side of the midrib was cut along one-quarter to one-third the distance of the blade, but did not cause the side to collapse. Only 2 of the 19 complete tents were located directly in the sun. Bat tents were found in clump sizes ranging from 1 to 23 (mean = 6.2). The average height of all plants (N = 232) was 2.3 m. The average height of plants with tents was 2.0 m (N = 16).

One adult male with enlarged testes and six pregnant females were watched over a 3-day period in one of the roosts. Additionally, two solitary nonpregnant female *Artibeus phaeotis* roosted under separate leaves (fig. 3).

Artibeus phaeotis appears to select banana leaves with specific characteristics. Usually these are the oldest fully expanded leaves, just over 2 m above the ground, with the center of the midrib nearly horizontal to the ground and positioned far enough from nearby stems and branches to limit access by predators. Roost sites generally are located in the shade of surrounding forest trees where apparently they are protected by the forest overstory from wind, blowing rain, and sunlight.

Although bananas are not native to the New World, they are now common throughout the range of *Artibeus phaeotis* and probably provide roost sites in other localities. *Heliconia, Calathea,* and broad-leafed palms are uncommon at Palo Verde, hence are not readily available to *A. phaeotis* for tent sites there. *Artibeus phaeotis* used only banana leaves for tent making at Palo Verde, but constructed similar tents in *Heliconia imbricata* at La Selva. *Heliconia imbricata* is the largest species of *Heliconia* in Costa Rica, and its leaves are similar in size and shape to banana. The tents in *Heliconia* were similar in all respects to those in banana leaves. An adult male and adult female were found roosting together under a single *Heliconia* leaf in late June.

Villa-R. (1967) found a single specimen roosting near the mouth of a small cave in Mexico.

Artibeus toltecus (Saussure, 1860)

Artibeus toltecus, the lowland fruit-eating bat, is found along the coasts of eastern and western Mexico from Nuevo León and Sinaloa south through Central America and perhaps to extreme northwestern Colombia. This species appears to be restricted to midelevational slopes, and in Costa Rica

FIG. 3. Photograph of an adult female *Artibeus phaeotis* roosting in a banana leaf tent. Details of the cut side nerves can be seen along the midrib of the leaf.

I have found it only from 650 to 1500 m in elevation.

In April 1986 I observed a single *Artibeus toltecus* roosting under a cut leaf of *Anthurium caperatum* in Braulio Carrillo National Park, northeastern Costa Rica. The *Anthurium* was growing as an epiphyte on a small tree at 800 m (2½ km S, 11 km E of San Miguel, 10°17′N, 84°05′W). One leaf on the plant was altered; it was 2½ m off the ground and the midrib hung parallel to the ground. Four or five lateral nerves were cut basally on each side, causing the sides of the leaf to fold down around the midrib. A break of the midrib at its midpoint caused the distal half of the leaf to droop down (fig. 4). Seven additional tents of this style were observed on *Anthurium* in this area, ranging in elevation from 700 to 1400 m. It is assumed that they were made by *A. toltecus,* the only small species of *Artibeus* we netted there, although these tents were not occupied. Six tents were observed in a 3-km transect ranging from 700 to 1100 m in elevation.

Davis (1944) reported that *Artibeus toltecus* roosts under banana leaves, although he did not indicate that the bats were modifying the leaves. Davis (1944, p. 378) stated:

> . . . they had regularly established roosts under the large, drooping leaves of the banana trees, each one easily recognized by the manner in which the vane of the leaf hung limply suspended from the midrib. The closely appressed vanes of the leaf, plus the natural darkness within the depths of the grove, afforded good concealment. These bats, too, were wary and that feature coupled with the nature of their retreat caused considerable difficulty in procuring specimens.

In light of Davis's description of the roost sites of *A. toltecus* in banana leaves and my own observations on *A. toltecus,* I suspect that this species was creating tents similar to those I observed for *A. phaeotis* in Costa Rica. The tents formed by *A. phaeotis* in *Musa* (see fig. 2) are similar in ap-

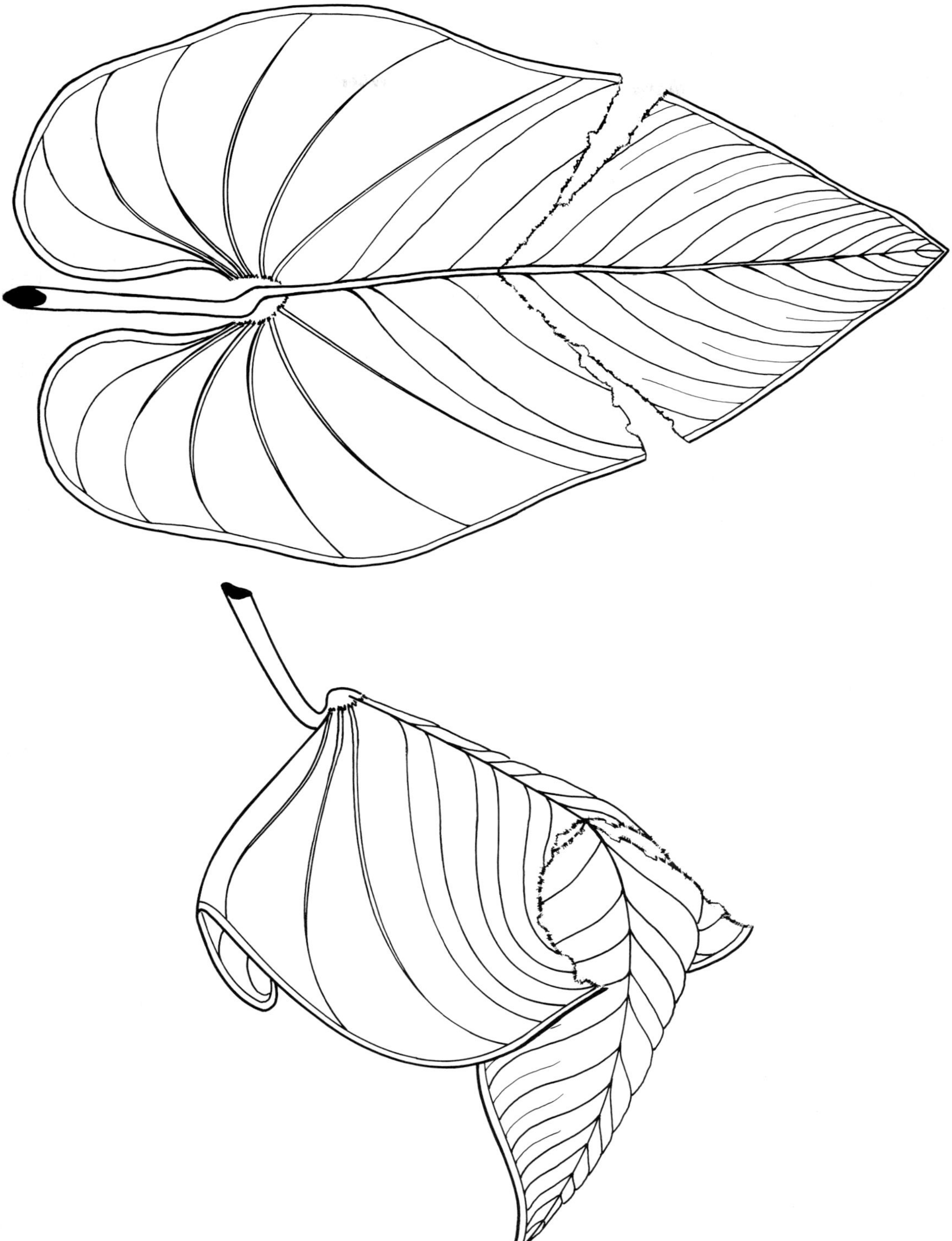

FIG. 4. **Top,** Ventral view of an *Anthurium caperatum* leaf showing the cut nerves along the base of the leaf and the broken midrib; **bottom,** tent of *Artibeus toltecus.*

pearance to those described by Davis (1944) for *A. toltecus*.

Artibeus toltecus has also been reported in caves (Davis et al., 1964; Jones, 1966; Jones & Alvarez, 1964), and Goodwin (1934, p. 12) reported a single specimen collected "in one of the buildings (church?) at San Lucas . . . [the] rest of the congregation seemed to be *Glossophaga.*"

Artibeus watsoni Thomas, 1901

Artibeus watsoni, Thomas's fruit-eating bat, is one of the smaller members of the genus *Artibeus* and found from southern Veracruz south through Central America to northern South America. It appears to be restricted to lowland and midelevation humid forests.

During the summers of 1974, 1982, 1984, and 1986, numerous individuals of *Artibeus watsoni* were seen roosting under 19 different species of broadleafed plants at several localities in Costa Rica. At Parque Nacional de Corcovado, Costa Rica, trails were surveyed on three separate occasions for the presence of tents made by *Artibeus watsoni,* in June and August 1982 and again in August 1984. In mid-June 1982, the following groupings of *Artibeus watsoni* were observed: two (both adults, a male and pregnant female), two (pregnant female and one not captured), two (not captured), and six hanging singly (of which two were captured and found to be adult males). Additionally, several tents on banana and *Heliconia* were marked for relocation later in the summer. A tent marked on *Heliconia imbricata* was relocated 60 days later. The tent was still intact, although it was beginning to break down; a single *A. watsoni* was using it. All other marked tents had decomposed.

On 10 August 1982 I found 90 tents constructed by *Artibeus watsoni* along the trail through Corcovado's "Monkey Woods." These tents were made from the following species of plants: *Musa* × *paradisiaca* (49, 54%), *Anthurium ravenii* (13, 14%), unidentified aroid (11, 12%), *Heliconia imbricata* (9, 10%), *Heliconia latispatha* (1, 1%), *Heliconia* sp. (3, 3%), and *Calathea insignis* (4, 4%). Tents located on *Anthurium ravenii* were most often found clumped, with an average of 2.6 tents per plant, whereas in the other species of plants it was uncommon to find more than one tent per individual plant. Bats were found singly (five) or in three groupings of four, three, and two individuals. Four of the single bats were all adult males.

A trail running up to a ridge top was surveyed from 9 through 11 August 1982, with the following results: 25 tents found of which 16 were on *Heliconia imbricata* (64%), 8 on *Calathea insignis* (32%), and 1 on *Carludovica palmata* (4%). Three tents were occupied by two (sexes unknown), one male, and one female. Near the mouth of the Río Llorona on 8 and 9 August I counted the following groups of bats: eight (three adult females, three young, and two not caught), two (adult female with volant young), two (sexes unknown), and three singles (one a nonreproductive adult female). Additional tents were observed in banana, coconut palm (*Cocos nucifera*), *Calathea insignis,* and *Carludovica* cf. *drudei*.

In August 1984, I found 63 tents constructed by *Artibeus watsoni*. These were distributed on the following plants: *Anthurium ravenii* (36, 57%), *Heliconia* sp. (14, 22%), *Musa* × *paradisiaca* (7, 11%), *Calathea insignis* (3, 5%), *Carludovica palmata* (1, 1.6%), *Welfia georgii* (1, 1.6%), and *Geonoma* sp. (1, 1.6%). Only 3 of the 63 tents were occupied; one had two bats and two each had single bats. As I noted in 1982, tents on *Anthurium ravenii* were often clumped on the same plant with an average of 2.5 tents per plant.

At La Selva in July 1982, 43 *Artibeus watsoni* tents were located over a 5-day period in the following species of plants: *Asterogyne martiana* (33, 77%), *Geonoma congesta* (6, 14%), *Geonoma cuneata* (2, 5%), and an unidentified species of Cyclanthaceae (2, 5%). One adult male *A. watsoni* was found under an *Asterogyne martiana* tent on the first day. On the fifth day an adult female with young was found under another *A. martiana* tent that had been unoccupied for the previous four days, as was a third adult (not captured) under another *A. martiana* tent. All other tents were unoccupied.

In 1974 I surveyed approximately 10 km of trails at La Selva and found 29 tents on the following species of palms: *Asterogyne martiana* (19, 66%), *Bactris wendlandiana* (1, 3%), *Geonoma congesta* (2, 7%) and *Geonoma cuneata* (7, 24%); all were unoccupied. Foster and Timm (1976) reported tents in these palms, although they were not able to associate bats with the tents. My recent studies at La Selva have confirmed that these tents were made by *A. watsoni*.

At Finca Las Cruces in mid-August 1982, 13 tents constructed by *Artibeus watsoni* were located;

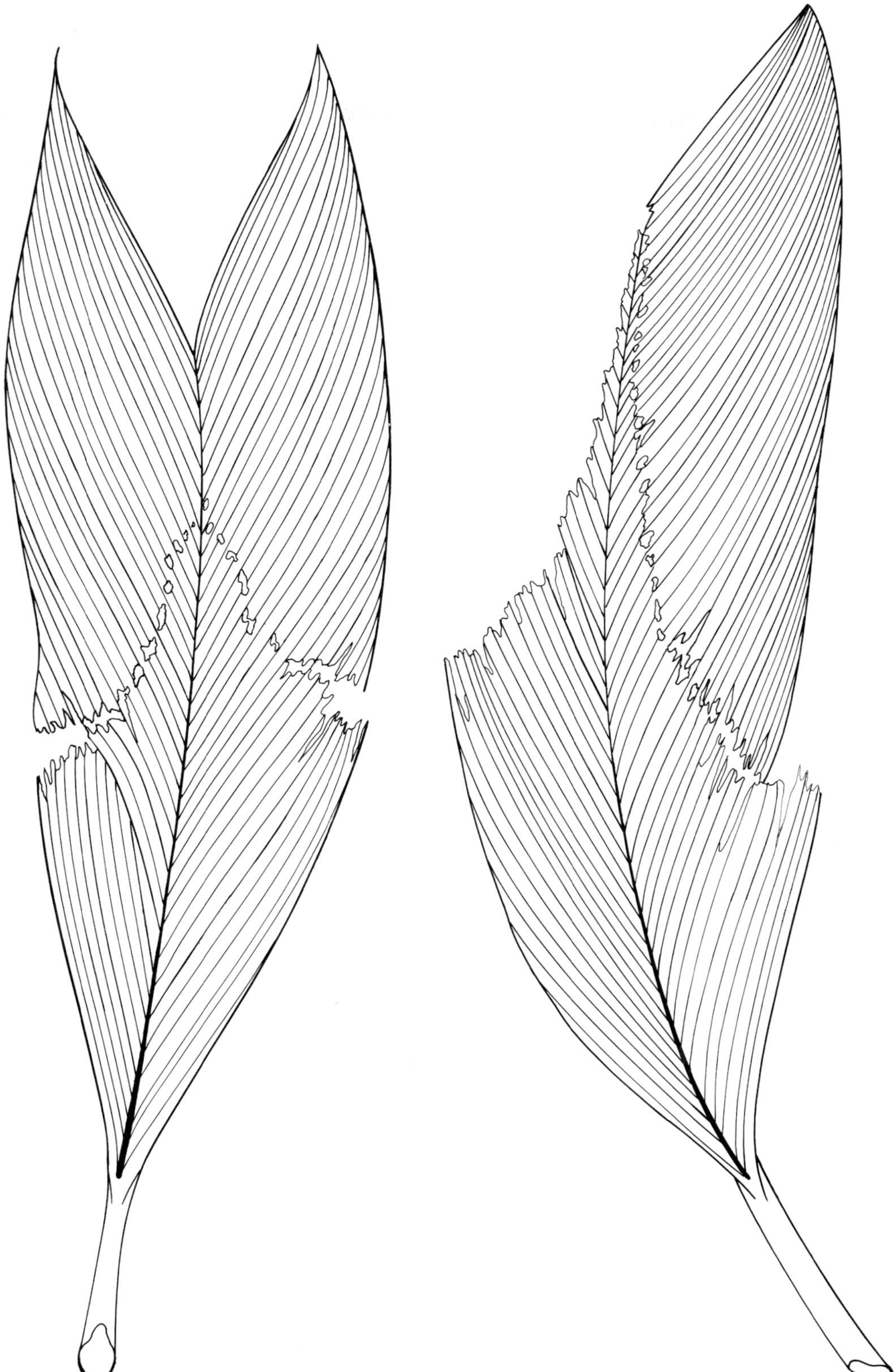

FIG. 5. **Top,** Dorsal view of the ground palm, *Asterogyne martiana*, showing the cut nerves running along the midrib and swinging out to the sides about midway in the leaf (described in the text as the J-shaped cut); **bottom,** tent of *Artibeus watsoni* on *A. martiana.*

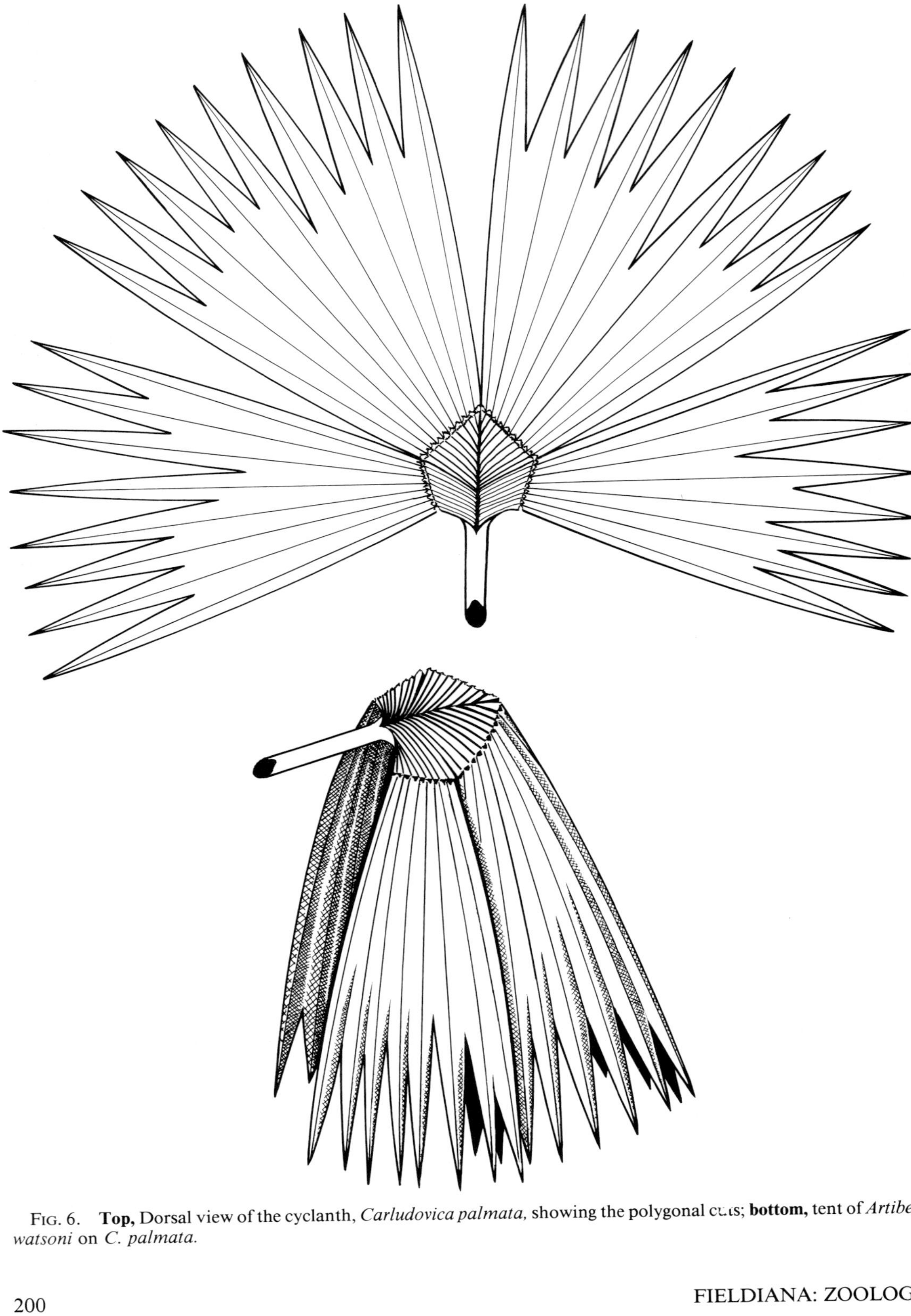

FIG. 6. **Top,** Dorsal view of the cyclanth, *Carludovica palmata,* showing the polygonal cuts; **bottom,** tent of *Artibeus watsoni* on *C. palmata.*

FIG. 7. Photograph of adult female *Artibeus watsoni* and her subadult offspring roosting in a *Carludovica* leaf tent. Details of the polygonal cuts and folds can be seen in the background.

one was occupied by an adult male. The tents were distributed on the following species of cyclanths: *Asplundia euryspatha* (6, 46%), *Carludovica drudei* (4, 31%), and *Cyclanthus bipartitus* (3, 23%).

Artibeus watsoni uses a variety of species of plants and a wide array of leaf shapes for diurnal roosts. I have found four distinct styles of tents at a single locality (Corcovado). These styles include the simple V-shaped cuts on bifurcated palms (fig. 5), cutting a few side veins on aroids to produce a rounded pyramid, the elongated J-shaped cuts on banana and *Heliconia* leaves, and the polygonal cuts on *Carludovica* (figs. 6–7). For each distinct leaf shape, the cuts create a well-concealed diurnal roost. *Artibeus watsoni* probably is an obligate tent-rooster, as it has only been found roosting under cut leaves.

On several instances a bat occupied the same tent, or tents in close proximity, for two to three days in succession. Those tents might then remain unoccupied for several days in succession. Disturbed bats generally flew directly to another tent from 20 to 50 m away, or attempted to return to the tent where originally found.

Tents generally are found clumped, both on a single plant if leaf morphology and age are appropriate, and in restricted areas. Up to five tents have been found on a single *Anthurium ravenii,* and when present the mean number of tents was 2.5.

At Parque Nacional de Corcovado, Choe and Timm (1985) found that *Artibeus watsoni* showed strong preference for *Anthurium ravenii* leaves that were medium sized, low within the plant, and grew closer to the ground than average *A. ravenii* leaves. Also at this site, Boinski and Timm (1985) documented that squirrel monkeys (*Saimiri oerstedi*) were major predators on *A. watsoni,* with the adult male monkeys being the most successful at capturing bats. Additionally, double-toothed kites (*Harpagus bidentatus*) followed troops of foraging squirrel monkeys, using them as "beaters." When tent-making bats were flushed by the monkeys and escaped, they were routinely captured and consumed by the attending double-toothed kites.

Artibeus watsoni has long been known to cut palm tents for diurnal roosts, although prior to this study little had been published on roosting

FIG. 8. Dorsal view of *Scheelea rostrata* showing the leaflets cut by *Uroderma bilobatum* to form a tent.

behavior of this species. The elongate J-shaped cuts made on bifurcated palms were first described and illustrated by Chapman (1932, p. 555). He stated that "both vanes of the leaf whence the bat flew were cut diagonally to the midrib of the leaf, so that their terminal portions drooped downward to form a tentlike shelter." Chapman appropriately termed these three-sided diurnal roosts "tents," and I have expanded the use of the word tents to include all modified leaves by bats.

Barbour's (1932) original description of *Uroderma bilobatum* cutting palm leaves for roosts also provides a secondhand report (p. 308) by H. C. Clark stating that "Clark has just found for the first time a youngish coconut palm, a single leaf of which was being cut by bats of the genus Uroderma in a very similar way." The common use of young coconut palms (*Cocos nucifera*) by *Artibeus watsoni* in Costa Rica, coupled with the total lack of evidence that *Uroderma bilobatum* uses juveniles of this palm, leaves that small, or roosts that close to the ground, suggests that the tents seen by Clark were in fact made by *A. watsoni. Artibeus watsoni* also is abundant on Barro Colorado Island, Panama. Apparently no voucher specimens of the bats were preserved at the time. Allen (1939, p. 69) reported "a specimen of *A. watsoni* that was hanging by day from the underside of a banana leaf." Perhaps the natural-looking folds caused by the cuts running parallel to the midrib were not noticed at the time. Ingles (1953) reported on tents of *A. watsoni* in two species of *Geonoma* on Barro Colorado Island; one tent was occupied by three individuals. Thomas's fruit-eating bat has been found roosting in an artificial tent, an inverted hanging box. Wilson (1970) reported that several females raised young in the corner of a suspended box on Barro Colorado Island.

Uroderma

Uroderma bilobatum Peters, 1865

Uroderma bilobatum has been given the distinctive "common" name of Peters's tent-making bat. Tents constructed by *Uroderma bilobatum* were seen at three separate localities in Costa Rica during the summers of 1982, 1984, and again in 1986, the first at Bosque Brrancia near Cuarto Cruces in the Pacific lowlands of northwestern Costa Rica, the second at Corcovado on the Osa Peninsula, and the third at La Selva in the Caribbean lowlands.

On 25 August 1982 a colony of *Uroderma bilobatum* was roosting under a modified frond of the palm *Scheelea rostrata* at Bosque Brrancia. The colony included an adult male with enlarged testes and four adult females.

The *Scheelea rostrata* frond in which the colony of *Uroderma bilobatum* roosted was a mature leaf, approximately 6.5 m in length. The bats were hanging approximately 4.5 m off the ground; most were clustered together, although a few were spread out over 50 cm of the frond. The cut leaflets started at about 3.5 m off the ground and proceeded up the frond for the next 2.5 m (fig. 8). The general pattern of the cut leaflets was a tapering effect, with the cuts on the lowermost leaflets being farthest from the midrib. Leaflets along the proximal 2 m and the distal 50 cm were unmodified. Only the midrib of the leaflets was cut. Each leaflet had a distinctive V-shaped fold at its base where it was attached to the midrib. The bats were hanging from the leaflets rather than the midrib.

From the dorsal aspect of the leaf, the proximal portion of the tent (cut leaflets) extended 50 cm further down on the right side than on the left to include 10 basal leaflets whose opposites on the left were unaltered. The basalmost cut leaflet was cut 34 cm from the midrib. Proceeding distally, the length of the unmodified basal portion of each leaflet decreases. The basalmost cut leaflet on the left was cut 19 cm from the midrib. The overall appearance of the tent was a sharp, convergent taper for the next meter. Following this section, there was a 75-cm section in which the cuts were close to the midrib (within 3 cm). On the distalmost 30 cm of the tent, the leaflets were cut closer to the midrib on the left side than on the right. Similar tents, each housing a colony of *Uroderma bilobatum,* were found in a large stand of *Scheelea rostrata* at Corcovado in 1984, and William A. Haber (pers. comm.) informed me that he has seen similarly cut leaves in the same species of palm at Cahuita (09°44′N, 82°49′W) in the Caribbean lowlands of southeastern Costa Rica in 1984.

In June of 1982 and again in March of 1986, I found numerous banana leaves cut by *Uroderma bilobatum* just to the north of the field station at La Selva. The midrib on vertical leaves was cut to the extent that the distal portion of the leaf collapsed downward to form a two-sided tent (fig. 9). Severing the midrib on vertical leaves had the effect of folding the leaf back upon itself creating a tight, dark crevice at the fold where the bats roosted (fig. 10). In addition to severing the midrib, the bat cut a large V-shaped pattern running

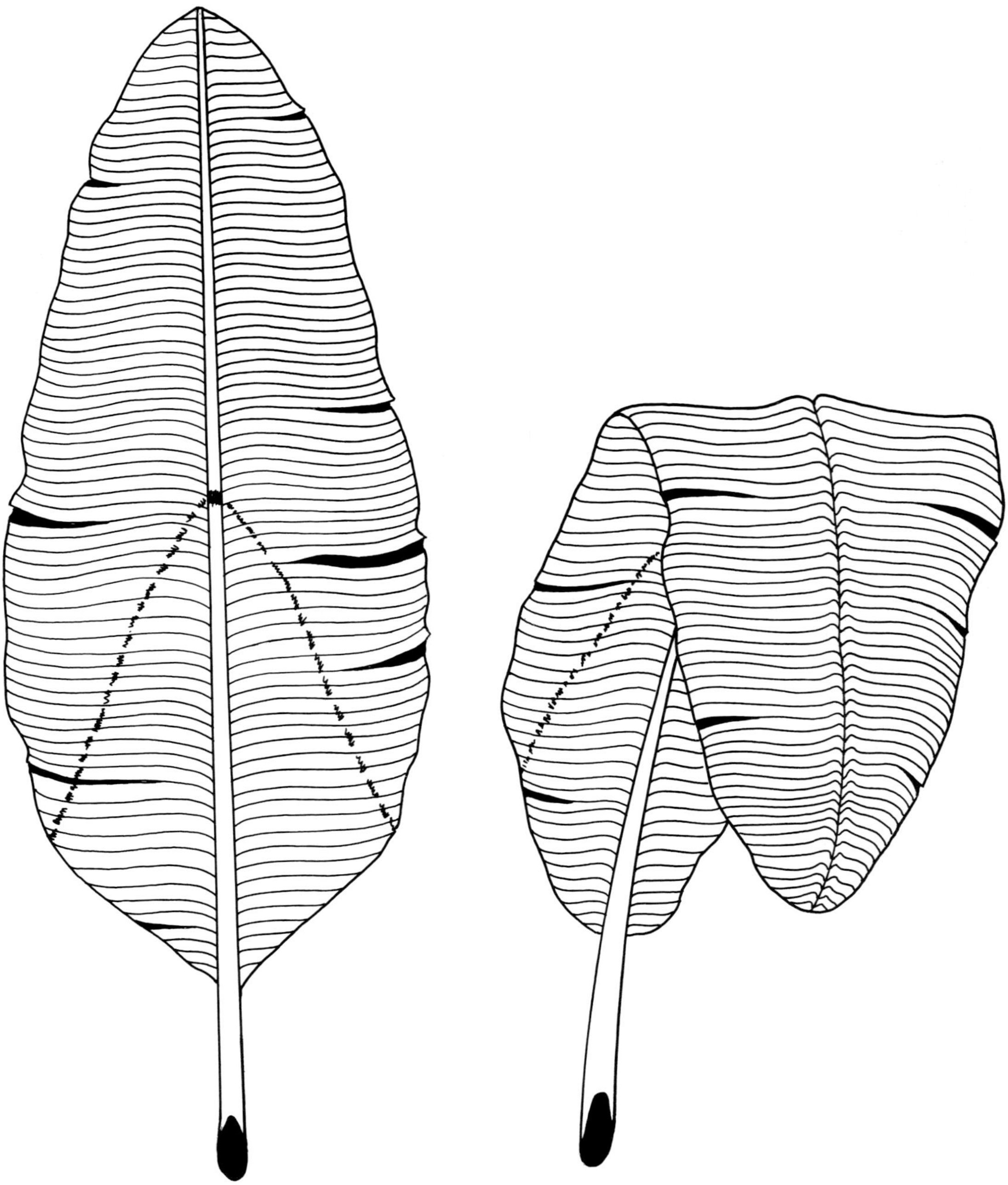

FIG. 9. **Left,** Dorsal view of a banana leaf showing the cut midrib and the large V-shaped cuts through the side nerves; **right,** tent of *Uroderma bilobatum* made from a banana leaf.

from the midrib to the base of the leaf. The side veins and interconnected tissues were partly severed. However, because the leaf stood nearly vertical, these V-shaped cuts did not cause further folding of the leaf. The only cut creating the tent was that of the midrib.

In 1982 five tents in widely separated banana leaves were located (table 4). One was occupied by eight *Uroderma bilobatum,* which included one adult male with enlarged testes and seven females. In 1986 eight tents were observed in the same banana patch. On this occasion eight *U. bilobatum*

FIG. 10. Photograph of a colony of *Uroderma bilobatum* roosting in a banana leaf tent. The bats are roosting in the fold of a leaf that was created by the distal half of the leaf collapsing downward at the severed midrib.

tents were clustered in three clumps of bananas. Only one tent was occupied; it contained 13 bats.

This folded broad-leaved style of tent is undoubtedly the tent style illustrated by Walker (1960, p. 30) in his photograph of roosting *Uroderma bilobatum,* although he did not describe it nor mention where it was observed. Interestingly, I have searched several dozen banana and larger *Heliconia* groves throughout Costa Rica and Ecuador specifically looking for this style of tent, and none were observed. *Uroderma bilobatum* is an abundant and widespread species in the lowlands, but employment of this particular style of tent appears spotty as I have not observed it elsewhere.

In Panama, Barbour (1932) found that colony size under a single cut leaf of *Prichardia pacifica* varied from a few bats to 56. Prior to Barbour's discovery that *Uroderma* altered leaves, Goldman (1920, p. 199) stated of these bats in Panama.

> In the forest near Gatun *Uroderma bilobatum* was located several times, a few in a place, clinging during the day in clusters to the midribs on the under sides of large palm leaves. They usually choose darkened spots where the leaf was folded over, or overhanging pinnae shut out much of the light.

Burt and Stirton (1961) reported *U. bilobatum* in El Salvador roosting beneath the leaves of bananas and coconut palms. I strongly suspect that the bats had altered these leaves and that these authors had failed to notice it or failed to associate the hanging bats with the damage to the leaves. Bloedel (1955, p. 234) stated of *Uroderma bilobatum* in Panama that:

> I observed these bats only in their palm-leaf tents. . . . In the latter part of March most females have nursing young, and are roosting in clusters of 20 to 40, while the males are separated from them, usually solitary or in small groups of from 2–5.

In Trinidad, Goodwin and Greenhall (1961, p. 254) found them

> on the under side of the fan-shaped leaves of certain palm trees, especially the carat palm (*Sabal glaucescens*). . . . The bat makes a series of cuts across the pleated surface of a leaf, causing half of the leaf to bend at an angle to form a protected retreat.

TABLE 4. Individual measurements (in cm or degrees) collected from four tents constructed by *Uroderma bilobatum* on banana, *Musa × paradisiaca,* in Costa Rica.

Blade			Cut	
Length	Width	Petiole angle	Cut midrib from base	Remarks
210	95	40	117	Not occupied
220	90	70	90	8 *U. bilobatum*
...	...	80	...	1 *U. bilobatum*
...	...	80	...	Not occupied
210	95	...	...	Not occupied

They were found roosting in colonies of 10 or more individuals. In Suriname, Husson (1962, p. 161; 1978, p. 143) collected three pregnant female *U. bilobatum* "in a plantation where they were found hanging on the under side of a leaf of [a] so-called 'paloeloe,' *Ravenala guyanensis.*" In Nicaragua, Jones (1964, p. 507) collected four female *U. bilobatum* that "hung together about 10 feet above the ground in the 'tent' formed by a cut palm frond. Each was pregnant with a single embryo." In Guatemala, Dickerman et al. (1981, p. 409) reported, "Palm leaf tents were frequently found occupied by one to seven individuals, but nursing females were usually found alone or with juveniles."

Leaves selected by *Uroderma bilobatum* for tents are all large, and of a variety of shapes. The large V-shaped pattern cut into the leaves is a characteristic of *Uroderma* tents. *Artibeus watsoni* also uses a variety of leaf shapes for tents; however, the nature of the cuts and style of tent created vary with leaf shape. *Uroderma bilobatum,* on the other hand, makes patterned cuts that appear to be an innate response to large leaves, regardless of the shape. As noted in the banana tents, a single cut across the midrib creates the tent, and the V-shaped pattern had no effect upon tent shape; this perhaps represents wasted effort by the bats.

There are a few records of *Uroderma bilobatum* being found roosting in a hollow tree and one "under the eave of house" (Davis, 1968, p. 695). In all cases these have been of single individuals, and I suspect represent either recently dispersed young that have yet to join a breeding colony or bachelor males. *Uroderma bilobatum* roosting under a cut *Prichardia* leaf was illustrated by Kunz (1982). A colony of *U. bilobatum* roosting under a banana leaf tent was illustrated by Keller (1986). Macdonald's (1984, p. 806) photograph of two tent-making bats under a *Heliconia* tent is erroneously labeled *Uroderma bilobatum.* These bats are ac-

FIG. 11. Tent of *Uroderma magnirostrum* on the pinnately leafed palm *Astrocaryum murumuru.*

tually a small species of *Artibeus,* probably *A. phaeotis*; the tent style and size and coloration of the bats are typical of *A. phaeotis.*

Uroderma magnirostrum Davis, 1968

Although *Uroderma magnirostrum* is a widely distributed bat found from Mexico to Bolivia, it was not recognized as a species distinct from *U. bilobatum* until 1968 (Davis, 1968), and few specimens are represented in collections. When W. B. Davis described this new species he commented that although it was widely distributed only seven specimens had been collected prior to the widespread use of mist nets in the 1960s, and that all specimens available to him had come from localities less than 1000 feet in elevation. He stated that "These facts strongly suggest basic differences in the habits of the two species and that those bats with a deep rostrum are not 'tent-makers' as are members of the species *Uroderma bilobatum*" (Davis, 1968, p. 678). There have been no reports on the behavior or ecology of *U. magnirostrum.*

On 25 July 1985 Barbara L. Clauson discovered a colony of two male and three female *Uroderma magnirostrum* roosting under the cut leaflets of the pinnately leafed palm, *Astrocaryum murumuru.* The single occupied tent found was on the ridge above Hacienda Amazonia at 825 m in southeastern Peru. The entire colony was collected by John W. Fitzpatrick. On 3 November Clauson returned to the site to measure ..ie leaf and noted an additional cut leaf in the same plant.

When first observed, four bats were hanging together and one was hanging several centimeters away. When observed an hour later, all five were hanging together in a tight cluster approximately 7.5 m off the ground. The bats were hanging from the leaflets rather than the midrib, approximately 200 cm from the tip of the leaf. The colony included one adult male with enlarged testes, two adult females, and two subadults, one female and one male.

The roosting structure of *Uroderma magnirostrum* was in a pinnately leafed palm (fig. 11). The bats severed the leaflets along the upper two-thirds of the leaf; those along the lower third were unaltered, as were the leaflets at the very tip. As the leaflets proceeded up the tent they were severed closer to the midrib forming an elongate, convergently tapering tent (fig. 11). The general appearance of the *U. magnirostrum* tent is similar to that described herein for *U. bilobatum* on the pinnately leafed palm *Scheelea rostrata.*

The *Astrocaryum* frond in which the colony of *Uroderma magnirostrum* roosted was a mature leaf, approximately 6.1 m in length and 1.9 m in width at the widest point, with the petiole 1.1 m long. The leaf left the trunk (d.b.h. .4 m) at 3 m from the ground and hung at an angle of approximately 50°. The bats were hanging approximately 7.5 m off the ground. The cut leaflets started at 1.5 m from the lowest leaflet and proceeded up the frond for the next 2.9 m to nearly the tip (fig. 11). The cuts on the lowermost leaflets were furthest from the midrib. The lowest severed leaflets were cut up to 34 cm from the midrib, whereas the distal leaflets were severed only 2 cm from the midrib of the leaf. The midribs of the leaflets were cut causing the distal portion of the leaflets to fold downward. Leaflets along the proximal 1.5 m and the distal .5 m were unmodified. The trunk, petiole, and midrib of this palm were covered with sharp, penetrating spines several centimeters in length. After 14½ weeks this tent was still alive and green, most of the leaflets appearing as fresh in November as they did in July.

This *Astrocaryum* contained a second cut leaf that was unoccupied. This roost was also in a mature leaf which was an older leaf than the occupied tent, with many broken, yellowed, and brown leaflets. The leaf was approximately 6 m in length and 2 m in width at the widest point, with a petiole .8 m long. The leaf left the trunk at 2.9 m from the ground and hung at an angle of approximately 60°. Cuts were distributed asymmetrically along the length of the leaf. The cut leaflets on the left side of the leaf started 3.05 m from the lowest leaflet and proceeded up the frond for the next 1.3 m, to .88 m from the tip. The cut leaflets on the right side of the leaf started 3.25 m from the lowest leaflet and proceeded up the frond for the next 1.49 m, to .49 m from the tip. The cuts on the lower leaflets were farthest from the midrib. The lowest severed leaflets were cut up to 35 cm from the midrib, whereas the distal leaflets were severed as close as 1.5 cm from the midrib. Leaflets along the proximal 3.05 m and 3.25 m and the distal .88 m and .49 m were unmodified.

I propose the common name of Davis's tent-making bat for this species.

Conclusions

A review of the literature on tent-making bats contains some 32 primary references covering the 55-year period from 1932 through 1986. Surprisingly, we actually know very little about the biology of these bats.

As late as 1975 Eisentraut was yet doubting that bats were cutting leaves to make tents, stating:

> . . . observers maintain that the bats form these tentlike structures themselves, by making a series of holes running across the middle of a large palm leaf. The bats then supposedly bend the outer half of the leaf around, so they can then rest inside this 'tent'. . . . On the basis of personal observations in tropical regions in Africa, I tend to believe instead that these holes were made by insect larvae while the leaves were still rolled up. A storm can then easily break the leaf along the line of holes and form the tent roof which is so convenient for the bats (Eisentraut, 1975, p. 142).

Eisentraut, by his own admission, had never seen a bat tent. I believe that if he had, he would have come to the same conclusion Thomas Barbour did nearly a half century earlier, that the bats and not insects were making the cuts.

Although we have yet to actually observe bats cutting leaves to form the roosting structures described herein, I hope the volume of data presented here and in my other works establishes for a fact that many species of small and medium sized stenodermines are indeed tent-makers. The observations presented represent data collected from several hundred tents located over a 15-year period. Several facts consistently emerge between my observations and those independently corroborated by others. Bats of the genera *Artibeus* and *Uroderma* (as well as *Ectophylla, Mesophylla,* and *Vampyressa*) roost under cut leaves. These leaves may be on a wide variety of species of plants, but generally the shape of the leaves is similar. The shape of the cuts is very characteristic for each species of bat and the patterns and styles of tents created by the bat species are consistent.

The behavioral repertoire associated with tent-making in bats certainly evolved more than once, as evidenced by the patchy distribution of tent-making species with the chiropteran suborders Megachiroptera and Microchiroptera. Within the Megachiroptera, a single species of tent-maker is known, *Cynopterus sphinx.* Within the Microchiroptera, tent-makers are known only from one subfamily of the Phyllostomidae, the Stenoderminae. The Stenoderminae constitute an extremely speciose and diverse group of bats, with more than 30 species currently recognized. Tent construction within stenodermines may be a trait that evolved once, twice, or as many as three times. The *Artibeus-Uroderma* group are sister genera and form one clade of the tent-making repertoire. Secondly, the *Mesophylla-Vampyressa* group are sister genera (and perhaps should be considered congeneric) and would constitute the second clade. Finally, *Ectophylla* would constitute a third lineage. The relationship between these three lineages is uncertain and warrants further investigation.

Knowing that bats modify the leaves of several species of plants to produce diurnal roosting structures led to the following questions: (1) Are bats selecting specific species of plants for tents? (2) What styles of tents are cut by bats and do these differ between species? (3) Do bats select for a particular angle, size, or shape of leaf for diurnal tents? (4) Are leaves selected preferably in larger clumps or smaller clumps? (5) Are older or younger leaves selected? (6) Are leaves of a particular height class selected? (7) Are leaves that are not adjacent to solid objects selected? (8) What do typical tents look like? (9) How and why did tent construction evolve?

On occasion I have found "cheaters," species of bats roosting in a tent made by another species. Is cheating an evolved strategy of roost site selection of some bats?

Bats of the genera *Artibeus* and *Uroderma* actively modify leaves to produce diurnal roosting structures, but by biting the tissue between veins along the midrib and leaving the midrib and most veins intact, do not kill the leaves. The resulting tent is available for use as a roost for an extended period of time; one was observed in use for more than 60 days. Bats select for specific sizes and shapes of leaves. Tents provide concealment from predators and protection from the rain, wind, and sun. This type of roost offers the additional advantage that the bats are warned about the approach of a potential predator, because even slight movements of the leaf stem or the leaf itself are transmitted as magnified vibrations to the roosting bat. Tents may provide bats with suitable roosting sites that would not otherwise be available in close proximity to prime food resources.

One of the most productive areas for future research will be exploring aspects of the biology of these bats from an evolutionary perspective. Future subjects I will be addressing include the role of tent roosting in controlling ectoparasites and the correlation between complexity of tents and social systems in these bats. I believe that tent-making originated as an antipredation strategy and has since, secondarily, evolved to play a major role in controlling ectoparasites and in social behavior.

Many factors influence the choice of roost site selection by bats. Included among these are vulnerability to predation, physical stability of the site, proximity to food sources, and general appropriateness of the nest microenvironment for the rearing of young. It seems likely that tent construction requires considerable time and energy expenditure by bats, attesting to intense selection pressures involved.

Acknowledgments

I thank Eduardo López Pizarro and El Departamento de Vida Silvestre and Fernando Cortés and Servicio de Parques Nacionales of Costa Rica for making this study possible. The Organization for Tropical Studies (OTS), Rebecca Butterfield,

William A. Haber, Gary Hartshorn, Charles E. Schnell, and Joe M. Wunderle are gratefully acknowledged for assistance with logistics. Robert J. Izor assisted with field logistics in Costa Rica and Peru, and John W. Fitzpatrick assisted in Peru. In Ecuador I thank the Comandancia General del Ejército Ecuatoriano, the Corporación Estatal Petrolera Ecuatoriana, and the Ministerio de Agricultura y Ganadería for making our studies there possible. Luis Albuja, Ramiro Barriga, Angelitos Garrett, Myriam Ibarra, Gustavo Orcés, and Donald J. Stewart provided logistic assistance in Ecuador. Kerry A. Barringer, William C. Burger, Thomas B. Croat, Robin B. Foster, Barry Hammel, and Timothy Plowman provided identifications or confirmed identifications of the plants. Barbara L. Clauson, Alfred L. Gardner, Lawrence R. Heaney, Karl F. Koopman, Thomas H. Kunz, Bruce D. Patterson, and Timothy Plowman provided valuable suggestions on earlier drafts of the manuscript. My wife, Barbara, provided superb assistance with all aspects of this project, including providing several of the photographs used and all data on *Uroderma magnirostrum*. Rosanne Miezio prepared the illustrations. Nina Cummings, Ron Testa, and Diane White expeditiously and cheerfully executed my photography requests. This project was funded in part by grants from the Rice Foundation of Chicago, the National Science Foundation [INT-8303194], National Geographic Society, and Field Museum of Natural History. I especially thank Mr. and Mrs. Arthur A. Nolan, Jr. for their continuing support of my research.

This paper is dedicated to Philip Hershkovitz in recognition of his contributions to Neotropical mammalogy, and most especially for the friendship he has shown me.

Literature Cited

Allen, G. M. 1939. Bats. Harvard University Press, Cambridge, Massachusetts, 368 pp.

Barbour, T. 1932. A peculiar roosting habit of bats. Quarterly Review of Biology, **7:** 307–312.

Bloedel, P. 1955. Observations on the life histories of Panama bats. Journal of Mammalogy, **36:** 232–235.

Boinski, S., and R. M. Timm. 1985. Predation by squirrel monkeys and double-toothed kites on tent-making bats. American Journal of Primatology, **9:** 121–127.

Burt, W. H., and R. A. Stirton. 1961. The mammals of El Salvador. Miscellaneous Publications, Museum of Zoology, University of Michigan, **117:** 1–69.

Chapman, F. M. 1932. A home-making bat. Natural History, **32:** 555–556.

Choe, J. C., and R. M. Timm. 1985. Roosting site selection by *Artibeus watsoni* (Chiroptera: Phyllostomidae) on *Anthurium ravenii* (Araceae) in Costa Rica. Journal of Tropical Ecology, **1:** 241–247.

Davis, W. B. 1944. Notes on Mexican mammals. Journal of Mammalogy, **25:** 370–403.

———. 1968. Review of the genus *Uroderma* (Chiroptera). Journal of Mammalogy, **49:** 676–698.

———. 1970. A review of the small fruit bats (genus *Artibeus*) of Middle America. Part II. Southwestern Naturalist, **14:** 389–402.

Davis, W. B., D. C. Carter, and R. H. Pine. 1964. Noteworthy records of Mexican and Central American bats. Journal of Mammalogy, **45:** 375–387.

Dickerman, R. W., K. F. Koopman, and C. Seymour. 1981. Notes on bats from the Pacific lowlands of Guatemala. Journal of Mammalogy, **62:** 406–411.

Eisentraut, M. 1975. The insectivorous bats, pp. 111–148. *In* Grzimek, H. C. B., ed., Grzimek's Animal Life Encyclopedia. Vol. 11, Mammals II. Van Nostrand Reinhold Company, New York, 635 pp.

Foster, M. S., and R. M. Timm. 1976. Tent-making by *Artibeus jamaicensis* (Chiroptera: Phyllostomatidae) with comments on plants used by bats for tents. Biotropica, **8:** 265–269.

Goldman, E. A. 1920. Mammals of Panama. Smithsonian Miscellaneous Collections, **69**(5): 1–309.

Goodwin, G. G. 1934. Mammals collected by A. W. Anthony in Guatemala, 1924–1928. Bulletin of the American Museum of Natural History, **68:** 1–60.

Goodwin, G. G., and A. M. Greenhall. 1961. A review of the bats of Trinidad and Tobago: Descriptions, rabies infection, and ecology. Bulletin of the American Museum of Natural History, **122:** 187–302.

Goodwin, R. E. 1979. The bats of Timor: Systematics and ecology. Bulletin of the American Museum of Natural History, **163:** 73–122.

Handley, C. O., Jr. 1987. New species of mammals from northern South America: Fruit-eating bats, genus *Artibeus* Leach, pp. 163–172. *In* Patterson, B. D., and R. M. Timm, eds., Studies in Neotropical mammalogy: Essays in honor of Philip Hershkovitz. Fieldiana: Zoology, n.s., **39:** 1–506.

Hartshorn, G. S. 1983. Plants, pp. 118–157. *In* Janzen, D. H., ed., Costa Rican Natural History. University of Chicago Press, Chicago, xi + 816 pp.

Herwitz, S. R. 1981. Regeneration of selected tropical tree species in Corcovado National Park, Costa Rica. University of California Publications in Geography, **24:** 1–109.

Holdridge, L. R. 1967. Life Zone Ecology. Tropical Science Center, San José, Costa Rica, 206 pp.

Holdridge, L. R., W. C. Grenke, W. H. Hatheway, T. Liang, and J. A. Tosi, Jr. 1971. Forest Environments in Tropical Life Zones: A Pilot Study. Pergamon Press, New York, 747 pp.

Husson, A. M. 1954. On Vampyrodes caracciolae (Thomas) and some other bats from the island of Tobago (British West Indies). Zoologische Mededelingen, **33:** 63–67.

———. 1962. The bats of Suriname. Zoologische Verhandelingen, No. **58:** 1–282.

———. 1978. The mammals of Suriname. Zoölogische monographieën van het Rijksmuseum van Natuurlijke Historie, No. **2:** 1–569.

Ingles, L. G. 1953. Observations on Barro Colorado Island mammals. Journal of Mammalogy, **34:** 266–268.

Janzen, D. H. 1971. The fate of Scheelea rostrata fruits beneath the parent tree: Predispersal attack by bruchids. Principes, **15:** 89–101.

Jones, J. K., Jr. 1964. Bats new to the fauna of Nicaragua. Transactions of the Kansas Academy of Sciences, **67:** 506–508.

———. 1966. Bats from Guatemala. University of Kansas Publications, Museum of Natural History, **16:** 439–472.

Jones, J. K., Jr., and T. Alvarez. 1964. Additional records of mammals from the Mexican state of San Luis Potosí. Journal of Mammalogy, **45:** 302–303.

Keller, W. E. 1986. Phenomena, comment and notes. Smithsonian, **17:** 28, 30–31.

Koepcke, J. 1984. "Blattzelte" als Schlafplätze der Fledermaus *Ectophylla macconnelli* (Thomas, 1901) (Phyllostomidae) im tropischen Regenwald von Peru. Säugetierkundliche Mitteilungen, **31:** 123–126.

Koopman, K. F. 1978. Zoogeography of Peruvian bats with special emphasis on the role of the Andes. American Museum Novitates, **2651:** 1–33.

Kunz, T. H. 1982. Roosting ecology of bats, pp. 1–55. *In* Kunz, T. H., ed., Ecology of Bats. Plenum Press, New York, 425 pp.

Kunz, T. H., P. V. August, and C. D. Burnett. 1983. Harem social organization in cave roosting *Artibeus jamaicensis* (Chiroptera: Phyllostomidae). Biotropica, **15:** 133–138.

Macdonald, D. 1984. The Encyclopedia of Mammals. Facts on File Publications, New York, 895 pp.

Morrison, D. W. 1979. Apparent male defense of tree hollows in the fruit bat, *Artibeus jamaicensis*. Journal of Mammalogy, **60:** 11–15.

Ramirez-Pulido, J., A. Martinez, and G. Urbano. 1977. Mamíferos de la Costa Grande de Guerrero, México. Anales del Instituto de Biología, Universidad Nacional Autonoma de México, Serie Zoología, **48:** 243–292.

Sawyer, J. O., and A. A. Lindsey. 1971. Vegetation of the life zones in Costa Rica. Indiana Academy of Science Monograph, **2:** 1–214.

Slud, P. 1960. The birds of Finca "La Selva," Costa Rica: A tropical wet forest locality. Bulletin of the American Museum of Natural History, **121:** 49–148.

———. 1980. The birds of Hacienda Palo Verde, Guanacaste, Costa Rica. Smithsonian Contributions to Zoology, **292:** 1–92.

Timm, R. M. 1984. Tent construction by *Vampyressa* in Costa Rica. Journal of Mammalogy, **65:** 166–167.

———. 1985. *Artibeus phaeotis*. Mammalian Species, **235:** 1–6.

Timm, R. M., and J. Mortimer. 1976. Selection of roost sites by Honduran white bats, *Ectophylla alba* (Chiroptera: Phyllostomatidae). Ecology, **57:** 385–389.

Tuttle, M. D. 1976. Collecting techniques, pp. 71–88. *In* Baker, R. J., J. K. Jones, Jr., and D. C. Carter, eds., Biology of Bats of the New World Family Phyllostomatidae. Part I. Special Publications, The Museum, Texas Tech University, **10:** 1–218.

Villa-R., B. 1967. Los murciélagos de México. Anales del Instituto de Biología, Universidad Nacional Autonoma de México, xvi + 491 pp.

Walker, E. P. 1960. Studying our Fellow Mammals. The Animal Welfare Institute, New York, 174 pp.

Watkins, L. C., J. K. Jones, Jr., and H. H. Genoways. 1972. Bats of Jalisco, México. Special Publications, The Museum, Texas Tech University, **1:** 1–44.

Wilson, D. E. 1970. An unusual roost of *Artibeus cinereus watsoni*. Journal of Mammalogy, **51:** 204–205.

Wilson, D. E., and D. H. Janzen. 1972. Predation on *Scheelea* palm seeds by bruchid beetles: Seed density and distance from the parent palm. Ecology, **53:** 954–959.

Appendix

List of scientific names of plants mentioned in the text and used by Neotropical bats for tent construction.

Anacardiaceae

Anacardium excelsum (Bertero & Balbis) Skeels

Araceae

Anthurium caperatum Croat & Baker
Anthurium ravenii Croat & Baker

Araceae (cont'd.)

Monstera lechleriana Schott
Xanthosoma sp.

Cyclanthaceae

Asplundia euryspatha Harl.
Carludovica drudei Masters

Cyclanthaceae (cont'd.)
Carludovica palmata R. & P.
Cyclanthus bipartitus Poit.

Marantaceae

Calathea insignis Petersen

Musaceae

Heliconia imbricata (Kuntze) Baker
Heliconia latispatha Benth.
Musa × *paradisiaca* L.
Plenakospermum guyanense Endl. (syn. *Ravenala guyanensis* Petersen)

Palmae

Asterogyne martiana (H. Wendl.) H. Wendl. ex Hemsley
Astrocaryum murumuru Mart.
Bactris wendlandiana Burret
Cocos nucifera L.
Geonoma congesta H. Wendl. ex Spruce
Geonoma cuneata H. Wendl. ex Spruce (syn. *G. decurrens* H. Wendl.)
Geonoma oxycarpa Martius (syn. *G. binervia* Oerst.)
Livistona chinensis (Jacq.) R. Br. ex. Mart.
Prichardia pacifica Seem. & H. Wendl.
Sabal mauritiiformis (Karsten) Griseb. & H. Wendl. ex Griseb. (syn. *S. glaucescens* Lodd. ex H. E. Moore)
Scheelea rostrata (Oersted) Burret
Welfia georgii H. Wendl. ex Burret

Comparative Ultrastructure and Evolutionary Patterns of Acinar Secretory Product of Parotid Salivary Glands in Neotropical Bats

Carleton J. Phillips, Toshikazu Nagato, and Bernard Tandler

ABSTRACTS

Secretory products produced by acinar cells of the parotid salivary glands of 15 species of Neotropical bats (*Pteronotus parnellii, Phyllostomus elongatus, P. latifolius, Tonatia bidens, T. sylvicola, Trachops cirrhosus, Glossophaga soricina, Leptonycteris sanborni, Sturnira lilium, Artibeus jamaicensis, Ariteus flavescens, Eptesicus lynni, E. brasiliensis, Tadarida brasiliensis,* and *Molossus molossus*) were compared by transmission electron microscopy. Extensive inter- and intrageneric differences were found in the ultrastructure of the mature acinar secretory granules. This variation in secretory cell product exceeded any previously reported intraordinal phenotypic variation at the cellular level, but was in keeping with previously reported biochemical data on salivary protein polymorphism in primates and rodents. Data from molecular biology and systematics lend support to the hypothesis that the microscopic variations are directly representative of genic differences among species. It also is postulated that intrageneric microscopic differences at least partly are due to neutral (nonfunctional) differences in molecular structure or charge (or both) rather than evolutionary selection. Among the phyllostomids, a general trend in parotid acinar cell product was found in *Artibeus* and *Ariteus,* in which a decrease in enzymatic content of the product could be correlated with ultrastructural differences. The secretory product in *Artibeus* and *Ariteus* also differed significantly from that of the genus *Sturnira,* and it is proposed that the phenotypic differences between *Sturnira* and the other two stenodermatines represent a major genetic difference of systematic importance. The ultrastructural appearance and substructure of the parotid acinar secretory granules could not be consistently correlated with diet alone, although insectivorous-animalivorous species have enzyme-rich, mostly electron-dense granules, whereas two fruit bats, *Artibeus* and *Ariteus,* have pale, enzyme-poor parotid granules.

Productos de secreción producidos por células acinares de las glándulas salivales de 15 especies de murciélagos neotropicales (*Pteronotus parnellii, Phyllostomus elongatus, P. latifolius, Tonatia bidens, T. sylvicola, Trachops cirrhosus, Glossophaga soricina, Leptonycteris sanborni, Sturnira lilium, Artibeus jamaicensis, Ariteus flavescens, Eptesicus lynni, E. brasiliensis, Tadarida brasiliensis* y *Molossus molossus*) fueron comparados mediante microscopio electrónico de transmisión. Extensivas diferencias inter- e intragenéricas fueron encontradas en la estructura de gránulos glandulares acinares maduros. Esta variación en productos de células secretoras excedió

From the Department of Biology, Hofstra University, Hempstead, NY 11550 (Phillips); and Department of Oral Biology, School of Dentistry, Case Western Reserve University, Cleveland, OH 44106 (Nagato and Tandler). Dr. Nagato's present address is Department of Oral and Maxillofacial Surgery, Ehime University School of Medicine, Shizukawa, Shigenobu, Onsen-gun, Eshime 791-02, Japan.

cualquier variación fenotípica intraordinal previamente reportada a nivel celular, pero estuvo en armonía con datos bioquímicos previamente reportados de polimorfismo en proteinas salivales de primates y roedores. Datos de biología molecular y sistemática proveen apoyo a la hipótesis de que las variaciones microscópicas son directamente representativas de diferencias genéticas entre especies. Es también postulado que diferencias microscópicas intragenéricas, al menos parcialmente, son debidas a diferencias neutrales (no funcionales) en estructura molecular o carga (o ambas) más bien que a selección evolutiva. Entre los filostómidos, una tendencia general en el producto parótido de las células acinares fue encontrado en *Artibeus* y *Ariteus*, en los cuales una disminución en contenido enzimático del producto podría estar relacionado con diferencias ultraestructurales. El producto secretado en *Artibeus* y *Ariteus* también se diferenció significativamente de aquel del genero *Sturnira* y es propuesto que las diferencias fenotípicas entre *Sturnira* y los otros dos stenodermatinos representan una mayor diferencia genética de importancia sistemática. La apariencia ultraestructural y subestructura de los gránulos secretorios parótidos acinares podría no estar consistentemente correlacionada sólo con la dieta, aunque especies insectívoras-animalívoras tienen gránulos ricos en enzimas mayormente densos en electrones, mientras que los murciélagos frugívoros, *Artibeus* y *Ariteus*, tienen gránulos parótidos pálidos y pobres en enzimas.

Produtos secretórios produzidos por ácinos das glandulas parótidas de 15 espécies de morcegos neotrópicos (*Pteronotus parnellii, Phyllostomus elongatus, P. latifolius, Tonatia bidens, T. sylvicola, Trachops cirrhosus, Glossophaga soricina, Leptonycteris sanborni, Sturnira lilium, Artibeus jamaicensis, Ariteus flavescens, Eptesicus lynni, E. brasiliensis, Tadarida brasiliensis,* e *Molossus molossus*) foram comparados através da microscopia de transmissão eletrônica. Vastas diferenças intra- e intergenéricas foram encontradas na ultra-estrutura dos granulos maduros dos ácinos secretórios. Esta variação no produto das células secretórias supera qualquer variação fenotípica intraordinal prèviamente relatada para o nível celular, porém concorda com as relações publicadas sobre dados bioquímicos do polimorfismo de proteínas salivares em primatas e em roedores. Dados sistemáticos, e de biologia molecular, apoiam a hipótese que variações microscópicas são diretamente representativas das diferenças genéticas entre espécies. Propõese também, que estas diferenças microscópicas intragenéricas são ao menos parcialmente causadas por diferenças neutras (i.e., nãofuncionais) nas estruturas moleculares ou nas suas cargas elétricas (ou ambas), ao invés de serem consequencias da seleção evolutiva. Entre os morcegos da família Phillostomidae, foi encontrado em *Artibeus* e em *Ariteus* um padrão geral nos produtos dos ácinos parotídeos, onde uma redução do conteúdo enzimático é correlacionado à diferencas nas ultraestruturas dos granulos produzidos. Os produtos secretórios em *Artibeus* e em *Ariteus* são significantemente diferentes dos produtos do genero *Sturnira*, e propõe-se que as diferenças fenotípicas entre *Sturnira* e os outros dois stenodermatinos representam uma grande diferença genética, de importancia sistemática. A aparencia ultraestrutural, e a subestrutura dos granulos secretórios dos ácinos parotídeos, não se correlacionam consistentemente com a dieta por si, apesar de que as espécies insetívoras-animalívoras possuem granulos ricos em enzimas e densos em elétrons, equanto que dois morcegos frugivoros, *Artibeus* e *Ariteus*, possuem granulos parotídeos pálidos e com poucas enzimas.

Introduction

Neotropical bats are extremely diversified; existing species represent perhaps the most outstanding mammalian example of ecomorphological adaptation. Dentitions, jaw morphology, brains, kidneys, tongues, and digestive tracts are only a few examples among the anatomical features that have been investigated in recent years (e.g., Phillips, 1971; Forman, 1972; Phillips et al., 1977, 1984; Freeman, 1979, 1981; Eisenberg & Wilson, 1978; Studier et al., 1983). The major salivary glands are yet another system that has attracted attention, primarily because histological, ultrastructural, and histochemical investigations have consistently revealed striking interspecific differences and unusual histological and secretory features (Wimsatt, 1956; DiSanto, 1960; Junqueira & Fava de Morães, 1965; Junqueira et al., 1967, 1973; Phillips et al., 1977; Mineda, 1977, 1978;

Pinkstaff et al., 1982; Tandler & Cohan, 1984; Nagato et al., 1984; Tandler & Phillips, 1985; Phillips & Tandler, 1985, 1987; Tandler et al., in press).

Mammalian salivary glands are highly complex organs that not only secrete digestive enzymes but also can secrete hormones, antibodies, and toxins, to name but a few known products (e.g., Tandler, 1972; Hand, 1980b). Data are available on the biochemistry of saliva in humans and common laboratory species; but very little is known about the specific biochemistry of the parotid acinar secretory granules except for laboratory rats in which some of the proteins have been characterized (Robinovitch & Sreebny, 1969; Ball, 1974; Wallach et al., 1975). The complex structure and function of salivary glands is underscored by data from studies of Neotropical bats, which recently have been analyzed by both transmission electron microscopy and histochemistry. For example, the accessory submandibular gland of *Trachops cirrhosus* has been shown to differ histologically from any known mammalian salivary gland, with the exception of the same gland in *Megaderma lyra* and *M. spasma*. *Megaderma lyra* is an Old World ecological equivalent of *Trachops* (Phillips & Tandler, 1985, 1987; Phillips et al., 1987). Both of these unrelated species feed on frogs (Lekagul & McNeely, 1977; Tuttle & Ryan, 1981), which possibly has been a factor in the convergent evolution of their submandibular glands. A previously unknown cellular organelle has been described in another Neotropical bat, *Tonatia sylvicola* (Nagato et al., 1984). In this species, the presence of unique crystalloid smooth endoplasmic reticulum in seromucous acinar cells is sex-linked (being found only in submandibular acinar cells in males); a steroid product produced by this organelle possibly serves as a species-isolating mechanism or as part of a chemo-behavioral system, or both. Lastly, a comparative investigation of the secretory product in seromucous acinar cells in submandibular glands of five species of *Artibeus* has revealed that the ultrastructural characteristics of secretory products can have systematic implications (Tandler et al., 1983, 1986). This study was of particular interest because the salivary gland data matched genic data independently derived by Koop and Baker (1983).

For the present investigation we surveyed parotid acinar cell secretory products in a selected group of 15 species of Neotropical bats. Our group comprised four families (Mormoopidae, Phyllostomidae, Vespertillionidae, and Molossidae) and included a group of species in which dietary habits ranged from insectivory and animalivory to omnivory and frugivory. This investigation is the first comprehensive interspecific survey of secretory product ultrastructure, and addresses the following questions: (1) what is the range of variation in secretory products; (2) what are the evolutionary patterns in parotid secretory product; and (3) what systematic conclusions can be reached by comparative ultrastructural analysis?

Materials and Methods

Numbers and sex of specimens used in the present study are given in the Appendix. All of these bats were collected during fieldwork in Mexico, Jamaica, and Suriname. Voucher specimens for all species and collecting localities are deposited in the mammal collections of either the Carnegie Museum of Natural History or The Museum, Texas Tech University. Bats typically were collected at night with mist nets, and were kept overnight without food until they were killed between 0900 and 1200 the following morning. The animals were anesthetized with 0.25 ml of sodium pentobarbital (50 mg/ml, intraperitoneally) and the salivary glands removed, placed on dental wax, flooded with freshly mixed fixative, and diced into pieces measuring approximately 1 mm^3.

Two fixation protocols were used at different times during the project. Specimens collected in Mexico and Jamaica were fixed in 2% glutaraldehyde in 0.1 M phosphate buffer and then stored unrefrigerated in fresh fixative. The specimens collected in Suriname were fixed in a modified trialdehyde-dimethylsulfoxide (DMSO) mixture, first described by Kalt and Tandler (1971) and later modified slightly for fieldwork (Phillips, 1985). The trialdehyde fixative, consisting of 3% glutaraldehyde, 1% paraformaldehyde, 0.5% acrolein, 2.5% DMSO, and 1 mM $CaCl_2$ in a 0.05 M cacodylate buffer and sucrose at pH 7.2, proved superior to the simple glutaraldehyde fixative in that (1) a higher percentage of tissues proved to be adequately fixed for study, and (2) the mitochondria tended to remain intact instead of being disrupted. Although the composition of the fixatives was different, we have not found any evidence that these differences introduced undesirable artifacts that would influence our analysis. Additional details about the techniques can be found in both Phillips (1985) and Nagato et al. (1984).

To remove unbound aldehydes, the tissue blocks were subjected to prolonged washing in phos-

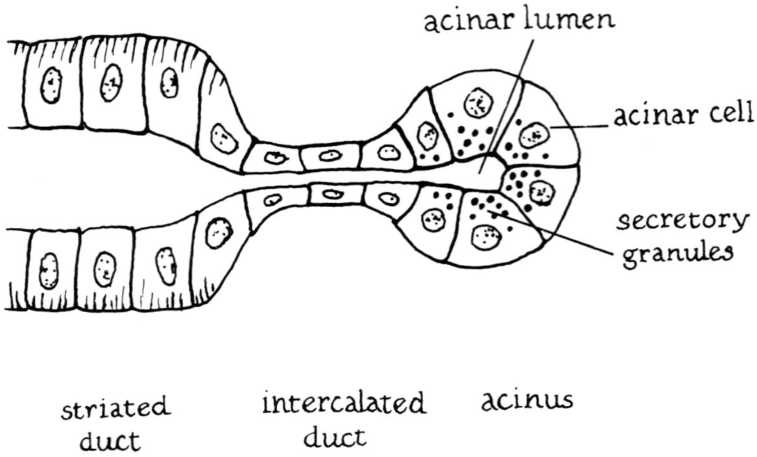

FIG. 1. Schematic diagram of a secretory unit in a typical parotid gland. A cluster of secretory cells, arranged around a lumen to form an acinus, drains into an intercalated duct that in turn empties into a striated duct. In reality, more than one acinus may be associated with an intercalated duct, which itself may be branched. The acinar cells elaborate the primary saliva, which has plasma-like concentrations of electrolytes, and add their organic secretory products to it. The precise role of the intercalated duct cells has not been established, but they probably play a role in secretion and transport. The principal function of the striated ducts is resorption of electrolytes, mainly sodium, from the primary saliva, rendering it hypotonic.

phate-buffered sucrose. The blocks were postfixed for two hours in phosphate-buffered 2% OsO_4, rinsed in distilled water, soaked overnight in cold aqueous 0.25 uranyl acetate, rinsed again in distilled water, dehydrated in ascending concentrations of ethanol, passed through propylene oxide, and embedded in Epon-Maraglas (Tandler & Walter, 1977). Thin sections were doubly stained with methanolic uranyl acetate (Stempak & Ward, 1964) and lead citrate (Venable & Coggeshall, 1965). All sections were examined in a Siemens Elmiskop 1a transmission electron microscope (TEM). Semithin sections (1 μm) used for orientation were stained with toluidine blue (Björkman, 1962) and examined in a Zeiss Ultraphot.

Tissue samples used for comparisons were selected carefully from among available tissue blocks. We generally avoided edges of tissues where mechanical trauma often affects not only the cellular ultrastructure but also the microscopic appearance of the secretory granules. In selecting representative "mature" granules we took into account the full array of inter- and intracellular variability as well as the often complex substructural geometry of the granules. Selection of the most representative granules admittedly was subjective, but based on our experience, four principal criteria were applied: (1) cells selected for analysis showed no signs of shrinkage or swelling, and sensitive organelles, such as mitochondria, were not distorted; (2) the appearance of the granule was not altered in any meaningful way by variations in fixation; (3) the appearance of the secretory granules had to be unrelated to their location within the specimen block; and (4) the development of the granules could be traced (from Golgi complex to apical cytoplasm) without any major breaks in developmental sequence.

Results

The basic histology of the parotid salivary gland was similar in all 15 species examined; in each species the acinus was formed by a cluster of cells connected to the striated duct by intercalated ducts of varying lengths (fig. 1). At the TEM level, acinar cell secretory products were found to differ in all species examined (figs. 2–5). These differences could not be related to granule ontogeny by comparing "immature" Golgi-GERL (Golgi-endoplasmic reticulum-lysosomes)-associated granules with "mature" granules in the apical cytoplasm. In our samples that consisted of both males and females we found no evidence of sexual dimorphism in secretory granule substructure, and in our largest samples (5 to 10 specimens) we found no example of individual or geographic variation that could not be attributed to typical inter- or intracellular variation or to granule geometry. Although vari

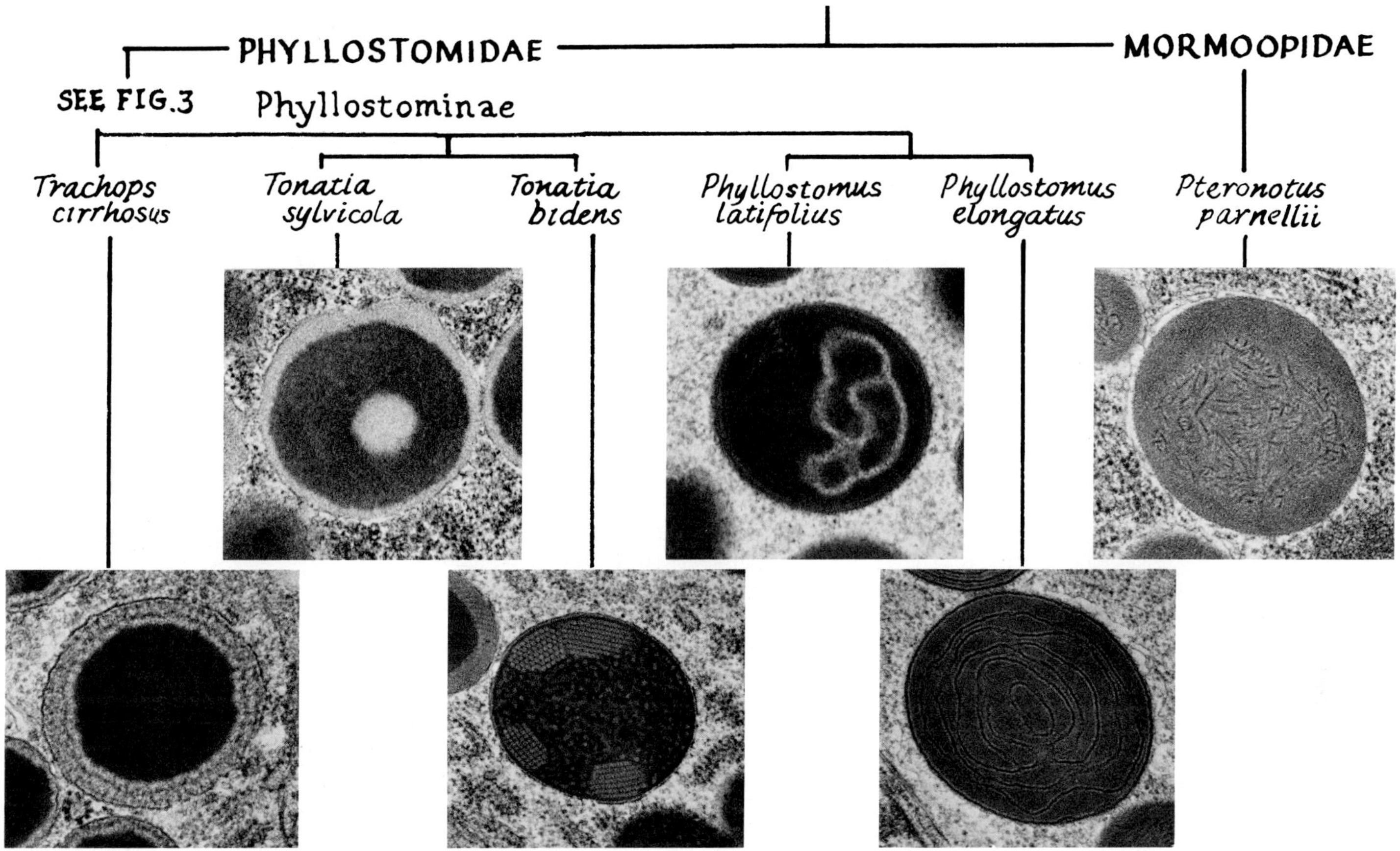

FIG. 2. Comparative ultrastructural views of representative "mature" secretory granules in parotid acinar cells from a mormoopid and five phyllostomine species. Although interspecifically variable, note that each contains a substantial electron-dense component. *Pteronotus parnellii,* × 30,200; *Phyllostomus elongatus,* × 25,900; *P. latifolius,* × 32,400; *Tonatia bidens,* × 24,500; *T. sylvicola,* × 30,000; *Trachops cirrhosus,* × 30,000.

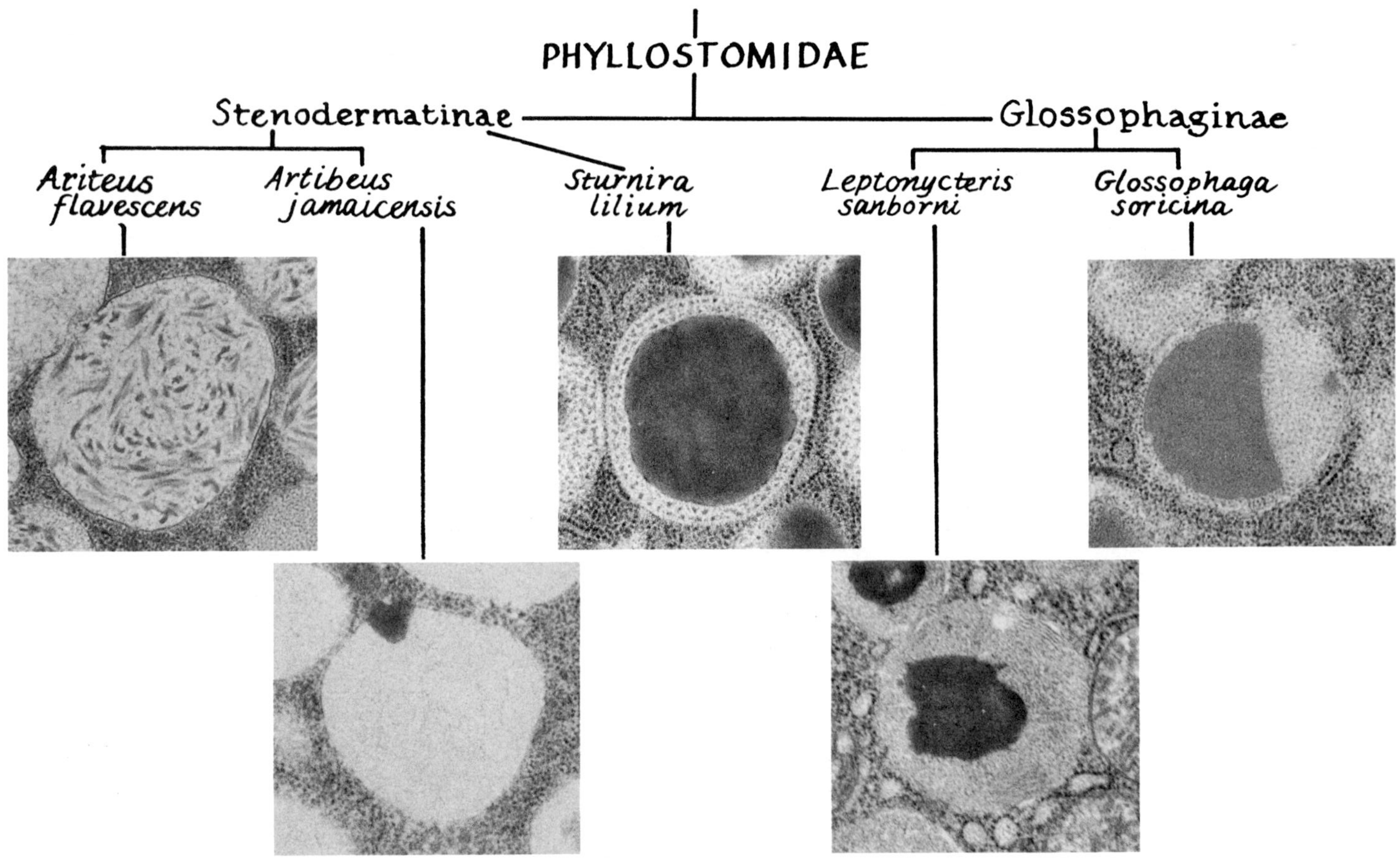

FIG. 3. Comparative ultrastructural views of representative "mature" secretory granules illustrating the evolutionary trends in the Glossophaginae and Stenodermatinae. In particular note the distinct difference between *Sturnira* and the other two stenodermatine genera. *Glossophaga soricina,* × 19,400; *Leptonycteris sanborni,* × 17,800; *Sturnira lilium,* × 26,400; *Artibeus jamaicensis,* × 19,200; *Artibeus flavescens,* × 17,300.

ation may occur at the individual level, we found no unequivocal evidence that it is detectable by transmission electron microscopy. Last, the examples selected to illustrate each species (figs. 2–5) should be regarded as "average" or representative of the most common, well-fixed granule morphology.

PTERONOTUS PARNELLII—The secretory granules consisted of a moderately dense, homogeneous, variably thick outer zone and a slightly lighter interior in which were embedded a collection of linear densities, usually randomly disposed but sometimes showing a degree of orderliness. Each of these densities was outlined by a thin, barely discernible, lucent layer (fig. 2).

PHYLLOSTOMUS ELONGATUS—The matrix of the secretory granules was uniformly dense and contained a series of irregular shell-like densities, which in thin section appeared as ribbons consisting of a dense line flanked by thin lucent plies. In certain granules, the ribbons were aligned in concentric fashion, imparting a layered appearance to the periphery of these cell structures.

PHYLLOSTOMUS LATIFOLIUS—These granules bore some resemblance to those in *P. elongatus* in that some had one or two ribbon-like layers immediately subjacent to their limiting membrane, but many granules lacked these layers. Instead, the prevailing inclusion was a lucent, coreless strand disposed in a complex skeinlike or twisted pretzel conformation (fig. 2).

TONATIA BIDENS—These secretory granules contained several groups of hexagonally-packed tubules just beneath the limiting membrane. In favorable transverse sections, such packets appeared as honeycomb structures, whereas in longitudinal section they appeared as parallel linear densities. The dense matrix shows a vermiculate pattern (fig. 2).

TONATIA SYLVICOLA—The secretory granules in this species differed from those in *T. bidens*. The granules consisted of light and dark material that displayed two principal patterns. Either the dark material was centrally placed, where it often contained one or several lucent spaces, or the light and dark material each occupied one hemisphere of the granule. In either case, crystalloid tubules of the sort found in *T. bidens* never were present (fig. 2).

TRACHOPS CIRRHOSUS—In this species, the granules consisted of a large, central, very dense spherule surrounded by a lucent cortex. The cortical material contained linearly arranged punctate densities that, at low to moderate magnifications, gave the appearance of several continuous dense layers in the outer portion of the granules (fig. 2).

GLOSSOPHAGA SORICINA—The bulk of each granule was occupied by homogeneously dense material, which formed a smooth interface with the remainder of the granule. The other portion of each granule was pale and contained minute, dense specks (fig. 3). The lighter material occasionally intruded into the dense zone in the form of a tortuous ribbon.

LEPTONYCTERIS SANBORNI—The secretory granules contained an amorphous, dense, central inclusion in a paler matrix. A small bundle of short filaments was occasionally present in the matrix (fig. 3).

STURNIRA LILIUM—The secretory granules in this species resembled those in *Trachops cirrhosus*. They consisted of a large, central, dense core surrounded by a rim of lucent material in which were embedded distinct punctate densities often disposed in layers (fig. 3).

ARTIBEUS JAMAICENSIS—The granules were large and pale and contained some faintly discernible, twisted fibrils. A few granules possessed a small, dense, usually peripheral inclusion (fig. 3).

ARITEUS FLAVESCENS—These secretory granules had a pale matrix in which were numerous, flat, narrow, moderately dense lamellae in a random arrangement. Viewed on edge, the lamellae appeared as dense lines; face on, as gray, irregular structures (as seen in fig. 3). In a few granules, there was a dense polygonal plate that lacked periodicity.

EPTESICUS LYNNI—These granules had a complex substructure. The limiting membrane was underlaid by a layer of moderately dense material that made incursions, arcuate and anfractuous, into a dense matrix, producing a variety of patterns. These extensions of the peripheral light material had a layered structure (fig. 4, top).

EPTESICUS BRASILIENSIS—As in *E. lynni*, the limiting membrane was subtended by layered material, usually in several plies, that extended into the dense matrix. These extensions were less tortuous than those in *E. lynni* and appeared to subdivide the granule interior (fig. 4, bottom).

TADARIDA BRASILIENSIS—The secretory granules in this species exhibited a spectrum of patterns. The most common was relatively simple, with either short, flat, dense prisms or dense dots being suspended in a slightly less dense matrix. A few granules contained dense hollow spheres, whereas others had a mazelike configuration based on light and dark laminations (fig. 5, top).

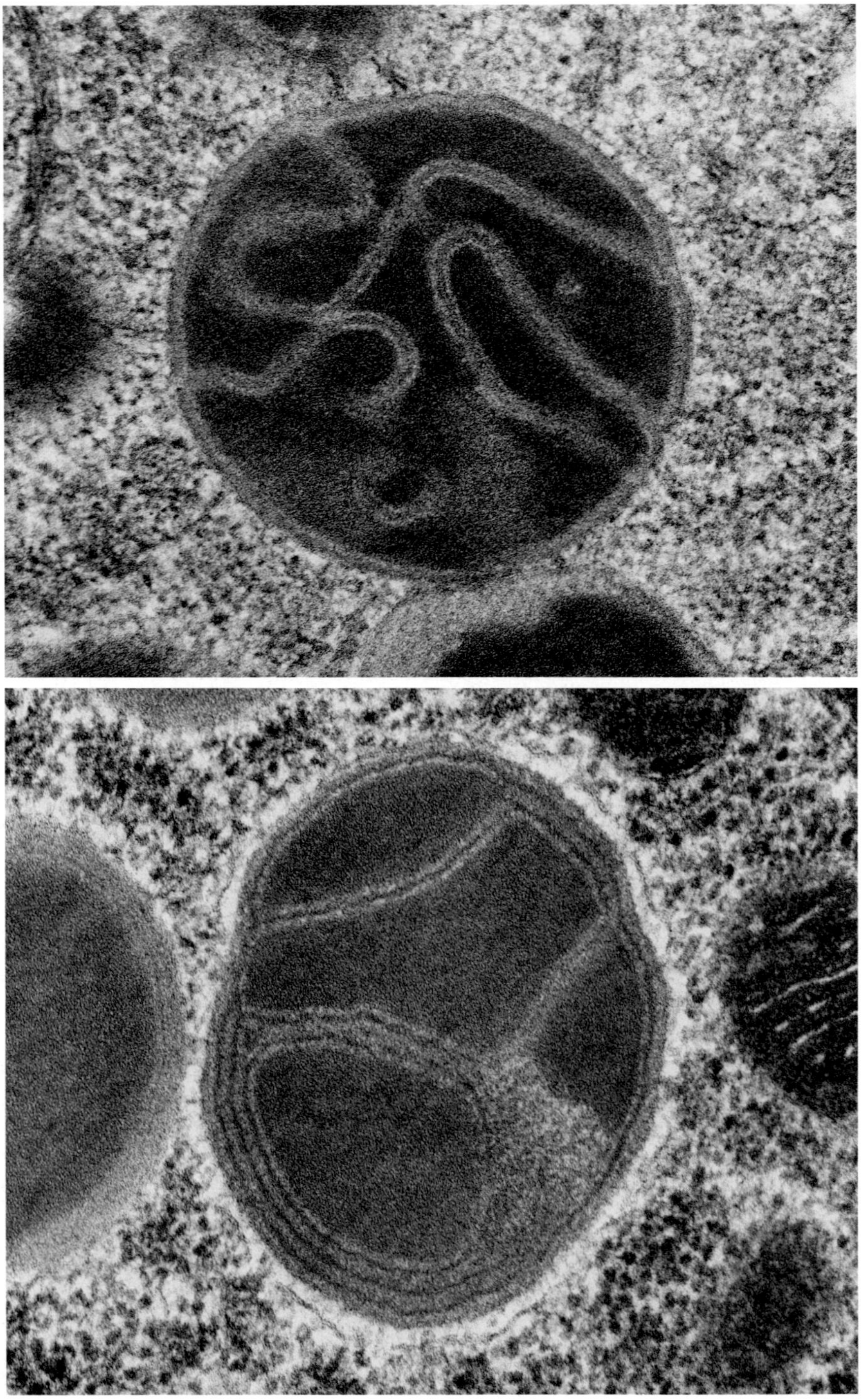

FIG. 4. Higher magnification of mature secretory granules in parotid acinar cells in *Eptesicus lynni* (top) and *E. brasiliensis* (bottom). Note the subtle but consistent differences in the granule substructure. *Eptesicus lynni,* × 63,000; *E. brasiliensis,* × 76,800.

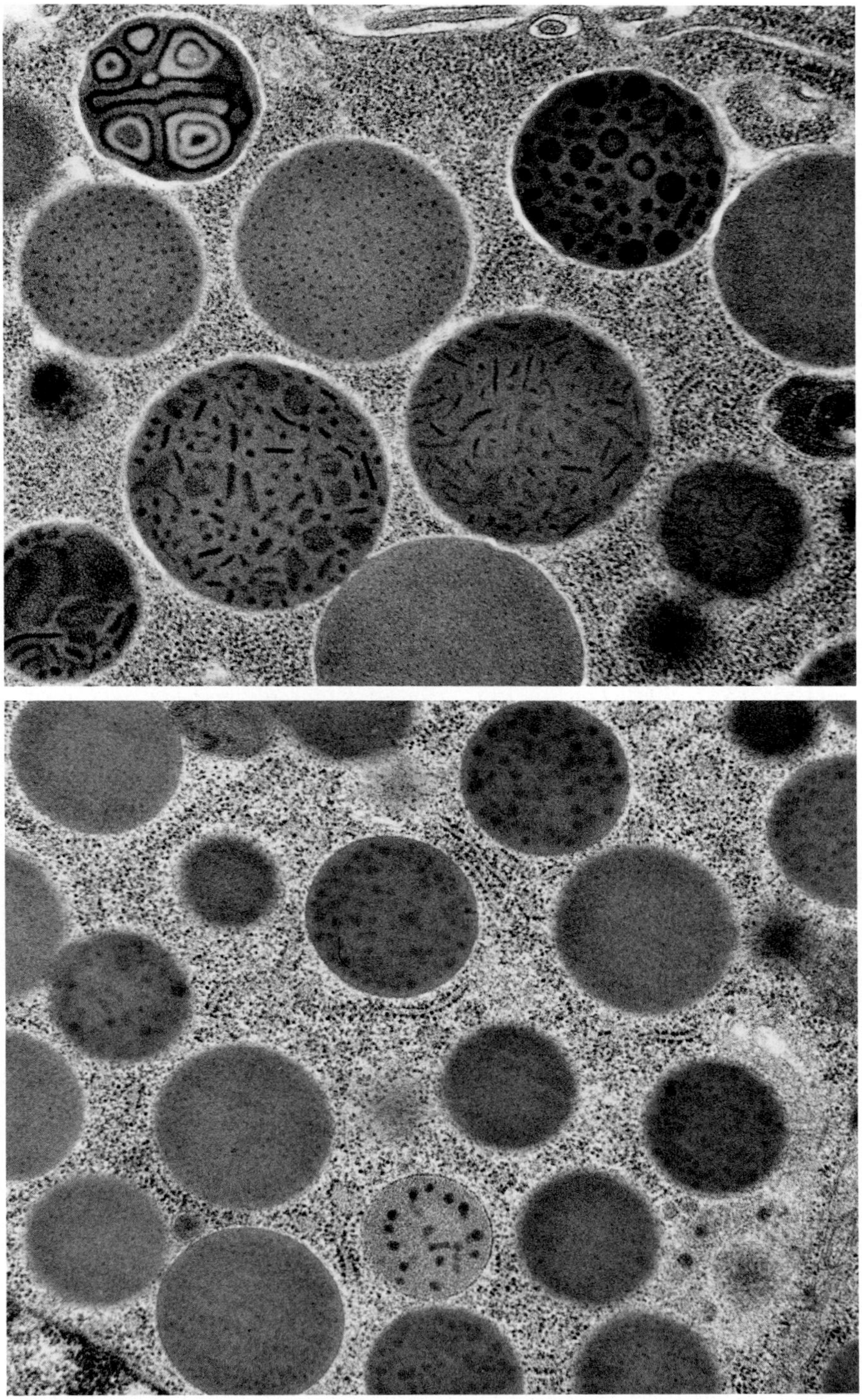

FIG. 5. Parotid acinar cell granules in *Tadarida brasiliensis* (top) are compared to those found in *Molossus molossus* (bottom). Note the range of variation in granule substructure in *T. brasiliensis*. *Tadarida brasiliensis,* × 26,000; *Molossus molossus,* × 24,000.

MOLOSSUS MOLOSSUS—The variable morphology of the granules in this species appeared to depend on their stage of maturation. Early granules were small, with a moderately dense matrix in which were some prominent dense particles. As the granules matured (based on size, density, and spatial relationship to the Golgi complex), they enlarged and the particles decreased somewhat in density. The mature granules had a farinaceous matrix probably resulting from comminution of the dense particles (fig. 5, bottom).

Discussion

The idea of using comparative ultrastructural analysis of homologous, morphologically differentiated cells to study evolutionary pathways and to explore systematic relationships is new to the study of mammalian orders, but has been used in broader studies of vertebrates and invertebrates (Eakin, 1968; Rieger & Tyler, 1979; Phillips & Tandler, 1987). Nevertheless, the potential value of ultrastructural comparisons has been demonstrated recently by studies of gastric mucosa, retina, and submandibular salivary glands (Phillips et al., 1984; Feldman & Phillips, 1984; Tandler et al., 1983, 1986; Phillips & Tandler, 1987). The discovery of patterns of interspecific differences in cellular architecture and cellular secretory products is in keeping with previously successful use of histology in evolutionary and systematic mammalogy (e.g., Quay, 1954; Forman, 1972; Phillips & Oxberry, 1972; Sands et al., 1977; Naumova, 1981; Hood & Smith, 1982, 1983).

Secretory cells, such as the parotid acinar cells used in the present study, seem to hold special promise for comparative investigation. Firstly, the entire secretory process—from nuclear DNA to synthesis of proteins and complex carbohydrates and packaging of materials into secretory granules and their subsequent discharge—has been studied intensively over the past several decades (e.g., Jamieson & Palade, 1971; Castle et al., 1975; Palade, 1975). Secondly, the secretory process seems to be relatively conservative—in the sense that basic pathways are the same in virtually all secretory cells—and, therefore, interspecific comparisons are facilitated and interpretation is somewhat simplified (Phillips & Tandler, 1987).

In the present study we limited our descriptive comparisons to "mature" secretory granules. These are the secretory granules that accumulate within the apical cytoplasm of the acinar cell; this product either is discharged from the cell into the acinar lumen (thus becoming part of the formative saliva) or, after an unknown storage interval, is broken down through autophagy and recycled within the cell. The glycoprotein components of the acinar secretory product are elaborated by the rough endoplasmic reticulum acting in concert with the Golgi complex and are perhaps the most interesting feature from a comparative point of view. Synthesis of these glycoproteins begins with transcription of a very small segment of the genome into MRNA. The exportable proteins of the secretory granules could be regarded as providing a "window" on the genome because, although all somatic cells have the same genome (Briggs & King, 1952, 1957), only a small portion actually is operational in any given fully differentiated cell. A protein synthesized for export thus directly reflects a portion of the operational genome. Any polysaccharide components of the secretory granules probably are at least one more step removed from the genome because synthesis of complex carbohydrates is enzymatically determined and usually takes place within saccules in the Golgi complex (Tandler, 1978). Even so, the polysaccharides also have considerable potential for comparative analysis.

Production of secretory granules can be considered in terms of ontogeny. The first "granules" typically are seen in direct proximity to the Golgi complex, where carbohydrates are linked to proteins and the membrane that will encase the granule is synthesized in a process involving both the GERL and the Golgi complex itself (Tandler, 1978; Hand, 1980a; Hand & Oliver, 1984). Newly formed ("immature") granules differ greatly from "mature" granules (see Castle et al., 1975) and were not used by us in describing the product for each of our species. Nevertheless, in view of species differences found by us, it is interesting to ask whether or not immature granules of one species might resemble mature granules in another species. Such similarities might be expected if heterochronic differences accounted for differences in mature product. However, because no such cross-species similarities were found by us between immature and mature secretory granules, heterochrony does not seem to account for species differences, at least among the 15 bats examined here.

No two Neotropical species examined here exhibited the same mature secretory granules in their parotid gland acinar cells, although in a few cases there was a degree of resemblance. This extreme variability is easily the greatest ever reported for homologous cells within an order or, as in the case of the Phyllostomidae, within a family of mammals (or any other vertebrates). To what can we attribute this striking finding?

One consideration is fixation, which certainly affects the appearance of any cellular feature as viewed with transmission electron microscopy. Indeed, it can be said that the appearance of cells and their products essentially is the consequence of their intrinsic chemistry combined with the chemistry of the fixative at the moment in time when fixation occurred. Different fixatives and tissue processing can have profound effects on the microscopic appearance of secretory granules in salivary glands (Simson et al., 1978). However, we used consistent processing techniques and two similar fixatives. Our specimens of *Ariteus* and *Leptonycteris* both were fixed with 2% glutaraldehyde, whereas all of our other specimens were fixed with a trialdehyde-DMSO fixative (Kalt & Tandler, 1971; Phillips, 1985). We were able to eliminate fixation as a source of variation because we also have examined specimens of *Artibeus* that had been fixed in both solutions, coincidentally with *Leptonycteris, Ariteus,* and all of the other species examined (see *A. phaeotis* parotid in Phillips et al., 1977).

Generally speaking, given consistent preparation techniques, microscopic differences in secretory granule substructure can be ascribed to biochemical differences among the granules. Microscopically detectable sequestration of individual types of macromolecules within secretory granules has been demonstrated only rarely (Ravazzola & Orci, 1980; Kousvelari et al., 1982) but clearly is the best available explanation of intragranule substructure. Based on fundamental principles of biochemistry, it thus can be concluded that macromolecules packaged in the secretory granules most likely sort themselves out according to charge and steric effects, as well as chemical interactions, to yield a characteristic pattern for each species. However, it also should be noted that a homogeneous appearance of intragranular substance does not in itself preclude sequestration of different enzymes within the granule. Separate localization of different enzymes (α-amylase and chymotrypsinogen A or α-amylase and trypsinogen) within pancreatic cell zymogen granules has been demonstrated with a combination of HRP (horseradish-peroxidase)-labeled and ferritin-labeled antibodies (Ono et al., 1980).

At present it is impossible to correlate exactly secretory granule substructure with particular chemical components such as certain enzymes or mucosubstances, so from microscopic images alone we cannot say precisely how the granules in our species differ chemically from one another. However, some conclusions can be inferred from the literature. For example, an electron-dense image (see, for example, fig. 2), labeled classically as "serous," can be associated with granules rich in enzymes. Such an image is typical in species such as laboratory rodents and primates, for which some data are available on the biochemistry of parotid saliva (e.g., Jacobsen & Hensten-Pettersen, 1974). On the other hand, the presence of electron-dense "serous" granules does not preclude the presence of mucosubstances within the secretory granules. Pinkstaff et al. (1982) reported that, although the parotid product in the little brown bat, *Myotis lucifugus,* was "serous" with standard histological techniques, both neutral and acidic mucosubstances could be demonstrated histochemically. The parotid granules in *Artibeus* and *Ariteus* are interesting in this regard because electron-dense material is scarce (especially in *Artibeus,* fig. 3) and enzyme production is extremely low, at least in *Artibeus* (Junqueira et al., 1973). By way of contrast, *Sturnira lilium* has largely electron-dense granules (fig. 3) and thus differs considerably from the other two stenodermatines; *Sturnira* produces saliva rich in enzymes (Junqueira et al., 1973).

Although the parotid granules in *Artibeus* (and, by extension, *Ariteus*) are low in enzyme content, to what can we attribute their TEM image? This is an intriguing question because Wimsatt (1956) reported that the gland was negative for mucosubstances, whereas Radtke (1972) reported the presence of sialomucins in parotid acinar cells. In part, this apparent disagreement is the result of differences in techniques that cannot be resolved by transmission electron microscopy.

What can be determined about the parotid salivary glands in Neotropical bats that relates to their diets, evolutionary history, or systematic relationships? Clearly the ultrastructure of parotid secretory granules has systematic significance, because no two genera (or species either) are exactly alike. This finding is in keeping with our studies of the submandibular gland in five species of *Artibeus* (Phillips et al., 1977; Tandler et al., 1983, 1986). The seromucous cells in this gland were found to contain granules that allowed for three groupings of *Artibeus* species (*A. cinereus-A. phaeotis, A. jamaicensis-A. lituratus,* and *A. concolor*) that matched the independently derived genic data (from isozyme analysis) reported by

Koop and Baker (1983). The differences in parotid acinar product in *Eptesicus lynni* and *E. brasiliensis* (fig. 4) also are interesting from this perspective because independent genic data from 19 presumptive loci show that *E. lynni* probably originated from the *E. fuscus* species complex, separate from the origin of *E. brasiliensis* (Arnold et al., 1980). *Eptesicus lynni* was found to share only 62% of analyzed alleles with *E. brasiliensis.*

It might seem surprising that the microscopic images of salivary gland secretory products are generically (and often specifically) distinctive, but variation of such a fine resolution actually is in keeping with numerous genetic studies of mammalian saliva. Several genetic markers have been found in human saliva (Azen & Oppenheim, 1973; Ashton & Balakrishinan, 1974; Tan & Ashton, 1976), and salivary proteins in particular tend to be polymorphic (Azen, 1972, 1973). Additionally, both sex and strain differences in salivary proteins have been reported in laboratory mice (Ikemoto & Matsushima, 1984). Although none of these differences has been demonstrated microscopically (all are based on biochemical analysis alone), their occurrence nevertheless is significant to comparisons among mammalian species.

Although the sometimes subtle but consistent microscopic differences between species within the genera *Phyllostomus, Tonatia,* and *Eptesicus* (figs. 2, 4) are in keeping with our hypothesis of the sensitivity of secretory product to genic differences, they cannot readily be related to any known ecological differences between species within each genus. It is altogether possible that such species differences do not represent direct evolutionary selection. Instead, the differing images might represent species-specific protein polymorphisms of a type that would not significantly affect "performance" of the saliva even though differences in primary molecular structure or surface charges (or both) could be indirectly detected by transmission electron microscopy (Phillips & Tandler, 1987). The existence of such "nonfunctional" (and presumably nonselected, i.e., "neutral") interspecific variation in a protein molecule has been demonstrated previously in the otherwise conservative hemoglobin molecule (Perutz, 1983). In this example the tertiary and quaternary structures apparently are conserved regardless of large numbers of functionally neutral amino acid substitutions in the primary structure. Although these substitutions in themselves most often have no effect on the functional capacity of the molecule, they nevertheless are known sometimes to change molecular surface charges (Perutz, 1983). For the present we postulate that similar nonselected, nonfunctional variations could account for the microscopically detectable intrageneric differences in salivary proteins in *Tonatia bidens* and *T. sylvicola, Phyllostomus latifolius* and *P. elongatus,* and *Eptesicus lynni* and *E. brasiliensis.* This explanation seems more parsimonious than the alternative, which would be to assume that the acinar cell component of the parotid saliva is "functionally" different in closely related, ecologically similar species. The extent or degree of microscopic differences in the mature acinar cell product ultimately might tell us more about relative times of divergence than about ecological differences among congeneric species of bats.

Microscopic comparisons of salivary gland secretory products could be valuable to cladistic studies of chiropteran families, at least to judge from our data. The phyllostomines generally are regarded as the most primitive (least derived) of the Phyllostomidae (Smith, 1976), and their parotid acinar products thus would qualify as plesiomorphic (following Henning, 1966). The products in all five species examined (*Phyllostomus elongatus, P. latifolius, Tonatia bidens, T. sylvicola,* and *Trachops cirrhosus*) contained large amounts of electron-dense material (fig. 2), as did the secretory product in *Pteronotus parnellii* (a closely related mormoopid), *Eptesicus lynni* and *E. brasiliensis* (Vespertilionidae), and *Tadarida brasiliensis* and *Molossus molossus* (Molossidae), all of which serve as "outgroups."

If the electron-dense, enzyme-rich secretory granules are regarded as plesiomorphic in microchiropteran bats, then it would be reasonable to regard a granule with less electron-dense material as apomorphic (derived). The glossophagine genera examined (*Glossophaga* and *Leptonycteris*) are representative of a phyllostomid evolutionary trend in which dentition, tongues, and associated musculature were modified for feeding on fruit, pollen, and nectar (Park & Hall, 1951; Phillips, 1971; Greenbaum & Phillips, 1974; Griffiths, 1982, 1983; Smith & Hood, 1984). In these genera the electron-dense component has been reduced in comparison to the phyllostomines. *Carollia perspicillata,* which is omnivorous, fits into this category in that the parotid secretory granules contain relatively little electron-dense material. In this bat the secretory granules are very distinctive because the electron-dense inclusions are often in the form of cagelike geodesic structures (Phillips & Tandler, 1987; Tandler et al., in press).

Perhaps the most interesting systematic finding in our data lies within the nominal subfamily Stenodermatinae. These bats represent an evolutionary trend toward frugivory that includes extreme gastric adaptation at the gross, histological, histochemical, and cellular levels (Forman, 1972; Forman et al., 1979; Phillips & Studholme, 1982; Phillips et al., 1984). The pale parotid secretory granules in *Artibeus* and *Ariteus* are synapomorphous, whereas the parotid granules in *Sturnira* are electron-dense, enzyme-rich, and more nearly like the plesiomorphic granules of the phyllostomines, in particular *Trachops cirrhosus* (figs. 2–3). This example is interesting because *Sturnira* always has been something of an enigma. Although de la Torre (1961) allied this genus with *Vampyrops*-like stenodermatines and Smith (1976) included it with the "long-faced" stenodermatines, others (e.g., Walton & Walton, 1968) previously had placed the genus in a separate subfamily (Sturnirinae) based on a variety of morphological features that seemed inconsistent with the other, more traditional, stenodermatines such as *Artibeus*. Indeed, Slaughter (1970) pointed out that *Sturnira* has some distinctive dental features that possibly link the genus to the glossophagines. The differences in the parotid secretory granules thus are in keeping with a variety of other phenotypic differences. While it is reasonable to suggest that secretory granule differences of this magnitude represent a major interspecific difference in the operational segment of the genomes of homologous parotid acinar cells, the eventual systematic value of such data will await availability of data about still other phyllostomid genera.

Although our comparative data clearly document great microscopically detectable plasticity in the secretory product of parotid acinar cells, many questions about their evolution remain unanswered. For example, what has been the role of diet? At first glance, our data suggest that insectivorous and animalivorous species have enzyme-rich, electron-dense granules, whereas frugivores have enzyme-poor, pale secretory granules. Yet, what about *Sturnira lilium,* which certainly includes large amounts of fruit in its diet (Gardner, 1977)?

At least three factors interfere with any effort to correlate parotid acinar cell product with diet. First, acinar cells are but one cell type among several that influence the biochemistry of parotid saliva. Second, and possibly more important, the parotid is but one salivary gland among a host of glands (the submandibular and sublingual and minor glands) that are located throughout the oral region and that contribute substances to the saliva. Some data suggest that different glands in different species might have been more responsive, in an evolutionary sense, to changes in diet. For example, it is the accessory submandibular glands of *Trachops* and *Megaderma* that are unique in histology and might correlate with feeding on frogs (Phillips & Tandler, 1985; Phillips et al., 1987); the parotid in *Trachops* is similar to that of other phyllostomines.

A third factor, for which we presently have no data, is the possibility that parotid acinar cells can respond in some way to diet at the individual level. Our sample sizes are large enough to convince us of the near uniformity of secretory granule morphology within a population. However, Schick et al. (1984) have recently published the first report of a secretory cell responding to dietary intake by a shift in enzyme production. This is the first such detailed molecular data known to us and, although pancreatic acinar cells in laboratory rats were the source of the data, the implications for comparative studies of salivary glands are worth noting.

A final factor for consideration is the complex role that salivary glands play in the lives of mammals. Salivary glands in *Artibeus, Chiroderma,* and *Ametrida* might contribute to gastric cytoprotection (Studier et al., 1983; Phillips et al., 1984). In *Tonatia sylvicola* an unusual organelle found in submandibular seromucous cells of males might be related to species isolation or sex recognition, or both (Nagato et al., 1984). If these examples are typical for bats, then digestion is but one of several major functions of salivary glands.

In conclusion, this first systematic microscopic survey of a secretory cell product has demonstrated a previously unknown, extraordinary degree of variation within a group of related species. Based on our findings, one might conclude that Neotropical microchiropteran bats will serve as a significant model for study of how secretory cells have evolved in mammals; meanwhile, data from comparative investigations will contribute to our knowledge of genic relationships among these animals.

Acknowledgments

Financial support for field and laboratory research that led to the data reported here came from a variety of sources, which we are pleased to acknowledge: Research Corporation Grant C-1251 (to Phillips); NIH Grant DE 03455-02 (to Phillips); NSF Grant CDP-801-8653 (to Phillips); Hofstra University HCLAS Executive Committee grants (to Phillips); NIH Grant AM-08305 (to Tandler); NIH Grant RO DE 07648-01A1 (to Tandler & Phillips); the Alcoa Foundation (through Dr. Hugh H. Genoways, Carnegie Museum of Natural History); and the Graham Netting Research Fund via a grant from the Cordelia S. May Charitable Trust (through Genoways). In addition to recognizing our financial support, we also wish to thank several colleagues whose field companionship in such places as Suriname and Jamaica, hard work, and willingness to share time and ideas in their laboratories played a major part in our study. These are: Hugh H. Genoways, Carnegie Museum of Natural History; Gary W. Grimes and Dorothy E. Pumo, Hofstra University; Robert J. Baker, Texas Tech University; and Henry A. Reichart, formerly of STINASU, in Suriname. Field assistance was provided by K. M. Studholme, S. L. Williams, N. M. Sposito, J. Groen, R. L. Honeycutt, B. Koop, M. Arnold, B. A. Oxberry, P. Billeter, J. Bickham, and J. Patton. We also acknowledge technical assistance of Carol Ayala, Case Western Reserve University, and typing by Linda Cossen, Special Secretarial Services, Hofstra University. Artwork and lettering were done by Helen Tandler. All specimens were collected and imported under proper permits issued by the relevant authorities in Mexico, Jamaica, Suriname, and the United States.

Literature Cited

ARNOLD, M. L., R. J. BAKER, AND H. H. GENOWAYS. 1980. Evolutionary origin of *Eptesicus lynni*. Journal of Mammalogy, **61**: 319–322.

ASHTON, G. C., AND C. R. BALAKRISHINAN. 1974. Polymorphisms of human salivary proteins. American Journal of Human Genetics, **26**: 145–153.

AZEN, E. A. 1972. Genetic polymorphism of basic proteins from parotid saliva. Science, **176**: 673–674.

———. 1973. Properties of salivary basic proteins showing polymorphism. Biochemical Genetics, **9**: 69–85.

AZEN, E. A., AND F. G. OPPENHEIM. 1973. Genetic polymorphism of proline-rich human salivary proteins. Science, **180**: 1067–1069.

BALL, W. D. 1974. Isolation of salivary ribonuclease, deoxyribonuclease and amylase from the parotid gland of the rat, *Rattus norvegicus albinus*. Biochimica et Biophysica Acta, **341**: 305–317.

BJÖRKMAN, N. 1962. Low magnification electron microscopy in histological work. Acta Morphologica Neerlando-Scandinavica, **4**: 344–348.

BRIGGS, R., AND T. J. KING. 1952. Transplantation of living nuclei from blastula cells into enucleated frogs' eggs. Proceedings of the National Academy of Sciences, USA, **38**: 455–463.

———. 1957. Changes in the nuclei of differentiating endoderm cells as revealed by nuclear transplantation. Journal of Morphology, **100**: 269–312.

CASTLE, J. D., J. D. JAMIESON, AND G. E. PALADE. 1975. Secretion granules in the rabbit parotid gland. Isolation, subfractionation, and characterization of the membrane and content subfractions. Journal of Cell Biology, **64**: 182–210.

DE LA TORRE, L. 1961. The evolution, variation, and systematics of the Neotropical bats of the genus *Sturnira*. Ph.D. diss., University of Illinois, Urbana.

DISANTO, P. E. 1960. Anatomy and histochemistry of the salivary glands of the vampire bat, *Desmodus rotundus murinus*. Journal of Morphology, **106**: 301–335.

EAKIN, R. M. 1968. Evolution of photoreceptors, pp. 194–242. *In* Dobzhansky, T., M. K. Hecht, and W. C. Steere, eds., Evolutionary Biology, vol. 2. Appleton-Century-Crofts, New York, 452 pp.

EISENBERG, J. F., AND D. E. WILSON. 1978. Relative brain size and feeding strategies in the Chiroptera. Evolution, **32**: 740–751.

FELDMAN, J. L., AND C. J. PHILLIPS. 1984. Comparative retinal pigment epithelium and photoreceptor ultrastructure in nocturnal and fossorial rodents: The eastern woodrat, *Neotoma floridana*, and the plains pocket gopher, *Geomys bursarius*. Journal of Mammalogy, **65**: 231–245.

FORMAN, G. L. 1972. Comparative morphological and histochemical studies of stomachs of selected American bats. University of Kansas Science Bulletin, **59**: 591–729.

FORMAN, G. L., C. J. PHILLIPS, AND C. S. ROUK. 1979. Alimentary tract, pp. 205–227. *In* Baker, R. J., J. K. Jones, Jr., and D. C. Carter, eds., Biology of bats of the New World family Phyllostomatidae. Part III. Special Publications, The Museum, Texas Tech University, **16**: 1–441.

FREEMAN, P. W. 1979. Specialized insectivory: Beetle-eating and moth-eating molossid bats. Journal of Mammalogy, **60**: 467–479.

———. 1981. Correspondence of food habits and morphology in insectivorous bats. Journal of Mammalogy, **62**: 166–173.

GARDNER, A. L. 1977. Feeding habits, pp. 293–350. *In* Baker, R. J., J. K. Jones, Jr., and D. C. Carter, eds., Biology of bats of the New World family Phyllostomatidae. Part II. Special Publications, The Museum, Texas Tech University, **13**: 1–364.

GREENBAUM, I. F., AND C. J. PHILLIPS. 1974. Comparative anatomy and general histology of tongues of long-nosed bats (*Leptonycteris sanborni* and *L. nivalis*) with reference to infestation of oral mites. Journal of Mammalogy, **55**: 489–504.

GRIFFITHS, T. A. 1982. Systematics of the New World nectar-feeding bats (Mammalia, Phyllostomidae), based on the morphology of the hyoid and lingual regions. American Museum Novitates, **2742**: 1–45.

———. 1983. On the phylogeny of the Glossophaginae and the proper use of outgroup analysis. Systematic Zoology, **32:** 283–285.

Hand, A. R. 1980a. Cytochemical differentiation of the Golgi apparatus from GERL. Journal of Histochemistry and Cytochemistry, **28:** 82–86.

———. 1980b. Salivary glands, pp. 336–370. *In* Bhaskar, S. N., ed., Orban's Oral Histology and Embryology. C. V. Mosby Co., St. Louis, Mo., 482 pp.

Hand, A. R., and C. Oliver. 1984. Effects of secretory stimulation on the Golgi apparatus and GERL of rat parotid acinar cells. Journal of Histochemistry and Cytochemistry, **32:** 403–412.

Henning, W. 1966. Phylogenetic Systematics. Translated by D. D. Davis and R. Zanger. University of Illinois Press, Urbana, 263 pp.

Hood, C. S., and J. D. Smith. 1982. Cladistical analysis of female reproductive histomorphology in phyllostomatoid bats. Systematic Zoology, **31:** 241–251.

———. 1983. Histomorphology of the female reproductive tract in phyllostomoid bats. Occasional Papers, The Museum, Texas Tech University, **86:** 1–38.

Ikemoto, S., and Y. Matsushima. 1984. Sex and strain differences of mouse variant salivary proteins. Differentiation, **26:** 55–58.

Jacobsen, N., and A. Hensten-Pettersen. 1974. Some biochemical characteristics of the salivary gland products of *Macaca irus* and *Cercopithecus aethiops*. Comparative Biochemistry and Physiology, **49B:** 605–614.

Jamieson, J. D., and G. E. Palade. 1971. Synthesis, intracellular transport, and discharge of secretory proteins in stimulated pancreatic exocrine cells. Journal of Cell Biology, **50:** 135–158.

Junqueira, L. C. U., and F. Fava de Morães. 1965. Comparative aspects of the vertebrate major salivary glands biology, pp. 36–48. *In* Wohlfarth-Bottermann, K. E., ed., Funktionelle und Morphologische Organisation der Zelle. Sekretion und Exkretion. Springer-Verlag, Berlin.

Junqueira, L. C. U., F. Fava de Morães, and A. M. Toledo. 1967. Sialic acids in vertebrate salivary glands, saliva and pancreas. Archives of Oral Biology, **12:** 151–157.

Junqueira, L. C. U., A. M. S. Toledo, and A. I. Doine. 1973. Digestive enzymes in the parotid and submandibular glands of mammals. Anais da Academia Brasileira de Ciencias, **45:** 629–643.

Kalt, M. R., and B. Tandler. 1971. A study of fixation of early amphibian embryos for electron microscopy. Journal of Ultrastructure Research, **36:** 633–645.

Koop, B. F., and R. J. Baker. 1983. Electrophoretic studies of relationships of six species of *Artibeus* (Chiroptera: Phyllostomidae). Occasional Papers, The Museum, Texas Tech University, **83:** 1–12.

Kousvelari, E. E., F. G. Oppenheim, and L. S. Cutler. 1982. Ultrastructural localization of salivary acidic proline-rich proteins from *Macaca fascicularis*. Journal of Histochemistry and Cytochemistry, **30:** 274–278.

Lekagul, B., and J. A. McNeely. 1977. Mammals of Thailand. Kurusapha Ladprao Press, Bangkok, 758 pp.

Mineda, T. 1977. An electron microscopic investigation on the submandibular gland of the Japanese horseshoe-bat (*Rhinolopus* (sic) *ferrumequinum nippon*). Histological and histochemical studies. Aichi-Gakuin Journal of Dental Science, **15:** 35–47.

———. 1978. An electron microscopic study of the parotid gland of the Japanese horseshoe-bat (*Rhinolopus* [sic] *ferrumequinum nippon*) in the functional and hibernating period. Aichi-Gakuin Journal of Dental Science, **16:** 139–145.

Nagato, T., B. Tandler, and C. J. Phillips. 1984. Unusual smooth endoplasmic reticulum in submandibular acinar cells of the male round-eared bat, *Tonatia sylvicola*. Journal of Ultrastructure Research, **87:** 275–284.

Naumova, E. I. 1981. Functional morphology of the digestive system of rodents and lagomorphs. USSR Academy of Sciences, Moscow, 261 pp. (In Russian.)

Ono, T., Y. Sakai, N. Yamamoto, and K. Yasuda. 1980. Double staining for immunohistochemistry: Simultaneous demonstration of two enzymes in an identical zymogen granule of an ultra-thin frozen section. Acta Histochemica et Cytochemica, **13:** 436–448.

Palade, G. 1975. Intracellular aspects of the process of protein synthesis. Science, **189:** 347–358.

Park, H., and E. R. Hall. 1951. The gross anatomy of the tongues and stomachs of eight New World bats. Transactions of the Kansas Academy of Sciences, **54:** 64–72.

Perutz, M. F. 1983. Species adaptation in a protein molecule. Molecular Biology and Evolution, **1:** 1–28.

Phillips, C. J. 1971. The dentition of glossophagine bats: Development, morphological characteristics, variation, pathology, and evolution. Miscellaneous Publications, University of Kansas, Museum of Natural History, **54:** 1–138.

———. 1985. Field fixation and storage of museum tissue collections suitable for electron microscopy. Acta Zoologica Fennica, **170:** 87–90.

Phillips, C. J., G. W. Grimes, and G. L. Forman. 1977. Oral biology, pp. 121–246. *In* Baker, R. J., J. K. Jones, Jr., and D. C. Carter, eds., Biology of bats of the New World family Phyllostomatidae. Part II. Special Publications, The Museum, Texas Tech University, **13:** 1–364.

Phillips, C. J., and B. Oxberry. 1972. Comparative histology of molar dentitions of *Microtus* and *Clethrionomys,* with comments on dental evolution in microtine rodents. Journal of Mammalogy, **53:** 1–20.

Phillips, C. J., and K. Studholme. 1982. Comparative ultrastructure of gastric chief cells in carnivorous and frugivorous bats. IRCS Medical Science, Cell and Membrane Biology, **10:** 573–574.

Phillips, C. J., K. M. Studholme, and G. L. Forman. 1984. Results of the Alcoa Foundation Suriname expeditions. VIII. Comparative ultrastructure of gastric mucosae in four genera of bats (Mammalia: Chiroptera), with comments on gastric evolution. Annals of Carnegie Museum, **53:** 71–117.

Phillips, C. J., and B. Tandler. 1985. Unique salivary gland structure in two genera of tropical bats. Anatomical Record, **211:** 152A.

———. 1987. Mammalian evolution at the cellular level, pp. 1–66. *In* Genoways, H. H., ed., Current Mammalogy. Plenum Press, New York, 519 pp.

Phillips, C. J., B. Tandler, and C. A. Pinkstaff. 1987. Unique salivary glands in two genera of tropical microchiropteran bats: An example of evolutionary convergence in histology and histochemistry. Journal of Mammalogy, **68:** 235–242.

Pinkstaff, C. A., B. Tandler, and R. P. Cohan. 1982. Histology and histochemistry of the parotid and the principal and accessory submandibular glands of the little brown bat. Journal of Morphology, **172:** 271–285.

Quay, W. B. 1954. The meibomian glands of voles and lemmings (Microtinae). Miscellaneous Publications, Museum of Zoology, University of Michigan, **82:** 1–24.

Radtke, B. M. 1972. Observacões estruturais, histoquímicas e bioquímicas das glándulas salivares, parotida e submandibular, de morcegos (*Chiroptera,* Mammalia). Ph.D. diss., Institute of Biomedical Science, University of São Paulo, Brazil.

Ravazzola, M., and L. Orci. 1980. Glucagon and glicentin immunoreactivity are topologically segregated in the α granule of the human pancreatic A cell. Nature, **284:** 66–67.

Rieger, R., and S. Tyler. 1979. The homology theorem in ultrastructural research. American Zoologist, **19:** 655–664.

Robinovitch, M. R., and L. M. Sreebny. 1969. Separation and identification of some of the protein components of rat parotid saliva. Archives of Oral Biology, **14:** 935–949.

Sands, M. W., R. P. Coppinger, and C. J. Phillips. 1977. Comparisons of thermal sweating and histology of sweat glands of selected canids. Journal of Mammalogy, **58:** 74–78.

Schick, J., R. Verspohl, H. Kern, and G. Scheele. 1984. Two distinct adaptive responses in the synthesis of exocrine pancreatic enzymes to inverse changes in protein and carbohydrate in the diet. American Journal of Physiology, **247** (Gastrointestine and Liver Physiology, 10): G611–G616.

Simson, J. A., R. M. Dom, P. L. Sannes, and S. S. Spicer. 1978. Morphology and cytochemistry of acinar secretory granules in normal and isoproterenol-treated rat submandibular glands. Journal of Microscopy, **113:** 185–203.

Slaughter, B. H. 1970. Evolutionary trends of chiropteran dentitions, pp. 51–83. *In* Slaughter, B. H., and D. W. Walton, eds., About Bats. Southern Methodist University Press, Dallas, Texas, 339 pp.

Smith, J. D. 1976. Chiropteran evolution, pp. 49–69. *In* Baker, R. J., J. K. Jones, Jr., and D. C. Carter, eds., Biology of bats of the New World family Phyllostomatidae. Part I. Special Publications, The Museum, Texas Tech University, **10:** 1–218.

Smith, J. D., and C. S. Hood. 1984. Genealogy of the New World nectar-feeding bats reexamined: A reply to Griffiths. Systematic Zoology, **33:** 435–460.

Stempak, J. G., and R. T. Ward. 1964. An improved staining method for electron microscopy. Journal of Cell Biology, **22:** 697–701.

Studier, E. H., B. C. Boyd, A. T. Feldman, R. W. Dapson, and D. E. Wilson. 1983. Renal function in the Neotropical bat, *Artibeus jamaicensis.* Comparative Biochemistry and Physiology, **74A:** 199–209.

Tan, S. G., and G. C. Ashton. 1976. An autosomal glucose-6-phosphate dehydrogenase (hexose-6-phosphate dehydrogenase) polymorphism in human saliva. Human Heredity, **26:** 113–123.

Tandler, B. 1972. Microstructure of salivary glands, pp. 8–21. *In* Rowe, N. H., ed., A Symposium on Salivary Glands and Their Secretion. University of Michigan Press, Ann Arbor, 125 pp.

———. 1978. Salivary glands and the secretory process, pp. 547–592. *In* Shaw, J. H., E. A. Sweeney, C. C. Cappuccino, and S. M. Meller, eds., Textbook of Oral Biology. W. B. Saunders Co., Philadelphia, 1178 pp.

Tandler, B., and R. P. Cohan. 1984. Ultrastructure of the parotid gland in the little brown bat. Anatomical Record, **210:** 491–502.

Tandler, B., T. Nagato, and C. J. Phillips. 1983. Ultrastructure of the submandibular gland in four species of fruit bats (*Artibeus* sp.). Anatomical Record, **205:** 198A.

———. 1986. Systematic implications of comparative ultrastructure of secretory acini in the submandibular salivary gland in *Artibeus* (Chiroptera: Phyllostomidae). Journal of Mammalogy, **67:** 81–90.

Tandler, B., and C. J. Phillips. 1985. Unusual secretory granules in the accessory submandibular gland of long-winged bats. Anatomical Record, **211:** 192A.

Tandler, B., C. J. Phillips, and T. Nagato. In press. Parotid salivary gland ultrastructure in an omnivorous Neotropical bat: Evolutionary diversity at the cellular level. Zoologica Scripta.

Tandler, B., and R. J. Walter. 1977. Epon-Maraglas embedment for electron microscopy. Stain Technology, **52:** 238–239.

Tuttle, M. D., and M. J. Ryan. 1981. Bat predation and the evolution of frog vocalizations in the Neotropics. Science, **214:** 677–678.

Venable, J. H., and R. Coggeshall. 1965. A simplified lead citrate stain for use in electron microscopy. Journal of Cell Biology, **25**(no. 2, pt. 1): 407–408.

Wallach, D., R. Tessler, and M. Schramm. 1975. The proteins of the content of the secretory granules of the rat parotid gland. Biochimica et Biophysica Acta, **382:** 552–564.

Walton, D. W., and G. W. Walton. 1968. Comparative osteology of the pelvic and pectoral girdles of the Phyllostomatidae (Chiroptera; Mammalia). Journal of the Graduate Research Center, Southern Methodist University, **37:** 1–35.

Wimsatt, W. A. 1956. Histological and histochemical observations on the parotid, submaxillary and sublingual glands of the tropical American fruit bat *Artibeus jamaicensis* Leach. Journal of Morphology, **99:** 169–209.

Appendix

Specimens Examined

Voucher specimens of each species and each locality are deposited in the collections of the Carnegie Museum of Natural History and The Museum, Texas Tech University. All specimens except those of *Leptonycteris sanborni* (Jalisco, Mexico) and *Ariteus flavescens* and *Eptesicus lynni* (Jamaica) were collected in Brokopondo Province, Suriname.

Pteronotus parnellii, 10 (5 ♂♂, 5 ♀♀); *Tonatia bidens,* 3 (2 ♂♂, 1 ♀); *T. sylvicola,* 2 (♂♂); *Trachops cirrhosus,* 2 (1 ♂, 1 ♀); *Phyllostomus elongatus,* 3 (1 ♂, 2 ♀♀); *P. latifolius,* 1 (♀); *Glossophaga soricina,* 3 (2 ♂♂, 1 ♀); *Leptonycteris sanborni,* 2 (♂♂); *Sturnira lilium,* 5 (2 ♂♂, 3 ♀♀); *Artibeus jamaicensis,* 3 (♀♀); *Ariteus flavescens,* 1 (♀); *Eptesicus lynni,* 2 (♂♂); *E. brasiliensis,* 3 (♀♀); *Tadarida brasiliensis,* 1 (♂); *Molossus molossus,* 8 (4 ♂♂, 4 ♀♀).

Distribution of the Species and Subspecies of Cebids in Venezuela

Roberta Bodini and Roger Pérez-Hernàndez

ABSTRACTS

Thirteen species of Primates representing nine genera in the family Cebidae are found in Venezuela. The geographic distribution of these species exhibits four main patterns. *Alouatta* and *Cebus* are widely distributed in all parts of the country. Four genera, *Saimiri, Callicebus, Cacajao,* and *Chiropotes* are restricted to south-central Venezuela; their distribution is centered in the Amazonian lowlands. *Aotus* and *Ateles* are each represented by one species or subspecies in south-central Venezuela and another in northwestern Venezuela. *Pithecia* is mainly restricted to the Guianan highlands in eastern Venezuela, with a single known outlying locality in south-central Venezuela. Of the nine cebid genera in Venezuela, all occur in south-central Venezuela, four occur in northwestern and north-central Venezuela, and only three occur in eastern Venezuela. Distribution maps for the 13 species in Venezuela are presented with exact localities of specimens.

Nueve géneros de cébidos representados por trece especies se hallan en Venezuela. La repartición geográfica de las especies se resuelve en cuatro patrones geográficos principales. De este modo, *Alouatta* y *Cebus* son extensamente repartido por todo el pais. La distribución de los géneros, *Saimiri, Callicebus, Cacajao* y *Chiropotes* está restringida al sur-central de Venezuela con concentración en las tierras bajas Amazónicas. *Aotus* y *Ateles* son representatados, cada cual, por una especie (o subespecie), en el sur-central, y otra especie in el noroeste del pais. *Pithecia* se reparte en las alturas guayanas venezolanas del este, y es conocido, a la véz, por un solo dato de captura del sur-central venezolano. En resumen, la totalidad de los nueve géneros está presente en el sur-central de Venezuela, cuatro de ellos en el noroeste y norte-central, y tres en la Venezuela oriental. La repartición de las trece especies está documentado por mapas.

Treze espécies de Primatas, representando nove gêneros da família Cebidae, são encontradas na Venezuela. A distribuição geográfica destas espécies exibem quatro padrões principais. *Alouatta* e *Cebus* espalham-se por todas partes do país. Quatro gêneros, *Saimiri, Callicebus, Cacajao,* e *Chiropotes,* limitam-se ao centro-sul de Venezuela; a Bacia Amazonica sendo seus focos de distribuições. *Aotus* e *Ateles* são representados por uma espécie ou subespécie cada no centro-sul da Venezuela, e por outra espécie no noroeste do país. *Pithecia* limita-se ao planalto Guianense, no leste da Venezuela, com apenas uma localidade conhecida no centro-sul do País. Em suma, todos nove gêneros de Cebidae ocorrem no centro-sul da Venezuela, quatro ocorrem no centro-norte e noroeste do País, e apenas três ocorrem no região leste da Venezuela. Apresentam-se mapas de distribuições das 13 espécies na Venezuela, com localidades exatas dos espécimes colecionados.

From the Instituto de Zoología Tropical, Universidad Central de Venezuela, Apartado 47058, Los Chaguaramos, Caracas 1041-A, Venezuela.

Introduction

Primates of the family Cebidae are some of the most conspicuous mammals in the Neotropics, as they are diurnal, often forage in large troops, and may be quite vocal. However, we actually know very little about the exact distributions of most species of cebids. In Venezuela, several authors have discussed cebids as part of generic revisions (see Elliot, 1912; Hershkovitz, 1949, and later works; Hill, 1960, 1962; Kellogg & Goldman, 1944) and in works on the status or collections of specific species (i.e., Bodini, 1983; Handley, 1976; Mondolfi & Eisenberg, 1979; Rudran & Eisenberg, 1982).

Herein we summarize the distribution of cebids in Venezuela based on recent collections, specimen records from several museums, and the literature. Specimens reported on are housed in the following collections: Estación Biológica Rancho Grande (EBRG); Museo de Biología, Universidad Central de Venezuela (MBUCV); Museo de Ciencias Naturales (MCN); and Museo de Historia Natural La Salle (MHNLS). Our objectives are to provide accurate locality records and distribution maps that may be utilized in future investigations.

Generic Distributions

The nine genera of cebids that inhabit Venezuela exhibit four main patterns of geographic distribution (figs. 1–7). (1) Two genera—*Alouatta* and *Cebus*—are widely distributed in all parts of the country. (2) Four genera—*Saimiri, Callicebus, Cacajao,* and *Chiropotes*—are restricted to south-central Venezuela, centering on the Amazonian lowlands (see Eisenberg & Redford, 1979). (3) Two genera—*Aotus* and *Ateles*—are each represented by one species or subspecies in south-central Venezuela and another in northwestern Venezuela or northwestern and north-central Venezuela, which is the region of the northeastern spurs of the Andean chain and the enclosed Maracaibo basin. (4) One genus—*Pithecia*—is mainly restricted to the Guianan highlands in eastern Venezuela, with a single known outlying locality in south-central Venezuela. In summary, of the nine cebid genera in Venezuela, all occur in south-central Venezuela (including the outlier record of *Pithecia*), four occur in northwestern and north-central Venezuela, and only three occur in eastern Venezuela.

Specific and Subspecific Distributions

Saimiri

In a recent revision of the squirrel monkeys, Hershkovitz (1984) recognized four species: *Saimiri boliviensis, S. oerstedi, S. sciureus,* and *S. ustus.* All those squirrel monkeys found in northern South America he referred to the single species *S. sciureus* (fig. 1). Squirrel monkeys from Amazonian Venezuela and adjacent Brazil and Colombia he regarded as *S. sciureus cassiquiarensis.* Hershkovitz (1984) plotted several localities in Territorio Federal Amazonas, but listed only a single specific locality: "Casiquiare, Río (mouth), 2°01′N, 67°07′W." Hill (1960) previously had mapped the distribution of *S. sciureus* as occurring throughout all of Venezuela, although he listed only a single locality in the state of Bolivar (Camp Canaracuni, 4°36′N, 64°10′W). The name *cassiquiarensis* Lesson is based on Humboldt's description of a captive female from the banks of the Río Casiquiare, Amazonas, Venezuela.

Specimens Examined—Total 29. **Bolivar:** Canaracuni (4°36′N, 64°10′W). **Territorio Federal Amazonas:** Campo Cacuri (4°49′N, 65°26′W); Caño Yureba, Río Ventuari (3°35′N, 66°46′W); Río Puruname, 40 km from union with Río Orinoco (3°19′N, 65°15′W); Río Ventuari (3°59′N, 67°02′W); San Fernando de Atabapo (4°02′N, 67°37′W); San Juan de Manapiare (5°14′N, 66°02′W).

Aotus

Night monkeys are found from Panama throughout much of Amazonian South America to Paraguay. Historically, it has generally been assumed that all night monkeys represented the single species *Aotus trivirgatus.* However, in a recent revision of the genus, Hershkovitz (1983) recognized nine allopatric species. A tenth species, *Aotus hershkovitzi* from Colombia, has recently been proposed by Rámirez-Cerquera (1983). Two species of *Aotus, A. lemurinus* and *A. trivirgatus,* have been reported from Venezuela (fig. 2).

Aotus lemurinus griseimembra Elliot is known in Venezuela only from the extreme northwestern region, the states of Cojedes and Mérida (Hershkovitz, 1983), and the states of Táchira, Trujillo, and Zulia (Handley, 1976).

Specimens Examined—Total 10. **Zulia:** Campo El Rosario, Distrito Catatumbo (1°44′N, 67°03′W);

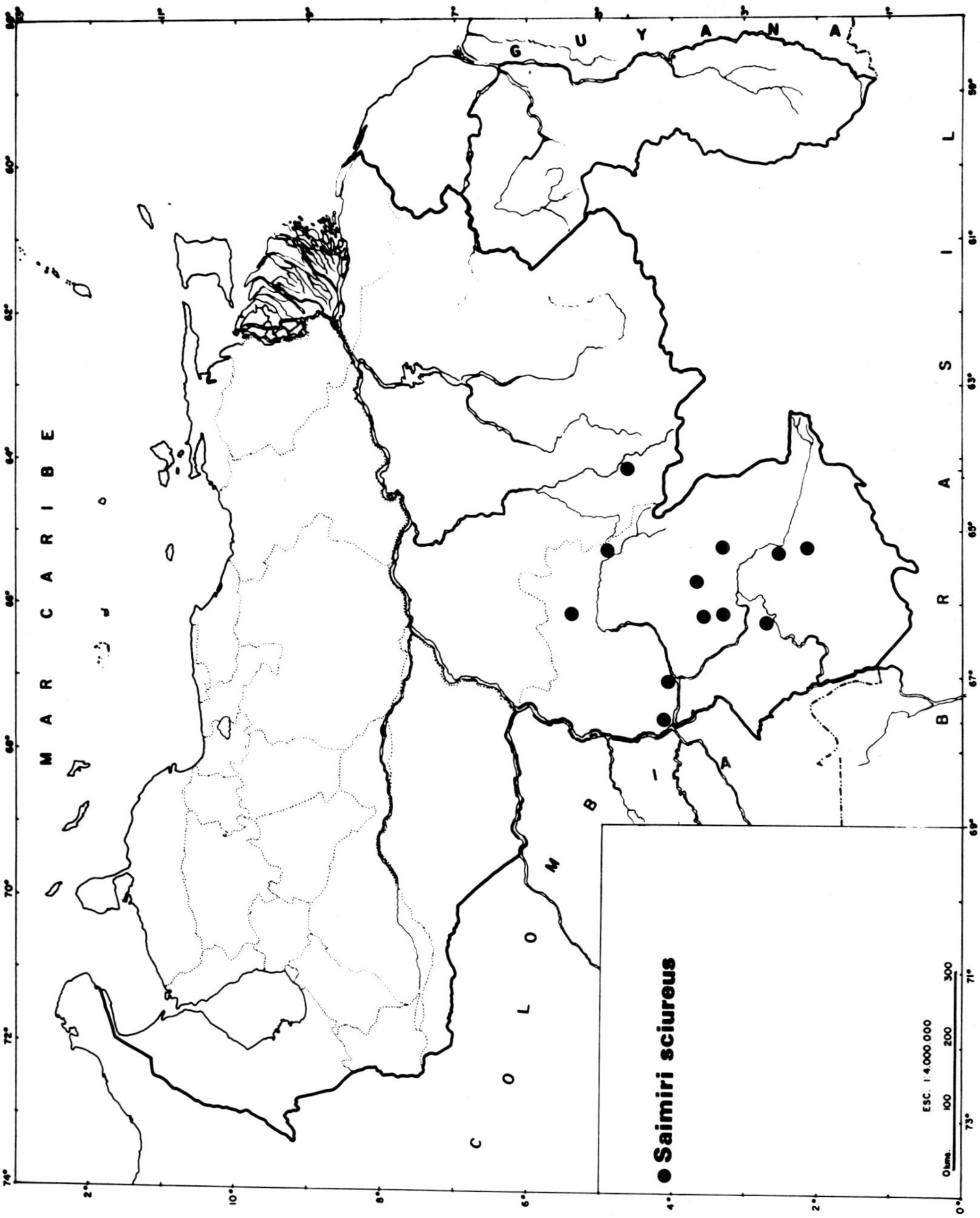

FIG. 1. Distribution of subfamily Saimirinae in Venezuela.

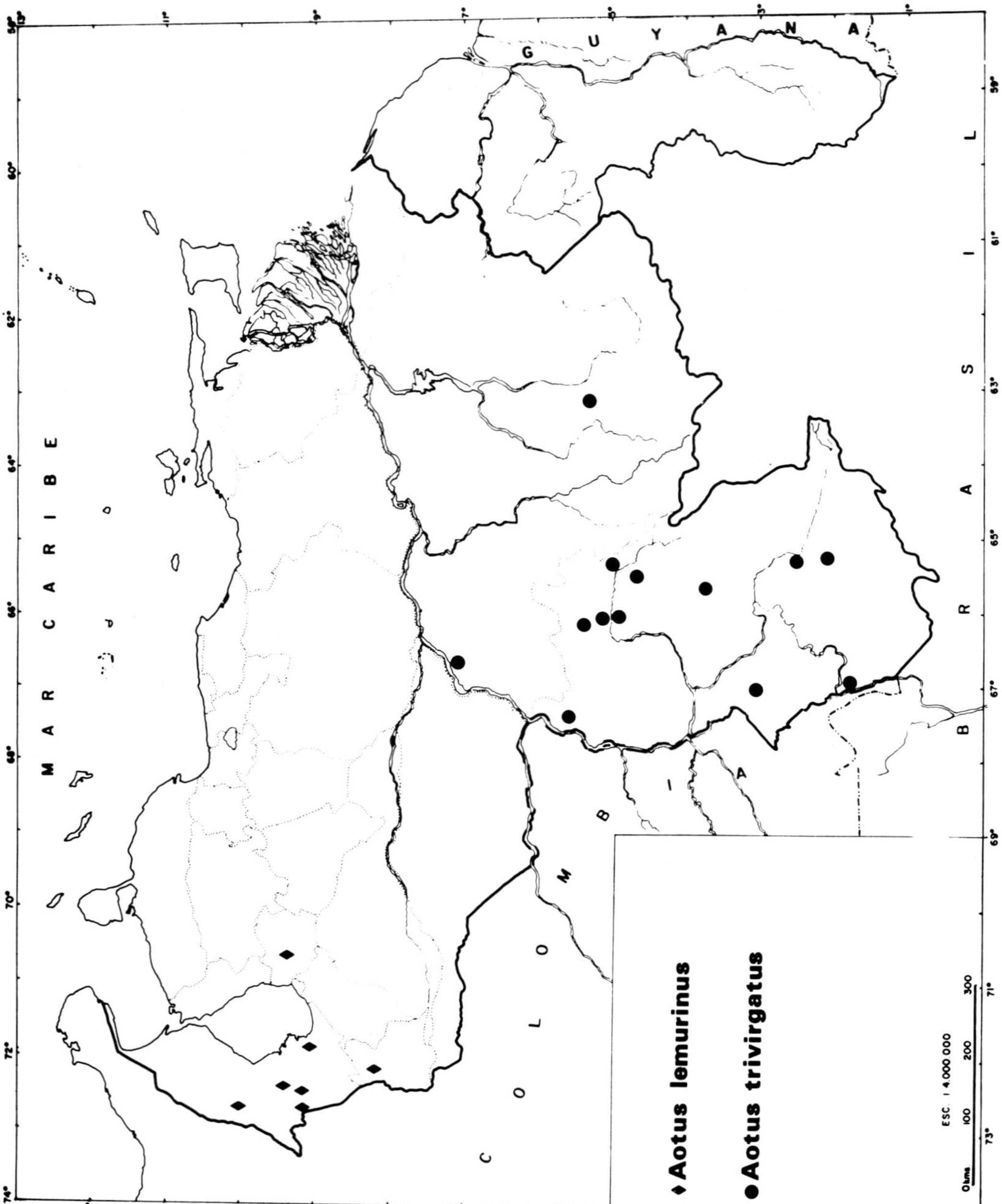

FIG. 2. Distribution of subfamily Aotinae in Venezuela.

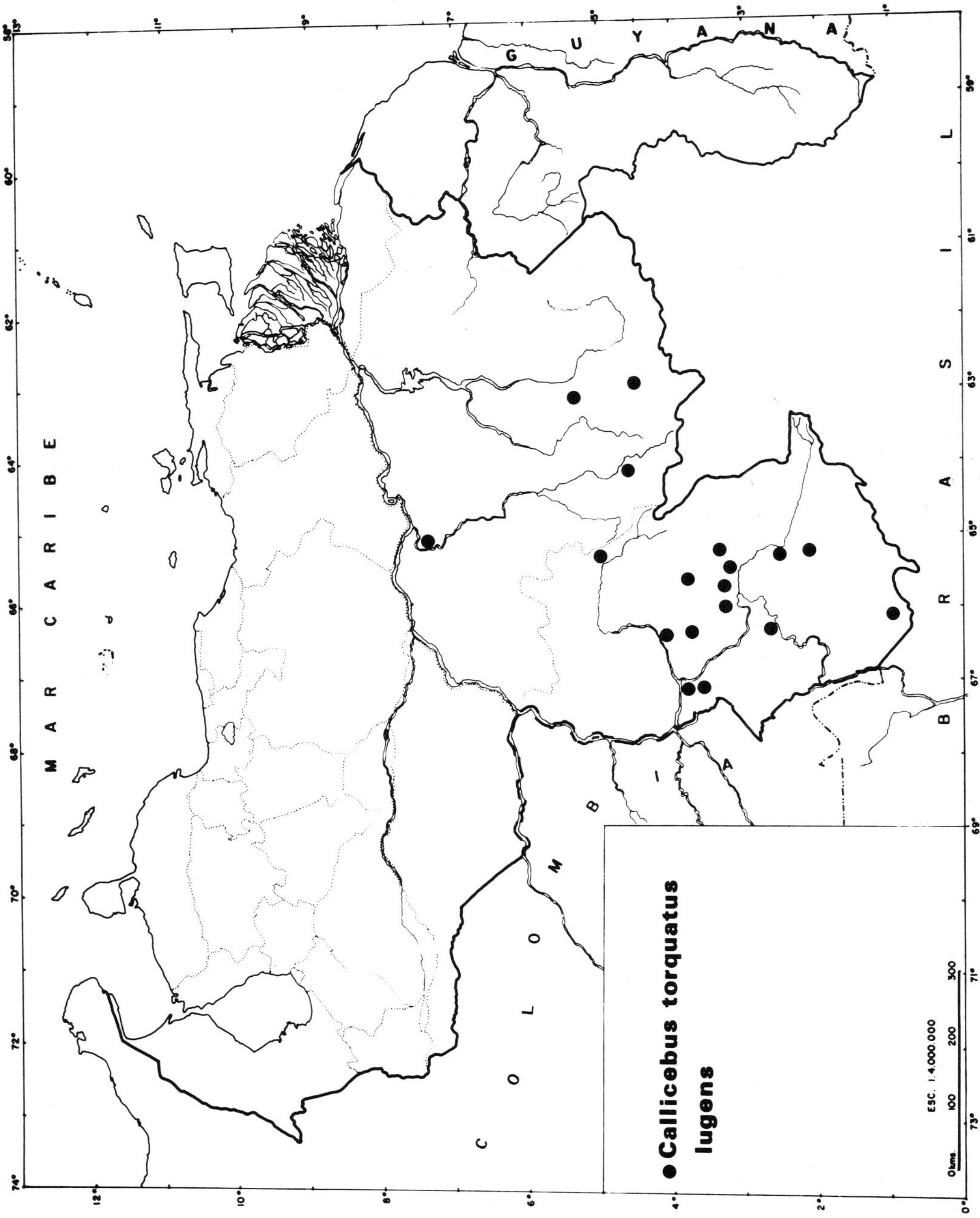

FIG. 3. Distribution of subfamily Callicebinae in Venezuela.

Reserva San Manuel, Los Cañaguatos (9°04′N, 71°56′W); Kunana, Río Negro, Sierra de Perijá (9°25′N, 72°25′W); Río Tocuco (72°25′W).

Aotus trivirgatus trivirgatus Humboldt is known from the region south of the Río Orinoco in Territorio Federal Amazonas and the state of Bolivar. The Venezuelan national collections contain specimens from Caño Manapiare and Río Antavari, and two literature records report specimens from the right bank of the Río Caroní (INPARQUES, 1982; MARNR-DGS-POA-SFS, 1982).

Our recent discovery of *Aotus trivirgatus* in Guyana to the east of the Río Caura, well outside of the Río Orinoco basin, led to an investigation of geographic variation in the species. Preliminary results suggest that these *A. trivirgatus* represent a new geographic race (Bodini & Ferreira, in prep.).

SPECIMENS EXAMINED—Total 11. **Bolivar:** Caño Manapiare (7°11′N, 66°40′W); Río Antavari (5°20′N, 63°10′W). **Territorio Federal Amazonas:** Alto Manapiare (5°03′N, 66°02′W); San Juan de Manapiare (5°14′N, 66°02′W); near Isla Cigarrón, Río Negro (1°44′N, 67°03′W); recently collected in La Neblina, east of Río Varía (0°59′N, 66°10′W).

Callicebus

The genus *Callicebus* is the sole representative of the subfamily Callicebinae and is found from northern South America south to northern Paraguay. Hershkovitz (1963) and Kinzey (1982) recognized three species, *Callicebus moloch, C. personatus,* and *C. torquatus,* of which only *C. torquatus* is found in Venezuela.

Callicebus torquatus lugens Humboldt is known in Venezuela south of the Río Orinoco (fig. 3). Hill (1960) originally suggested that *C. torquatus* is found as far east as Guyana, although this was questioned by both Handley (1976) and Hershkovitz (1963), who reported Venezuelan specimens only from Territorio Federal Amazonas. Handley (1976) reported 31 specimens from southern Territorio Federal Amazonas. Bodini (1981) reported specimens from Maripa, Camp Canaracuni, and the Río Antavari in Venezuela. Recent reports by Kinzey (1982) and Bodini (1981) confirm the presence of *C. torquatus* in Guiana Region.

SPECIMENS EXAMINED—Total 26. **Bolivar:** Canaracuni (4°36′N, 64°10′W); Maripa, 150 km from Ciudad Bolivar (7°22′N, 65°09′W); Río Antavari (5°20′N, 63°10′W); Alto Paragua (4°30′N, 63°00′W). **Territorio Federal Amazonas:** Alto Caño Caname, Departamento Atabapo (3°33′N, 67°06′W); Alto Ventuari (3°50′N, 67°04′W); Boca Caño Maica, Río Ventuari (66°30′W); Caño Yagua, Cerro Cucurito, Departamento Atabapo (3°38′N, 66°25′W); Cacuri, Alto Ventuari (4°49′N, 65°26′W); La Esmeralda (3°08′N, 65°32′W); Río Cunucunuma (3°10′N, 66°01′W); Río Puruname (3°19′N, 65°15′W); La Neblina, east of Río Varía (0°59′N, 66°10′W).

Alouatta

Alouatta is the only genus within the subfamily Alouattinae and is represented by some six species widely distributed in Central and South America.

A single species, *Alouatta seniculus,* is abundant and widely distributed in Venezuela. Hill (1962) recognized three subspecies of *A. seniculus* as occurring in Venezuela. *Alouatta seniculus seniculus* Linnaeus is found in extreme northwestern Venezuela, primarily in the states of Apure, Táchira, and Zulia, with a single record from Barinas (Ticoporo Forest, 8°06′N, 70°40′W). *Alouatta seniculus arctoidea* Cabrera, called the "Caracas howler monkey" by Humboldt, inhabits all the coastal region from Falcón to the state of Miranda. Cabrera (1958) proposed Caracas as the type locality. A third subspecies, *A. seniculus stramineus* Humboldt, inhabits all of Venezuela south of the Río Orinoco in Territorio Federal Amazonas and the state of Bolivar (fig. 4).

In addition to these three subspecies, we believe a fourth, previously unrecognized, form exists and is widely distributed throughout the Venezuelan *llanos.* This undescribed subspecies is characterized by coloration and size. The southern limit of its distribution is clearly defined by the Río Orinoco, but its northern and western limits in the Andean piedmont are as yet undetermined.

Howler monkeys are extremely adaptable to a wide array of environments, and we believe they are found throughout Venezuela. Their apparent absence in certain regions probably reflects lack of collecting rather than true distributional gaps, as our records demonstrate for the state of Anzoategui. The problem of current and historical distribution and systematic relationships of the various populations in Venezuela is in need of study.

SPECIMENS EXAMINED—Total 80. **Anzoategui:** Los Cocos, Río Caris (8°30′N, 64°05′W). **Apure:** Caño San Agustín, Selva de San Camilo (7°19′N, 71°57′W); Hato El Frío (7°44′N, 68°54′W); Las Raicitas, El Samán (7°55′N, 68°40′W). **Aragua:** Asentamiento Los Castillos, Turagua (10°09′N, 67°31′W); Cumbre de Guacamaya (10°21′N, 67°40′W); Los Picachos, Rancho Grande (10°21′N,

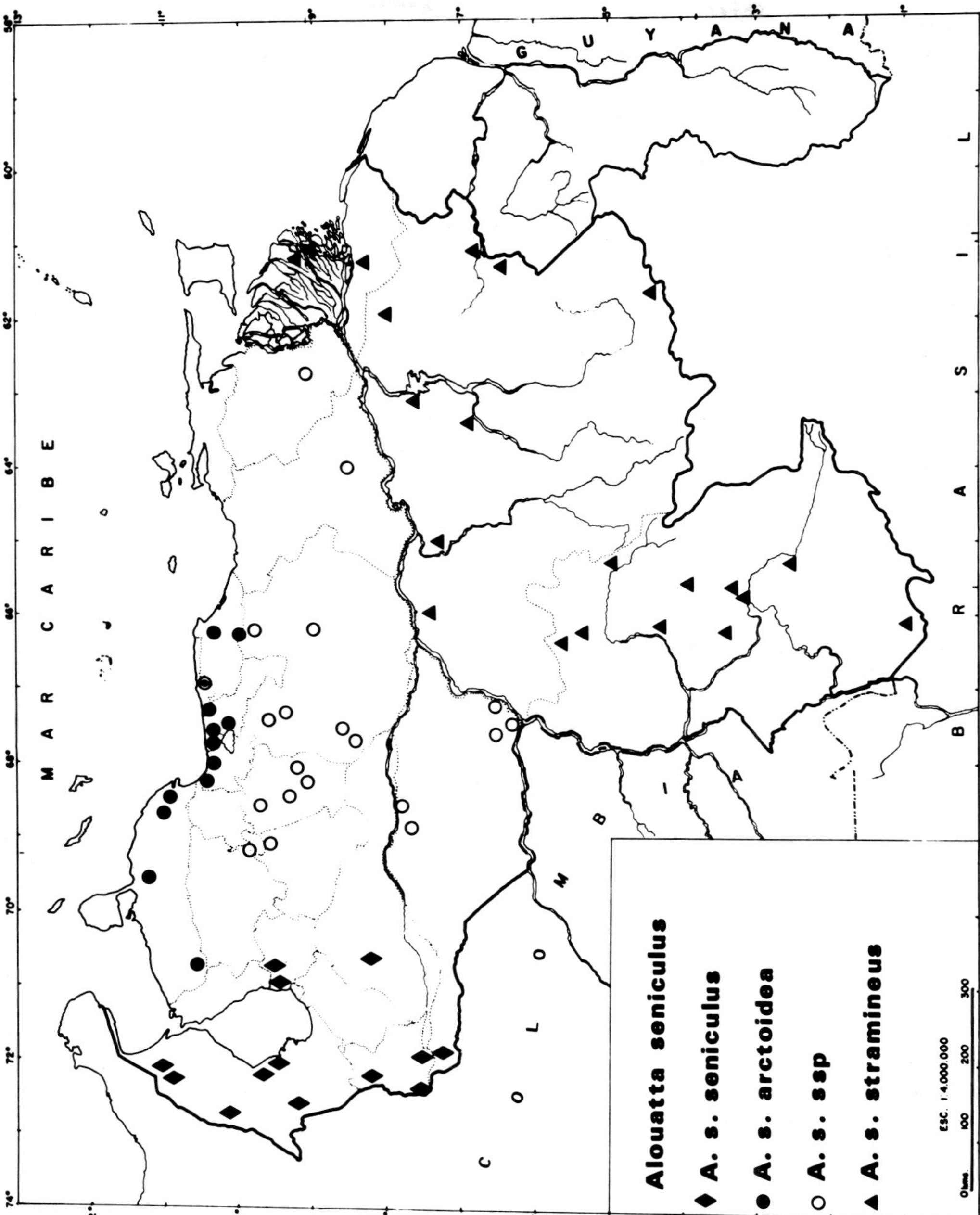

FIG. 4. Distribution of subfamily Alouattinae in Venezuela.

67°41′W). **Barinas:** Reserva Forestal Ticoporo (8°09′N, 70°40′W). **Bolivar:** Guayoba, Río Caura, 4 km from Maripa (7°20′N, 65°10′W); Hato Bella Vista, El Palmar (8°00′N, 62°00′W); Represa del Guri, Operación Rescate (7°40′N, 63°00′W); San Martín de Turumban, Anacoco-Cuyuní (6°42′N, 61°02′W). **Carabobo:** Las Quiguas, San Esteban (10°25′N, 68°01′W). **Cojedes:** Hato Barbasco (9°04′N, 68°08′W); Hato Itabana, 38 km from Las Vegas (9°17′N, 68°13′W); Las Queseras, El Baul (8°25′N, 68°17′W); Montaña Las Lomas, San Carlos (9°38′N, 68°34′W). **Distrito Federal:** Hacienda El Limón (10°28′N, 67°17′W). **Falcon:** Sanare, Departamento Silva (10°23′N, 68°25′W); Sierra San Luis (11°15′N, 69°10′W). **Guarico:** Corozo Pando (8°14′N, 67°17′W); Hato Mapurite, 40 km N of Calabozo (9°17′N, 67°24′W); Manapire, near San Antonio (9°17′N, 66°11′W); Río Tiznado (8°18′N, 67°48′W). **Lara:** Cumbre de Las Trojas, 45 km S of Cabudare (9°45′N, 69°07′W). **Miranda:** Río Negro (10°20′N, 66°15′W); La Guzmanera, Guatopo (10°00′N, 66°15′W). **Portuguesa:** Agua Blanca (9°40′N, 69°07′W). **Táchira:** Cerro El Teteo, Buragua (7°30′N, 71°57′W); La Fría (8°13′N, 72°14′W); Páramo Tamá (7°27′N, 72°26′W). **Territorio Federal Amazonas:** Cacuri, Alto Ventuari (4°49′N, 65°26′W); Caño Yureba, Ventuari (3°35′N, 66°46′W); Río Hacha, Alto Ventuari (3°47′N, 65°38′W); Río Puriname, 40 km from confluence of Río Orinoco (3°19′N, 65°15′W); San Juan de Manapiare (9°05′N, 66°02′W); Cerro La Neblina, E of Río Varía (0°59′N, 66°10′W). **Territorio Federal Delta Amacuro:** Caño Caneima (9°05′N, 60°55′W); Güiniquina (9°10′N, 61°06′W). **Zulia:** Laguna de Manaties, Departamento Catatumbo (9°27′N, 72°02′W); La Victoria, Río Negro (9°36′N, 72°15′W); Río Guasare (11°02′N, 72°05′W); Sierra de Perijá (9°00′N, 72°00′W).

Chiropotes

The bearded sakis comprise two species limited to northern South America (Cabrera, 1958; Hershkovitz, 1977; Mittermeier & Coimbra-Filho, 1981). One species is found in southern Venezuela represented by a single subspecies, *Chiropotes satanas chiropotes* Humboldt (fig. 5). All of our records and those reported by Handley (1976) are from Territorio Federal Amazonas. Cruz Lima (1945) and Rudran and Eisenberg (1982) proposed the occurrence of this species in the state of Bolivar on the basis of Humboldt's description; Mondolfi (1976) reported observations of *Chiropotes* at Maripa and Caño Maniapure (Bolivar), but we are aware of no specimens from this region.

SPECIMENS EXAMINED—Total 29. **Territorio Federal Amazonas:** Cacuri, Alto Ventuari (4°49′N, 65°26′W); Caño Yureba, Departamento Atabapo (3°31′N, 66°44′W); Caño Yagua, Cerro Cucurito (3°31′N, 66°44′W); Laguna de Chiripo, Caño Blanco (3°27′N, 66°40′W); Río Ocamo, Alto Orinoco (2°44′N, 65°11′W); Puerto Ayacucho (5°36′N, 67°35′W); Río Orinoco, S of San Fernando de Atabapo (4°00′N, 67°38′W); San Fernando de Atabapo (4°02′N, 67°37′W); San Juan de Manapiare (5°14′N, 66°02′W).

Cacajao

The genus *Cacajao,* or uakaris, contains two species which are found in northern South America (Cabrera, 1958; Hershkovitz, 1972; Mittermeier & Coimbra-Filho, 1981). One species, *Cacajao melanocephalus,* is found in Venezuela restricted to the upper Orinoco region of southern Territorio Federal Amazonas (fig. 5). Although few specimens exist in collections, they appear to be abundant.

SPECIMENS EXAMINED—Total 2. **Territorio Federal Amazonas:** Alto Caño Atacavi, Departamento Casiquiare (3°05′N, 67°02′W); La Neblina, E of Río Varía (0°59′N, 66°10′W).

Cebus

The capuchin monkeys are found from Honduras south through Central America and the northern two-thirds of South America. Four species currently are recognized (Cabrera, 1958; Hershkovitz, 1972). Three species of *Cebus* are represented in Venezuela: *Cebus albifrons, C. apella,* and *C. nigrivittatus* (Hershkovitz, 1949, 1958).

Three subspecies of *Cebus albifrons* are found in Venezuela (fig. 6). *Cebus albifrons adustus* Hershkovitz was described on the basis of three specimens from "near head of Río Cogollo (Apón) at eastern base of Sierra de Perijá, about 5 kilometers northwest of Machiques, Lake Maracaibo region, Zulia" (Hershkovitz, 1949, p. 369). We report additional specimens from Río Guasare and Kasmera. This subspecies is restricted to the Sierra de Perijá of extreme northwestern Venezuela and adjacent Colombia. *Cebus albifrons leucocephalus* Gray is found in extreme western Venezuela, in the region south of the Lago de Maracaibo basin,

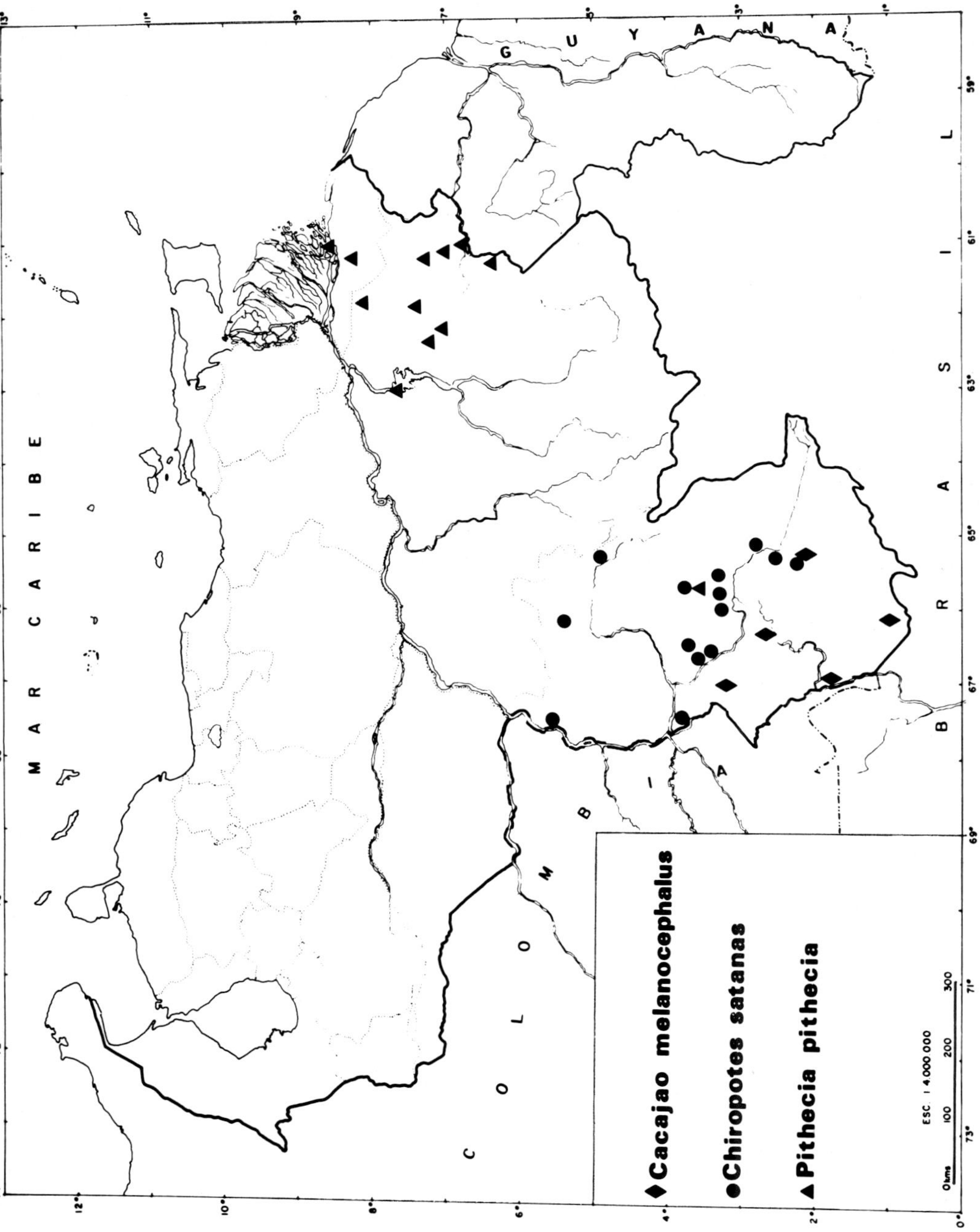

FIG. 5. Distribution of subfamily Pithecinae in Venezuela.

and the states of Apure, Mérida, Táchira, and Zulia. *Cebus albifrons unicolor* Spix is found in extreme southern Venezuela in Territorio Federal Amazonas. Hershkovitz (1949) reported specimens from Marimonda, Río Orinoco, and from Solano, Río Casiquiare. Handley (1976) reported specimens from Río Mavaca and Tamatama. All locality records for this subspecies in Venezuela are south of the Río Ventuari.

SPECIMENS EXAMINED—Total 15. **Apure:** Caño San Augustín, Selva de San Camilo (7°19′N, 71°57′W). **Mérida:** Palmichoso, S of Las Virtudes (9°09′N, 70°57′W). **Tachira:** La Fría (8°13′N, 72°14′W). **Territorio Federal Amazonas:** Caño Yagua, Cerro Cucurito, Departamento Atabapo (3°38′N, 66°25′W); near Boca Padamo, left side of Río Orinoco (3°02′N, 65°13′W). **Zulia:** Kasmera, Perijá (10°05′N, 72°45′W); Kunana (9°36′N, 72°15′W); Río Bravo, Distrito Catatumbo (9°05′N, 72°22′W); Río Guasare (11°02′N, 72°05′W).

Cebus apella is represented in Venezuela by two subspecies (fig. 6). *Cebus apella apella* Linnaeus is restricted to Amazonian Venezuela, the Territorio Federal Amazonas, and is found along both banks of the upper Orinoco. *Cebus apella margaritae* Hollister is endemic to and restricted to Margarita Island. The 800-km gap between the ranges of these two subspecies is striking and unexplained.

SPECIMENS EXAMINED—Total 10. **Nueva Esparta:** Sierra de Copey (11°03′N, 63°56′W). **Territorio Federal Amazonas:** Alto Caño Caname (3°22′N, 67°08′W); Caño Yapacana (3°30′N, 66°45′W); San Fernando de Atabapo (4°02′N, 67°37′W).

Cebus nigrivittatus is widely distributed in Venezuela and represented by perhaps five subspecies (fig. 6) (Hershkovitz, 1949; Cabrera, 1958). *Cebus nigrivittatus apiculatus* was described by Elliot (1912) on the basis of specimens from La Unión, Río Caura, near its confluence with the Río Orinoco. It is distributed throughout central Venezuela south of the Orinoco between the Río Caroní and the Río Ventuari.

Cebus nigrivittatus brunneus was described by J. A. Allen (1914) from specimens from Aroa, a station on the Bolívar Railway, Yaracuy, northwestern Venezuela. Hershkovitz (1949) reported an additional specimen from the Paria Peninsula. These records plus our specimens suggest that *C. n. brunneus* is continuously distributed throughout the Cordillera de la Costa of extreme northern Venezuela. *Cebus nigrivittatus nigrivittatus* Wagner is restricted in Venezuela to the Amazonian region of Territorio Federal Amazonas. *Cebus nigrivittatus olivaceus* Schömburg is found in southeastern Venezuela. Hershkovitz (1949, p. 348) reports the type locality as "Vicinity of 'Our Village,' said to be situated at latitude 4°57′N., 61°1′W., altitude 3,100 feet above sea level, southern foot of Mount Roraima."

We believe that an undescribed subspecies, *Cebus nigrivittatus* subsp., is widely distributed throughout central and northern Venezuela north of the Orinoco. The status of this population is currently under study.

The Orinoco Delta region, which Eisenberg and Redford (1979) excluded from their consideration of biogeographic regions due to insufficient data, is now shown by the distributions of *Alouatta, Cebus,* and *Pithecia* to be clearly aligned with the Guyana highlands. The Llanos region now extends up to the western edge of the delta, but does not include it, as demonstrated by the distributions of *Alouatta* and *Cebus.*

SPECIMENS EXAMINED—Total 75. **Anzoategui:** Los Cocos, Río Caris (8°30′N, 64°05′W); 10 km W of Laguna de Unare (10°02′N, 65°12′W); Morichal Largo [between Anzoategi and Monagas] (8°18′N, 63°15′W). **Aragua:** Rancho Grande (10°10′N, 67°19′W). **Barinas:** Reserva Forestal Ticoporo, on Barinas-Pedregal road (8°03′N, 70°18′W). **Bolivar:** Caño La Urbina (7°15′N, 66°25′W); Carretera Caicara-S. Juan de Manapiare, km 175 (6°02′N, 66°29′W); Carretera El Dorado-Santa Elena, km 33 (6°12′N, 61°14′W); Curaima, El Palmar (8°01′N, 61°26′W); El Dorado-Santa Elena, km 121 (5°18′N, 61°11′W); Guayopo, Río Caura, 14 km from Maripa (7°09′N, 65°10′W); Canaracuni (4°17′N, 64°05′W); Guri, Operación Rescate (7°18′N, 63°00′W); Río Antavari (5°09′N, 63°05′W); Río Marajano, Meseta de Jaua (4°08′N, 64°11′W); Río Villacoa, 4 km N of mouth (6°16′N, 67°05′W); San Martin de Turumban, Río Cuyuní (6°19′N, 61°09′W). **Carabobo:** Bahia de Patanemo (10°12′N, 67°26′W); Urama (10°12′N, 68°08′W). **Cojedes:** Cerro Azul, La Blanquera (8°26′N, 68°07′W); Montaña Las Loma, San Carlos (9°17′N, 68°16′W); Pica, Las Vegas (9°15′N, 68°17′W). **Distrito Federal:** El Avila, Caracas (10°14′N, 66°13′W); Hacienda El Limón (10°13′N, 67°08′W). **Falcon:** Sanare, Distrito Silva (8°23′N, 68°12′W). **Guarico:** Hato Flores Moradas, Calabozo (8°23′N, 67°13′W); Hato Mapurite, 40 km N of Calabozo (9°08′N, 67°11′W); Parmana (7°28′N, 65°18′W); San José de Tiznados (9°16′N, 67°16′W). **Lara:** La Pastora, 11 km SSW of Sanare (9°21′N, 70°07′W). **Miranda:** Río Negro (10°20′N, 66°17′W); La Guzmanera, Guatopo (10°00′N, 66°15′W). **Territorio Federal Amazonas:** Alto Manapiare (5°13′N, 66°01′W); Alto Ventuari (4°45′N, 65°20′W); Caño Yureba, Río Ventuari (3°16′N, 66°21′W); Caño

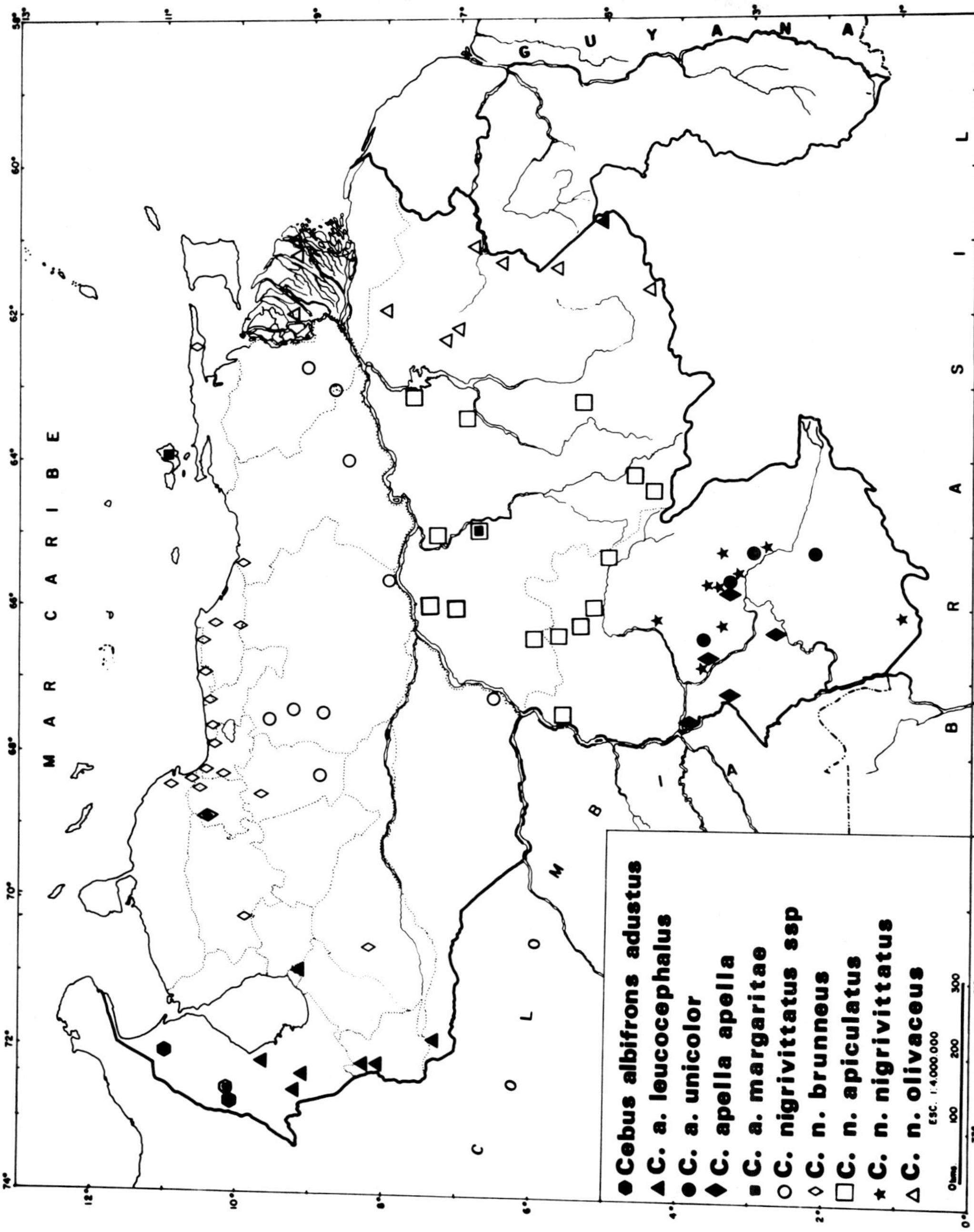

FIG. 6. Distribution of subfamily Cebinae in Venezuela.

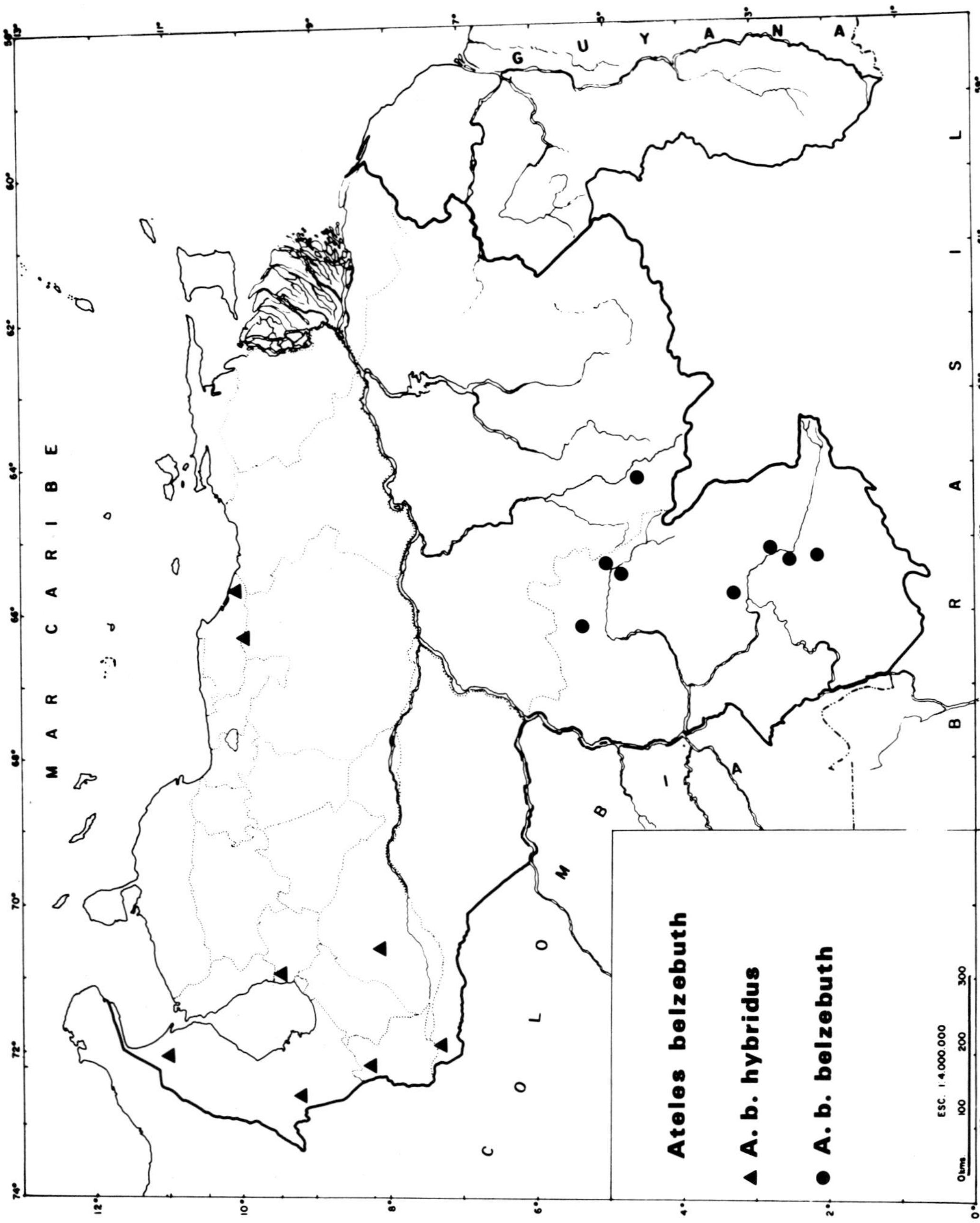

FIG. 7. Distribution of subfamily Atelinae in Venezuela.

Mayaba, Río Ventuari (4°07′N, 66°16′W); Caño Morrocoy, Alto Ventuari (5°08′N, 66°02′W); La Esmeralda (3°08′N, 65°32′W); Ocamo, Río Ocamo (2°20′N, 65°15′W); Puruname, 40 km from Río Orinoco (3°19′N, 65°15′W); La Neblina, E of Río Varía (0°59′N, 66°10′W). **Territorio Federal Delta Amacuro:** Caño Araguabisi (9°12′N, 60°27′W); Güiniquina (9°10′N, 61°03′W); Tobejuba, Guayo (9°09′N, 61°25′W). **Yaracuy:** Agua Negra (10°14′N, 68°14′W); Carretera Boca de Aroa, 20 km from Palmasola (10°01′N, 69°27′W).

Ateles

Spider monkeys are widely distributed from northeastern Mexico throughout tropical South America. Four species are recognized, of which only *Ateles belzebuth* occurs in Venezuela (fig. 7).

Ateles belzebuth belzebuth Geoffroy is found in southern Venezuela, south of the Río Orinoco; most records are from Territorio Federal Amazonas. Kellogg and Goldman (1944) indicate a wide distribution in Guyana and report the Venezuelan localities of La Unión, Río Mato, and El Llagual (on both banks of the Río Caura).

Ateles belzebuth hybridus Geoffroy is known from northern and western Venezuela (Hershkovitz, 1949; Cabrera, 1958; Kellogg & Goldman, 1944). Handley (1976) reported specimens from Apure and Trujillo in western Venezuela. Our specimens are from the states of Barinas, Táchira, and Zulia. Mondolfi and Eisenberg (1979) reported it from Cúpira and Guatopo, state of Miranda, suggesting a discontinuous distribution on the coast.

Specimens Examined—Total 16. **Barinas:** Reserva Forestal de Ticoporo, Sabana de Anare (8°06′N, 70°40′W). **Bolivar:** Canaracuni (4°06′N, 64°10′W). **Miranda:** Cúpira (10°10′N, 65°44′W). **Táchira:** La Fría (8°13′N, 72°14′W). **Territorio Federal Amazonas:** Cacuri, Río Ventuari (4°49′N, 65°26′W); Río Ocamo, Alto Orinoco (2°44′N, 65°11′W); Salto del Oso, Alto Ventuari (4°55′N, 65°25′W); San Juan de Manapiare (5°14′N, 66°02′W). **Zulia:** Río Guasare (11°02′N, 72°05′W).

Lagothrix

Woolly monkeys have not been collected in Venezuela; however two subspecies of *Lagothrix lagothricha* are to be expected: *L. lagothricha lagothricha* should be found in Territorio Federal Amazonas south of the Río Ventuari, and *L. l. lugens*, in the Selva de San Camilo, state of Apure (Fooden, 1963; Hernández-Camacho & Cooper, 1976).

Pithecia

Sakis are found only in northern South America; Hershkovitz (1979) recognized four monotypic species, of which only *Pithecia pithecia* Linnaeus is found in Venezuela. *P. pithecia* is found in Venezuela south of the Río Orinoco and throughout the Guianas and northeastern Brazil. In Venezuela specimens have been reported primarily from the extreme northeastern region, the state of Bolivar and Territorio Federal Delta Amacuro (fig. 5), with a single outlier locality record for Belén, Río Cunucunuma, Territorio Federal Amazonas (3°39′N, 65°46′W) (Handley, 1976). If the distribution proposed by Hershkovitz (1979) and Mittermeier and Coimbra-Filho (1981) is correct, *P. pithecia* should inhabit the entire region between the upper Río Orinoco and the Río Caroní, an enormous area for which no specimens have been recorded.

Specimens Examined—Total 24. **Bolivar:** Río Curumo (7°15′N, 61°20′W); Río Grande (8°16′N, 61°17′W); Gurisoco, El Palmar (8°62′N, 61°26′W); La Trinidad, El Palmar (7°12′N, 61°23′W); Guri, Operación Rescate (7°18′N, 63°00′W); Río Botonamo, near Río Cuyuní (6°59′N, 61°11′W); San Martín de Turumban, Río Cuyuní (6°59′N, 61°02′W). **Territorio Federal Delta Amacuro:** Altiplanicie de Nuria (7°50′N, 61°18′W); Yotacuay, SW of Cupiare (8°30′N, 61°00′W).

Literature Cited

Allen, J. A. 1914. New South American Monkeys. Bulletin of the American Museum of Natural History, **33:** 653.

Bodini, R. 1981. Musculatura locomotora de la viudita (*Callicebus torquatus lugens*). Sus implicaciones funcionales y filogenéticas. Memoria de la Sociedad de Ciencias Naturales La Salle, **16**(XL): 1–165.

———. 1983. Distribución, status de la investigación y conservación de los cebidos en Venezuela, p. 139. *In* Symposio sobre Primatología en Latinoamerica. IX Congreso Latinoamericano de Zoología, Arequipa, Perú.

Cabrera, A. 1957 [1958]. Catálogo de los mamíferos de America del sur. Revista Museo Argentino de Ciencias Naturales, "Bernardino Rivadavia," **4**(1): xviii + 138 pp.

Cruz Lima, E. 1945. Mammals of Amazonia. I. General introduction and Primates. Contribution from the

Museu Paraense Emilio Goeldi de Historia Natural e Etnografia, Belém do Pará, Rio de Janeiro, Brasil, 274 pp., 42 pls.

Eisenberg, J. F., and K. Redford. 1979. A biogeographic analysis of the mammalian fauna of Venezuela, pp. 31–36. *In* Eisenberg, J. F., ed., Vertebrate Ecology in the Northern Neotropics. Smithsonian Institution Press, Washington, D.C., 271 pp.

Elliot, D. G. 1912. A Review of the Primates. Vols. I and II. American Museum of Natural History, New York.

Fooden, J. 1963. A revision of the woolly monkey (genus *Lagothrix*). Journal of Mammalogy, **44**: 213–247.

Handley, C. O., Jr. 1976. Mammals of the Smithsonian Venezuelan project. Brigham Young University Sciences Bulletin, Biological Series, **20**(5): 1–91.

Hernandez-Camacho, J., and R. W. Cooper. 1976. The nonhuman Primates of Colombia, pp. 35–69. *In* Thorington, R. W., Jr., and P. G. Heltne, eds., Neotropical Primates: Field Studies and Conservation. National Academy of Sciences, Washington, D.C.

Hershkovitz, P. 1949. Mammals of northern Colombia. Preliminary report No. 4: Monkeys (Primates), with taxonomic revisions of some forms. Proceedings of the United States National Museum, **98**(3232): 323–427.

———. 1958. Type localities and nomenclature of some American primates, with remarks on secondary homonyms. Proceedings of the Biological Society of Washington, **71**: 53–56.

———. 1963. A systematic and zoogeographic account of the monkeys of the genus *Callicebus* (Cebidae) of the Amazonas and Orinoco river basins. Mammalia, **27**: 1–79.

———. 1972. The Recent mammals of the Neotropical region: A zoogeographic and ecological review, pp. 311–431. *In* Keast, A., F. C. Erk, and B. Glass, eds., Evolution, Mammals, and Southern Continents. State University of New York Press, Albany, N.Y. 543 pp.

———. 1977. Living New World Monkeys (Platyrrhini) with an Introduction to Primates. University of Chicago Press, Chicago, 1117 pp.

———. 1979. The species of sakis, genus *Pithecia* (Cebidae, Primates), with notes on sexual dichromatism. Folia Primatologica, **31**: 1–22.

———. 1983. Two new species of night monkeys, genus *Aotus* (Cebidae, Platyrrhini): A preliminary report on *Aotus* taxonomy. American Journal of Primatology, **4**: 209–243.

———. 1984. Taxonomy of squirrel monkeys genus *Saimiri* (Cebidae, Platyrrhini): A preliminary report with description of a hitherto unnamed form. American Journal of Primatology, **7**: 155–210.

Hill, W. C. O. 1960. Primates. Comparative Anatomy and Taxonomy. IV. Cebidae, Part A. Edinburgh University Press, London, 523 pp.

———. 1962. Primates. Comparative Anatomy and Taxonomy. V. Cebidae, Part B. Edinburgh University Press, London, 537 pp.

INPARQUES. 1982. Guía de los Parques Nacionales y Monumentos Naturales de Venezuela. Ediciones Fundación de Educación Ambiental, Caracas.

Kellogg, R., and E. A. Goldman. 1944. Review of the spider monkeys. Proceedings of the United States National Museum, **96**(3186): 1–45.

Kinzey, W. G. 1982. Distribution of primates and forest refuges, pp. 455–482. *In* Prance, G. T., ed., Biological Diversification in the Tropics. Columbia University Press, New York, 714 pp.

MARNR-DGS-POA-SFS. 1982. Informe Nacional de Fauna Silvestre. Ministerio del Ambiente y de los Recursos Naturales Renovable, Caracas.

Mittermeier, R. A., and A. Coimbra-Filho. 1981. Systematic: Species and subspecies, pp. 29–109. *In* Coimbra-Filho, A. F., and R. A. Mittermeier, eds., Ecology and Behavior of Neotropical Primates. Academia Brasileira de Ciẽncias, Rio de Janeiro.

Mondolfi, E. 1976. Fauna silvestre de los bosques humedos tropicales de Venezuela, pp. 113–181. *In* Hamilton, L. S., J. Steyermark, J. P. Veillon, and E. Mondolfi, eds., Conservación de los bosques húmedos de Venezuela. Sierra Club—Consejo de Bienestar Rural, Caracas.

Mondolfi, E., and J. F. Eisenberg. 1979. New records for *Ateles belzebuth hybridus* in northern Venezuela, pp. 93–96. *In* Eisenberg, J. F., Vertebrate Ecology in Northern Neotropics. Smithsonian Institution Press, Washington, D.C., 271 pp.

Ramirez-Cerquera, J. 1983. Reporte de una nueva especie de Primates del género *Aotus* de Colombia, p. 146. Symposio sobre Primatología en Latinoamerica. IX Congreso Latinoamericano de Zoología. Arequipa, Perú.

Rudran, R., and J. F. Eisenberg. 1982. Conservation and status of wild primates in Venezuela, pp. 52–59. *In* Olney, P. J. S., ed., International Zoo Yearbook, Vol. 22. The Zoological Society of London, London, 488 pp.

Host Associations and Coevolutionary Relationships of Astigmatid Mite Parasites of New World Primates I. Families Psoroptidae and Audycoptidae

Barry M. OConnor

ABSTRACTS

Coevolutionary patterns among mites of the families Psoroptidae and Audycoptidae and New World Primates are reviewed. Host records for primate parasites originally described from artificial situations are compared with field collections from Peruvian primates, with most host associations verified. A new species of Audycoptidae, **Saimirioptes hershkovitzi**, is described from *Cebus apella.* The psoroptid subfamily Cebalginae is hypothesized to be a monophyletic group based upon 10 synapomorphies. Phylogenetic relationships within the Cebalginae are reviewed, with cospeciational histories supported for the genera *Alouattalges* and *Schizopodalges* and their hosts, and a more diffuse cospeciational pattern observed for the genera *Cebalgoides, Cebalges,* and *Fonsecalges* and their hosts. Historical relationships of the genus *Procebalges* remain problematical.

Patrones coevolutivos entre ácaros de las familias Psoroptidae y Audycoptidae y primates del Nuevo Mundo son revisadas. Registros de huéspedes para parásitos de primates originariamente descritos de situaciones artificiales son comparados con colecciones de campo de primates peruanos, con la mayoría de asociaciones de huésped verificadas. Una nueva especie de Audycoptidae, **Saimirioptes hershkovitzi**, es descrita de *Cebus apella.* Los psoróptidos de la subfamilia Cebalginae son hipotetizados ser un grupo monofilético basado en 10 sinapomorfías. Relaciones filogenéticas entre los Cebalginae son revisadas, con historias coespeciacionales soportadas por el género *Alouattalges* y *Schizopodalges* y sus huéspedes, y un patrón coespeciacional más difuso observado en los géneros *Cebalgoides, Cebalges* y *Fonsecalges* y sus huéspedes. Relaciones históricas del género *Procebalges* permanecen problemáticas.

Padrões coevolucionários entre os acarinos das famílias Psoroptidae e Audycoptidae, e os primatas do Novo Mundo, são revisados. Os registros de hóspedes primatas de parasitas, que foram descritos em situações artificiais (cativeiros), são comparados com coleções de campo de primatas peruanos, e a maioria das associações hospedeiras atualmente registradas foram averiguadas. Uma nova espécie de Audycoptidae, **Saimirioptes hershkovitzi**, encontrada em *Cebus apella* é descrita. Baseando-se num estudo de 10 sinapomorfias, propõe-se ser a subfamília psoroptídea, Cebalginae, um grupo monofilético. As relações filogenéticas entre os Cebalginae são revisadas, e as histórias de coespecialização entre os generos *Alouattalges* e *Schizopodalges,* e seus respectivos hóspedes, são confirmadas. Os padrões de coespecialização entre os generos *Cebalgoides, Cebalges* e *Fonsecalges,* e seus hóspedes, são mais difusos. Os relacionamentos históricos do genero *Procebalges* permanecem incertos.

From the Museum of Zoology and Department of Biology, The University of Michigan, Ann Arbor, MI 48109.

Introduction

Coevolutionary patterns among hosts and parasites have been the focus of much recent study and discussion (Brooks, 1979, 1981, 1985; Brooks & Glen, 1982; Futuyma & Slatkin, 1983; Nitecki, 1983). Phylogenetic analysis of evolutionary relationships among parasite groups or host groups can provide additional data sets (e.g., treating parasite distributions as character states of their hosts), or such analyses can be used as tests for hypotheses regarding evolutionary relationships of the other lineages of associated organisms. These phylogenetic analyses are especially useful when the parasite groups are host specific and indicate little history of secondary colonization of new hosts. The associations between mites and primates are particularly amenable to such analysis because a number of acarine groups are specifically associated with primates, and these lineages exhibit enough within-group diversity to allow the construction and comparison of phylogenetic hypotheses. I have previously detailed such hypotheses for several groups of astigmatid mites parasitic on primates (OConnor, 1984).

Formulation of hypotheses regarding the history of associations between primates and their associated mites requires three steps. First, the parasite taxa must be described and their natural host and geographic ranges discovered. Second, taxa above the species level in classically derived classifications must be tested for naturalness (i.e., monophyly). Finally, phylogenetic relationships among all taxa must be elucidated. In the present paper, each of these questions will be addressed for the associations among certain groups of astigmatid mites and New World Primates. For the reader interested in summary information regarding known host-parasite relationships for New World primates, an exhaustive list of literature records may be found in Hershkovitz (1977).

One major difficulty in the application of this methodology to the study of the history of mite-primate associations is the scanty knowledge of the distribution and identity of the parasite species and their natural host ranges. Although many acarine parasites of primates have been described, much of the material has originally come from zoos and primate research centers where the possibility of unnatural interspecific contact between host species makes the transfer of parasites a real problem. Many other species have been described from preserved host specimens in museum collections where contamination in the field or in the museum may also have been a problem. Finally, the actual field locality from which either parasites or hosts were collected is known for extremely few of the known primate parasites. The uncertainties involved in the host range and geographic distribution of so many of the known species of primate parasites render the phylogenetic analyses proposed earlier (OConnor, 1984) subject to some doubt.

In 1981, I was invited by Philip Hershkovitz to collect parasites from specimens of a number of primate species which had been collected or obtained during field studies in Peru in 1980. A very large number of mites was collected, providing a unique survey of the primate parasites in a small area in Peru and a test for previously reported host-parasite associations. These collections yielded specimens belonging to three families of astigmatid mites: Psoroptidae, Audycoptidae, and Atopomelidae. In the present paper, the collections of Psoroptidae and Audycoptidae will be discussed. The collections of Atopomelidae, consisting of a number of described and undescribed species of the genus *Listrocarpus,* will be studied separately.

Materials and Methods

Primate specimens were collected in the field by Hershkovitz or obtained from local individuals. These specimens were prepared by removing and simply drying the skins at the time of collection. Upon their arrival in the United States, one or more skins of each species collected were made available to me for parasite removal before the skins were sent for tanning. The following species, all identified by Hershkovitz, were examined (number examined in parentheses): *Cebus apella* (3); *C. albifrons* (1); *Lagothrix lagothricha* (1); *L. flavicauda* (1); *Alouatta seniculus* (1); *Pithecia hirsuta* (3); *Callicebus moloch* (1); *Aotus nancymai* (4); and *Saimiri sciureus* (5).

Parasites were removed in two ways. First, all skins examined were vigorously brushed over white paper, with the dislodged parasites collected under a dissecting microscope. Finally, one skin of each species was soaked in water and mild soap until soft (the single specimen of *L. flavicauda* was not soaked). These skins were then gently washed, the wash water filtered through a 200-mesh sieve (mesh openings 75 micrometers), and the residue examined under a dissecting microscope. Mites collected were preserved in 70% ethanol for subsequent study.

In the laboratory, mites were cleared in lactophenol and mounted in Hoyer's medium, with some specimens retained in alcohol in the cases of large series. Voucher specimens of nominal species are deposited in Field Museum of Natural History, Chicago, and, when available, will be placed in the following institutions: Museum of Zoology, The University of Michigan, Ann Arbor; The United States National Museum of Natural History, Washington, D.C.; L'Institut Royale des Sciences Naturelles, Brussels, Belgium; and the collection of F. S. Lukoschus, Katholieke Universiteit, Nijmegen, Netherlands.

Species Accounts

Family PSOROPTIDAE

Seven species of mites in the family Psoroptidae, subfamily Cebalginae, are known to parasitize New World primates. These species, with their known hosts and distributions, are listed below, with new records from the Peruvian collections of Hershkovitz indicated under "Material Examined." Keys to most of these species may be found in Fain (1963c).

Cebalgoides cebi Fain, 1963

Cebalgoides cebi Fain, 1963, Bull. Ann. Soc. Roy. Ent. Belg., 99: 331. Fain, 1963, Bull. Inst. Roy. Sci. Nat. Belg., 39(32): 91.

This species was briefly diagnosed by Fain (1963a) from specimens collected from *Cebus albifrons* which originated in "Amerique du Sud" and died in the Antwerp (Belgium) Zoo. It was more thoroughly described and illustrated by Fain (1963c), who listed specimens from *Cebus albifrons* from Venezuela (type collection), *C. apella* from "Amerique du Sud," and *Leontocebus* (*Oedipomidas*) *oedipus* (= *Saguinus oedipus*) from Colombia. All these hosts had died in the Antwerp Zoo.

Fain (1963c) noted sexual dimorphism in tritonymphs of *C. cebi*. The two protonymphs he examined exhibited the characteristics of the male tritonymphs. Among 11 protonymphs in the present collection, eight exhibit characteristics similar to the male tritonymphs (i.e., posterior opisthosomal lobes sclerotized; coxal fields III well sclerotized) while three exhibit characteristics of female tritonymphs (posterior lobes unsclerotized; sclerotization of coxal fields III much weaker). I interpret these differences as evidence for sexual dimorphism at the protonymphal instar in this species.

MATERIAL EXAMINED—Total 133. Twenty-nine females, 43 males (of which 31 were in tandem with female tritonymphs), 38 female tritonymphs, 8 male tritonymphs, 3 female protonymphs, 8 male protonymphs, 4 larvae from *Cebus albifrons*. **PERU, Loreto:** Nauta, Río Tigre, 6 km above Río Tigrillo; 18 December 1980; P. Hershkovitz (9264). Host now a tanned skin (FMNH 122795). Mites labeled BMOC 81-0811-3. No specimens were recovered from *Cebus apella.*

Alouattalges corbeti Fain, 1963

Alouattalges corbeti Fain, 1963, Bull. Inst. Roy. Sci. Nat. Belg., 39(32): 122. Fain, 1966, Acarologia 8: 103.
Rosalialges cruciformis Lavoipierre, 1964, Acarologia, 6: 348.

This species was briefly described from the holotype female collected from a preserved specimen of *Alouatta seniculus macconnelli* which had been collected at Paramaribo, Surinam and preserved in the British Museum (Natural History) (Fain, 1963c). Although only the holotype was described, several specimens were apparently recovered from this host. Almost simultaneously, Lavoipierre (1964a) described and figured the female based upon two specimens collected from an "*Aotes*" (*sic*) sp. which had died in San Francisco, California, after its importation from Peru. Fain (1966) provided illustrations of the female and illustrated but did not describe the male beyond length and width measurements.

MATERIAL EXAMINED—Total 15. Five females, 8 males, 2 female tritonymphs from *Alouatta seniculus.* **PERU, Loreto:** Nauta, Río Samiria; 18 November 1980; P. Hershkovitz (9050). Host now a tanned skin (FMNH 122789). Mites labeled BMOC 81-0809-5. No specimens were recovered from *Aotus nancymai,* suggesting the possibility that the two specimens collected by Lavoipierre (1964a) represent contamination.

Schizopodalges lagothricola Fain, 1963

Schizopodalges lagothricola Fain, 1963, Bull. Ann. Soc. Roy. Ent. Belg., 99: 469. Fain, 1963, Bull. Inst. Roy. Sci. Nat. Belg., 39(32): 100.

This species was described from numerous specimens from two juvenile *Lagothrix lagothricha* which died in the Antwerp Zoo. The origin of the hosts was stated as "Amerique du Sud" (Fain, 1963b). Fain (1963c) provided illustrations of male and female and indicated that the hosts originated in Colombia.

MATERIAL EXAMINED—Total 30. Eleven females, 11 males, 3 tritonymphs, 2 protonymphs, 3 larvae from *Lagothrix lagothricha.* **PERU, Loreto:** Nauta, Río Samiria; 15 November 1980; P. Hershkovitz (9032). Host now a tanned skin (FMNH 122790). Mites labeled BMOC 81-0809-10.

Cebalges gaudi Fain, 1962

Cebalges gaudi Fain, 1962, Rev. Zool. Bot. Afr., 66: 160. Fain, 1963, Bull. Inst. Roy. Sci. Nat. Belg., 39(32): 81.

This species was briefly described from numerous specimens collected from a preserved specimen of *Cebus capucinus* with no locality information (Fain, 1962). Full descriptions and figures were provided later (Fain, 1963c).

MATERIAL EXAMINED—Four females from *Cebus apella.* **PERU, Loreto:** Nauta, Río Samiria; 18 November 1980; P. Hershkovitz (9049). Host now a tanned skin (FMNH 122792). Mites labeled BMOC 81-0811-2.

Fonsecalges johnjadini Fain, 1962

Fonsecalges johnjadini Fain, 1962, Rev. Zool. Bot. Afr., 66: 161. Fain, 1963, Bull. Ann. Soc. Roy. Ent. Belg., 99: 468. Fain, 1963, Bull. Inst. Roy. Sci. Nat. Belg., 39(32): 86.

This species was briefly described from several *Callithrix jacchus* which had died in captivity (Fain, 1962). Fain (1963b) provided an illustration of the male, and Fain (1963c) gave a complete description with figures of both male and female. The hosts, which were listed as eight *Hapale jacchus* (= *Callithrix jacchus*) imported from the "bassin de l'Amazone," died in Antwerp. Additional specimens were recorded from a specimen of *Hapale jacchus leucocephalus* (= *Callithrix jacchus geoffroyi*) collected in Bahia, Brazil, and preserved in Brussels.

MATERIAL EXAMINED—None.

Fonsecalges saimirii Fain, 1963

Fonsecalges saimirii Fain, 1963, Bull. Ann. Soc. Roy. Ent. Belg., 99: 330. Fain, 1963, Bull. Inst. Roy. Sci. Nat. Belg., 39(32): 90. Fain, 1966, Acarologia, 8: 107.

Dunnalges lambrechti Lavoipierre, 1964, Acarologia, 6: 343.

This species was briefly described from numerous specimens collected from two *Saimiri sciureus* from "Amerique du Sud" and which died in the Antwerp Zoo in 1959 and 1963 (Fain, 1963a). Fain (1963c) provided further descriptive information and indicated the original hosts had come from "Amazonie" and had died shortly after their arrival at the Antwerp Zoo. He also mentioned additional specimens from "Tamarins spp. originaires d'Amazonie." The inexact locality information leaves the actual identity of the type host in doubt following the revision of the genus *Saimiri* by Hershkovitz (1984). Lavoipierre (1964a) described and illustrated this species as *Dunnalges lambrechti,* from "a long series of specimens comprising all stages" from several *Tamarinus nigricollis* (= *Saguinus nigricollis*) from eastern Peru which died in San Francisco, California. Fain (1966) illustrated parts of this species from the type specimens.

Fain (1963c) did not observe sexual dimorphism in the nymphs of this species, as only two tritonymphs and some shed tritonymphal cuticles were examined. Sexual dimorphism was noted in the tritonymphs during the present study. All larvae and protonymphs examined bear three pairs of sclerotized apophyses in the ventrolateral region: a rounded apophysis immediately posterior to trochanter II, a pointed apophysis lateral to the base of leg III, and a pointed apophysis posteriolaterally near the posterior lobes. Tritonymphs in which these apophyses are retained are here interpreted as males, while those in which all apophyses are lost are interpreted as females.

MATERIAL EXAMINED—Total 59. Three females from *Saimiri sciureus macrodon.* **PERU, Loreto:** Nauta, Río Tigre, 5 km above Río Tigrillo; 17 December 1980; P. Hershkovitz (9257). Host now a tanned skin (FMNH 122810). Mites labeled BMOC 81-0809-15. Two females, 1 male from same host species and locality; P. Hershkovitz (9258, FMNH 122811, BMOC 81-0809-16). Twenty-four females, 8 males, 5 female tritonymphs, 8 male tritonymphs, 5 protonymphs, 1 larva from same host species. **PERU, Loreto:** Nauta, Río Tigre, 6 km above Río Tigrillo; same date; P. Hershkovitz (9268, FMNH 122816, BMOC 81-0809-18). Two females from same host species and locality; P. Hershkovitz (9267, FMNH 122815, BMOC 81-0809-19).

Procebalges pitheciae Fain, 1963

Procebalges pitheciae Fain, 1963, Bull. Ann. Soc. Roy. Ent. Belg., 99: 332. Fain, 1963, Bull. Inst. Roy. Sci. Nat. Belg., 39(32): 96.

This species was briefly described from specimens collected from a *Pithecia monachus* imported from "Amerique du Sud" and which died in the Antwerp Zoo (Fain, 1963a). The descriptions were completed and illustrations provided later (Fain, 1963c).

MATERIAL EXAMINED—Total 242. Nineteen females, 28 males (7 in tandem with female tritonymphs), 15 female tritonymphs, 1 protonymph from *Pithecia hirsuta*. **PERU, Loreto:** Nauta, Río Samiria; 30 November 1980; P. Hershkovitz (9115). Host now preserved as a tanned skin (FMNH 122797). Mites labeled BMOC 81-0809-7. Forty-seven females, 81 males (17 in tandem with female tritonymphs), 22 female tritonymphs, 8 male tritonymphs, 13 protonymphs, 8 larvae from same host and locality; P. Hershkovitz (9088, FMNH 122796, BMOC 81-0809-9).

Family AUDYCOPTIDAE

Three species of hair follicle inhabiting mites of the family Audycoptidae have been previously described from New World primates, all from squirrel monkeys identified as *Saimiri sciureus.* With the recent recognition of several valid species in the genus *Saimiri* (Hershkovitz, 1984), the exact identification of the reported hosts is problematical. Specimens representing a fourth species were recovered from *Cebus apella* from the present collections.

Audycoptes greeri Lavoipierre, 1964

Audycoptes greeri Lavoipierre, 1964, Ann. Natal Mus., 16: 194.

This species was described from females collected from the sinus-hair follicles of *Saimiri sciureus* collected in eastern Peru and kept in captivity in California (Lavoipierre, 1964b). In the absence of more detailed collection information, the true host may have been either *S. sciureus* or *S. boliviensis,* as both occur in eastern Peru (Hershkovitz, 1984).

MATERIAL EXAMINED—Five females from "*Saimuri sciurea*" (*sic*), without further collection data, from the Lavoipierre collection, University of California, Davis.

Audycoptes lawrencei Lavoipierre, 1964

Audycoptes lawrencei Lavoipierre, 1964, Ann. Natal Mus., 16: 199.

This species was described from the same hosts and habitat as *Audycoptes greeri* (Lavoipierre, 1964b), so the actual specific identity of the host remains uncertain, as indicated for *greeri.*

MATERIAL EXAMINED—Two females from "*Saimuri sciurea*" (*sic*), without further collection data, from the Lavoipierre collection, University of California, Davis.

Saimirioptes paradoxus Fain, 1968

Saimirioptes paradoxus Fain, 1968, Acarologia, 10: 286.

This species was described from a single female, containing a larva, collected from a *Saimiri sciureus* which had died in the Antwerp Zoo (Fain, 1968). No information concerning the geographic origin of the host was given, making the specific identity of the host uncertain.

MATERIAL EXAMINED—None.

Saimirioptes hershkovitzi, new species

In the following description, all measurements are given in micrometers and are presented as holotype (range of three measured specimens).

FEMALE (FIGS. 1–2)—Body elongate, cylindrical

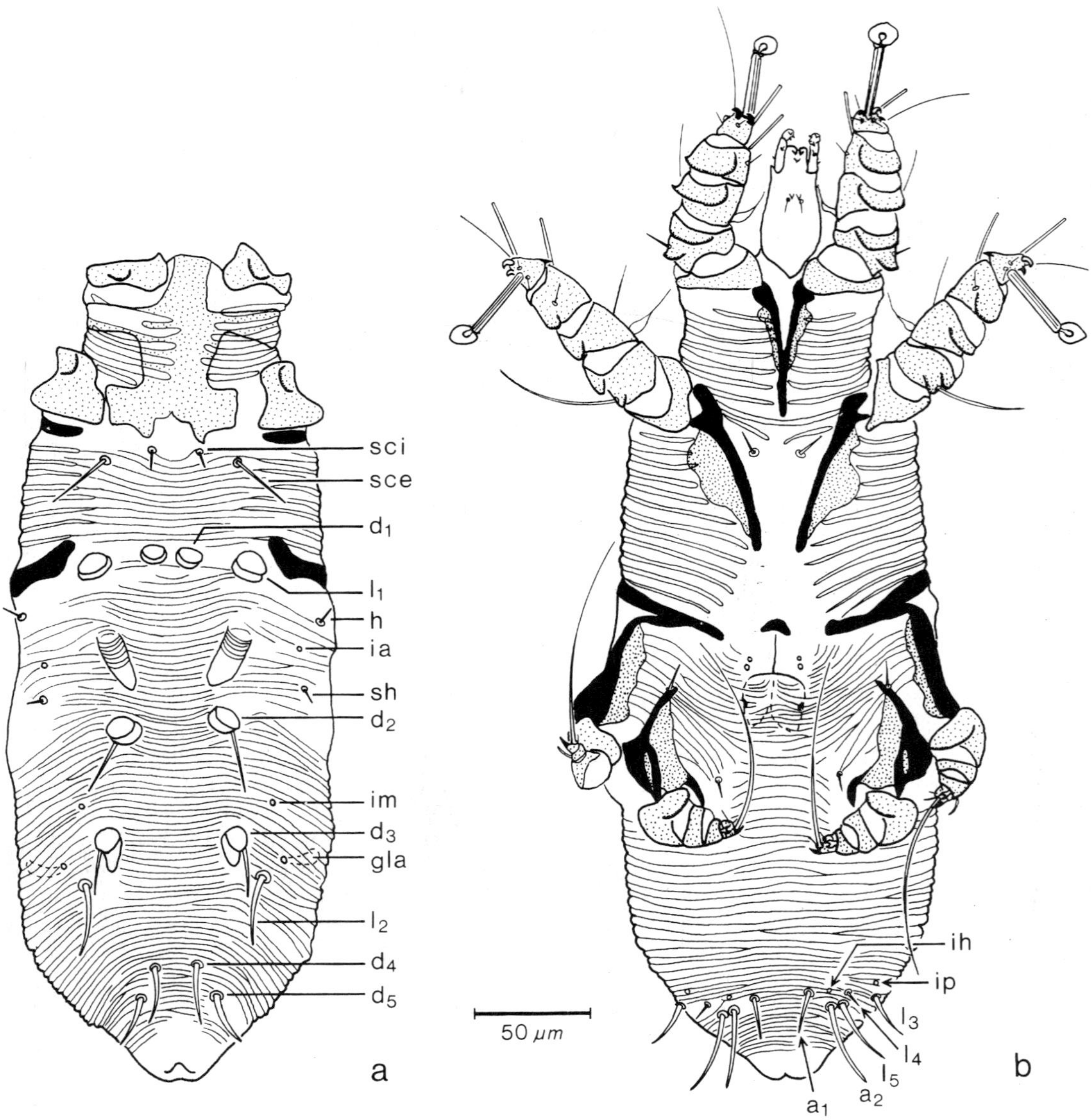

FIG. 1. *Saimirioptes hershkovitzi,* female. **A,** dorsum; **b,** venter.

to somewhat flattened dorsoventrally; length including gnathosoma 415 (415–421), width at level of transverse coxal apodemes III 146 (135–146); entire body with transverse striations which are very thick anteriorly, thin posteriorly.

Dorsum (fig. 1a)—Prodorsal sclerite narrow anteriorly, much widened medially, narrowed posteriorly and fused internally with apodemes of coxal fields II in most posterior region. Sclerite traversed by anterior transverse striations laterally. Length of sclerite 70 (70–77). Paired dorsal protuberances present mesal to setae *h* and *sh,* with striations encircling protuberances along most of their length; lengths 29 (28–29). Idiosomal chaetotaxy as follows: scapular setae filiform, in a transverse line posterior to prodorsal sclerite, lengths *sci* 4 (4–5), *sce* 39 (29–39); setae d_1 and l_1 very short and rounded, subdivided into two distinct lobes; setae *h* and *sh* filiform, lengths 9 (6–9); setae d_2 subdivided into three parts, a ventral filiform part, length 26 (23–26), and two dorsal rounded lobes; setae d_3 similar to d_2 but filiform part shorter, length 19 (14–19), and middle lobe more elongate; remaining setae filiform; l_2 29 (28–

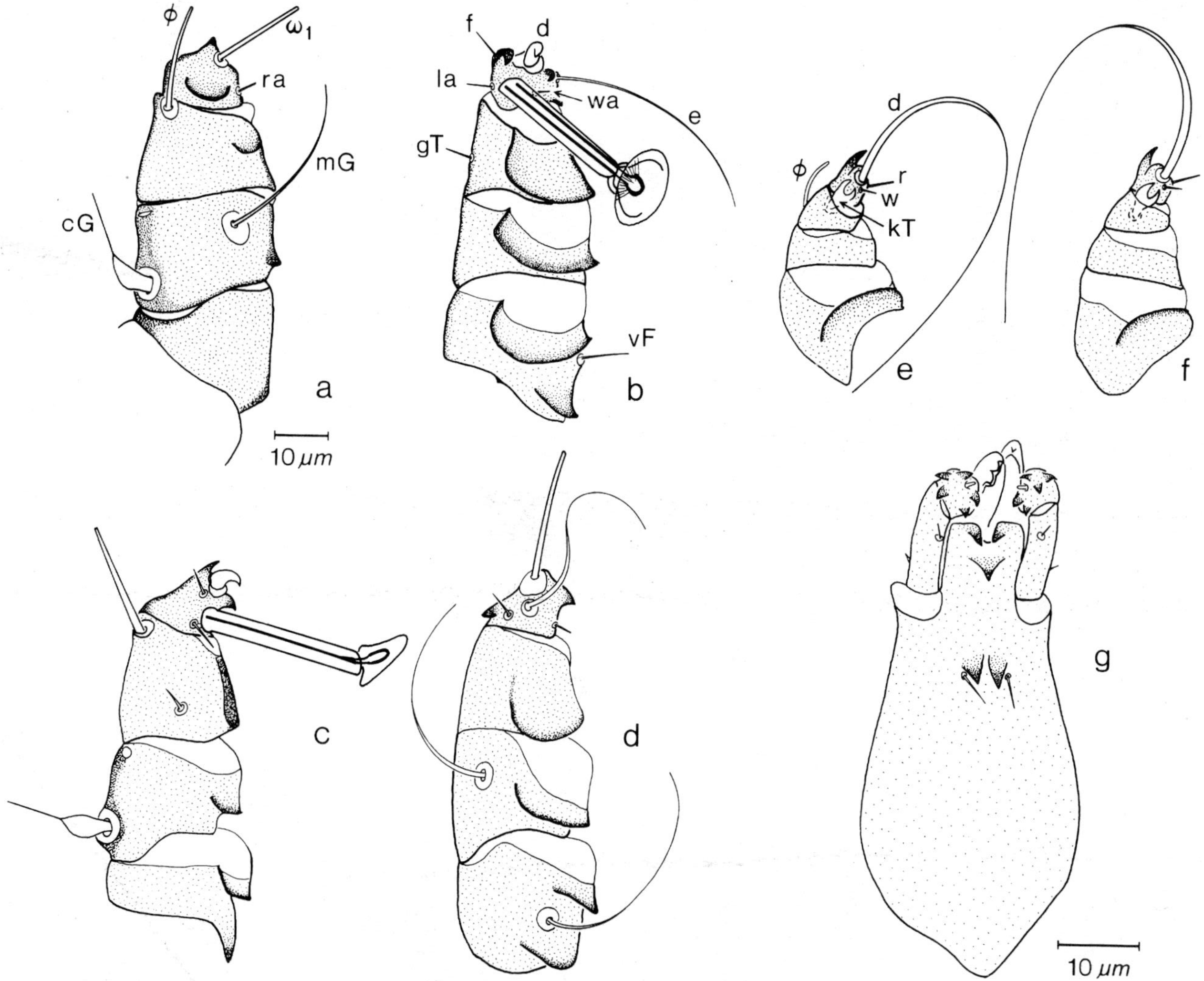

FIG. 2. *Saimirioptes hershkovitzi,* female. **A,** leg I, dorsal view; **b,** leg I, ventral view; **c,** leg II, dorso-antiaxial view; **d,** leg II, ventro-paraxial view; **e,** leg III, ventral view; **f,** leg IV, ventral view; **g,** gnathosoma, ventral view.

29); l_3 ventrally positioned, 18 (18–19); d_4 24 (19–24); d_5 21 (no variation); l_4 ventral, 5 (4–5); l_5 ventral, 30 (27–32); a_1 ventral, 14 (no variation); a_2 30 (27–30). Four pairs of idiosomal cupules present: *ia* dorsal between setae *h* and *sh*; *im* dorsal between and lateral to setae d_2 and d_3; *ip* ventral slightly anterior to seta l_3; *ih* ventral, between and mesal to setae l_4 and l_5. Opisthosomal glands present, small, opening between and lateral to setae d_3 and l_2. Anus terminal. Opening to bursa copulatrix terminal, above anal opening.

Venter (fig. 1b)—Coxal apodemes well developed. Anterior apodemes of coxal fields I straight, length 66 (57–66), fused medially to form a sternum of length 23 (23–24). Anterior apodemes of coxal fields II curving mesally, length 75 (75–79); posterior apodemes of coxal fields II transverse, fused ventromedially with anterior apodemes of coxal fields III, extending dorsally, length of ventral portion 50 (46–50), length of dorsal portion 31 (28–31). Anterior apodemes of coxal fields III extending almost longitudinally from trochanters III to posterior apodemes of coxal fields II, then bending at right angles to fuse with the latter, length of longitudinal portion 49 (49–51), transverse portion 39 (39–44). Anterior apodemes of coxal fields IV very wide, anteriomedially directed, length 50 (50–53). Ovipore located between coxal fields III and IV; a crescentic epigynal apodeme positioned anterior to ovipore; genital valve elongate, disappearing posteriorly under transverse flap; a pair of small thickenings along posterior, internal portion of valves; two pairs of vestigial genital papillae lateral to genital opening. Three pairs of filiform coxal setae present: *cxI* at posterior end of coxal fields I between coxal apodemes II, length 12 (9–12); *cxIII* on median apex of coxal apodemes IV, length 9 (6–9); *cxIV* mesad of the base of leg IV, length 7 (6–7). Genital setae absent.

Gnathosoma (fig. 2g)—Gnathosoma elongate, somewhat widened posteriorly, then tapering to a point ventrally, length (excluding chelicerae) 62 (59–62), maximum width 26 (24–26). Chelicerae elongate, with small, toothed digits, length 46 (44–46). Palps apparently two-segmented; basal segment bearing two very short, filiform setae, 1 proximodorsal, 1 medioventral; distal segment with short solenidion and 5 sclerotized, pointed projections. Subcapitulum with pair of short, filiform setae medioventrally; with pair of sclerotized, pointed projections positioned slightly anterior to subcapitular setae, and unpaired projection anteriomedially; median rutellar lobes with sclerotized points medially.

Legs (figs. 2a-f)—Legs I–II similar in structure, with all segments free, lengths 84 (80–90). Femora I–II bearing two ventral apophyses, one filiform seta *vF*, lengths: I, 9 (8–9); II, 33 (30–33). Genua I–II each with large ventral apophysis; setae *cG* expanded basally, then tapering, lengths 22 (20–25), setae *mG* filiform, lengths 43 (40–45). Tibiae I–II each with large ventral apophysis; setae *gT* short, filiform, lengths 6 (5–6); solenidia ϕ apicodorsal, lengths: I, 20 (18–20); II, 22 (21–23). Tarsi I–II each with pointed dorsal apophysis, two distal hooked apophyses; chaetotaxy of tarsi I–II similar; setae *d* as strongly hooked, apical claws, lengths 5 (no variation); *e* filiform, lengths 40 (43–50); *f* filiform, lengths 7 (6–7); *la* filiform, lengths 6 (5–6); *wa* very short, not measurable; *ra* filiform, lengths 6 (5–6); solenidia ω_1 blunt apically, lengths: I, 14 (no variation); II, 26 (24–26). Pretarsi I–II similar, total lengths 40 (37–40); pretarsi divided into long ambulacral stalk and rounded ambulacral disc; ambulacral disc with condylophore guide and ventral "rays" (fig. 2b); condylophores attenuate, appearing to divide at base of ambulacral disc.

Legs III–IV similar, each with fused trochanter-femur bearing a large, ventral apophysis. Genua III–IV glabrous. Tibiae III–IV each with setae *kT* enlarged, bifurcate, and deeply rooted; tibia III with solenidion ϕ, length 6 (6–7), ϕ IV absent. Tarsi III–IV each with apical, pointed apophysis, three setae: *d* apical, lengths: III, 100 (91–100); IV, 97 (86–97); *r* filiform, lengths 7 (6–7); *w* filiform, lengths 3 (no measurable variation).

TYPES—Holotype and two paratype females from *Cebus apella*. **PERU, Loreto:** Nauta, Río Samiria; 18 November 1980; P. Hershkovitz (9049). Host now a tanned skin (FMNH 122792). Mites labeled BMOC 81-0811-2. Holotype and 1 paratype deposited in Field Museum of Natural History, Chicago; 1 paratype in Museum of Zoology, University of Michigan, Ann Arbor.

SYSTEMATIC POSITION—*Saimirioptes hershkovitzi* shares with *S. paradoxus*, the type-species and only other member of the genus, the presence of dorsal lobes between setae d_1 and d_2, the similar form of setae d_1 and l_1 (rounded, bifurcate, and without filiform part), and the expansion of the bases of setae *cG* of genua I–II. The new species differs from the type-species in the relative lengths of the scapular setae (*sce:sci* 5–6:1 in *S. hershkovitzi*, 1.5:1 in *S. paradoxus*); the shorter filiform portion of setae d_2 (not reaching d_3 in *S. hershkovitzi*, extending past d_3 in *S. paradoxus*); the shorter lengths of setae l_2 (29 in *S. hershkovitzi*,

67 in *S. paradoxus*); the presence of setae l_4 (absent in *S. paradoxus*); the longer pretarsi I–II (37–40 in *S. hershkovitzi,* 17 in *S. paradoxus*); and the greater lengths of setae *d* on tarsi III–IV (at least twice as long as the entire leg in *S. hershkovitzi,* less than half the leg length in *S. paradoxus*).

Discussion

The Peruvian collections reported upon here support the hypothesis that most of the records of primate-Cebalginae associations previously reported reflect natural host-parasite associations. The association between *Alouattalges corbeti* and *Aotus* species was not verified and remains questionable. As there were no field collected representatives of the Callitrichidae examined during this study, the occurrence of cebalgine mites on these hosts remains generally untested in natural situations, with only *Fonsecalges johnjadini* recorded from noncaptive callitrichids (Fain, 1963c).

Having established that most records of primate-cebalgine associations reflect natural associations, two additional hypotheses must be proposed and tested before any coevolutionary hypotheses may be tested. These preliminary hypotheses concern the monophyly of the Cebalginae and the phylogenetic relationships of the taxa within the group. I have previously discussed these questions (OConnor, 1984) but, due to space constraints, was unable to detail the reasoning behind my conclusions.

Monophyly of the Cebalginae

Before any hypothesis of historical associations may be tested, at least one of the lineages must be hypothesized to be monophyletic (Brooks, 1981). In this study, the psoroptid subfamily Cebalginae, which comprises all psoroptid mites parasitizing New World Primates, must be tested for monophyly. In last defining this subfamily, Fain (1963c) listed 21 character states for the group. Because the comprehensive morphological studies of Fain (1963c) were carried out before the methods of phylogenetic systematics became widely discussed and utilized, no distinction between ancestral and derived states was made in the diagnosis of the Cebalginae. In order to test whether the Cebalginae represents a monophyletic group, I have examined the characters listed by Fain (1963c) using outgroup comparison to polarize the states. I have previously hypothesized that the taxa comprising the four subfamilies of Psoroptidae which parasitize the Primates form a monophyletic group (OConnor, 1984). These taxa include the Makialginae (*sensu* OConnor, 1984; i.e., including the Cheirogalalginae and Galagalgidae of Fain), parasites of the Strepsirrhini; the Paracoroptinae, parasites of African Cercopithecidae and Hominidae; the Nasalialginae, parasites of Asian Cercopithecidae; and the Cebalginae, from New World primates. This grouping of taxa is regarded as the ingroup in the following analysis. Outgroups used in defining the character state polarities were the other subfamilies of Psoroptidae and earlier derivative groups in the Astigmata.

Character Analysis

1. Presence of retrograde apophyses on coxal fields III. This state is unique to the cebalgine genera, with such apophyses not occurring in other taxa in the ingroup or outgroup.
2. Male body size much smaller than female. In the outgroups and other ingroups, males are similar in size to females or somewhat larger or smaller. The substantial reduction in body size of male Cebalginae is unique.
3. Female opisthosoma more or less squared posteriorly and laterally, reinforced with sclerotized areas. This condition is exhibited by all Cebalginae, but not in any outgroup. Among the ingroups, the squared body is unique to the Cebalginae, but sclerotized reinforcement is also found in the Nasalialginae and some Paracoroptinae (*Pangorillalges*). Among outgroup taxa, sclerotized reinforcement is present in some Psoralginae (*Edentalges*), but again without the squared body form.
4. Male with legs III very modified, with 3–4 terminal segments fused and bearing medially directed projections. The modification of the third pair of legs in male Cebalginae is unique and present in all taxa. No similar modification occurs anywhere in ingroup or outgroup.
5. Reduction or loss of paranal suckers in males. Paranal suckers are present in males in the outgroups and ingroups. In the Cebalginae, the suckers are very reduced or absent.
6. Tarsus I with 2 apical solenidia. I regard the apical displacement of solenidion ω_1 in the Cebalginae as derived. In the other ingroups and most outgroups, this solenidion is median or basal on

the tarsus. Apical displacement of this solenidion also occurs in the subfamily Psoroptinae and the monobasic Marsupialginae, conditions I regard as convergent.

7. Loss of dorsal seta d_1. Seta d_1 is absent in all Cebalginae and retained in the other ingroups and most outgroups. This seta is also lost in the Listropsoralginae and the psoroptid parasites of rodents (*Echimyalges, Myoproctalges,* and *Coendalges*), conditions I regard as convergent.

8. Reduction or loss of apophyses from the posterior tarsi of the female. Apophyses are present on the posterior tarsi of some of the outgroups such as the Listropsoralginae and are retained in some ingroups, the Makialginae and Paracoroptinae. I consider the presence of these apophyses to be plesiomorphic for the primate-associated psoroptid lineage. In the Cebalginae, taxa in which females have well-developed posterior legs retain vestiges of these apophyses, while taxa in which the legs are reduced retain no traces. I consider this reduction to be a derived state for the Cebalginae and regard the loss of these structures in the Cebalginae and Nasalialginae as convergent.

9. Posterior edge of female opisthosoma with 2 pairs of long, strong setae. In most ingroups and psoroptid outgroups, seta l_5 is long and strong while seta d_5 is shorter and thinner. Of the ingroups, setae l_5 and d_5 are equally well developed in the Cebalginae and Nasalialginae. However, in the Cebalginae, setae l_5 and d_5 are closely associated, usually on a single projection, while in the Nasalialginae these setae are separate, on distinctly different projections. I regard these two conditions as convergent. I also regard as convergence the elongation of seta d_5 in some Psoralginae.

10. Dorsal seta l_4 sometimes absent. Seta l_4 is absent in all Cebalginae except *Cebalgoides cebi.* The seta is retained in that species, the other ingroups, and most outgroups. The loss of seta l_4 could be regarded as a synapomorphy defining a group containing all Cebalginae except *Cebalgoides.* However, this hypothesis conflicts with groupings suggested by all other characters (see page 256). At this point, it is more parsimonious to regard the loss of seta l_4 as a synapomorphy for the Cebalginae, with a reversal in *Cebalgoides.*

11. Loss of retrograde apophyses on the anterior legs. Retrograde apophyses are present on the anterior legs in some of the outgroups, notably the Listropsoralginae and Marsupialginae in the Psoroptidae and in the related families Audycoptidae, Rhyncoptidae, and Myocoptidae. Among the ingroups, these apophyses are retained in the Makialginae but lost in the Cebalginae, Paracoroptinae, and Nasalialginae. I regard this loss as characterizing these latter three groups as a natural unit and thus plesiomorphic for the Cebalginae.

12. Female bursa copulatrix subterminal or ventral. In the outgroups, the female bursa copulatrix is terminal, as it is in the Cebalginae, Makialginae, and Nasalialginae among ingroups. I regard the dorsal position of the bursa in the Paracoroptinae as derived. Thus, this character retains the plesiomorphic condition in the Cebalginae.

13. Base of the gnathosoma with retrograde apophyses. In some outgroups as well as in the Cebalginae, Makialginae and some Paracoroptinae (i.e., *Pangorillalges*), the base of the gnathosoma bears retrograde apophyses. These are absent in certain outgroups (e.g., the Psoroptinae) and in *Paracoroptes* and *Nasalialges* among ingroup taxa. I regard the retention of these apophyses in the Cebalginae as plesiomorphic.

14. Absence of retrograde apophyses on coxae I–II. Retrograde apophyses occur on coxae I–II only in the Makialginae. They do not occur on both coxal fields in any other ingroup or outgroup taxa, although the Listropsoralginae and Myocoptidae have apophyses on coxae II. I regard these structures as synapomorphies characterizing the Makialginae, and their absence in the Cebalginae as plesiomorphic.

15. Well-developed "claws" on tarsi I–II. Clawlike apophyses on the apices of tarsi I–II are present in many outgroup taxa, including both mammal and bird parasites. I regard the presence of this character state as ancestrally characterizing all Psoroptidia, and thus plesiomorphic for the Cebalginae.

16. Tarsus II with solenidion apically displaced. Solenidion ω is apically displaced in the Cebalginae, Paracoroptinae, and Nasalialginae. In most outgroups and in the Makialginae, the solenidion retains its ancestral, basal position. I regard this displacement as diagnosing a natural group consisting of the above three primate-associated subfamilies and thus plesiomorphic for the Cebalginae.

17. Female tarsal chaetotaxy 7–7–6–4 or –5. Possession of 7 setae on tarsi I–II involves the loss of seta *ba* in the Paracoroptinae and Cebalginae and probably the Nasalialginae. Retention of 8 setae on these segments (including *ba*) is characteristic for most outgroup taxa and the Makialginae. I regard this state as diagnosing a natural group comprising the Cebalginae, Paracoroptinae,

and Nasalialginae and thus plesiomorphic for the Cebalginae.

18. Angles of female opisthosoma generally with strong hooks. Such hooks are found in the cebalgine genera *Cebalges, Cebalgoides,* and *Fonsecalges,* but not in other taxa in the outgroup or ingroup. I regard the presence of these structures as characterizing a monophyletic unit within the Cebalginae but not the group as a whole.

19. Posterior legs of female normal or atrophied. Legs III–IV of the female are reduced in size and may exhibit fusion of segments in the genera *Cebalges* and *Fonsecalges* but not in other Cebalginae nor other members of the ingroup. Well-developed legs are characteristic of most outgroup taxa, although similar reduction of the posterior legs occurs in some Psoroptinae and Psoralginae. I regard the atrophied legs of some Cebalginae as characterizing a smaller monophyletic unit within the group and convergent with the outgroup taxa noted above.

20. Posterior legs of nymphal stages normal or short and atrophied and bearing a long seta. The reduction of the legs in the nymphs is found in all cebalgine genera except *Procebalges.* It is also found in a few outgroup taxa (e.g., the Psoroptinae and Psoralginae), but not in most outgroups or in other ingroup taxa. I regard this state as diagnosing a group within the Cebalginae and not characterizing the group as a whole.

21. Dorsal seta d_4 sometimes absent. Seta d_4 is absent in *Schizopodalges lagothricola* and *Fonsecalges johnjadini* (but not *F. saimirii*), and present in other Cebalginae, other ingroups, and most outgroups. The distribution of this derived state within the Cebalginae strongly suggests the independent loss of this seta in the two species sharing the state. This character at best diagnoses a group within the Cebalginae, but is more likely a case of convergence. In no way does this character diagnose the group as a whole.

Conclusions

On the basis of the above analysis, I conclude that, of the 21 character-states listed by Fain (1963c) as diagnosing the Cebalginae, five represent unique synapomorphies which diagnose the Cebalginae and occur in no other group (character-states 1–5), five are synapomorphies which diagnose the Cebalginae but which also occur as convergent states in certain other taxa (character-states 6–10), seven are symplesiomorphies diagnosing larger groups which include the Cebalginae (character-states 11–17), and four represent within-group apomorphies diagnosing smaller groups within the Cebalginae (character-states 18–21). The ten synapomorphies diagnosing the Cebalginae leave no doubt that the group is a natural one.

Phylogenetic Relationships Within the Cebalginae

I have previously presented a hypothesis of phylogenetic relationships among genera in the Cebalginae based upon 17 characters (OConnor, 1984). This cladogram is reproduced here (fig. 3), with numbers on the cladogram referring to the derived states of the characters listed below. In the earlier study, space limitations prevented discussion of the different states of these characters and the reasons for interpreting their polarity. Additionally, an error in character 12 appeared in the list of character-states. As I have hypothesized that the Cebalginae form a monophyletic group within a larger lineage comprising the Makialginae, Paracoroptinae, and Nasalialginae, taxa in these three lineages were used as outgroups to polarize the character-states within the Cebalginae.

1. Female with coxal apodemes III–IV fused. In all taxa in the outgroups, coxal apodemes III–IV end freely. Within the Cebalginae, these apodemes are fused together on either side in the genera *Procebalges, Schizopodalges,* and *Fonsecalges.* Although I regard the fused condition to be derived, conflicts with many other characters suggest that these apodemes have fused independently in the three genera.

2. Female epigynum fused with coxal apodemes I. In most outgroup taxa, the female ovipore is located between coxal fields II and III. There has been a trend toward anterior displacement of the ovipore in a number of psoroptid lineages (e.g., Psoralginae, Listropsoralginae, Psoroptinae), and the more anterior position occurs in certain taxa among the primate associated lineage as well (e.g., *Lemuralges* in the Makialginae). Within the Cebalginae, this derived state occurs in the genera *Procebalges* and *Schizopodalges.* Because this character-state distribution conflicts with many other characters, I regard the presence of this state in the two genera as convergence.

3. Male with paranal suckers lost. In all outgroups, the male retains paranal suckers. Within the Cebalginae, the suckers are retained only in

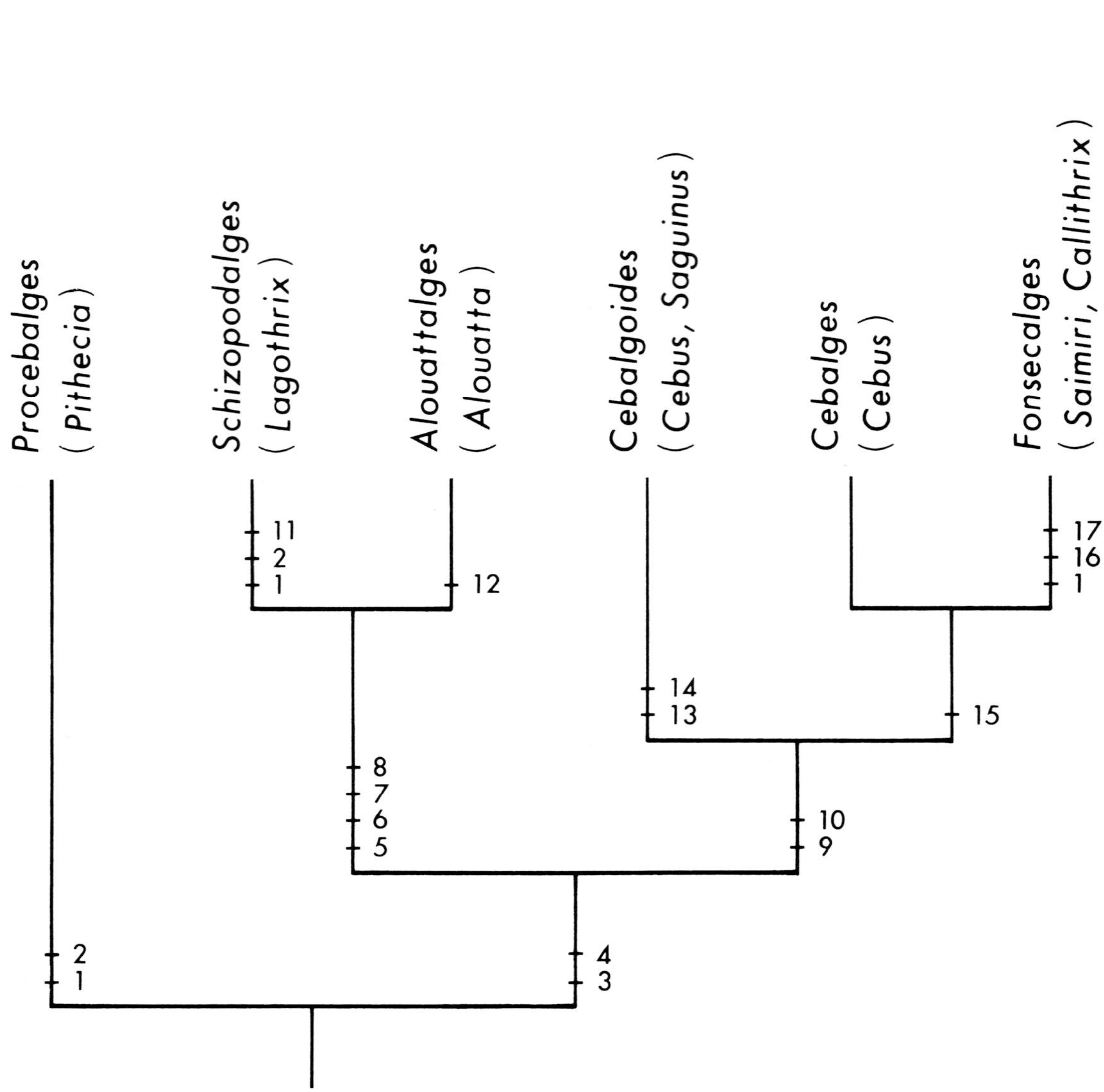

FIG. 3. Phylogenetic relationships among genera in the subfamily Cebalginae. Numbers refer to derived states of characters discussed in the text. Hosts of each mite taxon given in parentheses.

Procebalges. I regard the loss of these structures as derived and characterizing the sister group of *Procebalges.*

4. Immature stages with pretarsi III–IV lost. Described immature stages of outgroup taxa all retain the pretarsus on legs III–IV. Within the Cebalginae, only *Procebalges* retains this ancestral state. In all other Cebalginae, pretarsi III–IV are lost in the immature stages, a condition I regard as derived.

5. Gnathosomal apophyses lost in both sexes. As discussed in the previous section, I regard the presence of gnathosomal apophyses as ancestral in the Cebalginae. These structures are lost in both sexes in the genera *Schizopodalges* and *Alouattalges* within the Cebalginae, a condition I regard as derived. I regard this state as a synapomorphy for these two genera and convergent with the similar state occurring in the genera *Lemuralges* (Makialginae), *Paracoroptes* (Paracoroptinae), and *Nasalialges* (Nasalialginae). I have treated loss of these apophyses in the female only as a separate character (16, below).

6. Female with seta *s* of tarsi I–II in the form of a strongly hooked claw. In the outgroup taxa, seta *s* is simple and filiform. Within the Cebalginae, this state is retained in all taxa except *Schizopodalges* and *Alouattalges* where the seta is enlarged and hooked. I regard the latter state as derived and a synapomorphy for the two genera.

7. Male with a large spur on tarsus III. This condition is unique to the cebalgine genera *Schi-*

zopodalges and *Alouattalges.* In the outgroups and other ingroup taxa, such a spur does not exist. I regard the presence of this character-state as indicative of common ancestry of these two genera.

8. Male with ridges on leg III restricted to tibial element. Ridges on the distal portion of leg III of the male characterize the Cebalginae (see above). Such ridges do not occur in any outgroup. I regard the restriction of the ridges to a spur extending from the tibial portion of the leg as a derived state defining the group *Schizopodalges* + *Alouattalges* because this condition forms a functional complex with the spur on tarsus III (character 7) and a dorsal spur on femur IV. This latter character was not considered in the previous analysis (OConnor, 1984). The conjunction of the three spurs serves to lock legs III and IV together into a unit.

9. Female with recurved hooks on posterior border of opisthosoma. Strong hooks are present in this position in females of *Cebalgoides, Cebalges,* and *Fonsecalges.* Weaker development in this area is characteristic of other cebalgine genera, while in the outgroups, no projections exist. I regard the possession of strong hooks as a derived condition defining a lineage comprising the three genera noted above.

10. Male with pretarsus III lost. Males retain a pretarsus on leg III in most outgroup taxa. This pretarsus is lost in the genera *Galagalges* and *Cheirogalalges* (Makialginae), and also in the cebalgine genera *Cebalgoides, Cebalges,* and *Fonsecalges.* I regard the loss of pretarsus III as a synapomorphy for the latter three genera within the Cebalginae, with independent loss in the makialgine lineage comprising the former two genera.

11. Female without opisthosomal sclerite. A median sclerite is present in females of most outgroup taxa and is retained in all Cebalginae except *Schizopodalges.* I regard this loss as derived in the latter genus. Convergence in this character occurs with some outgroup taxa. This sclerotization is also lost in *Gaudalges caparti,* and *Lemuralges* (Makialginae), and *Nasalialges* (Nasalialginae).

12. Female with seta *s* of tarsi III–IV enlarged and clawlike. In the previous study (OConnor, 1984), a lapsus occurred in that the state "seta *s* of tarsi III–IV reduced" was listed as a derived condition for the genus *Alouattalges.* In fact, possession of a small, filiform seta *s* on these tarsi must be regarded as the ancestral condition, as it occurs in all outgroup taxa as well as in most Cebalginae. In the genus *Alouattalges,* seta *s* is enlarged and recurved on tarsi III–IV, a condition I regard as the true derived state.

13. Male with opisthosomal lobes widely spaced. In most outgroup taxa, the opisthosomal lobes of the male are close together, a condition also found in most Cebalginae. These lobes are widely spaced in the genus *Cebalgoides* within the Cebalginae, a condition I regard as derived. Convergence occurs, with this state also occurring in *Nasalialges* (Nasalialginae).

14. Male with apodeme between genital and anal region strongly reduced. In the outgroup taxa, males do not possess a transverse apodeme between the genital and anal regions. A large apodeme is present in this position in most male Cebalginae. In the prior analysis (OConnor, 1984), I stated that this apodeme was absent in *Cebalgoides* as indicated by Fain (1963c) and regarded this as a reversal. Closer examination of a number of specimens indicates that a very small apodeme is present in some males of this taxon. I regard the presence of the apodeme as a derived condition for the Cebalginae, and its reduction or loss in *Cebalgoides* as a further derived state.

15. Female with tibiae-tarsi III–IV fused. In the outgroups, the tibia and tarsus of legs III–IV are freely articulated. This condition is present in most Cebalginae as well. In the genera *Cebalges* and *Fonsecalges,* these segments are fused in the female, a condition I regard as derived within the Cebalginae. Similar fusions occur in some other psoroptid subfamilies but not within the primate-associated lineage.

16. Female with gnathosomal apophyses lost. Gnathosomal apophyses are retained in females of most outgroup taxa and are retained in some Cebalginae. These apophyses are lost in the female but retained in the male in the genus *Fonsecalges.* I regard this loss as independent of the loss of apophyses in the *Schizopodalges-Alouattalges* lineage, where the apophyses are lost in both sexes (character 5).

17. Female with pretarsi III–IV lost. Pretarsi III–IV are retained in the female in all outgroup taxa and in all Cebalginae except *Fonsecalges.* I regard this loss as derived and convergent with the similar loss in females in other psoroptid groups (e.g., some Psoroptinae, Psoralginae).

History of Primate-Cebalgine Evolution

The present knowledge of the diversity and host associations of the family Audycoptidae is not sufficient for formulating hypotheses concerning the history of their host associations. However, given the phylogenetic relationships among cebalgine taxa presented in the cladogram (fig. 3), two methods exist for using these relationships to make hypotheses regarding the history of the associations between the New World primates and the Cebalginae. The first method would involve comparing the phylogenetic hypothesis for the parasite group with a similar hypothesis for the hosts. The hypothesis tested by this comparison is whether the current pattern of host-parasite associations directly results from strict cospeciation between the hosts and their parasites. Given a phylogenetic hypothesis for each group, such a hypothesis is easily tested. The second type of conclusion which might be drawn from the parasite cladogram is a phylogenetic hypothesis concerning host relationships. This method is dependent upon the assumption of cospeciation, or at least noncolonization, between the two lineages.

The current state of knowledge concerning the phylogenetic relationships among New World primates is somewhat confused. Many early hypotheses were based upon classical methodology in which ancestral and derived character-states were not differentiated in proposing hypotheses of relationships. Schwartz et al. (1978) summarized the state of knowledge only a few years ago by saying that "Platyrrhini appears in and of itself to be a natural group, although both its wider relationships and the relationships among its members remain unclear" (p. 128). Much new information relating to this question was presented in symposia in 1978–1979 and published in a volume edited by Ciochon and Chiarelli (1980). In the following discussion, I refer to phylogenetic hypotheses for New World primates suggested by Rosenberger (1977) and contributors to the Ciochon and Chiarelli volume as representing the most modern phylogenetic thinking regarding Platyrrhine relationships.

Returning to the question of cospeciation between primates and cebalgine mites, the first difficulty encountered is the lack of a consensus concerning phylogenetic relationships among all New World primates. Phylogenetic hypotheses based on the following types of data sets are in strong disagreement: dentition (Rosenberger, 1977; 3 separate hypotheses); integumentary characters (Perkins & Meyer, 1980); karyology (Chiarelli, 1980); immunological methods (Sarich & Cronin, 1980); and other immunological data and protein sequence data (Baba et al., 1980). Despite the strong disagreement among these data sets, some patterns are common to several or all of these. These may be compared with the cladogram of mite relationships to test the cospeciational hypothesis.

One host relationship which is supported by almost all data sets is the relationship between *Alouatta* and the Atelinae (including *Lagothrix*). Of the hypotheses cited above, only the karyological evidence did not support this relationship, and then only because the karyotype of *Lagothrix* is so derived that Chiarelli (1980) made no hypothesis as to its relationships. This relationship between *Alouatta* and *Lagothrix* is mirrored by the sister group relationship between their parasites, *Alouattalges* and *Schizopodalges,* which is among the most strongly supported relationships among the Cebalginae.

A second host relationship which is partially mirrored in the mite phylogeny is the relationship between the genera *Cebus* and *Saimiri,* which was supported by all data sets except the integumentary data (Perkins & Meyer, 1980). In this case, the mite phylogeny suggests a cospeciational process, but some additional hypotheses are required to explain the distributions of the genera *Cebalgoides, Cebalges,* and *Fonsecalges.* The relationships among these three parasite genera are compatible with a cospeciational scenario if an early speciation event between the *Cebalgoides* and the *Cebalges-Fonsecalges* lineages occurred in conjunction with a common ancestor of *Cebus* and *Saimiri.* Extinction of the *Cebalgoides* lineage on the *Saimiri* line, with cospeciation of the *Cebalges-Fonsecalges* lineage in both host lines, leads to the present distribution on the Cebidae. In all cases, colonization events are required to explain the presence of both of these lineages on the Callitrichidae, given the probable monophyly of that taxon.

Certain other relationships among the parasite taxa are not mirrored by host phylogenies. The sister group relationship between *Procebalges* and all other cebalgines is not reflected in any hypothesis of host relationships. The relationship of *Pithecia* to the *Alouatta-Lagothrix* lineage was suggested by both dental data (Rosenberger, 1977) and karyology (Chiarelli, 1980). Interestingly, the integumentary data (Perkins & Meyer, 1980) suggest that *Pithecia* retains the most plesiomorphic skin characteristics of any mite-bearing New World primate (*Aotus* was regarded as even more plesiomorphic). As cebalgine mites are skin inhabitants, it might be suggested that plesiomorphic skin retains plesiomorphic mites, a hypothesis which would require the ancestral possession of a *Procebalges* lineage on other cebids with subsequent extinction on all but *Pithecia* in order to retain a basically cospeciational history. Discovery of cebalgine mites on hosts related to *Pithecia* (*Chiropotes, Cacajao*) will provide a test of these hypotheses.

The utility of using the proposed mite phylogeny to infer a host phylogeny at higher levels than mentioned above depends upon the degree to which cospeciational patterns outweigh colonizations or extinctions in the hypothesized historical relationships. Among the entire psoroptid lineage parasitizing all Primates, cospeciational patterns appear to be supported in a large majority of cases (OConnor, 1984). However, noncospeciational patterns such as the distribution of the genus *Lemuralges* in the Makialginae, or the cebalgine parasites of the Callitrichidae, make the use of these mites as consistent indicators of host phylogeny at least somewhat suspect. Given the large disparity in phylogenetic hypotheses generated from subsets of the overall character matrix for the New World Primates, this parasite data deserves to be at least considered by future workers in this area.

Acknowledgments

I would like to thank Philip Hershkovitz, without whose cooperation and interest this study could not have been attempted. I also thank Robert Timm and Bruce Patterson, Field Museum of Natural History, for their hospitality and cooperation during the processing of host specimens and for their critical review of the manuscript. I thank J. H. S. Klompen, University of Michigan, for his comments on the manuscript. I thank Margaret van Bolt, Museum of Zoology, University of Michigan, for assistance with the illustrations.

Literature Cited

BABA, M., L. DARGA, AND M. GOODMAN. 1980. Biochemical evidence on the phylogeny of Anthropoidea, pp. 423–443. *In* Ciochon, R. L., and A. B. Chiarelli, eds., Evolutionary Biology of the New World Monkeys and Continental Drift. Plenum Press, New York.

BROOKS, D. R. 1979. Testing the context and extent of host-parasite coevolution. Systematic Zoology, **28:** 299–307.

———. 1981. Hennig's parasitological method: A proposed solution. Systematic Zoology, **30:** 229–249.

———. 1985. Historical ecology: A new approach to studying the evolution of ecological associations. Annals of the Missouri Botanical Garden, **72:** 660–681.

BROOKS, D. R., AND D. R. GLEN. 1982. Pinworms and Primates: A case study in coevolution. Proceedings of the Helminthological Society of Washington, **49:** 76–85.

CHIARELLI, A. B. 1980. The karyology of South American Primates and their relationship to African and Asian species, pp. 387–398. *In* Ciochon, R. L., and A. B. Chiarelli, eds., Evolutionary Biology of the New World Monkeys and Continental Drift. Plenum Press, New York.

CIOCHON, R. L., AND A. B. CHIARELLI, EDS. 1980. Evolutionary Biology of the New World Monkeys and Continental Drift. Plenum Press, New York, 528 pp.

FAIN, A. 1962. Diagnoses d'acariens nouveaux. Revue de Zoologie et de Botanique Africaines, **66:** 154–162.

———. 1963a. Nouveaux acariens psoriques parasites de marsupiaux et de singes sud-américains (Psoralgidae: Sarcoptiformes). Bulletin et Annales de Société Royale d'Entomologie de Belgique, **99:** 322–332.

———. 1963b. Un nouvel acarien producteur de gale chez un singe sud-américain. Bulletin et Annales de Société Royale d'Entomologie de Belgique, **99:** 467–470.

———. 1963c. Les acariens producteurs de gale chez les lémuriens et les singes avec une étude des Psoroptidae (Sarcoptiformes). Bulletin du Institut Royal des Sciences Naturelles de Belgique, **39**(32): 1–125.

———. 1966. Les acariens producteurs de gale chez les lémuriens et les singes II. Nouvelles observations avec description d'une espèce nouvelle. Acarologia, **8:** 95–114.

———. 1968. Notes sur trois acariens remarquables (Sarcoptiformes). Acarologia, **10:** 276–291.

FUTUYMA, D. J., AND M. SLATKIN, EDS. 1983. Coevolution. Sinauer Associates Inc., Sunderland, Mass., 555 pp.

HERSHKOVITZ, P. 1977. Living New World Monkeys (Platyrrhini), vol. 1. University of Chicago Press, Chicago, 1117 pp.

———. 1984. Taxonomy of squirrel monkeys genus *Saimiri* (Cebidae, Platyrrhini): A preliminary report with description of a hitherto unnamed form. American Journal of Primatology, **7:** 155–210.

LAVOIPIERRE, M. M. J. 1964a. A note on the family Psoralgidae (Acari: Sarcoptiformes) together with a description of two new genera and two new species parasitic on primates. Acarologia, **6:** 342–352.

———. 1964b. A new family of acarines belonging to the Suborder Sarcoptiformes parasitic in the hair follicles of Primates. Annals of the Natal Museum, **16:** 191–208.

NITECKI, M. H., ED. 1983. Coevolution. University of Chicago Press, Chicago, 392 pp.

OCONNOR, B. M. 1984. Co-evolutionary patterns between astigmatid mites and primates, pp. 186–195. *In* Griffiths, D. E., and C. E. Bowman, eds., Acarology VI, vol. 1. Ellis Horwood Ltd., Chichester, England.

PERKINS, E. M., AND W. C. MEYER. 1980. The phylogenetic significance of the skin of primates: Implications for the origin of New World monkeys, pp.

331–346. *In* Ciochon, R. L., and A. B. Chiarelli, eds., Evolutionary Biology of the New World Monkeys and Continental Drift. Plenum Press, New York.

Rosenberger, A. L. 1977. *Xenothrix* and ceboid phylogeny. Journal of Human Evolution, **6**: 461–481.

Sarich, V. M., and J. E. Cronin. 1980. South American mammal molecular systematics, evolutionary clocks, and continental drift, pp. 399–422. *In* Ciochon, R. L., and A. B. Chiarelli, eds., Evolutionary Biology of the New World Monkeys and Continental Drift. Plenum Press, New York.

Schwartz, J. H., I. Tattersall, and N. Eldredge. 1978. Phylogeny and classification of the Primates revisited. Yearbook of Physical Anthropology, **21**: 95–133.

Notes on Bolivian Mammals
2. Taxonomy and Distribution of Rice Rats of the Subgenus *Oligoryzomys*

Nancy Olds and Sydney Anderson

ABSTRACTS

There are at least three kinds of small, long-tailed rice rats of the subgenus *Oligoryzomys* (genus *Oryzomys*) in Bolivia. We use the names *Oryzomys microtis* (including *O. fornesi*), *O. chacoensis,* and *O. longicaudatus* for these taxa. The correctness of these names is less certain and resolution of the nomenclatorial questions awaits study of specimens from outside Bolivia. The ranges of *O. chacoensis* and *O. microtis* are at low elevations and probably overlap to some degree geographically. The range of *O. longicaudatus* is in the highlands. No sample from one locality includes specimens of more than one species; therefore, ecological and microgeographic differences at places where any two of the three species meet are unknown. Further study may reveal other and more cryptic species within Bolivia. No one measurement or other characteristic that we have studied will unequivocally distinguish all adult specimens of any one of the three kinds from Bolivia. Geographic variation probably occurs within as well as beyond Bolivia in at least two of the three species (the most uniform seems to be *O. chacoensis*), but more material is needed to describe such patterns. We refrain from using subspecies names in consideration of an ignorance of both geographic patterns of variation and the status of available names. The only name in the subgenus with a Bolivian type locality, *O. chaparensis,* is tentatively considered a synonym of *O. microtis.*

Existen en Bolivia por lo menos tres clases de pequeñas ratas arroceras de cola larga del subgénero *Oligoryzomys* (género *Oryzomys*). Nosotros usamos los nombres de *Oryzomys microtis* (incluyendo *O. fornesi*), *O. chacoensis,* y *O. longicaudatus* para éstos taxa. La exactitud de éstos nombres es menos cierta y la resolución de problemas de nomenclatura necesita estudio de especímenes de fuera de Bolivia. Los rangos de distribución de *O. chacoensis* y *O. microtis* están en bajas elevaciones y probablemente se superponen geográficamente en algún grado. Ninguna muestra de una localidad contiene especímenes de más de una especie. Diferencias ecológicas y microgeográficas son desconocidas en lugares donde cualquiera de las dos o tres espécies se encuentran. Ulterior estudio puede revelar otras y más crípticas espécies en el interior de Bolivia. Ninguna medida u otra característica que nosotros hemos estudiado separara claramente todos los especímenes adultos de Bolivia. Variación geográfica ocurre probablemente dentro así como fuera de Bolivia en al menos dos de las tres espécies (la más uniforme parece ser *O. chacoensis*), pero más material es necesario para describir tales patrones de variación. Nos abstenemos de usar nombres subespecíficos en consideración de la ignorancia de los patrones de variación y el estado de los nombres disponibles. El único nombre en el subgénero con una localidad típica en Bolivia, *O. chaparensis,* es tentativamente considerado como *O. microtis.*

From the Department of Mammalogy, American Museum of Natural History, New York, NY 10024.

Existem no mínimo três tipos de pequenos ratos-de-arroz, de cauda longa, pertencentes ao subgênero *Oligoryzomys* (gênero *Oryzomys*) na Bolívia. A estes taxa damos os nomes de *Oryzomys microtis* (incluindo *O. fornesi*), *O. chacoensis*, e *O. longicaudatus*. A precisão destes nomes é incerta, e resoluções da nomenclatura aguardam estudos de espécimes não bolivianos. *O. chacoensis* e *O. microtis* ocorrem em elevações baixas, e as duas espécies provàvelmente coocorrem em algumas áreas. *Oryzomys longicaudatus* ocorre em áreas montanhosas. Amostras de um só local não incluem mais do que uma espécie, e não se conhecem diferenças ecológicas ou microgeográficas em áreas onde duas ou mais espécies possam coocorrer. Com futuros estudos, novas espécies, mais ocultas, poderão ser encontradas na Bolívia. Nenhum único caráter que estudamos pode inequìvocamente separar espécimes adultos dos três tipos de *Oligoryzomys* na Bolívia. Variações geográficas provàvelmente ocorrem na Bolívia, bem como em outras áreas, mas mais material é necessário para poder documentar quaisquer padrões geográficos (*O. chacoensis* parece ser a espécie mais uniforme em aparência). Dada a falta de conhecimento, ambos dos padrões geográficos, e da disponibilidade dos nomes, não usamos nomes de subespécies. O único subgênero com localidade de tipo na Bolívia, *O. chaparensis*, é aqui tentativamente considerado como sinônimo de *O. microtis*.

Introduction

Bolivian landscapes range from less than 300 m to more than 4000 m, and habitats range from the humid lowland Amazonian tropical forests and subtropical savannahs to the high barren plains and snow-capped peaks of the Andean altiplano (fig. 1). Habitats may change abruptly, often within only a few kilometers. The mammals of Bolivia are also diverse, and provide excellent opportunities for ecological and taxonomic studies of broad scope. However, the animals are poorly known (Mares & Genoways, 1982); before satisfactory general conclusions can be reached, the Bolivian species need to be clearly delimited, both morphologically and geographically.

Mice of the genus *Oryzomys* occur throughout South America and are important members of small mammal communities (Myers & Carleton, 1981; Mares et al., 1981; Alho, 1982; O'Connell, 1982; Streilein, 1982a–c; Viega-Borgeaud, 1982). We examined critically one subgenus of *Oryzomys, Oligoryzomys,* in Bolivia to determine how many species are present and where they occur. For a general description of the subgenus, see Myers and Carleton (1981, pp. 9–12).

The subgenus *Oligoryzomys* needs revision. There are few published studies of the more than 45 named forms (Tate, 1932; Ellerman, 1941; Cabrera, 1961; Myers & Carleton, 1981; Honacki et al., 1982). Myers and Carleton (1981) studied *Oligoryzomys* from Paraguay, where they recognized three species: *Oryzomys nigripes, O. chacoensis,* and *O. fornesi.* They also clarified nomenclatorial questions relating to the name *Oryzomys nigripes.* Since Paraguay borders Bolivia, this recent study was used as our starting point. We used the same measurements and comparable analyses. We assumed that *O. chacoensis* and *O. fornesi* occurred also in southeastern Bolivia, near the Paraguayan border. If *O. nigripes* occurs in Bolivia, the most probable place for it is in eastern Santa Cruz, from which no specimens are now available.

At least six names have been used in the literature or in museum collections for Bolivian specimens of the subgenus *Oligoryzomys*: *Oryzomys longicaudatus* and *O. stolzmanni* (of the highlands); *O. chaparensis* (from the lowlands of Cochabamba); *O. nigripes* (used for all forms); *O. delicatus* (used for a few specimens at middle elevations); and *O. flavescens* (lowland). These names have been used with considerable uncertainty in the past (see summaries in Tate, 1932, and especially Myers & Carleton, 1981). One cause of this problem is that the original descriptions are vague or apply equally well to more than one species of *Oligoryzomys.*

The taxonomic confusion can be resolved by detailed study of adequate numbers of museum specimens, and the nomenclatorial confusion resolved by comparisons with type specimens. Renewed interest in South American mammals has resulted in more specimens, which will help in these tasks.

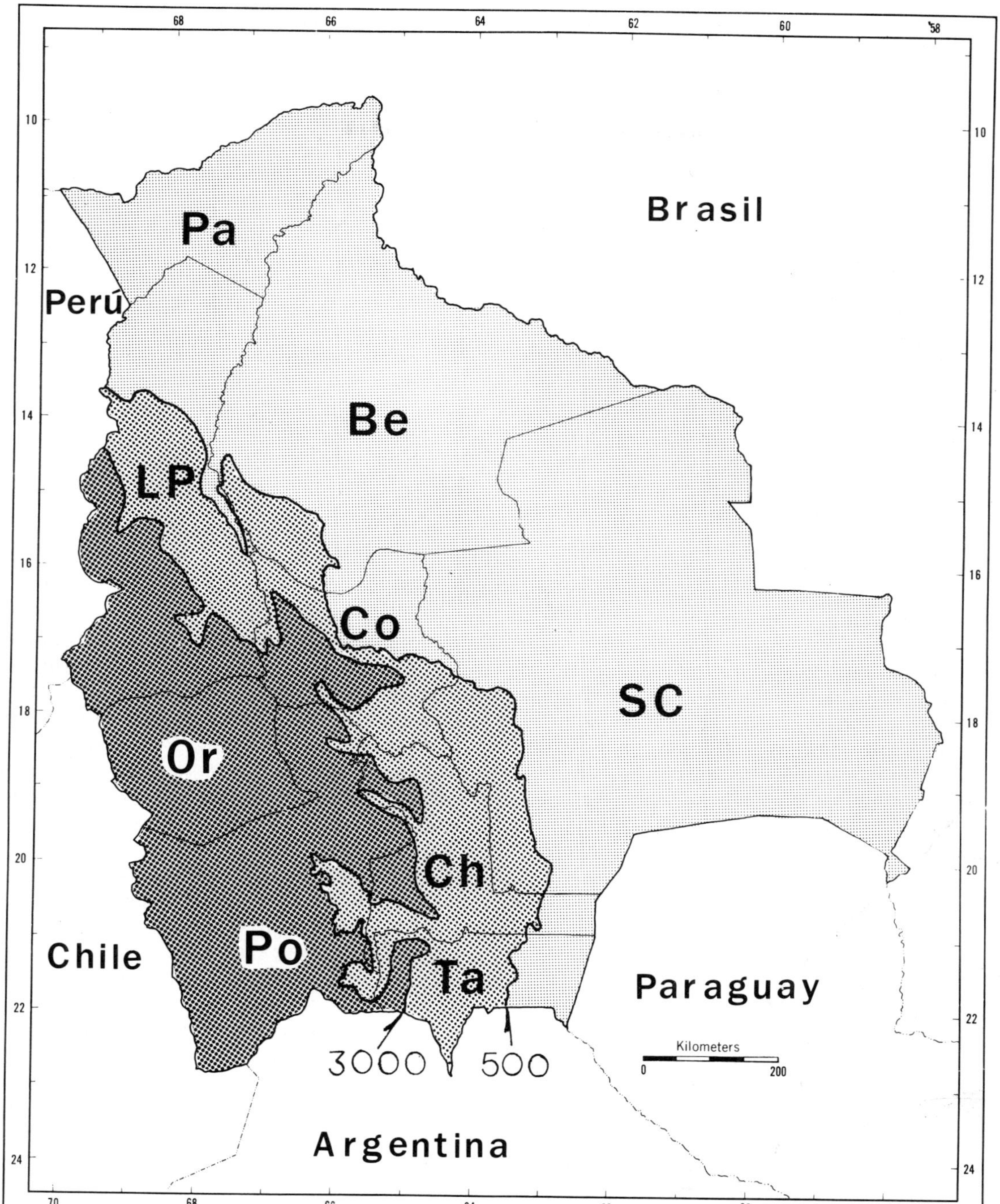

FIG. 1. Map of Bolivia showing the departments and the 500- and 3000-m contour lines. Be, Beni; Ch, Chuquisaca; Co, Cochabamba; LP, La Paz; Or, Oruro; Pa, Pando; Po, Potosí; SC, Santa Cruz; Ta, Tarija.

Methods

We examined specimens in the collections of the American Museum of Natural History (AMNH); Academy of Natural Sciences in Philadelphia (ANSP); California Academy of Sciences (CAL); Field Museum of Natural History (FMNH); Museum of Vertebrate Zoology, University of California (MVZ); Museum of Zoology, University of Michigan (UMMZ); and United States National Museum of Natural History (USNM).

Measurements were obtained as follows: external measurements are from the original labels or field notes, when available, or remeasured (fluid-preserved specimens only). The following cranial measurements were taken to the nearest 0.01 mm on a craniometer (see Anderson, 1968) following Myers and Carleton (1981) and Musser (1979):

ZN = depth of zygomatic notch
LR = length of rostrum from tip of nasals to posterior edge of zygomatic notch
GLS = greatest length of skull
ZB = zygomatic breadth
BB = breadth of braincase
BIC = breadth of interorbital constriction
LIF = length of incisive foramen
LPB = length of palatal bridge
LM1, LM2, LM3 = crown lengths of upper molars
WM1, WM2, WM3 = crown widths of upper molars
MM = greatest breadth across molars (labial edges)
LMX = crown length of upper toothrow
LNP = length of nasal projection
LD = length of diastema
LB = length of bulla
HB = height of bulla
TL = total length, including tail
T = tail length
HF = length of hind foot, including claw
E = length of ear, from notch

Specimens were sorted by age, using dental criteria as outlined here (following Myers & Carleton, 1981):

Age Class I: M^3 not erupted or newly erupted, M^2 unworn.
Age Class II: M^3 slightly to moderately worn, but not flat; M^2 slightly worn; enamel island formed by the isolation of the internal part of the mesoflexus of M^2.
Age Class III: M^3 flat or slightly concave; enamel island of above well isolated; M^1 and M^2 substantially worn.
Age Class IV: M^3 concave; enamel island obliterated; teeth well worn, but main cusps still discernible.
Age Class V: M^1 and M^2 flat or concave; folding pattern obliterated.

Individuals in age class I were examined, measured, and included in mapping geographic ranges, but were excluded from statistical treatment of character variation.

Specimens examined are listed by locality and museum catalogue number in the Appendix. Localities are plotted in Figure 2. Statistical analyses were done using the computer facilities of the City University of New York (CUNY) and programs from SAS (Statistical Analysis System) Institute, Inc. (1982).

Taxonomy

Our knowledge of geographic variation and the status of some of the names is sketchy at best, so we do not use subspecific names. The name *Oryzomys* (*Oligoryzomys*) *longicaudatus stolzmanni* was first used by Hershkovitz (1940, p. 81), by inference for Bolivian populations, but we are not certain that *O. stolzmanni* (Thomas, 1894; type locality Huambo, 3700 ft, department of Amazonas, Peru) and *O. longicaudatus* (type locality restricted to Valparaiso, Chile) are conspecific or that the Bolivian specimens are conspecific with either. The resolution of these problems awaits further study beyond Bolivia.

We noticed no difference between Bolivian specimens from Beni and Brazilian specimens from the vicinity of the type locality of *O. microtis* (Lower Rio Solimões, 50 mi above mouth), and thus the subspecific name *Oryzomys microtis microtis* might be applied to Bolivian specimens. However, we need to know more about geographic variation in the species (see Remarks under *O. microtis*). Some other names that may refer to conspecific populations and thus be relevant as possible subspecific epithets are as follows:

1. *Oryzomys destructor* from lowland eastern Peru has been assigned to *longicaudatus* but may prove to be conspecific with *microtis*; if so, *destructor* is the senior synonym. If *destructor* and *microtis* are conspecific, the northern Bolivian mice

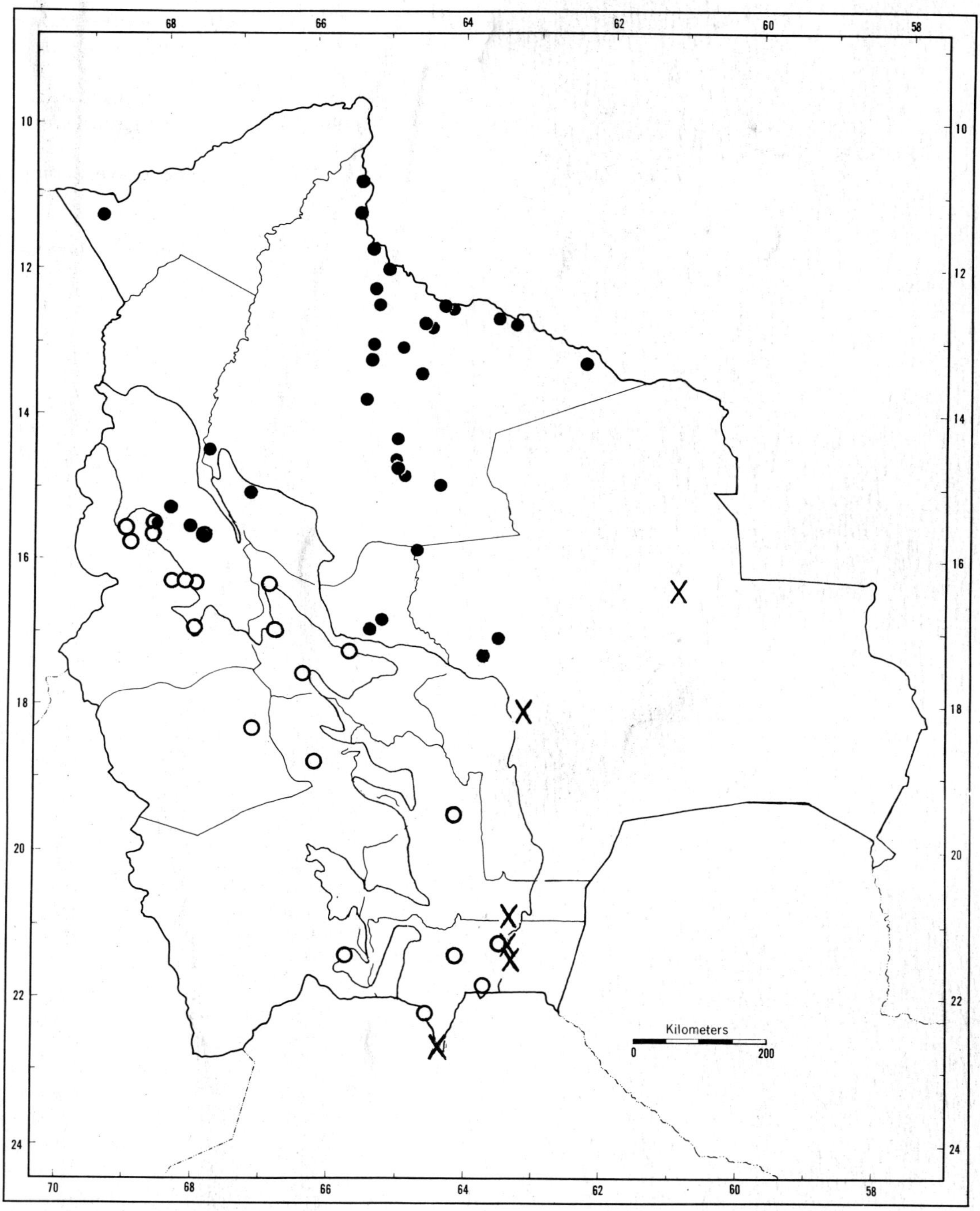

FIG. 2. Distribution of Bolivian *Oligoryzomys*: ● = *Oryzomys microtis*; ○ = *O. longicaudatus*; X = *O. chacoensis*.

might be known as *Oryzomys destructor destructor*, or perhaps *O. destructor microtis*.

2. *Oryzomys fornesi* from northern Argentina might also be expected to occur in southeastern Bolivia on geographic and ecological grounds, although no specimens are presently available. Myers and Carleton (1981) referred specimens from San Joaquín, Beni, to *O. fornesi*.

3. *Oryzomys mattogrossae* from Brazil just to the east of Bolivia is conspecific with *O. microtis* and has its type locality nearest the department of Beni, where most of the Bolivian specimens have been taken; thus, for geographic reasons, it must be considered in any future subspecific determinations. The uncertain specific status of *utiaritensis* from the same type locality as *mattogrossae* is noted below. If it is conspecific with *mattogrossae, utiaritensis* should be regarded as a synonym thereof at the subspecies level also.

We concur with Myers and Carleton (1981) in the recognition of two species groups of *Oligoryzomys*. These are the smaller-bodied, small-toothed *Oryzomys microtis, O. flavescens,* and *O. delicatus,* and the larger-bodied, larger-toothed *O. chacoensis, O. longicaudatus,* and *O. nigripes.*

In the Species Accounts section, synonymies cover only Bolivian records. Other relevant names are discussed elsewhere in the text.

Results of Statistical Analyses

Sex and Age Variation

We examined sex and age variation in *Oryzomys microtis, O. longicaudatus, O. chacoensis,* and *O. flavescens* (from Uruguay). Males are slightly larger in general; the average size differences, considering all characters for each species, are 2.7%, 1.6%, 2.5%, and 2.1%, respectively. We examined sexual dimorphism in all measurements for each species by one-way and two-way analyses of variance of sex and of sex and age for each species. Results of two-way analyses of variance of sex and age (using all specimens of tooth wear class II or greater) on specimens of *O. microtis* from the department of Beni, Bolivia, are presented in Table 1. Our results are roughly comparable to those of Myers and Carleton (1981). We did not separate sexes in further statistical analyses, although we watched for unbalanced sex ratios in samples when interpreting results.

Age variation is more difficult to assess, as no detailed study has been published. Mice of this subgenus probably continue to grow for most of their lives (see Myers & Carleton, 1981), although the rate slows with age. We performed least-squares regression analyses (General Linear Models procedures of SAS) on five variables: greatest length of skull (GLS), total length (TL), zygomatic breadth (ZB), length of the first upper molar (LM1), and length of the hind foot (HF), by species, to plot graphically the relationship between relative age and size. Based on study of these graphs, we decided to include age classes II–V in further analyses, as did Myers and Carleton (1981). Most specimens were juveniles (age class I) or young adults (age classes II–III). Few specimens of age classes IV–V were present among the species we examined. See Tables 2 and 3 for the mean adult ages of the specimens studied.

Principal Components

We analyzed principal components (with the Princomp procedure of SAS on a correlation matrix) using measurements taken on individuals of all species (the approximate numbers of specimens are in tables 2 and 3, some specimens were excluded because of missing measurements). When plotted, the first principal component tends to separate the smaller species (*O. microtis* and *O. flavescens*) from the larger species (*O. chacoensis* and *O. longicaudatus*). The percentage of the total variance accounted for by the first three components is 68.7%.

Discriminant Analysis

We performed several different discriminant analyses (using the SAS programs Discrim, Stepdisc, and Candisc). A stepwise discriminant analysis chose the following characters (in order of selection): tail length, breadth of interorbital constriction, diastema length, length of incisive foramina, rostral length, bullar length, length of hind foot, molar breadth, length of nasal projection, length of zygomatic notch, total length, zygomatic breadth, length of palatal bridge, and length of maxillary toothrow.

Using the SAS program Discrim, we were able to test the posterior probability of group membership. In all analyses, we used as "known" groups samples of the species *Oryzomys microtis* from Beni, Bolivia; *O. flavescens* from Uruguay; *O. longicaudatus* from the department of La Paz, Bolivia; and *O. chacoensis* from Paraguay. Plots of the individual mice on the first two canonical correlates showed little overlap among the four species. We assumed that the following "unknowns" belonged to one of the four "known" groups. We submitted as "unknowns" the holotypes of *O. mi-*

crotis, O. mattogrossae, and O. chaparensis; the paratype of O. chaparensis; and two specimens from Beni assigned by Myers and Carleton (1981) to *O. chacoensis* (see Remarks under *O. microtis*). The holotypes of *O. microtis* and *O. mattogrossae* were assigned to the *O. microtis* species sample, with a posterior probability of more than 95%. The holotype of *O. chaparensis* was also assigned to *O. microtis* ($P = 0.987$). The paratype of *O. chaparensis* was assigned to *O. microtis* ($P = 0.985$). Several other specimens that we wanted to test, including the type of *O. delicatus*, had missing values and could not be used in the analysis. We recognize that the validity of the taxonomic conclusions based on these analyses depends on the correctness of the initial assumption, and that it needs further testing. Not all measurements were available for all skulls, but in each comparison the largest possible subset was used.

TABLE 1. Nongeographic variation in *Oryzomys microtis*. Results of analysis of variance of Bolivian *O. microtis*.

Character	Error d.f.	F (sex)	F (age)	F (interaction)
ZN	98	3.75	0.44	0.52
LR	89	26.31***	1.87	1.35
GLS	81	22.69***	1.77	0.75
ZB	86	13.65***	1.19	1.33
BB	92	6.57*	0.41	2.55*
BIC	96	0.74	1.79	1.22
LIF	95	4.64*	1.71	0.51
LPB	93	4.32*	0.81	1.01
LM1	98	0.08	2.38*	0.76
LM2	99	1.38	5.37***	5.57***
LM3	95	4.23*	1.93*	1.82
WM1	98	0.51	1.38	1.68
WM2	99	0.82	1.43	1.06
WM3	95	0.17	1.90*	1.24
MM	95	3.07	3.08**	1.81
LMX	94	1.53	2.38*	1.36
LNP	87	15.05***	1.26	1.73
LD	96	10.87**	1.23	1.58
LB	92	5.45*	0.48	0.71
HB	91	1.77	0.41	0.83
TL	91	5.36*	1.38	1.21
T	91	0.02	1.65	2.34*
HF	92	6.41*	0.95	0.67
E	48	0.02	1.53	0.84

* $P \leq 0.05$; ** $P \leq 0.01$; *** $P \leq 0.001$.

Species Accounts

Subgenus **Oligoryzomys** Bangs, 1900

DIAGNOSIS—Within the genus *Oryzomys*, the subgenus *Oligoryzomys* is distinguished by small size and delicate structure throughout, tail relatively long, hind foot long and slender; skull small, delicate, interorbital region narrow, outer edges of frontals squarish but unbeaded, braincase smooth and unridged, zygomatic plate narrow and with slight forward projection, molar teeth and incisors small and delicate but with cusp pattern like other *Oryzomys* (description adapted from Bangs, 1900). *Oryzomys* of the subgenus *Microryzomys* are also small and delicate, but they differ from *Oligoryzomys* in having a more slender rostrum, shallower zygomatic notch, shorter and more rounded braincase, sphenofrontal foramen and squamosoalisphenoid groove present, and karyotype with a low FN/2*n* ratio (Myers & Carleton, 1981, p. 12).

Oryzomys chacoensis Myers and Carleton, 1981

Oryzomys chacoensis Myers and Carleton, 1981, p. 19 (type locality "419 km by road NW Villa Hayes [alongside the Trans Chaco Highway], Dept. Boquerón, Paraguay").

DIAGNOSIS—*Oryzomys chacoensis* Myers and Carleton (1981, p. 20) was diagnosed as "A medium-sized species of the subgenus *Oligoryzomys* unique in its whitish underside with hair white to the base on the chin and throat, relatively long ears having hairs on inner surface with unusually short or absent dark basal bands, small but distinctive tufts of orangish hairs anterior to the ears, and karyotype with 2n = 58, FN = 74."

DISTRIBUTION IN BOLIVIA—We examined specimens (see Appendix) from the departments of Cochabamba, Santa Cruz, and Tarija (fig. 2). Myers and Carleton (1981) reported two specimens from Beni. These specimens have been restudied and are here reassigned to *Oryzomys microtis* (see Remarks under that species). Habitat of *O. chacoensis* is grassland and thornscrub. The highest known elevation is 640 m (Río Lipeo).

GENERAL DESCRIPTION AND COMPARISONS—Cranial and external measurements are listed in Table 2. The dorsal pelage is rufous and heavily lined with black hairs. The venter is white and sharply distinguished from sides, which are clearer than dorsum. Frequently a thin orange line separates the sides from the belly. The cheeks are slightly paler than the rest of the face. The tail is dark gray, weakly bicolored, and long relative to the body. The hind feet are whitish above. Juve-

TABLE 2. Mean, standard deviation, and range (in mm) for *Oryzomys chacoensis* and *O. longicaudatus*.

Character	***O. chacoensis*** Bolivia (N = 15)	Paraguay (N = 16)	Argentina (N = 5)	***O. longicaudatus*** Bolivia (N = 34)
ZN	1.15 ± 0.28 (0.88–1.59)	1.30 ± 0.20 (0.90–1.61)	1.15 ± 0.12 (0.94–1.24)	1.02 ± 0.20 (0.57–1.36)
LR	7.27 ± 0.45 (6.62–7.94)	7.39 ± 0.57 (6.57–8.86)	7.67 ± 0.34 (7.25–8.07)	7.14 ± 0.72 (4.49–8.15)
GLS	24.90 ± 0.82 (23.80–26.69)	24.52 ± 1.09 (22.52–27.12)	26.12 ± 0.32 (25.79–26.43)	25.01 ± 1.39 (21.06–27.29)
ZB	13.10 ± 0.59 (11.94–13.98)	12.76 ± 0.62 (11.91–14.27)	13.64 ± 0.09 (13.54–13.71)	13.17 ± 0.65 (11.55–13.98)
BB	11.33 ± 0.41 (10.42–11.87)	11.16–0.34 (10.59–11.59)	11.39 ± 0.06 (11.35–11.45)	11.25 ± 0.37 (10.58–11.96)
BIC	3.89 ± 0.20 (3.52–4.27)	3.76 ± 0.10 (3.64–3.95)	3.86 ± 0.13 (3.76–4.04)	3.63 ± 0.22 (3.35–4.17)
LIF	4.62 ± 0.26 (4.25–5.20)	4.71 ± 0.39 (4.17–5.56)	5.07 ± 0.29 (4.64–5.31)	4.68 ± 0.27 (4.23–5.26)
LPB	4.14 ± 0.22 (3.65–4.54)	3.98 ± 0.26 (3.50–4.65)	4.22 ± 0.31 (3.75–4.52)	3.97 ± 0.21 (3.49–4.41)
LM1	1.68 ± 0.10 (1.47–1.84)	1.63 ± 0.07 (1.54–1.79)	1.74 ± 0.5 (1.70–1.83)	1.59 ± 0.09 (1.42–1.81)
LM2	1.09 ± 0.06 (0.96–1.18)	1.10 ± 0.07 (1.00–1.28)	1.14 ± 0.07 (1.08–1.26)	1.05 ± 0.04 (0.97–1.15)
LM3	0.84 ± 0.05 (0.74–0.92)	0.80 ± 0.04 (0.73–0.88)	0.84 ± 0.04 (0.78–0.88)	0.81 ± 0.05 (0.67–0.92)
WM1	1.07 ± 0.07 (0.93–1.24)	1.03 ± 0.03 (0.96–1.08)	1.11 ± 0.05 (1.05–1.17)	1.04 ± 0.06 (0.91–1.15)
WM2	1.02 ± 0.07 (0.86–1.14)	1.00 ± 0.04 (0.93–1.06)	1.04 ± 0.06 (0.97–1.11)	0.99 ± 0.06 (0.87–1.09)
WM3	0.87 ± 0.05 (0.78–0.97)	0.85 ± 0.03 (0.78–0.92)	0.90 ± 0.05 (0.84–0.94)	0.83 ± 0.06 (0.67–0.95)
MM	4.66 ± 0.20 (4.32–4.99)	4.57 ± 0.17 (4.34–5.09)	4.70 ± 0.13 (4.58–4.87)	4.61 ± 0.27 (4.03–5.08)
LMX	3.60 ± 0.12 (3.29–3.78)	3.50 ± 0.13 (3.23–3.68)	3.71 ± 0.15 (3.61–3.98)	3.46 ± 0.14 (3.20–3.77)
LNP	1.43 ± 0.24 (1.04–1.77)	1.59 ± 0.17 (1.35–2.07)	1.49 ± 0.17 (1.29–1.64)	1.40 ± 0.25 (0.99–2.04)
LD	5.66 ± 0.34 (5.17–6.33)	5.63 ± 0.44 (5.02–6.44)	5.89 ± 0.24 (5.63–6.18)	6.04 ± 0.47 (4.96–7.07)
LB	3.54 ± 0.14 (3.28–3.81)	3.60 ± 0.14 (3.28–3.82)	3.75 ± 0.03 (3.72–3.78)	3.47 ± 0.20 (3.07–3.84)
HB	2.75 ± 0.18 (2.43–3.04)	2.73 ± 0.15 (2.40–2.96)	2.72 ± 0.06 (2.66–2.76)	2.76 ± 0.32 (2.23–3.32)
TL	227.93 ± 12.45 (204–252)	219.50 ± 12.29 (194–241)	229.00 ± 8.34 (215–237)	217.68 ± 16.88 (172–252)
T	134.27 ± 10.26 (112–150)	126.06 ± 8.46 (112–139)	138.00 ± 5.70 (130–145)	124.15 ± 12.18 (87–145)
HF	25.33 ± 1.33 (23–28)	24.38 ± 1.54 (23–29)	23.40 ± 2.30 (21–26)	25.13 ± 1.20 (22–28)
E	16.70 ± 1.33 (14–18)	15.07 ± 1.69 (12–17)	18.40 ± 1.14 (17–20)	15.03 ± 1.72 (12–18)
AGE	2.65 ± 0.82 (2–4)	2.31 ± 0.68 (2–4)	2.55 ± 0.87 (2–4)	3.04 ± 0.88 (2–5)

niles are grayer, as are all young *Oligoryzomys* in Bolivia, but this is especially noticeable on the venter, which has mixed gray and white hairs. Incisive foramina extend posteriorly to the anterior edge of M^1 or slightly beyond. Alisphenoid strut (a strut of the alisphenoid bone that covers the lateral part of the alisphenoid canal, see Musser, 1982, p. 29) is generally absent (table 4). The sides of the interorbital constriction are divergent posteriorly (less parallel-sided than in *O. longicaudatus*) (fig. 3). The following quotes are from Myers and Carleton (1981, first and second quotes, p. 21, third quote, p. 24):

> *Oryzomys chacoensis* differs from *O. chaparensis* Osgood (1916) primarily in color pattern: the type of *chaparensis* is much darker and less hispid dorsally, yellowish ventrally, lacks the orange tufts anterior to the ears, and has a grayish throat. The distal portions of the nasals of the holotype flare laterally to an extent not seen in *chacoensis.*
>
> *Oryzomys chacoensis* can be distinguished from *fornesi,* with which it occurs sympatrically, by its larger size (maxillary toothrow usually > 3.3mm, ears usually > 15mm), characteristic karyotype, lack of preputial glands, and in most specimens by its lack of buff on the belly.
>
> . . . the hind feet of *chacoensis* are relatively short compared to those of the more terrestrial *fornesi.*

In reference to the last point, however, our calculations of length of hind foot relative to length of head and body (using data from tables 2 and 3) are about 26% for both *O. chacoensis* and *O. microtis* (including *fornesi*). Comparison of specimens also reveals no noticeable difference.

REMARKS—This species is clearly distinct from *O. microtis* (here including *O. fornesi*). In most morphometric characters, *O. chacoensis* grossly resembles *O. longicaudatus* from Bolivia and *O. nigripes* from Brazil.

Our specimens from Santa Cruz show some variation in coloration: two specimens (AMNH 247772–247773) are pale. The specimens from Tarija resemble the Paraguayan samples more closely than do the specimens from Santa Cruz.

A series of 13 specimens from Río Lipeo includes two with darker pelage that resemble *O. longicaudatus.* We measured these mice and examined the two darker ones. When skulls are compared, the two darker mice are among the older and larger individuals in the series. They seem to have relatively broader braincases than most, but there is no character shared by these two that is not also seen in one or more of the others. In a discriminant analysis (SAS program Discrim) the posterior probability of membership (in the four reference species) allies these two mice with *O. chacoensis* (ANSP 18187, $P = 0.997$, and ANSP 18188, $P = 0.954$ with *chacoensis*). We have assigned these mice to *O. chacoensis.* Since Río Lipeo is in the area where these two species meet, further study there should reveal whether sympatry exists and if species differences remain distinct.

Specimens from Argentina (listed in the Appendix) are similar in coloration to both Paraguayan and Bolivian specimens, but are woolier. This sample extends the known range of *O. chacoensis* into the department of Jujuy, northwestern Argentina.

Oryzomys longicaudatus (Bennett, 1832)

Mus longicaudatus Bennett, 1832, p. 2 (type locality "In trees in Chile," restricted to Valparaiso by Cabrera, 1961, p. 391).

Oryzomys longicaudatus: Thomas, 1898, p. 3 (Aguairenda Missión, San Francisco, perhaps not *O. longicaudatus*); Thomas, 1926, p. 194 (Tupiza).

Oryzomys Stolzmanni: Thomas, 1902, p. 130 (Charuplaya, Choro); Neveu-Lemaire and Grandidier, 1911, p. 9 (Charuplaya, Choro).

Oryzomys sp., *flavescens* group: Thomas, 1925, p. 578 (Carapari, perhaps not *O. longicaudatus*).

Oryzomys stolzmanni stolzmanni: Sanborn, 1950, p. 2 (Río Aceramarca, Cocapunco, Ñequejahuira, Okara, Pitiguaya, Pongo).

DIAGNOSIS—Bolivian specimens referred to this species differ from other Bolivian *Oligoryzomys* in that they are larger (especially in the size of the teeth) than *O. microtis* and lack the buffy wash on the venter; are grayer-bellied, darker, and less grizzled dorsally than *O. chacoensis*; and occur at generally higher elevations (at least in northern Bolivia).

DISTRIBUTION IN BOLIVIA—*O. longicaudatus* is found in the valleys and mountains of the Andes from at least middle elevations (1200 m, Entre Ríos) up to 3720 m (Poopó). Specimens from near Camiri and Cuyambuyo in southern Bolivia are from elevations of 780 to 1000 m. Specimens (see Appendix) have been reported or examined from

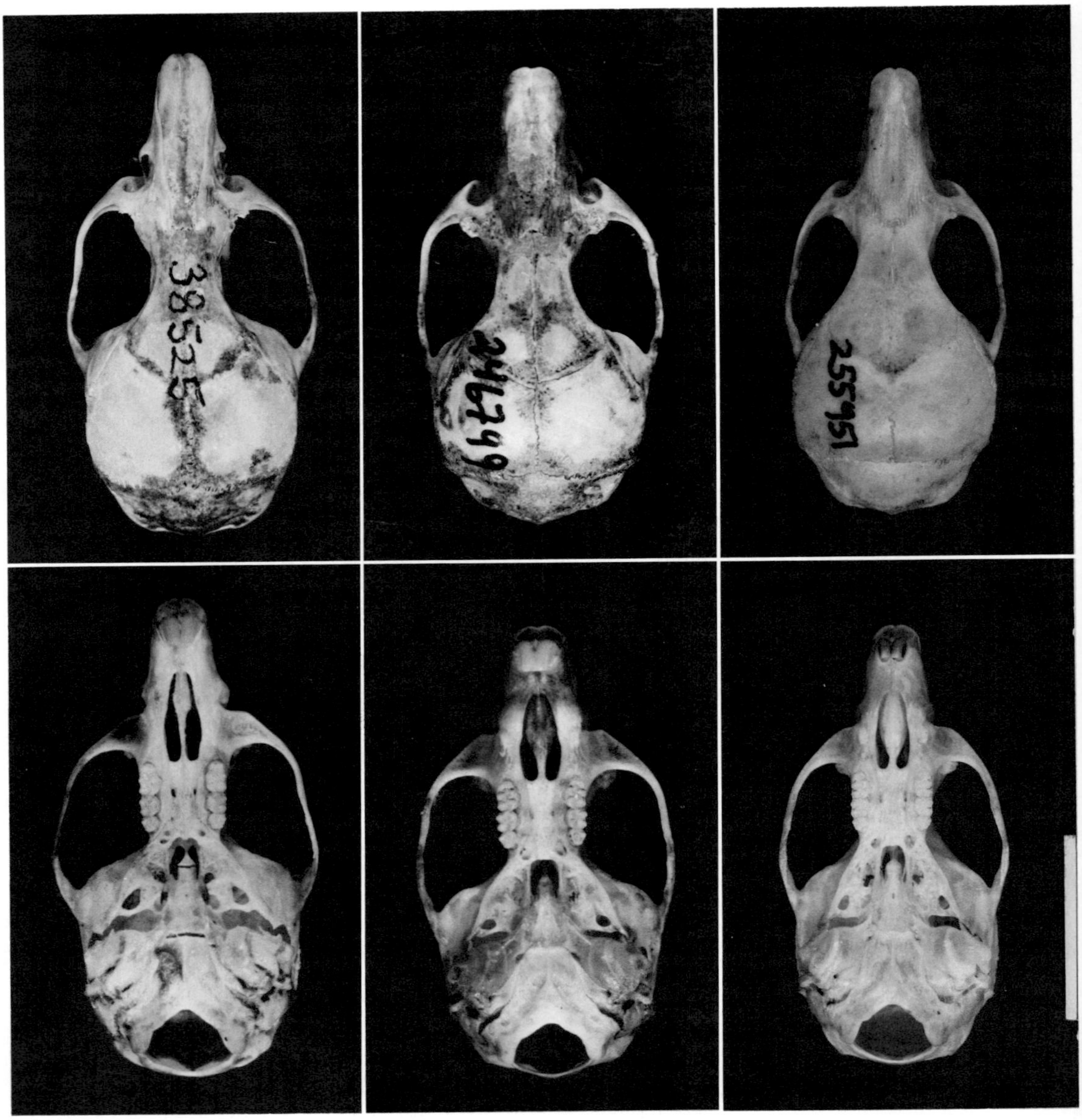

FIG. 3. Dorsal and ventral views of skull of **(left),** *Oryzomys longicaudatus stolzmanni*; **(middle),** *O. chacoensis,* and **(right),** *O. microtis microtis.* Specimens in the American Museum of Natural History. Scale at lower right represents 10 mm.

the departments of Chuquisaca, Cochabamba, La Paz, Oruro, Potosí, Santa Cruz, and Tarija (see map, fig. 2).

GENERAL DESCRIPTION AND COMPARISONS—Cranial and external measurements are presented in Table 2. The belly is gray, rarely washed with buff; upper parts are a dark brown lined with black hairs, and often the sides are more rufous. A thin orange lateral line may be present. Nose and face are dark with paler cheeks. Often there are pale spots just behind the ears. Hind feet are pale above. The tail is brown above, weakly to strongly bicolored, and relatively long compared to that of any other Bolivian *Oligoryzomys.* Incisive foramina generally extend to the anterior edge of M^1 or slightly behind. Alisphenoid strut is generally absent (table 4). The interorbital area is slightly more constricted on the average than that of *O. cha-*

coensis, and has relatively parallel sides (see fig. 3).

REMARKS—*Oryzomys longicaudatus* may be a composite of more than one species. *Oryzomys destructor* and *O. stolzmanni* were considered subspecies of *O. longicaudatus* by Cabrera (1961) and full species by Soukup (1961). Gardner and Patton (1976) found four karyotypic variants in specimens they assigned to *O. longicaudatus,* and commented that these may represent four separate species. One of their karyotypic variants, no. 2, is represented by 17 specimens from Balta, department of Loreto (now Acayali), Peru, at an elevation of about 300 m. Myers and Carleton (1981, p. 26) mentioned that the karyotype of these specimens as reported by Gardner and Patton (1976) agreed better with results they obtained for *O. fornesi* (= *O. microtis*) than for either *O. chacoensis* or *O. longicaudatus.* We have examined these specimens and judge that they are clearly different from *O. longicaudatus* and belong with either *O. microtis* or *O. destructor* (Balta is near the type locality of *O. destructor*). The inclusion of karyotype no. 3 under *O. longicaudatus,* represented by one female from Venezuela, also needs verification.

We have five specimens from Caracato, department of La Paz (AMNH 248977–248981), that are quite different from other Bolivian *Oryzomys longicaudatus.* The pelage closely resembles that of *O. nigripes* from Paraguay in color and texture; it is paler and more obviously lined with black hairs dorsally than the pelage of typical *O. longicaudatus.* They are also slightly larger bodied, but four of the five are old individuals (age classes IV–V) and this may account for the larger size. Because of the small sample size and the single locality, we chose not to distinguish this population taxonomically.

It is possible that Andean populations may occupy areas that are rather widely scattered and physically isolated in different valleys or side canyons. If so, there may be considerable geographic variation. More specimens are needed to detect and describe such patterns.

The specimens from near Cuyambuyo are at comparatively low elevations (980 to 1000 m) for *O. longicaudatus,* although the species occurs at progressively lower elevations farther south (Osgood, 1916). The two localities near Cuyambuyo are only about 60 km from Río Lipeo, where *O. chacoensis* occurs. The SAS Discrim program analysis allied the largest adult specimen from near Cuyambuyo (UMMZ 155891) with *O. longicaudatus* ($P = 0.956$). There are problems in distinguishing *O. longicaudatus* and *O. chacoensis* solely by cranial morphology in all areas.

Oryzomys microtis Allen, 1916

Oryzomys (*Oligoryzomys*) *microtis* Allen, 1916, p. 525 (type locality Lower Rio Solimões, 50 mi above mouth [80 km from its confluence with Rio Negro, Amazonas, Brazil]).

Oryzomys chaparensis Osgood, 1916, p. 205 (holotype and paratype, type locality Todos Santos, on Río Chaparé, Department of Cochabamba, Bolivia); Gyldenstolpe, 1932, p. 25 (Todos Santos); Myers and Carleton, 1981, p. 38 (Todos Santos); all in reference to the same specimens.

Oryzomys fornesi Massoia, 1973, p. 22 (type locality Naineck, Dept. Río Pilcomayo, province of Formosa, Argentina); as used by Myers and Carleton, 1981, p. 25 for five specimens from San Joaquín, Beni, Bolivia.

DIAGNOSIS—Allen described *Oryzomys microtis* as being readily distinguished from other *Oligoryzomys* by its pale coloration, relatively small ears, and tail less than half total length. However, the tail of the holotype is more than half of the total length, a discrepancy noted by Goodwin (1953). Massoia (1973) diagnosed *O. fornesi,* here considered a junior synonym of *O. microtis,* as the smallest species of the subgenus in Argentina, having the ears covered with short ochraceous hairs, the pterygoids short (shorter than the molar series), the interorbital constriction narrow, and the incisive foramina not extending to the molar series. In Bolivia, it can be distinguished by its small body size, toothrow generally less than 3.3, gray-buffy color of the belly, and relatively short tail in proportion to body.

DISTRIBUTION IN BOLIVIA—Specimens (see Appendix) from the lowlands of Bolivia, in the departments of Beni, Cochabamba, Pando, La Paz, and Santa Cruz have been examined. Habitat includes marshes and wet forests up to an elevation of 1800 m (Guanay, see fig. 2).

GENERAL DESCRIPTION AND COMPARISONS—Smallest of the Bolivian *Oligoryzomys,* with small teeth, toothrow generally less than 3.3 (averaging 3.10). Massoia (1973) gave general measurements useful in distinguishing this species from other Argentine species of *Oryzomys* (*Oligoryzomys*) as follows: length of hind foot generally less than 24, length of ear generally less than 13, breadth of braincase less than 10.8, and length of incisive foramina generally less than 4.5. All of his mea-

TABLE 3. Mean, standard deviation, and range (in mm) for *Oryzomys microtis* and *O. flavescens*.

Character	*O. microtis* Bolivia (N = 67)	*O. microtis* Brazil (N = 20)	*O. microtis* Peru (N = 26)	*O. flavescens* Uruguay (N = 20)
ZN	1.01 ± 0.17 (0.53–1.46)	0.98 ± 0.18 (0.78–1.48)	0.84 ± 0.13 (0.61–1.18)	1.07 ± 0.15 (0.79–1.32)
LR	6.82 ± 0.51 (5.17–8.16)	6.82 ± 0.42 (6.16–7.46)	6.60 ± 0.41 (5.96–7.36)	6.75 ± 0.43 (5.89–7.39)
GLS	23.54 ± 1.06 (21.24–25.78)	23.79 ± 0.88 (22.47–25.04)	23.27 ± 0.82 (21.75–24.53)	23.41 ± 0.94 (21.66–24.81)
ZB	12.28 ± 0.51 (11.42–13.73)	12.40 ± 0.44 (11.59–13.32)	11.86 ± 0.53 (11.01–12.93)	12.05 ± 0.49 (11.25–12.76)
BB	10.78 ± 0.36 (9.38–11.78)	10.71 ± 0.23 (10.35–11.06)	10.47 ± 0.32 (9.49–10.87)	10.70 ± 0.26 (10.29–11.22)
BIC	3.76 ± 0.17 (3.38–4.19)	3.72 ± 0.14 (3.44–3.93)	3.63 ± 0.11 (3.45–3.94)	3.35 ± 0.17 (2.95–3.77)
LIF	3.96 ± 0.27 (3.38–4.53)	4.01 ± 0.22 (3.65–4.37)	4.00 ± 0.18 (3.73–4.33)	4.59 ± 0.35 (3.54–5.20)
LPB	4.04 ± 0.32 (3.14–4.76)	4.06 ± 0.22 (3.64–4.38)	3.98 ± 0.26 (3.42–4.51)	3.77 ± 0.19 (3.51–4.06)
LM1	1.46 ± 0.10 (1.22–1.64)	1.43 ± 0.07 (1.32–1.56)	1.51 ± 0.06 (1.39–1.66)	1.47 ± 0.06 (1.37–1.60)
LM2	0.95 ± 0.06 (0.81–1.09)	0.95 ± 0.05 (0.84–1.06)	0.93 ± 0.08 (0.67–1.04)	1.01 ± 0.06 (0.83–1.09)
LM3	0.70 ± 0.07 (0.54–0.85)	0.68 ± 0.05 (0.59–0.78)	0.66 ± 0.05 (0.55–0.77)	0.73 ± 0.04 (0.66–0.80)
WM1	0.94 ± 0.06 (0.78–1.08)	0.93 ± 0.06 (0.80–1.04)	0.99 ± 0.04 (0.91–1.10)	0.97 ± 0.04 (0.93–1.07)
WM2	0.88 ± 0.06 (0.70–1.01)	0.89 ± 0.06 (0.78–0.99)	0.93 ± 0.04 (0.82–1.00)	0.93 ± 0.05 (0.86–1.03)
WM3	0.73 ± 0.06 (0.55–0.85)	0.74 ± 0.06 (0.58–0.85)	0.80 ± 0.05 (0.71–0.93)	0.78 ± 0.04 (0.70–0.85)
MM	4.33 ± 0.21 (3.83–4.93)	4.39 ± 0.23 (3.99–4.81)	4.33 ± 0.18 (4.07–4.70)	4.26 ± 0.15 (3.98–4.50)
LMX	3.10 ± 0.17 (2.64–3.43)	3.07 ± 0.16 (2.75–3.34)	3.14 ± 0.13 (2.93–3.39)	3.22 ± 0.12 (2.96–3.47)
LNP	1.32 ± 0.22 (0.91–1.74)	1.25 ± 0.32 (0.75–1.70)	1.20 ± 0.20 (0.81–1.62)	1.33 ± 0.14 (1.16–1.68)
LD	5.75 ± 0.41 (4.82–6.62)	5.78 ± 0.36 (5.13–6.51)	5.32 ± 0.40 (4.49–6.01)	5.35 ± 0.35 (4.67–5.90)
LB	3.29 ± 0.14 (2.94–3.57)	3.18 ± 0.19 (2.90–3.63)	3.01 ± 0.12 (2.75–3.27)	3.30 ± 0.15 (2.91–3.53)
HB	2.46 ± 0.20 (1.96–2.87)	2.10 ± 0.23 (1.72–2.45)	2.22 ± 0.12 (1.96–2.38)	2.64 ± 0.23 (2.02–2.94)
TL	185.88 ± 11.12 (165–214)	185.00 ± 10.60 (164–202)	179.77 ± 9.16 (160–196)	198.56 ± 17.66 (160–223)
T	101.00 ± 6.67 (87–116)	95.44 ± 7.88 (75–105)	101.69 ± 8.13 (77–115)	110.37 ± 11.05 (85–127)
HF	22.81 ± 1.21 (19–25)	21.68 ± 1.34 (20–25)	21.37 ± 1.60 (18–25)	25.21 ± 1.85 (20–28)
E	14.49 ± 1.04 (12–16)	... ...	... ...	13.97 ± 1.05 (12–16)
AGE	2.57 ± 0.73 (2–5)	2.35 ± 0.59 (2–4)	2.97 ± 0.88 (2–5)	2.18 ± 0.35 (2–3)

surements fall within the extremes of our values for *O. microtis,* except for measurements of the toothrow (LMX), in which his are on the upper end of our range. Values of cranial and external measurements are listed in Table 3. The venter is white mixed with gray to gray-buffy and clear buffy. The dorsum is rufous brown, lined with black hairs, and has paler sides. Often there is no clear demarcation between sides and belly. There are no white spots behind ears. Juveniles are similar to adults, but show a grayer belly. The tail is dark above and weakly bicolored. Occasionally orange-tipped hairs lie anterior to the ear. Hind feet are white above. Incisive foramina generally extend posteriorly to the anterior edge of M^1 and not beyond. In comparison with adults of *O. chacoensis* and *O. longicaudatus, O. microtis* has a shorter skull, longer braincase, shorter incisive foramina, a relatively broader interorbital constriction with divergent sides, and smaller, narrower teeth; also, an alisphenoid strut is more often present (table 4).

REMARKS—This species is similar to *Oryzomys flavescens* of Uruguay and Argentina (see Myers & Carleton, 1981). Specimens of both species were reported from the same locality, Capitan Solari, in the province of Chaco, northern Argentina, by Contreras and Berry (1983). Measurements of *O. flavescens* from Uruguay are listed in Table 3 (see also Langguth, 1963). In comparison with *O. microtis,* specimens of *O. flavescens* have slightly larger teeth, larger bullae, longer incisive foramina, and are generally larger. *Oryzomys flavescens* needs to be more clearly defined. Massoia and Fornes (1967) once synonymized *O. flavescens* with *O. nigripes,* but according to the later analyses of Massoia (1973) and Myers and Carleton (1981) these taxa are not conspecific. Uruguayan specimens of *O. flavescens* are clearly different from *O. nigripes* from eastern Paraguay. Future studies may reveal additional relationships, including the possibility that *O. microtis* and *O. fulvescens* of Central America and northern South America are conspecific (Handley, 1976, referred Venezuelan specimens to *O. fulvescens*).

The name *Oryzomys microtis fornesi* might be used to reflect the conspecific status of *fornesi* and *microtis.* This subspecific name would apply to at least the five specimens identified by Massoia (1973), pending more detailed studies that would test whether geographic differences warrant continued subspecific recognition.

Until the relationships of the Andean "*longicaudatus*-group" are clarified, it is also possible

TABLE 4. Presence or absence of alisphenoid strut in five species of *Oryzomys* (*Oligoryzomys*) expressed as a percentage; number of specimens in parentheses.

Species	Alisphenoid strut: Present	Alisphenoid strut: Absent
O. chacoensis	2.4 (1)	97.6 (41)
O. longicaudatus	5.6 (2)	94.4 (34)
O. microtis	35.8 (54)	64.2 (97)
O. flavescens	3.2 (1)	96.8 (30)
O. nigripes	8.3 (3)	91.7 (33)

that one or more of the earlier names currently assigned to this group actually belong with *Oryzomys microtis,* and have priority. For example: *O. stolzmanni* (named in 1894, type locality listed under synonymy) and *O. destructor* (named in 1844, type locality: "the house mouse of the 'Plantagen at the border of the forest,'' eastern Peru, above 6000 ft, according to Tate, 1932, p. 9; restricted by Cabrera, 1961, to the haciendas along the Río Chinchao in the department of Huánuco, between 900 and 1000 m) have been associated with *O. longicaudatus.* These type localities are at low elevations, which suggests the possible affinity of these forms with *microtis* rather than with *longicaudatus.* We have examined a series from low elevation in the department of Pasco, Peru, that is clearly *O. microtis.* This locality is near the type locality of *destructor* (Cole, 1984). If future study of the holotype of *destructor* proves it to be conspecific with *O. microtis,* the name of the species should be *O. destructor* instead of *O. microtis.*

We have examined the type specimens of *Oryzomys microtis* (AMNH 37091, type locality Lower Rio Solimões, 50 mi above mouth, Brazil), *O. mattogrossae* (AMNH 37542, type locality Utiarity, Mato Grosso, Brazil), *O. utiaritensis* (AMNH 37541, type locality Utiarity, Rio Papagaia, Mato Grosso, Brazil), and *O. delicatus* (AMNH 7317/5925, type locality Trinidad, West Indies). We judge that the small lowland Bolivian specimens are *O. microtis* and also represent the same species as specimens referred by Myers and Carleton to *O. fornesi* (*O. microtis* has priority). The Bolivian specimens differ slightly from Massoia's description of *O. fornesi* in having a shorter average toothrow, otherwise in being slightly larger. If *O. fornesi* and *O. microtis* are conspecific, it is possible that this difference in measurements reflects his small sample size (N = 5), or that there may be significant geographic variation between Argentine and Bolivian populations. We further treat *O. mattogrossae* as a synonym of *O. microtis* (*O. mattogrossae* is not included in the synonymy above because the name has never been used in print for a Bolivian specimen). As mentioned by Myers and Carleton (1981), *O. utiaritensis* may belong with *O. nigripes.* More specimens from Mato Grosso are needed to establish the characteristics of the species there. *Oryzomys microtis* was considered a synonym of *O. delicatus* by Alho (1982) and Pine (1973); however, because of the slightly damaged condition of the skin and skull of the holotype of *O. delicatus* and the remoteness of its type locality (on the island of Trinidad, Allen & Chapman, 1897), we defer any decision on this question.

We have reexamined the two specimens from Beni that Myers and Carleton (1981) referred to *Oryzomys chacoensis* (USNM 391297, 460741) and have assigned them to *O. microtis.* These specimens are young adults (age class II), have external measurements that fit well with *O. microtis* and that are rather small for *O. chacoensis* (TL 200, 200; T 111, 110; HF 19, 24, respectively), and have toothrows measuring 3.34 and 3.40, respectively. The toothrows fall on the upper end of the range for *O. microtis* and the lower end of the range for *O. chacoensis* (tables 2–3). We then compared their measurements through a discriminant analysis with the samples of *O. flavescens, O. microtis, O. chacoensis,* and *O. longicaudatus.* The posterior probability of membership for both was greatest for *O. microtis* (USNM 391297, $P = 0.937$; USNM 460741, $P = 0.953$).

We refer the holotype (FMNH 21330) and the one paratype (AMNH 40787) of *Oryzomys chaparensis* to *Oryzomys microtis.* The holotype is the largest and one of the oldest individuals we have identified as *O. microtis.* Its greatest skull length is 26.8, zygomatic breadth 13.4, breadth of braincase 11.2, crown length of maxillary toothrow 3.5, tail length 111, and length of hind foot 26. The skull length, the toothrow, and hind foot length are larger than other Bolivian *O. microtis* (see table 3). The pelage coloration of this specimen falls within the range of variation in *O. microtis.* For a comparison of pelage coloration between *O. chaparensis* and *O. chacoensis,* see the account of *O. chacoensis* under General Description and Comparisons.

GEOGRAPHIC VARIATION—We were unable to detect significant differences in pelage coloration between our samples from Brazil, Bolivia, and Peru. The Brazilian specimens are much like the Bolivian specimens. A *t* test showed only three variables that differ significantly: bullar height ($t = 5.45$, $P \leq 0.001$), tail length ($t = 2.61$, $P \leq 0.05$), and length of hind foot ($t = 3.24$, $P \leq 0.01$); all have greater average values in Bolivian specimens, but the ranges of values overlap. One possible confounding factor is that the Brazilian sample is composed of mostly males (ca. 95%) and has a lower average adult age than the Bolivian or Peruvian samples (see tables 2–3). If the degree of sexual dimorphism in this species is considered (2.72%, see p. 266), the difference may actually be less than noted in our tests. For these reasons the subspecies name *Oryzomys microtis microtis* might be appropriate for specimens from northern Bolivia.

More differences exist between Peruvian and Bolivian samples. The Bolivian specimens are generally larger; the zygomatic notch is deeper, diastema longer, braincase broader, zygomatic breadth greater, and interorbital constriction broader. However, Bolivian specimens have significantly narrower teeth. Brazilian specimens, as noted above, are similar to Bolivian specimens and differ from the Peruvian material about as much as Bolivian specimens do. The Peruvian mice used in this comparison come from a lowland site in Pasco, the department just south of Huánuco, which is near the type locality of *Oryzomys destructor.*

One skin (AMNH 247776) from near Villa Tunari in Cochabamba is dorsally paler and ventrally whiter (not buffy) than any skin from Pando or Beni. A skin (AMNH 246809) from near Buena Vista in Santa Cruz is comparable to the darker specimens from Beni, but is ventrally less buffy.

Concluding Remarks

Given the considerable pelage differences noted in reference to the few available specimens from Cochabamba and Santa Cruz in the accounts of *Oryzomys chacoensis* and *O. microtis,* the problems noted in reference to specimens from southern Tarija in the accounts of *O. chacoensis* and *O. longicaudatus,* the difficulty of identification that results from overlapping ranges of measurements, the sketchy knowledge of geographic variation, and the small samples from most critical areas, the taxonomy of Bolivian *Oligoryzomys* will probably prove to be more complicated than that presented here. We hope our summary will provide information useful in further taxonomic work on the subgenus. All of our data and detailed analyses are deposited in the Department of Mammalogy,

American Museum of Natural History, and are available for the use of other interested workers.

More than 400 additional specimens, many including karyological and biochemical preparations, have been collected in 1984, 1985, and 1986 since this paper was written. Study of these will certainly help to clarify remaining taxonomic problems.

Acknowledgments

This study was partially funded by a grant from the Undergraduate-Graduate Research Fund Program, which is supported by the Susan Greenwall Foundation and administered by the American Museum of Natural History. We thank Dr. Leslie F. Marcus for his statistical advice. Masaaki Yoneda kindly provided us with a list of specimens in the Museo Nacional de Historia Natural, La Paz, Bolivia (MNHN). We are grateful to the curators of the collections referred to here, especially Dr. Philip Myers, Dr. Guy G. Musser, and Dr. Michael D. Carleton, for their generous advice and assistance in various ways. We also thank Dr. Robert S. Voss, Dr. James L. Patton, and Dr. Robert M. Timm for reviewing the manuscript.

Literature Cited

ALHO, C. J. R. 1982. Brazilian rodents: Their habitats and habits, pp. 143–166. *In* Mares, M. A., and H. H. Genoways, eds., Mammalian Biology in South America. Special Publication Series, Pymatuning Laboratory of Ecology, University of Pittsburgh, **6:** 1–539.

ALLEN, J. A. 1916. New mammals collected on the Roosevelt Brazilian Expedition. Bulletin of the American Museum of Natural History, **35:** 523–530.

ALLEN, J. A., AND F. CHAPMAN. 1897. On a second collection of mammals from the island of Trinidad, with descriptions of new species, and a note on some mammals from the island of Dominica, W.I. Bulletin of the American Museum of Natural History, **9:** 13–30.

ANDERSON, S. 1968. A new craniometer and suggestions for craniometry. Journal of Mammalogy, **49:** 221–228.

BANGS, O. 1900. List of the mammals collected in the Santa Marta region of Colombia by W. W. Brown, Jr. Proceedings of the New England Zoological Club, **1:** 87–102.

BENNETT, E. T. 1832. Characters of a new species of *Otter* (*Lutra,* Erxl.), and of a new species of *Mouse* (*Mus,* L.), collected in Chili by Mr. Cuming. Proceedings of the Zoological Society of London, part **2:** 1–2.

CABRERA, A. 1961. Catálogo de los mamíferos de América del Sur. Revista del Museo Argentino de Ciencias Naturales "Bernardino Rivadavia," Ciencias Zoologicas, **4:** 309–732.

COLE, Y. I. 1984. Genetic structure of a wild population of a Neotropical rodent (*Oryzomys* [*Oligoryzomys*] *longicaudatus*; Muridae, Cricetinae). M.A. thesis, New York University, New York, 165 pp.

CONTRERAS, J. R., AND L. M. BERRY. 1983. Notas acerca de los roedores del genero *Oligoryzomys* de la Provincia del Chaco, Republica Argentina (Rodentia, Cricetidae). Historia Natural, **3**(15): 145–148.

ELLERMAN, J. R. 1941. The Families and Genera of Living Rodents. Vol. 2, Family Muridae. British Museum (Natural History), London, xii + 690 pp.

GARDNER, A. L., AND J. L. PATTON. 1976. Karyotypic variation in oryzomyine rodents (Cricetinae) with comments on chromosomal evolution in the Neotropical cricetine complex. Occasional Papers of the Museum of Zoology, Louisiana State University, **49:** 1–48.

GOODWIN, G. G. 1953. Catalogue of type specimens of Recent mammals in the American Museum of Natural History. Bulletin of the American Museum of Natural History, **102:** 207–412.

GYLDENSTOLPE, N. 1932. A manual of Neotropical sigmodont rodents. Kungl. Svenska Vetenskapsakademiens Handlingar. Tredje Serien, Bd. **11**, no. 3: 1–164, 17 plates.

HANDLEY, C. O., JR. 1976. Mammals of the Smithsonian Venezuelan Project. Brigham Young University Science Bulletin, Biological Series, **20:** 1–91.

HERSHKOVITZ, P. 1940. Four new oryzomyine rodents from Ecuador. Journal of Mammalogy, **21:** 78–84.

HONACKI, J. H., K. E. KINMAN, AND J. W. KOEPPL, EDS. 1982. Mammal Species of the World. A Taxonomic and Geographic Reference. Allen Press, Inc. and the Association of Systematics Collections, Lawrence, Kansas, 694 pp.

LANGGUTH, A. 1963. Las especies Uruguayas del genero *Oryzomys.* Comunicaciones zoologicas del Museo de Historia Natural de Montevideo, v. **7**, no. 99: 1–19.

MARES, M. A., AND H. H. GENOWAYS, EDS. 1982. Mammalian Biology in South America. Special Publication Series, Pymatuning Laboratory of Ecology, University of Pittsburgh, **6:** 1–539.

MARES, M. A., R. A. OJEDA, AND M. P. KOSCO. 1981. Observations on the distribution and ecology of the mammals of Salta Province, Argentina. Annals of Carnegie Museum, **50:** 151–206.

MASSOIA, E. 1973. Descripción de *Oryzomys fornesi,* nueva especie y nuevos datos sobre algunas especies y subespecies argentinas del subgénero *Oryzomys* (*Oligoryzomys*) (Mammalia-Rodentia-Cricetidae). Revista de Investigaciones Agropecuarias, INTA, Buenos Aires, Serie 1, Biología y Producción Animal, **10:** 21–37.

MASSOIA, E., AND A. FORNES. 1967. El estado sistemático, distribución geográfica y datos etoecológicos de algunos mamíferos neotropicales (Marsupialia y Rodentia) con la descripción de *Cabreramys,* género nuevo (Cricetidae). Acta Zoologica Lilloana, **23:** 407–430.

Musser, G. G. 1979. Results of the Archbold Expeditions. No. 102. The species of *Chiropodomys,* arboreal mice of Indochina and the Malay Archipelago. Bulletin of the American Museum of Natural History, **162:** 379–445.

———. 1982. Results of the Archbold Expeditions. No. 108. The definition of *Apomys,* a native rat of the Philippine Islands. American Museum Novitates, **2746:** 1–43.

Myers, P., and M. D. Carleton. 1981. The species of *Oryzomys* (*Oligoryzomys*) in Paraguay and the identity of Azara's "Rat sixième ou Rat à Tarse Noir." Miscellaneous Publications of the Museum of Zoology, University of Michigan, **161:** 1–41.

Neveu-Lemaire, M., and G. Grandidier. 1911. Notes sur les mammifères des hauts plateaux de l'Amerique du Sud. Le Sondier, Paris, viii + 127 pp.

O'Connell, M. A. 1982. Population biology of North and South American grassland rodents: A comparative study, pp. 167–185. *In* Mares, M. A., and H. H. Genoways, eds., Mammalian Biology in South America. Special Publication Series, Pymatuning Laboratory of Ecology, University of Pittsburgh, **6:** 1–539.

Osgood, W. H. 1916. Mammals of the Collins-Day South American Expedition. Field Museum of Natural History, Zoological Series, **10**(14): 199–216.

Pine, R. H. 1973. Mammals (exclusive of bats) of Belém, Pará, Brazil. Acta Amazonica, **3:** 47–79.

Sanborn, C. C. 1959. Small rodents from Peru and Bolivia. Publicaciones del Museo de Historia Natural "Javier Prado," serie A, Zoologica, **5:** 1–16.

SAS Institute, Inc. 1982. SAS User's Guide: Statistics. Cary, North Carolina, 584 pp.

Soukup, J. 1961. Materiales para el catálogo de los mamíferos peruanos. Biota, **3:** 240–276.

Streilein, K. E. 1982a. Ecology of small mammals in the semiarid Brazilian caatinga. I. Climate and faunal composition. Annals of Carnegie Museum, **51:** 79–107.

———. 1982b. Ecology of small mammals in the semiarid Brazilian caatinga. II. Water relations. Annals of Carnegie Museum, **51:** 109–126.

———. 1982c. Ecology of small mammals in the semiarid Brazilian caatinga. III. Reproductive biology and population ecology. Annals of Carnegie Museum, **51:** 251–269.

Tate, G. H. H. 1932. The taxonomic history of the South and Central American cricetid rodents of the genus *Oryzomys.* Part 2. Subgenera *Oligoryzomys, Thallomyscus,* and *Melanomys.* American Museum Novitates, **580:** 1–17.

Thomas, O. 1894. Descriptions of some new Neotropical *Muridae.* Annals and Magazine of Natural History, (6), **14:** 346–366.

———. 1898. On the small mammals collected by Dr. Borelli in Bolivia and northern Argentina. Bolletino dri Musei di Zoologia ed Anatomia Comparata, R. Universita di Torino, **13:** 1–4.

———. 1902. On mammals from Cochabamba, Bolivia, and the region north of that place. Annals and Magazine of Natural History, (7), **9:** 125–143.

———. 1925. The Spedan Lewis South American Exploration. I. On mammals from southern Bolivia. Annals and Magazine of Natural History, (9), **15:** 575–582.

———. 1926. The Spedan Lewis South American Exploration. IV. List of mammals obtained by Sr. Budin on the boundary between Jujuy and Bolivia. Annals and Magazine of Natural History, (9), **18:** 193–195.

Viega-Borgeaud, T. 1982. Données écologiques sur *Oryzomys nigripes* (Desmarest, 1819) (Rongeurs; Cricétidés) dans le foyer natural de peste de Barracao des Mendes (Etat de Rio de Janeiro, Brésil). Mammalia, **46:** 335–359.

Appendix

Specimens Reported or Examined

The latitude and longitude are abbreviated; 1657/6523 means 16°57′S, 65°23′W, for example. Where references are given (to Thomas), the specimens have not been examined by us. As noted, there are a few other specimens that we have not seen.

O. chacoensis

Specimens: 88.

BOLIVIA

SANTA CRUZ

1623/6059, San Ignacio, province of Velasco, 10 USNM (390120–390121, 390664–390666, 391522–391526).
1808/6312, 7 km E and 3 km N Ingeniero Mora, 13 AMNH (247758–247761, 247765–247773).

TARIJA

2056/6321, 2 km S and 10 km E of Tiquipa, 4 AMNH (246799–246801, 246822).
2119/6325, 8 km S and 10 km E of Villa Montes, 8 AMNH (246802–246806, 246810, 146912, 246914).
2128/6317, 35 km by road SE Villa Montes, Taringuiti, 5 UMMZ (155937–155938, 156332–156334).
2241/6426, Río Lipeo, 13 ANSP (18176–18188).

PARAGUAY

PRESIDENTE HAYES

Río Pilcomayo, 15 mi W Río Paraguay, 3 AMNH (143892–143894).

NUEVA ASUNCIÓN

19 km by road WSW, km 588 Trans Chaco, 8 AMNH (249255–249262).

BOQUERÓN

419 km by road NW Villa Hayes, 1 AMNH (248397).

CHACO

50 km WNW Fortín Madrejón, 9 AMNH (248398–248406).

ARGENTINA

JUJUY

Yuto, 10 AMNH (167855, 179976, 179980, 182570–182571, 182738–182739, 183312, 185226–185227, 185269).
Sta. Barbara, 4 AMNH (185224–185225, 185228, 186954).

O. longicaudatus

Specimens: 134.

BOLIVIA

CHUQUISACA

1927/6407, Tola Orko, Tomina Province, 6 USNM (271588–271590, 545226–545227, 545229).
1929/6433, Horcus, 80 km SE Sucal, 3 MVZ (134654–134656).
1931/6409, Monte Canto, 1 USNM (271591).
?, Chuyayacu, 1 FMNH (72889).

Cochabamba

1620/6645, Yungas, 1 CM (5276).
1654/6642, El Choro, 1 AMNH (119405), 12 FMNH (74902–74913), 1 BM (Thomas, 1902).
1714/6541, Incachaca, 19 AMNH (38525–38542, 38550, not seen), 2 CM (5081–5082).
1734/6621, Parotani, 2 AMNH (38543–38544), 1 FMNH (21668).
1742/6509, 20 mi E Totora, 1 MVZ (119916).
1807/6509, Aiquile, 2 FMNH (50970–50971).

La Paz

1530/6824, Cocapunco, 1 AMNH (72644).
1535/6843, Tacacoma, 1 AMNH (91540).
1535/6843, Tacacoma-Sorata, 2 AMNH (91541–91542).
1539/6824, Okara, 2 AMNH (72704–72705).
1543/6840, 10 km by road N of Sorata, 3 UMMZ (156301–156303).
1547/6840, Sorata, at base of Mt. Sorata, 4 AMNH (91536–91539).
1600/6516, Charuplaya, 16 BM (Thomas, 1902).
1618/6753, Río Aceramarca, 2 AMNH (72693–72694).
1619/6752, Ñequejahuira, 3 AMNH (72722, 72724–72725).
1620/6808, Mt. Chacaltaya, 1 UMMZ (115711).
1620/6756, Pongo, 15 AMNH (72702–72703, 72706–72715, 72726, 81283, 241612).
1621/6747, Pitiguaya, 4 AMNH (72716, 72729–72731).
1659/6749, Caracato, 5 AMNH (248977–248981).
1823/6659, Poopo, 1 MNLP (not seen).

Potosí

1844/6609, 3 km SE of Pocoata, 1 AMNH (255946).
2127/6543, Tupiza, 1 BM (Thomas, 1926).

Santa Cruz

1754/6429, 5 mi W Comarapa, 1 MVZ (119917).
2005/6334, nr. Camiri, 1 CAL (13803).

Tarija

2131/6445, 10 mi NW Tarija, 1 MVZ (119918).
2132/6412, Entre Ríos, 3 USNM (271411–271412, 271432).
2212/6436, 8 km by road N of Cuyambuyo, 2 UMMZ (156326–156327).
2213/6436, 4 km by road N of Cuyambuyo, 10 UMMZ (155889–155891, 156312–156313, 156315–156317, 156319–156320).

Department Uncertain

?, Zapial, 1 USNM (270911).

O. microtis

Specimens: 230.

BOLIVIA

Beni

1049/6525, Guayaramerin, 2 AMNH (210050–210051).
1110/6522, 4 km below Santa Cruz, 1 AMNH (211727).
1142/6516, 4 km S Santa Rosa, 4 AMNH (211729–211730, 211791–211792).
1200/6502, Río Iténez, 20 km from mouth, 2 AMNH (211756–211757).
1200/6506, Puerto More, Río Iténez, 3 AMNH (211758–211760).
1213/6513, Mamoré River on bank opposite Cascajal, 16 AMNH (211724–211725, 211754, not seen, 211762–211774).
1226/6511, Mamoré River, 2 AMNH (211721, 211761).
1229/6417, Río Iténez, opposite Costa Marques, Brazil, 11 AMNH (210122, 210038–210045, 210365–210366).
1229/6418, Río Iténez, 1.5 km below Costa Marques, Brazil, 1 AMNH (210364).
1230/6415, Pampa de Meio, 2 AMNH (210046–210047).
1230/6418, Baures River mouth, 6 AMNH (210028–210031, 210382, 210383, not seen).
1240/6330, Curiche River mouth, 6 AMNH (210032–210037).
1241/6432, 15 km above Horquilla on Río Machupo, 1 AMNH (210053).
1244/6318, Versalles, 1 AMNH (210052).
1244/6428, Las Peñas, 3 FMNH (117079–117081), 1 USNM (460741).

1300/6515, Mamoré River, 1 AMNH (211722).
1304/6449, San Joaquín, 1 FMNH (117075), 6 USNM (364735, 364738, 391299, 460273, 460742–460743).
1312/6410, Cachuelita, 1 USNM (460739).
1312/6515, 8 km N Exaltación, 6 AMNH (211775–211780).
1313/6221, 20 km W Larangeira, Bahia de los Casara, 2 AMNH (210048–210049).
1313/6409, Boroica, 1 USNM (460740).
1322/6520, Palacios, province of Yacuma, 1 USNM (461082).
1324/6442, Chaco Lejo, 20 km SE San Ramón, 2 USNM (391295–391296).
1325/6435, Tacuaral, 1 USNM (391297).
1343/6521, Puerto Caballo, 2 AMNH (211785, 214597).
1420/6455, 10 km W San Pedro, on Mamoré River, 5 AMNH (211786–211790).
1428/6734, Rurrenabaque, 2 AMNH (247774–247775).
1434/6455, 23 km W San Javier, 1 AMNH (214760).
1437/6457, Ibare River mouth, 2 AMNH (211783–211784).
1446/6451, Ibare River, 24 km from mouth, 1 AMNH (211755, not seen).
1454/6422, 6 km W of Casarabe, 7 AMNH (255947–255953).
1503/6658, 1 km E of La Embocada, 2 UMMZ (155940, 156292).
1515/6415, El Triunfo, 1 USNM (391298).
1550/6441, 5 km NW Grande River mouth, 2 AMNH (211781–211782).
?, Centenela, 1 FMNH (117074).
?, "Beni" only, 1 AMNH (232699).

COCHABAMBA

1648/6508, Todos Santos, 1 AMNH (paratype of *O. chaparensis,* 40787), 1 FMNH (holotype, 21330).
1657/6523, 2 km E of Villa Tunari, 4 AMNH (247662–247664, 247776).

LA PAZ

1515/6810, Mapiri, 5 AMNH (72721, 73011, 72732, 72697–72698).
1528/6752, Guanay, 1 AMNH (72701).
1528/6818, Ticunhuaya, 1 AMNH (72700).
1540/6742, 4 km NW Alcoche, 2 UMMZ (126777, 127167).

PANDO

1117/6855, Río Nareuda, 11 AMNH (248982–248990, 248993–248994).

SANTA CRUZ

1703/6335, 7 km N Santa Rosa, 1 AMNH (254601).
1724/6344, 7 km N and 17 km W of Buena Vista, 3 AMNH (246809, 246820, 246935).

BRAZIL

AMAZONAS

Solimões, 10 AMNH (37088–37097, 37091 = holotype of *O. microtis*).
Guatsue, 1 AMNH (37100).
Lower Solimões, 1 AMNH (37157).

PARÁ

Capim, 150 mi SE Belém, 1 AMNH (188964).
Capim, Est. B.R. 14, km 97, 1 AMNH (203400).
Rio Xingu; Porto de Moz, 1 AMNH (95983).
Rio Xingu; Villarinho do Monte, 4 AMNH (95984–95986, 95997).

GOIÁS: RIO MADEIRA

Auara Igarapé, 5 AMNH (91874, 91876–91878, 94245).
Sta. Antonio du Uayara, 4 AMNH (92258–92261).
Rosarinho, Lago Miguel, 11 AMNH (92705–92715).
Rosarinho, Lago Sampaio, 2 AMNH (92716–92717).

MATO GROSSO

Utiarity, Papagaio River, 3 AMNH (37540–37542; 37541 = type of *O. utiaritensis,* 37542 = type of *O. mattogrossae*).

? AMAZONAS

S bank R. Amazon: Villa Bella Imperatriz, 2 AMNH (91899–91900).

PARAGUAY

CAAGUAZÚ

24 km NNW Carayaó, 1 UMMZ (133821).

PRESIDENTE HAYES

24 km NW Villa Hayes, 3 UMMZ (133830–133831, 133833).

PERU

AYACUCHO

Hda. Luisiana on Río Apurímac, 1 AMNH (242484).

PASCO

Ca. 10 km N Puerto Bermúdez, 38 AMNH (245551–245588).

O. flavescens

Specimens: 33.

URUGUAY

ARTIGAS

6 km NNW Belén, 2 AMNH (205986–205987).

CERRO LARGO

Sierra de Vaz, Río Tacuari, 20 km SE Melo, 1 AMNH (205988).

LAVALLEJA

12 km WSW Zapicán, 4 AMNH (205989–205905).

MALDONADO

Barra del Arrojo Maldonado, 9 km ENE Punta del Este, 1 AMNH (205994).

MONTEVIDEO

Río Santa Lucia, 1 km SE Santiago Vásquez, 2 AMNH (205995–205996).

ROCHA

22 km SE Lascano, 18 AMNH (205997–206009, 206011–206015).

SAN JOSÉ

Dept. only, 3 AMNH (232216–232218).

SORIANO

3 km E Cardona, 1 AMNH (206016).

TREINTA Y TRES

16 km SSW Boca del Río Tacuarí, 1 AMNH (206017).

O. nigripes

Specimens: 31.

BRAZIL

GOIÁS

Anápolis, 7 AMNH (134528–134530, 134532, 134534, 134538, 134540).

MINAS GERAIS

(Serra do Caparão), Rio Caparão, 8 AMNH (80369–80370, 80372, 80375–80379).
(Serra do Caparão), Boa Espera, 1 AMNH (80391).

MATO GROSSO

Maracaju, 9 AMNH (134541–134542, 134544–134547, 134551, 134838, 134902).

SÃO PAULO

Piquete, 1 AMNH (36496).

PARAGUAY

AINAMBAY

4 km by road SW Cerro Cora, 1 UMMZ (125523).

CAAGUAZÚ

N of Coronel Oviedo, 1 UMMZ (124212).
24 km NNW Carayao, 3 UMMZ (133835–133837).

New Records and Current Status of *Euneomys* (Cricetidae) in Southern South America

José L. Yáñez, Juan C. Torres-Mura, Jaime R. Rau, and Luis C. Contreras

ABSTRACTS

New records of *Euneomys* from central and southern Chile are given. Specific assignment of this material is difficult, owing to the lack of good diagnostic characters, small number of specimens, and spotty distribution. Absence of clear-cut differences between *Euneomys* localities suggests that there is only a single species, *E. chinchilloides*. Subspecific assignments are deferred, because at present they can only be made geographically, and the gaps between localities seem to be more apparent than real.

Se citan nuevos registros de *Euneomys* para Chile Central y Sur. La determinación específica de este material es difícil debido a la carencia de buenos caracteres diagnósticos, al bajo número de especímenes y a la distribución localizada con grandes hiatos intermedios. La ausencia de claras diferencias entre *Euneomys* de distintas localidades sugiere la existencia de tan solo una especie, *E. chinchilloides*. Las nominaciones subespecíficas se posponen, ya que al momento estás sólo se pueden realizar basadas en las localidades geográficas y los hiatos parecen ser más aparentes que reales.

São citados novos registros de *Euneomys* para o Sul e Centro do Chile. Dada a falta de bons caráteres diagnósticos, o baixo número de espécimes, e a distribuição localizada com hiatos intermediários, é difícil a determinação específica deste material. A ausência de características distintas entre *Euneomys* de localidades diferentes sugere a existencia de uma espécie apenas: *E. chinchilloides*. Não foram designadas nominações subespecíficas já que, no momento, estas só poderiam ser realizadas na base das localidades geográficas, e os hiatos podem ser mais aparentes do que reais.

Introduction

The genus *Euneomys* is one of the least studied cricetid rodents in southern South America. These herbivorous, volelike rodents have a relatively large, heavy body and a short tail and ears (Osgood, 1943). *Euneomys* has close affinities with the phyllotine and sigmodontine groups (Hershkovitz, 1962; Mann, 1978). The genus has a known distribution from 33°00′S to Cape Horn (fig. 1), mainly along and to the east of the Andes (Hershkovitz, 1962; and subsequent reports).

In Chile *Euneomys* is known from Lago Fagnano and Hermite Island (Tierra del Fuego), Punta Arenas (Magallanes), Torres del Paine and Laguna Lazo (Ultima Esperanza), Puerto Ibáñez (General Carrera), Campo Bandera (Aysén), Pino Hachado (Malleco), and Lo Valdés and Farellones (Santiago) (Osgood, 1943; Mann, 1944, 1978; Markham, 1971; Greer, 1965; Reise & Venegas, 1974; Pine et al., 1979). In Argentina *Euneomys* is certainly known from Bahía del Buen Suceso (Tierra del Fuego), upper Río Chico (Santa Cruz), Paso de

From Museo Nacional de Historia Natural, Casilla 787, Santiago, Chile (Yáñez); Departamento Biología y Química, Facultad de Ciencias, Universidad de Talca, Casilla 747, Talca, Chile (Torres-Mura and Contreras); and Estación Biológica de 'Doñana' (Sevilla), Apartado 1056, Sevilla 41013, España (Rau).

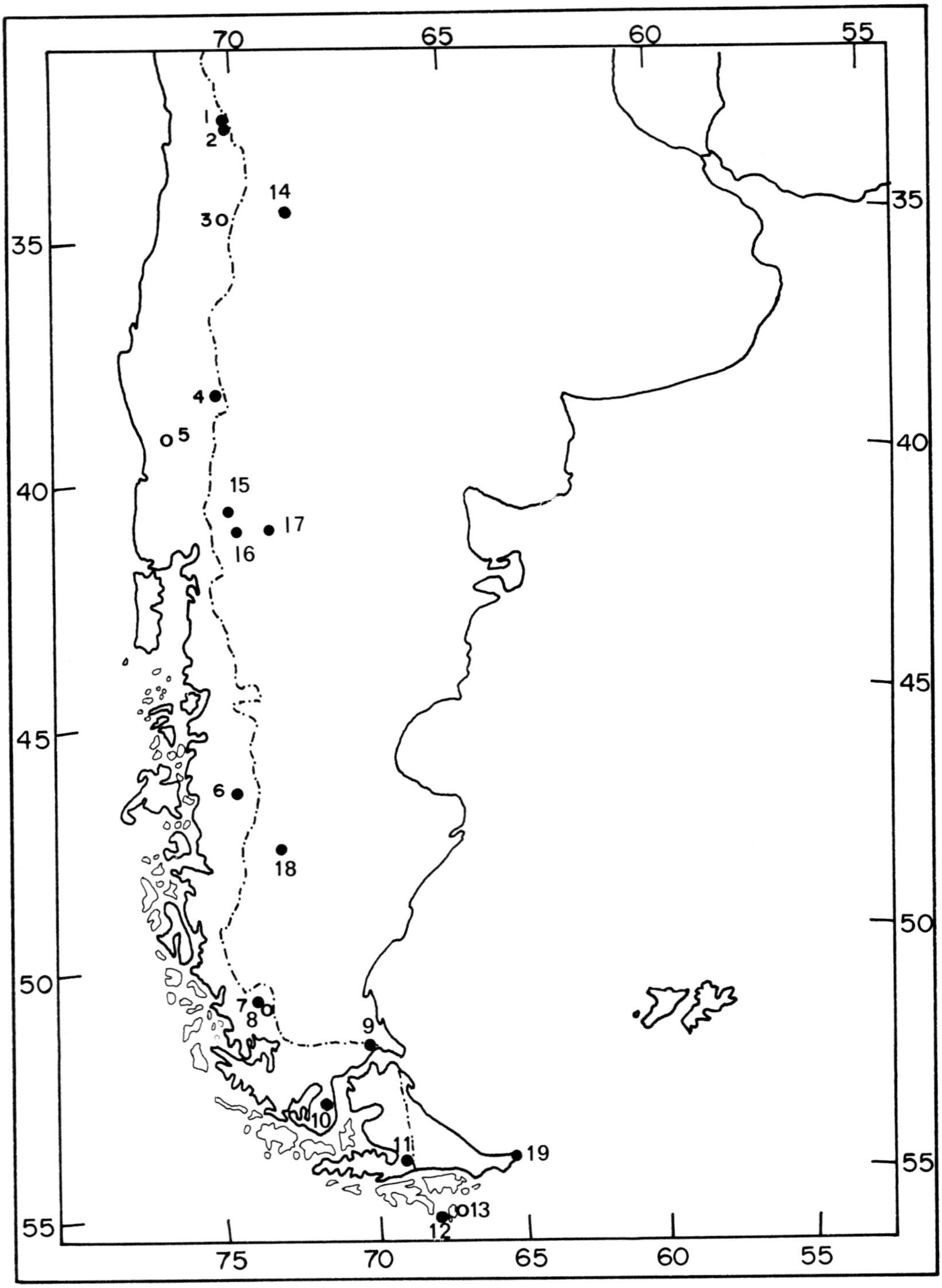

FIG. 1. Collecting localities of *Euneomys*. **1,** Farellones; **2,** Lo Valdés; **3,** Baños del Flaco; **4,** Pino Hachado; **5,** Lastarria; **6,** Puerto Ibáñez; **7,** Torres del Paine; **8,** Río Baguales and Cueva del Milodón; **9,** Estrecho de Magallanes and Palli Aike; **10,** Punta Arenas; **11,** Lago Fagnano; **12,** Hermite Island; **13,** Grevy Island; **14,** San Rafael; **15,** Cueva Traful; **16,** Cerro Leones; **17,** Paso de los Molles, Pilcaniyeu; **18,** Upper Río Chico, Santa Cruz; **19,** Bahía del Buen Suceso. Open circles represent new localities.

TABLE 1. Mandibular measurements of *Euneomys* found in barn owl pellets from Lastarria and Baños del Flaco and from animals trapped in Farellones near Santiago.

Character	Farellones*	Baños del Flaco	Lastarria
MH-1	4.07 ± 0.12 (3.8–4.1)	4.1, 3.9	2.6,† 4.0, 3.1†
MHC	8.8 ± 0.44 (8.1–9.1)	9.2, 9.1	7.9, 6.1
MAL	6.07 ± 0.08 (5.9–6.2)	6.1, 6.1	5.9, 6.0, 6.1
MDL	4.03 ± 0.24 (3.6–4.4)	3.6, 4.1	3.6, 3.9, 3.1
SL	6.24 ± 0.16 (6.1–6.5)	6.2, 6.3	5.8, 6.0, 5.5

MH-1 = Mandibular height at the first molar; MHC = mandibular height from condyle to angular process; MAL = mandibular alveolar length; MDL = mandibular diastema length; SL = symphysis length.

Values from Farellones near Santiago are $\bar{X} \pm 2$ SE (and range; N = 7). Individual measurements are given for Baños del Flaco and Lastarria.

* LCM collection. † Juvenile, the last cheektooth is not fully erupted.

Los Molles, Pilcaniyeu, and Cerro Leones (Río Negro), Cueva Traful (Neuquén), and San Rafael (Mendoza) (Hershkovitz, 1962; Pine et al., 1978; Pearson & Pearson, 1982; Massoia, 1982). In this paper we report new records of *Euneomys* in Chile and review the scant literature on the taxonomy and geographic distribution of this poorly known genus.

New Records and Current Status

Some new records are from barn owl (*Tyto alba*) pellets from Lastarria (near Temuco), 39°14′S, 70°40′W, and from Baños del Flaco (near San Fernando), 34°57′S, 70°26′W. In addition, six other specimens of *Euneomys* were recently captured on Grevy Island (Tierra del Fuego; 55°32′S, 67°37′W) and at Palli Aike (Magallanes; 52°25′ S, 69°42′W) and Río Baguales (Ultima Esperanza; 51°02′S, 72°30′W); all are deposited in the Museo del Instituto de la Patagonia (MIP), Punta Arenas. Data obtained from these new specimens have been incorporated by locality into Table 2.

Determination of this material to genus was not difficult, using Reise's (1973) key and studies by Hershkovitz (1962) and Mann (1978). However, determination to species is difficult, owing to a lack of good diagnostic characters, the small number of specimens, and their spotty distribution.

Osgood recognized two species of *Euneomys* in Chile: *E. chinchilloides* (with two subspecies) in Tierra del Fuego and Magallanes and *E. petersoni* in Ultima Esperanza and Aysén. Subsequently, Mann (1944) described *E. noei* from Lo Valdés in the Andes near Santiago. Hershkovitz (1962) considered *E. noei* as doubtfully separable from *E. mordax,* with type locality at San Rafael (Mendoza Province, Argentina), and thought that *mordax* was probably a subspecies of *E. chinchilloides.* Greer (1965) referred his four Malleco specimens to *E. chinchilloides petersoni,* despite their greater similarity to *noei* or *mordax* in measured characters. Miller and Rottmann (1976) identified all their *Euneomys* from near Santiago as *E. mordax.* Mann (1978) thought all Chilean specimens represented a single polytypic species: *E. chinchilloides chinchilloides* from Grande de Tierra del Fuego Island and the adjacent mainland, *E. c. petersoni* from the eastern Andes of Ultima Esperanza and Aysén, and *E. c. noei* from the Andes outside Santiago. Subsequently, Pine et al. (1979) suggested that two sympatric species of *Euneomys* occur in Farellones near Santiago, but declined to identify the other species until comparisons could be made with specimens representing the entire range of the genus. Karyotypes of nine specimens from Farellones show insignificant differences (Spotorno, pers. comm.).

Thus, Hershkovitz (1962) and Mann (1978) recognized only one polytypic species, whereas Pine et al. (1979) and Tamayo and Frassinetti (1980) regarded specimens from different areas as different species. Tamayo and Frassinetti (1980) thought Greer's Malleco specimens also represented *Euneomys* sp. Unfortunately, this taxonomic uncertainty is difficult to dispel, as only seven intact skulls of *Euneomys* from the Andes outside of Santiago are deposited in Chilean collections (Laboratorio Citogenética de Mamíferos [LCM], Fac. Medicina, Universidad de Chile), and none is deposited in collections in Mendoza or San Rafael, Argentina (R. A. Ojeda, pers. comm.; H. A. Lagiglia, pers. comm.).

Euneomys material recently obtained from owl pellets and deposited in the Museo Nacional de Historia Natural de Chile is incomplete, consisting

TABLE 2. Range and sample size (N) of measurements of characters in adult *Euneomys* from different localities (the name given to a form by certain authors is given below the locality).

Locality (and name)	Total length	Tail length	Hind foot	Ear	Greatest skull length	Basal length	Maxillary alveolar length	Zygomatic width	Rostral width	Maxillary diastema	Interorbital constriction	Source
Tierra del Fuego and adjacent mainland (*chinchilloides*)	186–237 (5)	59–83 (5)	25.4–32 (5)	12.0–23 (5)	29.6–34.8 (5)	26.2–28.4 (4)	5.6–6.2 (5)	18.0–21.1 (5)	5.6–6.2 (3)	7.4–9.0 (4)	4.0–4.4 (4)	Osgood, 1943; Hershkovitz, 1962; Pine et al., 1978; MIP collection
Ultima Esperanza (*petersoni*)	166–201 (6)	56–68 (6)	24.2–28 (6)	17–21 (5)	28.2–30.7 (5)	23.3–25.8 (2)	5.2–6.0 (5)	16.6–18.8 (5)	4.7–5.0 (2)	6.5–8.5 (4)	3.5–4.0 (4)	Osgood, 1943; MIP collection
Malleco (*petersoni*)	199–258 (4)	74–103 (4)	30–33 (4)	24–27 (4)	31.4–32.3 (2)	...	5.8–6.8 (4)	18.0–20.6 (4)	5.5–7.0 (4)	7.3–9.6 (4)	3.6–4.6 (4)	Greer, 1965
Cordillera Santiago (*noei*)	200–268 (10)	61–88 (10)	24–32 (10)	13–27 (10)	29.6–35.1 (10)	26.8–31.1 (3)	5.0–6.3 (10)	16.2–19.7 (11)	6.9 (2)	6.8–9.5 (11)	3.8–4.7 (11)	Mann, 1944; Pine et al., 1979; LCM collection
(*sp*)	194–210 (3)	65–76 (3)	26–28 (3)	19–25 (3)	30.8–30.9 (2)	26.8 (2)	5.9–6.0 (3)	17.3–17.7 (2)	5.5–5.6 (3)	7.8 (3)	3.7–3.9 (3)	Pine et al., 1979
Mendoza, Argentina (*mordax*)	225	78	28	31.0	31.0	...	6.1	20.0	...	...	...	Hershkovitz, 1962

mainly of mandibles and maxillae. The limited samples and their disintegration make a statistical analysis impossible. Comparisons of the owl pellet material with the seven specimens collected near Santiago (table 1) show no differences. The same results are obtained when comparisons are extended to include all known Chilean specimens (table 2). Using these criteria it is impossible for us to assign specimens from Lastarria and Baños del Flaco to either *E. c. mordax* or *E. c. petersoni.*

Conclusion

The absence of clear-cut differences between localities in Table 2 suggests that there is only a single species represented, *E. chinchilloides.* Because subspecific assignments at present can only be made geographically, treatment of subspecies is deferred until an adequate sample of specimens is available. The fact that *Euneomys* are not taken in traps at the same localities where they are found in owl pellets (e.g., Reise & Venegas, 1976; Pearson & Pearson, 1982; present study) suggests that current gaps in the distribution of *Euneomys* might be attributable to sampling error rather than to truly disjunct distributions. The new records reported here fill both morphological and geographical gaps.

Acknowledgments

We thank Angel Spotorno and Claudio Venegas for making specimens of *Euneomys* available to us. Bruce D. Patterson, Ronald H. Pine, and an anonymous reviewer made valuable comments. Mrs. Verónica Aguirre typed the manuscript, and Luz Uribe assisted with the figure. This report is dedicated to P. Hershkovitz in recognition of his pioneering contributions to our understanding of Neotropical mammals.

Literature Cited

Greer, J. K. 1965. Mammals of Malleco Province, Chile. Publications of The Museum, Michigan State University, Biological Series, **3:** 51–151.

Hershkovitz, P. 1962. Evolution of Neotropical cricetine rodents (Muridae). Fieldiana: Zoology, **46:** 1–524.

Mann, G. 1944. Dos nuevas especies de roedores. Biológica (Santiago), **1:** 95–126.

———. 1978. Los pequeños mamíferos de Chile. Gayana: Zoología, **40:** 1–342.

Markham, B. 1971. Catálogo de los anfibios, reptiles, aves y mamíferos de la Provincia de Magallanes, Chile. Publicaciones Instituto de la Patagonia, Serie Monografías N$^{\underline{o}}$. 3, 64 pp.

Massoia, E. 1982. Restos de mamíferos recolectados en el paraje Paso de Los Molles, Pilcaniyeu, Río Negro. Revista Investigaciones Agropecuarias INTA, Buenos Aires, **17:** 39–53.

Miller, S. D., and J. Rottmann. 1976. Guía para el reconocimiento de mamíferos chilenos. Ed. Gabriela Mistral, Santiago, 200 pp.

Osgood, W. H. 1943. The mammals of Chile. Field Museum of Natural History, Zoological Series, **30:** 1–268.

Pearson, O. P., and A. K. Pearson. 1982. Ecology and biogeography of the southern rainforests of Argentina, pp. 129–142. *In* Mares, M. A., and H. H. Genoways, eds., Mammalian Biology in South America. A Symposium Held at the Pymatuning Laboratory of Ecology, May 10–14, 1981. Special Publication Series, Pymatuning Laboratory of Ecology, University of Pittsburgh, **6:** 1–539.

Pine, R. H., J. P. Angle, and D. Bridge. 1978. Mammals from the sea, mainland, and islands at the southern tip of South America. Mammalia, **42:** 105–114.

Pine, R. H., S. D. Miller, and M. L. S. Schamberger. 1979. Contributions to the mammalogy of Chile. Mammalia, **43:** 339–376.

Reise, D. 1973. Clave para la determinación de los cráneos de marsupiales y roedores chilenos. Gayana: Zoología, **27:** 1–20.

Reise, D. L., and W. Venegas. 1974. Observaciones sobre el comportamiento de la fauna de micromamíferos en la Región de Puerto Ibáñez (Lago General Carrera), Aysén, Chile. Boletín Sociedad de Biología, Concepción, Chile, **47:** 71–85.

Tamayo, M., and D. Frassinetti. 1980. Catálogo de los mamíferos fósiles y vivientes de Chile. Boletín Museo Nacional de Historia Natural, Chile, **37:** 323–399.

Morphological Variation, Karyology, and Systematic Relationships of *Heteromys gaumeri* (Rodentia: Heteromyidae)

Mark D. Engstrom, Hugh H. Genoways, and Priscilla K. Tucker

ABSTRACTS

Morphological variation was assessed within and among populations of *Heteromys gaumeri* using univariate and multivariate statistical analyses of external and cranial measurements. Although patterns and amount of nongeographic variation in *H. gaumeri* were similar to other heteromyines, geographic variation was relatively conservative. Mean values of most characters were statistically homogeneous among localities and spatially unpatterned. Consequently, no association was found between levels of within- and among-sample variation for individual characters (the "Kluge-Kerfoot phenomenon"). Populations of *H. gaumeri* were chromosomally monomorphic. The lack of morphological and chromosomal variation in *H. gaumeri* contrasts sharply with patterns in other heteromyines. *Heteromys gaumeri* is morphologically and chromosomally distinct from the *H. desmarestianus* species group (to which it is currently assigned) and appears to share some primitive characters with *Liomys* (the sister group of *Heteromys*). We recommend that *H. gaumeri* be removed from the *H. desmarestianus* group.

La variación morfológica intra e interpoblacional de *Heteromys gaumeri* fue evaluada usando análisis estadísticos univariados y multivariados de medidas externas y craneales. A pesar de que los patrones y cantidad de variación intrapoblacional en *H. gaumeri* fue similar a la de otros heterómінos, la variación geográfica fue relativamente conservadora. Los valores promedio de la mayoría de los caracteres fueron estadisticamente homogeneos entre las localidades, sin mostrar ningún patrón de variación espacial. En conservencia, no se encontró asociación alguna entre los niveles de variación intra e interpoblacional para caracteres individuales ("fenómeno Kluge-Kerfoot"). Las poblaciones de *H. gaumeri* fueron monomórficas cromosómicamente. La falta de variación tanto morfológica como cromosómica en *H. gaumeri* contrasta marcadamente con los patrones encontrados anteriormente para otros heterómінos. *Heteromys gaumeri* es morfológica y cromosómicamente distinguible del grupo *H. desmarestianus* (al cual se asigna actualmente) y aparentemente comparte algunos caracteres primitivos con *Liomys* (el grupo hermano de *Heteromys*). Nosotros recomendamos que se remueva a *H. gaumeri* del grupo *H. desmarestianus*.

Avalia-se a variação morfológica intra- e interpopulacional de *Heteromys gaumeri*, através de análises estatísticas uni- e multivariadas de medidas externas e craniais. Apesar dos padrões, e da quantidade de variação intrapopulacional em *H. gaumeri* serem similares aos de outros heteromídeos, a variação geográfica é relativamente conservadora. Os valores médios da maior parte dos caráteres examinados são estatìsticamente homogeneos entre as localidades, e não

From Department of Biology, Angelo State University, San Angelo, TX 76909 (Engstrom); Section of Mammals, Carnegie Museum of Natural History, 4400 Forbes Avenue, Pittsburgh, PA 15206 (Genoways; present address, University of Nebraska State Museum, 212 Morrill Hall, Lincoln, NE 68588); and Department of Wildlife and Fisheries Sciences, Texas A&M University, College Station, TX 77843 (Tucker; present address, The Jackson Laboratory, Bar Harbor, ME 04609).

surgiu nenhum padrão de variações locais. Consequentemente, não foram encontradas associações entre os níveis de varaiações intra- e interpopulacionais para caráteres individuais (o "fenômeno Kluge-Kerfoot"). Populações de *H. gaumeri* mostraram-se cromossômicamente monomórficas. A falta de variação morfológica ou cromossômica em *H. gaumeri* é altamente contrastante aos padrões encontrados em outros heteromídeos. *Heteromys gaumeri* distingue-se tanto morfológica quanto cromossômicamente do grupo *H. desmarestianus,* ao qual está atualmente designado, e aparentemente possue caráteres primitivos em comum com *Liomys*—grupo irmão de *Heteromys.* Recomendamos que *H. gaumeri* seja removido do grupo *H. desmarestianus.*

Introduction

Spiny mice of the genus *Heteromys* (Heteromyidae: Heteromyinae) are Neotropical rodents that typically occur in rainforest and cloudforest habitats from east-central Mexico south to northern South America. Currently, the genus is partitioned into two subgenera (*Xylomys* and *Heteromys*), and two species groups (*desmarestianus* and *anomalus* groups) are recognized in the nominate subgenus (Goldman, 1911; Hall, 1981; Rogers & Schmidly, 1982). Hall and Kelson (1959) and Hall (1981) noted that taxonomic relationships within *Heteromys* and in particular the *H. desmarestianus* species group were problematical and in need of revision. Recently, Rogers and Schmidly (1982) reviewed morphological variation among representatives of the *H. desmarestianus* species group from northern Middle America, exclusive of *H. gaumeri,* and concluded that only two species (*H. desmarestianus* and *H. goldmani*) were represented in the material they examined.

Heteromys gaumeri, the third species currently recognized in the *desmarestianus* group, is endemic to the Yucatán Peninsula. Ecologically this species characteristically occurs in relatively dry deciduous and subdeciduous-subperennial forest. Systematic relationships of *H. gaumeri* are enigmatic. In his review of the subfamily Heteromyinae, Goldman (1911) placed *H. gaumeri* in his *desmarestianus* group for convenience, but noted (p. 29) that it was "not closely related to any known species." In this paper, we review morphological and chromosomal variation within *H. gaumeri* and comment on the systematic relationships of this species to the *H. desmarestianus* species group and the subgenus *Heteromys.*

Materials and Methods

A total of 322 specimens of *Heteromys gaumeri* was examined in the morphological analyses. Specimens examined are listed in the Appendix, and institutions housing those specimens are as follows (abbreviations for North American collections follow Choate & Genoways, 1975): American Museum of Natural History (AMNH); British Museum (Natural History), London (BMNH); Field Museum of Natural History (FMNH); Museum of Natural History, University of Kansas (KU); James Ford Bell Museum of Natural History, University of Minnesota (MMNH); Royal Ontario Museum (ROM); Texas Cooperative Wildlife Collections, Texas A&M University (TCWC); The Museum, Texas Tech University (TTU); Museum of Zoology, University of Michigan (UMMZ); Universidad Nacional Autónoma de México (UNAM); and National Museum of Natural History (USNM).

Each specimen was assigned to one of five presumptive age classes (I–V, from youngest to oldest) based on pelage characteristics and state of eruption and relative wear of the maxillary toothrow, following Genoways (1973). For each adult and selected subadult specimens, the following four external (recorded from specimen labels) and 10 cranial measurements (measured to the nearest 0.1 mm using dial calipers) were taken, as defined by Genoways (1973): total length (TL); length of tail (TV); length of hind foot (HF); length of ear (LE); greatest length of skull (GLS); zygomatic breadth (ZB); interorbital constriction (IOC); mastoid breadth (MB); length of nasals (LN); length of rostrum (LR); length of maxillary toothrow (MTR); depth of braincase (DBC); interparietal width (IW); and interparietal length (IL). Each measurement was chosen because it was examined in previous studies of variation in other heteromyine rodents and was geographically variable in some taxa (Goldman, 1911; Genoways, 1973; Rogers & Schmidly, 1982).

Initially, four qualitative cranial characters and dorsal coloration also were examined, as described by Genoways (1973) for *Liomys* (the sister group of *Heteromys*). Unlike patterns found in species of *Liomys,* each of these characters was equally or more variable within populations of *H. gaumeri* as among populations, and the characters varied in no apparent geographic pattern. Consequently, qualitative cranial characters and dorsal coloration were not analyzed further.

Nongeographic variation in the 14 mensural characters was examined in one sample of 94 specimens collected near Campo Experimental Forestal "El Tormento," 7.5 km W Escárcega, Campeche. These specimens were collected within a two square kilometer area of transitional tropical evergreen-tropical deciduous forest and were here considered to represent a single population. All calculations were performed using subprograms of the Statistical Analysis System (SAS; SAS Institute, Inc., 1982). Standard statistics (mean, range, standard deviation, standard error, coefficient of variation, skewness, and kurtosis) were calculated for each variable within each subgroup (MEANS and UNIVARIATE procedures). Student's *t* test (or an approximation of variances were unequal) was used to test for significant differences between sexes within each age class (TTEST procedure). A model I, one-way analysis of variance (ANOVA) was used to test for significant differences among age classes with sexes pooled (GLM procedure). Subsequently, a Duncan's multiple range test (DUNCAN) was used to determine maximally nonsignificant subsets of age classes.

Straney (1978) criticized the use of *F* tests in unbalanced ANOVAs to determine patterns of nongeographic variation. To augment hypothesis testing, variance partitioning of a model II, two-way ANOVA (VARCOMP procedure) was used to estimate the relative contributions of sex, age, sex by age interaction, and error to within sample variation. Age and sex were considered random factors (see discussion in Leamy, 1983) and the percent contribution of each factor was estimated from variance components. The main effects (sex and age) in the two-way ANOVAs were not independent because the data were unbalanced (Searle, 1971). Consequently, ANOVAs were run with sex entered into the model first, then again with age entered first.

For analysis of geographic variation of mensural data, adult specimens were assigned to one of 11 grouped localities to increase sample size (fig. 1). In no instance did a grouped locality cross a major physiographic or previously recognized taxonomic boundary. The specific geographic composition of samples is as follows: Group 1—Chuntuquí and Laguna de Sotz (= Zotz), Petén, Guatemala; 103 km SE Escárcega (= Francisco Escárcega), Campeche, Mexico. Group 2—7.5 km W Escárcega, Campeche, Mexico. Group 3—Apazote, 7 km N, 51 km E Escárcega, and La Tuxpeña, Campeche, Mexico. Group 4—Dzibalchén and San José Carpizo, Campeche, Mexico. Group 5—Esmeralda and Santa Rosa, Yucatán, Mexico. Group 6—Chichén Itzá and Pisté, Yucatán, Mexico. Group 7—Tizimín and Tunkás, Yucatán, Mexico. Group 8—La Vega, Pueblo Nuevo Xcan, and Puerto Morelos, Quintana Roo, Mexico. Group 9—Felipe Carrillo Puerto, Quintana Roo, Mexico. Group 10—Bacalar, Quintana Roo, Mexico. Group 11—Kate's Lagoon and Rockstone Pond, Belize, Belize.

For each mensural character, we analyzed two aspects of geographic variation: (1) statistical heterogeneity of mean values among geographic samples, and (2) significant departure of means from spatial randomness (see Sokal & Oden, 1978). The null hypothesis of homogeneity of means among grouped localities for each character was tested using a model I, one-way ANOVA (SAS:GLM). By ANOVA, the variance of each character was partitioned into among and within (error) locality effects, and the percentage of variation attributable to each effect was estimated from variance components (SAS:VARCOMP). Homogeneity among grouped localities across all characters was examined using a multivariate analysis of variance (SAS:GLM, MANOVA).

The null hypothesis of no geographic pattern among grouped locality means was examined by testing for significant association between geographic and phenetic distance matrices. For each character, a phenetic distance matrix was constructed in which the elements were calculated as the absolute differences between means for all pairs of localities (Sokal, 1979). Multivariate phenetic matrices of taxonomic distance (Sneath & Sokal, 1973) were calculated for all 14 mensural characters and a set restricted to those showing significant heterogeneity, using the NT-SYS library of computer programs (Rohlf et al., 1982). Elements of the geographic distance matrix were the actual map distances (in km) between all pairs of grouped localities (taken from the center of each grouped locality); all connections between pairs of localities were maintained because there are no obvious physiographic or ecological barriers to gene flow among populations of *H. gaumeri* on the Yucatán Peninsula. Three test statistics (Mantel's Z; Spearman's *rho,* R; and a component of Kendall's *tau,* K_c) were used to test for significant association between each phenetic distance matrix and the

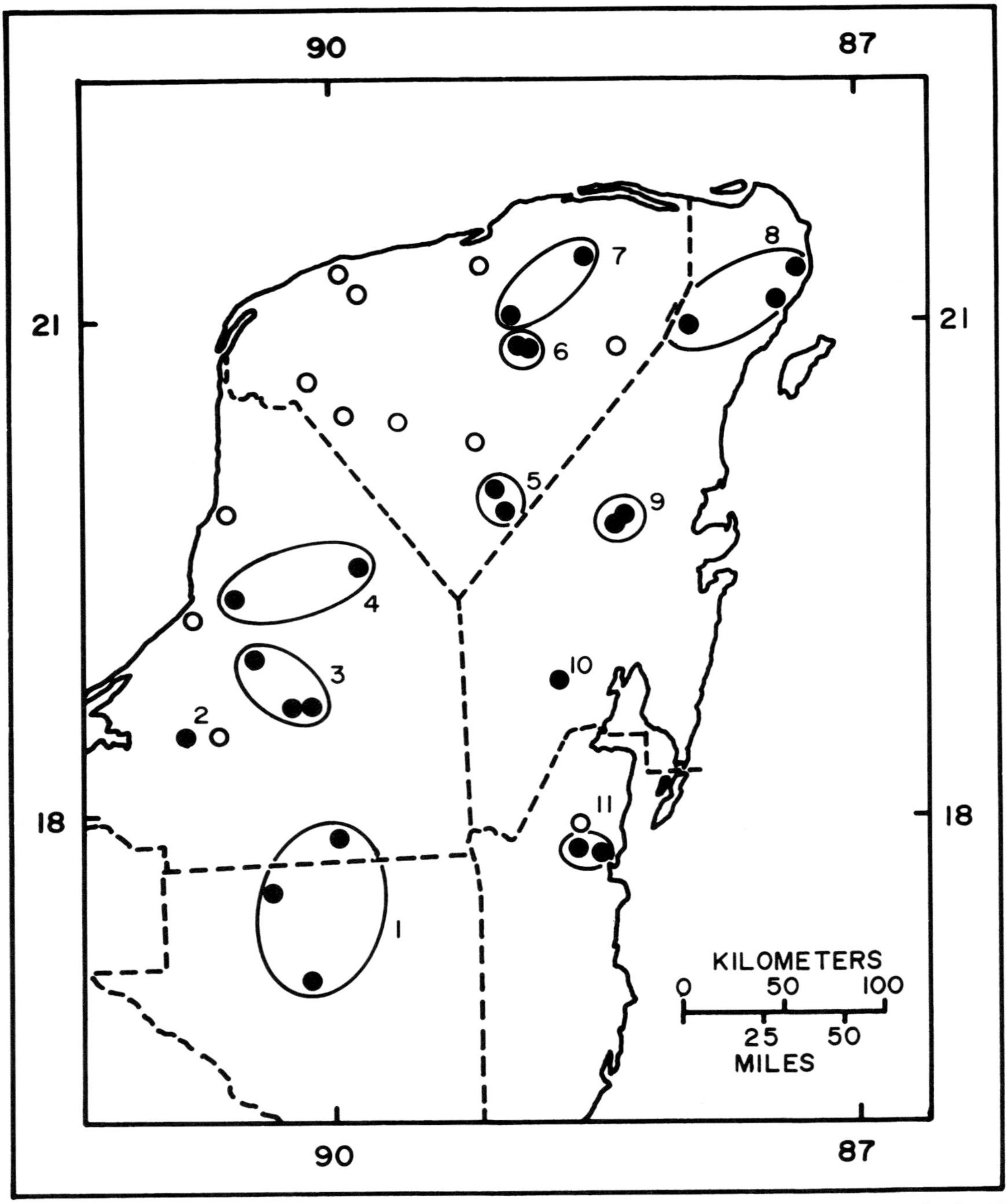

FIG. 1. Geographic distribution of *Heteromys gaumeri* and approximate geographic areas included in the 11 grouped localities used in analyses of morphological variation. Closed circles denote specimens used in analyses of geographic variation and open circles denote specimens examined but not included in statistical analyses.

geographic distance matrix (Dietz, 1983). Values of *P* associated with each statistic were estimated from 2,000 random permutations using a FORTRAN program supplied by E. J. Dietz.

In the chromosomal analysis, standard karyotypes were examined for 26 specimens of *Heteromys gaumeri.* Additionally, karyotypes of one specimen of *H. desmarestianus* and 20 specimens of *H. anomalus* were examined for comparison. Standard karyotypes were prepared from specimens sampled from natural populations using the *in vivo* bone marrow technique of Patton (1967), as modified by Lee (1969). Terminology regarding relative chromosome arm ratios is that of Patton (1967). In the calculation of Fundamental Numbers (FN), relative chromosome arm ratios were scored conservatively. Several chromosomes recorded as acrocentric here had telomeric knobs of chromatin in elongated preparations. Only chromosomes which consistently displayed second arms of chromatin, regardless of state of contraction of the preparation, were scored as biarmed. The following voucher specimens are deposited in the Texas Cooperative Wildlife Collection, Texas A&M University, and the Carnegie Museum of Natural History (sample size in parentheses): *Heteromys gaumeri* (total 26)—**MEXICO. Campeche:** 7.5 km W Escárcega (8). **Quintana Roo:** 2 km N, 8 km W Bacalar (8); 8 km NNE Felipe Carrillo Puerto (3); 2.5 km NNE Felipe Carrillo Puerto (1). **Yucatán:** Chichén Itzá (1); Cenote Seco, 2 km E Chichén Itzá (5). *Heteromys desmarestianus* (total 1)—**MEXICO. Chiapas:** 9.4 km S Palenque (1). *Heteromys anomalus* (total 20)—**VENEZUELA. Miranda:** 25 km N Altagracia de Oricuto (6); 24 km N Altagracia de Oricuto (1). **Monagas:** Caripito (4). **Sucre:** 40 km SW Caripito (9).

Results

Morphological Variation

NONGEOGRAPHIC VARIATION—The sample of *Heteromys gaumeri* (N = 94) from 7.5 km W Escárcega, Campeche, was used to estimate within-sample variation of the 14 mensural characters. Two approaches were taken, hypothesis testing and estimation of variance components.

Initially, *t* tests were used to test for significant differences between sexes in each of age classes I–IV. Males average larger than females for most characters in most age classes; however, significant differences between sexes were found for only one measurement in age class I (length of rostrum) and seven measurements in age class III (table 1).

Results of ANOVA among age classes with sexes pooled are given in Table 1. Separate analyses of the age effect within each sex gave results similar to those in Table 1 and are not presented. Significant variation with age was found in each of the 14 measurements. Each measurement tended to increase with age, although this pattern was not pronounced for length of maxillary toothrow. In the *a posteriori* DUNCAN analysis, age class V averaged significantly larger than the other age classes for two characters (interorbital constriction and interparietal length), whereas age classes IV–V formed a homogeneous subset for the remaining 12 characters. Patterns of variation among age classes I–III were less consistent, although age classes I–III differed significantly from each other and age classes IV–V in six of 10 cranial measurements.

To complement hypothesis testing, the relative contributions of sex, age, sex by age interaction, and error (residual variation) to a two-way ANOVA were estimated from variance components (table 2). Separate analyses with either sex or age entered into the model first generally yielded results that differed by 1% to 4%. Because of the similarity of these estimates (and because sex most often is analyzed before age in studies of nongeographic variation), only the results in which sex was entered into the model first are presented.

Most of the variation (average 97%) in the ANOVA was attributable to the effects of age and error. Age contributed the largest proportion of variance for most characters (average 53%); error was nearly as important, contributing an average of 44%. For length of hind foot, length of maxillary toothrow, interparietal width, and interparietal length, age contributed a relatively small proportion of the variance, and variation was mostly attributable to error. Homogeneity of estimates across characters was examined using *z*-transformations and a subsequent *chi*-square test (Sokal & Rohlf, 1981). Estimates of age and error were significantly heterogeneous across characters.

Sex and interaction were relatively unimportant factors in the ANOVA for most characters, contributing an average of 1% and 2% of the variance, respectively. The only noteworthy exception to this pattern was interparietal length, for which interaction accounted for 23% of the variance. Despite this exception, estimates of the effects of sex and interaction were statistically homogeneous across characters. Given that only a small proportion of the variance of each character was attributable to the effect of sex, the significant sexual dimorphism found for seven of 14 characters in age class III in preliminary *t* tests (table 1) probably was due to trivial differences accentuated by large sample size (N = 47). For most characters, the largest proportion of variance was attributable to age and

TABLE 1. Age variation in 14 external and cranial measurements of *Heteromys gaumeri* from 7.5 km W Escárcega, Campeche, Mexico.

Age class	N	Mean (range) ± 2 SE	CV
	TOTAL LENGTH		
V	2	265.0 (237.0–272.0) ± 10.00	2.7
IV	14	261.2 (235.0–288.0) ± 9.82	7.0
III*	42	242.6 (186.0–275.0) ± 5.87	7.8
II	17	220.6 (200.0–252.0) ± 6.70	6.3
I	4	190.3 (179.0–210.0) ± 13.60	7.1
	$F = 19.45$***		
	LENGTH OF TAIL		
IV	14	145.5 (137.0–159.0) ± 4.58	5.9
V	2	139.5 (139.0–150.0) ± 1.00	0.5
III*	42	132.5 (87.0–152.0) ± 3.65	8.9
II	17	120.1 (101.0–136.0) ± 4.62	7.9
I	4	99.8 (96.0–117.0) ± 6.85	6.9
	$F = 20.27$***		
	LENGTH OF HIND FOOT		
V	3	35.0	...
IV	16	34.2 (32.0–36.0) ± 0.66	3.9
III*	47	33.2 (30.0–35.0) ± 0.40	4.1
II	19	33.0 (30.0–35.0) ± 0.70	4.6
I	6	31.0 (29.0–34.0) ± 1.71	6.8
	$F = 6.76$***		
	LENGTH OF EAR		
V	3	17.7 (17.0–18.0) ± 0.67	3.3
IV	16	17.4 (17.0–19.0) ± 0.31	3.6
III	47	16.1 (14.0–18.0) ± 0.26	5.6
II	19	15.6 (14.0–18.0) ± 0.44	6.1
I	6	14.8 (14.0–16.0) ± 0.61	5.1
	$F = 15.77$***		
	GREATEST LENGTH OF SKULL		
V	2	35.8 (35.5–36.0) ± 0.50	1.0
IV	15	35.2 (33.9–37.5) ± 0.49	2.7
III*	47	32.9 (30.2–34.5) ± 0.26	2.7
II	19	31.2 (29.6–34.2) ± 0.50	3.5
I	6	28.8 (27.8–29.8) ± 0.65	2.7
	$F = 71.20$***		
	ZYGOMATIC BREADTH		
V	3	16.3 (16.1–16.5) ± 0.23	1.2
IV	16	16.1 (15.3–16.7) ± 0.19	2.4
III*	44	15.1 (14.0–16.3) ± 0.14	3.1
II	19	14.4 (13.7–15.1) ± 0.20	3.0
I	6	13.5 (12.9–13.8) ± 0.33	3.0
	$F = 59.28$***		
	INTERORBITAL CONSTRICTION		
V	3	9.3 (9.2–9.5) ± 0.20	1.9
IV	16	8.7 (8.3–9.0) ± 0.10	2.4
III*	47	8.2 (7.5–9.3) ± 0.12	4.8
II	19	7.8 (7.2–8.6) ± 0.16	4.3
I	6	7.5 (7.0–7.7) ± 0.20	3.3
	$F = 29.71$***		
	MASTOID BREADTH		
V	3	15.5 (15.3–15.8) ± 0.30	1.7
IV	15	15.3 (14.8–15.9) ± 0.18	2.3
III	47	14.7 (14.0–15.4) ± 0.10	2.3

Table 1. *Continued.*

Age class	N	Mean (range) ± 2 SE			CV
II	19	14.4	(13.4–15.2)	± 0.21	3.2
I	6	13.7	(13.1–14.1)	± 0.29	2.6
	$F = 27.94$***				
	Length of Nasals				
V	2	14.5	(14.0–15.0)	± 1.00	4.9
IV	16	14.3	(13.2–15.6)	± 0.37	5.1
III	47	13.0	(11.8–14.2)	± 0.18	4.7
II	19	11.9	(10.8–13.4)	± 0.29	5.3
I	6	10.5	(10.0–10.9)	± 0.26	3.0
	$F = 59.06$***				
	Length of Rostrum				
V	2	15.6	(15.1–16.0)	± 0.90	4.1
IV	16	15.2	(14.2–16.3)	± 0.27	3.6
III*	47	14.0	(12.7–15.3)	± 0.18	4.3
II	19	12.9	(11.9–13.9)	± 0.23	3.9
I*	6	11.6	(11.0–12.1)	± 0.37	3.9
	$F = 65.77$***				
	Length of Maxillary Toothrow				
IV	16	4.9	(4.7–5.2)	± 0.07	3.0
II	17	4.8	(4.5–5.1)	± 0.08	3.3
V	3	4.8	(4.7–4.8)	± 0.07	1.2
III	46	4.7	(4.3–5.2)	± 0.05	3.4
I	5	4.7	(4.5–4.8)	± 0.17	2.8
	$F = 3.72$**				
	Depth of Braincase				
V	3	9.1	(8.9–9.4)	± 0.29	2.8
IV	16	9.0	(8.5–9.5)	± 0.13	3.0
III	46	8.8	(8.3–9.7)	± 0.08	3.3
II	19	8.5	(7.9–8.8)	± 0.13	3.3
I	6	8.5	(8.3–8.8)	± 0.16	2.3
	$F = 12.68$***				
	Interparietal Width				
V	3	9.3	(8.9–9.8)	± 0.54	5.1
IV	15	8.8	(8.0–10.0)	± 0.31	6.8
II	19	8.6	(7.6–9.4)	± 0.21	5.3
III	45	8.5	(7.3–9.7)	± 0.17	6.7
I	6	8.0	(7.3–8.6)	± 0.36	5.6
	$F = 4.11$**				
	Interparietal Length				
V	3	5.5	(5.1–6.1)	± 0.61	9.6
IV	15	5.1	(4.2–5.8)	± 0.21	8.0
III	45	5.0	(4.2–5.7)	± 0.10	6.8
II	19	4.9	(4.2–5.3)	± 0.14	6.3
I	6	4.7	(3.8–5.3)	± 0.45	11.9
	$F = 3.23$*				

Vertical lines alongside age classes denote nonsignificant subsets. Asterisks after F statistics indicate levels of significance (* $P < 0.05$; ** $P < 0.01$; *** $P < 0.001$). Males and females in age classes I–IV were used to test for significant mean differences due to sex (t test). Asterisks following age classes indicate significant sexual dimorphism ($P < 0.05$).

TABLE 2. Percentage of total variation attributable to sex (S), age (A), sex by age interaction (S × A), and error (E) for 14 external and cranial measurements of *Heteromys gaumeri* from 7.5 km W Escárcega, Campeche, Mexico.

Characters*	Variance components			
	S	A	S × A	E
TL	0	61.3	1.5	37.2
TV	0	63.6	0	36.4
HF	4.8	29.9	0	65.3
LE	0	50.3	0	49.7
GLS	0	84.1	0	15.9
ZB	0	77.5	4.6	17.9
IOC	0	66.7	0.5	32.8
MB	0	64.1	2.0	33.9
LN	0	79.7	1.4	18.9
LR	0	82.1	0.8	17.1
MTR	2.2	21.1	0	76.7
DBC	1.2	42.4	0.6	55.8
IW	3.9	19.1	0	77.0
IL	0	1.1	22.7	76.2
Mean	0.9	53.1**	2.4	43.6**

* Abbreviations of characters are defined in text. Estimates were calculated from variance components (considering sex as a random factor).

** Significant heterogeneity ($P < 0.01$) of estimates among characters.

error, and the main effects in the model (sex and age) were independent.

Variance components also were estimated from a two-way ANOVA restricted to age classes IV and V (results available on request). In this analysis, the pattern of variance partitioning changed considerably. For each character, error contributed the largest proportion of variance (average 64%), whereas sex, age, and interaction had smaller average contributions (10%, 12%, and 14%, respectively). These estimates, however, should only be regarded as approximations because of the small sample size of age class V (N = 3). Estimates derived from larger subsets of age classes included a large average contribution (> 40%) attributable to the effect of age. Based on these results, geographic analyses were restricted to age classes IV–V with sexes pooled.

GEOGRAPHIC VARIATION—Standard statistics were calculated for each mensural character in each grouped locality. By ANOVA, the variance of each character was partitioned into among and within (error) locality effects. Only six of 14 characters were significantly heterogeneous among grouped localities (table 3). For seven characters (HF, ZB, IOC, MB, DBC, IW, IL), the among-locality variance component was zero and for the remaining characters, locality accounted for 6% (LE) to 30% (TL) of the total variance. For those characters that were significantly heterogeneous, an average of 17% of the variance was attributable to locality. In a MANOVA across all mensural characters, localities were marginally heterogeneous ($P < 0.05$) by two test statistics (Hotelling-Lawley Trace and Wilk's Criterion), but homogeneous by a third test statistic (Pillai's Trace), suggesting an overall lack of mensural differentiation among localities.

Because significant spatial patterning is theoretically possible (although unlikely) even with statistically homogeneous means (Sokal & Oden, 1978), each character was tested for significant departures from spatial randomness. We found no significant association between matrices of phenetic and geographic distance for any character by any of three test statistics. Multivariate taxonomic distance matrices calculated by using all characters and using only those characters that were significantly heterogeneous among localities were also incongruent with geographic distance. Accordingly, mean values of most mensural characters were homogeneous among grouped localities and not spatially patterned as a simple function of geographic distance.

Karyology

HETEROMYS GAUMERI (2n = 56, FN = 76; fig. 2a)—The autosomal complement comprises graded series of 11 pairs of large- to small-sized metacentric and submetacentric chromosomes and 16 pairs of large- to small-sized acrocentric elements. The X chromosome is large and submetacentric and the Y is medium-sized and subtelocentric. No variation was noted among individuals.

HETEROMYS DESMARESTIANUS (2n = 60, FN = 66; fig. 2b)—The autosomal complement comprises four pairs of submetacentric and metacentric chromosomes, two large, one medium-sized, and one small, and 25 acrocentric pairs graded from large to small. The X chromosomes are presumed to be large and submetacentric (no males were examined).

HETEROMYS ANOMALUS (2n = 60, FN = 68; fig. 2c)—The autosomal complement comprises two large, one medium-sized, and two small pairs of submetacentric and metacentric chromosomes and 24 acrocentric pairs graded from large to small. The X chromosome is large and submetacentric and the Y is medium-sized and subtelocentric. No variation was noted among individuals.

Discussion

Patterns of nongeographic variation in mensural characters of *Heteromys gaumeri* are generally concordant with those observed in other heteromyine rodents (Genoways, 1973; Rogers & Schmidly, 1982; see also Straney, 1978). In all species examined, age contributes a large proportion of within-sample variance for most characters; only a minor component of total variance is attributable to sex. Male heteromyines generally average larger than females, however, and with large sample size secondary sexual dimorphism often appears significant for some characters (Genoways, 1973; age class III, this study).

Relative levels of variability of individual characters among species of heteromyines were compared using coefficients of variation of adults for the 12 external and cranial measurements common to each study of intralocality variation (total length, length of tail, and the 10 cranial measurements included in this study; see Genoways, 1973; Rogers & Schmidly, 1982). Coefficients of variation (CVs) appeared congruent among species for individual characters, although CVs appeared heterogeneous among characters within species (external measurements and interparietal width and length consistently were more variable than other characters). There was no indication of a reduction of within-sample variation for any character in *H. gaumeri* relative to other heteromyines.

Although patterns and level of intralocality variation in *Heteromys gaumeri* appear similar to other heteromyines, geographic variation in *H. gaumeri* is relatively conservative. In *H. gaumeri*, mean values for most characters were homogeneous among localities and geographically unpatterned; only a small proportion of variance (average 8%) was attributable to interlocality variation. We found no positive statistical correlation between levels of within- and among-sample variance (the "Kluge-Kerfoot phenomenon," Kluge & Kerfoot, 1973) for mensural characters in *H. gaumeri* (tested using Kendall's rank correlation between the W_s and A_s statistics suggested by Sokal, 1976; but see Rohlf et al., 1983). These data contrast with studies of geographic variation within species of *Liomys* (Genoways, 1973) and other members of the *H. desmarestianus* species group (Rogers & Schmidly, 1982), in which population samples were statistically heterogeneous and appeared spatially patterned.

The relative lack of interlocality variance in *Heteromys gaumeri* might be attributable to a restricted geographic distribution, to relative environmental homogeneity on the Yucatán Peninsula, and/or to a lack of genetic divergence among populations. Compared to wide-ranging species of *Liomys* and *Heteromys*, the geographic area occupied by *H. gaumeri* is relatively small, with little topographic or climatic relief. Plant communities grade gradually from lowland tropical evergreen forest in the south and east, to deciduous forest in the northwestern and north-central portions of the peninsula, to a scrub zone bordering the northern coast (Paynter, 1955). For statistically heterogeneous but spatially unpatterned characters, character distributions might be determined mainly by stochastic factors (e.g., genetic drift) in the absence of strong migration or selective gradients (Sokal & Oden, 1978). *Peromyscus yucatanicus*, which has a similar geographic range and occupies similar habitats to *H. gaumeri*, is geographically variable in color and cranial size (Lawlor, 1965; Huckaby, 1980), and at least color closely tracks vegetational changes on the peninsula. Consequently, lack of interlocality differentiation in color, qualitative, and mensural characters in *H. gaumeri* might not result solely from selective responses or stochastic processes in a homogeneous environment.

The reduced level of geographic variation in *Heteromys gaumeri* also is consistent with a hypothesis of little genetic divergence among populations. If populations of *H. gaumeri* are genetically similar, similarity probably is not a product of panmixia, but more likely of a reduction of genetic variation through a genetically depauperate founding population (see Johnston, 1976; Johnston & Klitz, 1977; Baker, 1980) or genetic bottleneck. Although we have little direct evidence of genetic variation among populations of *H. gaumeri*, individuals sampled appear chromosomally monomorphic. In *H. desmarestianus*, interlocality polymorphism in fundamental number is pronounced (D. S. Rogers, pers. comm.). More sensitive estimation of the level of interpopulational genetic divergence of *H. gaumeri* awaits study of genic variation.

Systematic Relationships of *H. gaumeri*

Morphology—Allen and Chapman (1897) described *Heteromys gaumeri* from seven individuals from Chichén Itzá, Yucatán. In his review of

TABLE 3. Geographic variation in six external and cranial measurements of *Heteromys gaumeri*.

Grouped locality	N	Mean (range) ± 2 SE	Percentage of variation: Locality	Percentage of variation: Error	*F*
		TOTAL LENGTH			
1	3	288.7 (276.0–298.0) ± 13.13	23.7	76.3	3.75***
2	16	261.7 (235.0–288.0) ± 8.63			
3	10	276.9 (245.0–302.0) ± 10.98			
4	5	272.0 (262.0–286.0) ± 8.83			
5	9	280.0 (263.0–295.0) ± 7.51			
6	17	286.8 (265.0–324.0) ± 8.29			
7	4	294.5 (288.0–300.0) ± 4.93			
8	3	281.7 (275.0–290.0) ± 8.82			
9	3	263.3 (250.0–280.0) ± 17.64			
10	2	256.5 (253.0–260.0) ± 7.00			
11	4	270.5 (250.0–290.0) ± 18.65			
		LENGTH OF TAIL			
1	3	154.7 (148.0–166.0) ± 11.39	13.6	86.4	2.02*
2	16	144.8 (129.0–159.0) ± 4.12			
3	10	150.4 (124.0–173.0) ± 9.73			
4	5	154.4 (141.0–166.0) ± 9.48			
5	9	161.0 (144.0–193.0) ± 9.22			
6	17	154.4 (135.0–183.0) ± 5.53			
7	4	162.8 (160.0–166.0) ± 2.75			
8	3	152.7 (151.0–155.0) ± 2.40			
9	3	145.7 (140.0–150.0) ± 5.93			
10	2	147.5 (147.0–148.0) ± 1.00			
11	4	146.5 (129.0–160.0) ± 12.92			
		GREATEST LENGTH OF SKULL			
1	3	35.8 (35.2–36.2) ± 0.61	15.6	84.4	2.30*
2	17	35.3 (33.9–37.5) ± 0.44			
3	9	35.0 (33.5–36.3) ± 0.57			
4	5	34.9 (33.6–36.2) ± 0.88			
5	9	35.7 (34.3–36.7) ± 0.52			
6	23	35.9 (34.1–38.2) ± 0.41			
7	6	36.4 (35.8–37.0) ± 0.41			
8	3	35.2 (33.8–36.6) ± 1.62			
9	4	35.5 (34.2–36.6) ± 1.10			
10	3	34.4 (34.0–34.7) ± 0.40			
11	2	34.6 (34.3–34.8) ± 0.50			
		LENGTH OF NASALS			
1	3	15.2 (14.7–15.7) ± 0.58	13.8	86.2	2.19*
2	18	14.3 (13.2–15.6) ± 0.34			
3	10	13.9 (12.8–14.5) ± 0.40			
4	5	14.2 (13.6–15.0) ± 0.61			
5	9	14.6 (13.7–15.5) ± 0.39			
6	25	14.8 (13.6–15.8) ± 0.27			
7	6	14.8 (14.0–15.5) ± 0.47			
8	4	14.6 (13.7–15.8) ± 0.87			
9	4	14.6 (13.9–15.2) ± 0.54			
10	3	13.8 (13.2–14.5) ± 0.75			
11	2	13.9 (13.3–14.5) ± 1.20			
		LENGTH OF ROSTRUM			
1	3	15.4 (14.8–16.3) ± 0.92	10.5	89.5	2.16*
2	18	15.4 (14.2–16.3) ± 0.25			
3	10	15.1 (14.1–15.8) ± 0.38			
4	5	15.2 (14.2–16.0) ± 0.61			
5	8	15.5 (14.8–16.0) ± 0.33			
6	25	15.6 (14.6–16.6) ± 0.23			
7	6	16.0 (15.1–16.7) ± 0.62			
8	4	15.4 (14.4–16.4) ± 0.84			

TABLE 3. *Continued.*

Grouped locality	N	Mean (range) ± 2 SE	Percentage of variation		F
			Locality	Error	
9	4	15.6 (15.2–16.0) ± 0.37			
10	3	14.8 (14.2–15.5) ± 0.75			
11	2	14.6 (14.4–14.8) ± 0.40			
		LENGTH OF MAXILLARY TOOTHROW			
1	3	4.9 (4.7–5.1) ± 0.23	18.9	81.1	2.76**
2	19	4.9 (4.7–5.2) ± 0.07			
3	10	4.9 (4.6–5.2) ± 0.13			
4	5	5.0 (4.9–5.1) ± 0.06			
5	8	4.8 (4.6–5.0) ± 0.08			
6	26	5.0 (4.4–5.3) ± 0.08			
7	6	5.2 (4.9–5.6) ± 0.21			
8	4	4.9 (4.7–5.1) ± 0.17			
9	4	4.8 (4.6–5.0) ± 0.17			
10	4	4.9 (4.8–5.1) ± 0.13			
11	2	5.2 (5.0–5.4) ± 0.40			

Asterisks following *F* statistics indicate levels of significance (see footnote to Table 1). Percentages of variation are based on variance components estimated from the entire data set. Grouped localities are defined in text and outlined in Figure 1.

the Heteromyinae, Goldman (1911) placed *H. gaumeri* in his *H. desmarestianus* group, but noted that it was aberrant and not closely related to any known species. Although *H. gaumeri* possesses characters diagnostic for *Heteromys,* including upper and lower molars with three lophs and lower permanent premolars with three lophs, it also shares some characters with the sister group *Liomys* not shared with other *Heteromys,* including posterior sole of hind foot haired as in *Liomys* rather than naked as in other *Heteromys* and early disappearance of the enamel island between cingulum and metaloph(id) of upper and lower molars, a condition somewhat intermediate between the genera (Goldman, 1911; Genoways, 1973). Assuming *Heteromys* is monophyletic, the characters shared between *H. gaumeri* and *Liomys* probably are primitive (based on the "operational rule" outgroup procedure outlined by Watrous & Wheeler, 1981). Dorsal coloration of *H. gaumeri* also is unusual in that the dorsum has a definite ochraceous cast with a bright ochraceous lateral line resembling *Liomys pictus* and *L. spectabilis* more than other, darker *Heteromys.* However, dorsal coloration in *H. gaumeri* might be convergent on *Liomys* through their common occupation of xeric forest and thorn scrub habitats rather than more mesic forest characteristic of most other species of *Heteromys.*

KARYOLOGY—Compared to the nominate species in the *Heteromys desmarestianus* and *H. anomalus* species groups, the karyotype of *H. gaumeri* also is divergent (see fig. 2). Superficially, the karyotypes of *H. desmarestianus* (2n = 60, FN = 66) and *H. anomalus* (2n = 60, FN = 68) appear similar and might differ by a single rearrangement (heterochromatic addition/deletion or pericentric inversion), whereas the karyotype of *H. gaumeri* (2n = 56, FN = 76) differs by a minimum of seven rearrangements from *H. desmarestianus* and six from *H. anomalus.* The karyotype reported herein for *H. desmarestianus* differs from that reported by Genoways (1973; 2n = 60, FN = 82) in having a larger number of acrocentric elements. Populations of *H. desmarestianus* are variable in fundamental number and the karyotype reported herein is among the lowest FN karyotypes known for the species (D. S. Rogers, pers. comm.).

CONCLUSIONS—We agree with Goldman (1911, p. 30) that *Heteromys gaumeri* "is a somewhat aberrant species, presenting characters which set it off from all the others [species of the *H. desmarestianus* group]." The distinctive morphological, ecological, and karyotypic features of *H. gaumeri* distinguish it from other members of the *H. desmarestianus* group (including *H. goldmani*; D. S. Rogers, pers. comm.). Shared (probably primitive) characters with *Liomys* suggest that *H. gaumeri* might represent an early branch of the lineage leading to other *Heteromys* (which share probable derived states for these characters). At present, we believe the distinctness and unresolved phyloge-

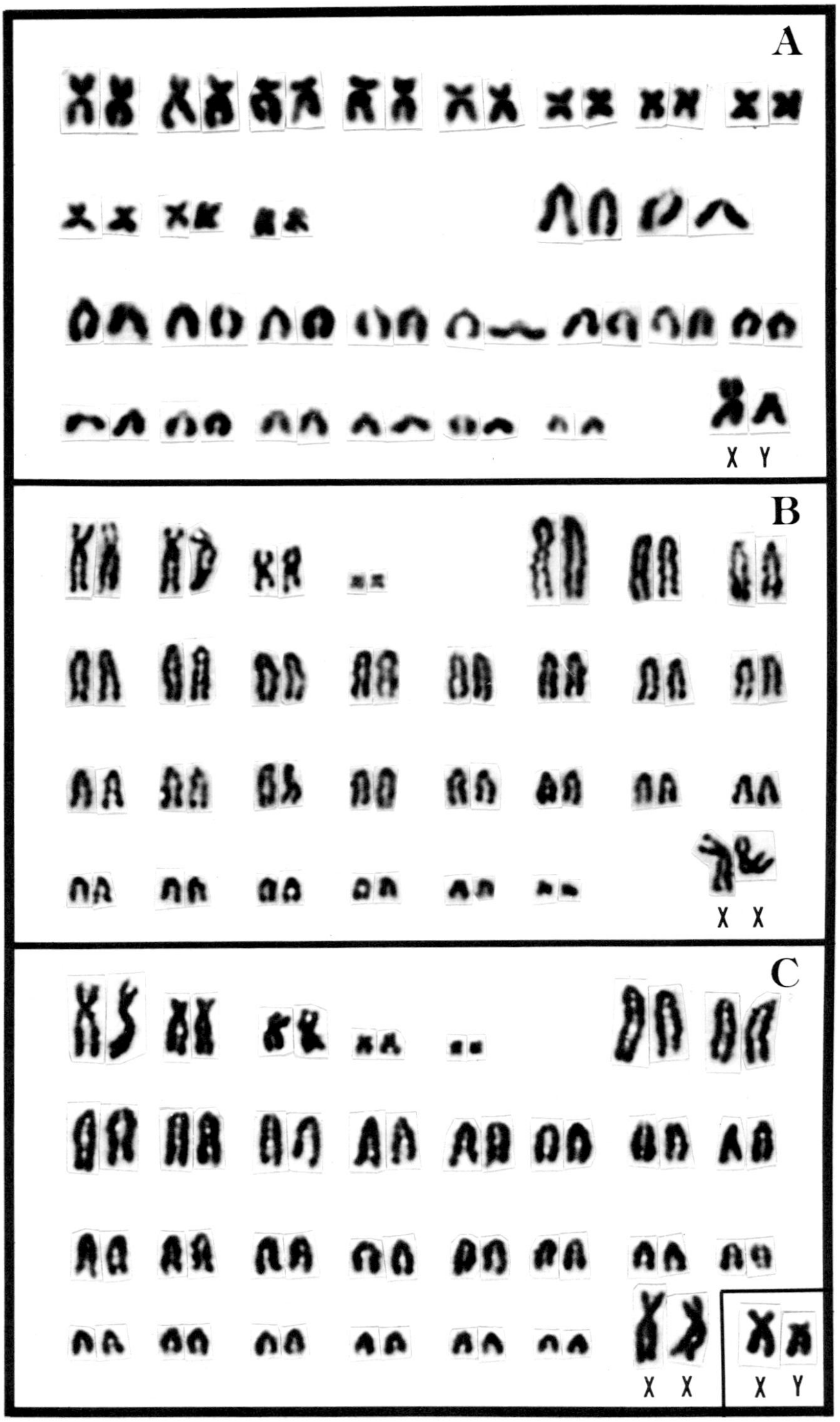

FIG. 2. Representative karyotypes of *Heteromys*: **A,** karyotype of a male *H. gaumeri* (2n = 56, FN = 76) from 7.5 km W Escárcega, Campeche, Mexico; **B,** karyotype of a female *H. desmarestianus* (2n = 60, FN = 66) from 9.4 km S Palenque, Chiapas, Mexico; **C,** karyotype of a female *H. anomalus* (2n = 60, FN = 68) from Caripito, Monagas, Venezuela. Insert shows sex chromosomes of a male.

netic position of *H. gaumeri* would best be emphasized by removing it from the *H. desmarestianus* group, and recognizing it as a divergent species at the same cladogenic level as the species groups in the subgenus *Heteromys*. Further investigation of heteromyines might indicate that *H. gaumeri* warrants subgeneric recognition.

Species Account

Heteromys gaumeri Allen and Chapman, 1897

Heteromys gaumeri Allen and Chapman, 1897, February. Bull. Amer. Mus. Nat. Hist., 9: 9.

HOLOTYPE—Adult male, skin and skull, AMNH 12028/10461; from Chichén Itzá, Yucatán, Mexico. Type examined.

DISTRIBUTION—Northern Belize; Petén, Guatemala; and Campeche, Quintana Roo, and Yucatán, Mexico (see fig. 1); subdeciduous-subperennial tropical rain forest, tropical deciduous forest, and thorn scrub forest from sea level to 100 m.

DESCRIPTION—Dorsal coloration of adults ranges from dark to medium gray, with heavy admixture of orange buff hairs lending an overall ochraceous cast to the otherwise gray dorsum (adults in worn pelage appear more ochraceous and molting individuals often have a 'salt and pepper' appearance); rich orange buff lateral line, usually broad and conspicuous, extending from cheeks to base of tail, ochraceous hairs often extending onto dorsal and ventral surfaces of ankle; margins of forearm orange buff, interrupted on dorsal surface by white line; venter and feet white; tail well haired, grayish brown above, dull white below, with conspicuous terminal tuft of hairs; ears dusky, lightly edged with dull white; sole of hind foot haired, posteriorly. Subadults medium to dark gray above, dorsum without ochraceous hairs; ochraceous lateral line faint and narrow. Juvenile pelage similar to that of subadults, but spiny hairs on dorsum absent. Tail longer than head and body; soles of hind feet with six tubercules; body size medium for the genus. Skull size medium, with relatively large auditory bullae; lower permanent premolar with three lophs; upper and lower molars with three lophs, enamel island formed between metaloph(id) and cingulum disappearing quickly with wear. $2n = 56$, $FN = 76$.

COMPARISONS—*Heteromys gaumeri* is geographically isolated from all other heteromyines except *H. desmarestianus,* which it might contact at the southern edge of the Yucatán Peninsula (see Jones et al., 1974). *Heteromys gaumeri* differs from *H. desmarestianus* in averaging smaller in most external and cranial measurements; having a broad, bright ochraceous lateral line, extending onto cheeks and ankles (a narrow, pale ochraceous lateral line often is present in *H. desmarestianus,* but seldom extends onto cheeks or ankles); having soles of hind feet haired posteriorly (this area is naked in *H. desmarestianus* and all other *Heteromys*); having a relatively well-haired tail with terminal tuft (the tail is sparsely haired in *H. desmarestianus,* without a conspicuous terminal tuft); having relatively large auditory bullae; and in having a diploid number of 56 chromosomes (compared to 60 in *H. desmarestianus*). For additional comparisons with other heteromyines, see Goldman (1911) and Genoways (1973).

REMARKS—Laurie (1957, p. 387) assigned eight specimens from three localities in the state of Yucatán, Mexico (Tekom, 2; X-Cala-Koop, 1; Chichén Itzá, 5) to *Heteromys desmarestianus.* One of us (HHG) reexamined her material, which is stored in alcohol; based on size, coloration, and a haired posterior sole of the hind foot, all are assignable to *H. gaumeri.* Consequently, the northernmost locality for *H. desmarestianus* on the Yucatán Peninsula is 85 km W Chetumal, Quintana Roo, Mexico (Jones et al., 1974). Two other specimens, from Kate's Lagoon, Belize, identified by Laurie (1957, p. 387) as *H. desmarestianus,* are referable to *H. gaumeri.* These specimens, along with additional material from Honey Camp Lagoon (reported by Izor & McCarthy, 1984) and Rockstone Pond reported here, suggest that *H. gaumeri* is distributed throughout northern Belize.

Acknowledgments

Permits to collect specimens in Mexico were kindly provided by the Director General of the Departmento de Conservación de la Fauna Silvestre. Especial thanks are extended to Biol. Francisco Rodriguez Gallegos, Ing. Antonio Sánchez, and personnel of Campo Experimental Forestal "El Tormento" for their hospitality during fieldwork near Escárcega, Campeche. We also thank the many curators who permitted examination of specimens in their care. E. Jaquelin Dietz provided the computer program for testing the association between distance matrices. Kenneth Schoenly prepared the final copy of Figure 1. D. S. Rogers provided information and helpful discussion on character-states within the Heteromyinae. Portions of this study were funded by the Texas Agricultural Experiment Station (project H-1977 to D. J. Schmidly) and a Faculty Organized Research Grant, Angelo State University. Fieldwork in Venezuela was supported by NSF Grant DEB 79-21519 to J. W. Bickham and the Carnegie Museum of Natural History through the M. Graham Netting Research Fund, established by a grant from the Cordelia S. May Charitable Trust.

Literature Cited

ALLEN, J. A., AND F. M. CHAPMAN. 1897. On mammals from Yucatan, with descriptions of new species. Bulletin of the American Museum of Natural History, **9**: 1–12.

BAKER, A. J. 1980. Morphometric differentiation in New Zealand populations of the house sparrow (*Passer domesticus*). Evolution, **34**: 638–653.

CHOATE, J. R., AND H. H. GENOWAYS. 1975. Collections of Recent mammals in North America. Journal of Mammalogy, **56**: 452–502.

DIETZ, E. J. 1983. Permutation tests for association between two distance matrices. Systematic Zoology, **32**: 21–26.

GAUMER, G. F. 1917. Monografia de los mamíferos de Yucatán. Departmento de Taleres Gráficos de la Secretaria de Fomento, México, 331 pp.

GENOWAYS, H. H. 1973. Systematics and evolutionary relationships of the spiny pocket mice, genus Liomys. Special Publications, The Museum, Texas Tech University, **5**: 1–368.

GOLDMAN, E. A. 1911. Revision of the spiny pocket mice (genera Heteromys and Liomys). North American Fauna, **34**: 1–70.

HALL, E. R. 1981. The Mammals of North America, vol. 1. John Wiley and Sons, New York, 600 + 90 pp.

HALL, E. R., AND K. R. KELSON. 1959. The Mammals of North America, vol. 1. Ronald Press, New York, 546 + 79 pp.

HATT, R. T., H. I. FISHER, D. A. LANGEBARTEL, AND G. W. BRAINERD. 1953. Faunal and archeological researches in Yucatan caves. Bulletin of the Cranbrook Institute of Science, **33**: 1–119.

HUCKABY, D. G. 1980. Species limits in the *Peromyscus mexicanus* group (Mammalia: Rodentia: Muroidea). Contributions in Science, Natural History Museum of Los Angeles County, **326**: 1–24.

IZOR, R. J., AND T. J. MCCARTHY. 1984. *Heteromys gaumeri* (Rodentia: Heteromyidae) in the Northern Plain of Belize. Mammalia, **48**: 465–467.

JOHNSTON, R. F. 1976. Estimating variation in bony characters and a comment on the Kluge-Kerfoot effect. Occasional Papers of the Museum of Natural History, University of Kansas, **56**: 1–8.

JOHNSTON, R. F., AND W. J. KLITZ. 1977. Variation and evolution in a granivorous bird: The house sparrow, pp. 15–51. *In* Pinowski, J., and S. C. Kendeigh, eds., Granivorous Birds in Ecosystems, Their Evolution, Populations, Energetics, Adaptations, Impact and Control. Cambridge University Press, London.

JONES, J. K., JR., H. H. GENOWAYS, AND T. E. LAWLOR. 1974. Annotated checklist of mammals of the Yucatan Peninsula, Mexico II. Rodentia. Occasional Papers, The Museum, Texas Tech University, **22**: 1–24.

KLUGE, A. G., AND W. C. KERFOOT. 1973. The predictability and regularity of character divergence. American Naturalist, **107**: 426–442.

LAURIE, E. M. O. 1957. Rodents from British Honduras, Mexico, Trinidad, Haiti, and Jamaica collected by Mr. I. T. Sanderson. Annals of the Magazine of Natural History, series 12, **6**: 382–394.

LAWLOR, T. E. 1965. The Yucatan deer mouse, Peromyscus yucatanicus. University of Kansas Publications, Museum of Natural History, **16**: 421–438.

LEAMY, L. 1983. Variance partitioning and effects of sex and age on morphometric traits in randombred house mice. Journal of Mammalogy, **64**: 55–61.

LEE, M. R. 1969. A widely applicable technique for direct processing of bone marrow for chromosomes of vertebrates. Stain Technology, **44**: 155–158.

PATTON, J. L. 1967. Chromosome studies of certain pocket mice, genus *Perognathus* (Rodentia: Heteromyidae). Journal of Mammalogy, **48**: 27–37.

PAYNTER, R. A., JR. 1955. The ornithogeography of the Yucatán Peninsula. Bulletin of the Peabody Museum of Natural History, Yale University, **9**: 1–347.

ROGERS, D. S., AND D. J. SCHMIDLY. 1982. Systematics of spiny pocket mice (genus *Heteromys*) of the *desmarestianus* species group from México and northern Central America. Journal of Mammalogy, **63**: 375–386.

ROHLF, F. J., A. J. GILMARTIN, AND G. HART. 1983. The Kluge-Kerfoot phenomenon—A statistical artifact. Evolution, **37**: 180–202.

ROHLF, F. J., J. KISHPAUGH, AND D. KIRK. 1982. Numerical Taxonomy System of Multivariate Statistical Programs. State University of New York at Stony Brook, Stony Brook, New York, 98 pp.

RYAN, R. M. 1960. Mamíferos colectados en Guatemala en 1954. Acta Zoologica Mexicana, **4**: 1–19.

SAS INSTITUTE, INC. 1982. SAS User's Guide: Statistics, 1982 ed. SAS Institute, Cary, North Carolina, 584 pp.

SEARLE, R. S. 1971. Linear Models. John Wiley and Sons, New York, 535 pp.

SNEATH, P. H. A., AND R. R. SOKAL. 1973. Numerical Taxonomy. W. H. Freeman, San Francisco, 573 pp.

SOKAL, R. R. 1976. The Kluge-Kerfoot phenomenon reexamined. American Naturalist, **110**: 1077–1091.

———. 1979. Testing statistical significance of geographic variation patterns. Systematic Zoology, **28**: 232–238.

SOKAL, R. R., AND N. L. ODEN. 1978. Spatial autocorrelation in biology. 2. Some biological implications and four applications of evolutionary and ecological interest. Biological Journal of the Linnean Society, **10**: 229–249.

SOKAL, R. R., AND F. J. ROHLF. 1981. Biometry. W. H. Freeman, San Francisco, 859 pp.

STRANEY, D. O. 1978. Variance partitioning and nongeographic variation. Journal of Mammalogy, **59**: 1–11.

WATROUS, L. E., AND Q. D. WHEELER. 1981. The outgroup comparison method of character analysis. Systematic Zoology, **30**: 1–11.

Appendix

Specimens Examined

MEXICO: Campeche: Apazote, near Yohaltún, 6 (USNM); Campeche, 1 (USNM); 5 km S Champotón, 10 m, 3 (KU); Dzibalchén, 1 (KU); 7 km N, 51 km E Escárcega (= Francisco Escárcega), 4 (KU); 7.5 km W Escárcega, 65 m, 94 (15 KU, 79 TCWC); 7 km E Escárcega, 4 (KU); 103 km SE Escárcega, 2 (KU); La Tuxpeña, Champotón, 12 (USNM); San José Carpizo, 3 (FMNH); San José Carpizo, 45 km S Campeche, 19 (UMMZ); San Juan, 4 (FMNH). **Quintana Roo:** 2 km N, 8 km W Bacalar, 11 (TCWC); 8 mi NNE Felipe Carrillo Puerto, 4 (TCWC); 4 km NNE Felipe Carrillo Puerto, 30 m, 9 (KU); 2.5 mi NNE Felipe Carrillo Puerto, 1 (TCWC); La Vega, 5 (USNM); Pueblo Nuevo Xcan, 10 m, 1 (KU); 1.5 km S, 1 km E Pueblo Nuevo Xcan, 1 (KU); Puerto Morelos, 4 (USNM). **Yucatán:** Calcehtok, 55 km SW Mérida, 2 (1 KU, 1 UMMZ); Chichén Itzá, 10 m, 36 (15 AMNH, 11 BMNH, 2 FMNH, 1 KU, 7 USNM); Chichén Itzá, Cenote Xtoloc, 1 (TCWC); 2 km E Chichén Itzá, Cenote Seco, 5 (TCWC); 4 km E Dzitya, 1 (MMNH); Esmeralda, Laguna de Chichancanab, 2 (1 KU, 1 UMMZ); Esmeralda, 45 km S Peto, 4 (UMMZ); 1 km SW Casa de la Esmeralda, Laguna de Chichancanab, 2 (1 KU, 1 UNAM); 2 km SSW Casa de la Esmeralda, Laguna de Chichancanab, 3 (1 AMNH, 2 UNAM); La Ceiba, 2 km SE Laguna de Chichancanab, 1 (UNAM); 66 km NE Mérida, 1 (KU); 14 km SW Muna, 1 (TTU); Oxkutzcab, Ebiztic (cave deposit), 6 (USNM); Peto, 3 (KU); 3 km N Pisté, 16 (7 KU, 6 MMNH, 3 TTU); 2 km N Pisté, 2 (KU); Pisté, 10 m, 2 (KU); Progreso, 1 (USNM); Santa Rosa, 1 (UMMZ); Santa Rosa, 20 km S Peto, 2 (UMMZ); Santa Rosa, 25 km S Peto, 1 (UMMZ); 1 km SSW Santa Rosa, 3 (1 KU, 1 UMMZ, 1 UNAM); Tekom, 2 (BMNH); 6 km N Tizimín, 1 (KU); Tunkás, 7 (USNM). Buena Vista, 1 (FMNH) and X-Cala-Koop, 1 (BMNH) were not located exactly. **GUATEMALA: Petén:** Chuntuquí, 2 (USNM); Laguna de Sotz [= Zotz], 2 (USNM). **BELIZE: Belize:** Kate's Lagoon, 2 (BMNH); Rockstone Pond, 11 (ROM). **Orange Walk:** Honey Camp Lagoon, 2 (FMNH).

ADDITIONAL RECORDS (Hatt et al., 1953, p. 64, unless noted otherwise)—**MEXICO: Yucatán:** Actun Chacaljas, 3 km SSW Calcehtok (cave deposit); Actun Coyok [= Coyoc], 3.5 km SSE Oxkutzcab (cave deposit); Actun Has, 3.5 km WSW Yokat (cave deposit); Actun Lara, 3 km SW Yokat (cave deposit); Actun Oxkintok, 3 km SW Santa Cruz (cave deposit); Actun Spukil, 4.5 km SSW Calcehtok (cave deposit); Loltun, 5 km SW Oxkutzcab (cave deposit); Xbac (Gaumer, 1917, p. 13); Yaxcach (Gaumer, 1917, p. 13). **GUATEMALA: Petén:** 11 km NE Flores, 3.2 km inland SE shore Laguna Petén Itzá, ca. 100 m (Ryan, 1960, p. 11).

Species Groups of Spiny Rats, Genus *Proechimys* (Rodentia: Echimyidae)

James L. Patton

ABSTRACTS

Nine species groups of the spiny rat subgenus *Proechimys,* family Echimyidae, are defined on the basis of bacular characters and qualitative features of the cranium. The latter include the structure of the incisive and mesopterygoid foramina, temporal ridge and infraorbital canal development, and counterfold pattern of the cheekteeth.

Three groups are apparently monotypic, including the *decumanus*-group of southwestern Ecuador and adjacent Peru, the *canicollis*-group of northeastern Colombia and adjacent Venezuela, and the *simonsi*-group of the western Amazon Basin from Colombia to northern Bolivia. The remaining six groups are polytypic, but the number of species in each remains unclear. The *semispinosus*-group ranges from Central America south to southwestern Ecuador in the Pacific lowlands; its only Amazonian representative is *P. oconnelli* from east-central Colombia. The *longicaudatus*-group ranges from southeastern Colombia through the western Amazon Basin into the northern Paraná Basin of Brazil and northern Paraguay. The *goeldii*-group ranges throughout the Amazon Basin from eastern Peru to eastern Brazil. The *guyannensis*-group occurs from the coastal Guianan region through the Río Negro and eastern half of the Amazon Basin in Brazil, with an isolate in Goiás and Minas Gerais states. The *cuvieri*-group has a similar distribution, but extends further up the Amazon into northern Peru, with one isolate in east-central Peru. Finally, the *trinitatus*-group is found from north-central Colombia eastward across northern Venezuela to Trinidad.

Nueve grupos de especies de la rata espinosa subgénero *Proechimys,* familia Echimyidae, son definidos primariamente en las bases de caracteres baculares que son soportados por razgos cualitativos del cráneo. El último incluye la estructura del foramen incisivo y mesopterigoideo, arista temporal y desarrollo del canal infraorbital y patrón de contraplegamiento de los dientes postcaninos.

Tres grupos son aparentemente monotípicos, incluyendo el grupo *decumanus* del suroccidente del Ecuador y adyacente Perú, el grupo *canicollis* del nororiente de Colombia y adyacente Venezuela, y el grupo *simonsi* del occidente de la cuenca amazónica desde Colombia hasta el norte de Bolivia. Los seis grupos remanentes son politípicos, pero el número de especies en cada uno permanece obscuro. El grupo *semispinosus* se extiende desde el sur de Centroamérica hasta el suroccidente ecuatoriano en las tierras bajas del Pacífico; su solo representante de la cuenca amazónica el *P. oconnelli* del centroriente colombiano. El grupo *longicaudatus* se distribuye desde el suroriente de Colombia a través del occidente de la cuenca amazónica hasta el norte de la cuenca del Paraná en Brazil y norte de Paraguay. El grupo *goeldii* se distribuye a través de la cuenca amazónica desde el Perú oriental hasta el Brazil oriental. El grupo *guyannensis* ocurre desde la región costera guayanesa a través de río Negro y la mitad oriental de la cuenca amazónica en Brazil, con un aislamiento de poblaciones en los estados de Goiás

From Museum of Vertebrate Zoology, University of California, Berkeley, CA 94720.

y Minas Gerais. El grupo *cuvieri* tiene una distribución similar, pero se extiende más arriba del Amazonas en el interior del norte del Perú, con un aislamiento de poblaciones en el centroriente peruano. Finalmente, el grupo *trinitatus* es encontrado desde el centronorte de Colombia hacia el oriente a través del norte de Venezuela hasta Trinidad.

Nove grupos de espécies de ratos-de-espinho, do subgênero *Proechimys,* família Echimyidae, são definidos principalmente na base de caráteres baculares que concordam também com carácteres qualitativos do crânio. Estes incluem: a estrutura dos incisivos e do fôrame mesopterigóideo, o desenvolvimento das têmporas e do canal infraorbital, e o padrão dos molares.

Três grupos são aparentemente monotípicos, incluindo o grupo *decumanus* do sudoeste do Equador, o grupo *canicollis* do nordeste da Colombia e das áreas adjacentes na Venezuela, e o grupo *simonsi* da Bacia Amazônica ocidental, desde a Colombia até o norte da Bolívia. Os seis grupos restantes são politípicos, mas o número de espécies em cada continua incerto. O grupo *semispinosus* estende-se da America Central ao sudoeste do Equador, nas planícies do Pacífico. Seu único representante na Bacia Amazônica é *P. oconnelli,* do Centro-leste da Colombia. O grupo *longicaudatus* estende-se do sudeste da Colombia, através da Amazonia ocidental, até o norte da Bacia do Paraná no Brasil e no norte do Paraguai. O grupo *goeldii* ocorre na Bacia Amazônica, do leste do Peru ao leste do Brasil. O grupo *guyannensis* ocorre da costa guianense, até o Rio Negro e à região oriental da Bacia Amazônica no Brasil, com uma espécie isolada nos estados de Goiás e de Minas Gerais. A distribuição do grupo *cuvieri* é parecida, mas este ocorre também até o norte do Perú, com uma espécie isolada no centro-leste do País. Por final, o grupo *trinitatus* é encontrado do centro-norte da Colombia, através da Venezuela, até Trinidade.

Introduction

Spiny rats of the genus *Proechimys* represent one of the most diverse groups of Neotropical rodents; with the possible exception of tuco-tucos, *Ctenomys,* the number of taxa of spiny rats is unmatched by any other caviomorph (Woods, 1984). The genus extends throughout lowland forests from Nicaragua to northern Paraguay and the coastal regions of Brazil. Despite this diversity and large geographic range, however, the group is taxonomically one of the most poorly understood among all of the Neotropical mammals. Only a few studies have succeeded in recognizing the number of taxa sympatric at any single locality (e.g., Moojen, 1948; Patton & Gardner, 1972), and no study has been able to follow geographic character trends within a clearly defined taxon over any but the shortest distances. Diagnosis of species and hence definition of natural units in *Proechimys* have been severely hampered by the often extreme level of variability within and between populations for most morphological characters that have been examined (Moojen, 1948; Hershkovitz, 1948; Patton & Gardner, 1972). Even karyotypes, which have proven useful in differentiating sympatric taxa of spiny rats (Patton & Gardner, 1972), are often highly variable geographically (Reig & Useche, 1976; Reig et al., 1980; Gardner & Emmons, 1984).

In this report I will challenge some of these precepts of character instability (see Thomas, 1928, p. 262) by using the structure of the baculum and specific characters of the cranium to define major taxonomic groups of spiny rats. Taxa can be diagnosed, despite both within- and between-population variation, and the patterns of character variation over geography are coherent, permitting a consistent view of these taxa throughout their range. Some of the more traditional characters that have been used to make taxonomic decisions in spiny rats (e.g., counterfold patterns on the cheekteeth) are not chaotic in their variation patterns, but are quite helpful in defining units.

This paper will consider only members of the subgenus *Proechimys,* excluding entirely the group of species found along the Atlantic highlands of Brazil that compose the subgenus *Trinomys.*

Species Groups of *Proechimys*

In the section below I provide the basic species groups of spiny rats, subgenus *Proechimys,* listing

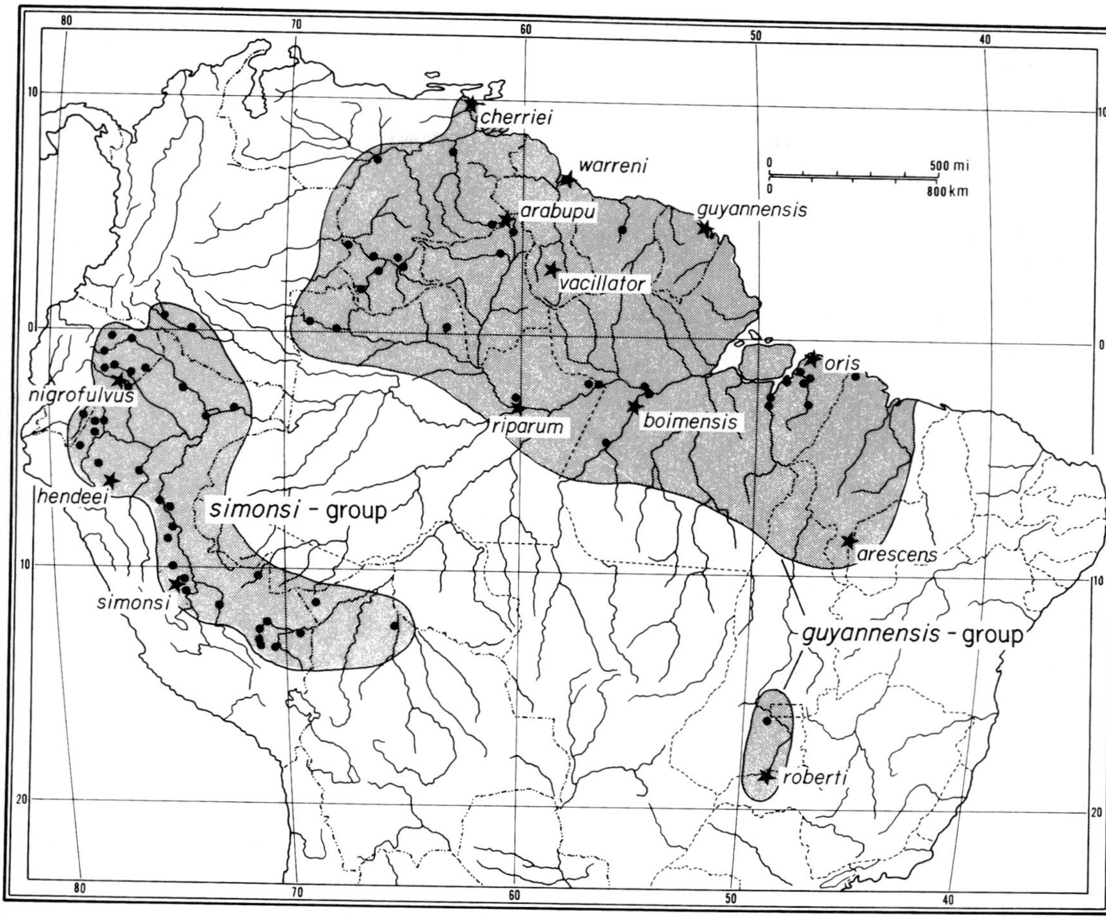

FIG. 1. Geographic distribution of taxa of the *simonsi*- and *guyannensis*-groups. Type localities for each included taxon are indicated by stars; dots represent other localities listed in the Appendix.

those named forms I consider as component parts. In recognizing these groups and their membership I make no conclusions here as to the specific, subspecific, or other status of these names. Because of the plethora of names available and the confusion with which each has been applied to the genus over the past century, this synopsis is provided first to allow for coherent discussion; the documentation upon which these decisions are based follows.

I recognize nine species groups within the subgenus *Proechimys.* Five of these are widespread, while the remaining ones are more restricted in their ranges. Maps of the distribution of each group, with localities of included holotypes, are presented in Figures 1–4 (see Specimens Examined for lists of localities). Unless otherwise stated, allocation of any given holotype to a specific species-group is based on my examination of that specimen. The groups are defined by a combination of palatal (particularly incisive foramina and mesopterygoid fossa) characters, counterfold patterns of the cheekteeth, temporal ridge development, infraorbital notch development, and bacular characters (see below). In each case, the group name is taken from the oldest assignable name for that unit.

guyannensis-group

Named forms in this group include:

guyannensis (E. Geoffroy, 1803)
cherriei Thomas, 1899
roberti Thomas, 1901
vacillator Thomas, 1903
oris Thomas, 1904
warreni Thomas, 1905

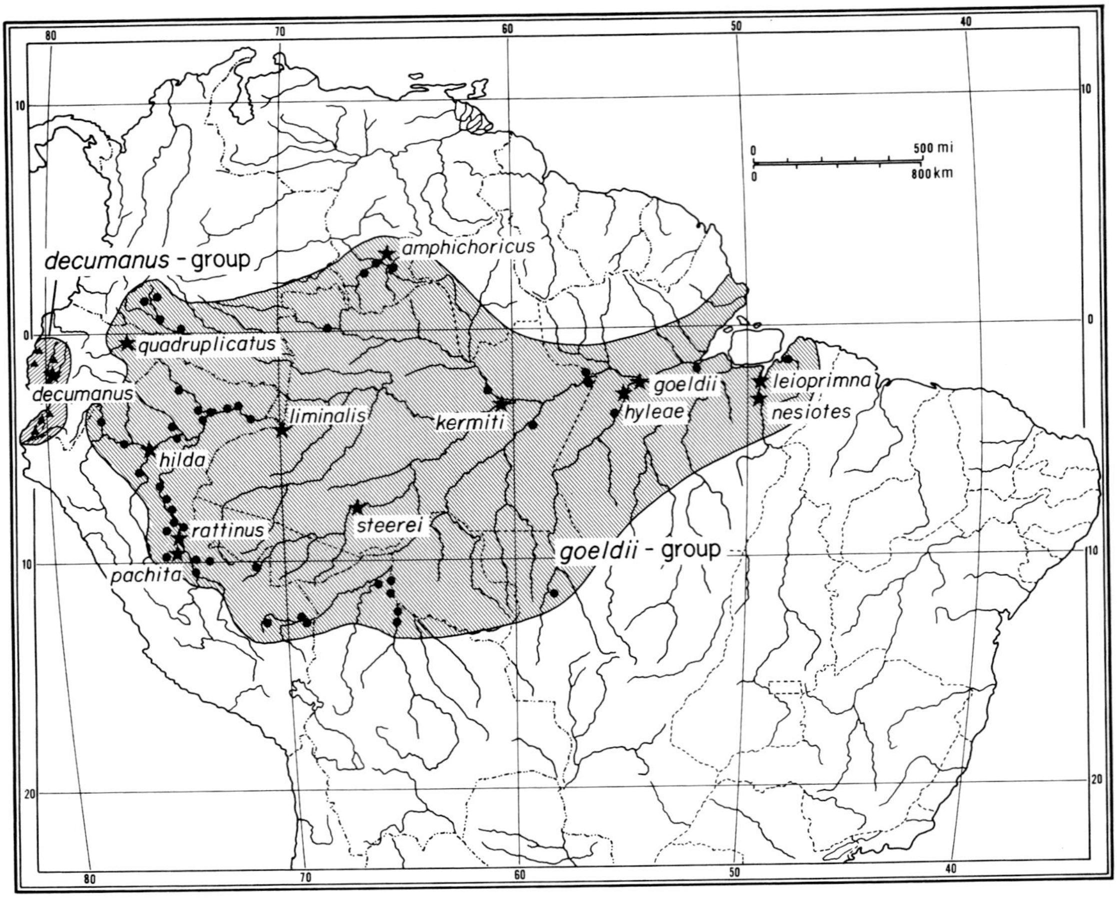

FIG. 2. Geographic distribution of taxa of the *goeldii*-group (dots) and *decumanus*-group (triangles). Type localities of taxa are indicated by stars.

boimensis Allen, 1916
arescens Osgood, 1944
riparum Moojen, 1948
arabupu Moojen, 1948

COMMENTS—On the basis of septal patterns in the bullae, Gardner and Emmons (1984) included these taxa in their *brevicauda*-group, an all-inclusive unit combining taxa that are here allocated to six separate groups. As will be apparent below, my *guyannensis*-group only shows close similarity to the taxa included in the *simonsi*-group. Members of these two groups share virtually no characters with the remaining taxa listed by Gardner and Emmons (1984) in their *brevicauda*-group.

This group is confined in its distribution to the Guianan region and southern Venezuela through the central Amazon Basin of Brazil, with an isolate (*roberti*) in Minas Gerais and Goiás states in Brazil (see map, fig. 1). It is sympatric with members of the *cuvieri*-group in the Guianan region (see Petter, 1978) and with those of both the *cuvieri*- and *goeldii*-groups in the central Amazon Basin.

goeldii-group

Included are the following named forms:

goeldii Thomas, 1905
steerei Goldman, 1911
kermiti Allen, 1915
pachita Thomas, 1923
hilda Thomas, 1924
rattinus Thomas, 1926
quadruplicatus Hershkovitz, 1948
liminalis Moojen, 1948
amphichoricus Moojen, 1948
hyleae Moojen, 1948
nesiotes Moojen, 1948
leioprimna Moojen, 1948

COMMENTS—I have not examined the holotypes of *liminalis* Moojen or *hyleae* Moojen; their inclusion here is based on the original descriptions (Moojen, 1948).

This group is distributed throughout the Amazon Basin, from the most western margins in northern Bolivia, eastern Peru, Ecuador, and southeastern Colombia to southern Venezuela east along the central Amazon to the lower Rio Tapajós in Pará state, Brazil (see map, fig. 2). Members of this group are sympatric with those of the *guyannensis*-group and *cuvieri*-group in the central and eastern Amazon Basin, and with those of the *cuvieri*-, *longicaudatus*-, and *simonsi*-groups in the western parts of the Basin.

longicaudatus-group

Named forms in this group include:

longicaudatus (Rengger, 1830)
brevicauda (Gunther, 1877)
boliviensis Thomas, 1901
securus Thomas, 1902
gularis Thomas, 1911
leucomystax Ribeiro, 1914
elassopus Osgood, 1944
villacauda Moojen, 1948
ribeiroi Moojen, 1948

COMMENTS—The taxa *leucomystax* Ribeiro, *villacauda* Moojen, and *ribeiroi* Moojen are included on the basis of descriptions given in Moojen (1948); I have not examined the holotypes.

The *longicaudatus*-group is confined to the western and southwestern parts of the Amazon Basin and northern Paraná Basin, from northern Paraguay and adjacent Brazil west and northwest through Bolivia, eastern Peru, eastern Ecuador, and southeastern Colombia (see map, fig. 3). In this region it is sympatric with members of the *goeldii*-, *cuvieri*-, and *simonsi*-groups.

simonsi-group

Included members are:

simonsi Thomas, 1900
hendeei Thomas, 1926
nigrofulvus Osgood, 1944

COMMENTS—This is perhaps the most readily recognizable of all groups of *Proechimys*; the level of differences with sympatric taxa of the *goeldii*-, *longicaudatus*-, or *cuvieri*-groups is geographically consistent and quite sharp.

The *simonsi*-group is geographically restricted to the western margins of the Amazon Basin from northern Bolivia through eastern Peru and Ecuador to southeastern Colombia (see map, fig. 1). In this region, it ranges to higher elevations than any other species in the genus, occurring as high as 2000 m.

cuvieri-group

This group includes only the nominate form:

cuvieri Petter, 1978

COMMENTS—Specimens assigned to this species are relatively few in number and are known from localities scattered from the coastal Guianan region and along the Amazon River from near its mouth to northern Peru (see map, fig. 4). In the Guianas, *cuvieri* is sympatric with *guyannensis*; in central Brazil, with *guyannensis*- and *goeldii*-group taxa; and in northern Peru, with *simonsi*-, *longicaudatus*-, and *goeldii*-group members. Despite the paucity of widely scattered locality records, these specimens share common bacular, palatal, counterfold, and karyotypic characters, the former being particularly divergent from other taxa in the genus *Proechimys*.

Because of similar bacular (but not karyotypic) features, the form from eastern Peru referred to *P. guyannensis* by Patton and Gardner (1972) is included here. This form is an enigma at the moment, and its placement must be considered provisional; it is not known with certainty from any locality other than Balta, Río Curanja, Ucayali, Peru (see map, fig. 4, and Patton & Gardner, 1972). Gardner and Emmons (1984) consider it closely related to *P. guyannensis,* perhaps even conspecific, but it does not share bacular or incisive foraminal characters with members of that group (see below).

trinitatus-group

Named forms in this group include:

trinitatus (Allen and Chapman, 1893)
chrysaeolus (Thomas, 1898)
mincae (Allen, 1899)
urichi (Allen, 1899)

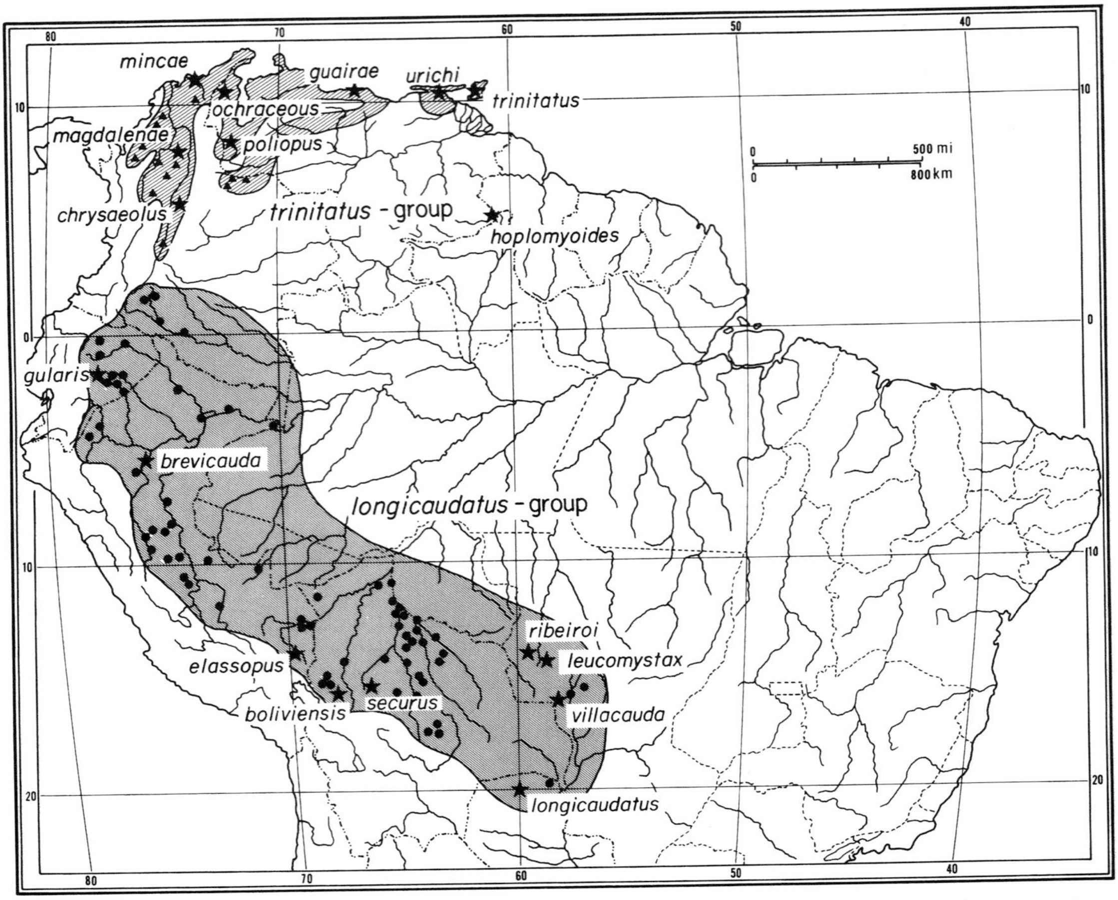

FIG. 3. Geographic distribution of taxa of the *longicaudatus*-group (dots) and *trinitatus*-group (triangles). Type localities are indicated by stars.

guairae Thomas, 1901
ochraceous Osgood, 1912
poliopus Osgood, 1914
hoplomyoides Tate, 1939
magdalenae Hershkovitz, 1948

COMMENTS—Gardner and Emmons (1984) included *magdalenae* in their *brevicauda*-group and *chrysaeolus* in their *semispinosus*-group based on similarities of bullar septal patterns. Bacular and other characters, however, align these two taxa with others of the *trinitatus*-group as here defined. This group includes all members referred to as the "*guiarae* complex" by Reig and co-workers, based on karyotypic data (see Benado et al., 1979; Reig et al., 1980; Reig, 1981), or as the *guairae*-group by Gardner and Emmons (1984).

Members of this group are distributed across northern South America, from Trinidad through the coastal mountains and upper llanos of Venezuela and Colombia, and including the isolated northern Andean valleys of Colombia (see map, fig. 3). Only *hoplomyoides* occurs south of the Río Orinoco in southeastern and southern Venezuela (see Gardner & Emmons, 1984). The included taxa are largely allopatric, or parapatric (see Reig et al., 1980; Reig, 1981), and are separated by rivers (upper llanos of Venezuela) or mountain ridges (northwestern Venezuela and northern Colombia). Sympatric contact between members of this group and other *Proechimys* occurs in several areas in northern Colombia: *chrysaeolus* with *semispinosus* and both *chrysaeolus* and *mincae* with *canicollis*.

***semispinosus*-group**

Membership in this group includes the following named forms:

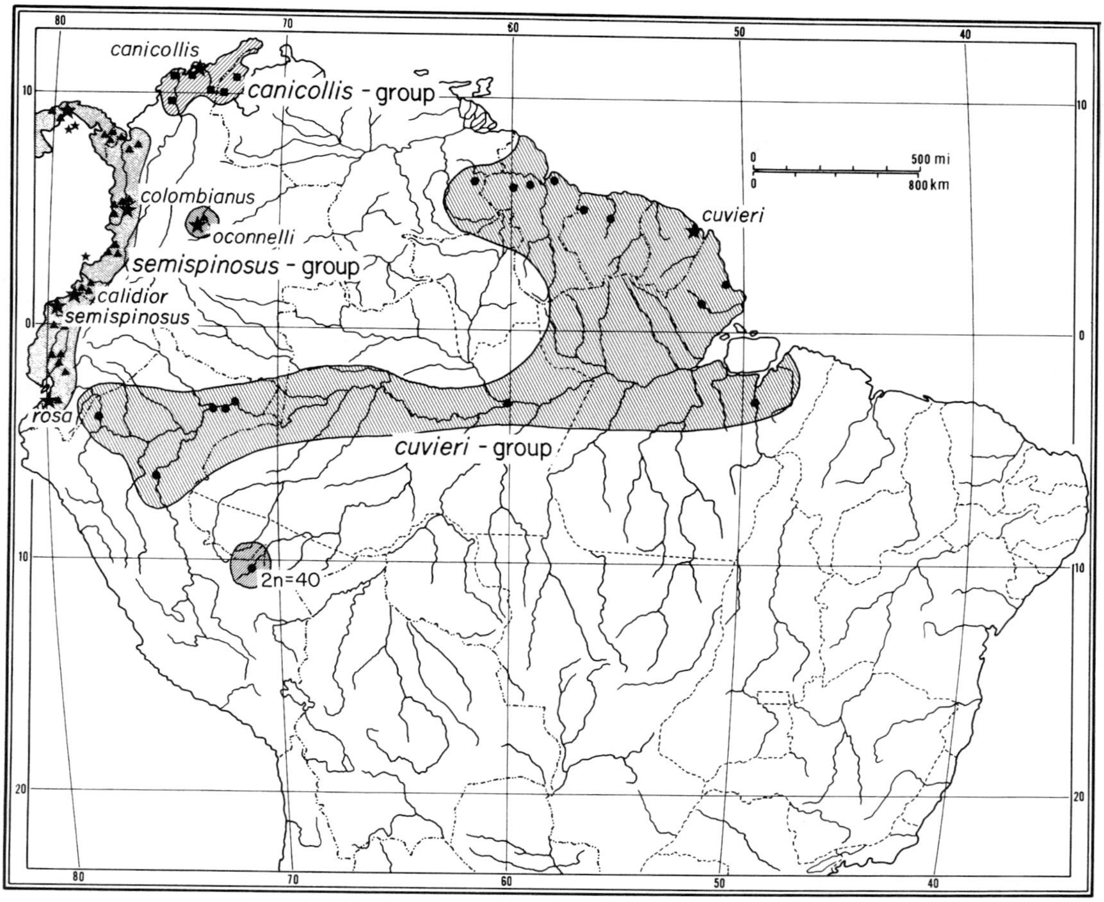

FIG. 4. Geographic distribution of taxa of the *cuvieri*-group (dots), *semispinosus*-group (triangles), and *canicollis*-group (squares). Type localities are indicated by stars.

semispinosus (Tomes, 1860)
centralis (Thomas, 1896)
rosa Thomas, 1900
chiriquinus Thomas, 1900
panamensis Thomas, 1900
burrus Bangs, 1901
gorgonae Bangs, 1905
calidior Thomas, 1911
oconnelli Allen, 1913
rubellus Hollister, 1914
colombianus Thomas, 1914
goldmani Bole, 1937
ignotus Kellogg, 1946

COMMENTS—Gardner (1983) reviewed the membership, distribution, and taxonomic history of this species group, and I concur with him. Gardner and Emmons (1984) expanded the group to include *oconnelli* and *chrysaeolus* based on common bullar septal patterns. I treat *oconnelli* as a component part of the *semispinosus*-group, but place *chrysaeolus* in the *trinitatus*-group based on bacular, palatal, and counterfold patterns (see below).

Members of the *semispinosus*-group are largely restricted in distribution to Central America and the Pacific lowlands of Colombia and Ecuador (see map, fig. 4). The only Amazonian representative is *oconnelli*, which is restricted to the limited area of Villavicencio in east-central Colombia, midway between the northernmost distributional extensions of members of the *goeldii*-, *simonsi*-, and *longicaudatus*-groups and the westernmost extension of the *trinitatus*-group.

***canicollis*-group**

This group is limited to the nominate species:

canicollis (Allen, 1899)

COMMENTS—The species *P. canicollis* is one of the more readily recognizable in the entire genus *Proechimys* (see below), although on the basis of bullar septal patterns, Gardner and Emmons (1984) included it within their *brevicauda*-group. It is limited in its distribution to the coastal forested foothills from northern Bolívar, Colombia, to northwestern Zulia, Venezuela (see map, fig. 4). It is sympatric with *mincae* and *chrysaeolus* of the *trinitatus*-group.

decumanus-group

This group includes only the nominate species:

decumanus (Thomas, 1899)

COMMENTS—I consider this taxon to represent a separate species group, although Gardner and Emmons (1984) placed it in their *brevicauda*-group based on bullar septal patterns and karyotypic similarities. It is, however, readily distinguishable on external, palatal, and bacular grounds from other members of that group.

Proechimys decumanus is restricted to the Pacific lowland forests of extreme southwestern Ecuador and adjacent northwestern Peru (see map, fig. 2), where it is sympatric with the named form *rosa* of the *semispinosus*-group.

Bacular Structure and Characteristics

Didier (1962) described and figured bacular variants of echimyid rodents, with an emphasis on variation within the genus *Proechimys*. Bacular variants of *Proechimys* were also described by Martin (1970) and Patton and Gardner (1972). In the latter paper, bacular characters and karyotypes permitted the delineation of taxa sympatric at several localities in eastern Peru, suggesting that such structures could be of use in distinguishing taxa within the genus as a whole. Didier's (1962) material and the names he used were supplied by Philip Hershkovitz, from specimens in Field Museum collections. I have examined each of the bacula discussed by Didier, as well as the associated skulls and skins. Below, as I describe bacular variation in the context of the species groups recognized in this report, I will emend Didier's groupings according to this reexamination of materials.

A basic bacular type characterizes most species in the genus *Proechimys*, as well as other echimyid genera, despite differences in overall size. This baculum is an elongated, narrow structure, with a rather rounded and broadened base and a shaft tapering distally. The distal tip shows only a weakly developed median depression, if any at all. This bacular type is characteristic of the dactylomyine genera *Thrinacodus, Kannabateomys,* and *Dactylomys,* species in the genera *Echimys, Makalata, Mesomys, Diplomys,* and *Isothrix* (see Didier, 1962; Patton & Emmons, 1985), and most species of *Proechimys*. It is also characteristic of most other caviomorphs examined to date (e.g., *Cavia, Abrocoma, Ctenomys, Agouti, Dasyprocta* [Didier, 1962; Hooper, 1961]). There are, however, both subtle as well as more marked differences among bacula of this general form, and these will be detailed below in the discussion of variation in *Proechimys.*

Overall size of the baculum of *Proechimys* depends strongly on age, although age does not noticeably affect shape. Hence, in the descriptions and summarized measurements given below, analyses are restricted to those individuals considered adults on the basis of tooth wear patterns (age classes 8 through 10 of Patton & Rogers, 1983). Bacula of each species group of *Proechimys* are illustrated in Figures 5–11, and measurements for geographic representatives of each group are summarized in Table 1. Figure 12 illustrates differences among the species groups in proportions of length and width. It is clear from this figure that there are two major classes of bacular variants in the subgenus *Proechimys.* Members of the *guyannensis-, simonsi-, trinitatus-, goeldii-, decumanus-*, and *canicollis*-groups have long and narrow bacula (despite differences in relative size and other characters), while taxa of the *semispinosus-, cuvieri-*, and *longicaudatus*-groups have massively long and broad bacula with well-developed distal apical wings or extensions.

guyannensis-group (Figure 5a–g)

Despite Didier's (1962) use of the name *guyannensis,* no specimen he examined can be referred to this group as defined here or as recognized by Gardner and Emmons (1984). The four bacular types Didier (1962, pp. 408–415) referred to *Proechimys guyannensis* in fact represent members of five different species groups, as follows: Type I (part *semispinosus*-group, part *cuvieri*-group), Type II (*longicaudatus*-group), Type III (*P.

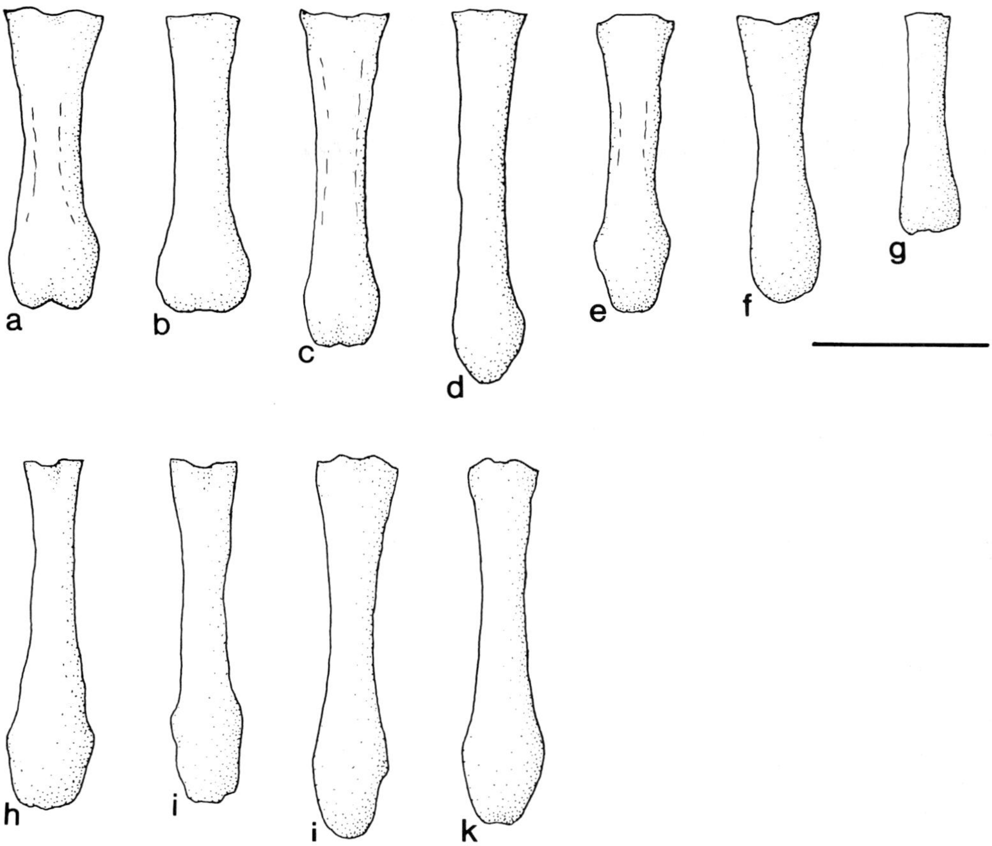

FIG. 5. Representative bacula of members of the *guyannensis*-group (**a**–**g**) and the *simonsi*-group (**h**–**k**); scale = 5 mm. **a,** BM[NH] 52.1124—Suriname: Zanderig. **b,** FMNH 95726—Suriname: Brokopando; Saramacca River, Loksie Hatti. **c,** AMNH 75803—Venezuela: Bolivár; Arabupu, Mt. Roraima (topotype of *arabupu* Moojen). **d,** AMNH 75451—Brazil: Roraima; Rio Cotingo, Limão. **e,** USNM 554847—Brazil: Amazonas; 72 km N Manaus. **f,** USNM 555653—Brazil: Pará; Altar da Chao, Rio Tapajós. **g,** BM[NH] 1.11.3.64—Brazil: Minas Gerais; Araguay, Rio Jordão (topotype of *roberti* Thomas). **h,** AMNH 71866—Peru: Loreto; Boca Río Curaray. **i,** USNM 461305—Peru: Ucayali; 59 km W Pucallpa. **j,** AMNH 213487—Peru: Pasco; Bermudas de Loma Linda. **k,** FMNH 84261—Peru: Madre de Dios; Itahuanía.

chrysaeolus of the *trinitatus*-group), and Type IV (part *goeldii*-group and part *semispinosus*-group).

The baculum is relatively long and narrow, averaging in adult specimens nearly 8 mm long and approximately 2 mm wide at both the proximal and distal extremities (see fig. 5a–g; table 1). The shaft is rather straight, with little dorsoventral curvature and only slightly tapered lateral indentations near mid-shaft. The proximal end is usually evenly rounded and paddle-shaped, although samples from every examined locality include bacula with a basal median notch of varying depth. The distal tip shows only slight development of apical wings and a moderate median depression. Except for topotypes of *P. roberti* from Minas Gerais, Brazil, there is no demonstrable geographic variation in length and width measurements among bacula of the same age class from different geographic regions (see table 1). The few bacula examined of *P. roberti* are smaller in length and width relative to others of the group. However, *roberti* is a rather small animal (see Thomas, 1901), and the proportions of its baculum are similar to those of other members of the group (see fig. 12).

simonsi-group (Figure 5h–k)

Bacula of specimens referred to this group were described and figured by Didier (1962, pp. 416–417, 419, 422) as *Proechimys guyannensis brevicauda* and *P. hendeei,* and by Patton and Gardner (1972) as *P. hendeei.* Didier's supposed specimen of *guyannensis brevicauda* (FMNH 62095) is clearly

TABLE 1. Measurements of length, distal width, and proximal width of bacula (mean ± SD) of spiny rats, subgenus *Proechimys*. Data are presented only for adult individuals (age classes 8 through 10 of Patton & Rogers, 1983).

Species group/ region or taxon	**Age**	**N**	**Length**	**Distal width**	**Proximal width**
guyannensis-group					
Venezuela	8	2	6.38 ± 1.21	1.99 ± 0.12	2.17 ± 0.60
Guianas	8	8	7.44 ± 0.85	2.00 ± 0.30	2.06 ± 0.41
	9	12	8.09 ± 0.58	2.17 ± 0.35	1.94 ± 0.51
	10	9	8.87 ± 1.04	2.38 ± 0.36	1.83 ± 0.32
Central Brazil	8	8	6.81 ± 1.45	1.89 ± 0.23	1.87 ± 0.39
	9	4	6.91 ± 1.35	2.28 ± 0.18	2.29 ± 0.31
	10	4	8.12 ± 0.86	2.16 ± 0.16	2.37 ± 0.24
Southern Brazil	8	1	6.08	1.28	1.74
[*roberti*]	10	1	5.43	1.79	2.27
goeldii-group					
Colombia–Bolivia	8	24	7.21 ± 1.05	2.46 ± 0.41	2.45 ± 0.43
	9	31	7.11 ± 1.18	2.79 ± 0.45	2.57 ± 0.41
	10	21	8.17 ± 1.02	3.11 ± 0.42	2.93 ± 0.45
Venezuela–Brazil	8	5	6.58 ± 1.01	2.08 ± 0.20	2.06 ± 0.44
	9	9	7.44 ± 0.55	2.36 ± 0.44	2.58 ± 0.47
	10	3	8.36 ± 0.81	2.48 ± 0.68	2.62 ± 0.20
longicaudatus-group					
Colombia–Northern	8	33	10.81 ± 1.73	5.13 ± 0.76	4.28 ± 0.62
Peru	9	18	11.43 ± 1.61	5.75 ± 0.80	4.68 ± 0.69
	10	11	11.67 ± 1.90	5.87 ± 1.02	5.12 ± 0.87
Central Peru	8	8	7.79 ± 0.65	4.57 ± 0.55	3.72 ± 0.92
	9	11	9.30 ± 1.25	4.61 ± 0.45	4.10 ± 0.80
	10	7	10.11 ± 0.66	5.02 ± 0.68	4.08 ± 0.63
Southern Peru–Bolivia	8	4	8.66 ± 1.45	4.85 ± 0.50	4.41 ± 0.55
	9	6	9.60 ± 1.12	4.74 ± 0.79	4.46 ± 0.62
SE Bolivia–Brazil	8	2	10.17 ± 0.84	5.22 ± 0.82	4.43 ± 0.72
	9	2	11.01 ± 0.49	5.13 ± 0.40	3.20 ± 0.30
	10	4	11.08 ± 1.02	5.10 ± 0.32	5.02 ± 0.47
cuvieri-group					
Guianas	8	7	6.27 ± 0.49	5.51 ± 0.56	5.30 ± 0.50
	9	6	7.21 ± 0.84	5.94 ± 0.82	5.72 ± 0.84
	10	6	8.61 ± 0.80	7.09 ± 0.45	6.70 ± 0.58
Central Brazil	8	3	5.95 ± 1.29	5.18 ± 1.24	4.69 ± 1.08
	9	3	7.26 ± 0.41	6.02 ± 0.54	5.54 ± 0.45
	10	2	8.09 ± 1.45	6.14 ± 1.39	5.31 ± 1.05
Northern Peru	8	5	6.13 ± 0.85	5.82 ± 0.74	5.38 ± 0.50
	9	3	6.96 ± 0.45	6.88 ± 1.29	6.29 ± 0.45
	10	3	7.41 ± 0.75	7.40 ± 0.05	6.53 ± 0.49
2n = 40, Balta	8	1	7.27	5.76	4.45
	9	1	7.39	4.99	4.71
simonsi-group	8	8	8.42 ± 1.44	1.51 ± 0.21	1.93 ± 0.23
	9	11	9.02 ± 1.06	1.89 ± 0.38	2.05 ± 0.27
	10	11	10.09 ± 1.18	1.77 ± 0.22	2.50 ± 0.32
semispinosus-group					
semispinosus	8	9	8.54 ± 0.95	5.11 ± 0.91	4.21 ± 1.11
	9	11	9.16 ± 1.26	6.11 ± 0.67	4.67 ± 0.54
	10	9	9.66 ± 1.01	6.19 ± 0.31	4.94 ± 0.67
oconnelli	9	1	8.43	5.26	4.94
	10	1	8.16	4.63	4.93
trinitatus-group					
trinitatus	9	1	9.97	3.14	3.10
	10	1	10.46	3.32	3.15
magdalenae	8	2	8.07 ± 0.81	2.37 ± 0.41	2.56 ± 0.22
	10	2	10.47 ± 0.16	2.62 ± 0.30	2.88 ± 0.07

TABLE 1. *Continued.*

Species group/ region or taxon	Age	N	Length	Distal width	Proximal width
guairae	8	1	8.62	2.79	2.93
	9	1	9.25	2.78	3.57
	10	1	10.01	3.34	3.50
hoplomyoides	10	1	11.64	1.79	2.47
mincae	8	1	7.29	2.42	2.04
	9	3	9.98 ± 0.99	2.94 ± 0.18	2.17 ± 0.25
chrysaeolus	8	6	9.01 ± 0.94	2.34 ± 0.27	2.74 ± 0.23
	9	7	10.03 ± 0.73	2.34 ± 0.32	2.97 ± 0.44
	10	3	10.97 ± 1.35	2.63 ± 0.66	3.11 ± 0.31
decumanus-group	8	1	10.63	3.84	3.41
	9	2	9.16 ± 1.44	2.93 ± 0.64	2.80 ± 0.52
	10	3	11.83 ± 1.30	3.08 ± 0.34	2.92 ± 0.42
canicollis-group	9	1	7.74	2.80	2.55
	10	1	8.11	2.64	2.78

referable to *P. simonsi* by the cranial characters that distinguish this group from *P. brevicauda* (see below).

The baculum is elongate and narrow, with a rounded and slightly broadened base. In older specimens the base is often laterally expanded with thin wings of bone. The weakly expanded distal end is usually characterized by a small lateral platform on each side separated by a shallow median depression. In general aspects, the baculum of this group is similar to that described for the *guyannensis*-group, although it averages longer and narrower (see table 1). No geographic variation is apparent, but the samples are not adequate to document this.

goeldii-group (Figure 6)

Included in this group are bacula described and figured by Didier (1962) as *Proechimys guyannensis* Type IV and as *P. quadruplicatus,* both from Caquetá, Colombia. Specimens from Peru referred to *P. brevicauda* by Patton and Gardner (1972) also belong to this unit (see Gardner & Emmons, 1984, for the correct allocation of these specimens to *P. steerei*). One specimen figured by Martin (1970, p. 8; fig. 4d), from Riberalta, El Beni, Bolivia, is also referable to this group.

There is virtually no distinguishable geographic variation in the baculum of *goeldii*-group members among samples which range from southern Venezuela to Bolivia and from the Rio Tapajós to eastern Colombia and Peru (see fig. 6; table 1). In general form, this baculum is similar to that described for members of the *guyannensis*-group. At similar cranial ages, it is nearly the same length but slightly wider both basally and distally, giving the baculum a somewhat stouter appearance. The base varies in shape from rounded to bilobed with a median notch, the sides are parallel to only slightly concave, and the tip shows only faint development of apical wings and a median depression. In lateral view, the baculum is straight to slightly convex dorsally and concave ventrally.

trinitatus-group (Figure 7)

Didier (1962, pp. 412, 417–418) described bacula of specimens referred to this group as *Proechimys guyannensis* Type III (which represent *P. chrysaeolus*) and as *P. guyannensis mincae.*

The baculum is long (averaging over 10 mm in adults; table 1) and narrow, but except for *hoplomyoides* it is considerably stouter than that of *simonsi*-group members which resemble them in overall length (see fig. 12). The lateral margins are only slightly concave; the base is broadened and rounded, usually with a distinct median notch; and the distal tip has only slightly developed apical wings and a median depression (fig. 7).

decumanus-group (Figure 8a–b)

The baculum of *Proechimys decumanus* has apparently not been described before. It is similar in general size and shape to that of the *trinitatus*-group (fig. 12; table 1), being elongate yet stout, and with rather parallel sides. The base is somewhat rounded and the distal tip only slightly ex-

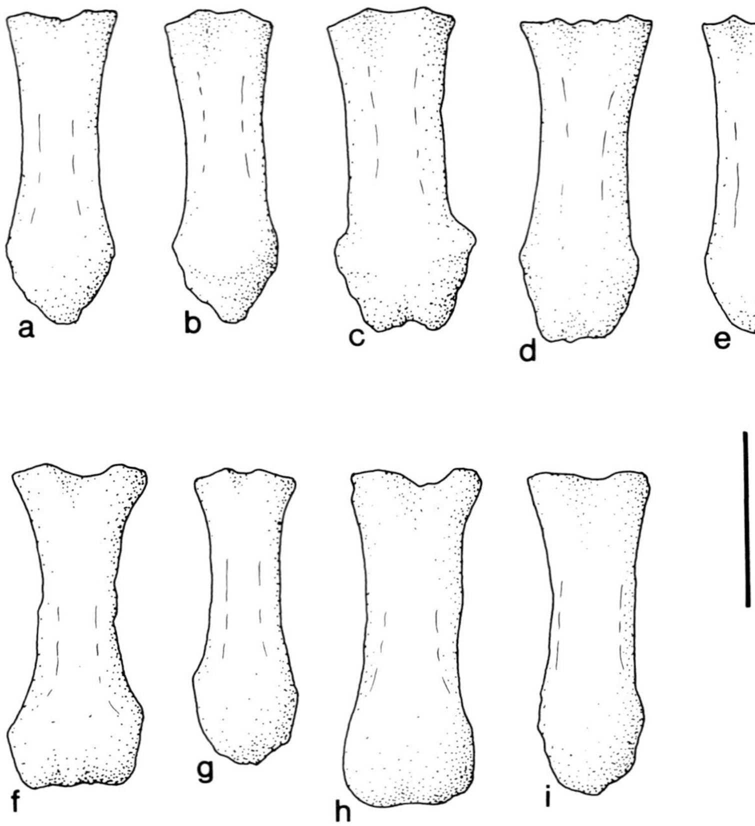

FIG. 6. Representative bacula of members of the *goeldii*-group; scale = 5 mm. **a,** FMNH 71178—Colombia: Caquetá; Florencia, Mantanito. **b,** MVZ 157955—Peru: Amazonas; La Poza, Río Santiago. **c,** AMNH 71592—Peru: Loreto; Boca Río Curaray. **d,** AMNH 76268—Peru: Loreto; Sarayacu. **e,** USNM 530931—Peru: Madre de Dios; Río Manu. **f,** AMNH 97017—Venezuela: Amazonas; Esmeralda. **g,** USNM 415117—Venezuela: Amazonas; Capibara, Casiquiare Canal. **h,** AMNH 92272—Brazil: Amazonas; Rosarinho, Río Madiera. **i,** AMNH 96828—Brazil: Pará; Ilha do Taiuno, Río Tocantins.

panded, with a weak median depression (fig. 8a–b).

canicollis-group (Figure 8c–d)

The baculum of *Proechimys canicollis* was described and figured by both Didier (1962, p. 419) and Martin (1970, p. 8). It is most similar to that of the *goeldii*-group in both shape and size, being relatively short and stout with a rounded base, weakly concave sides, and a rather flat distal tip with only weakly developed apical wings (see fig. 8c–d; table 1).

longicaudatus-group (Figure 9)

Bacula of this group were figured by Didier (1962, p. 410, fig. 2, p. 412) as *Proechimys guyannensis* Type II (specimens from Caquetá, Colombia, and Santa Cruz, Bolivia), and by Martin (1970, p. 8, fig. 4c,e–k), also as *P. guyannensis,* from southeastern Peru, Bolivia, and southwestern Brazil. Patton and Gardner (1972) described and figured the bacula of Peruvian specimens of this group as *P. longicaudatus* (now referred to *P. brevicauda* [see Patton & Rogers, 1983; Gardner & Emmons, 1984]).

In general aspect, the baculum is elongate and broad, with well-developed apical wings (see fig. 9). The margins are concave and the proximal and distal ends are usually about equal in width. In some specimens, the proximal end bears a median indentation of variable depth; in others, the proximal base is evenly rounded. The shaft is arched dorsally from base to tip and transversely concave along its entire ventral length. While overall length varies considerably, as do width measures to a lesser extent (table 1), the uniform and character-

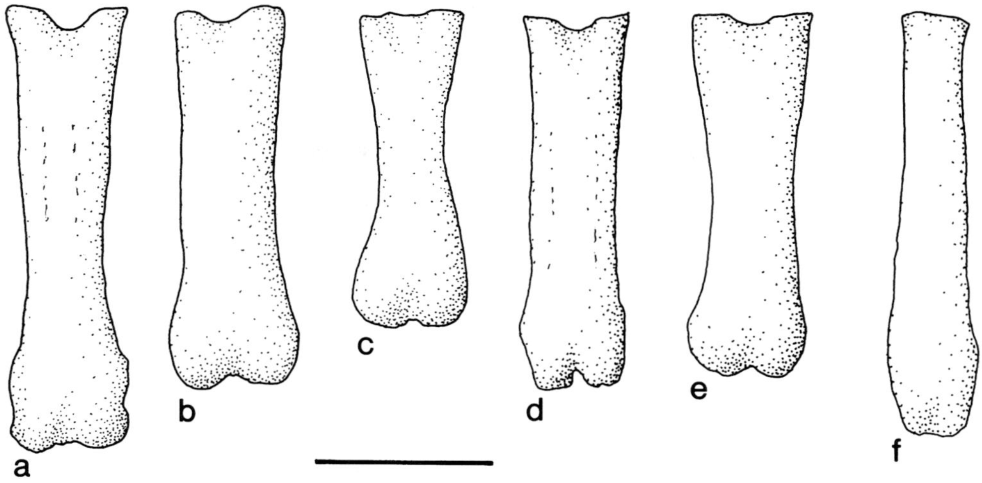

FIG. 7. Representative bacula of members of the *trinitatus*-group; scale = 5 mm. **a,** *P. chrysaeolus,* BM[NH] 12.4.2.4—Colombia: Santander; Margarita. **b,** *P. chrysaeolus,* FMNH 69109—Colombia: Bolivár; San Juan Nepomuceno. **c,** *P. guairae,* MVZ 168945—Venezuela: Portuguesa; Sto. Domingo. **d,** *P. magdalenae,* USNM 499719—Colombia: Antióquia; 22 km S and 22 km W Zaragoza. **e,** *P. trinitatus,* MVZ 168946—Trinidad: Chaguaramas. **f,** *P. hoplomyoides,* AMNH 75827—Venezuela: Bolivár; Arabupu, Mt. Roraima.

istic shape of the baculum renders members of this group easily identifiable. Geographically, samples allocated to *P. brevicauda* average larger and broader in the northern (e.g., southern Colombia, Ecuador, and northern Peru) than in more southern localities (e.g., southeastern Peru and adjacent Bolivia); samples from southeastern Bolivia and Brazil referred to *P. longicaudatus* approach the general size of northern samples of *P. brevicauda* (see table 1).

cuvieri-group (Figure 10)

Didier (1962, p. 411) recorded one specimen (FMNH 18198) of *Proechimys cuvieri* from Guyana in his *P. guyannensis* Type I bacular group. This group is otherwise made up of specimens of *P. semispinosus* (see below), although the bacula of *P. cuvieri* and some populations of *semispinosus* have similarities of shape in common (see figs. 10–11). Patton and Gardner (1972) figured and de-

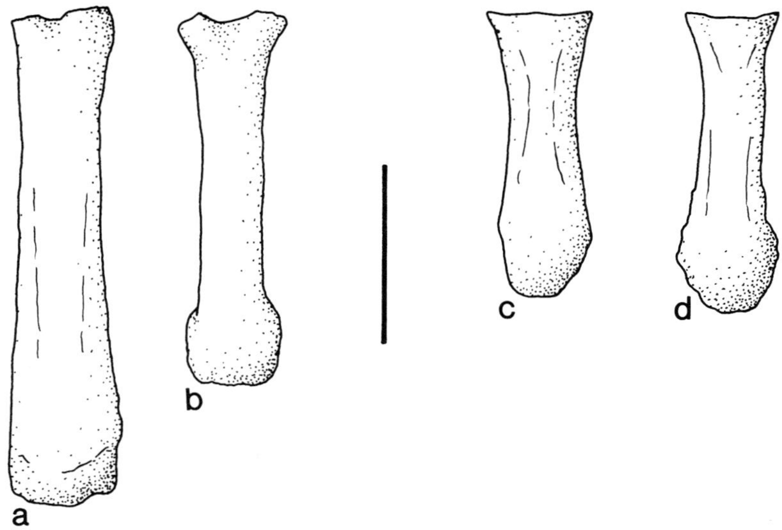

FIG. 8. Representative bacula of the *decumanus*-group (**a**–**b**) and *canicollis*-group (**c**–**d**); scale = 5 mm. **a,** FMNH 82023—Peru: Piura; Laguna Lamadero. **b,** FMNH 81199—Peru: Tumbes; Matapalo. **c,** USNM 280113—Colombia: Magdalena; Río Cesar, opposite El Orinoco. **d,** USNM 280114—Colombia: Magdalena; Río Cesar, opposite El Orinoco.

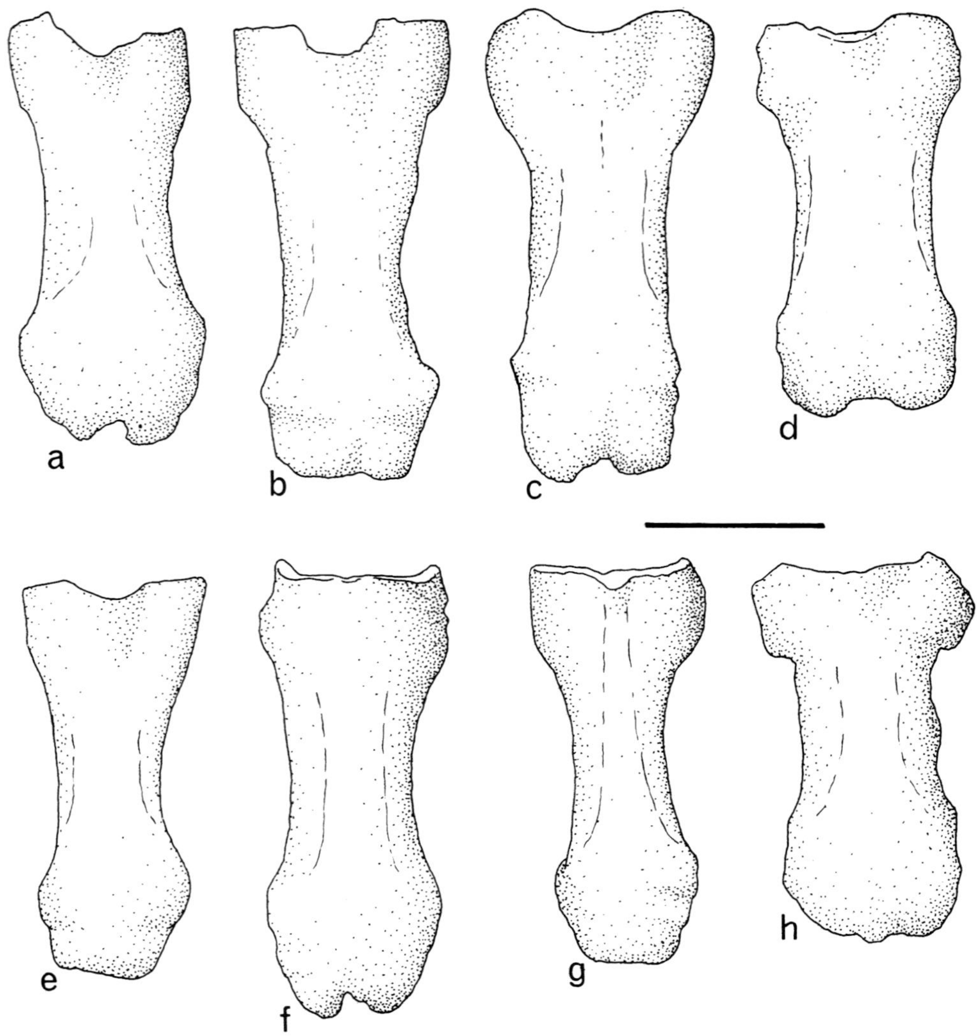

FIG. 9. Representative bacula of members of the *longicaudatus*-group; scale = 5 mm. **a,** FMNH 71174—Colombia: Caquetá; Florencia, Mantanita. **b,** MVZ 155034—Peru: Amazonas; Huampami, Río Cenepa. **c,** MVZ 157934—Peru: Amazonas; La Poza, Río Santiago. **d,** AMNH 71877—Peru: Loreto; Boca Río Curaray. **e,** MVZ 157854—Peru: Madre de Dios; Lago Sandoval. **f,** FMNH 119356—Bolivia: El Beni, San Pedro. **g,** BM[NH] 28.2.9.48—Bolivia: Santa Cruz; Buenavista. **h,** BM[NH] 3.7.7.94—Brazil: Mato Grosso; Serra de Chapada.

scribed bacula of specimens from eastern Peru with 2n = 40 karyotype that they referred to *P. guyannensis,* and that have this type of baculum.

The baculum is massive, with a broad shaft and a thickened and expanded base (see fig. 10; table 1). In cross section, the proximal two-thirds is convex dorsally and deeply concave ventrally. The distal end has a pair of diverging apical extensions separated by a wide median depression of varying depth. This is the most distinctive of any of the bacular types in *Proechimys.* It characterizes specimens from widely scattered localities in the Guianas and along the entire length of the Amazon River, as well as the karyotypically differentiated 2n = 40 form from Balta, Río Curanjá, Ucayali, Peru (see Patton & Gardner, 1972). These latter specimens were considered to be conspecific with *P. guyannensis* by Gardner and Emmons (1984) on karyotypic grounds. In bacular characters, however, they are clearly different from *guyannensis*-group members and are placed here solely because of these uniquely shared bacula. The true taxo-

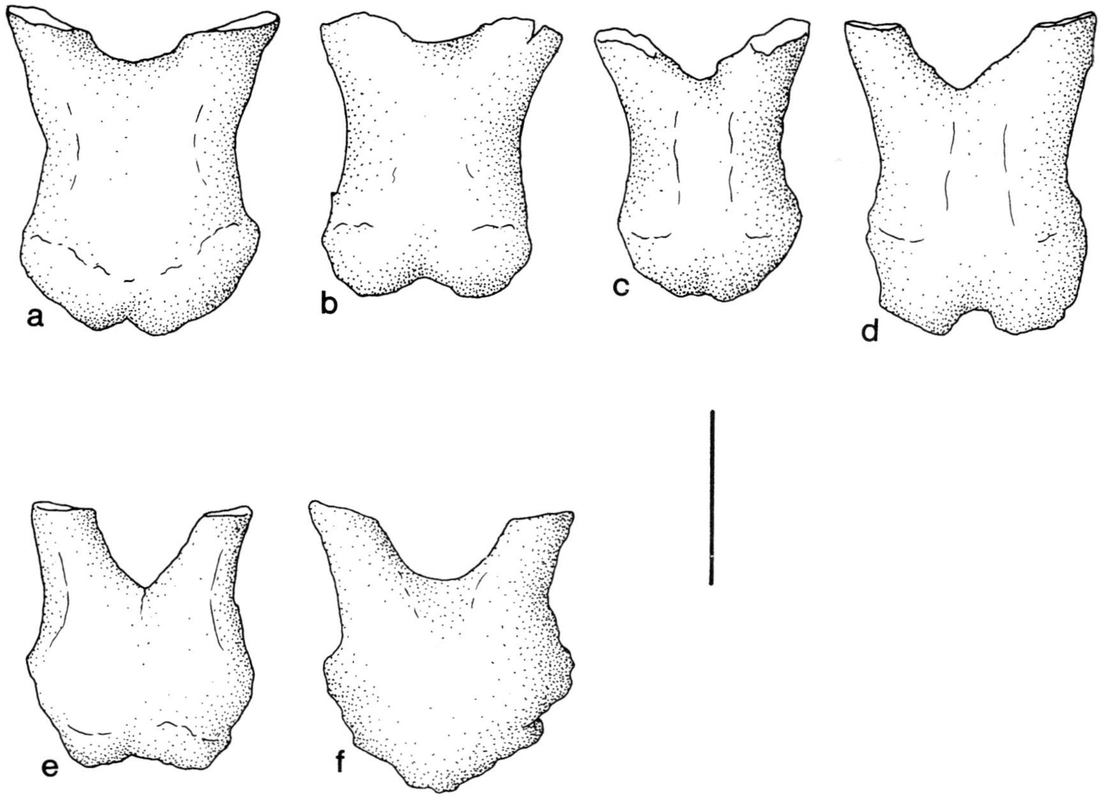

FIG. 10. Representative bacula of members of the *cuvieri*-group; scale = 5 mm. **a,** BM[NH] 7.6.20.5—Guyana: Hyde Park, Demarara River. **b,** AMNH 140540—Guyana: Kamakusa. **c,** AMNH 96844—Brazil: Pará; Ilha do Taiuno, Río Tocantins. **d,** AMNH 96867—Brazil: Pará; Ilha do Taiuno, Río Tocantins. **e,** MVZ 157874—Peru: Amazonas; La Poza, Río Santiago. **f,** AMNH 74091—Peru: Loreto; Orosa, Río Amazonas.

nomic position of these specimens remains an enigma, as material clearly assignable to this form has not been found elsewhere.

In the meager samples available, there appears to be little geographic variation in bacular size within *P. cuvieri,* although specimens from the Guianas average slightly larger than those from Brazil or Peru (table 1).

semispinosus-group (Figure 11)

Didier (1962, pp. 409–411) defined his *Proechimys guyannensis* Type I baculum based largely on specimens from western Colombia which represent the *semispinosus*-group as defined herein. Interestingly, three of the individuals he included in this group (FMNH 69063, 69064, 69071) were also listed as members of his *guyannensis* Type IV complex (Didier, 1962, p. 415), which is otherwise composed of specimens here referred to the *goeldii*-group. All cranial characters, as well as bacular ones (see fig. 11b), show that these three specimens represent *P. semispinosus.* Patton and Gardner (1972, pp. 16–17) also described and figured specimens of *P. semispinosus* from Costa Rica. The additional specimens examined here do not differ importantly from the descriptions provided in these two papers.

In general aspects, the baculum is intermediate between the *longicaudatus* and *cuvieri* groups. The shaft is long and massive, with deeply concave margins, a broadly expanded and thickened base, and a wide distal portion with well-developed apical wings separated by a median depression. In cross section, the proximal and distal portions are convex dorsally and deeply concave ventrally. The width across the distal portion of the baculum typically exceeds that of the proximal portion (table 1), which is also characteristic of the *cuvieri*-

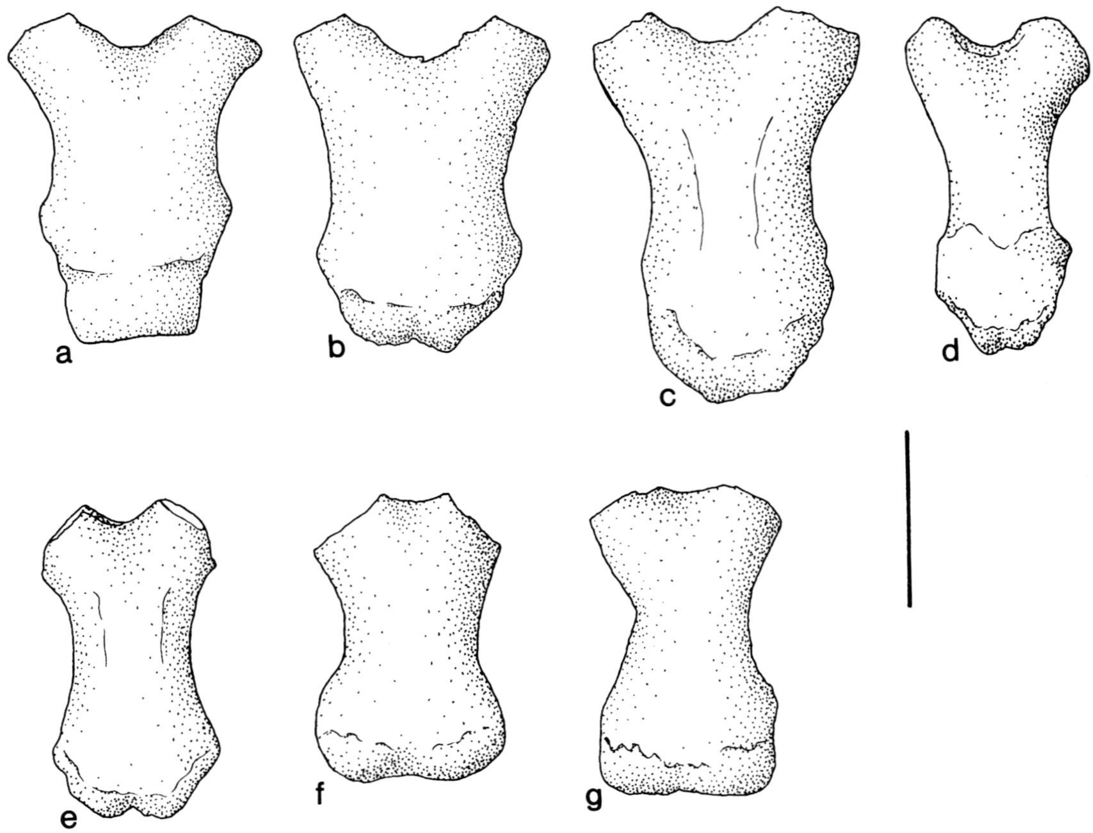

FIG. 11. Representative bacula of members of the *semispinosus*-group (*P. semispinosus* [**a**–**e**] and *P. oconnelli* [**f**–**g**]); scale = 5 mm. **a,** USNM 592694—Panama: Canal Zone. **b,** FMNH 69063—Colombia: Bolivár; Socorré, upper Río Sinu. **c,** FMNH 70085—Colombia: Choco; Unguía. **d,** FMNH 89525—Colombia: Nariño; La Guayacana. **e,** FMNH 90150—Colombia: Cauca; La Boca. **f,** MVZ 99680—Colombia: Meta; Villavicencio. **g,** FMNH 88050—Colombia: Meta; Los Micos, San Juan de Aramas.

group. However, specimens from the northern (Costa Rica and western Panama) and southern limits (Cauca, Colombia, and southward) of the range of *P. semispinosus,* as well as those of *P. oconnelli* (table 1), tend to be more symmetrical, those from central Panama south through Chocó, Colombia, more expanded distally (compare fig. 11a–d with 11e–g).

Qualitative Cranial Characters

The usual morphometric approach to specific and infraspecific taxonomy of small mammals has met with little success in studies of *Proechimys.* In part, this is due to the large age-related component of character variation within localities that obscures any geographic patterns and species differences (see Patton & Rogers, 1983). Use of qualitative characters has proven more successful, but only Patton and Gardner (1972) and Gardner and Emmons (1984) have marshalled such features as palatal structure, bullar septal pattern, temporal ridge development, and counterfold pattern on the cheekteeth into coherent patterns that identify geographically overlapping forms. Characters of the palate, in particular, proved to be concordant with bacular and karyotypic differences in delineating taxa in Peru (Patton & Gardner, 1972).

Here, I focus on the qualitative description of five cranial features that prove useful in the discrimination of sympatric taxa and in the definition of homogeneous regional units. These include (see Patton & Gardner, 1972; Gardner & Emmons,

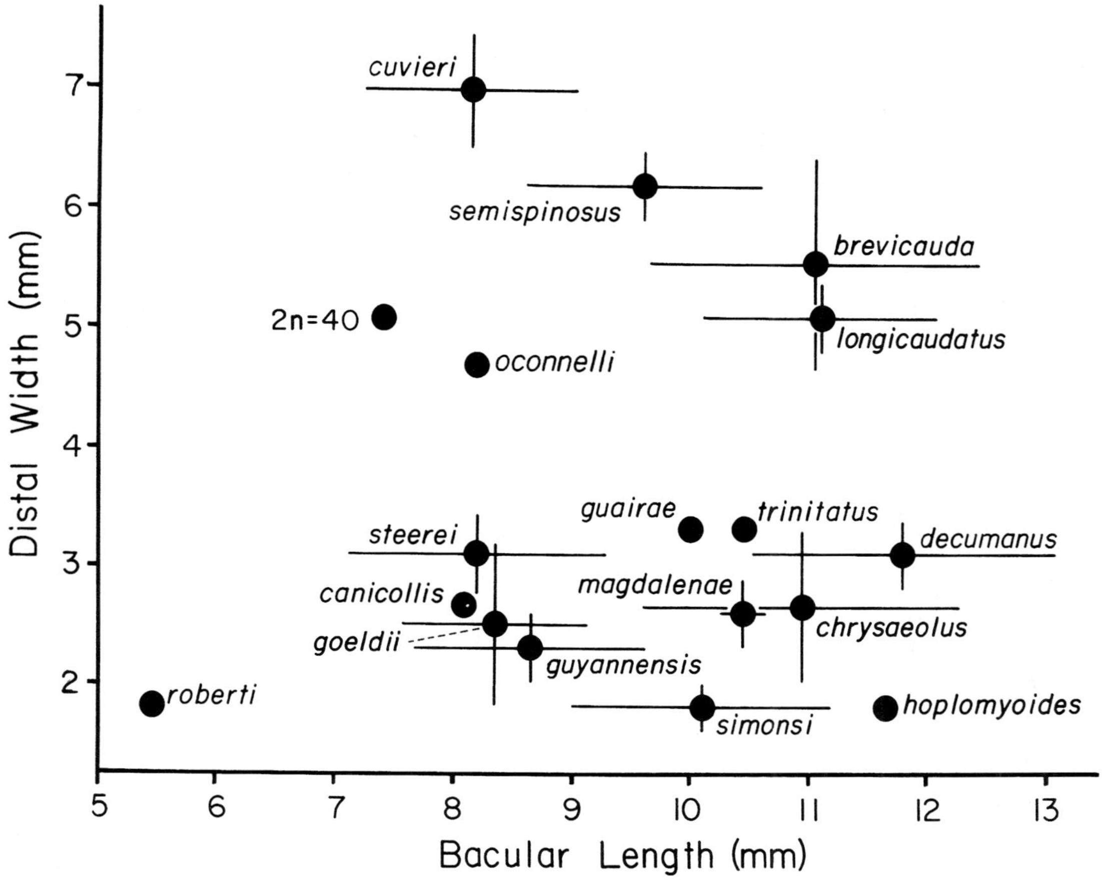

FIG. 12. Bivariate plot of distal bacular width and bacular length measurements for taxa of spiny rats, subgenus *Proechimys*. All individuals represent age class 10 of Patton and Rogers (1983). Circles indicate means, vertical and horizontal lines indicate standard deviations. See text for the allocation of taxa into species groups.

1984): (1) shape and structure of the incisive foramen; (2) angle and depth of mesopterygoid fossa; (3) degree of development of a bony groove in the floor of the infraorbital foramen; (4) degree of development of the temporal ridges across the parietals; and (5) the counterfold pattern of the upper and lower cheekteeth. These structures were used by Moojen (1948) with limited success to segregate sympatric taxa in Brazil, but he failed to use them to group regional samples into consistently defined morphological entities. Hence, his analysis appears to show that these features are much more variable, and thus of less utility, than actually proves to be the case.

Other qualitative features of the cranium, such as the size and shape of the bullae, the size and position of the hamular processes of the pterygoids, and the degree of lateral indentation of the paroccipital processes, need further examination (see, e.g., Patton & Gardner, 1972). My preliminary analysis of each of these features indicates strong concordance with those characters used here to define taxonomic limits.

Shape and Structure of the Incisive Foramen

Specimens of each recognized species group were analyzed for the following characteristics of the incisive foramen: (1) general shape and size [lyre-shaped, constricted posteriorly; oval; evenly tapered posteriorly or parallel sided]; (2) presence or absence of grooves extending onto the anterior portion of the palate; (3) flanged or flat postero-lateral margins of the foramen; (4) degree of development of the maxillary and premaxillary por-

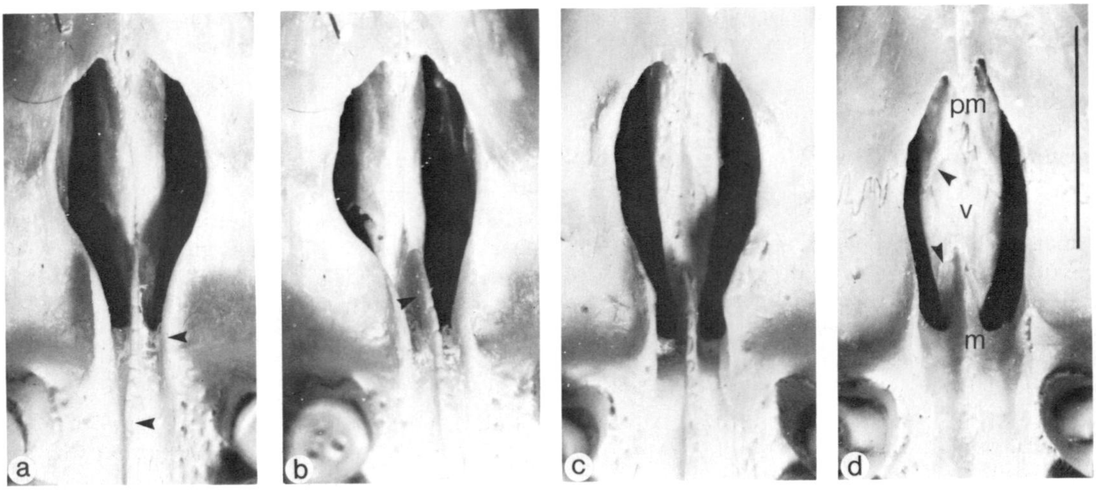

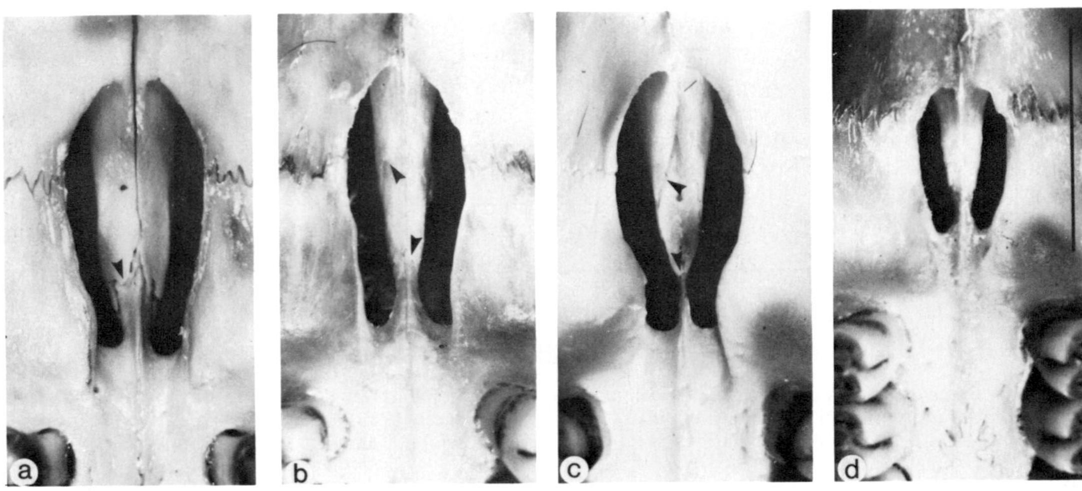

FIG. 13 (**top**). Representative incisive foramina of specimens of *P. brevicauda* of the *longicaudatus*-group; scale = 5 mm. **a,** MVZ 153596—Peru: Amazonas; Huampami, Río Cenepa. Arrows indicate median palatal ridge and elevated flange marking posterolateral foraminal margins. **b,** Same specimen as in **a,** emphasizing the strongly keeled maxillary portion of the foraminal septum (arrow). **c,** MVZ 153607—Peru: Amazonas; Huampami, Río Cenepa. **d,** MVZ 157855—Peru: Amazonas; La Poza, Río Santiago. The premaxillary (pm), vomerine (v), and maxillary (m) portions of the foraminal septum are identified; arrows indicate sutures between these elements.

FIG. 14 (**bottom**). Representative incisive foramina of specimens of the *cuvieri*-group (*P. cuvieri* [**a**-**c**] and the 2n = 40 karyotypic form from Balta, eastern Peru [**d**]); scale = 5 mm. **a,** FMNH 95720—Suriname: Brokopando; Saramacca River, Loksie Hattie. Arrow indicates direct contact between premaxillary and maxillary portions of the septum. Note that only a small part of the vomerine portion is visible ventrally. **b,** MVZ 160091—Venezuela: Bolivár; 69 km S Río Cuyuni. An expanded vomerine portion of the septum is evident (arrows identify vomerine contact with the premaxilla and maxilla). **c,** MVZ 157874—Peru: Amazonas; La Poza, Río Santiago. The vomerine portion of the septum is visible ventrally, widely separating the premaxillary and maxillary components (arrows). **d,** LSU 14425—Peru: Ucayali; Balta, Río Curanja. Note the small size of this specimen relative to the others, and the elongated premaxillary portion of the septum.

tions of septum; (5) whether the vomerine portion of septum is visible ventrally; (6) whether or not the maxillary portion of the septum is keeled; and (7) whether the anterior portion of the palate has a median ridge. Descriptions of the incisive foramina for representatives of each of the species groups are given below and are illustrated in Figures 13–20; examples were chosen to express the full range of character variation for each group, regardless of the specific localities from which specimens were collected.

longicaudatus-group (Figure 13a–d)

The general features of this type of incisive foramen were given by Patton and Gardner (1972, p. 10). Figure 13 illustrates the range of form typical for members of the group. The most diagnostic features include: a lyre-shaped foramen, usually with a strongly constricted posterior portion; the maxillary terminus of the foramen deeply grooved onto the anterior palate; the posterolateral margins of the foramen strongly flanged; an expanded, long premaxillary portion of the septum, usually extending more than one-half its length; a well-developed and strongly keeled maxillary portion of the septum (see fig. 13b), the maxillary keel extending onto the anterior palate resulting in a well-developed median ridge; and a vomerine portion of the septum exposed ventrally between the premaxillary and maxillary components.

This is one of the more consistently recognizable types of incisive foramina within the genus *Proechimys,* varying mostly in the degree of constriction at the posterior margins, hence in the degree of the general lyre-shape.

cuvieri-group (Figure 14a–d)

The incisive foramen of *Proechimys cuvieri* is most similar in structure to that of the *longicaudatus*-group. The general conformation is weakly to strongly lyre-shaped with strongly developed posterolateral flanges. The anterior palate, however, is only weakly to moderately grooved resulting in a slight median ridge. The premaxillary portion of the septum is strongly developed, extending more than one-half its length; the maxillary portion varies from stout to attenuate, but is always short and is only weakly keeled. The vomer is varyingly exposed ventrally (compare fig. 14a with 14b–c).

The 2n = 40 specimens from Balta which have bacula similar to *P. cuvieri* share only some foraminal characters with that taxon (fig. 14d). The foramen is weakly lyre-shaped and the posterolateral margins are only weakly flanged, hence the anterior palate is scarcely grooved. Nevertheless, the premaxillary portion of the septum is elongated and broad, similar in shape and structure to that of *P. cuvieri.* The maxillary portion is well developed and appears to contact directly the premaxillary portion.

goeldii-group (Figure 15a–d)

The general features of this foraminal type were provided by Patton and Gardner (1972, p. 4) under *P. brevicauda.* The foramen is usually only weakly lyre-shaped, or with margins tapering slightly posteriorly or parallel-sided. The premaxillary portion of the septum is short, usually one-half or less of the length of the foramen; the maxillary portion varies greatly, being usually rather weak and attenuate, often not in contact with the premaxillary portion (fig. 15d) but sometimes broadly spatulate and filling much of the foramen (fig. 15a). Nevertheless, the vomer is only rarely exposed ventrally, being completely enclosed in the premaxillary sheath. The maxillary portion of the septum often exhibits a median vacuity (fig. 15b); it may be slightly ridged, but is never strongly keeled, and seldom does this ridge extend onto the anterior palate (fig. 15b). Thus, there are only moderately developed grooves onto the anterior palate, and the posterolateral margins of the foramen are only moderately flanged.

semispinosus-group (Figure 16a–d)

Specimens of this group from Costa Rica were described by Patton and Gardner (1972, p. 15). Foraminal shape varies from rather evenly tapered margins to moderately lyre-shaped ones. The posterolateral margins are usually strongly flanged, creating deep grooves extending onto the anterior palate despite only moderate development of a medial ridge (compare fig. 16b with 16c). The premaxillary portion of the septum is dominant, usually broadly filling the foramen and extending well over one-half its length. The maxillary portion varies from moderately developed to attenuate, but is almost always in direct contact with the premaxillary portion. The vomer is completely encased within the premaxilla and thus is not visible in ventral aspect.

simonsi-group (Figure 17a–d)

Again, the incisive foramen of the *simonsi*-group was described fully by Patton and Gardner (1972,

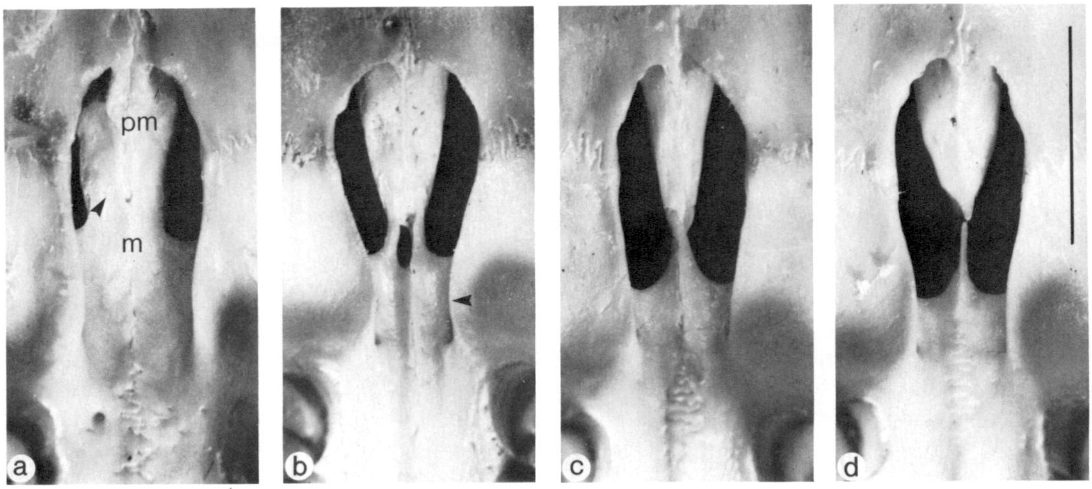

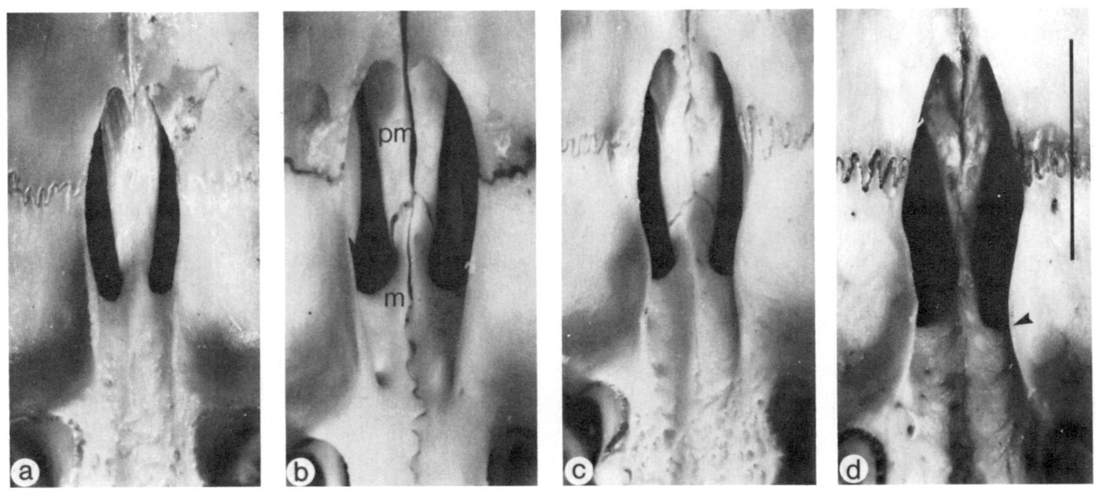

FIG. 15 (**top**). Incisive foramina of *P. steerei* of the *goeldii*-group. All are from La Poza, Río Santiago, Amazonas, Peru; scale = 5 mm. **a,** MVZ 157949. Note the enlarged maxillary portion of the septum (m) and the direct contact (arrow) between it and the premaxillary portion (pm). **b,** MVZ 157956. Note the moderately developed posterolateral flange (arrow). **c,** MVZ 157861. **d,** MVZ 157869. Note the attenuate maxillary portion of the septum and the lack of contact between it and the premaxilla portion.

FIG. 16 (**bottom**). Incisive foramina of *P. semispinosus* of the *semispinosus*-group; scale = 5 mm. **a,** MVZ 165794—Panama: Panamá; 0.8 km N Paraiso. **b,** FMNH 90169—Colombia: Chocó; Río Baudo. Note the direct contact between the premaxillary (pm) and maxillary (m) portions of the septum. **c,** FMNH 90177—Colombia: Chocó; Río Baudo. **d,** FMNH 70080—Colombia: Chocó; Unguía. Note the well-developed posterolateral flange (arrow).

p. 19). This is a distinctive foraminal type, as all specimens examined were consistent in most features despite variation in overall shape. The foramen is oval in general shape, although often asymmetrical in anteroposterior direction (fig. 17b). The premaxillary portion of the septum is rather short, usually no more than one-half the length of the foramen. The maxillary portion is usually weak and attenuate, only rarely in contact with the premaxillary portion. When the septum is complete (fig. 17a), the vomer is either completely enclosed by the premaxilla or barely visible (fig. 17c). The posterolateral margins are flat, not flanged, and no groove extends onto the anterior palate. Rather, the palate is noticeably flat and smooth, without a medial ridge.

guyannensis-group (Figure 18a–d)

This foraminal type is virtually indistinguishable from that described for the *simonsi*-group.

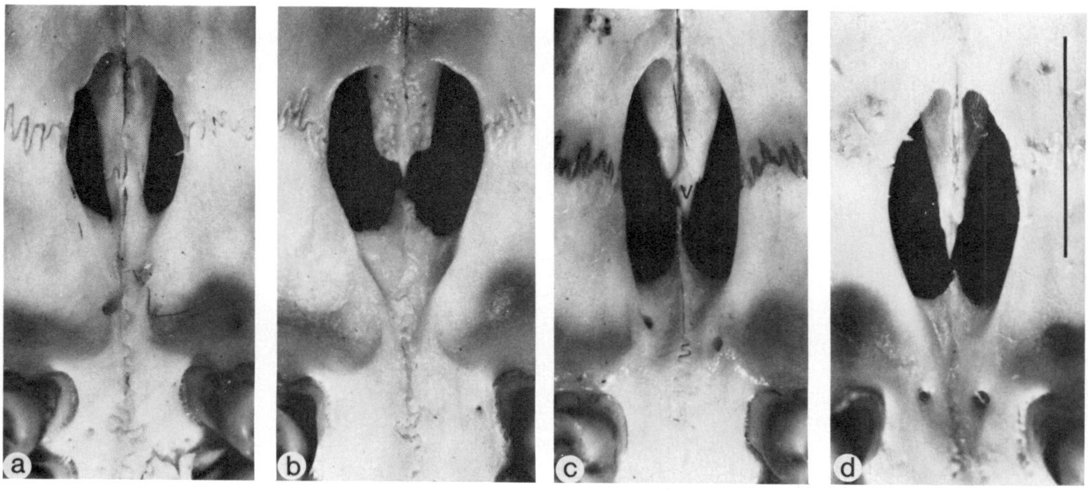

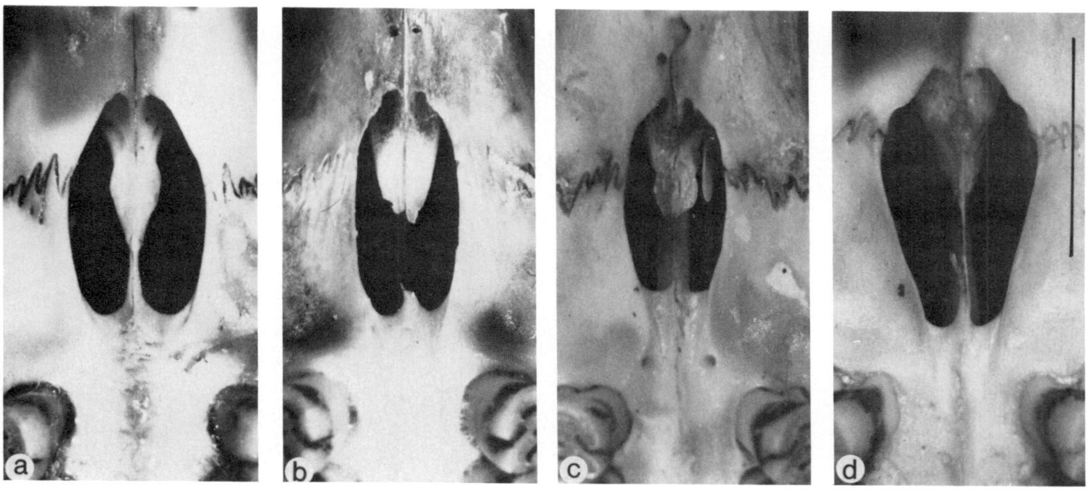

FIG. 17 (**top**). Incisive foramina of *P. simonsi* of the *simonsi*-group; scale = 5 mm. **a,** MVZ 155045—Peru: Amazonas; headwaters of Río Kagka. Note lack of posterolateral flanges or palatal grooves. **b,** MVZ 157914—Peru: Amazonas; La Poza, Río Santiago. **c,** MVZ 136654—Peru: Ucayali; Balta, Río Curanja. Note slightly exposed vomer (v) and attenuate maxillary portion of septum. **d,** MVZ 168955—Peru: Madre de Dios; Albergue, Río Madre de Dios.

FIG. 18 (**bottom**). Incisive foramina of the *guyannensis*-group (**a**–**b,** *arabupu*; **c,** *oris*; and **d,** *roberti*); scale = 5 mm. **a,** AMNH 139741—Venezuela: Bolivár; Auyantepui. Note lack of posterolateral flanges and anterior palatal grooves, and the attenuate maxillary portion of the septum. **b,** MVZ 160094—Venezuela: Bolivár; 4 km E El Paují. Note the lack of contact between the premaxillary and maxillary portions of the septum. **c,** AMNH 93997—Brazil: Pará; Faro, north bank Río Amazon. Note weakly developed posterolateral flanges. **d,** AMNH 134309—Brazil: Goiás; Anápolis.

The shape is oval, although often unequal (fig. 18d). The anterior palate is flat, without grooves or a median ridge, and the posterolateral margins of the foramen are not flanged, or only weakly flanged (fig. 18c shows maximal development of flanges). The premaxillary portion of the septum is relatively short, usually less than one-half the length of the foramen, and the maxillary portion is attenuate, usually not in contact with the premaxillary portion. The vomer generally does not contribute to the ventral aspect of the septum.

trinitatus-group (Figure 19a–d)

Members of this species group generally exhibit the most enlarged foramina within the subgenus *Proechimys*. Specimens referred to *mincae, polio-*

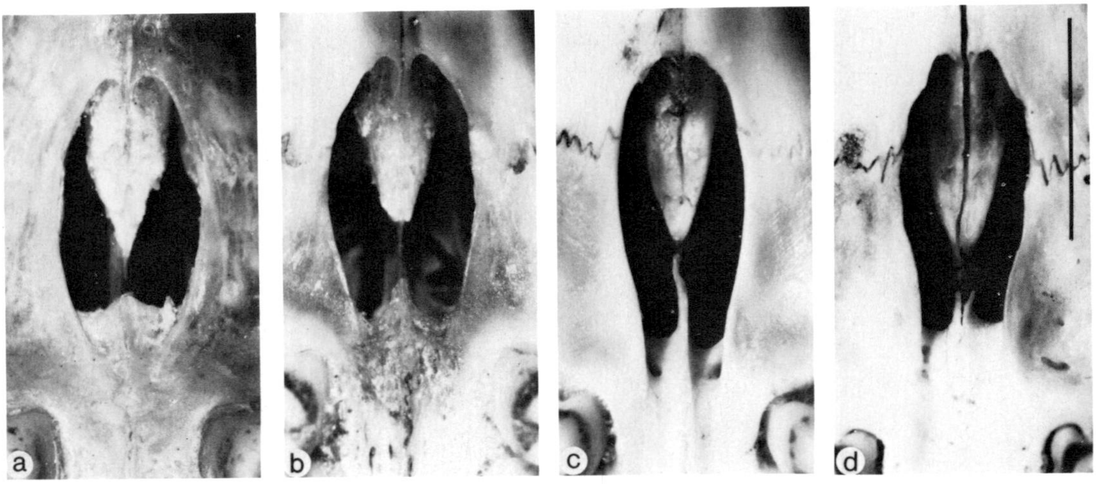

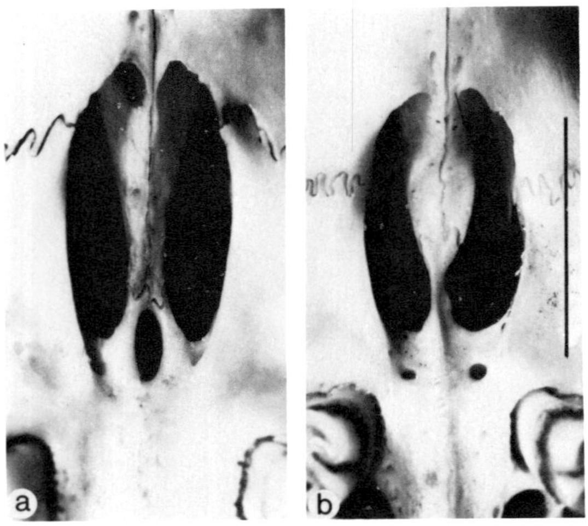

FIG. 19 (**top**). Incisive foramina of the *trinitatus*-group; scale = 5 mm. **a,** *P. mincae,* FMNH 13203—Colombia: Magdalena; Minca (topotype). **b,** *P. ochraceous,* FMNH 18688—Venezuela: Zulia; El Panorama, Río Aurare (topotype). **c,** *P. guairae,* FMNH 92588—Colombia: Arauca; Río Cobaria. **d,** *P. chrysaeolus,* FMNH 69037—Colombia: Bolivár; San Juan Nepumoceno.

FIG. 20 (**bottom**). Incisive foramina of the *decumanus*- and *canicollis*-groups; scale = 5 mm. **a,** *P. decumanus,* FMNH 82024—Peru: Piura; Laguna Lamadero. **b,** *P. canicollis,* FMNH 69111—Colombia: Bolivár; San Juan Nepumoceno.

pus, and *ochraceous* have smoother, less ridged palates and ovoid foramina lacking posterolateral flanges (fig. 19a–b); those referred to *guairae, trinitatus, urichi, hoplomyoides,* and *chrysaeolus* have somewhat more lyre-shaped foramina with weakly to moderately flanged posterolateral margins which define grooves extending onto the anterior palate (fig. 19c–d). In all forms the premaxillary portion of the septum is enlarged, usually extending one-half or more of the length of the foramen, while the maxillary portion is attenuate, most often not in direct contact with the premaxillary portion. Only in specimens referred to *guairae* and *trinitatus* does the maxillary portion of the septum show a medial ridge (fig. 19c).

decumanus-group (Figure 20a)

This foraminal type is oval in shape and large, with poorly defined posterolateral flanges and weak grooves extending onto the anterior palate. The premaxillary portion of the septum is long, but tapering posteriorly and in direct contact with a

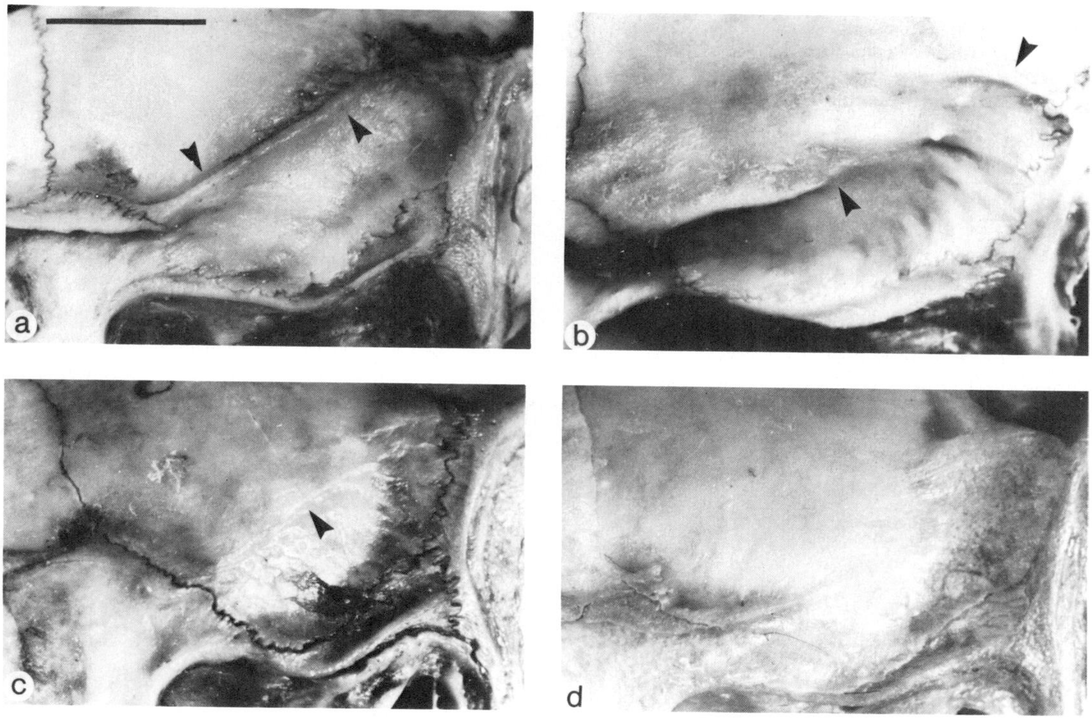

FIG. 21. Patterns of temporal ridge development in *Proechimys*; scale = 5 mm. **a,** *P. semispinosus,* FMNH 90145—Colombia: Cauca; Río Saija. Note extreme development of ridge across parietals continuous between the supraorbital ledge and the lambdoidal ridge (arrows). **b,** *P. steerei,* MVZ 157955—Peru: Amazonas; La Poza, Río Santiago. Note separation of temporal ridge into an anterior part distinctly ventral to a posterior section (arrows). **c,** *P. quadruplicatus,* UMMZ 80069—Ecuador: Napo; San Francisco, Río Napo. Note barely perceptible temporal ridge (arrow). **d,** *P. cuvieri,* MVZ 160091—Venezuela: Bolivár; 69 km S Río Cuyuní. Note relatively smooth parietal with only an anterior temporal ridge extension.

short but wide (and often perforated) maxillary portion. The vomer is not visible ventrally.

canicollis-group (Figure 20b)

The shape and structure of this foramen is similar to that described for *P. decumanus.* The opening is oval in shape, posterolateral flanges are weakly developed, and the anterior palate shows only faint grooves. The premaxillary portion of the septum is broad and extends to one-half the length of the foramen; the maxillary portion is moderately developed and in direct contact with the premaxilla. Hence, the vomer does not form part of the ventral aspect of the septum.

Development of the Temporal Ridge

Four conditions of temporal ridge development were recognized, as follows:

1. Ridge well developed, extending across the parietals from the supraorbital ledge to the lambdoidal ridge (fig. 21a).
2. Ridge moderately developed, but with an anterior parietal portion separated from and distinctly ventral to the posterior lambdoidal portion (fig. 21b).
3. Ridge continuous across parietals, but weakly developed, being a barely perceptible change in the lateral curvature of the parietals (fig. 21c).
4. No ridge development, or only a weak ridge extending from the supraorbital ledge onto the anterior parietals (fig. 21d).

Representatives of each species group were scored for these conditions, and the patterns are indicated in Table 2. Only *P. semispinosus* of the *semispinosus*-group displays complete and well-developed ridges, a characteristic noted by Gardner and Emmons (1984). *P. oconnelli* of this group, however, does not exhibit temporal ridges. The remainder of the species groups, however, show

TABLE 2. Frequency classes of the development of the temporal ridge of spiny rats, subgenus *Proechimys*. See text for explanation of character-states.

Species group/ region or taxon	Ridge score				
	N	1	2	3	4
semispinosus-group					
Central America	82	0.82	0.18	...	...
Northern Colombia	63	0.94	0.06	...	...
Central Colombia	31	1.00	...	...	...
Northern Ecuador	19	0.95	0.05	...	...
Southern Ecuador	29	1.00	...	...	...
oconnelli	20	...	...	...	1.00
longicaudatus-group					
Colombia–Ecuador	6	...	0.33	0.17	0.50
Northern Peru	145	...	0.53	0.27	0.20
Central Peru	39	...	0.58	0.19	0.23
SE Peru–Bolivia	35	...	0.46	0.36	0.18
Bolivia–Brazil	32	...	0.39	0.36	0.26
cuvieri-group					
Guianas	3	...	...	1.00	...
Central Brazil	37	...	0.35	0.50	0.15
Peru	4	...	0.25	0.50	0.25
2n = 40, Balta	5	...	...	0.40	0.60
goeldii-group					
Northern Peru	104	...	0.17	0.77	0.06
Central Peru	31	...	0.24	0.76	...
SE Peru–Bolivia	17	...	...	0.89	0.11
Venezuela–Brazil	21	...	...	0.61	0.39
Central Amazon	63	...	0.07	0.70	0.23
Eastern Amazon	123	...	0.07	0.81	0.12
guyannensis-group					
Goiás, Brazil	20	...	0.05	0.05	0.90
Pará, Brazil	50	...	0.13	0.12	0.75
Central Amazon	31	...	...	0.12	0.88
Venezuela–Brazil	129	...	0.01	...	0.99
Suriname	1	...	...	...	1.00
simonsi-group					
Colombia–Ecuador	5	...	...	...	1.00
Northern Peru	32	...	...	...	1.00
Central Peru	15	...	...	...	1.00
SE Peru–Bolivia	27	...	...	...	1.00
trinitatus-group					
chrysaeolus	5	...	...	0.25	0.75
mincae	19	...	...	0.05	0.95
trinitatus	12	...	...	0.08	0.92
urichi	8	...	...	...	1.00
guairae	3	...	...	...	1.00
ochraceous	1	...	...	...	1.00
poliopus	1	...	...	...	1.00
hoplomyoides	1	...	...	...	1.00
canicollis-group	13	...	...	...	1.00
decumanus-group	17	...	0.59	0.29	0.12

considerable overlap in the expression of this feature, although *decumanus*-, *longicaudatus*-, *goeldii*-, and *cuvieri*-group members consistently show some ridge development while taxa of the *guyannensis*-, *simonsi*-, *trinitatus*-, and *canicollis*-groups characteristically do not (table 2).

Ventral Canal of the Infraorbital Foramen

The infraorbital nerve courses near the medial floor of the infraorbital foramen, producing a canal of varying distinctness in many caviomorph rodents, including *Proechimys* (Woods, 1984). Both Moojen (1948) and Gardner and Emmons (1984) suggested that the presence and degree of development of this canal or groove has taxonomic significance in the genus. To evaluate this view, several grades of notch development were scored for representatives of each species group:

1. No groove present (fig. 22c)
2. Groove present, with moderately developed lateral flange (fig. 22b)
3. Groove present, with extreme development of a lateral flange (fig. 22a)

Intermediate levels between each of these classes were also recognized, providing five scores ranging from 1.0 to 3.0 in half increments.

The degree of groove development for geographic and taxonomic representatives of each species group is given in Table 3. While groove development varies widely within the genus, each species group displays a relatively narrow range. Members of the *longicaudatus*-, *trinitatus*-, *cuvieri*-, and *guyannensis*-groups exhibit the least development of a groove, the *longicaudatus*-group rarely showing any groove at all. The *goeldii*- and *simonsi*-groups show moderate development of the groove, while a notch is most strongly developed in the *semispinosus*-group. Specimens of *P. oconnelli* of the *semispinosus*-group consistently displayed the most extensive lateral flange development.

Geographic variation in the expression of groove development within each species group is virtually absent. The *longicaudatus*- and *semispinosus*-groups, with the lowest and highest mean scores, respectively, are the most uniform geographically; the *guyannensis*- and *goeldii*-groups are most variable (table 3).

Angle and Depth of the Mesopterygoid Fossa

The angle formed by the anterior margins of the mesopterygoid fossa was measured with a pro-

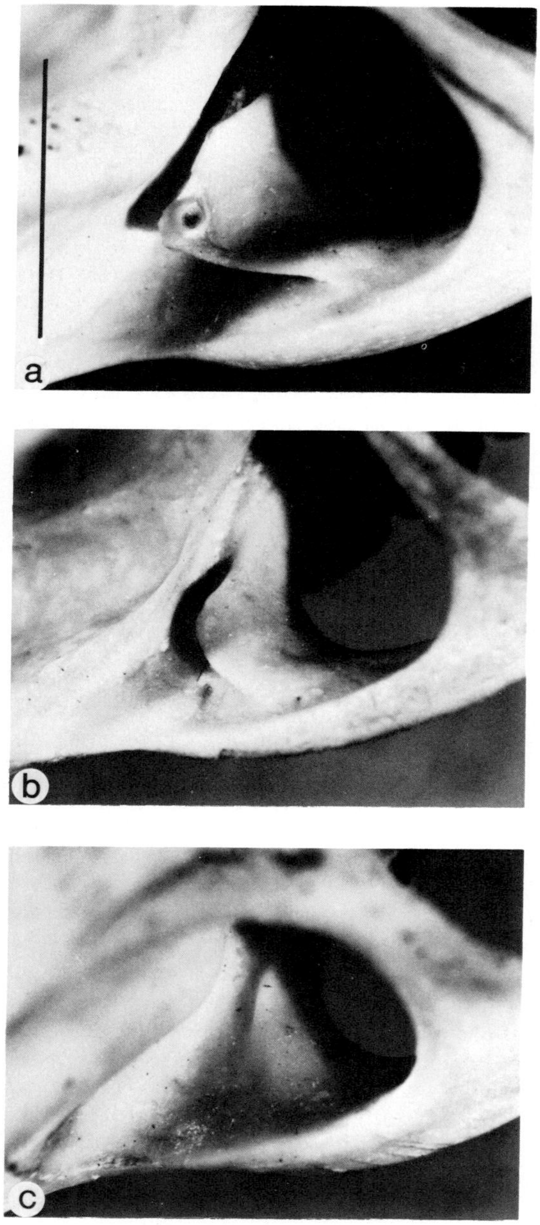

FIG. 22. Degrees of development of the canal or groove on the medial floor of the infraorbital foramen that accommodates the infraorbital nerve; see text for description of scoring system used; scale = 5 mm. **a,** Score = 3.0, *P. oconnelli,* MVZ 99685—Colombia: Meta; Villavicencio. **b,** Score = 2.0, *P. oconnelli,* MVZ 99684—Colombia: Meta; Villavicencio. **c,** Score = 1.0, *P. quadruplicatus,* UMMZ 80069—Ecuador: Napo; San Francisco, Río Napo.

TABLE 3. Mean scores and ranges for the development of the infraorbital foramen canal of spiny rats, subgenus *Proechimys.* See text for explanation of scoring system.

Species group/ region or taxon	N	Mean	Range
longicaudatus-group			
Colombia–Ecuador	6	1.1	1.0–1.5
Northern Peru	145	1.1	1.0–1.4
Central Peru	39	1.1	1.0–1.5
SE Peru–Bolivia	35	1.1	1.0–1.5
Southern Bolivia–Brazil	32	1.0	1.0–1.5
cuvieri-group			
Guianas	3	1.4	1.0–1.5
Central Amazon	37	1.3	1.0–2.0
Peru	4	1.4	1.0–2.0
2n = 40, Balta	5	1.4	1.0–2.0
semispinosus-group			
Central America	82	2.6	1.0–3.0
Northern Colombia	63	2.7	1.5–3.0
Central Colombia	31	2.6	2.0–3.0
Northern Ecuador	19	2.7	2.0–3.0
Southern Ecuador	29	2.7	1.5–3.0
oconnelli	20	2.8	2.0–3.0
goeldii-group			
Northern Peru	104	2.1	1.0–3.0
Central Peru	31	1.8	1.0–3.0
SE Peru–Bolivia	17	1.5	1.0–2.0
Venezuela–Brazil	21	1.8	1.0–3.0
Central Amazon	63	1.5	1.0–2.5
Eastern Amazon	123	1.9	1.0–3.0
guyannensis-group			
Goiás, Brazil	20	1.2	1.0–1.5
Pará, Brazil	50	1.1	1.0–1.5
Central Amazon	31	2.0	1.0–3.0
NW Brazil	34	2.0	1.5–3.0
SE Venezuela–Brazil	95	1.5	1.0–3.0
Suriname	1	2.5	. . .
simonsi-group			
Colombia–Ecuador	5	2.0	. . .
Northern Peru	32	2.1	1.5–3.0
Central Peru	15	1.6	1.0–2.0
SE Peru–Bolivia	27	1.8	1.0–2.5
trinitatus-group			
chrysaeolus	7	1.7	1.0–2.0
mincae	19	1.8	1.0–2.0
trinitatus	12	1.9	1.5–3.0
urichi	6	1.3	1.0–2.0
guairae	3	2.0	1.5–3.0
ochraceous	1	1.5	. . .
poliopus	1	1.5	. . .
hoplomyoides	1	1.0	. . .
canicollis-group	13	1.2	1.0–1.5
decumanus-group	17	1.4	1.0–2.0

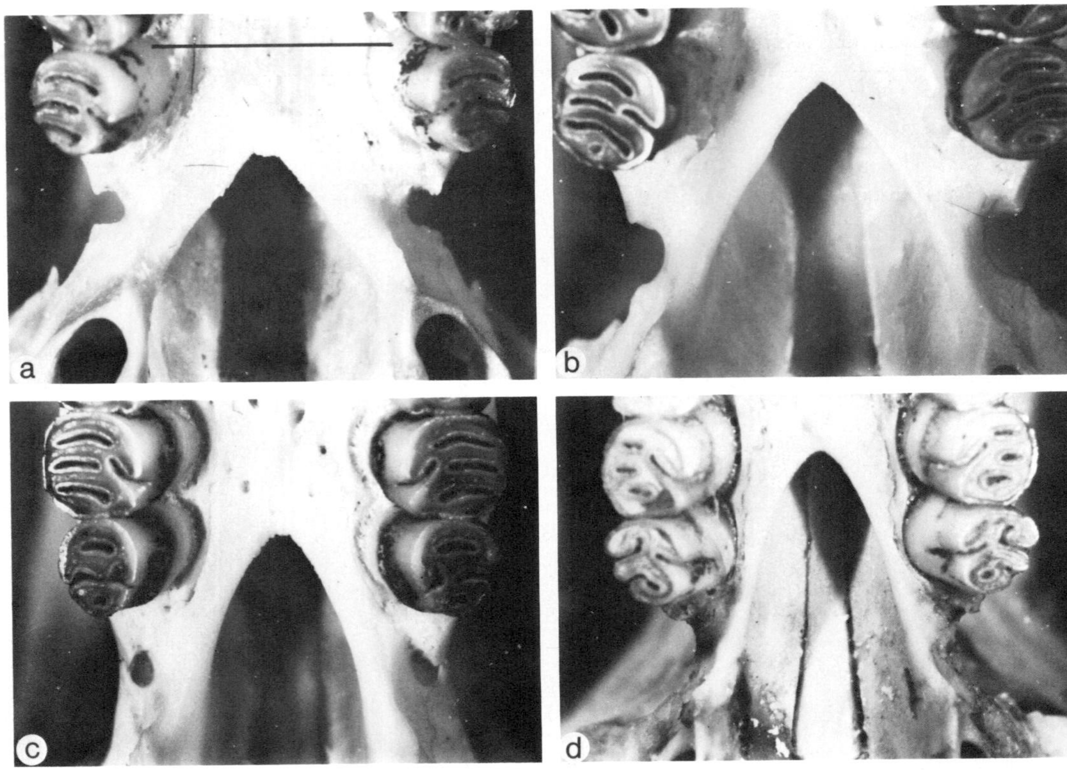

FIG. 23. Angle and extent of the mesopterygoid fossa for representative taxa of *Proechimys*; scale = 5 mm. **a,** *P. brevicauda,* MVZ 157854—Peru: Amazonas; La Poza, Río Santiago. **b,** *P. steerei,* MVZ 157888—Peru: Amazonas; La Poza, Río Santiago. **c,** *P. simonsi,* MVZ 157950—Peru: Amazonas; La Poza, Río Santiago. **d,** *P. guyannensis,* AMNH 75820—Venezuela: Bolivár; Arabupu, Mt. Roraima (topotype of *arabupu*).

tractor to the nearest degree, and the maximal penetration of the fossa into the palate was scored relative to the cheekteeth, as follows:

1. Not extending to the posterior margins of M3
2. Extending to the posterior one-half of M3
3. Extending to anterior one-half of M3
4. Extending to posterior one-half of M2
5. Extending to anterior one-half of M2

These two characters, angle and depth, are correlated in that the greater the depth usually the more acute the angle (fig. 23).

Table 4 provides data for the mesopterygoid fossa characters for representatives for each recognized species group of *Proechimys.* Members of the *longicaudatus*-group consistently have the broadest angle with the most shallow fossa (fig. 23a); those of the *simonsi*-group have the most acutely-angled fossa and, with members of the *guyannensis*-group, the deepest penetration into the palate (fig. 23c–d). Most other groups show moderate angles and degree of penetration. Except for the *guyannensis*-group, where samples referred to *P. oris* (Pará state, Brazil) and *P. roberti* (Goiás and Minas Gerais states, Brazil) are quite different from other samples examined (table 4), there is little geographic variation in the expression of mesopterygoid fossa characters within each species group.

Counterfold Pattern of the Cheekteeth

Early attempts to establish systematic relationships within *Proechimys* placed considerable emphasis on variation in counterfolds of the cheekteeth, both in number and pattern (e.g., Hershkovitz, 1948; Moojen, 1948). In general, these earlier studies indicated that fold number and pattern are quite variable geographically within taxa. As a result, counterfolds have been used primarily to recognize taxa sympatric at given localities rather than as a character complex capable of uniting distinct populations into cohesive and

TABLE 4. Mesopterygoid fossa (MPF) angle (in degrees) and depth scores of spiny rats, subgenus *Proechimys*. See text for detailed descriptions of character-states.

Species group/region or taxon	N	MPF angle Mean ± SD	MPF angle Range	MPF depth Mean	MPF depth Range
longicaudatus-group					
Colombia–Ecuador	6	79.0 ± 8.9	64–89	2.0	1–3
Northern Peru	145	79.6 ± 5.8	64–107	2.0	1–3
Central Peru	39	78.7 ± 7.1	64–93	1.9	1–3
SE Peru–Bolivia	35	72.7 ± 6.3	60–90	1.9	1–3
Southern Bolivia–Brazil	32	77.6 ± 8.5	60–91	2.2	1–3
cuvieri-group					
Guianas	3	69.7 ± 4.9	64–73	2.0	...
Central Brazil	37	72.8 ± 5.6	57–92	2.2	1–3
Peru	4	66.7 ± 5.7	59–71	2.0	1–3
2n = 40, Balta	5	56.4 ± 3.2	52–60	3.2	2–4
semispinosus-group					
Central America	82	56.4 ± 6.4	45–73	2.6	1–3
Northern Colombia	63	57.7 ± 6.1	45–72	2.5	2–4
Central Colombia	31	57.1 ± 5.6	46–62	2.8	2–4
Northern Ecuador	19	62.6 ± 6.8	53–75	3.0	...
Southern Ecuador	29	62.1 ± 4.8	53–71	2.8	2–4
oconnelli	20	63.3 ± 5.3	56–72	2.9	2–4
goeldii-group					
Northern Peru	104	60.6 ± 7.0	49–80	2.7	1–4
Central Peru	31	65.3 ± 7.1	45–76	2.6	2–3
SE Peru–Bolivia	17	67.9 ± 7.0	55–79	2.4	2–3
Venezuela–Brazil	21	68.2 ± 6.2	56–80	2.7	2–4
Central Brazil	63	62.1 ± 5.1	52–78	2.9	2–4
Eastern Brazil	123	65.3 ± 5.1	50–80	2.6	1–4
guyannensis-group					
Goiás, Brazil	20	67.7 ± 6.1	54–78	2.8	2–3
Pará, Brazil	50	64.2 ± 6.2	44–82	3.0	2–4
Central Brazil	31	56.1 ± 4.9	45–66	3.7	3–5
Venezuela–Brazil	129	47.5 ± 6.3	34–67	3.9	2–5
Suriname	1	58.0	...	4.0	...
simonsi-group					
Colombia–Ecuador	5	49.3 ± 6.9	39–53	3.6	3–4
Northern Peru	32	51.1 ± 3.3	47–62	4.1	2–5
Central Peru	15	53.3 ± 4.6	46–62	3.8	3–5
SE Peru–Bolivia	27	51.2 ± 4.5	45–64	3.2	2–4
trinitatus-group					
chrysaeolus	7	59.3 ± 9.9	53–67	2.3	1–3
mincae	19	56.7 ± 3.2	53–64	3.7	3–4
trinitatus	12	52.8 ± 4.9	48–65	3.4	3–4
urichi	6	56.3 ± 4.9	48–64	3.3	3–4
guairae	3	55.7 ± 3.1	53–58	3.3	3–4
ochraceous	2	48.5 ± 3.5	46–51	4.0	...
poliopus	2	47.0 ± 1.4	46–48	3.5	3–4
hoplomyoides	1	40.0	...	4.0	...
canicollis-group	13	54.2 ± 2.9	50–61	3.0	...
decumanus-group	17	52.9 ± 3.5	46–59	3.1	2–4

definable groups. While there is geographic variation in counterfold pattern and number within taxa of spiny rats, these characters can be used to form groups which are consistently separable from others. In this respect, the counterfold pattern and number reinforces group membership delineated by other morphological features, such as bacula and palatal characters.

Illustrations of counterfold patterns for both upper and lower toothrows for each species group

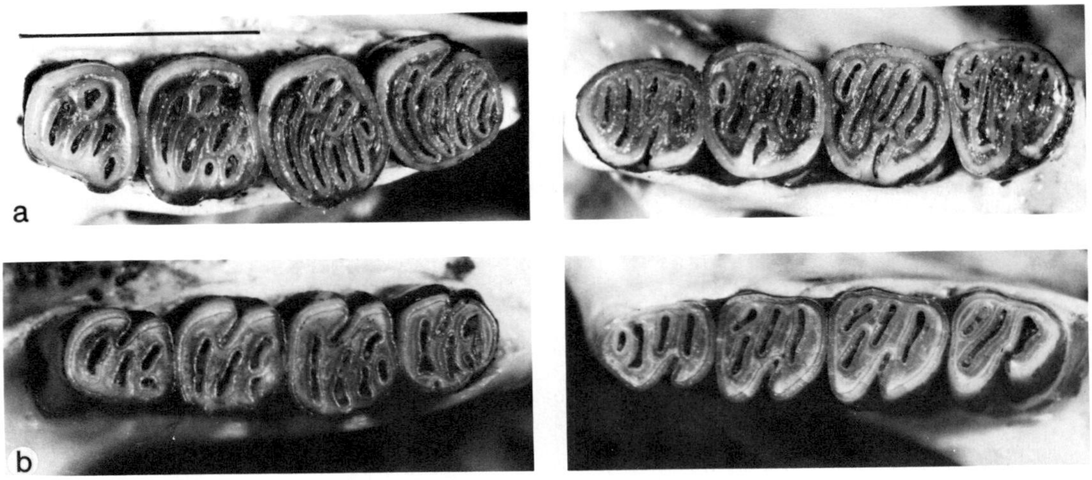

FIG. 24. Upper (left) and lower (right) toothrows of *P. semispinosus* of the *semispinosus*-group of *Proechimys*; scale = 5 mm. **a,** FMNH 90143—Colombia: Cauca: Río Saija. **b,** FMNH 70072—Colombia: Chocó; Unguía.

(with the exception of the *decumanus*-group) are presented in Figures 24–30. As with other character complexes I examined, these examples were chosen to illustrate the range of variation. Geographic variation within each group is summarized in Table 5.

Counterfold number and, to a lesser extent, pattern change with increasing age. Obviously, folds become obliterated in advanced age, but even in moderately aged individuals smaller folds easily can become lost, and coalescence or isolation of folds occurs (see Moojen, 1948). These age-related

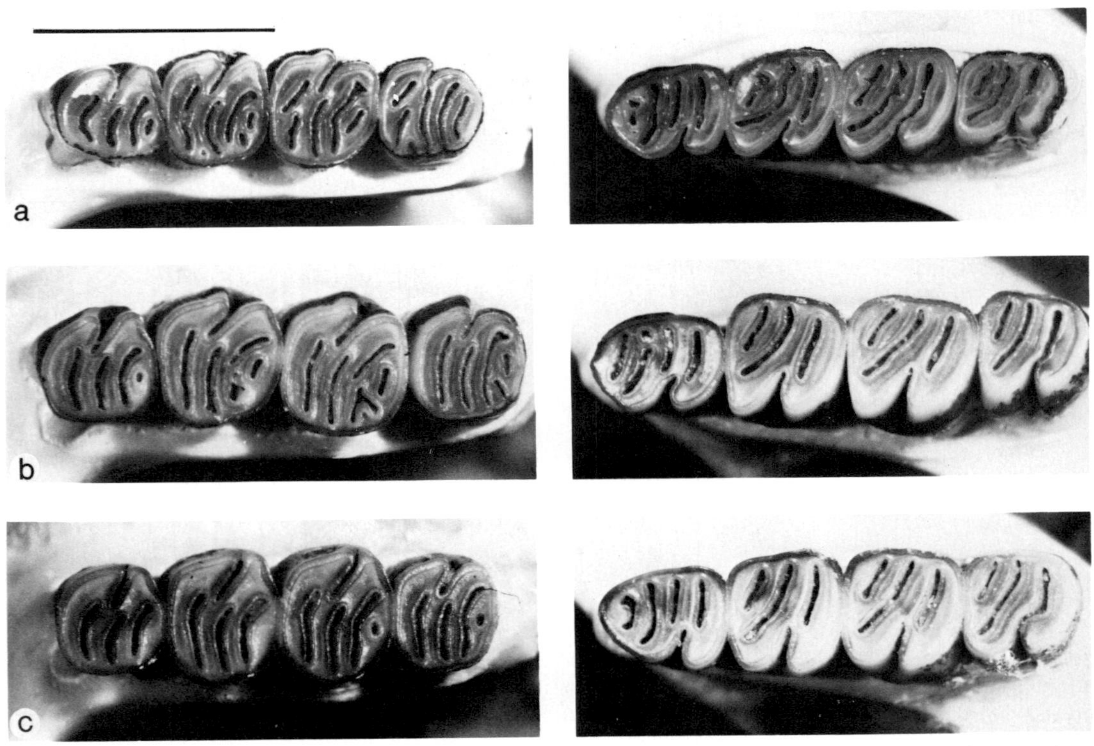

FIG. 25. Upper (left) and lower (right) toothrows of *P. steerei* of the *goeldii*-group; all specimens from La Poza, Río Santiago, Amazonas, Peru; scale = 5 mm. **a,** MVZ 157871. **b,** MVZ 157863. **c,** MVZ 157861.

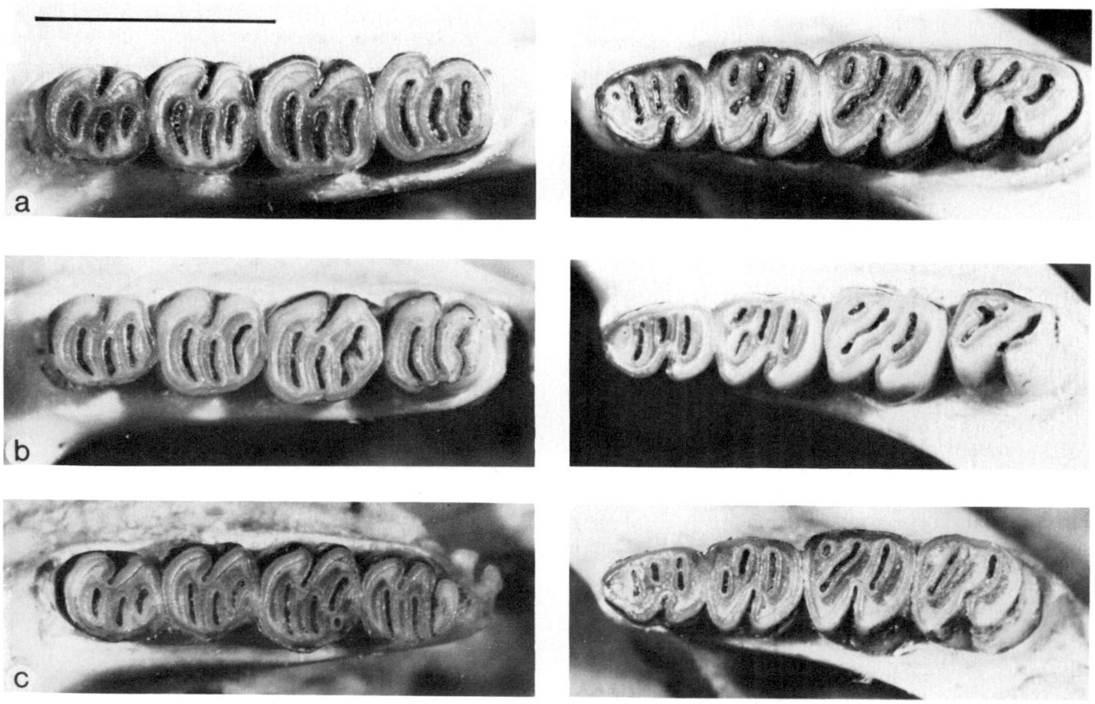

FIG. 26. Upper (left) and lower (right) toothrows of *P. brevicauda* (**a–b**) and *P. longicaudatus* (**c**) of the *longicaudatus*-group; scale = 5 mm. **a,** MVZ 157584—Peru: Amazonas; La Poza, Río Santiago. **b,** MVZ 157855—Peru: Amazonas; La Poza, Río Santiago. **c,** JRS 222—Paraguay: Chaco; 54 km E Agua Dulce (specimen to be deposited in National Museum, Asunción, Paraguay).

phenomena create difficulties in counting folds and clearly are partly responsible for some of the variability observed within and between samples. To minimize this extraneous variation, the data assembled here are based on individuals in age classes 8 or 9 (as defined by Patton & Rogers, 1983) where folds are still mostly confluent with the sides of each tooth.

Taxa of the *semispinosus*-group (fig. 24; table 5), followed by those of the *goeldii*-group (fig. 25),

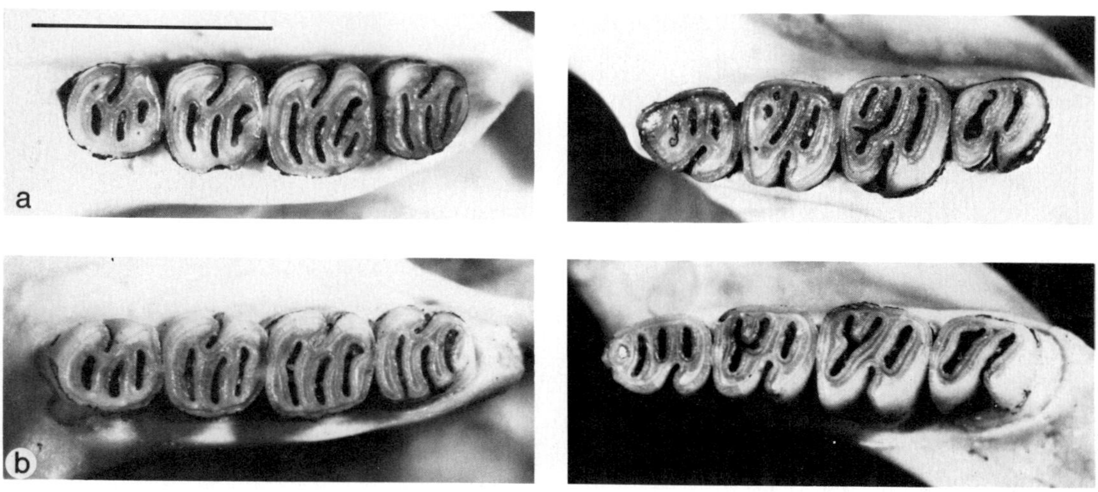

FIG. 27. Upper (left) and lower (right) toothrows of *P. cuvieri* of the *cuvieri*-group; scale = 5 mm. **a,** MVZ 157874—Peru: Amazonas; La Poza, Río Santiago. **b,** MVZ 160091—Venezuela: Bolivár; 69 km S Río Cuyuní.

TABLE 5. Counterfold patterns of the cheekteeth of spiny rats, subgenus *Proechimys*. Folds are given as frequencies for each tooth, based largely on specimens of the age classes 8 and 9 (Patton & Rogers, 1983).

Species group/ region or taxon	**N**	**dPM^4**			**M^1**			**M^2**		
		2	**3**	**4**	**2**	**3**	**4**	**2**	**3**	**4**
semispinosus-group										
Central America	76	...	1.00	...	...	0.92	0.08	...	0.70	0.30
Northern Colombia	36	...	1.00	...	...	0.89	0.11	...	0.71	0.29
Central Colombia	12	...	0.67	0.33	...	0.58	0.42	...	0.30	0.70
Northern Ecuador	10	...	0.80	0.20	...	0.30	0.70	...	...	1.00
Southern Ecuador	29	...	1.00	...	...	0.97	0.03	0.05	0.85	0.10
oconnelli	19	...	1.00	...	...	1.00	...	...	1.00	...
goeldii-group										
Colombia–Ecuador	9	...	0.25	0.75	...	0.12	0.88	...	0.08	0.92
Northern Peru	91	...	0.36	0.64	...	0.30	0.70	...	0.10	0.90
Central Peru	13	...	0.85	0.15	...	0.75	0.25	...	0.55	0.45
SE Peru–Bolivia	18	...	0.92	0.12	...	0.78	0.22	...	0.22	0.78
Brazil–Venezuela	21	...	0.79	0.21	...	0.58	0.42	...	0.25	0.75
Central Amazon	61	...	0.86	0.14	...	0.72	0.28	...	0.57	0.43
Eastern Amazon	108	...	1.00	...	...	1.00	...	...	0.98	0.02
longicaudatus-group										
Colombia–Ecuador	7	...	1.00	...	...	1.00	...	...	1.00	...
Northern Peru	134	...	1.00	...	...	1.00	...	...	0.97	0.03
Central Peru	19	...	1.00	...	...	1.00	...	...	1.00	...
SE Peru–Bolivia	36	...	1.00	...	...	1.00	...	...	0.95	0.05
Southern Bolivia–Brazil	37	...	1.00	...	...	1.00	...	...	1.00	...
cuvieri-group										
Guianas	3	...	1.00	...	...	1.00	...	...	1.00	...
Brazil	24	...	1.00	...	...	1.00	...	...	1.00	...
Peru	20	...	1.00	...	...	1.00	...	...	0.80	0.20
2n = 40, Balta	5	...	1.00	...	...	1.00	...	...	1.00	...
simonsi-group										
Colombia–Ecuador	12	...	1.00	...	...	1.00	...	...	1.00	...
Northern Peru	39	...	1.00	...	...	1.00	...	...	0.88	0.12
Central Peru	24	...	1.00	...	...	0.98	0.02	...	0.76	0.24
SE Peru–Bolivia	31	...	1.00	...	...	1.00	...	...	0.68	0.32
guyannensis-group										
Goiás	20	...	1.00	...	...	1.00	...	...	1.00	...
Pará	50	...	1.00	...	...	1.00	...	...	0.98	0.02
Central Amazon	31	...	1.00	...	...	1.00	...	...	1.00	...
NW Brazil–Venezuela	34	...	1.00	...	...	1.00	...	...	1.00	...
SE Venezuela	95	...	1.00	...	0.02	0.98	...	0.03	0.97	...
Suriname	17	...	1.00	...	...	1.00	...	...	1.00	...
trinitatus-group										
chrysaeolus	12	...	1.00	...	...	1.00	...	...	1.00	...
mincae	19	...	1.00	...	...	1.00	...	0.16	0.84	...
trinitatus	12	...	1.00	...	...	1.00	...	...	1.00	...
urichi	6	...	1.00	...	...	1.00	...	...	1.00	...
guairae	2	...	1.00	...	...	1.00	...	...	1.00	...
hoplomyoides	1	...	1.00	...	...	1.00	...	...	1.00	...
decumanus-group	17	...	1.00	...	...	1.00	...	...	1.00	...
canicollis-group	13	...	1.00	...	1.00	...	...	1.00	...	...

* Five folds occur on dpm_4 with frequencies of 0.02 and 0.01, respectively.

exhibit the most complex counterfold patterns. Specimens of *P. semispinosus* from central Colombia south into northern Ecuador have the highest average number of folds per tooth, particularly of the maxillary row (table 5); this number decreases slightly to the north and south. *Proechimys oconnelli* shows the lowest counterfold number for this group. In the *goeldii*-group, specimens from Co-

TABLE 5. *Continued.*

M³			dpm₄			m₁			m₂			m₃		
2	3	4	2	3	4	2	3	4	2	3	4	2	3	4
···	0.91	0.09	···	0.62	0.38	0.01	0.99	···	0.05	0.90	0.05	0.18	0.81	0.01
···	0.77	0.23	···	0.44	0.56	···	1.00	···	···	0.97	0.03	0.20	0.73	0.07
···	0.83	0.17	···	···	1.00	···	1.00	···	···	1.00	···	···	0.83	0.17
···	···	1.00	···	···	1.00	···	1.00	···	···	0.90	0.10	···	1.00	···
0.05	0.85	0.10	···	0.25	0.75	···	1.00	···	···	1.00	···	0.48	0.52	···
0.04	0.96	···	···	1.00	···	0.18	0.82	···	0.22	0.78	···	0.85	0.15	···
···	0.28	0.72	···	···	1.00	···	1.00	···	···	1.00	···	···	1.00	···
···	0.14	0.86	···	···	0.98*	···	0.95	0.05	···	0.93	0.07	···	1.00	···
···	0.23	0.77	···	···	1.00	···	1.00	···	···	1.00	···	···	1.00	···
···	0.58	0.42	···	···	1.00	···	1.00	···	···	1.00	···	···	1.00	···
···	0.54	0.46	···	···	1.00	···	1.00	···	···	1.00	···	0.07	0.93	···
···	0.70	0.30	···	0.06	0.93*	···	1.00	···	0.02	0.98	···	0.05	0.95	···
0.06	0.92	0.02	···	0.33	0.67	0.01	0.99	···	0.10	0.90	···	0.50	0.50	···
···	1.00	···	···	0.70	0.30	···	1.00	···	···	1.00	···	0.23	0.77	···
···	0.98	0.02	···	0.74	0.26	···	1.00	···	0.02	0.98	···	0.26	0.74	···
···	1.00	···	···	0.75	0.25	···	1.00	···	0.11	0.89	···	0.46	0.54	···
···	0.98	0.02	···	0.71	0.29	0.02	0.98	···	0.02	0.98	···	0.43	0.57	···
···	1.00	···	···	0.98	0.02	0.14	0.86	···	0.10	0.90	···	0.69	0.31	···
···	1.00	···	···	0.67	0.33	···	1.00	···	0.18	0.82	···	0.33	0.67	···
···	1.00	···	···	0.94	0.06	0.02	0.98	···	0.24	0.76	···	0.33	0.67	···
···	0.98	0.02	···	0.25	0.75	···	1.00	···	···	1.00	···	0.95	0.05	···
···	1.00	···	···	0.79	0.21	···	1.00	···	···	1.00	···	1.00	···	···
0.04	0.96	···	···	0.75	0.25	···	1.00	···	···	1.00	···	···	1.00	···
···	0.97	0.03	···	0.70	0.30	···	1.00	···	0.01	0.99	···	0.15	0.84	0.01
···	1.00	···	···	0.42	0.58	0.03	0.97	···	···	1.00	···	0.21	0.79	···
···	0.79	0.21	···	0.29	0.71	···	1.00	···	···	1.00	···	0.11	0.89	···
0.08	0.92	···	···	1.00	···	0.52	0.48	···	0.75	0.25	···	1.00	···	···
0.11	0.89	···	···	0.85	0.15	0.33	0.67	···	0.39	0.61	···	0.96	0.04	···
0.11	0.89	···	···	0.91	0.09	0.75	0.25	···	0.75	0.25	···	0.92	0.08	···
0.12	0.88	···	···	0.90	0.10	0.67	0.33	···	0.62	0.38	···	0.84	0.16	···
0.36	0.64	···	···	1.00	···	0.78	0.22	···	0.85	0.15	···	0.97	0.03	···
···	1.00	···	···	1.00	···	1.00	···	···	1.00	···	···	1.00	···	···
···	1.00	···	···	0.97	0.03	···	1.00	···	0.35	0.65	···	0.50	0.50	···
0.62	0.38	···	···	1.00	···	0.98	0.02	···	0.98	0.02	···	0.97	0.03	···
0.05	0.95	···	···	1.00	···	0.18	0.82	···	0.22	0.78	···	0.30	0.70	···
0.21	0.95	···	···	1.00	···	0.25	0.75	···	0.30	0.70	···	0.71	0.29	···
1.00	···	···	···	1.00	···	1.00	···	···	1.00	···	···	1.00	···	···
···	1.00	···	···	1.00	···	···	1.00	···	···	1.00	···	···	1.00	···
0.04	0.96	···	···	1.00	···	0.04	0.96	···	0.63	0.37	···	0.98	0.02	···
1.00	···	···	1.00	···	···	1.00	···	···	1.00	···	···	1.00	···	···

lombia south to northern Peru and east into Venezuela and adjacent Brazil have the highest fold counts (table 5); those from Ecuador represent *quadruplicatus,* which was named by Hershkovitz (1948) for its high fold number. Counts decrease slightly to the south into Bolivia and central and eastern Brazil (table 5).

The *longicaudatus*-group members have a con-

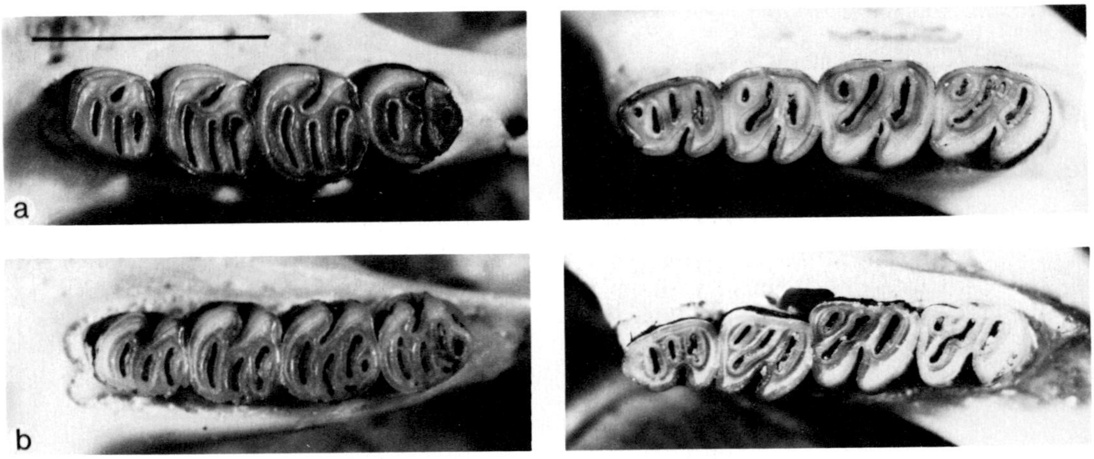

FIG. 28. Upper (left) and lower (right) toothrows of *P. simonsi* of the *simonsi*-group; scale = 5 mm. **a,** MVZ 157950—Peru: Amazonas; La Poza, Río Santiago. **b,** MVZ 166035—Peru: Madre de Dios; Albergue, Río Madre de Dios.

sistent fold count for the upper teeth throughout their ranges (fig. 26; table 5). The lower cheekteeth, however, show a decrease in fold number from north to south: samples referred to *P. longicaudatus* are characterized by only two folds on the last molar; specimens referred to *P. brevicauda* typically have the first and second medial folds on m_3 displaying degrees of coalescence (see fig. 26a–b). The *cuvieri*-group shows a fold count similar to that of the *longicaudatus*-group (table 5), and these two groups cannot be distinguished in pattern (compare fig. 26 with fig. 27). Taxa of the *simonsi*-group are somewhat intermediate between the *goeldii*- and *longicaudatus*-groups in fold number (fig. 28). The upper cheekteeth, particularly M^2, show a slight increase in number of folds from north to south, as does dpm_4 (table 5). This geographic pattern is the reverse of that seen in both the *goeldii*- and *longicaudatus*-groups over the same part of the western Amazon Basin.

Taxa of the *guyannensis*-group are uniform in number and pattern of counterfolds throughout their range (fig. 29; table 5). They are characterized by having three folds on most teeth, often with two folds on the lower molars, and only rarely four folds on dpm_4. Members of the *trinitatus*-, *decumanus*-, and *canicollis*-groups display the lowest counterfold number (table 5) and, hence, the simplest pattern (fig. 30). Specimens of *P. canicollis* have the least complex cheekteeth of any taxon in the subgenus *Proechimys,* with two folds on each tooth the general rule.

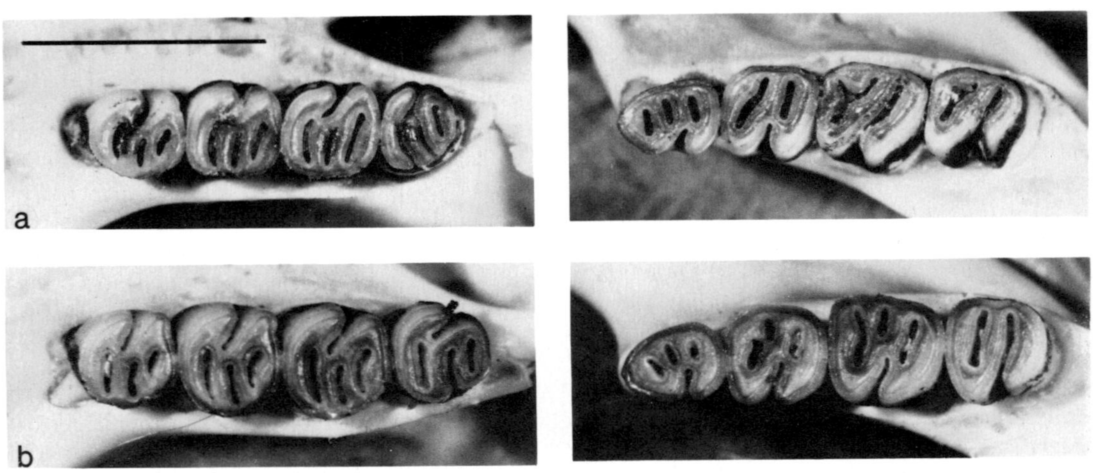

FIG. 29. Upper (left) and lower (right) toothrows of *P. guyannensis* (**a**) and *P. roberti* (**b**) of the *guyannensis*-group; scale = 5 mm. **a,** AMNH 130737—Venezuela: Bolivár; Auyantepui. **b,** AMNH 134309—Brazil: Goiás; Anápolis.

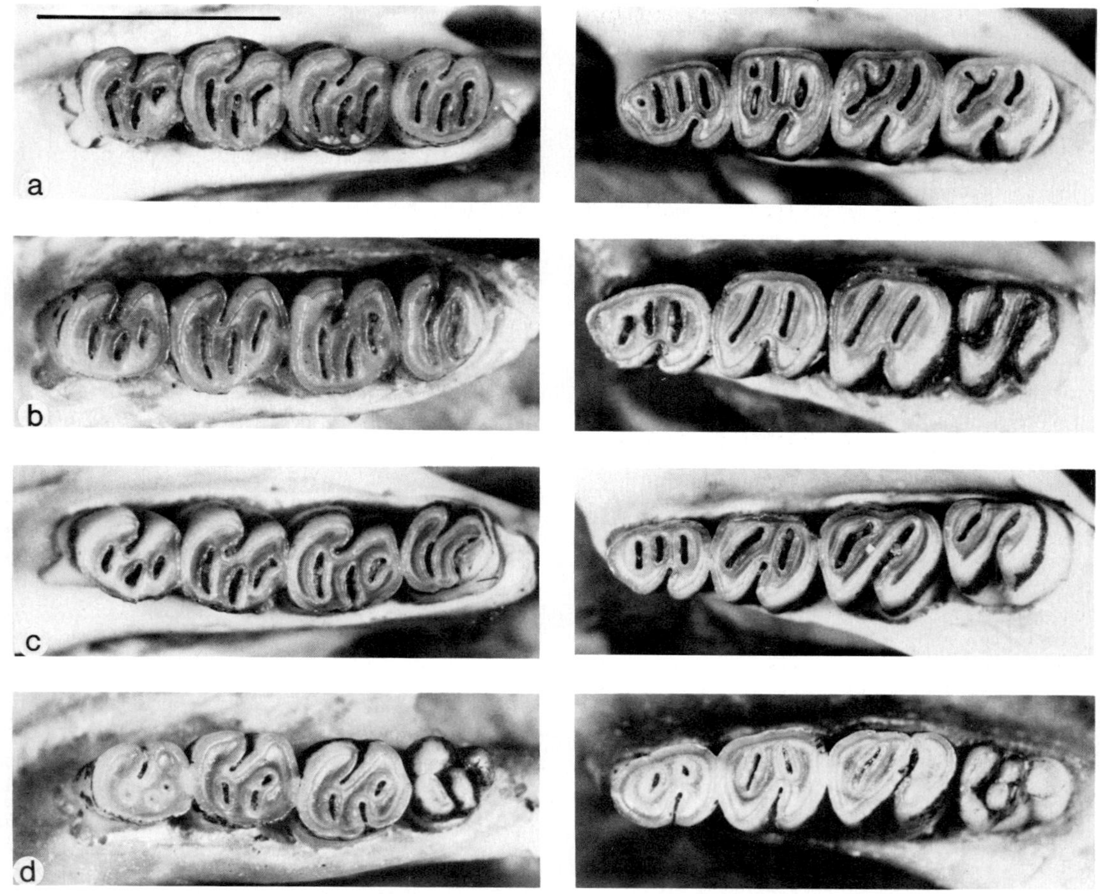

FIG. 30. Upper (left) and lower (right) toothrows of representatives of the *trinitatus* (**a–c**) and *canicollis* (**d**) groups; scale = 5 mm. **a,** *P. chrysaeolus,* FMNH 69039—Colombia: Bolivár; San Juan Nepumoceno. **b,** *P. mincae,* FMNH 13203—Colombia: Magdalena; Minca (topotype). **c,** *P. guairae,* FMNH 92588—Colombia: Arauca; Río Cobaria. **d,** *P. canicollis,* FMNH 69109—Colombia: Bolivár; San Juan Nepumoceno.

Remarks and Prospectus

Nine species groups of spiny rats (subgenus *Proechimys*) are defined herein, and 59 of the 67 names which have been proposed are allocated to one or another of these groups. Although the defined groups differ from those proposed recently by Gardner and Emmons (1984), it is reassuring that we have grouped taxa similarly, with only minor exceptions, despite our use of different suites of characters. Our common conclusions indicate that characters are not hopelessly chaotic geographically (see, for example, Thomas, 1928), but that consistent patterns are recognizable.

The preceding discussion, however, neither suggests the number of species that are likely present in each of the groups defined, nor comments on the phyletic relationships among them. I would like here to summarize my opinions as to the likely number and distribution of the species contained within each group. It remains for future work, both in the field and in the museum, to verify the accuracy of these hypotheses.

The *decumanus*-, *canicollis*-, and *simonsi*-groups are considered monotypic; certainly the restricted ranges and uniform character distributions of both *Proechimys decumanus* and *P. canicollis* support this view. Despite a much broader geographic range, character variation among populations assigned to the *simonsi*-group is either negligible or clinal in nature. Indeed, this is perhaps the most consistently recognizable group of spiny rats because of this character uniformity (see also Gardner & Emmons, 1984). Even the karyotype is invariant throughout the species range, based on samples available from southern Colombia (Reig

& Useche, 1976) to southern Peru (Patton & Gardner, 1972; Gardner & Emmons, 1984). By seniority, the single species in the *simonsi*-group should be recognized as *P. simonsi* Thomas, with the names *hendeei* Thomas and *nigrofulvus* Osgood considered synonyms.

The *guyannensis*-group members are close to *P. simonsi* in most salient features described here. Their bacula are nearly indistinguishable, as are characters of the incisive foramina and mesopterygoid fossa. Differences exist in counterfold pattern, but these could represent nothing more than geographic variation. Only the analysis of samples from the hiatus in western Brazil between the known distributions of these taxa (see fig. 1) will permit such a determination. Certainly, of the groups defined herein, the *simonsi*- and *guyannensis*-groups are more similar in examined characters than any of the others. Future work may indicate that these groups should be condensed into one.

More than one species is likely present, however, in my *guyannensis*-group. Karyotypic variation is large, as diploid number ranges from 44 to 30 among the limited geographic samples (see Gardner & Emmons, 1984). Moreover, there are some seemingly striking geographic differences in some of the characters examined, although not in all. For example, specimens from eastern Pará, Goiás, and Minas Gerais states of Brazil (referred to *oris* Thomas and *roberti* Thomas) have broader and shallower mesopterygoid fossae and a less developed infraorbital canal than do those from elsewhere in the group's range. Clearly, a more refined and critical examination of detailed geographic variation in these characters is needed; the view provided here is simply too general to judge adequately the significance of this variation.

The *goeldii*-group varies more over its geographic range than any other; nevertheless, much of the variation in counterfold pattern, for example, appears clinal, and abrupt character shifts which might signal species-level demarcations are not readily apparent in the characters I examined. Known karyotypic variation also appears limited, with samples examined from southern Venezuela (2n = 26, FN = 42; Reig & Useche, 1976), Ecuador and northern Peru (2n = 28, FN = 42–44; Gardner & Emmons, 1984), and central and southern Peru (2n = 24, FN = 42; Patton & Gardner, 1972; Gardner & Emmons, 1984). Specimens from the western Amazon Basin and from the Casiquiare region of southern Venezuela appear fairly homogeneous in counterfold and pelage color characters. There does, however, appear to be relatively sharp transition of some characters, particularly those of the pelage (not examined in this report), in the central Amazon Basin between the lower Río Negro and the Río Tapajós. It is probable that at least two species are present in this group, a western one to which the name *steerei* Goldman would apply, and an eastern one to which the senior name *goeldii* Thomas applies.

The character summaries provided in this paper tend to minimize the difficulties that I had in assigning individual specimens to species groups, and thus may provide a sense of false security. This is particularly true for specimens of the *goeldii* and *guyannensis* groups from the eastern parts of both ranges, primarily in Pará state, Brazil. There are more individual question marks regarding group assignments for specimens from this region than for any other area or group, and much more detailed effort is necessary to confirm the character differences described herein.

Within the *longicaudatus*-group there appears to be at least two species, *Proechimys longicaudatus* Rengger from eastern Bolivia east through adjacent Brazil into northern Paraguay; and *P. brevicauda* Gunther, which occupies the remainder of the group range as depicted in Figure 3. An area of rather sharp character transition, particularly in pelage color and color pattern but also in bacular measurements, for example, occurs in the upper Río Iténez and Río Mamoré of southern El Beni and Santa Cruz in Bolivia. The limited samples of *P. longicaudatus* examined show little variation throughout its range. Samples of *P. brevicauda* from southern Colombia to northern Bolivia are, however, quite variable, and more than one species may be represented here. Gardner and Emmons (1984) suggested that the Ecuadoran populations referred to *gularis* are specifically distinct from northern Peruvian *brevicauda* based on karyotypic differences (2n = 30, FN = 48 without large subtelocentric autosomes, versus 2n = 30, FN = 48 with two pairs of large subtelocentrics). They also suggested that the central and southern Peruvian populations might represent a valid subspecies of *P. brevicauda,* to which the name *elassopus* would apply, based both on karyotypic (2n = 28, FN = 50) and color pattern differences. A thorough analysis of geographic variation within this group is certainly warranted.

No more than two species appear to be represented in the *semispinosus*-group; *P. semispinosus* (Tomes) is distributed from Nicaragua south along the Pacific lowlands to southern Ecuador, and *P.*

oconnelli is restricted to the western llanos in the vicinity of Villavicencio, Colombia. Karyotypic variation occurs in *P. semispinosus,* but such is minimal (2n = 30, FN = 50–54); *P. oconnelli* differs only by a single fusion/fission (Gardner & Emmons, 1984). A more detailed examination of geographic variation in *P. semispinosus* is needed before the intraspecific status of the large number of named forms referred to this group can be properly evaluated.

As mentioned previously, the *cuvieri*-group is clearly divisible into two biological units. The widespread species *P. cuvieri* Petter is uniform in its characters, including karyotype, from the Guianas to northern Peru. The status of the 2n = 40 karyotypic form from Balta in eastern Peru, however, is an enigma at present. It is clearly specifically distinct from *cuvieri* and is only placed in this group because of similar bacular design. I do not believe that it is a relative of *guyannensis*-group taxa, as suggested by both Patton and Gardner (1972) and Gardner and Emmons (1984). I have also not been able to identify this form with certainty anywhere except at Balta.

Finally, I have treated the various taxa assigned to my *trinitatus*-group as though they were species, primarily because I have examined relatively few specimens from only scattered localities. Nevertheless, it is likely that a number of species exist in this group. Karyotypic variation is extensive (see reviews by Reig et al., 1980; Reig, 1981), and both karyotypic and electromorphic data (Benado et al., 1979) differentiate a *guairae* superspecies (including *guairae, poliopus, ochraceous,* and *mincae* [see Gardner & Emmons, 1984]) and a *trinitatus* superspecies (composed of *trinitatus* and *urichi*). *Proechimys hoplomyoides* is clearly a species distinct from the above, as indicated by Gardner and Emmons (1984). These latter authors included *chrysaeolus* in their *semispinosus*-group and *magdalenae* in their *brevicauda*-group, positions which are not supported by the bacular characters covered here. If their true relationships do lie with the *trinitatus*-group, they too are probably separate species. Certainly, *chrysaeolus* is the most distinctive member of my *trinitatus*-group in incisive foraminal and counterfold characters.

Acknowledgments

Alfred Gardner introduced me to *Proechimys* in the field 19 years ago. My continued interest in these animals results solely from his own infectious curiosity and our long-lasting friendship; I value both immeasurably. I am grateful to J. E. Hill and P. D. Jenkins of the British Museum (Natural History); G. G. Musser and S. Anderson of the American Museum of Natural History; A. L. Gardner and C. O. Handley, Jr. of the National Museum of Natural History; P. W. Freeman, R. M. Timm, and B. D. Patterson of Field Museum of Natural History; M. S. Hafner and J. P. O'Neill of the Museum of Zoology, Louisiana State University; and P. Myers of the Museum of Zoology, University of Michigan for the opportunity to examine materials in their respective collections. Special appreciation is extended to A. L. Gardner and L. H. Emmons for their constant willingness to share information and ideas on *Proechimys* and thus to work toward a common understanding. In the same vein, I also thank O. A. Reig for cooperative interactions over the past decade. Aid in the field has been generously provided by C. P. Patton, A. L. Gardner, J. E. Cadle, M. A. Barros, M. D. Robinson, J. P. O'Neill, P. Myers, and O. B. Berlin. This research has been supported by the National Science Foundation (BNS 76-17485), the National Geographic Society, and the Museum of Vertebrate Zoology. Fieldwork has been facilitated by the Dirección General Forestal y de Fauna, Ministerio de Agricultura, Lima, Peru.

Literature Cited

BENADO, M., M. AGUILERA, O. A. REIG, AND F. J. AYALA. 1979. Biochemical genetics of chromosome forms of Venezuelan spiny rats of the *Proechimys guairae* and *Proechimys trinitatus* superspecies. Genetica, **50:** 89–97.

DIDIER, R. 1962. Note sur l'os penien de quelques rongeurs de l'Amerique du Sud. Mammalia, **26**(3): 408–430.

GARDNER, A. L. 1983. *Proechimys semispinosus* (Rodentia: Echimyidae): Distribution, type locality, and taxonomic history. Proceedings of the Biological Society of Washington, **96:** 134–144.

GARDNER, A. L., AND L. H. EMMONS. 1984. Species groups in *Proechimys* (Rodentia, Echimyidae) as indicated by karyology and bullar morphology. Journal of Mammalogy, **65**(1): 10–25.

HERSHKOVITZ, P. 1948. Mammals of northern Colombia. Preliminary report no. 2: Spiny rats (Echimyidae), with supplemental notes on related forms. Proceedings of the United States National Museum, **97:** 125–140.

HOOPER, E. T. 1961. The glans penis in *Proechimys* and other caviomorph rodents. Occasional Papers of the Museum of Zoology, University of Michigan, no. 623, 18 pp.

Martin, R. E. 1970. Cranial and bacular variation in populations of spiny rats of the genus *Proechimys* (Rodentia: Echimyidae) from South America. Smithsonian Contributions to Zoology, **35:** 1–19.

Moojen, J. 1948. Speciation in the Brazilian spiny rats (genus *Proechimys,* family Echimyidae). University of Kansas Publications, Museum of Natural History, **1**(19): 301–406.

Patton, J. L., and L. H. Emmons. 1985. A review of the genus *Isothrix* (Rodentia, Echimyidae). American Museum Novitates, **2817:** 1–14.

Patton, J. L., and A. L. Gardner. 1972. Notes on the systematics of *Proechimys* (Rodentia: Echimyidae), with emphasis on Peruvian forms. Occasional Papers, Museum of Zoology, Louisiana State University, **44:** 1–30.

Patton, J. L., and M. A. Rogers. 1983. Systematic implications of non-geographic variation in the spiny rat genus *Proechimys* (Echimyidae). Zeitschrift für Säugetierkunde, **48:** 363–370.

Petter, F. 1978. Épidémiologie de la leishmaniose en Guyane française, en relation avec l'existence d'une espèce nouvelle de Rongeurs Échimyidés, *Proechimys cuvieri* sp. n. Comptes Rendus Hebdomadaires des Séances de l'Académie des Sciences, séries D, Sciences Naturelles, **287:** 261–264.

Reig, O. A. 1981. Modelos de especiación cromosómica en las casiraguas (género *Proechimys*), pp. 149–190. *In* Reig, O. A., ed., Ecology and Genetics of Animal Speciation. Equinoccio, Caracas, Venezuela, 295 pp.

Reig, O. A., M. Aguilera, M. A. Barros, and M. Useche. 1980. Chromosomal speciation in a Rassenkreis of Venezuelan spiny rats (genus *Proechimys,* Rodentia, Echimyidae), pp. 291–312. *In* Vorontsov, N. N., and J. M. Van Brink, eds., Animal Genetics and Evolution. Dr. W. Junk B. V., The Hague, Netherlands, 383 pp.

Reig, O. A., and M. Useche. 1976. Diversidad cariotípica y sistemática en poblaciones venezolanas de *Proechimys* (Rodentia, Echimyidae), con datos adicionales sobre poblaciones de Peru y Colombia. Acta Científica Venezolana, **27:** 132–140.

Thomas, O. 1901. On mammals obtained by Mr. Alphonse Robert on the Rio Jordão, S.W. Minas Geraes. Annals and Magazine of Natural History, ser. 7, **8:** 526–536.

———. 1928. The Godman-Thomas expedition to Peru. VII. The mammals of the Rio Ucayali. Annals and Magazine of Natural History, ser. 10, **2:** 249–265.

Woods, C. A. 1984. Histricognath rodents, pp. 389–446. *In* Anderson, S., and J. K. Jones, Jr., eds., Orders and Families of Recent Mammals of the World. John Wiley & Sons, New York, 686 pp.

Appendix: Specimens Examined

Repositories for specimens examined in this study are as follows: American Museum of Natural History (AMNH), British Museum (Natural History) (BM[NH]), Field Museum of Natural History (FMNH), Louisiana State University Museum of Zoology (LSU), University of California Museum of Vertebrate Zoology (MVZ), University of Michigan Museum of Zoology (UMMZ), and National Museum of Natural History (USNM).

canicollis-group

COLOMBIA: Atlántico: Cienaga de Guajaro, Sabana Larga (1, USNM). **Bolívar:** San Juan Nepumoceno (9, FMNH). **Magdalena:** Bonda (29, AMNH; 1, FMNH; 8, USNM; 2, BM[NH]); Bonda, Finca Veracruz (2, USNM); Mamatoca (11, AMNH; 7, FMNH); Minca Road (1, AMNH); El Líbano plantation (6, AMNH); Santa Marta (11, AMNH; 1, FMNH); Río Guaimaral, Valledupar (13, USNM); Aguas Verdes, Valledupar (26, USNM); Parmarito, Valledupar (2, USNM); El Orinoco, Río Cesar, Valledupar (14, USNM); Villanueva, Valledupar (33, USNM).

VENEZUELA: Zulia: Perijá, Río Cogollo (7, FMNH); Río Cachiri (1, MVZ).

cuvieri-group

P. CUVIERI

BRAZIL: Pará: Ilha do Taiuno, Rio Tocantins (25, AMNH); Vila Bela Imperatriz, Serra de Parintins, Rio Amazonas (1, AMNH).

GUYANA: Kamakusa (1, AMNH); Kalacoon (1, AMNH); Kartabo (8, AMNH); Minehaha Creek (1, AMNH); Samin Island, Mazarani River (6, AMNH); Maracai Creek, Demarara River (1, BM[NH]); Demarara River (2, BM[NH]); Supinaam River (2, BM[NH]).

PERU: Amazonas: La Poza, Río Santiago (1, MVZ). **Loreto:** Pebas, Río Amazonas (5, BM[NH]);

Orosa, Río Amazonas (14, AMNH); Pto. Indiana, Río Amazonas (1, AMNH); Santa Luisa, Río Nanay (3, FMNH); Sarayacu, Río Ucayali (3, AMNH).

SURINAME: Carolinakreek (5, FMNH); Wilhelmina Mts., West River (1, FMNH); Finisanti, Saramacca River (1, FMNH); Lelydorpplan (2, FMNH); La Poule (2, FMNH); Dirkshoop (3, FMNH).

2n = 40

PERU: Ucayali: Balta, Río Curanja (5, LSU).

***decumanus*-group**

ECUADOR: El Oro: Santa Rosa (2, AMNH). **Guayas:** Chongón (4, BM[NH]); Chongoncito (10, AMNH); Cerro Manglar Alto (3, AMNH); Cerro Baja Verde (1, AMNH); Los Pozos (25, AMNH). **Los Ríos:** Vinces, Hda. Pijigual (2, AMNH). **Manabí:** Bahia de Caraques, Río Briseno (7, AMNH).

PERU: Piura: Quebrada Bandarrango (1, FMNH); Laguna Lamadero (2, FMNH). **Tumbes:** Matapalo (6, FMNH).

***goeldii*-group**

BOLIVIA: El Beni: Riberalta, Vaca Diez (13, USNM); 13 km W Riberalta (11, USNM); 3.5 km NW Riberalta (2, USNM); 5 km NW Riberalta (3, USNM); Río Mamoré, 4 km below Santa Cruz (2, AMNH); 6 km S Buena Hora (1, AMNH); Río Mamoré, 7 km N Lagionha (2, AMNH); Río Mamoré, 5 km S Guayaramarin (1, AMNH); Río Mamoré, 5 km S Guayaramarin (1, AMNH); Río Mamoré (2, AMNH); Río Mamoré, opposite Cascajal (1, AMNH); Río Mamoré, 17 km NNW Nuevo Berlín (1, AMNH).

BRAZIL: Acre: Sena Madureira, Mandel Urbano (1, USNM); Rio Branco, 3–4 km S Rio Branco (2, USNM). **Amazonas:** Faro, Rio Yurnunda [= Nhamunda] (2, BM[NH]); Faro, Paraiso (2, AMNH); Acajutuba, Rio Negro (2, BM[NH]); Mirapinima, Rio Negro (5, AMNH); Cacauo Pereira Igarapé, Rio Negro (9, AMNH); Yucabi, Rio Negro (1, AMNH); Tatu, Rio Uaupés (3, AMNH); Itamarati, Rio Uaupés (1, AMNH); Tahuapunta, Rio Uaupés (2, AMNH); Manacapurú, Rio Solimoes (2, BM[NH]); Maturaca Mission, northern Amazonia (1, USNM); Humaitá, km 886–990, Br 230 (3, USNM); São Antonio de Amatari (1, AMNH); Borba, Rio Madeira (2, AMNH); Auara Igarapé, Rio Madeira (15, AMNH); São Antonio de Uayara (9, AMNH); Rosarinho, Rio Madeira (11, AMNH); Ipixuna, Rio Purus (1, USNM); Rio Purus, Hyutanahan (4, USNM). **Mato Grosso:** Serra da Chapada (4, BM[NH]); Utiariti, Rio Papagaio (1, AMNH). **Pará:** Cametá, Rio Tocantins (2, BM[NH]); Manapiri Island, Rio Tocantins (2, BM[NH]); Ilha do Taiuno, Rio Tocantins (57, AMNH); Mocajuba, Rio Tocantins (1, AMNH); Baião, Rio Tocantins (3, AMNH); Urucum de Corumbá (1, FMNH); Tuary, Rio Tapajós (1, FMNH; 11, AMNH); Aramanay, Rio Tapajós (6, AMNH); Piquiatuba, Rio Tapajós (7, AMNH); Igarapé Amorim, Rio Tapajós (30, AMNH); Igarapé Brabo, Rio Tapajós (2, AMNH); Farinicatuba, Rio Tapajós (1, AMNH); Aquiatuba, Rio Tapajós (1, AMNH); Limoal, Rio Tapajós (1, AMNH); Inajatuba, Rio Tapajós (1, AMNH); Fordlandia, Rio Tapajós (5, AMNH); Santarém (1, BM[NH]); km 84, Santarém-Cuiabá hwy (27, USNM); km 212, Santarém-Cuiabá hwy (4, USNM); km 216, Santarém-Cuiabá hwy (3, USNM); Itaituba, Rio Tapacurazinho (9, USNM); Mojui Dos Campos (15, USNM); km 19, Itaituba-Jacaréacanga hwy (12, USNM); km 25, Itaituba-Altamira hwy (4, USNM); Itaituba (6, USNM); 54 km S, 150 km W Altamira (3, USNM); Agrovila, km 43 Itaituba-Altamira rd (2, USNM); Vila Bela Imperatriz, south bank Rio Amazonas (3, AMNH); Serra de Parintins, Vila Bela Imperatriz, south bank Rio Amazonas (7, AMNH); Porto de Moz, Rio Xingu (2, AMNH); Vilarinho do Monte, Rio Xingu (4, AMNH). **Rondônia:** Pista Nova, 8 km N Porto Vêlho (4, USNM).

COLOMBIA: Caquetá: La Tagua, Tres Troncos, Río Caquetá (13, FMNH); Río Mecaya (1, FMNH); Florencia, Mantanito (3, FMNH); La Murelia, Río Bodoquera (2, FMNH).

ECUADOR: Napo: San Francisco, Río Napo (8, UMMZ); Llunchi, Río Napo (5, UMMZ).

PERU: Amazonas: La Poza, Río Santiago (128, MVZ). **Huánuco:** San Antonio, Río Pachitea (1, BM[NH]); Port Leguia, Río Pachitea (1, BM[NH]). **Loreto:** Yurimaguas (1, FMNH); Orosa, Río Amazonas (14, AMNH); Boca Río Curaray (2, AMNH); Pto. Indiana, Río Amazonas (27, AMNH); Río Mazán (2, AMNH); Río Panduro (6, AMNH); Pampa Chica, Iquitos (1, AMNH); Santa Rita, Iquitos (5, FMNH); Santa Luisa, Río Nanay (9, FMNH); Santa Elena, Río Samiria (11, FMNH); Río Samiria (10, FMNH); San Lorenzo, Río Marañón (1, FMNH; 4, BM[NH]); Boca Río Peruate, Río Amazonas (2, FMNH); Lagunas (10, FMNH; 1, BM[NH]); Quistacocha, Maynas (4, FMNH); Río Tigre, 1 km above Río Tigrillo (7, FMNH); Pebas, Río Amazonas (1,

BM[NH]); Sarayacu, Río Ucayali (16, AMNH); San Jeronimo, Río Ucayali (1, BM[NH]); Cantamana (3, BM[NH]); Lago Mirano, Río Napo (6, BM[NH]). **Madre de Dios:** La Pastora, Maldonado (1, FMNH); Tambopata, Puerto Maldonado (3, USNM); Itahuania (1, FMNH). **Ucayali:** Yarinacocha (6, FMNH; 9, LSU); Chicosa, upper Río Ucayali (6, BM[NH]); 59 km W Pucallpa (1, USNM); Pucallpa (2, AMNH); Santa Rosa, Río Ucayali (12, AMNH); Fernando Stahl Mission (14, AMNH); Cumaria (5, BM[NH]); Tushemo, Masisea, Río Ucayali (1, BM[NH]); Balta, Río Curanja (1, LSU).

VENEZUELA: Amazonas: 68 km SE Esmeralda (9, USNM); Río Orinoco, Tamatama (13, USNM); Casiquiare Canal, Capibara (14, USNM); 30 km SSE Puerto Ayacucho, Coromoto (8, USNM); 18 km SSE Puerto Ayacucho (2, USNM); Mt. Duida, Río Casiquiare, Quemapure (2, AMNH); Mt. Duida, 8 mi from Río Orinoco (2, AMNH); Mt. Duida, Esmeralda (11, AMNH); Mt. Duida, Caño Seco (1, AMNH); Mt. Duida, El Merey (2, AMNH); Río Orinoco, Paripari (1, AMNH); Río Orinoco, Boca del Río Ocamo (3, AMNH).

***guyannensis*-group**

BRAZIL: Amapá: Serra do Navio (3, USNM); Calicoene (1, USNM); Capoeira (1, USNM). **Amazonas:** Hd. Rio Tucaro (1, USNM); Serra de Neblina (1, USNM); Rio Uaupés, Tauarate (1, AMNH); Rio Uaupés, Tahuapunta (10, AMNH); Rio Negro, Tatu (1, AMNH); Rio Negro, Uacara (1, AMNH); Rio Negro, Camanaos (1, AMNH); Rio Negro, Pira-pocú (1, AMNH); Rio Negro, Manaus (1, AMNH); Faro, north bank Rio Amazonas (10, AMNH); Rio Paratucu (2, AMNH); Rio Nhamundá, Castanhal (20, AMNH); Rio Nhamundá, São José (3, AMNH). **Goiás:** Fazenda Cangalha (1, USNM); Anápolis (79, AMNH). **Minas Gerais:** Rio Jordão, Araguari (1, FMNH; 10, BM[NH]). **Pará:** Providência (1, FMNH); Cametá, Rio Tocantins (1, FMNH; 1, AMNH); Ilha do Taiuno, Rio Tocantins (1, AMNH); Baião, Rio Tocantins (1, AMNH); Manapiri Island, Rio Tocantins (1, AMNH); Maranhão, Alto Parnaíba (1, FMNH); km 84, Santarém-Cuiabá hwy (53, USNM); km 212, Santarém-Cuiabá hwy (1, USNM); km 217, Santarém-Cuiabá hwy (4, USNM); km 19, Itaituba-Jacaréacanga hwy (2, USNM); Rio Tapacurazinho (10, USNM); Agrovila, Altamira (6, USNM); km 43, Itaituba-Altamira hwy (5, USNM); Marabá, Serra Norte (7, USNM); Jatobal (11, USNM); Itupiranga (1, USNM); Belém (46, USNM; 1, AMNH); Igarapé-Açu (13, AMNH); Tury-Açu, Maranhão (1, BM[NH]); Abaeté (8, BM[NH]); Patagônia (15, AMNH); Capim (10, AMNH); Igarapé Amorim, Rio Tapajós (17, AMNH); Inajatuba, Rio Tapajós (1, AMNH); Igarapé Brabo, Rio Tapajós (4, AMNH); Limoal, Rio Tapajós (1, AMNH); Vila Bela Imperatriz, south bank Rio Amazonas, Lago Andina (1, AMNH); Vila Bela Imperatriz, south bank Rio Amazonas, Boca Rio Andina (3, AMNH); Vila Bela Imperatriz, south bank Rio Amazonas, Serra de Parintins (3, AMNH). **Roraima:** Uaico, Rio Uraricoera (3, AMNH); Rio Cotingo, Limão (64, AMNH).

SURINAME: Brokopando: Carolinakreek (4, FMNH); Lawa Mission, Lawa River (1, AMNH); Loksie Hattie, Saramacca River (4, FMNH); Finisanti, Saramacca River (8, FMNH).

VENEZUELA: Amazonas: 68 km SE Esmeralda, Mavaca (2, USNM); 68 km SE Esmeralda, Boca Masiaca (1, USNM); Río Canucunuma, Belén (26, USNM); Casiquiare Canal, Capibara (3, USNM); San Juan, Río Manapiari (17, USNM); Río Orinoco (2, USNM); Río Orinoco, Boca del Río Ocamo (5, AMNH); Río Casiquiare, El Merey (6, AMNH); Río Casiquiare, Buena Vista (4, AMNH); Río Casiquiare, Solano (2, AMNH); Río Casiquiare (1, AMNH); Río Casiquiare, 2 mi W Tamasu (1, AMNH); Mt. Duida, foothills camp (1, AMNH); Mt. Duida, middle camp (7, AMNH); Mt. Duida, Valle de los Monos (2, AMNH); Mt. Duida, Playa del Río Base (6, AMNH); Mt. Duida, Caño Seco (1, AMNH); Mt. Duida, Pie del Cerro (1, AMNH); Mt. Duida, La Lajo, Río Orinoco (1, AMNH).

***longicaudatus*-group**

BOLIVIA: El Beni: 6 km S Buena Hora (1, AMNH); Río Machupo, 15 km above Horquilla (1, AMNH); San Ignacio (47, USNM); 3.6 km NNE San Ignacio (9, USNM); Riberalta (2, FMNH); Riberalta, Vaca Diez (4, USNM); Fortaleza (2, USNM); San Marco, 3.2 km SW San Joaquín (3, USNM; 22, FMNH); San Joaquín (7, USNM; 90, FMNH; 4, AMNH); 20 km S San Joaquín, Est. Yutiole (1, AMNH); Cafetal, 20 km SE San Ramón (4, USNM); Río Iténez, opposite Príncipe da Beira (3, AMNH); Río Mamoré, 5 km NE Río Grande mouth (5, AMNH); Río Mamoré, 1 mi NW Guayaramarin (5, AMNH); Guayaramarin (1, AMNH); Río Mamoré, 5 km S Guayaramarin (2, AMNH); Río Mamoré (1, AMNH); Río Mamoré, opposite Cascajal (2, AMNH); Río Mamoré, 2 km SE Puerto Siles (5, AMNH); 10 km E San Antonio (1, AMNH); Rurrenabaque (1, AMNH);

Lago Victoria (1, USNM; 14, FMNH); La Esperanza, 42 km NE San Joaquín (1, USNM); Est. Barranquita, 20 km S San Joaquín (3, FMNH); El Carmen (20, FMNH); Azunta (27, FMNH); Santo Dios (10, FMNH); San Pedro (2, FMNH); Caravana (17, FMNH); San Pablo (8, FMNH); Filadelfia (6, FMNH); Acapulco (1, FMNH); Buena Vista (1, FMNH); Arruda (2, FMNH); Centenela (1, FMNH); Cinco (4, FMNH); Las Pavas (6, FMNH); Providencia (1, FMNH); Puerto Siles (1, FMNH); San Andrés (2, FMNH); San Juan (3, FMNH); Tapera Jorillo (3, FMNH); Veintedos (1, FMNH); Huchulu (1, USNM); Las Penas (1, USNM); Pampitas (1, USNM). **Cochabamba:** El Mojón (4, FMNH); San Rafael, 19 km SW Villa Tunari (2, USNM); 4 km SE Villa Tunari (1, USNM); 2 km E Villa Tunari (2, AMNH); Todos Santos (3, FMNH; 13, AMNH); Mission San Antonio, Río Chimore (8, AMNH); El Palmar (5, FMNH); Charuplaya, upper Río Secure (5, BM[NH]). **La Paz:** 5 km SE Guanay, Río Challana (2, UMMZ); Caranavi (4, UMMZ); Mapiri (4, AMNH; 3, BM[NH]); Ticunhuaya (5, AMNH); San Ernesto (2, BM[NH]). **Pando:** Río Nareuda (2, AMNH). **Santa Cruz:** Buenavista (4, FMNH; 10, BM[NH]); Ascención de Guarayos (12, FMNH); Río Surutu (1, BM[NH]); Río Ichilo, 54 km S Boca Río Chaparé (12, AMNH); Río Ichilo, 52 km S Boca Río Chaparé (2, AMNH); Río Ichilo, 34 km S Boca Río Chaparé (4, AMNH); Río Ichilo, 30 km S Boca Río Chaparé (1, AMNH); Río Mamoré, 2 km from Boca Río Chaparé (2, AMNH); Warnes (2, USNM); 1.3 km NE Warnes (8, USNM); 1 km NNW Warnes (9, USNM); 3 km SW Warnes, Santa Rosita (6, USNM); Florida, near Floripondio (2, FMNH); Cerro Hosana (1, FMNH).

BRAZIL: Acre: Sena Madureira, Mandel Urbano (1, AMNH); Rio Branco, 3–4 km S Rio Branco (2, USNM). **Mato Grosso:** Tapirapuã, Rio Siputuba (2, AMNH); Urucum (2, AMNH); Serra da Chapada (6, BM[NH]); Fazenda Acurizal (2, USNM); Aripuanã, Humboldt-Aripaunã (8, USNM); Corumbá (7, USNM); 7 km SE Corumbá (1, USNM); 22 km S Corumbá (2, USNM); Sta. Theresa, 7 km WSW Urucum (7, USNM); Cuiabá, 10 km N Cuiabá (1, USNM); Limão, 48 km W Cáceres, Rio Jauru (15, USNM). **Rondônia:** Pista Nova, 8 km N Porto Vêlho (6, USNM); Porto Vêlho (1, FMNH).

COLOMBIA: Caquetá: Río Mecaya (2, FMNH); Florencia, Mantanito (17, FMNH); Florencia (17, AMNH); La Murelia, Río Bodoquera (15, AMNH).

ECUADOR: Napo: San Francisco, Río Napo (17, UMMZ); Intillama, Río Napo (2, UMMZ); Llunchi, Río Napo (2, UMMZ); near Río Napo, Oriente (4, BM[NH]); San José Abajo (6, AMNH); Río Suno Abajo (2, AMNH). **Pastaza:** Canelos, Río Bobonaza (2, BM[NH]); Río Pastaza (2, BM[NH]); Río Pindo Yacu (2, FMNH); Río Yana Rumi (1, FMNH); Río Capihuara (3, FMNH); Río Copataza (3, FMNH); Río Lipuno (1, AMNH); Sarayacu (4, AMNH); Canelos (1, AMNH).

PERU: Amazonas: Huampami, Río Cenepa (173, MVZ); La Poza, Río Santiago (38, MVZ). **Huánuco:** Port Leguia, Río Pachitea (3, BM[NH]); San Antonio, Río Pachitea (1, BM[NH]); 35 km NE Tingo María, Sta. Elena (2, LSU); Tingo María (5, BM[NH]; 1, LSU; 9, FMNH); Chinchavita (10, BM[NH]). **Loreto:** Yurimaguas (2, USNM; 2, BM[NH]; 25, FMNH); Pebas, Río Amazonas (10, BM[NH]); Boca Río Curaray (32, AMNH); Iquitos, Río Amazonas (1, BM[NH]); Santa Luisa, Río Nanay (1, FMNH); San Fernando, Río Yavari (1, FMNH); Cantamana (3, BM[NH]). **Madre de Dios:** Tambopata, Puerto Maldonado (23, USNM); Lago Sandoval, Río Madre de Dios (3, MVZ); La Pastora, Puerto Maldonado (5, FMNH); Albergue, Río Madre de Dios (8, MVZ). **Pasco:** Nevati Mission (54, AMNH); San Pablo (32, AMNH). **Puno:** Santo Domingo [= Inca Mines] (1, BM[NH]; 5, AMNH; 6, FMNH). **San Martín:** Achinamiza, Río Huallaga (1, AMNH). **Ucayali:** Yarinacocha (1, FMNH); Pucallpa (1, FMNH); 59 km W Pucallpa (21, USNM); Fernando Stahl Mission (2, AMNH); Balta, Río Curanja (18, LSU; 10, MVZ).

semispinosus-group

P. SEMISPINOSUS

COLOMBIA: Antióquia: Urabá, Villa Arteaga, Río Curulao (23, FMNH). **Cauca:** Río Saija (16, FMNH); El Papayo, Río Saija (3, FMNH); La Boca, Río Saija (3, FMNH); San José (1, FMNH; 11, AMNH); Novita trail, western Andes (1, AMNH). **Chocó:** Condoto (1, BM[NH]); Río Docampado (12, FMNH); Río Saudo (16, FMNH); Unguía (24, FMNH); Bagado (4, AMNH); Andaqueda (1, AMNH). **Córdoba:** Socorre, upper Río Sinú (4, FMNH). **Nariño:** La Guayacana (15, FMNH); La Candelilla (4, FMNH); Isla Gorgona (5, BM[NH]; 2, FMNH); Barbacoas (8, AMNH). **Valle de Cauca:** Sabaleta 2, FMNH).

COSTA RICA: Alajuela: San Carlos (2, FMNH). **Limón:** Cariari (5, LSU); Finca La Lola (1, LSU); 4.6 km W Limón (2, MVZ). **Puntarenas:** San Geronimo (2, FMNH; 2, AMNH); Rincón de Osa (1, LSU); Palmar Sur (2, LSU); Palmar (28, AMNH). **San José:** 16.3 km SE San Isidro (2, MVZ); 34.7 km SE San Isidro (2, MVZ); 1.6 km W Villa Colón (2, MVZ);

2.8 km W Villa Colón (1, MVZ); 14.5 km N Quepos, Río Damitas (1, LSU); Caspirola (1, LSU).

ECUADOR: Esmeraldas: San Javier (6, BM[NH]; 1, FMNH); Esmeraldas (3, AMNH). **Manabí:** Río Mongaya (2, FMNH). **Pichincha:** Santo Domingo (9, BM[NH]). **El Oro:** Santa Rosa (3, BM[NH]; 4, AMNH); Pasaje (4, AMNH). **Los Ríos:** Bucay (1, AMNH); Puente de Chimbo, Bucay (2, AMNH); Cague, El Destino (6, AMNH); Limón, Balsapampa to Babahoyo road (7, AMNH); Ventura (1, AMNH).

NICARAGUA: Rivas: Río Grande (11, AMNH). **Zelaya:** Toro Rapids (2, AMNH); Bluefields (5, MVZ).

PANAMA: Canal Zone: Barro Colorado Island (3, AMNH); Gatun (20, AMNH); Maxim Ranch (3, AMNH); Buena Vista Peninsula (6, LSU); Río Chagres (2, AMNH); Balboa (1, AMNH). **Chiriquí:** Boquerón (1, BM[NH]; 12, FMNH; 56, AMNH); Bugaba (6, BM[NH]). **Darién:** Cituro (4, AMNH); El Real (8, AMNH); Boca de Cupe (4, AMNH); Tapaliza (3, AMNH); Tacarcuna (2, AMNH). **Panamá:** Tocumen (3, BM[NH]); Gobernador Island (7, BM[NH]); Savanna near Panamá (3, BM[NH]); 0.8 km N Paraíso (4, MVZ); Cebaco Island (5, BM[NH]); San Miguel Island (3, FMNH; 5, AMNH).

P. OCONNELLI

COLOMBIA: Meta: Quaicaramo (3, AMNH; 15, USNM); Mambita (2, USNM); La Aguadita (1, AMNH); Barrigona (2, AMNH); Restrepo (12, AMNH); Villavicencio (26, AMNH; 2, USNM; 10, UMMZ; 4, MVZ); 3 km N Villavicencio (1, USNM); Los Micos, San Juan de Arama (16, FMNH).

***simonsi*-group**

BOLIVIA: El Beni: Río Mamoré (1, AMNH). **Cochabamba:** Yungas (1, AMNH). **Pando:** Río Nareuda (1, AMNH).

COLOMBIA: Caquetá: Río Mecaya (9, FMNH); La Murelia, Río Bodoquera (1, AMNH).

ECUADOR: Napo: Intillana, Río Napo (5, UMMZ); near Río Napo (1, BM[NH]); San José Abajo (3, AMNH); Río Suno Abajo (1, AMNH). **Pastaza:** Río Pindo Yacu (4, FMNH); Río Bobonaza, Montalvo (3, FMNH); Río Bobonaza (2, BM[NH]); Río Yana Rumi (1, FMNH); Río Capihuara (2, FMNH); Río Pastaza (5, BM[NH]); Río Tigre (4, BM[NH]). **Zamora:** Gualaquiza (1, BM[NH]).

PERU: Amazonas: Huampami, Río Cenepa (3, MVZ); headwaters Río Kagka (2, MVZ); La Poza, Río Santiago (3, MVZ); Yambrasbamba (1, FMNH; 5, BM[NH]). **Cajamarca:** Huarandosa (1, AMNH). **Cuzco:** 40 km E Quincemil above Río Marcapata (2, LSU); Cosñipata, Hda. Villa Carmen (4, FMNH); Urubamba (1, BM[NH]). **Junín:** Río Perené (1, BM[NH]). **Loreto:** Yurimaguas (3, FMNH); Santa Luisa, Río Nanay (5, FMNH); Boca Río Curaray (22, AMNH); Orosa, Río Amazonas (1, AMNH); Cerro Azul, Cantamana, Río Ucayali (4, BM[NH]); Cantamana (1, BM[NH]). **Madre de Dios:** Itahuanía (4, FMNH); Tambopata, Puerto Maldonado (4, USNM); Aguas Calientes, Río Alto Madre de Dios (10, MVZ); Hda. Erika, Río Alto Madre de Dios (3, MVZ); Albergue, Río Madre de Dios (7, MVZ). **Pasco:** Mairo, Río Palcazu (2, BM[NH]); San Juan (1, USNM); Bermudas de Loma Linda (13, AMNH); San Pablo (3, AMNH). **San Martín:** Puca Tambo, 50 mi E Chachapoyas (10, BM[NH]). **Ucayali:** Balta, Río Curanja (2, MVZ; 17, LSU); Yarinacocha (1, FMNH); 59 km W Pucallpa (33, USNM).

***trinitatus*-group**

P. CHRYSAEOLUS

COLOMBIA: Antióquia: Puri, above Cáseres (8, FMNH); Medellín (1, BM[NH]). **Bolívar:** San Juan Nepumoceno (26, FMNH); Coloso (20, FMNH); Margarita (3, BM[NH]). **Boyacá:** Muzo (3, FMNH; 2, BM[NH]). **Cauca:** Río Chili (2, BM[NH]). **Córdoba:** Catival, upper Río San Jorge (4, FMNH); Socorre, upper Río Sinu (18, FMNH). **Tolima:** Santana (3, BM[NH]).

VENEZUELA: Táchira: San Cristóbal (1, BM[NH]).

P. GUAIRAE

COLOMBIA: Arauca: Río Cobaria (22, FMNH); Río Bojaba (5, FMNH); Río Arauca (18, FMNH); Fatima, Río Cobaria (12, FMNH). **Boyacá:** La Argentina, Río Cubugón (4, FMNH); El Porvenir, Río Cubugón (3, FMNH).

VENEZUELA: Barinas: Guaquitas (1, MVZ). **Portuguesa:** Sto. Domingo (1, MVZ).

P. HOPLOMYOIDES

VENEZUELA: Bolívar: Mt. Roraima, Arabupu (1, AMNH); Mt. Roraima, Rondon camp (1, AMNH).

P. MAGDALENAE

COLOMBIA: Antióquia: near La Providencia, SW Zaragoza (1, USNM); 25 km S and 22 km W Zaragoza (48, USNM). **Bolívar:** Norosi, Mompos, Río San Pedro (18, USNM).

P. MINCAE

COLOMBIA: Magdalena: Minca (2, BM[NH]; 6, USNM; 78, AMNH); Bonda (12, AMNH; 2, USNM); Onaca (4, AMNH); Buritaca (1, AMNH); Don Dago (1, AMNH); Cuaco (1, AMNH); Masinga Vieja (1, AMNH); Manzanares (2, USNM); Colonia Agrícola de Caracolicito (1, USNM); El Salado (5, USNM).

P. OCHRACEOUS

VENEZUELA: Zulia: El Panorama, Río Aurare (2, FMNH).

P. POLIOPUS

VENEZUELA: Táchira: San Juan de Colón (1, FMNH). **Zulia:** Kasmera (1, MVZ).

P. TRINITATUS

TRINIDAD: Caparo (2, FMNH); Princestown (2, FMNH; 1, AMNH); Turure Forest (2, AMNH); Cumaca (1, AMNH); Oropuche Heights (2, FMNH); Chaguaramas (1, MVZ).

VENEZUELA: Monagas: 2 km N and 4 km W Caripe (1, USNM). **Sucre:** 5 km S and 25 km E Carúpano (1, USNM).

P. URICHI

VENEZUELA: Sucre: San Estéban (2, FMNH); Quebrada Seca (1, FMNH; 1, AMNH); Campo Alegre (1, AMNH); Los Palmales (1, AMNH).

An Assessment of the Systematics and Evolution of the Akodontini, with the Description of New Fossil Species of *Akodon* (Cricetidae: Sigmodontinae)

Osvaldo A. Reig

ABSTRACTS

The taxonomy and systematics of the tribe Akodontini (Cricetidae: Sigmodontinae) are discussed and revised. *Bolomys* is distinguished from *Akodon,* and *Cabreramys* is considered a synonym of *Bolomys*; a diagnosis of *Bolomys* and a list of its species are given. *Akodon* is thought to comprise five subgenera, namely, *Akodon, Abrothrix, Chroeomys, Deltamys,* and *Hypsimys*; *Thaptomys* is included within the subgenus *Akodon,* and *Thalpomys* is considered a synonym of *Bolomys.* Distinctive character states of each subgenus are described, including a detailed description of cranial and dental features of *Akodon* and *Abrothrix. Microxus* is distinguished from *Abrothrix* and is retained as a genus. *Blarinomys, Oxymycterus, Lenoxus, Juscelinomys,* and *Podoxymys* are all accorded generic status within the Akodontini. The status of the long-clawed fossorial akodontines of southern South America is discussed; *Chelemys, Geoxus,* and *Notiomys* each deserve generic recognition and are diagnosed. New species of *Akodon* from the Plio-Pleistocene of Buenos Aires Province, Argentina, are described, and their significance for the evolution of the tribe is noted. The Pliocene *Akodon* (*Abrothrix*) *kermacki* Reig, 1978 is described in detail, with the new species **A. (Abrothrix) magnus** and **A. (Akodon) lorenzinii** from the Lower Pleistocene, and **A. (Ak.) johannis** from the Middle Pleistocene. Also described are remains of *A.* (*Ak.*) cf. *cursor* and *A.* (*Ak.*) cf. *iniscatus* from the Middle and Upper Pleistocene, respectively. A tentative scenario of the origin and evolutionary deployment of the Akodontini is presented, including an origin in the Puna by Middle Miocene times and an indication of probable dispersal corridors.

Se discute aqui y se revisa la situación taxonómica y las relaciones sistemáticas de los roedores cricétidos sudamericanos de la tribu Akodontini (Cricetidae: Sigmodontinae). Se distingue a *Bolomys* de *Akodon* como un género pleno, y se considera a *Cabreramys* sinónimo del primero, del que se proporciona una diagnosis y la lista de sus especies. Se reconocen dentro del género *Akodon* cinco subgéneros: *Akodon, Abrothrix, Chroeomys, Deltamys* e *Hypsimys*; a *Thaptomys* se lo incluye dentro del subgénero *Akodon,* y a *Thalpomys* se lo considera como un sinónimo de *Bolomys.* Se describen los estados de caracteres distintivos de cada uno de los subgéneros de *Akodon* y se proporciona una descripción detallada de los rasgos craneanos y dentarios de *Akodon* y *Abrothrix.* Se plantea la necesidad de distinguir a *Microxus* de este último y de conferirle rango pleno de género. Se reconoce también rango genérico pleno dentro de los Akodontini a *Blarinomys, Oxymycterus, Lenoxus, Juscelinomys* y *Podoxymys.* Se discute tambien la situación taxonómica de los akodontinos hipogéicos de uñas largas de la región austral de América del Sur; se reconoce la distinción genérica de *Chelemys, Geoxus* y *Notiomys* y se proporcionan diagnosis de los tres. Se describen nuevas especies de *Akodon* del Plio-Pleistoceno

From the Departamento de Ciencias Biológicas, F.C.E. y N., Universidad de Buenos Aires, Pabellon 2, Ciudad Universitaria Nunez, (1428) Buenos Aires, Argentina. Dr. Reig is a CONICET Career Investigator.

de la provincia de Buenos Aires, Argentina, comentándose su significado para la comprensión de la evolución de la tribu. Se describe en detalle la especie del Plioceno *Akodon* (*Abrothrix*) *kermacki* Reig, 1978, y las nuevas especies **A. (Abrothrix) magnus** y **A. (Akodon) lorenzinii** del Pleistoceno inferior, y **A. (Ak.) johannis** del Pleistoceno medio. Se incluye asi mismo la descripción de restos de *A.* (*Ak*) cf. *cursor* y *A.* (*Ak.*) cf. *iniscatus* del Pleistoceno medio y superior, respectivamente. Se esboza también un panorama interpretativo del origen y de la historia evolutiva de los Akodontini en su conjunto, en el que se sostiene su origen en la Puna hacia el Mioceno medio, y se indican las posibles rutas de la dispersión de sus distintos taxones.

A taxonomia e a sistemática de tribo Akodontini (Cricetidae: Sigmodontinae) é discutida. Justificam-se a separação de *Bolomys* do gênero *Akodon* e a sinonomia de *Cabreramys* com *Bolomys.* Uma diagnose do gênero *Bolomys* e oferecida, com una lista de suas espécies. O gênero politípico *Akodon* inclue 5 subgêneros: *Akodon, Abrothrix, Chroeomys, Deltamys* e *Hypsimys. Thaptomys* não e reconhecido como gênero ou subgênero, e sua espécie única, *nigrita,* é referida a *Akodon* s.s. Considera-se *Thalpomys* como sinonimo menor de *Bolomys.* Providenciam-se os carácteres distintos de cada subgênero, com uma descrição detalhada do crânio e das estruturas dentais de *Akodon* e de *Abrothrix.* Justifica-se o rango de gênero a *Microxus* e salientam-se as diferenças entre ele e *Akodon* (*Abrothrix*). *Blarinomys, Oxymycterus, Lenoxus, Juscelinomys,* e *Podoxymys* são gêneros akodontinos que não requerem agrupamento especial, ou colocacão numa tribo separada. Discutese o rango dos akodontes fossoriais com longas unhas que habitam a região sul do continente; *Chelemys, Geoxus,* e *Notiomys* são reconhecidos como gêneros válidos. Novas espécies de *Akodon* são descritas do Plio-Pleistoceno da Província de Buenos Aires na Argentina, e sua importância na história evolutiva é discutida. *Akodon* (*Abrothrix*) *kermacki* Reig, 1978, do Plioceno, é descrito em detalhe, e **A. (Abrothrix) magnus** e **A. (Ak.) johannis** (do Pleistoceno médio) são descritos como espécies novas. Descreve-se também os restos de *A.* (*Ak.*) cf. *cursor* e *A.* (*Ak.*) cf. *iniscatus* do Pleistoceno médio e superior, respetivamente. Um possível cenário da evolução dos Akodontinos é apresentado. Justifica-se o origem dos Akodontini na Puna, durante o Mioceno médio, e as possíveis rotas de dispersão são reconstruídas.

Introduction

In a recent paper (Reig, 1984), I advanced some interesting features of the frequency distribution of diversity among living South American mammals. One of 12 orders of living terrestrial mammals, the Rodentia, comprises 43.2% of the total extant mammalian species of this continent. Most rodent species belong either to the Myomorpha (21.9%) or to the Caviomorpha (19.1%). However, species of the latter are distributed among 11 different families, whereas South American living species of myomorphs are all grouped within a single family, Cricetidae. The extant South American cricetids include about 250 species and 53 or 54 genera. Except for two species belonging to two genera of North American affinities, all South American cricetids belong to the subfamily Sigmodontinae. This subfamily may itself be subdivided into seven well-defined tribes, namely Oryzomyini, Ichthyomyini, Akodontini, Scapteromyini, Phyllotini, Wiedomyini, and Sigmodontini (Reig, 1980, 1981). Reinforcing the observed tendency for a log-normal distribution of species diversities among South American mammals (which seems to be a general pattern of diversity in nature; see Williams, 1964), an overwhelming number of sigmodontine species (44.2%) belong to one tribe, the Oryzomyini. The other two tribes showing high species frequencies are the Akodontini (24.9%) and the Phyllotini (18.1%). Thus, three of seven tribes of sigmodontine rodents include 87.1% of the species.

The Akodontini comprise around 63 species distributed in 11 different genera (see later). They are predominately Andean in distribution, although many species are widely distributed in temperate, subtropical, and to a lesser degree, tropical lowlands (Reig, 1984). Despite extensive cytogenetic studies (see particularly Bianchi et al., 1971; Bianchi & Merani, 1984; Rodriguez et al., 1983; Vitullo et al., 1986), the systematics of Akodontini is still confusing in several important respects, including the taxonomic status and the rank of some supraspecific taxa; the generic allocation of several species; and the status, distribution, and geographical variation of many of their species. Moreover, little is known of their evolutionary history and fossil record, although their present diversity and distribution and a few fossil remains permit a plausible explanation of their age, place of origin and diversification, and phylogenetic relationships (Bianchi et al., 1971; Reig, 1978, 1980, 1981, 1984).

I started to revise the Akodontini at the British Museum about 15 years ago. Although the work remains uncompleted because of the pressure of other, more urgent duties, I gathered a considerable amount of information which is relevant to a better understanding of the systematics of this diverse tribe of sigmodontine rodents. Moreover, I have had the chance to study an important collection of fossil akodontine remains from different levels of the Argentinian Plio-Pleistocene stratigraphic column, partly described elsewhere (Reig & Linares, 1969; Reig, 1978). In this paper, I present and briefly document the more important conclusions of my revision in progress of the living Akondontini and describe all the known fossil representatives of the genus *Akodon.* I also present a tentative explanation of the origin and evolutionary and biogeographic deployment of these South American mice. The first purpose is treated unevenly, as some genera (i.e., *Bolomys* and *Akodon*) are covered in more detail than others. This is intentional, as clarification of the limits between these two genera is considered essential for an assessment of the whole tribe, and as an elucidation of the limits *Akodon* and its subgenera is a prerequisite to study its fossil representatives.

Materials and Methods

This systematic survey of the Akodontini was primarily based on direct examination of museum specimens, including in most cases the type specimens and original series. The collection of the British Museum (Natural History) in London (BMNH) was the main source of information for this study. However, specimens of other collections have been also extensively studied, particularly those of the American Museum of Natural History (AMNH), the United States National Museum (USNM), the Museum of Comparative Zoology, Harvard University (MCZ), and the Museum of Vertebrate Zoology, University of California (MVZ). Moreover, specimens from South American collections provided substantial information, especially those of the Museo Municipal de Ciencias Naturales de Mar del Plata "Lorenzo Scaglia" (MMP), the Universidad Central de Venezuela in Caracas (MBUCV), the Universidad Austral de Chile in Valdivia (UACH), and the Facultad de Ciencias Exactas y Naturales of Buenos Aires University (FCM). The fossil specimens belong primarily to the rich collection gathered by Galileo J. Scaglia and collaborators, which is deposited at the Museum of Mar del Plata, but also include material from the Museo de La Plata (MLP).

Skull and molar morphology has been the main source of information in the systematic revision of the living forms, and, indeed, the only source available in the study of the fossil forms. Skull measurements were taken from adult specimens with dial calipers graduated to 0.1 mm. The measurements of the teeth were taken through the reticule eyepiece of a Wild M-5 stereomicroscope. Only the greatest lengths and widths of the teeth are given. The names of the enameled components of the crown of the molar teeth follow my proposed unified nomenclature (Reig, 1977). Other aspects of nomenclature of tooth morphology follow Hershkovitz (1962, 1967). The drawings were made by me, with the aid of a drawing tube.

Although several conceptual tools of cladistic methodology are used, I follow a syncretic evolutionary approach within which paraphyletic taxa are not rejected (Reig et al., in press). Supraspecific taxa are treated as class-concepts and, as in previous papers (Reig, 1970, 1982), I use the logical terms "intension" and "extension" to refer to the set of attributes that determine the taxon-concepts and the set of subordinate taxa that are members of them, respectively. Taxa are also considered to be polythetic. Thus, their intension is defined by reference to a set of character-states which are not exclusive of the taxon, and no claim is made that membership to the taxon requires sharing all characters used in defining the taxon's intension (Sokal & Sneath, 1963).

Historical Shaping of the Concept of Akodontini

Before the recognition of the Akodontini as a formal taxon of tribal rank of South American cricetid rodents (Vorontzov, 1959; Reig, 1980, 1981), the assemblage of genera and subgenera making up this tribe was usually treated informally as the akodont rodents or the akodont group. The belief that those genera and subgenera formed a natural group arose from the studies of Thomas, but the concept of this taxonomic group long remained rather vague and controversial, both in intension and in extension. In 1916, Thomas wrote on the taxonomic status of a group of related species, which he allocated in seven different genera, namely *Akodon* (with *Chalcomys* as a subgenus), *Thalpomys, Thaptomys, Bolomys, Chroeomys, Abrothrix,* and *Zygodontomys.* Subsequently (Thomas, 1918), he added the genera *Hypsimys* and *Deltamys* to the same group. He recognized that other supraspecific taxa, such as *Blarinomys, Geoxus, Notiomys, Chelemys, Microxus, Oxymycterus,* and *Lenoxus,* were also related to the above group. Osgood (1925) was probably the first to revise Thomas's akodont genera, lumping *Notiomys, Chelemys,* and *Geoxus* into a single genus, *Notiomys,* a contention which was criticized by Thomas (1927). The monotypic genus *Podoxymys* was later added to the group by Anthony (1929).

Tate (1932b) presented a preliminary revision of the whole group, which introduced the designation "akodont rodents." As is true of his other revisions of South American rodents, Tate's work is a useful and careful historical and bibliographical review of the taxa involved rather than a revision based on actual collections, but it did not result in a clarification of the intension of the group. He recognized generic rank for most of Thomas's supraspecific groups, but followed Osgood regarding *Notiomys*; he also withdrew *Zygodontomys* from the akodont assemblage, placing it in the oryzomyine group (Tate, 1932a).

The next comprehensive study of the whole group was that of Gyldenstolpe (1932), which did not separate the group from other "sigmodont rodents," equivalent to my subfamily Sigmodontinae. Gyldenstolpe accorded generic rank to all taxa of akodont rodents created by Thomas, and he contributed to their definition. He also agreed with Thomas in recognizing the generic status of *Notiomys, Geoxus,* and *Chelemys.*

Ellerman (1941) made a thorough reappraisal of Thomas's akodont mice. He claimed that most of the genera created by Thomas in 1916, as well as the later *Hypsimys* and *Deltamys,* were best treated as subgenera of *Akodon.* He excluded *Zygodontomys* from this grouping, regarding it as a distinct genus related to *Akodon.* However, he retained *Microxus, Oxymycterus, Lenoxus,* and *Notiomys* as genera; following Osgood, the latter included *Geoxus* and *Chelemys.* Ellerman's balanced judgment and overall experience influenced the work of subsequent authors, and his concept of the akodont genera was essentially followed by Cabrera (1961).

Vorontzov (1959) coined the name Akodontini for the tribe containing *Akodon* (*sensu* Ellerman), *Zygodontomys, Microxus, Podoxymys, Lenoxus, Oxymycterus, Blarinomys,* and *Notiomys.* Hershkovitz (1962) later withdrew *Zygodontomys* from the akodonts, placing it in his phyllotine group, close to *Calomys.*

Hooper and Musser (1964), in their discussion of the bearing of phallic morphology on the interrelationships of cricetids and allied genera, concluded that *Oxymycterus* is distinct enough in the characters of the glans penis to be recognized as a group distinct from, but allied to the akodont group. Moreover, they stated that *Notiomys* is as distinct from *Akodon* as is *Oxymycterus,* that *Zygodontomys* is annectent between akodonts and oryzomyines (not closer to the phyllotines as claimed by Hershkovitz), and that *Calomys,* and especially *Eligmodontia,* are to be placed near *Akodon.* Many of the conclusions of these authors were criticized by Hershkovitz (1966) on methodological grounds, namely that phallic evidence coming from few taxa, each represented by a few individuals, should not be the basis for introducing major changes in a taxonomic arrangement based on a complex of character states from different organ systems and a large number of genera. Hershkovitz, however, followed Hooper and Musser in splitting an oxymycterine group (including *Oxymycterus, Podoxymys, Lenoxus,* and *Abrothrix*; he considered *Microxus* to be a synonym of *Abrothrix*) from the akodont group. A further basis for splitting *Oxymycterus* from the akodonts is suggested by specializations of its stomach and intestines for an insectivorous diet, as illustrated in the detailed studies of Vorontzov (1967; see also Tullberg, 1899; Echave Llanos & Vilches, 1964; Carleton, 1973). However, in view of the incomplete and scattered information on the anatomy of the digestive system in cricetids, it does not seem wise to put much weight on this sort of evidence. The complex stomach of *Oxymycterus* may well represent an exclusive autopomorphy.

To complete this picture of the Akodontini, Massoia and Fornes (1967) proposed a new genus, *Cabreramys,* for a group of species formerly referred to *Akodon* and *Zygodontomys.* They allege it to be transitional between akodontine and phyllotine cricetids. Additionally, Moojen (1965) described a new genus, *Juscelinomys,* which is clearly related to *Oxymycterus,* and Reig (1978) created the fossil *Dankomys,* which is related to *Bolomys.*

The Akodontini remain one of the most obscure groups of South American rodents. There is no agreement as to its extension (viz., *Zygodontomys*), to the rank of its supraspecific groupings, or even to its unity as a taxon. This situation is obviously due to the lack of comprehensive studies and to the failure of authors to draw conclusions from the partial evidence. Although we are still far from the goal of achieving an accurate knowledge of this group, I hope that recent advances in chromosomal systematics and certain conclusions about several critical genera and species can serve to improve our understanding of this complex array of rodents.

Various chromosome studies on several akodontine taxa have been undertaken by various authors in the last 15 years (Bárquez et al., 1980; Barros & Reig, 1979; Bianchi & Contreras, 1967; Bianchi & Merani, 1984; Bianchi et al., 1969, 1971, 1973, 1979; Gallardo, 1982; Gardner & Patton, 1976; Gentile de Fronza, 1970; Kiblisky et al., 1970, 1976; Lobato et al., 1982; Maia & Langguth, 1981; Pearson, 1984; Rodriquez et al., 1983; Sbalqueiro et al., 1984; Spotorno & Fernández, 1976; Vitullo et al., 1986; Yonenaga, 1972, 1975, 1979; Yonenaga et al., 1975). These studies demonstrate that species of *Akodon, Abrothrix, Bolomys, Microxus, Chroeomys, Thaptomys,* and *Oxymycterus* resemble one another in their karyotype, but that *Zygodontomys* is distinct from this group. The first group is characterized by karyotypes of no more than 54, mostly telocentric chromosomes, with several species showing diploid number reduction by Robertsonian fusions and tandem translocations. In contrast, *Zygodontomys* is exceptional among mammals, but resembles the Oryzomyini, in having karyotypes of $2n = 84–88$, mostly subtelocentric chromosomes. These results strongly suggest the unity of the oxymycterine and akodontine groups. An almost indistinguishable $2n = 52$ karyotype occurs in *Akodon* (*Ak.*) *olivaceus, A.* (*Ak.*) *xanthorhinus, A.* (*Ak.*) *andinus, A.* (*Abrothrix*) *sanborni, A.* (*Ak.*) *nigrita, A.* (*Chroeomys*) *jelskii, Chelemys macronyx,* and *Geoxus valdivianus.* The karyotype of *Oxymycterus* proved to be of $2n = 54$ chromosomes, quite similar to these in gross morphology. At the same time, these results distinguish *Zygodontomys* quite apart from the Akodontini. As suggested by Gardner and Patton (1976), it may represent a separate group or tribe of its own. Pending more complete evidence, I prefer to treat *Zygodontomys* as *incertae sedis* within the Sigmodontinae.

Once *Oxymycterus* and allied genera are included within the Akodontini, and *Zygodontomys* is excluded from this tribe, it is convenient to discuss the status of the taxa of generic and subgeneric rank which remain as its members.

Status of *Bolomys* and *Cabreramys*

By definition, the most typical akodontine is *Akodon.* It is also the most polytypic, the most complex, and the most problematic of the Akodontini. However, some recent conclusions help to understand its limits.

As I advanced and partially substantiated elsewhere (Reig, 1978), and in agreement with the independent results of Maia and Langguth (1981), I exclude from *Akodon* a series of species which have been variously grouped under *Akodon* proper, the "southern group" of species of *Zygodontomys* (Hershkovitz, 1962), or under the genus *Cabreramys* (Massoia & Fornes, 1967). Those species deserve separate generic status by virtue of clearcut morphological and karyological differences. Comparisons of character-states of this group with those of *Akodon amoenus* Thomas, 1900, which is the type species of *Bolomys* by original designation, show this species belongs to the same group, and therefore all these species must be grouped under the genus *Bolomys* Thomas, 1916. However, the matter is still confusing to some and deserves further substantiation.

This taxonomic confusion started from the very beginning; when he erected *Bolomys* as a generic name, Thomas (1916) selected *Akodon amoenus* as its type species, but included also within the same taxon *Akodon albiventer* and *A. berlepschii,* two species which proved to be quite inseparable from the typical *Akodon* (*A. boliviensis,* fig. 4A), or other allied species, such as *A. andinus* (fig. 1B). Afterward, Thomas (1926b) referred *lactens, negrito* (sic), and *orbus,* which are probably conspecific but different from *amoenus,* to *Bolomys.* However, these taxa share peculiar character-states with *amoenus* that strikingly differentiate the group from any other typical species of *Akodon.*

As clearly observed in the type species *A. amoenus* (figs. 1A, 2C–D), *Bolomys* shows characteristic cranial and dental features which can be

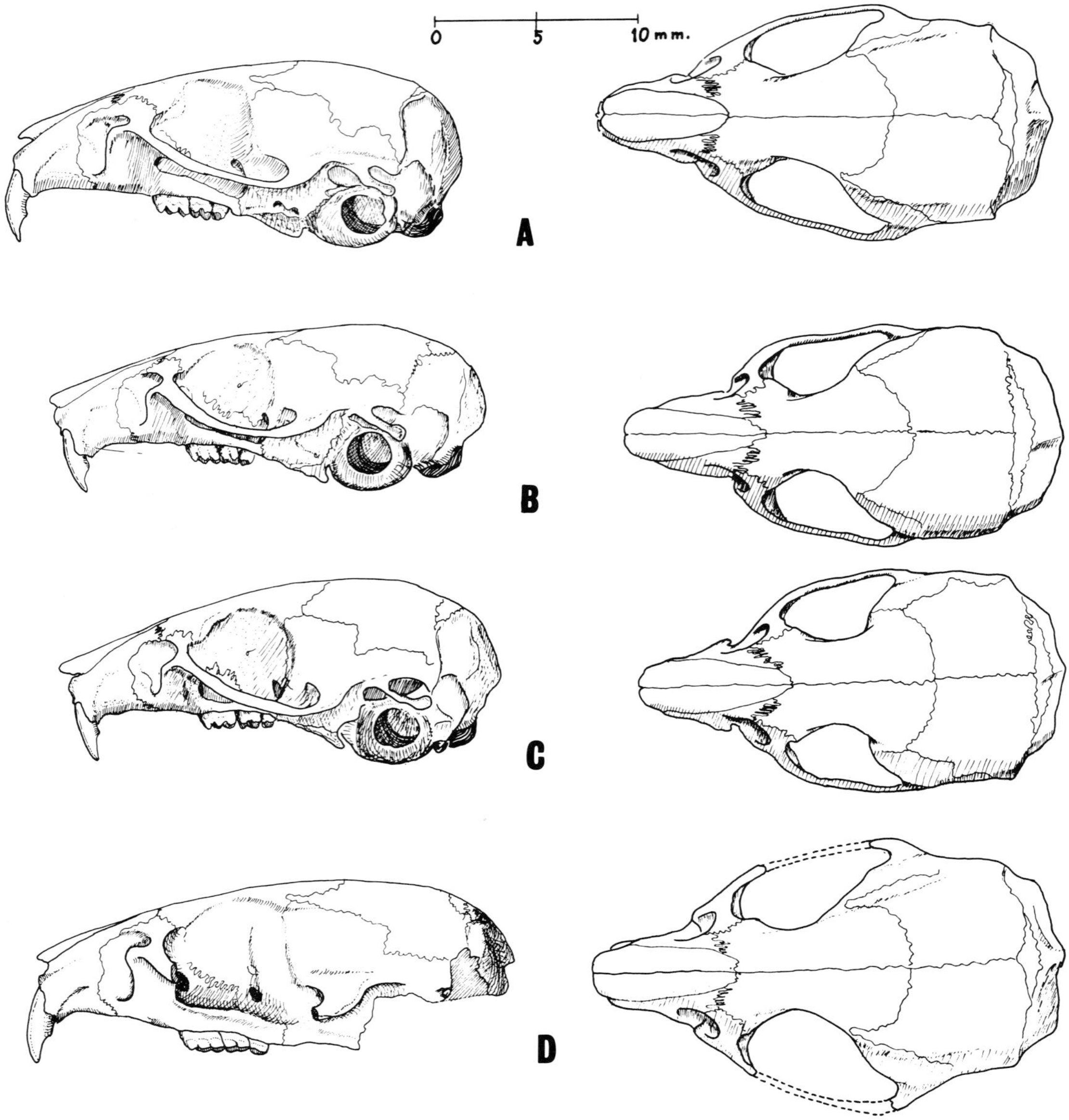

FIG. 1. Skulls of species of *Akodon* and *Bolomys*. Left row, lateral view; right row, dorsal view. **A,** *Bolomys amoenus* (Thomas); BMNH 1.1.1.12; Sanguero Puno, Peru. **B,** *Akodon* (*Akodon*) *andinus* (Philippi); female; type of *Akodon gossei* Thomas, BMNH 98.3.21.5; Puente del Inca, Mendoza, Argentina. **C,** *Akodon* (*Akodon*) *albiventer* Thomas; male; BMNH 21.11.1.51; Sierra de Zenta, Jujuy, Argentina. **D,** *Bolomys obscurus* (Waterhouse); sex unknown; lectotype, BMNH 55.12.24.161; Maldonado, Uruguay.

used in its diagnosis: braincase broad and deep; occipital region short; rostrum rather short and markedly tapering forward in lateral view; upper profile of skull gradually sloping forward from the middle of parietals; nasals short, with anterior borders well posterior to the level of the anterior border of incisors; frontals long, always longer than nasals; parietals short, less than half the length of frontals, and extending forward anterolaterally by means of narrow spines penetrating between frontals and temporals; interparietal noticeably reduced anteroposteriorly and transversely; occiput short and truncated; interorbital area with well-formed, anteriorly convergent borders; posterior palate moderately long and wide, the median posterior border of palatines behind the posterior bor-

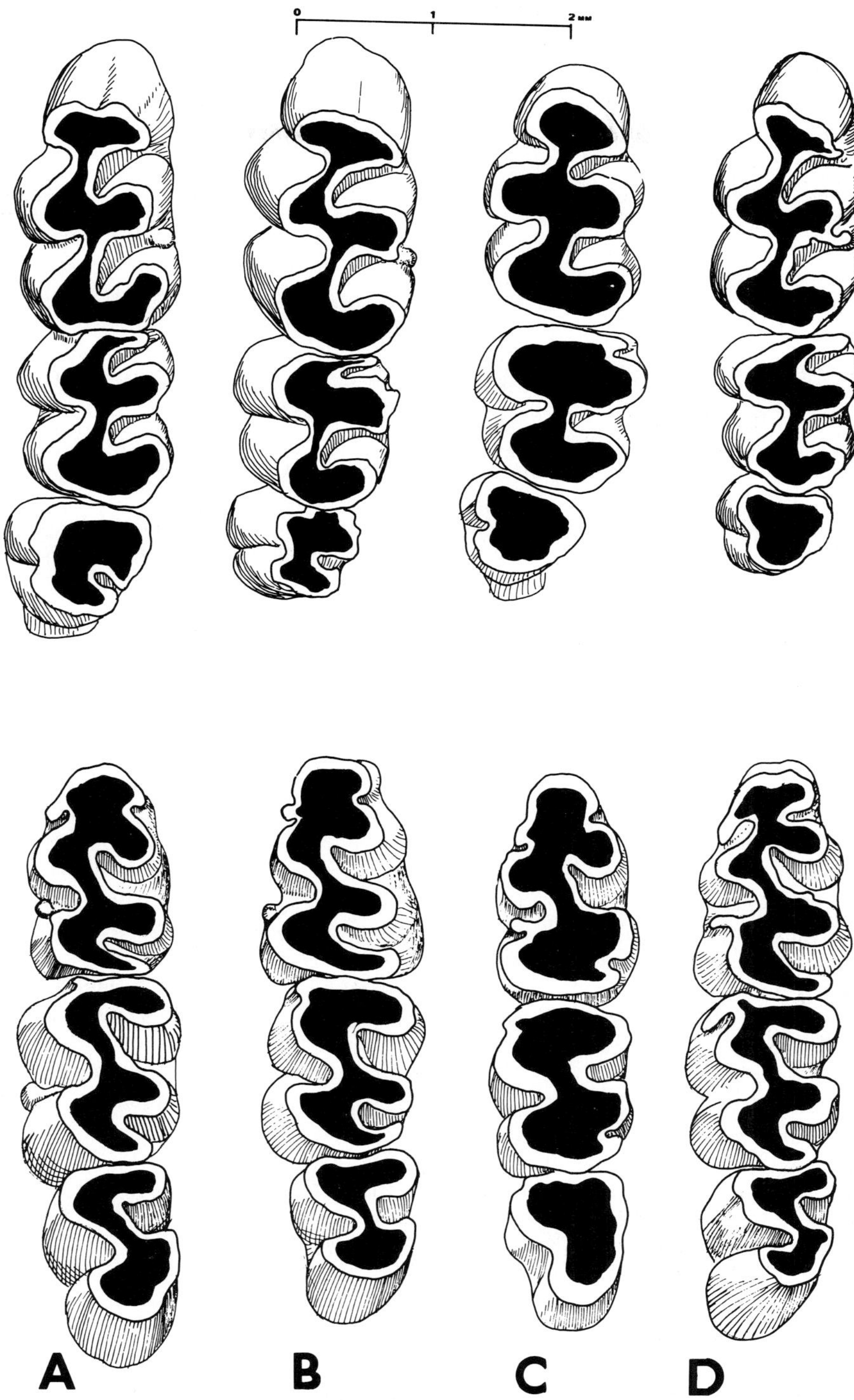

FIG. 2. Molar teeth of *Bolomys obscurus* and *Bolomys amoenus.* Upper row, left upper molar series; lower row, left lower molar series. **A,** *Bolomys obscurus* (Waterhouse); sex unknown; lectotype, BMNH 55.12.24.161; Maldonado, Uruguay. **B,** *Bolomys obscurus* (Waterhouse); male; holotype of *Akodon benefactus* Thomas, BMNH 16.10.3.35; Laguna Alsina (Bonifacio), Partido de Guaminí, Buenos Aires Province, Argentina. **C,** *Bolomys amoenus* (Thomas); male; holotype, BMNH 0.10.1.77; Río Colca, north of Sumbay, Peru (molars with advanced stage of wear). **D,** *Bolomys amoenus* (Thomas); male; BMNH 22.1.1.97; Huarconda, Peru.

der of M^3; zygomatic plate broad and strong, with anterior border straight or slightly concave, perpendicular to diastema; upper incisors orthodont or proodont; molars mesodont, terraced with moderate wear, broad and robust; upper molars with lophs almost completely transverse, and mesoloph usually completely coalesced with paraloph; procingulum of M^1 simple, with anteromedian flexus absent or only slightly developed; lower molars with lingual cusps somewhat anterior to the labial ones, with mesolophid remnants and mesostylids usually absent.

This combination of cranial and dental character-states is quite distinctive and clearly differentiates *Bolomys amoenus* and *B. lactens* (fig. 3I) from species of *Akodon* and *Zygodontomys* (restricted to *Z. brevicauda* and northern relatives) and from any other recognized genus of akodontine mice (fig. 3). Meaningfully, that set of character-states is also shared by *Akodon obscurus* (figs. 1D, 2A–B), which is the type species of *Cabreramys* Massoia & Fornes, 1967 by original designation. *Akodon benefactus* (fig. 3H) and *A. lenguarum* (fig. 3J), two other nominal species referred to *Cabreramys* by Massoia and Fornes (1967), also share that set of character-states. Therefore, there is no doubt that *Cabreramys* Massoia & Fornes, 1967 must be placed under the synonymy of *Bolomys* Thomas, 1916. Massoia and Fornes recognized the distinctiveness of the species they grouped in *Cabreramys* but, unfortunately, failed to realize that *A. amoenus* belonged to the same group. The latter being the type species of *Bolomys,* it was not necessary to propose a new generic name for this group of species.

It is also evident (see also Maia & Langguth, 1981) that the same set of character-states is also present in *Mus lasiurus* Lund, *Akodon fuscinus* Thomas (fig. 3G), *Zygodontomys pixuna* Moojen, *Zygodontomys tapirapoanus* Allen, and inferentially, "*Hesperomys*" *brachiurus* Wagner. All these nominal taxa, together with *A. lenguarum,* were placed by Hershkovitz (1962), probably inspired by Tate (1932a), in the "southern group" of *Zygodontomys* as subspecies of *Z. lasiurus.* In addition to the cogent arguments of Maia and Langguth (1981), the fact that *lasiurus* does not belong to *Zygodontomys* (fig. 3C–D) has been recently substantiated by Voss and Linzey (1981) on ventral prostate morphology as well as on some dental characters (and the chromosomal evidence discussed below). Similarities of *lasiurus* and species of *Akodon* in the presence of smaller medial than lateral prostates led Voss and Linzey to assign that species to *Akodon* and to withdraw it from *Zygodontomys.* This part of their conclusions agrees with the previous discussion. However, the evidence from prostate morphology should not be taken as an indication that *lasiurus* belongs to *Akodon.* The former shares enough derived character states in the skull, dentition, and chromosomes with *amoenus, obscurus, lactens,* and *benefactus* to make convincing its assignment to *Bolomys.* Derived prostate morphology in *lasiurus* and species of *Akodon* found by these authors is better interpreted as a suggestion that *Akodon* and *Bolomys* share a synapomorphous character-state which distinguishes them from the remaining genera of akodontines.

Hesperomys arviculoides Wagner also belongs to *Bolomys,* and it is a junior synonym of *Mus lasiurus* Lund. The type specimen of *arviculoides,* illustrated by Ximenez and Langguth (1970), is undoubtedly a *Bolomys* in all character-states visible in the illustration, and additionally, it cannot be differentiated from the type of *lasiurus* illustrated by Winge (1887) and in the Museum of Copenhagen. Moreover, it cannot be differentiated from a topotype of *lasiurus* (BMNH 88.1.9.4.) (fig. 3F), which I studied in the British Museum. Thus, I cannot agree with Maia and Langguth when they claim, without giving reasons, that *arviculoides* does not belong to *Bolomys.* The name *arviculoides* has been freely applied in recent cytogenetic literature to Brazilian akodontines with 2n = 14, 2n = 16, and 2n = 24 chromosomes (Cestari & Imada, 1968; Maia & Langguth, 1981; Yonenaga, 1972, 1979; Yonenaga et al., 1975) showing a complex system of intra- and interpopulational chromosomal variation. In fact, it has never been demonstrated that they agree phenotypically with the type of *arviculoides,* which seems inseparable from the type of *lasiurus* on morphological grounds. As specimens of *lasiurus* matching the character-states of Lund's type specimen have been shown to have a distinctive and different 2n = 34 karyotype, there are reasons to surmise that Brazilian forms with low chromosome numbers must be placed elsewhere. Thanks to the kindness of Dr. A. Langguth, I recently had the opportunity to examine specimens of so-called *arviculoides* from Pernambuco showing the 2n = 16 karyotype described by Maia and Langguth. I concluded that they belong to *Akodon s.s.,* probably to a new species related to *A. cursor.* The status of the São Paulo and Rio de Janeiro forms reported by Yonenaga and associates is still unsettled, although the chromosomal evidence indicates that they may belong to the

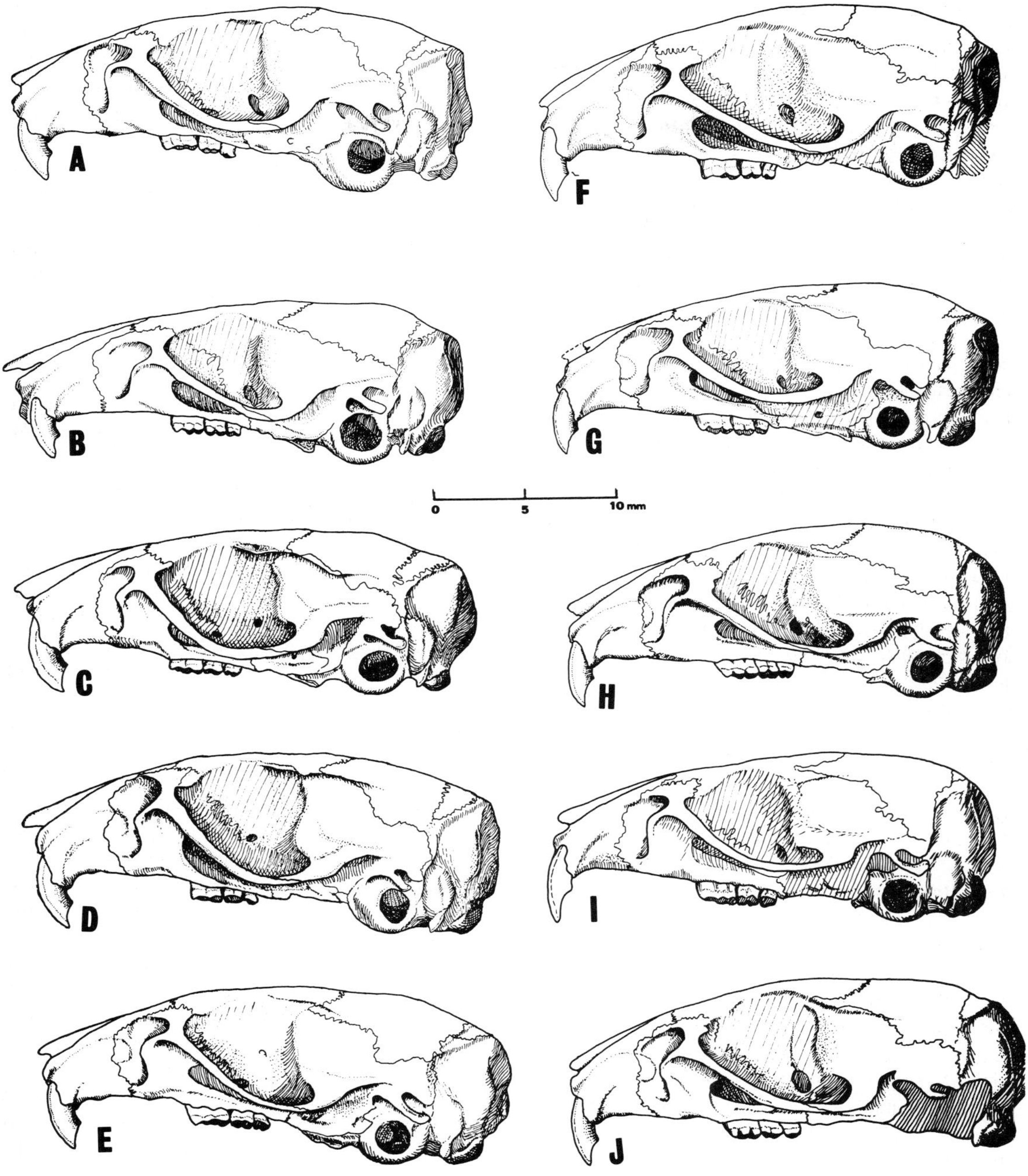

FIG. 3. Skulls in lateral view of species of *Bolomys, Akodon* (*Akodon*), and *Zygodontomys*. **A,** *Akodon* (*Ak.*) *dolores* Thomas; male; type specimen, BMNH 16.1.6.38; Villa Dolores, Córdoba, Argentina. **B,** *Akodon* (*Ak.*) *cursor* (Winge); female; BMNH 66.1874; Puerto Gisela, Misiones, Argentina. **C,** *Zygodontomys thomasi* Allen; female; BMNH 14.9.1.60; El Trompillo, Carabobo, Venezuela. **D,** *Zygodontomys microtinus* Thomas; female; holotype, BMNH 66.8.11.10; Surinam. **E,** *Akodon* (*Ak.*) *varius* Thomas; female; holotype, BMNH 2.1.1.67; Cochabamba, Bolivia. **F,** *Bolomys lasiurus lasiurus* (Lund); male; topotype, BMNH 88.1.9.4; Lagoa Santa, Minas Gerais, Brazil. **G,** *Bolomys lasiurus fuscinus* Thomas; male; holotype, BMNH 94.4.1.3; Soure, Ilha de Marajó, Pará, Brazil. **H,** *Bolomys obscurus benefactus* Thomas; male, holotype, BMNH 16.10.3.35; Laguna Alsina (Bonifacio), Partido de Guaminí, Buenos Aires Province, Argentina. **I,** *Bolomys lactens lactens* Thomas; female; holotype, BMNH 18.1.1.37; León, Jujuy Province, Argentina. **J,** *Bolomys lenguarum* (Thomas); sex unknown; holotype, BMNH 98.5.14.1; Waikthlatingmay-alwa, northern Chaco, Paraguay.

same group, and that at least the 2n = 24–25 polymorphic form described by Yonenaga as *Akodon* sp. from specimens captured in São Paulo may represent another, probably synmorphic new species.

To make this story even more complex, Langguth (1975) demonstrated that *Mus lasiotis* Lund, 1838 is based on a specimen which belongs to the same species as the holotype of *Mus lasiurus* Lund, 1837. Therefore, Langguth placed *lasiotis* under the synonymy of *lasiurus,* which seems correct. However, the specimens described by Winge (1887) as *Habrothrix lasiotis* belong certainly to a species distinct from *lasiurus.* One of the specimens identified by Winge as *lasiotis* was sent to the British Museum and was the basis of *lasiotis* Thomas, 1916, being selected as the type species of his genus *Thalpomys.* Langguth proposed a new name, *Akodon reinhardti,* for Winge's and Thomas's *lasiotis,* but as *Thalpomys* was based on the species *lasiotis* as described by Lund, and as *lasiotis* Lund is a junior synonym of *lasiurus* Lund, and as *lasiurus* belongs to *Bolomys, Thalpomys* becomes a subjective synonym of *Bolomys.* This nomenclatorial issue means that, if *lasiotis* as conceived by Winge and Thomas is really distinctive enough from *Akodon* to deserve generic or subgeneric distinction, a new name should be proposed to replace *Thalpomys* for the taxon containing *reinhardti* Langguth. However, this species is now too poorly known to justify this action, and I prefer to keep *reinhardti* in *Akodon s.s.*

Recently, Massoia (1982) described a new species, *Cabreramys temchuki* from Misiones, Argentina. Later, Contreras (1982) referred *temchuki* to *Bolomys* and described two new subspecies from northern Corrientes Province and northeastern Chaco Province, Argentina: *B. t. elioi* and *B. t. liciae.*

The previous conclusions on *Bolomys* and its species, based on morphological analysis, are also supported by cytogenetic data. Karyotypes of 2n = 34 identical in gross morphology have been described or recorded in *Akodon amoenus* (Gardner & Patton, 1976; O. Pearson, pers. comm.), *Akodon obscurus* (Bianchi et al., 1971), *Akodon arviculoides* (Yonenaga & Ricci, 1969), *Zygodontomys lasiurus* (Yonenaga, 1975), *Zygodontomys* sp. (probably *B. lenguarum*) (Gardner & Patton, 1976). More recently, banding patterns have been published for *Bolomys obscurus* (Bianchi et al., 1976) and *B. lasiurus* (Maia & Langguth, 1981), and we have recently confirmed that *B. temchuki* shares the same G-banded karyotype with the latter (Vitullo et al., 1986). As with *Oxymycterus, Bolomys* shows karyotypic homogeneity, although differences in banding pattern are not unexpected when more detailed comparative studies are completed. Species of *Akodon* are variable in karyotypes, but none of them have the 2n = 34 karyotype reported for *obscurus, lasiurus,* and allies. However, *B. obscurus* was found to be more closely related in karyotype and G-banding to some species of *Akodon* (e.g., *A. dolores* and *A. azarae*) than other species of *Akodon* are related to each other (Bianchi & Merani, 1984). This must be interpreted as an indication that *Bolomys* evolved amidst the radiation of *Akodon,* which is consistent with the results from prostate morphology. Once differentiated from *Akodon, Bolomys* diversified as a clade of its own by adaptive divergence that did not involve significant chromosomal evolution. Therefore, the karyotypic distinction and uniformity of *amoenus, lasiurus, lenguarum, obscurus,* and *temchuki* must be taken as additional evidence that this group of species (together with *lactens* for which we lack chromosomal information) belongs to a separate group for which the name *Bolomys* Thomas, 1916 is available and has priority. If, as it seems most likely, *Bolomys* got its identity as a genus within the cladogenesis of *Akodon,* the latter should be considered as a paraphyletic taxon, which I consciously admit.

Bolomys is polytypic and widespread, comprising several different species and subspecies distributed in the Central Andes, the Chacoan and Pampean regions, northeastern Argentina, eastern and southern Brazil, and Uruguay. It is not yet clear how many species must be recognized within this genus. The specific or subspecific status of the various nominal forms described are still problematical, and new forms may be recognized in poorly explored regions, as shown by the recent discoveries of Massoia (1982) and Contreras (1982). As a tentative arrangement, I propose the following:

Bolomys amoenus (Thomas, 1900). Living in highlands and Altiplano of southeastern Peru.

Bolomys bonapartei Reig, 1978. Fossil from the Pliocene Monte Hermoso Formation of southern Buenos Aires Province.

Bolomys lactens (Thomas, 1918) (fig. 3I) (including *Akodon orbus* Thomas, 1919 and *Bolomys negrito* Thomas, 1926 as junior synonyms; *Akodon leucolimnaeus* Cabrera 1926 may deserve recognition as a subspecies). Living in highlands and Pampean mountains of Jujuy, Catamarca, and Tucumán, Argentina.

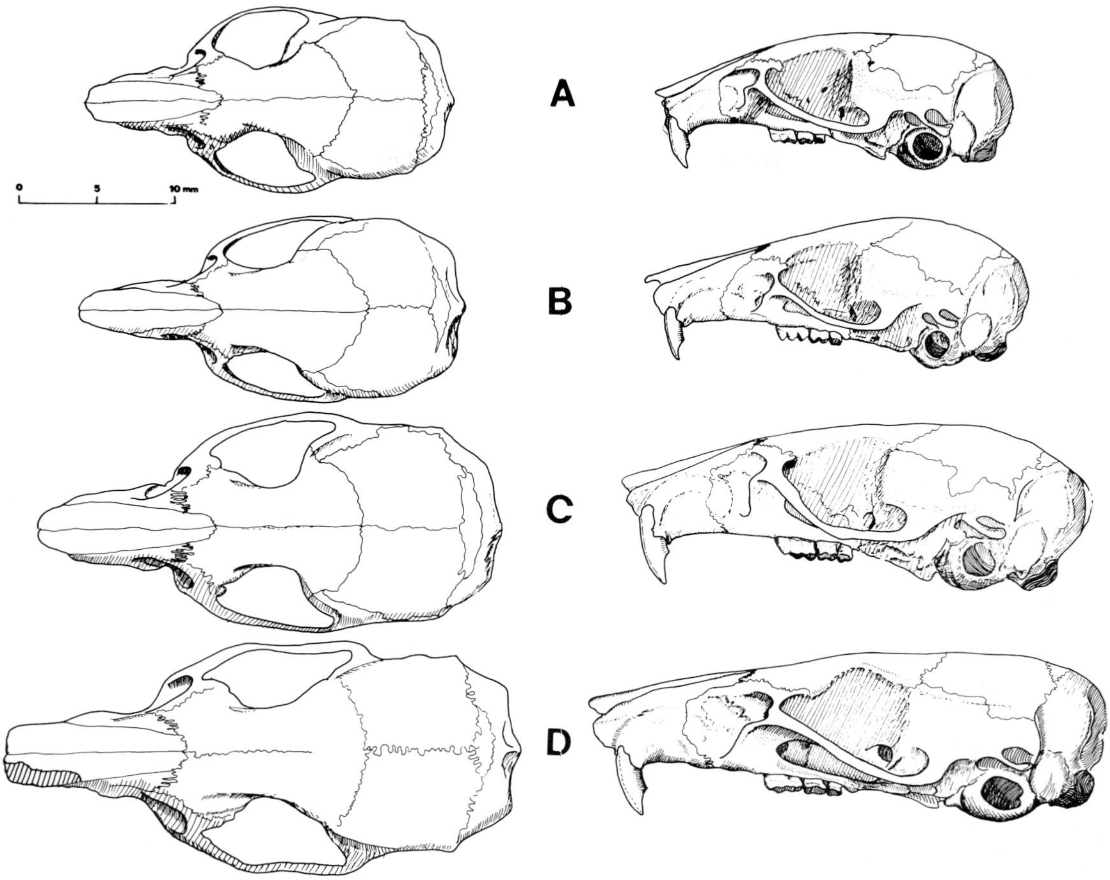

FIG. 4. Skulls of typical representatives of *Akodon* (*Akodon*), *Akodon* (*Abrothrix*), *Microxus,* and *Oxymycterus.* Left row, dorsal view; right row, lateral view. **A,** *Akodon* (*Ak.*) *boliviensis*; male; MBUCV 1.1889; 30 km N.W. Ollantaytambo, Cuzco, Peru. **B,** *Microxus mimus* Thomas; female; holotype, BMNH 1.1.1.48; Limbane, Puno, Peru. **C,** *Akodon* (*Ab.*) *longipilis* (Waterhouse); male; topotype, BMNH 98.2.2.2; Valparaiso, Chile. **D,** *Oxymycterus platensis* (= *Oxymycterus rufus platensis*?); male; holotype, BMNH 99.10.4.1; Ensenada, Río Santiago, La Plata, Buenos Aires Province, Argentina.

Bolomys lasiurus (Lund, 1838) (including *Mus lasiotis* Lund, 1838, *Hesperomys arviculoides* Wagner, 1842, and *Hesperomys brachyurus* Wagner, 1845, as junior synonyms; *Akodon fuscinus* Thomas, 1897, and *Zygodontomys pixuna* Moojen, 1943, may deserve recognition as subspecies). Living in eastern and southern Brazil, including the states of Pará, Ceará, Pernambuco, Paraíba, Sergipe, Bahia, Minas Gerais, São Paulo, and vicinities.

Bolomys lenguarum (Thomas, 1898) (probably including *Zygodontomys tapirapoanus* J. A. Allen, 1916 as a junior synonym). Living in Paraguayan, Bolivian, and (probably) Argentinian Chaco, and Planalto de Mato Grosso, Brazil. Contreras (pers. comm.) says that specimens from the Argentinian Chaco may belong to a different species.

Bolomys obscurus (Waterhouse, 1837) (including *Akodon benefactus* Thomas 1919 probably as a distinct subspecies). Living in southern Uruguay, northwest of Buenos Aires Province, south of Córdoba and Santa Fe, and east of La Pampa, Argentina.

Bolomys temchuki (Massoia, 1982) (including *Bolomys t. elioi* Contreras, 1982 and *B. t. liciae* Contreras, 1982, as subspecies). Living in Misiones, north of Corrientes and northeast of Chaco, Argentina.

Bolomys innom. sp. (reported as *Akodon obscurus* and *Zygodontomys obscurus*). Living in southern Buenos Aires Province. It is a larger and darker form found at Partidos de General Pueyrredón, Balcarce, and General Juan Madariaga and at Sierra de La Ventana, to be described in a forthcoming paper.

As advanced above, *Akodon albiventer* Thomas (fig. 1C) and *A. berlepschii* Thomas, originally included by Thomas (1916) in *Bolomys*, do not belong to that genus. In a previous paper (Bianchi et al., 1971), the former was considered to be a *Bolomys* by uncritical adoption of Thomas's opinion. The two species are probably conspecific (see Thomas, 1902). As already pointed out by Osgood (1943), *albiventer* shows several similarities with *Akodon andinus* (Philippi) (fig. 1B), and it does not exhibit any of the distinctive character-states of *Bolomys*. Moreover, *albiventer* has a 2n = 40 karyotype very similar to that of *A. boliviensis* and *A. varius* (Bianchi et al., 1971), which is easily distinguished from the 2n = 34 karyotype of species of *Bolomys*.

Akodon and Its Allied Supraspecific Taxa

Once *Bolomys* is excluded, the limits of *Akodon* as a genus appear more precise. However, it is still mostly a matter of sense and balance to decide whether *Abrothrix, Akodon s.s., Chalcomys, Chroeomys, Deltamys, Hypsimys, Microxus, Thalpomys*, and *Thaptomys* should be considered as separate genera, different subgenera within *Akodon*, or simply synonyms of *Akodon*.

Some time ago (Reig et al., 1971), following Cabrera (1961), I included *Chalcomys* as a synonym of *Akodon*. I also advocated (in Bianchi et. al., 1971) generic rank for the names listed above, excluding *Chalcomys* and *Deltamys*. Afterward (Reig, 1978, 1981), I reevaluated that view, as further study indicated more strongly that species referred to most of these names represent a complex array of allied forms, more closely related to each other than with other distinctive genera of akodontines, such as *Bolomys* and *Oxymycterus*. This is certainly a subjective feeling which badly needs to be substantiated by careful quantitative systematic analysis of morphological, chromosomal, and allozymic evidence. However, until such study is accomplished, the available evidence suggests that they are more likely to constitute a single unitary, although paraphyletic, taxon better treated as a large central genus within which some distinctive species or set of species may be distinguished as subgenera. However, as substantiated below, species referred to *Microxus* seem better excluded from *Akodon, Thalpomys* is a subjective synonym of *Bolomys* (as explained above), and *Chalcomys* seems not to deserve recognition. I also believe that separation of *Thaptomys* from *Akodon s.s.* is unwarranted. Actually, I could not find differences distinctly beyond the limits of variation within *Akodon s.s.* to keep *A. nigrita* Lichtenstein, the type and single species of *Thaptomys* (see Cabrera, 1961; Massoia, 1963a) as a separate subgenus. Cranially and dentally (figs. 10E,I), *nigrita* is a typical *Akodon*, and its alleged fossorial adaptations are too incipient to deserve any special taxonomic treatment. Its only distinction in male accessory glands is the presence of a single pair of ventral prostates in *A. nigrita* (Voss & Linzey, 1981), but the existing survey of species is still too small to give much weight to this isolated distinctive feature.

Thus, I tentatively recognize a central genus *Akodon* subdivided into five subgenera, namely *Abrothrix, Akodon s.s., Chroeomys, Deltamys*, and *Hypsimys*. I do not deny, however, that more detailed and comprehensive further studies could eventually elevate some of these taxa to generic level, but this action does not seem warranted by the present state of knowledge.

Recently, Massoia (1981b) and Contreras and Rosi (1981) advocated generic status for *Abrothrix*, whereas other modern authors accord it only subgeneric status (Yañez et al., 1978; Patterson et al., 1984). Actually, species of *Abrothrix* seem to be rather distinctive in skull and dental morphology (figs. 4–6). They differ from *Akodon s.s.* in the more robust and elongated skulls; long nasals which are clearly longer than frontals and that exceed backward the frontomaxillary suture and project also forward; long and rounded braincase; longer and more slender muzzle; slight inflation of the anterodorsal frontal sinuses; longer palate; low and elongated mandible with a projecting incisor capsule; broader molars, with obsolete or completely absent anteromedian flexus in M^1 and anteromedian flexid in the m_1; bulging of entoconid in m_1 and m_2; and total absence of ectolophids and rare presence of ectostylids. It also differs from typical *Akodon* in the woolly and longer pelage, in the relative sizes of ampullary glands and prostates, and in the presence of two pairs of preputial glands (Voss & Linzey, 1981). However, in overall resemblance, species of *Abrothrix* are clearly more closely related to species of *Akodon s.s.* than species of conventionally distinctive genera as *Oxymycterus, Bolomys*, or *Blarinomys*. Thus, it seems wiser to keep it as a subgenus of *Akodon. Abrothrix* is polytypic, but it is not clear how many species it contains. Typically, *longipilis, lanosus,* and *sanborni* and their subspecies (see Yañez et al., 1978) belong to *Abrothrix.* All inhabit the southern Andes and adjacent lowlands and show an identical 2n = 52 karyotype (Bianchi et al., 1971; Spotorno & Fernández, 1976; Gallardo, 1982; Reig, unpubl. data). *Akodon illutea* Thomas, 1925 (fig. 7D) is an *Abrothrix* in molar and skull morphology. However, living in the mountains of Tucumán in northwestern Argentina, it is isolated in distribution from the former group and was reported (Bianchi et al., 1971; Dulout et al., 1976) to have a polymorphic 2n = 41 karyotype. However, Bárquez et al. (1980) claimed that the identification of the specimens bearing those karyotypes was in error, and that they may represent *Akodon varius,* which seems quite probable. *Akodon xanthorhinus* has also been alleged to belong to *Abrothrix* (Bianchi et al., 1971), primarily on karyotypic and distributional grounds. It inhabits the same general region of *longipilis* and allies, and shows a 2n = 52 karyotype quite similar to that of *longipilis* (see Rodriguez et al., 1983). However, the same karyotype is present in *A. andinus, A. jelskii, A. nigrita, A. olivaceus, Chelemys macronyx,* and *Geoxus valdivianus* (Yonenaga, 1975; Gardner & Patton, 1976; Spotorno & Fernández, 1976; Pearson, 1984) and seems to represent a plesiomorphic character-state for akodontines as a group (Vitullo et al., 1986). Although the symplesiomorphy in karyotypic character-states hinders inclusion of *xanthorhinus* in *Abrothrix,* this conclusion is suggested by other attributes. Actually, *xanthorhinus* shares with species of *Abrothrix* an elongated rostrum, long nasals, rather inflated braincase, and similar molar patterns, including obliterated anteromedian flexi in the upper first molars. In this connection, it is quite relevant to take into account the recent parasitological results of Dolores del C. Castro. In her studies of the lice *Hoplopleura* (Anoplura), she erected the species *H. andina,* of the *travassosi* group (Castro, 1981). This group of species are typical ectoparasites of oryzomyines, whereas akodontines are more typically infected by lice of the *H. aikeni* species group. *Hoplopleura andina* was found on *Akodon olivaceus, A. andinus,* and *A. xanthorhinus* (Castro, 1982) and more recently also on *A. longipilis* (Castro, pers. comm.). Thus, *xanthorhinus* shares the same species of lice with a typical *Abrothrix* (*A. longipilis*) and with other southern species of *Akodon s.s.* having 2n = 52 chromosomes. Besides, the lice belong to a group which is more typical of the more primitive oryzomyines. Needless to say, a definite conclusion on the relationships of the southern akodonts needs more detailed studies, and especially an adequate evaluation of polarity in character-states. By overall resemblance, however, it seems that the available evidence should advise placing *xanthorhinus* in *Abrothrix.* Recently, Patterson et al. (1984) maintained *xanthorhinus* in *Akodon s.s.,* but they described a new species, *A. hershkovitzi,* which they claim to be closely related to *xanthorhinus.* From the illustrations and description of these authors, I am also inclined to place *hershkovitzi* in *Abrothrix,* in keeping with the view of Patterson and associates that *hershkovitzi* represents an island derivative of *xanthorhinus.* In addition to the species discussed above, *Akodon* (*Abrothrix*) *mansoensis* was recently described by de Santis and Justo (1980). The status of this name does not seem clear to me, and from the illustrations and description, I surmise that it may be a local form of *A. olivaceus,* a view which is also agreed upon by O. Pearson (pers. comm.). However, a comprehensive revision of the southern *Akodon* has not yet been conducted, and my conclusions on *xanthorhinus, hershkovitzi,* and *mansoensis* are only tentative. Meanwhile, and in the following treatment, the concept of *Abrothrix* is centered around the typical species *longipilis, sanborni, lanosus,* and *illutea.*

Chroeomys is certainly a well-differentiated offshoot from Andean akodonts. *Akodon jelskii,* its type species, can be easily distinguished from species of *Akodon s.s.* by the broadened braincase; enlarged and moderately swollen bullae; shorter palate; M^1 lacking a distinct anteromedian flexus, but with a well-developed and deeply infolded anteroflexus; reduction or absence of mesoloph remnants and mesostyles; deeply infolded anteromedian flexus of m_1; absence of mesolophids, ectostylids, and mesostylids. *Chroeomys* also differs from typical *Akodon* in having two pairs of preputial glands instead of one pair and in the large size of the median prostates, which are reduced in species of the subgenus *Akodon* (Voss & Linzey, 1981). Again, it is more like members of that genus in the sum of its characters than are members of other akodontine genera. Following the revision of Sanborn (1947), the five species recognized by Thomas are lumped into a single polytypic species, *A. jelskii.*

Hypsimys is also a well-differentiated subgenus of *Akodon.* Superficially, *A. budini,* its type species, may be confused with a *Bolomys,* but the confusion rapidly disappears when one examines the skull, which resembles that of *Akodon* in overall morphology. However, it differs neatly from *Akodon s.s.* in its long, round, and broad braincase; shorter palate; more elongated incisive foramina; more slender mandible and, especially, in the clearly more hypsodont molars, which are also more elongated and narrower than those in *Akodon, Chroeomys,* or *Abrothrix.* My associates and I have recently studied the karyotype of topotypical specimens of *A. budini,* finding that it has a peculiar 2n = 38 karyotype quite distinctive in its banding pattern (Vitullo et al., 1986). After comparison of the type specimens in the British Museum, I agree with Cabrera (1961) in considering *deceptor* Thomas, 1921 as a junior synonym of *A. budini* Thomas, 1918. Thus, *Akodon* (*Hypsimys*) is also monotypic.

Deltamys has also been recently accorded generic status (Massoia, 1981b; Gentile de Fronza et al., 1979). The absence of a small pair of metacentrics in its chromosome complement was given as an argument for its separation, but this is not convincing. *Hypsimys* also lacks this minute metacentric characteristic of akodontines, and *Zygodontomys microtinus* shows the character, being quite different from akodontines in many other respects. Cranially and dentally, *Deltamys* shares many more character-states with *Akodon s.s.* and other discussed subgenera than with any other genus. It is, however, quite distinctive as a subgenus in its elongated and narrow skull, with weak and low zygomatic plate; longer narrowed braincase; very reduced interparietal; shorter palate; low and elongated mandible with projecting incisor capsule; elongated and narrow, rather more hypsodont molars (less so than in *Hypsimys*); inclined molar flexi and flexids; M^1 with well-developed anteromedian flexus, but anteromedian flexid of m_1 obsolete; large m_3, almost as long as m_2. Up to now, only one species *A.* (*D.*) *kempi* is recognized, along with two subspecies (Massoia, 1981b).

Repeating the log-normal distribution of species mentioned in the Introduction section, *Akodon s.s.* is by far the most polytypic of all *Akodon.* Although the number of recognized species is a matter of debate, there are at least 23 species in this subgenus. It is probably the most generalized of the akodontines, representing the stock from which several episodes of akodontine diversification started in different places and probably at different times (see later). Both in morphology and in cytogenetics, it is a complex taxon, and it may prove convenient in the future to distinguish within it various species groups and superspecies. A formal description of its skull and dental characters and the present status of its species is considered below.

The situation of *Microxus* remains to be discussed. Described as a genus by Thomas (1909), it was accorded generic rank in his later synthesis (Thomas, 1916). It was also recognized as a genus by Tate (1932b) and Ellerman (1941). However, Cabrera (1961) treated *Microxus* as a subgenus of *Akodon,* a view followed by Arata (1967) and Voss and Linzey (1981). Hershkovitz (1966) included *Microxus* under *Abrothrix* on the argument that *M. mimus,* the type species, is actually an *Abrothrix,* a view echoed by Gardner and Patton (1976). I studied the type specimen of *M. mimus* (BMNH 1.1.1.48) and of *M. bogotensis* (BMNH 95.10.14.2) and several specimens referred to the latter from the Andes of Venezuela, and I concluded that *Microxus* belongs neither to *Akodon* nor to *Abrothrix* and that it is a distinctive genus of Akodontini. Including *M. mimus* (fig. 4B) and *M. bogotensis, Microxus* differs from *Akodon* and *Abrothrix* as well as from other akodontines by its inflated short and deep braincase; short and low zygomatic plate with anterior border gradually sloping forward; muzzle narrow and elongated; strongly reduced interparietal; long frontals; interorbital region rather broad; molars relatively large and broad, with a simplified enamel pattern and cusps well opposed to each other, and noticeably reduced upper and lower M3. *Microxus bogotensis* is also peculiar in possessing a single pair of ventral prostates (Voss & Linzey, 1981). Additionally, and as discussed above, species of *Akodon* (*Abrothrix*) share an identical 2n = 52 karyotype, whereas specimens assigned to *M. bogotensis* collected at Mucubaji, Mérida (Venezuela), showed polymorphic karyotypes of 35–37 chromosomes (FN = 48) (Barros & Reig, 1979). Further differences and distinctive character-states of *Microxus* in the skull and in different organ systems are provided by Vorontzov (1982). Thus, all lines of argument indicate that, at the present state of knowledge, this taxon is better kept as a genus within the Akodontini.

Other Genera Recognized Within the Akodontini

As discussed above, *Oxymycterus* is not separable from the Akodontini on chromosomal grounds. Moreover, it is clearly an akodont in cranial and tooth morphology and intestine and lung anatomy (Vorontzov, 1967, 1982). While there are no grounds to separate it from the Akodontini, it is indeed a distinct genus within the tribe. It sharply differs from the remaining akodont genera by its larger size, elongated skull with anteriorly expanded and trumpet-like nasals (fig. 4D), reduced zygomatic plate, simplified enamel pattern of molar teeth, diet (Kravetz, 1973), stomach morphology (Echave Llanos & Vilchez, 1964; Vorontzov, 1967, 1982; Carleton, 1973), phallic morphology (Hooper & Musser, 1964), and male accessory glands (Voss & Linzey, 1981).

Oxymycterus is rather polytypic, although the status and nomenclature of some of its species are still dubious. Cabrera (1961) recognized eight species and 13 subspecies. Honacki et al. (1982) accorded specific status to nine nominal forms. The genus needs revision. I examined the type specimens in the British Museum and can advance some results.

Following Massoia and Fornes (1969), I distinguish *O. nasutus* from *O. rufus*. At present, it is questionable whether *O. platensis* represents a subspecies of *O. rufus* or a distinct species. I am inclined to the former alternative; in any case, *platensis* is fully separated from *nasutus* (Massoia & Fornes, 1969). Regarding *rufus*, I follow Langguth (in Honacki et al., 1982, p. 459) in using Fischer's name instead of *rutilans* Olfers. It is at present unsettled whether *delator* is different from *rufus*. *Oxymycterus paramensis* and *O. akodontius* are reported to show some differences in skull and color characters, but it is unclear whether they are different species or merely phenotypic variants of a single species. *Oxymycterus incae* (including *doris, inca, iris,* and *juliacae* as subspecies) is quite distinct from *paramensis* or *akodontius*. As recognized by Massoia (1963b), *iheringi* is a good species of *Oxymycterus*, not a *Microxus*; this author is wrong, however, in allocating *Akodon sanborni* to *Oxymycterus*. I follow Cabrera (1961) in placing *judex, misionalis,* and *quaestor* under *O. hispidus*. However, this action is tentative until the type of *hispidus* is re-studied; but *judex* and *quaestor* are similar to each other and distinctive from *rufus* and the other mentioned species. *Oxymycterus angularis* and *O. roberti* are also tentatively accepted as valid species. The karyotypes of four studied species of *Oxymycterus* (Vitullo et al., 1986) are identical in number (2n = 54), morphology, and in the known cases, in G- and C-banding patterns of chromosomes. Thus, karyology may not afford good markers to clarify the status of the dubious species.

Lenoxus, known from the single species *L. apicalis* (with *boliviae* as a subspecies), is just an exaggerated *Oxymycterus* in size and skull morphology. The skull, as illustrated by Vorontzov (1982, fig. 112), is much like that of *Oxymycterus*, but can be distinguished by the broadened incisive foramina, larger palate, broader interorbital region, and greater development of the interparietal. The molar teeth show a less simplified enamel pattern, are more brachyodont, and have more sharply defined cusps; in the upper teeth, the lingual cusps are more displaced backward relative to the labial ones, and the anteromedian flexus and flexid are much more developed and penetrating than those in species of *Oxymycterus* (Vorontzov, 1982, p. 228, fig. 192). Additionally, *Lenoxus* can be distinguished from *Oxymycterus* in the less-developed foreclaws, the softer fur, and the longer tail.

Juscelinomys, which Moojen (1965) erected for *J. candango* from the "cerrado" near the city of Brasilia, and which probably also includes *Oxymycterus talpinus* Winge, 1888 (*fide* Moojen, 1965) from Minas Gerais, is certainly an akodontine closely related to *Oxymycterus* and *Lenoxus*. From its description and illustration, its generic distinction seems also warranted, as ensues from the diagnosis, which, taken with modifications from Moojen (1965, p. 283), is as follows:

> An akodontine cricetid close to *Oxymycterus* in dental character states, body form, and fossorial way of life. It differs from *Oxymycterus* in external characters in less elongated muzzle; tail very thick and densely covered with hairs that totally hide the tail scales; ventral region of fur, with rufous hairs whitish in their base. Skull stronger built; bullae swollen; rostrum short and broad; premaxillae not extending beyond the level of upper incisors; nasals reaching slightly in front of the incisors, but not trumpet-like; posterior border of infraorbital foramen in front of the level of first molars; zygomatic plate more anterior; diastema shorter; incisive foramina penetrating between the first upper molars beyond their protocone. Upper incisors slightly grooved, with the groove closer to their lateral borders; molars stronger; procingulum of first upper molar with a deep anteromedian flexus. Four pairs of mammae.

Podoxymys, erected by Anthony (1929) for the single species *P. roraimae* from the *tepuis* of the Guiana region, has also been alleged to stand near

Oxymycterus (Hershkovitz, 1966), although Ellerman (1941) pointed out that, from the description, it seems to be very close to *Microxus*. In fact, the description is rather vague, but doubtless it is a distinctive genus of the Akodontini. Its stomach is unilocular and hemiglandular (Carleton, 1973; Vorontzov, 1982) as in most akodontines, but differs in having a reduced area of glandular epithelium. Its molar teeth differ from those of *Microxus* and *Oxymycterus* in being much simplified, almost as much as in the otherwise quite different genus *Geoxus*.

Blarinomys is an akodontine highly modified for fossorial life and insectivorous diet (Vorontzov, 1982): extremely reduced eyes; tail short; ear very small; broad hind foot with fairly strong claws; hand with long claws. The skull is characterized by the absence of the interparietal, a broad interorbital region, and a very narrow zygomatic plate. The molars are highly modified in enamel pattern, and the m_3 is fairly reduced in size.

A point of clarification is required on the status of the fossorial southern akodonts currently referred to *Notiomys*. Thomas advocated that the burrowing, short-tailed mice of this group inhabiting southern Chile and Argentina represented three genera, namely *Notiomys* Thomas, 1890 (with *Hesperomys edwardsi* Thomas, 1890, as the type species), *Geoxus* Thomas, 1919 (with *Notoxus fossor* Thomas, 1919, as the type species), and *Chelemys* Thomas, 1903 (with *Hesperomys megalonyx* Waterhouse, 1844, as the type species). Later, Osgood (1925) proposed combining all three genera under *Notiomys*, a contention which, in spite of Thomas's (1927) criticisms, was accepted by most later authors, including, recently, Honacki et al. (1982). However, Gyldenstolpe (1932) accepted the generic rank of all three taxa. On the basis of the study of the types and series of specimens in the British Museum and the American Museum of Natural History, I concluded that two genera may be distinguished, namely *Notiomys* (including *Geoxus*) and *Chelemys* (Reig, 1981). More recently, Pearson (1983) treated *Chelemys* and *Geoxus* as different genera, and said that he considered it untenable to assign *Geoxus* to *Notiomys*. In a subsequent paper (Pearson, 1984), he supported his view with evidence from ecology, food habits, and external and craniodental morphology. With the new information of Pearson, and after a reconsideration of the information from my former studies, I am ready to accept the Thomas-Pearson view. It is in order, therefore, to afford an up-to-date definition of these three genera.

Notiomys Thomas, 1890

Size small: head and body less than 100 mm. Tail shorter than half the length of head and body, even shorter than in *Chelemys*; front claws long, less strong than in *Geoxus*; claws of the hind feet shorter than in *Geoxus*; fur not molelike, brightly colored, with whitish underparts and with bright rufous nose; margins of the hind feet with a shaggy fringe of hairs; ears with extremely thin and small pinnae, hidden in the fur; nose tipped with a dark leathery button. Skull rather heavy in shape, much shorter and wider than in *Geoxus*; muzzle shorter, wider and conical; incisive foramina short, not reaching the anterior borders of M^1; palate long, extending backward far behind the level of last molars, with small posterior palatal pits; pterygoid region shorter than in *Geoxus*, with rounded anterior border of mesopterygoid fossa; zygomatic plate reduced, moderately slanting to almost vertical; interparietal small and narrow; frontals not inflated, interorbital region broad, with sharply squared supraorbital ridges; mandible not as slender and proodont as in *Geoxus*, with long coronoid process, condyloid process not bent inward, and small capsular projection. Incisors slender, the upper ones orthodont; molars brachyodont, small, fairly narrow and elongated; M^1 with procingulum reduced, without anteroflexus but with a very shallow anteromedian flexus; upper molars with para- and metaflexus moderately oriented backward; mesoflexus present, deep and narrow on M^1. Mesostyle and enterostyle missing; M^3 simple and strongly reduced, less than ½ the length of M^2; procingulum of m_1 narrow, with a shallow anteromedian flexid; lower molars with meso- and posteroflexids not deeply infolded and lophids oriented rather transversely in position; mesolophid remnants visible on m_1; hypoflexid shallow; mesostylid and ectostylid absent; m_3 strongly reduced, rounded and simple, ½ the length of m_2; cecum absent; stomach and male glands unknown.

Included species: *edwardsi*.

Geoxus Thomas 1919

Size small: head and body length less than 100 mm; tail shorter than half the length of head and body, even shorter than in *Chelemys*; front claws stouter and longer than in *Notiomys* and *Chelemys*; hind foot with long claws, and without a shaggy fringe of hairs on the margins; fur molelike, uniform in color; ears small, but pinnae easily visible; nose normal. Skull slender and rather delicately built, longer and narrower than in *Notiomys*; muzzle long and narrow; incisive foramina long, surpassing backward the anterior borders of M^1; palate long, extending past the level of last molars, without posterior palatal pits; pterygoid region long; mesopterygoid fossa squarely open in front and nearly parallel-sided; zygomatic plate reduced, narrow and slanting; interparietal rather narrow and short, not extremely reduced; frontals rather inflated; interorbital region smooth and rounded, the supraorbital edges not square-shaped; mandible slender and proodont, with short coronoid process, condyloid process not bent inward, and small capsular projection. Incisors rather slender, the upper ones orthodont; molars small and narrow, simple and brachyodont; M^1 with reduced procingulum with a shallow anteroflexus and no anteromedian flexus; upper molars with transversely oriented para- and metaflexus, mesoflexus absent and para- and mesoloph coalesced; mesostyle and anterostyle missing; M^3 strongly reduced, less than ⅔ the length of M^2, ring-shaped; procingulum of m_1 narrow, without anteromedian flexid; lower molars with transverse lophids and flexids little infolded; m_3 reduced, less than ¾ the length of m_2, and with a simplified enamel pattern. Cecum missing; stomach unilocular-hemiglandular, with glandular epithelium more extended than in *Akodon*; medial ventral prostates large, and bulbourethral gland much larger than typical condition.

Included species: *valdivianus* (including *valdivianus, fossor, chiloensis, bullocki,* and *bicolor* as subspecies) and, probably, *michaelseni* (see Pearson, 1984, p. 233).

Chelemys Thomas, 1903

Size larger: head and body length longer than 120 mm; tail shorter than half the length of head and body; front claws stout and long, but comparatively less developed than those in *Geoxus*; hind foot with moderately long claws, without a shaggy fringe of hairs on their margins; fur short and dense, uniform in color; ears small, but pinnae visible; nose normal; skull robust and broadly built; muzzle broad and rather short; incisive foramina long, extending backward beyond the anterior borders of M^1; palate short, extending backward to the level of last molars, with small posterior palatal pits; mesopterygoid fossa deep, with rather squared front edge, and lateral borders slightly divergent backward; zygomatic plate comparatively strong and broad, its anterior border nearly vertical or slightly slanting forward; interparietal normally developed, not reduced, although relatively narrow; frontals not inflated, interorbital region smooth and rounded, with supraorbital edges not square-shaped; mandible rather strongly built; ramus much deeper than in *Geoxus* or *Notiomys,* with long coronoid process, condyloid process bent inward, and a medium-sized capsular projection. Incisors thick and robust, the upper orthodont, the lower less proodont than in *Notiomys* and *Geoxus*; molars rather large and broad, relatively hypsodont; M^1 with moderately reduced procingulum lacking any trace of anteromedian flexus and anteroflexus; upper molars with para- and metaflexus strongly oriented backward; mesoflexus completely reabsorbed by complete fusion of para- and mesoloph; mesostyle normally absent, enterostyle totally missing; M^3 relatively well developed, about ⅔ the length of M^2; m_1 with short procingulum without anteromedian flexid; lower molars with ento- and posteroflexids oblique, metaflexid more transverse; mesoloph and paraloph coalesced, mesostylid and ectostylid missing; m_3 sigmoid-shaped, not reduced, more than ¾ the length of m_2; cecum present; stomach and male glands unknown.

Included species: *macronyx* (including *macronyx, alleni, fumosus,* and *vestitus* as subspecies), *megalonyx* (including *megalonyx* and *microtis* as subspecies), and probably, *delfini,* a species from Punta Arenas whose status is dubious and which might represent a third subspecies of *megalonyx.* For removal of *angustus* from *Chelemys,* see Pearson (1984, p. 231).

Defining the Tribe Akodontini

As discussed in the previous sections, 11 extant genera are recognized within the Akodontini; namely, *Akodon, Blarinomys, Bolomys, Chelemys, Geoxus, Juscelinomys, Lenoxus, Microxus, Notiomys, Oxymycterus,* and *Podoxymys.* The extinct *Dankomys* is also included.

The various genera to which the term Akodontini is applied constitute in one sense its *meaning.* This is the referential approach to the assessment of the meaning of a concept (Alston, 1964). This sense of meaning is called the *extensional* or *denotative* meaning. Since defining is an operation by which we explain the meaning of a term, assessing in this way the meaning of the Akodontini is equivalent to formulating an extensional definition of that taxon-concept. However, this denotative definition is only one meaning of the taxon and is not convenient for further taxonomic purposes. Although frequently the elucidation of the extension of a taxon-concept precedes the assessment of the properties that are shared by the members of the concept (the *intensional* or *connotative* meaning of the term), a complete definition of concepts requires that the common properties or attributes which qualify a given taxon for membership be defined.

This definition of a taxon by its connotative or intensional meaning is an unavoidable aim in taxonomy, as it enables the zoologist to grasp the basic features which characterize the taxon as a natural product of evolution, and to decide whether as yet unknown or problematic taxa deserve membership in the taxon.

To arrive at this definition satisfactorily would require revision of the whole tribe and other related groups of the subfamily Sigmodontinae. However, we can agree on a definition of the term Akodontini by ascertaining from the present body of knowledge its *conventional intension* (Copi, 1953, p. 102).

This definition must be polythetic (Sokal & Sneath, 1963; Bechner, 1959 called the same kind of definition polytypic) in that the Akodontini are defined by reference, not to a set of attributes the common possession of all of which is both necessary and sufficient for membership in its extension, but to a set of attributes the common possession of a large (but unspecified) number of which is sufficient for belonging to its extension. Needless to say, the same approach is applicable to the definition of genera and subgenera within the tribe, and it permits defining the intensional meaning of various taxa by commonality of character-states, abjuring the typological claim of exclusive sharing of one or another of all the alternative states of the characters used in defining taxa of the same rank and which belong to the same taxon of immediate higher rank.

The intensional meaning of the polythetic concept of the Akodontini follows.

Akodontini Vorontzov, 1959

Sigmodontine cricetids of small to medium size, with omnivorous to insectivorous digestive system, without specializations for plant feeding; stomach normally of the unilocular-hemiglandular type, exceptionally of the unilocular-discoglandular type; large intestines short, usually less than 15% the length of small intestines; cecum small or absent, never enlarged; molar teeth subhypsodont to mesodont, rarely brachyodont, and crested, terraced, or secondarily planed; mesoloph and mesolophid reduced or vestigial when present, often fully or partially coalesced with paraloph or entolophid, and only shown as terminal remnants usually united with mesostyle or mesostylid; posteroloph coalesced with metaloph, and posteroflexus usually obsolete; skull with zygomatic plate little to moderately developed, never very high and strongly projecting before the antorbital bridge; palate broad, from short to moderately long; incisive foramina usually large and reaching backward to or beyond the anterior plane of M^1; preputial gland usually single; medial ventral prostate usually reduced or absent; glans penis complex; diploid karyotype never with more than 54 chromosomes.

The Akodontini are certainly more derived than the Oryzomyini in molar structure, skull, and male accessory glands (Voss & Linzey, 1981; Vorontzov, 1982); they are probably direct descendants of the latter. At the same time, they are more primitive than the Phyllotini in molar pattern and digestive system (Vorontzov, 1982) and may have been their ancestors. The Akodontini are predominantly Andean in distribution (Reig, 1984), and most of them are inhabitants of open land, although some of their representatives may dwell in forested tropical, subtropical or temperate habitats, and some of their genera—especially *Blarinomys, Chelemys, Geoxus,* and *Notiomys*—exploited the subterranean niche and might be considered the South American counterparts of the moles. Most species studied are quite omnivorous (Meserve, 1981; Meserve & Glanz, 1978; Pearson, 1983), but some became more specialized for an insectivorous diet, such as *Oxymycterus* (Kravetz, 1973), *Notiomys,* and *Geoxus* (Pearson, 1983, 1984). Although *Chelemys* eats mostly mushrooms, and species of *Bolomys* have been found to be partly vegetarians (Scaglia & Velazquez, pers. comm.), the remaining akodontines did not invade especially the herbivorous niches, which are heavily exploited by their probable derivatives, the phyllotines.

Description of New Plio-Pleistocene *Akodon*

The study of the fossil material described and discussed in this section necessitated evaluation of the character-states of the genus *Akodon* and its subgenera *Abrothrix* and *Akodon s.s.* in cranial, mandibular, and dental morphology. The full description of the character-states of these taxa is essential for the identification and description of fossil specimens. The following description of the morphological attributes of *Akodon* is based in the study of types and series of about 70% of the included species. It is intended, therefore, to apply to the whole genus. *Akodon* is, however, rather varied and, as already discussed, the distinction of various subgenera seems appropriate. Therefore, in all those cases where exceptions to the next description are known, the corresponding character-states have been qualified as "usually" present. Departures from those states in included subgenera are assessed in the description of morphological attributes of *Akodon s.s.* and *Abrothrix* provided afterward.

Genus **Akodon** Meyen, 1833

TYPE SPECIES—*Akodon boliviensis* Meyen, 1833, by original designation.

DISTRIBUTION OF LIVING SPECIES—Andean valleys, highlands, and fringes of mountain forests of Argentina, Bolivia, Chile, Colombia, Ecuador, Peru, and Venezuela; temperate-zone meadows, grasslands and brushlands of Argentina, Bolivia, SE Brazil, Chile, Paraguay, and Uruguay.

CRANIAL CHARACTER-STATES—Skull usually slender, with a typically fairly narrow and rounded braincase, a fairly elongated occipital region, and a rostrum of regular shape, not tapering forward in lateral view. Upper profile of the skull sloping forward and backward from the posterior part of the frontals. Zygomata slightly expanded and not markedly convergent anteriorly. Nasals usually longer than, or as long as, the frontals, their anterior border passing forward beyond the anterior plane of the incisors, but not projected or expanded to form a trumpet-shaped opening. Posterior borders of the nasals usually tapering backward and projecting beyond the fronto-premaxillary suture. Frontals long, usually with a narrow and transversely convex interorbital region with more or less sharply squared edges not defining a supraorbital ridge. Frontoparietal suture angular or crescentic in shape. Parietals relatively long, their length, in the midline, usually more than half the length of the frontals, not extending forward through lateral processes between frontals and temporals. Interparietal moderately reduced in width and length. Zygomatic plate moderately developed, relatively narrow and low, but usually with anterior border vertical, straight or slightly concave, not slanting gradually backward from its lower root to the upper border. Upper corner of the zygomatic plate rounded, not projecting forward. Incisive foramina wide and elongated, narrow behind, usually penetrating well beyond the anterior plane of the first molars to reach or slightly surpass the level of the protocone of M^1. Posterior palate long and wide, its median posterior border usually slightly behind the posterior plane of the M^3. Palatal surface relatively simple, without marked ridges and with shallow grooves. Mesopterygoid fossa narrow and less than width of parapterygoid fossa. Bullae usually small, less frequently moderately large. Mastoid not noticeably inflated. Occipital region somewhat elongated, its posterior border rounded and continuous with the line of the braincase when viewed from the lateral side.

Mandible somewhat slender, its height at m_1 usually shorter than diastema length. Lower masseteric crest present, but not strong, reaching forward to the level of the anterior half of m_1. Upper masseteric crest rather long and usually as strong as the lower one. Coronoid process rather short, with anterior border gradually slanting backward. Condyloid process relatively low, elongated, and projected backward. Articular surface of the condyle extending dorsally and slightly posteriorly. Capsule of incisor root normally not projected as a well-developed tubercle, lying on the anterior half of the sigmoid notch. Angular process longer than high.

DENTAL CHARACTER-STATES—Upper incisors usually opisthodont, less frequently orthodont, never proodont. Molar rows parallel-sided. Molars relatively small, usually with moderately developed hypsodonty, with crested and with bilevel occlusal surface in slightly worn teeth, terraced with advanced wear.

M^1 four-rooted. Upper molars with labial (paracone-metacone) main cusps only slightly posterior to the level of the lingual ones (protocone-hypocone), and main lophs transverse in position. M^1 and M^2 usually trilophodont in moderately worn teeth. Mesoloph reduced, usually only partially fused with the paraloph, so that a lingual remnant of it is usually present, forming a bifurcated broad median loph marked by a shallow mesoflexus in M^1 and M^2. Metaloph united with the posteroloph and not reaching the hypocone, almost completely coalesced with posteroloph, so that posteroflexus is very narrow or completely absent in moderately worn teeth. Paraflexus and metaflexus somewhat directed backward, hypoflexus and protoflexus somewhat forward, the opposite flexi slightly alternating. Enterostyle and enteroloph usually completely absent. Mesostyle often present in M^1 and M^2, usually united with the remnant of mesoloph. M^2 usually much longer than wide. M^3 much reduced, cylindriform in moderately worn teeth. Procingulum of M^1 moderately simple, slightly oblique in position, usually clearly biconulate by the presence of a well-developed anteromedian flexus. Procingulum united to protocone through an anteroposteriorly oriented anterior mure. Protoflexus absent or undivided by a protostyle, anteroflexus usually present but not deeply infolded. Anteroloph well defined and usually united to parastyle. Protoflexus of M^2 evident in moderately worn teeth, absent on M^3 of similar stage of wear.

Lower molars with lingual cusps (metaconid and entoconid) placed fairly anterior to the labial ones (protoconid-hypoconid), with metalophid, as well as entolophid and posterolophid usually oblique, directed slightly forward from the lingual border to the labial one. Mesolophid almost completely fused with entolophid, but a weak lingual remnant, usually united with a mesostyle, is often present in most species. Ectolophid and ectostylid frequently present in m_1, occasionally in m_2, rarely in m_3. In m_1 and m_2, hypoflexid broad and transverse, mesoflexid directed obliquely forward from outside. Posteroflexid well developed, oblique and parallel to the mesoflexid, absent in m_3. m_1 tetralophodont, with a somewhat complex procingulum, defined by a usually well-defined, well-infolded metaflexid, and protoflexid; anteroflexid normally absent. Protostylid and anterolabial cingulum usually present. Anteromedian flexid normally well developed in moderately worn teeth. m_2 trilophodont, well longer than wide, with protoflexid usually well marked in moderately worn teeth. m_3 relatively large, but even smaller than m_2, usually bilophodont and sigmoid-shaped in outline, with protoflexus frequently present in moderately worn teeth.

Subgenus **Akodon**, Meyen

Akodon Meyen, 1833, Nova Acta Leopold. 16, pt. 11: 599.
Hesperomys, Wagner, 1839, Schrebers Säugeth. Suppl. 3: 516 (in part); Leche, 1886, Zool. Jahrb. 1: 687 (in part).
Habrothrix, Thomas, 1884, Proc. Zool. Soc. London: 450.
Chalcomys, Thomas, 1916, Ann. Mag. Nat. Hist. (8)18: 338.
Thaptomys, Thomas, 1916, Ann. Mag. Nat. Hist. (8)18: 339.
Bolomys, Thomas, 1916, Ann. Mag. Nat. Hist. (8)18: 339 (in part).

TYPE SPECIES—*Akodon boliviensis* Meyen, by original designation.

KNOWN DISTRIBUTION—As for the genus.

INCLUDED SPECIES—*aerosus, albiventer, andinus, azarae, boliviensis, brachiotis, cursor, dolores, iniscatus, markhami, molinae, mollis, nucus, olivaceus, orophilus, pacificus, puer, reinhardti, serrensis, surdus, tolimae, urichi, varius,* and two unnamed species from Brazil, one with 2n = 14–16 chromosomes, the other with 2n = 24–25 chromosomes (Yonenaga et al., 1975; Yonenaga, 1979; Maia & Langguth, 1981). Additionally, two fossil species which are described below. [The number of living species is certainly larger, mostly because of the complex nature of *A. varius.* Philip Myers (pers. comm.) recognizes three distinct species in what is usually considered to be *A. varius*; namely, *A. varius, A. toba,* and *A. simulator.* He is also inclined to treat *A. neocenus* as a full species and to recognize *A. dayi* as a distinct species.]

CHARACTERS—Skull normally built and usually somewhat elongated behind. Nasals longer to slightly shorter than frontals. Zygomatic plate normal, with anterior border usually vertical in position. Braincase moderately long, usually not broadened, its breadth as large or slightly shorter than ½ the condylobasal length. Interorbital region usually fairly narrowed, without supraorbital ridges. Anterodorsal frontal sinuses not inflated. Interparietal normal to much reduced. Bullae usually not enlarged. Incisive foramina usually reaching the protocone of M^1 or slightly before it. Posterior border of palate behind the posterior border of M^3. Mandible relatively high and stout, more slender in the smaller species, with masseteric crest normally developed and reaching the middle of the m_1. Incisor capsule usually not projected as a definite tubercle. Upper incisors normally opisthodont, less frequently orthodont. Molars not markedly elongated and narrowed, with a moderately developed tubercular hypsodonty. Molar crowns usually bilevel, terraced to planate with wear, cusps neither noticeably tuberculate, nor with noticeably inclined enamel walls. M^1 usually with an anteromedian flexus and an anteroflexus. Paraflexus of M^2 directed lingually, the anteroloph in normal position. Mesoloph remnants usually united to mesostyle, typically on M^1 and M^2. Entoconid wide, but not noticeably bulging laterally in m_1 and m_2. Mesoflexid and posteroflexid of m_1 and m_2 normally inclined and well developed. Anteromedian flexid of m_1 frequently present, but only exceptionally deeply infolded, metaflexid moderately to scarcely infolded. Mesolophid remnants, ectolophids, and ecto- and mesostylids frequently present. m_3 long, but clearly shorter than m_2.

Subgenus **Abrothrix** Waterhouse

Mus (*Abrothrix*) Waterhouse, 1837, Proc. Zool. Soc. London: 21.
Habrothrix Wagner, 1843, Schrebers Säugeth. Suppl. 3 (in part).
Acodon Thomas, 1895, Ann. Mag. Nat. Hist. (6)14: 369 (in part).
Abrothrix Thomas, 1916, Ann. Mag. Nat. Hist. (8)18 (considered as a full genus).
Akodon (*Abrothrix*), Ellerman, 1941, The Families and Gen. of Living Rodents. Vol. 2: 409, 416; Osgood, 1943, Field Mus. Nat. Hist., Zool. Ser. 30: 184–198.

TYPE SPECIES—*Akodon* (*Abrothrix*) *longipilis* Waterhouse (by original designation).

DISTRIBUTION—Lowlands and mountain valleys of southern and central Chile; Andean slopes and low valleys of Argentina from Tierra del Fuego to Mendoza; mountain valleys of Tucumán, Argentina.

INCLUDED SPECIES—*hershkovitzi, illutea, lanosus, longipilis, sanborni, xanthorhinus.* (For the provisional assignment of *xanthorhinus* and *hershkovitzi* to *Abrothrix,* see above; see also Patterson et al. [1984], who specifically rejected this assignment.) Additionally, the fossil *A. kermacki* and a new fossil species as described below. Another species referred to *Abrothrix, A. mansoensis* (de Santis & Justo, 1980), is a dubious form. It is not evident that it belongs to *Abrothrix* or even that it is a valid species of the genus *Akodon.*

CHARACTERS—Skull strong and elongated, with a rather long and slender muzzle. Nasals well longer than frontals, exceeding backward the frontomaxillary suture and slightly projecting forward. Zygomatic plate relatively deep and short, with anterior border vertical or slightly inclined backward. Braincase relatively long, rounded and slightly broadened, its breadth as large as ½ the condylobasal length. Interorbital region of median breadth, without supraorbital ridges and with smoothly rounded edges. Anterodorsal frontal sinuses slightly inflated, its dorsal surface rounded. Interparietal normally reduced. Incisive foramina elongated but scarcely reaching the protocone of M^1. Posterior border of the palate well behind the posterior border of the M^3. Bullae not enlarged. Mandible moderately low and elongated, with upper masseteric crest better marked than the lower masseteric crest, slightly surpassing the middle of the m_1. Incisor capsule projected as a tubercle. Upper incisors orthodont, rather strong. Molars comparatively broad, with a moderately well-developed tubercular hypsodonty and a slight crown hypsodonty. Molar crowns bilevel, terraced to planate with advanced wear. Cusps not noticeably tuberculate, with somewhat inclined enamel walls. Anteromedian flexus of M^1 completely obsolete or barely noticeable. Anteroflexus present, but shallow. Paraflexus of M^2 directed lingually, anteroloph normal. Mesoloph remnants usually united to mesostyles in M^1 and M^2. Mesoflexid and posteroflexid of m_1 and m_2 well developed, the former scarcely inclined and the latter smaller and nearly transverse in position. In m_1 and m_2 entoconid typically bulging lingually. m_1 with anteromedian flexid obsolete or occasionally present in an incipient stage as a shallow and open notch, metaflexid little to moderately infolded. Mesolophid remnants constant on m_1–m_3, but poorly developed and projecting from the anterolateral border of the entoconid, usually connected with mesostylids. Ectolophids completely absent, ectostylids sometimes present on m_1, very rarely so in m_2. m_3 long, but smaller than m_2.

Akodon (Abrothrix) kermacki Reig, 1978

HOLOTYPE—MMP S-321, almost complete right and left lower jaws with the entire lower dentition; portion of the left maxilla including the three upper molars (figs. 5A,E; 6A). Collected by G. J. Scaglia in stratum IX of the Chapadmalal Formation (Kraglievich, 1952), 500 m north of "Bajada del Vivero" (Punta Lobería), Atlantic cliffs of the Partido de General Pueyrredón, SE of Buenos Aires Province, Argentina. Figured by Reig and Linares (1969) as *Akodon* sp.

HYPODIGM—The holotype and the following:

MMP S-222—Almost complete left mandibular ramus, with the entire dentition. Collected by G. J. Scaglia in stratum VIII or IX (Kraglievich, 1952) of the Chapadmalal Formation, 650 m north of "Bajada del Vivero" (Punta Lobería). Figured by Reig and Linares (1969) as *Akodon* sp.

MMP M-1067—Anterior part of left lower jaw with the whole dentition. Collected by G. J. Scaglia and Mr. Prima in Stratum IX of Chapadmalal Formation at "Bajada del Vivero" (Punta Lobería).

MMP M-1071—Anterior part of right lower jaw with the whole dentition. Collected by G. J. Scaglia

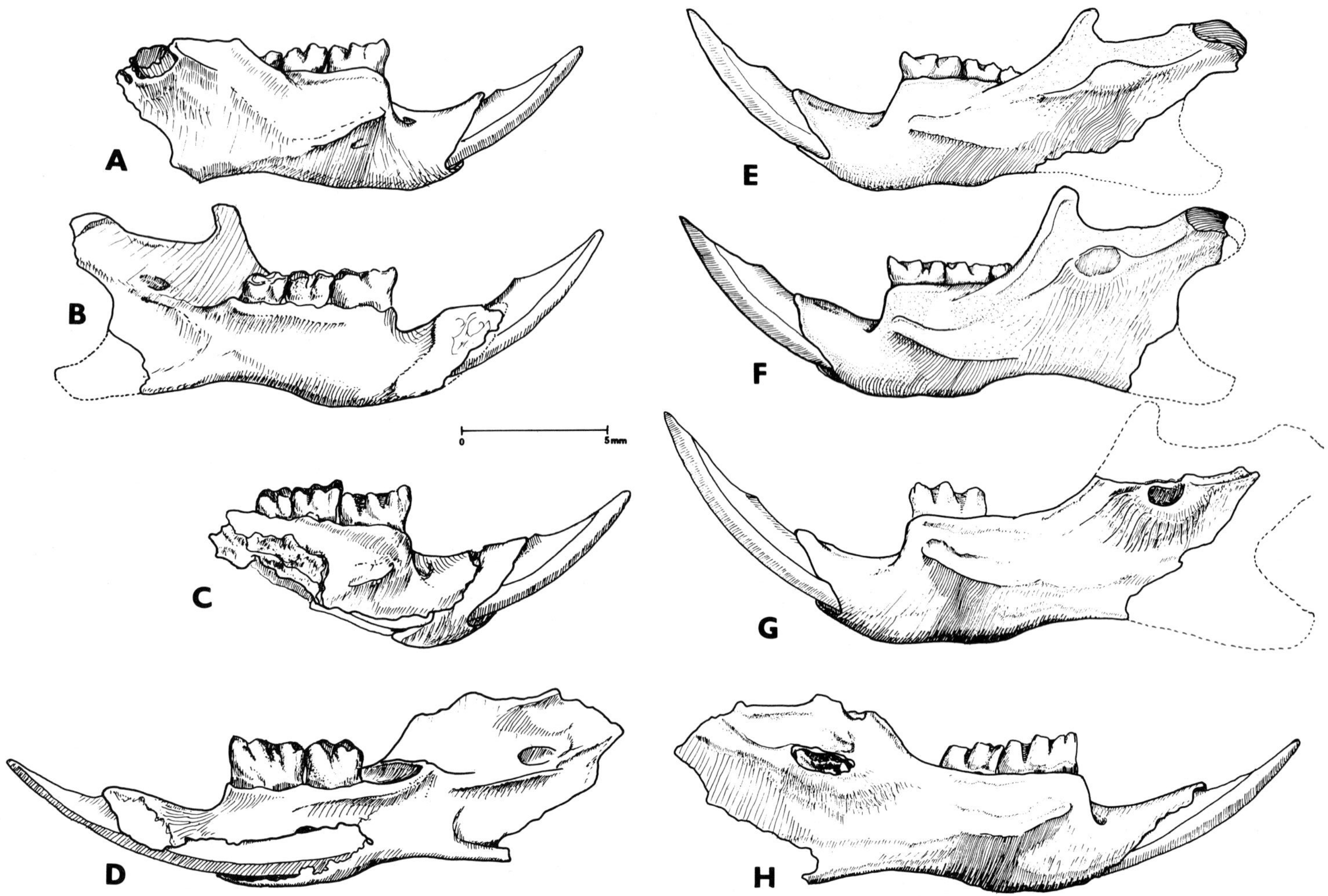

FIG. 5. Lower jaws of *Akodon* (*Abrothrix*) *kermacki* Reig, and *Akodon* (*Abrothrix*) *magnus* n. sp. **A,** External aspect of left lower jaw and **E,** external aspect of right lower jaw of *A.* (*Ab.*) *kermacki* Reig; holotype, MMP S-321. **B,** Internal aspect and **F,** external aspect of WA.E (*Ab.*) *kermacki*; MMP S-222. **C,** External view of *A.* (*Ab.*) *kermacki*; MMP M-1071. **D,** Internal view of right lower jaw, **H,** external view of right lower jaw, and **G,** external view of left mandible of *A.* (*Ab.*) *magnus*; type, MMP M-551.

in the lower levels of Barranca Lobos Formation (Kraglievich, 1952) in the cliffs north of "Bajada del Vivero" (Punta Lobería).

MLP 62.VII.27.84—Almost complete left lower jaw with the whole dentition. Collected by G. J. Scaglia in association with the holotype.

MMP M-1154—Incomplete right lower jaw with the whole dentition. Collected by G. J. Scaglia in association with the holotype. (For former numbers of the last two specimens, see Reig, 1978, p. 175.)

Known Distribution—Chapadmalalan (Upper Pliocene) and Uquian (Lowermost Pleistocene) ages, SE of Buenos Aires Province, Argentina (see Marshall et al., 1983, 1984).

Diagnosis—A species of *Abrothrix* close to *A. longipilis*; size larger than in *A. l. longipilis*. Incisor stronger and deeper, M^1 with a shorter and wider, noticeably oblique procingulum, with a visible, although weakly developed, anteromedian flexus. m_1 relatively shorter and m_3 larger than in *A. l. longipilis*. M^3 without any evidence of a metafossetus. Lower jaw with the capsular projection for the base of the incisor more developed than usual in the subgenus.

Description—The only known part of the skull, a piece of left maxilla including the cheekteeth which belongs to the type specimen, is too fragmentary to afford useful information about the structure of the palate. Moreover, it does not include any part of the usually diagnostic zygomatic plate. From the bone tissue preserved posterointernal to the M^3, it can be inferred that the posterior border of the palate was behind the posterior border of the last upper molar, as it is in *Abrothrix*.

The mandible (fig. 5) is very well preserved in the holotype, MLP 62.VII.27.84, MMP M-1154, and MMP S-222. It is relatively slender, moderately low and elongated, as in *Abrothrix* and some species of *Akodon s.s.* (e.g., *A. cursor*). The diastema has approximately the same length as the combined length of m_1 and m_2, and the depth of the horizontal ramus below m_1 is less than the diastema length in all specimens except MMP M-1071, in which it is slightly larger. The lower border of the ramus bends gently upward and backward behind the level of the middle of the m_2 and descends again behind the m_3, shaping a concave line, as is usual in *Abrothrix*. The border of the ramus immediately in front of the m_1 descends rather abruptly downward, making a slightly obtuse angle with the upper border of the symphysis. The symphysis is relatively long and moderately low, and the uppermost anterior point of the diastema is almost at a level with the alveolar row. The lower masseteric crest is smooth, but well marked, rather high in position, and better developed than is usual in *A. longipilis*; it reaches to a level anterior to the middle of the m_1, but behind its anterior border. The upper masseteric crest is not so well defined as the lower one, and is less developed than in *A. longipilis*. The mental foramen is normally developed and opens on the dorsolateral surface of the diastema. The anterior edge of the coronoid process originates at the level of the middle of the m_2 and slopes gradually upward, with most of the m_3 visible in lateral view when the mandible is seen perpendicular to the plane of the symphysis. The coronoid process is relatively short, and the condyloid process is low and elongated, slightly projected backward, resembling closely the situation found in *A. longipilis*. The capsular projection, which lies at the level of the anterior part of the sigmoid notch, is stronger than is usual in *Abrothrix*, reaching a development similar to that in *Deltamys*, but less developed than in *Bolomys*.

The greater development of the capsule of the incisor root is obviously a consequence of the more strongly developed lower incisor. This is unusually deep for *Akodon sensu lato*, and it is absolutely and proportionally deeper than in the living species of *Abrothrix*. In all the available specimens of *kermacki*, the mean depth of the incisor is almost as large as the length of the m_3 (length $m_3 \cdot 100$/depth incisor = 0.993), and in ½ of the available individuals, it surpasses the m_3 (see table 3, fig. 9). In a sample of 19 *A. l. longipilis* from Valparaiso, Chile, in the British Museum, the same index is 0.927, and the length of the m_3 is, in all the individual cases, longer than the depth of the incisor.

The molar teeth agree in all respects with the characters of *Abrothrix* as stated above in the diagnosis. In all six known specimens, the masticatory surface shows an appreciable, but not advanced, degree of wear, corresponding to wear stages 2–3 of *Akodon azarae* as described by Pearson (1967). Therefore, most of the details of the enameled structures of the crown can be observed. The upper dentition is only known from the type specimen. The total length of the upper molar row (4.91 mm in crown length) places *A. kermacki* among the largest species of *Akodon s.s.* (*A. urichi, A. varius*) and within the range of variation in the available sample of *A. l. longipilis* from Valparaiso, which is the largest subspecies of *longipilis*. In molar morphology, some differences are apparent, which might be of diagnostic value. In M^1 and M^2

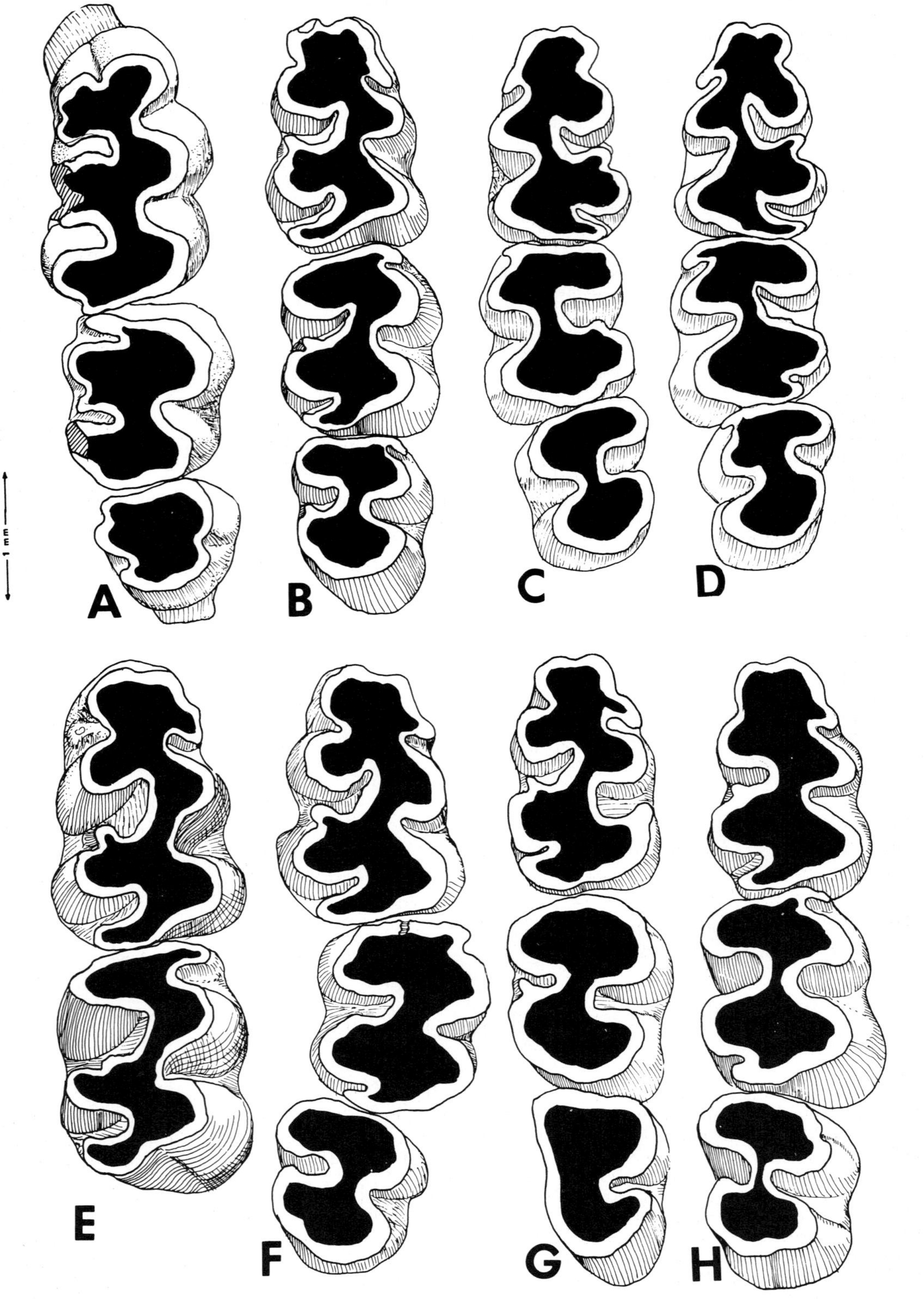
1 mm
A
B
C
D
E
F
G
H

TABLE 1. Measurement (in mm) of the known specimens of *Akodon* (*Abrothrix*) *magnus* n. sp. and of *Akodon* (*Abrothrix*) *kermacki* Reig (some of the specimens of *A. kermacki* were reported with slightly different values for some of the measurements in Reig & Linares, 1969; the new values given here result from the adoption of conventions used in the present study).

	***Akodon magnus*, Vorohué Formation**		***Akodon kermacki***					
			Chapadmalal Formation					**Barranca Lobos Formation**
Variate	**Type, MMP M-551**	**MMP S-407**	**Type, MMP S-321 (*r*)**	**MLP 62-7-25-84**	**MMP M-1154**	**MMP S-222**	**MMP M-1067**	**MMP M-1071**
Mandibular condyle-symphysis length	...	...	17.00	16.55	...	16.80	...	...
Lower diastema length	4.22	4.61	4.03	4.03	...	3.99	4.42	3.65
Ramus depth at middle m_1	3.96	4.48	3.71	3.78	...	3.82	4.03	3.84
m_1–m_3 (alveolar) length	6.40	6.14	5.06	5.04	5.57	5.25	5.69	5.57
m_1–m_3 (coronal) length	...	6.02	5.03	4.86	5.38	4.93	5.63	5.37
m_1 length	2.42	2.32	1.98	1.89	1.95	1.95	2.29	2.05
m_1 width	1.55	1.47	1.33	1.30	1.36	1.37	1.58	1.33
m_2 length	1.93	1.80	1.58	1.46	1.64	1.52	1.74	1.64
m_2 width	1.61	1.52	1.31	1.30	1.47	1.35	1.55	1.30
m_3 length	...	1.80	1.43	1.39	1.58	1.49	1.59	1.53
m_3 width	...	1.25	1.15	1.12	1.20	1.14	1.39	1.12
Lower incisor depth	1.36	1.34	1.43	1.47	1.44	1.54	1.70	1.41
Lower incisor width (thickness)	0.87	0.77	0.86	0.87	0.82	0.84	0.99	0.87
M^1–M^3 (alveolar) length	...	...	5.37	...	...	...	...	...
M^1–M^3 (coronal) length	...	...	4.91	...	...	...	...	...
M^1 length	...	...	2.45	...	...	...	...	...
M^1 width	...	...	1.39	...	...	...	...	...
M^2 length	...	...	1.49	...	...	...	...	...
M^2 width	...	...	1.30	...	...	...	...	...
M^3 length	...	...	1.05	...	...	...	...	...
M^3 width	...	...	1.12	...	...	...	...	...

the main cusps are nearly opposed, the paracone and metacone being only slightly posterior to the protocone and hypocone, respectively. As in *Abrothrix* and in *Akodon s.s.*, the M^2 is clearly longer than wide, and the M^1 is strong and comparatively broad. The procingulum of the M^1 is short and wide, more so than is usual in *A. longipilis* and *A. illutea,* and it is more strongly oblique in position, the anterolingual conule being more anterior than the anterolabial conule. The anteromedian flexus is distinct, although it is only very slightly infolded; its presence is also indicated in

Opposite Page:

FIG. 6. Upper and lower molar teeth of *Akodon* (*Abrothrix*) *kermacki* Reig, and *Akodon* (*Abrothrix*) *magnus* n. sp. **A,** Right upper molar series and **B,** Right lower molar series of *A.* (*Ab.*) *kermacki*; holotype, MMP S-321; Chapadmalal Formation, Partido de General Pueyrredón, Buenos Aires Province, Argentina (Upper Pliocene). **C,** Left lower molar series of *A.* (*Ab.*) *kermacki*; MMP M-1153; Chapadmalal Formation; found in association with S-321. **D,** Left lower molar series of *A.* (*Ab.*) *kermacki*; MMP S-222; Chapadmalal Formation, Partido de General Pueyrredón, Buenos Aires Province, Argentina. **E,** Right m_1 and m_2 of *A.* (*Ab.*) *magnus,* n. sp.; holotype, MMP M-551; Vorohué Formation (Lower Pleistocene), Chapadmalal region, Partido de General Pueyrredón, Buenos Aires Province, Argentina. **F,** Right lower molar series of *A.* (*Ab.*) *kermacki*; MMP M-1067; Chapadmalal Formation, Partido de General Pueyrredón, Buenos Aires Province, Argentina. **G,** Right lower molar series of *A.* (*Ab.*) *kermacki*; MMP M-1071; lower levels of Barranca Lobos Formation (lowermost Pleistocene), Partido de General Pueyrredón, Buenos Aires Province, Argentina. **H,** Right lower molar series of *A.* (*Ab.*) *kermacki*; MMP M-1154; Chapadmalal Formation; found in association with S-321 (**A–B**) and M-1153 (**C**).

TABLE 2. Length of upper molar row of different species and subspecies of *Akodon* (*Abrothrix*).

Species and subspecies	N	$\bar{x}$	Ref.*
Akodon (*Ab.*) *kermacki*	1	5.37	1
Akodon (*Ab.*) *longipilis longipilis*	19	5.13	1
Akodon (*Ab.*) *longipilis castaneus*	16	4.58	2
Akodon (*Ab.*) *longipilis francei*	11	4.18	2
Akodon (*Ab.*) *longipilis suffusa*	4	3.88	2
Akodon (*Ab.*) *longipilis apta*	27	4.27	2
Akodon (*Ab.*) *longipilis nubila*	14	4.07	2
Akodon (*Ab.*) *sanborni*	18	4.03	2
Akodon (*Ab.*) *lanosus*	4	3.63	2
Akodon (*Ab.*) *mansoensis*	14	3.48	3
Akodon (*Ab.*) *hershkovitzi*	1	3.90	4

* 1 = This paper; 2 = Yañez et al., 1978; 3 = de Santis & Justo, 1980; 4 = Patterson et al., 1984.

the anterior surface of the crown by a shallow groove descending to the alveolus. The anteroloph is barely defined by a very shallow anteroflexus, much as in the type specimens of *A. l. longipilis* and *A. illutea* (fig. 7C–D). As in them, the protoflexus is wide and moderately infolded to the center of the tooth, its innermost point reaching a level anterior to the level of the innermost point of the opposed paraflexus. The metaflexus is rather transverse in position, scarcely inclined backward, and it is at a level posterior to the main axis of the hypoflexus, which is wide and slightly oriented forward. The lingual surface of the crown is partially broken at the walls of the medial loph (paraloph + mesoloph), but the presence of a free lingual remnant of the mesoloph is clearly indicated, though it cannot be ascertained if a mesostyle was also present. In the M^2, the mesoloph remnant is evident, but there is no distinguishable mesostyle. The protoflexus is obsolete both in the M^2 and in the M^3. The M^2 is very similar to that in the types of *illutea* and *longipilis*, but the paraflexus is better indicated, probably because of less advanced wear, although the metaflexus is less reentrant than in those specimens. The M^3 is subcylindrical in outline, with traces of the lingual flexi. There is no trace, however, of a metafossetus in this tooth, whereas such an internal enamel island is present in the type specimens of *A. l. longipilis, A. l. hirtus, A. l. nubilus,* and *A. illutea,* and in all the specimens of the sample of *A. l. longipilis* from Valparaiso. Even in the most worn M^3 of observed *Abrothrix,* the presence of a metafossetus is constant; its absence in *A. kermacki* is a diagnostic feature.

The lower molars can be studied in the six available specimens. They are typical *Abrothrix* lower molars in the bulging of the entoconid and the small, anteriorly directed mesolophid remnant and the elongated and oblique median murid of m_1 and m_2. As it is usual in *Akodon s.l.,* the main cusps are disposed nearly in echelon, and the metaconid and entoconid are placed at a level anterior to the protoconid and hypoconid, respectively. The procingulum of the m_1 is wider than is usual in *Abrothrix* and bears a very shallow, but distinguishable, anteromedian flexid in five of six available individuals (~83%). In the studied sample of *A. l. longipilis* from Valparaiso, it was observed only in 25% of the cases, and then only as a very shallow notch. There is a well-developed anterolabial cingulum, similar to that in other species of *Abrothrix,* and, as in them, there is no evidence of a division of the protoflexid into an anterior and a posterior portion, although the lateral surface of the crown at the procingulum shows a pronounced concavity anterior to the anterolabial cingulum in some cases. As in other *Abrothrix,* the mesoflexid is obliquely directed anteriorly from the outside in m_1 and m_2; and the posteroflexid is less oblique, almost completely transverse with more advanced wear. In all the specimens, a well-defined, but weak mesolophid remnant connected with a mesostylid is apparent in the m_1. It grows out from the anterior border of the entolophid, and directs forward and outward, defining a very shallow entoflexid. In the m_2, this structure is even weaker, being obsolete in three of six known specimens. This development and disposition of

Opposite Page:

FIG. 7. Molar teeth (occlusal views) of representative species of living species of *Akodon* (*Abrothrix*). Upper row, right lower molar series; lower row, left upper molar series of same individuals. **A,** *Akodon* (*Ab.*) *longipilis longipilis* (Waterhouse); male; BMNH 97.5.1.6; Valparaiso, Chile. A rather young specimen of the sample of 20 individuals, showing internal remnants of the mesoloph in M^1 and M^2 in the form of a persisting mesofosettus and a persisting metaflexid in the m_3. **B,** *Akodon* (*Ab.*) *sanborni* Osgood; male; MBUCV 1-2025; Mehuin, Valdivia, southern Chile. Rather young specimen, showing persisting mesofosettus on M^1. **C,** *Akodon* (*Ab.*) *longipilis longipilis* (Waterhouse); holotype, BMNH 55.12.24.177; Coquimbo, Chile. **D,** *Akodon* (*Ab.*) *illutea* Thomas; female; type specimen, BMNH 28.10.14.2; Aconquija, Tucumán, Argentina.

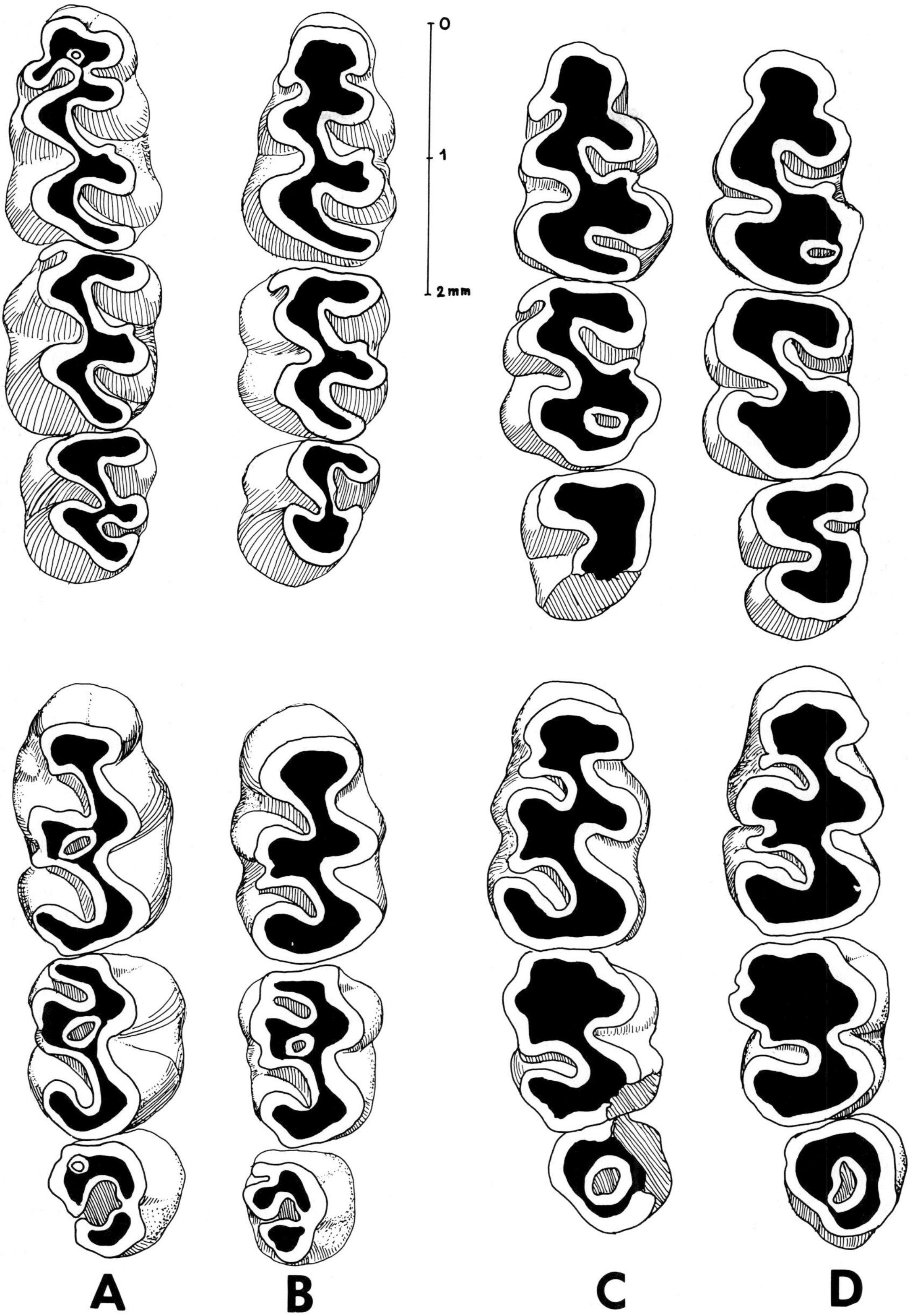
0
1
2 mm
A
B
C
D

TABLE 3. Statistics of the sample of *Akodon* (*Abrothrix*) *kermacki* Reig compared with a sample of *Akodon* (*Abrothrix*) *longipilis longipilis* from a living population at Valparaiso, Chile.

Variate	***Akodon* (*Ab.*) *kermacki*, Upper Pliocene and lowermost Pleistocene, SE Buenos Aires Prov. (N = 6)**			***Akodon* (*Ab.*) *longipilis*, BMNH sample from Valparaiso (N = 19)**		
	Range	$\bar{x}$	s	**Range**	$\bar{x}$	s
m_1–m_3 (alveolar) length	5.04–5.69	5.35	0.28	4.80–5.57	5.18	0.21
m_1–m_3 (coronal) length	4.86–5.63	5.19	0.30	4.73–5.31	5.02	0.19
m_1 length	1.89–2.29	2.02	0.14	1.95–2.23	2.09	0.08
m_1 width	1.30–1.58	1.38	0.10	1.27–1.39	1.32	0.04
m_2 length	1.46–1.74	1.60	0.10	1.43–1.62	1.53	0.05
m_2 width	1.30–1.55	1.38	0.10	1.21–1.39	1.30	0.05
m_3 length	1.39–1.59	1.50	0.08	1.27–1.49	1.37	0.05
Incisor depth	1.41–1.70	1.50	0.11	1.14–1.39	1.27	0.07
M^1–M^3 (alveolar) length	...	(5.37)	...	4.67–5.69	5.13	0.24
M^1–M^3 (coronal) length	...	(4.91)	...	4.35–5.18	4.79	0.20

the mesolophid remnant is also typical of other species of *Abrothrix.* As is also the case in other species of this subgenus, the ectolophid is absent in all the observed specimens, and the ectostylid is present in the m_1 in only one of the six specimens, but in none of them was there any evidence of it in the m_2. In the observed type specimens of living species of *Abrothrix,* there is no trace of ectostylid either on m_1 or m_2, and in the studied sample of *A. l. longipilis* from Valparaiso, an ectostylid was found in eight of 19 cases in the m_1, and in two of 19 cases in the m_2. However, no specimens showed any trace of an ectolophid. The ectolophid is usually present in the species of *Akodon s.s.* which are comparable in size to *kermacki.* Following our records, the presence of an ectolophid in the m_1 has a frequency of 92% in *Akodon tolimae* (N = 40), 83% in *Akodon urichi saturatus* (N = 48), 100% in *Akodon urichi venezuelensis* (N = 27), and 64% in *Akodon azarae* (N = 58). An ectolophid is completely absent, however, in some small species of *Akodon* as *A. iniscatus.*

In the m_2 the protoflexid is well defined, though it may be completely eroded by wear (cf. M-1071). It disappears earlier by wear than in the m_3. This tooth varies with wear from sigmoid-shaped to nearly eight-shaped. In one case (M-1071), the mesoflexid is completely obliterated by advanced wear. In size, the m_3 is relatively longer as regards the m_2 than in *A. l. longipilis* (fig. 7), and from what can be inferred in the type specimens, than in the other forms of the subgenus.

As regards metrical differences of *A. kermacki* in comparison with *A. l. longipilis* and large-sized species of *Akodon s.s.,* Figures 7 and 8 and Tables 1 through 4 show the corresponding data.

DISCUSSION—From the morphological characteristics of molar teeth and mandible, it seems mandatory to place *A. kermacki* in the subgenus *Abrothrix.* It matches all studied character-states of this subgenus, and the differences found are merely indicative of a clear-cut distinction at the species level. As regards the metrical analysis, the evidence is conclusive for statistical differences in some, but not in all, the studied variables, as expected in this kind of analysis. Metrically, *kermacki* is quite distinctive in the depth of the incisor, the relative shortness of the m_1, and the relatively longer m_3. These two differences are complementary, and the result is that the mean crown length of the molar row is not statistically different from living *A. l. longipilis.* This illustrates the care which must be taken in inferring systematic kinship or distance from overall statistical differences. Combining the results of the morphological studies and the metrical analysis, there can be little doubt that we are dealing with a distinctive species of *Akodon* (*Abrothrix*).

The fossil species is more similar in size to the living *A.* (*Ab.*) *longipilis* and *A.* (*Ab.*) *illutea* than to *A.* (*Ab.*) *sanborni, A.* (*Ab.*) *lanosus,* the dubious *A.* (*Ab.*) *mansoensis, A.* (*Ab.*) *xanthorhinus,* and *A.* (*Ab.*) *hershkovitzi* (for the assignment of the two last species to *Abrothrix,* see above). An examination of Table 2 clearly supports this assertion. The fossil species is slightly larger than the largest living subspecies of *A.* (*Ab.*) *longipilis,* namely, *A.* (*Ab.*) *l. longipilis* (tables 2–4), and it belongs, along with *A.* (*Ak.*) *varius simulator, A.* (*Ak.*) *urichi saturatus,* and *A.* (*Ab.*) *l. longipilis* (table 5, fig. 8), to the group of the largest species of *Akodon s.l.* Within the limits of this genus, its size is only surpassed by the new species described next.

In the larger size of the incisor and the correlated

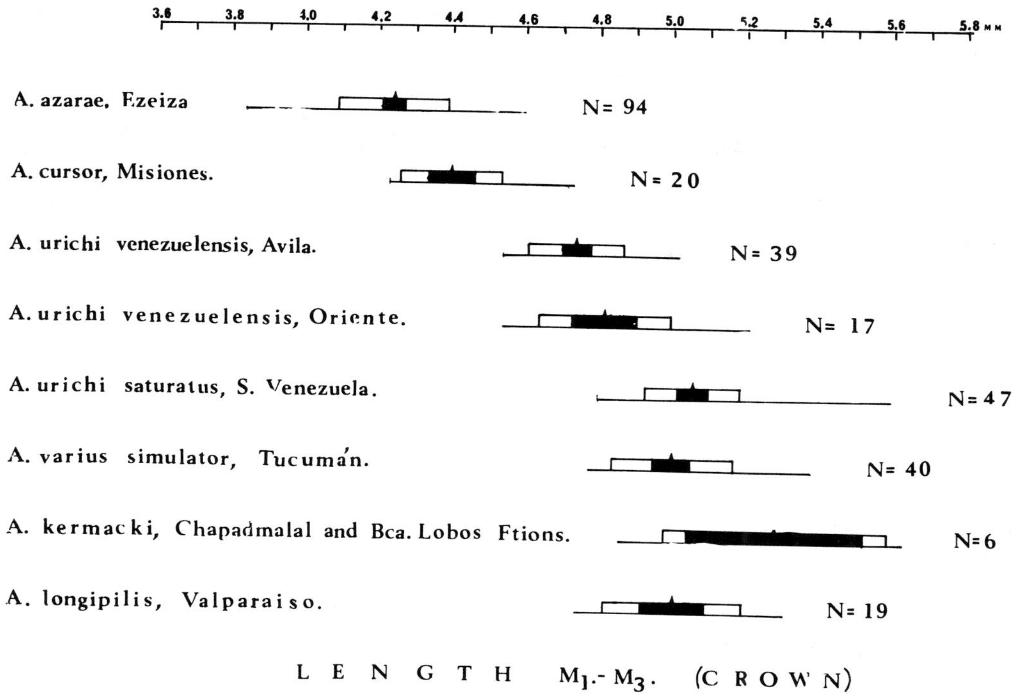

FIG. 8. Dice-grams of the variation in the length of the lower molar row (coronal) in various species of medium- and small-sized *Akodon* (*Akodon*) and *Akodon* (*Abrothrix*). The diagram shows the mean, the range, two standard errors to each side of the mean (black squares), and one standard deviation to each side of the mean (open squares + black squares).

greater development of the capsular projection of the base of the incisor, *kermacki* is more highly modified than the living species. If this is indicative of evolutionary divergence, *kermacki* could not be the ancestor of any living species of *Abrothrix*. This conclusion, however, is based on frail evidence, and more specimens and further study of other characters are necessary to evaluate its evolutionary significance.

In any case, *Abrothrix* in the Upper Pliocene and lowermost Pleistocene of SE Buenos Aires Province occurred almost 1,000 km eastward of the present distribution of this subgenus. Actually, the living representatives of *Abrothrix* are now limited to the lowlands and low valleys of central and southern Chile and the eastern Andean slopes of Mendoza and Patagonia, to Tierra del Fuego, plus the isolated *A. illutea* of Tucumán. The different distributions of living and fossil *Abrothrix* indicate a reduction of the range of this subgenus from a much more extended area to its present limits, a phenomenon which could have been caused by the climatic changes that occurred during the Pleistocene. The case of *Abrothrix* is not isolated, and other mammals presently restricted to Chilean or Andean distributions were also present in the Upper Pliocene of Buenos Aires Province. One is the fossil caviomorph *Pithanotomys*, which is hardly separable from the living *Aconaemys*, restricted now to a few isolated populations in southern and central Chile (Osgood, 1943) and high valleys of Mendoza and Neuquén, Argentina (Pearson, 1984), but abundant in the fossil de-

TABLE 4. Student's t test for differences between means in six selected variates of *Akodon* (*Abrothrix*) *longipilis* and *Akodon* (*Ab.*) *kermacki* (from data in tables 1 and 3).

Variate	***A. (Ab.) longipilis-A. (Ab.) kermacki* (df = 23)**	
	t	***P***
m_1–m_3 (alveolar) length	1.62	>0.05
m_1–m_3 (coronal) length	1.59	>0.05
m_1 length	1.36	>0.05
m_2 length	1.91	<0.05
m_3 length	3.94	<0.001
Incisor depth	5.41	<0.001

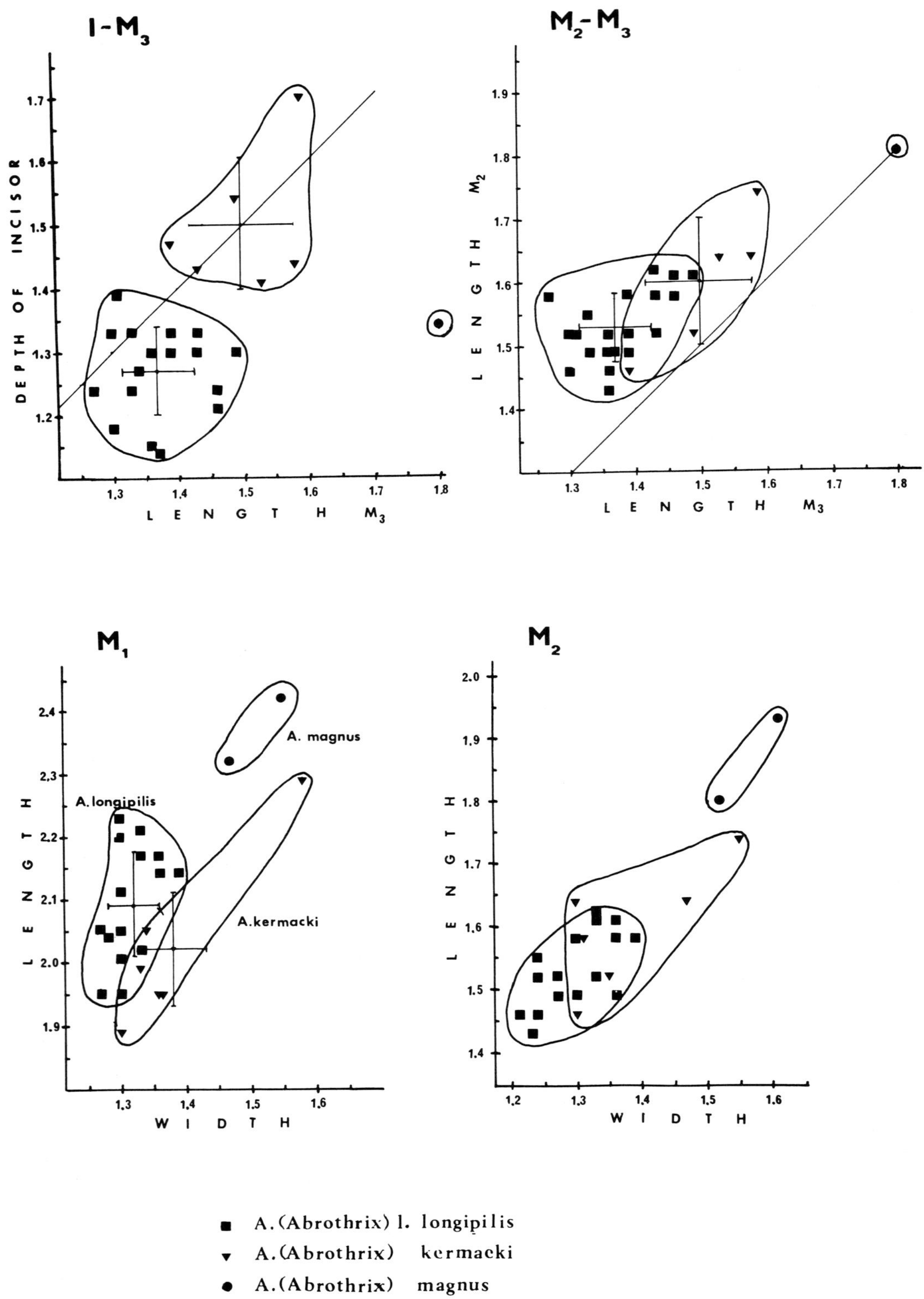

FIG. 9. Scattergrams of measurements of different teeth in living and fossil species of *Akodon* (*Abrothrix*).

posits of the Pliocene and Lower Pleistocene of Buenos Aires Province. The marsupial subfamily Caenolestinae, which is now restricted in southern South America to the south temperate forests of southern Chile and possibly Argentina, is also represented by the fossil genus *Pliolestes* in the Pliocene of Buenos Aires Province (Reig, 1955). Also the living genus *Abrocoma,* now restricted to the Andean region, had representatives in the Pliocene of Buenos Aires Province, and a specimen in the collection of the Museum of Mar del Plata (MMP-1059) from the Atlantic cliffs of the Chapadmalal region dates from the early Pleistocene.

Akodon (Abrothrix) magnus, n. sp.

HOLOTYPE—MMP M-551 (figs. 5D, G–H,; 6E): left lower jaw with incisor, m_1 and m_2, lacking m_3 and condyloid, coronoid and angular processes; right lower jaw with incisor and m_1, lacking m_2 and m_3, coronoid and condyloid processes; left femur broken in the middle of the shaft; right calcaneum (the postcranial bones are only tentatively associated with the mandibles). Found by G. J. Scaglia in Vorohué Formation (Kraglievich, 1952), at the Atlantic slopes of the Chapadmalal region near Baliza San Andrés, Partido de General Pueyrredón, SE of Buenos Aires Province, Argentina. These specimens were found associated with MMP M-869, the holotype of *Cholomys pearsoni* Reig, 1980, and MMP M-897 and M-868, the last two belonging to a new species of the subgenus *Akodon* described below.

HYPODIGM—The holotype and MMP S-407: right fragmentary lower jaw of an old individual, bearing the incisor and the three molar teeth, the latter greatly worn. Found by G. J. Scaglia in Vorohué Formation, at the sector of Atlantic slopes stretching from south of Arroyo Lobería, Chapadmalal region, Partido de General Pueyrredón, SE of Buenos Aires Province, Argentina.

DIAGNOSIS—A very large species of *Akodon* (*Abrothrix*), exceeding in size *A.* (*Ab.*) *kermacki*; mandible slender; incisor relatively much weaker; m_1 with a distinct metastylid.

KNOWN DISTRIBUTION—Vorohuean subage of the Uquian age (Lower Pleistocene), SE of Buenos Aires Province, Argentina (see Marshall et al., 1984).

DESCRIPTION—The description is based on the holotype, as specimen MMP S-407 is only tentatively included in the species.

The mandible is distinctly larger than that of *A. kermacki,* but more slender. The diastema is, however, a little shorter, its length being less than that of the two first molars. The horizontral ramus is shorter, with a height less than the diastema length, and its lower border bends slightly upward from the level of the anterior part of the m_2. The border of the ramus immediately in front of the m_1 is as in *A. kermacki,* but the upper border of the diastema is slightly more concave. The symphysis, while shorter, is lower and more slender. The lower masseteric crest is somewhat higher in position than in *A. kermacki.* The mental foramen is as in *kermacki.* The anterior border of the coronoid process is only partially preserved, but it is clear that it originates further posteriorly than in *kermacki,* at the level of the middle of the alveolus of m_3. The condyloid process is not preserved, but the remaining parts of the ascending ramus show a great deal of the sigmoid notch and suggest that the condyle was rather low and well projected backward. The capsular projection has broken walls in the two rami of the holotype, and it is slightly less pronounced than in *kermacki,* the root of the incisor lying further forward, between the coronoid process and the beginning of the sigmoid notch.

The incisor is much weaker than in *kermacki,* and its proportions are as normal in normal *A. longipilis.* Its absolute size in depth is less than in *kermacki,* even when *magnus* shows greater values for all the remaining measurements of the dentition.

The m_1 and m_2 are very similar to those of *kermacki,* the main distinction being absolute size. However, the m_1 clearly shows a metastylid, a character which has not been observed in any other specimen of *Abrothrix* examined by me, and which may be considered diagnostic. However, a larger sample is necessary to evaluate the constancy of this character-state. As is typical of *kermacki* and other *Abrothrix,* the mesolophid remnant is weak and grows out from the middle of the entolophid in the m_1, and the entoconid makes a noticeable bulge on the lingual face of the tooth. In the m_2 the mesolophid remnant is rudimentary, but the protoflexid is better marked than in specimens of *kermacki.* The roots of the m_3 show that this tooth was large, probably relatively larger than in *kermacki.* This is confirmed by the very worn m_3 of specimen MMP S-407.

DISCUSSION—Although the main distinction of *A. magnus* from *A. kermacki* is one of size, the slender mandible and weaker incisor confirm that they are different species. The difference in abso-

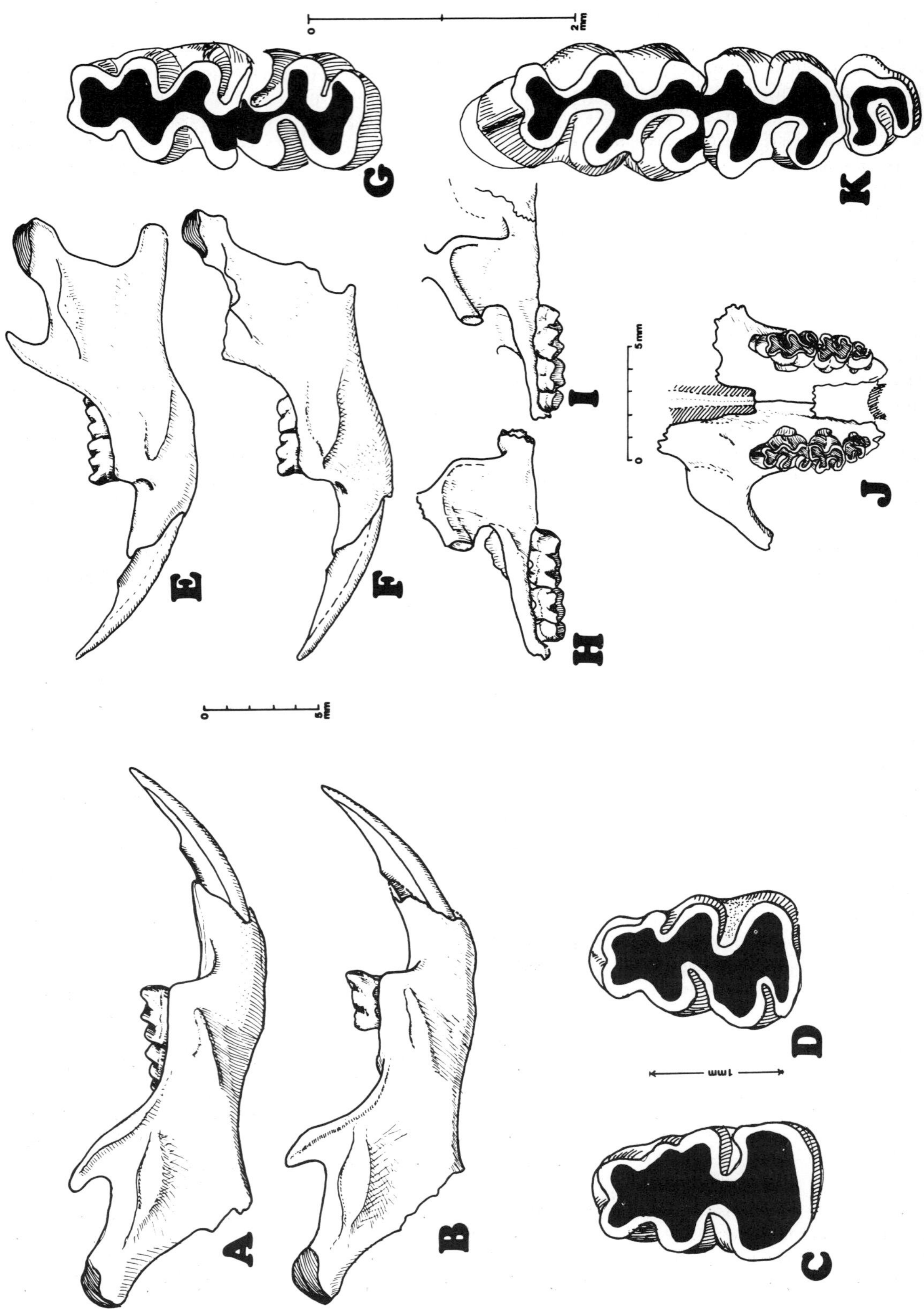
A
B
C
D
E
F
G
H
I
J
K
0
5 mm
0
2 mm
1 mm
0
5 mm

lute size is obvious at first sight (figs. 21–22), but one is tempted to wonder whether this specimen might not be an extremely large individual within the size range of *A. kermacki* (cf. fig. 21). One specimen of *A. kermacki* (MMP M-1067) from the Chapadmalal Formation has a first molar nearly as long as that of specimen MMP S-407, attributed to *A. magnus*. The Chapadmalal specimen, however, has a much shorter m_2, and agrees with *kermacki* in the mandible and relative size of the incisor (table 1). Another specimen of *kermacki* from the Chapadmalal Formation (MMP M-1154) has an m_2 which approaches the size of the m_2 in specimen MMP S-407, but its m_1 is much shorter and it also agrees with *kermacki* in incisor and mandible characters (table 1). In any case, there is no overlap in absolute size, and the specific distinction seems to be validated by the sum of all studied characters.

It could be alleged, however, that the transitions in size between organs in specimens of the two species might be a reflection of a real transition between the two taxa, and that a process of speciation by gradual transformation is involved here. Such alleged cases of phyletic speciation have been described in the echimyid rodent *Eumysops* from the same sequence (Kraglievich, 1965) and surmised in the case of the didelphoids *Thylatheridium* (Reig, 1959) and *Sparassocynus* (Reig & Simpson, 1972), also from the same sediments. Although this possible interpretation can only be substantiated by more material, I believe that the greater development of the incisor in *kermacki* precludes the possibility that this species is the direct ancestor of *magnus*. Moreover, the single specimen of *kermacki* known from the intermediate Barranca Lobos Formation (MMP M-1071), does not show intermediate character-states, but is closer to the holotype of *A. kermacki* in size than the other specimens referred to the same species. In any case, even if it were demonstrated that there is a direct phyletic link between the two species, there is still enough evidence to maintain *A. magnus* as a well-distinguished species; its size differences from *kermacki* are not compatible with the known range of size variation in species of the subgenus *Abrothrix* (figs. 8–9).

Akodon (Akodon) johannis, n. sp.

HOLOTYPE—MMP M-742 (fig. 10F,H,J): right lower jaw with incisor and m_1–m_2, lacking the coronoid and angular processes and the m_3; left lower jaw with m_1–m_2, with incisor broken, and lacking m_3 and the same processes; left maxilla with M^1–M^3; portion of right maxilla with M^1–M^3; the two tibiae, the right incomplete; incomplete right and left femora; right humerus and cubitus; portions of scapula and of left pelvis; two vertebrae. Found by G. J. Scaglia in Miramar Formation (Kraglievich, 1952) at the Atlantic slopes south of "Bajada de San Andrés," Chapadmalal region, Partido de General Pueyrredón, SE of Buenos Aires Province, Argentina.

ETYMOLOGY—The species name, *johannis,* is given for Juan Brkljacic, a close collaborator of G. J. Scaglia in the latter's work at the Museum of Mar del Plata. Brkljacic has been responsible for a great deal of progress by that institution in the recent past.

HYPODIGM—The holotype is the only known specimen.

DIAGNOSIS—A small species of *Akodon* (*Akodon*) the size of *A. andinus* or *A. nigrita*; moderately strong mandible with a low symphysis, a relatively deep incisor, and a fairly well-developed capsular projection. Incisive foramina almost level

---

Opposite Page:

FIG. 10. Lower jaws, maxillae, and molar teeth of fossil and living species of small *Akodon* (*Akodon*). **A,** Later view of left lower jaw of living *A.* (*Ak.*) *cursor montensis* Thomas; female; BMNH 1874; Puerto Gisela, Misiones, Argentina. **B,** Lateral view of left lower jaw of *A.* (*Ak.*) cf. *cursor* (Winge); MLP 66.VII.27.95 (a); Miramar Formation (Ensenadan Stage), vicinity of Camet, Partido de Mar Chiquita, SE of Buenos Aires Province, Argentina (Middle Pleistocene). **C,** Crown view of left m_1 of *A.* (*Ak.*) cf. *cursor* (Winge); MLP 66.VII.27.95. **D,** Crown view of left m_1 of living *A.* (*Ak.*) *cursor montensis* Thomas; type specimen, BMNH 4.1.5.3.36; Sapucay, Paraguay. **E,** Lateral view of right lower jaw of living *A. nigrita* Lichtenstein; male; BMNH 3.7.1.74; Roça Nova, Paraná, Brazil. **F,** Lateral view of right lower jaw of *A.* (*Ak.*) *johannis* n. sp.; holotype, MMP M-742; Miramar Formation (Ensenadan stage), Chapadmalal region, Partido de General Pueyrredón, Buenos Aires Province, Argentina (Middle Pleistocene). **G,** Crown view of right lower m_1 and m_2 of *A.* (*Ak.*) *johannis* n. sp.; holotype, MMP M-742. **H,** Lateral view of left maxilla of *A.* (*Ak.*) *johannis* n. sp.; holotype, MMP M-742. **I,** Lateral view of left maxilla of *A.* (*Ak.*) *nigrita,* BMNH 3.7.1.74. **J,** Palatal view of left and right maxillae of *A.* (*Ak.*) *johannis* n. sp.; holotype, MMP M-742. **K,** Crown view of upper molar series of *A.* (*Ak.*) *johannis* n. sp.; holotype, MMP M-742.

TABLE 5. Statistics of the coronal length of m_1–m_3 in 10 samples of 10 species and subspecies of the genus *Akodon.*

Species, subspecies, and locality	Statistic				
	N	Range	$\bar{x}$	SD	SE
A. puer (Peru and Bolivia)	14	3.58–4.16	3.76	0.176	0.040
A. iniscatus (Chubut)	20	3.39–4.22	3.97	0.186	0.041
A. azarae (Ezeiza, Buenos Aires)	94	3.84–4.60	4.24	0.154	0.014
A. cursor (Misiones, Argentina)	20	4.23–4.73	4.40	0.138	0.031
A. urichi venezuelensis (Avila)	39	4.54–5.02	4.74	0.128	0.019
A. urichi venezuelensis (Oriente)	17	4.54–5.21	4.82	0.179	0.043
A. urichi saturatus (Tepuyes)	47	4.80–5.50	5.06	0.135	0.019
A. varius simulator (Tucumán)	40	4.38–5.39	5.01	0.167	0.023
A. l. longipilis (Valparaiso)	19	4.73–5.31	5.01	0.186	0.042
A. kermacki (Pliocene, Buenos Aires)	6	4.86–5.63	5.19	0.299	0.118

$\bar{x}$ = arithmetic mean; SD = standard deviation; SE = standard error.

with the anterior borders of M^1. Zygomatic plate strong, wider than the length of the M^1, with a rounded and slightly forward-projecting anterior border. Molars relatively broad. M^1 with a wide and oblique procingulum showing a moderately developed anteromedian flexus, but without anteroflexus. Lower molars without mesolophid remnants and mesostylids; ectolophids and ectostylids also absent.

KNOWN DISTRIBUTION—Ensenadan age, Middle Pleistocene of SE of Buenos Aires Province, Argentina.

DESCRIPTION—Of the skull, only the maxilla and the middle palatal region can be studied. The palate is long and wide; the space between the internal borders of the crowns of the M^1 is greater than the length of the M^1. There is no direct evidence of the position of the posterior border of the palate, but the maxillary bone surrounding the M^3 clearly indicates that the border was slightly behind the posterior border of those molars. The posterior limits of the incisive foramina (fig. 10J) are clearly indicated in the two portions of maxillae. The foramina scarcely surpass the anterior border of the M^1, and they are even less expanded behind than in *Akodon nigrita,* a living species with rather short incisive foramina. In fact, the posterior position of these foramina resembles that in *Notiomys* and *Microxus* (in which they scarcely surpass the anterior border of the M^1) more than the usual condition in *Akodon s.s.,* in which they usually reach the level of the protocone of the M^1. In the zygomatic plate, however, *johannis* stands quite apart from *Notiomys* and *Microxus,* and shows an unusually strong and wide plate, with an anteroposterior length greater than the length of the M^1, as is also the case in *A. nigrita* and *A. andinus.* In most other species of small *Akodon,* the length of the M^1 either exceeds the anteroposterior diameter of the plate, as is the case in *A. puer, A. boliviensis,* and *A. azarae,* or the two measurements are roughly equivalent, as is the case in *A. iniscatus.* The anterior border of the zygomatic plate is quite upright, and it slightly projects forward at its rounded upper corner, and it is not sharply cut off above, the upper corner being very slightly turned into the anterior border.

The mandible is also characterized by its low symphysis and the very procumbent incisor. This is reflected in the anterior median point of the diastema, which is well below the level of the alveolar row, even more so than in *A. cursor* (fig. 10A), a species with a particularly low symphysis. In *A.* (*Deltamys*) *kempi* and in *A. nigrita* (fig. 10E) and *A. andinus* (fig. 11D), the symphysis is also low, but less markedly so than that in *A. johannis*; in most of the other species of *Akodon s.l.,* the symphysis is more upturned and, consequently, the incisor is less procumbent. The mandibular ramus is relatively deep: although the depth of the ramus at the m_1 is less than the diastema length, the ramus is higher than in species of similar size, such as *A. nigrita, A. iniscatus,* and *A. puer,* and it is longer than the combined m_1–m_2 length. The lower masseteric crest is high and moderately marked, more clearly so than the upper masseteric crest, and the two crests reach forward nearly to the anterior border of the m_1. The tip of the coronoid process is broken in the two mandibles, but its anterior border is partially preserved, and it slopes backward somewhat abruptly. The condyle is well posterior and fairly high in position, and the capsular projection is well developed, as compared to usual *Akodon s.s.*

The incisor is comparatively strong, markedly more so than in *nigrita, andinus,* and *puer,* and it

TABLE 6. Measurements of lower jaw and lower molars of fossil and living specimens of *Akodon* (*Akodon*) *cursor.*

	***Akodon* cf. *cursor,* Miramar Formation**			***Akodon c. montensis***				
						Sample of *Akodon c. montensis* in BMNH from Puerto Gisella, Misiones, Argentina		
Variate	**MLP 62-VII-27-95 (a)**	**MLP 62-VII-27-95 (b)**	**MLP 62-VII-27-95 (c)**	**Holotype, BMNH 4.1.5.35**	**Female, BMNH 66-1874, Misiones, Argentina**	**Range**	**$\bar{x}$**	**s**
Mandibular condyle-anterior border diastema length	14.20	13.60	...	14.12	14.41	...	...	...
Condyle-anterior border m_1 length	11.39	11.39	...	...	10.99	...	...	...
Diastema length	3.20	2.94	3.32	...	3.46	...	...	...
Condyle-posterior border alveolar m_3 length	6.46	5.82	...	...	6.21	...	...	...
Mandible depth at m_1	3.26	3.26	3.27	...	3.07	2.56–3.64	3.01	0.36
m_1–m_3 (alveolar) length	4.67	4.74	4.54	4.55	4.61	4.35–4.93	4.65	0.18
m_1 length	1.86	...	1.80	1.77	1.71	1.75–1.90	1.82	0.04
m_1 width	1.18	...	1.02	0.99	1.08	1.02–1.24	1.11	0.06
m_2 length	...	...	1.39	1.43	1.37	1.33–1.49	1.40	0.05
m_2 width	...	...	1.03	0.99	1.05	0.99–1.22	1.09	0.07
m_3 length	...	...	...	1.24	1.24	1.22–1.49	1.31	0.08
m_3 width	...	...	...	0.93	0.99	0.87–0.99	0.94	0.04
Lower incisor depth	1.19	1.05	...	1.08	1.04	...	...	...

is comparable in relative depth to the incisor of *iniscatus* (fig. 11A). As already indicated, the incisor is characteristically procumbent. The molar teeth (fig. 10G,K) are broad and rather short, and they look similar to, but are slightly more heavily built than, the molars of *nigrita.* They differ from molars of that species, however, in the lack of any indication of remnants of mesolophids, ectolophids, and ectostylids in the m_1 and m_2; the upper molars of the two species are more closely comparable in morphology and proportions and in the advanced reduction of the m_3. Both in the upper and in the lower series, the posterior border of the first molars is partially cut off, as is, although less markedly, the anterior border of the second molars. This is probably an individual anomaly.

DISCUSSION—*Akodon johannis* appears to be clearly distinctive from the living small-sized species of *Akodon s.l.* It is obviously distinct from the contemporary *A.* cf. *cursor,* which is described next. Among living species, it seems to be more closely related to *A. nigrita* than to any other compared species. Doubtless it should be allocated in the subgenus *Akodon s.s.,* and it probably represents an extinct lineage among the extensive diversification of the subgenus. The dubious *Necromys conifer* Ameghino, which Ameghino (1889) mentions as being represented in the coetaneous Ensenadan stage of the north of Buenos Aires Province, does not seem to have anything in common with *A. johannis.* Although the illustrations and the description are obscure, the drawings given by Ameghino (1889, Atlas, table IV, figs. 17–18) show a mandible with an upturned symphysis and a nonprocumbent incisor. Hershkovitz (1962) considers *Necromys* a mere synonym of *Calomys,* a contention that does not seem warranted by Ameghinos's data, even given the faulty nature of the illustrations.

Akodon (Akodon) cf. cursor (Winge)

REFERRED SPECIMENS—MLP 62.VII.27.95 (a) (fig. 10B–C): left lower with incisor and m_1 of an old individual; MLP 62.VII.27.95 (b): right lower jaw with the incisor and the alveoli of the molar teeth; MLP 62.VII.27.95 (c): fragment of right lower jaw with extremely worn m_1 and m_2. These three specimens were found in association with each other and with remains of *Reithrodon auritus, Nectomys squamipes,* and *Ctenomys* sp. in a bone conglomerate probably representing fossil owl pellets. The bone conglomerate was extracted from a rocky block from the Atlantic cliffs 5 km N of Colonia Camet (about 15 km N of the city of Mar del Plata), Partido de Mar Chiquita, SE of Buenos Aires Province, Argentina. The cliffs at this point expose sediments of the Miramar Formation (Kraglievich, 1952), and this geological provenance can be ascertained for the fossils found in the fallen block.

DESCRIPTION—The morphology and measurements (table 6) of the mandibles indicate that these represent *Akodon* closely allied, if not identical, to the living species *A.* (*Ak.*) *cursor* (Winge). As in the latter species, the mandible is elongate and slender relative to other species of intermediate-sized *Akodon.* The symphysis is elongated and low, the anterior median point of the diastema being at a level below that of the alveolar row. The length of the diastema is as long as m_1 and m_2, and the depth of the ramus below m_1 is slightly greater than the diastema length. The lower border of the ramus is gently concave behind the m_1. The lower masseteric ridge is relatively well marked and rather high in position, the upper masseteric ridge being scarcely noticeable and somewhat parallel to the alveolar border. The coronoid process is low, and its anterior border slopes upward very gently. The condyloid process is also low and projected well backward. There is a fairly well-developed capsular projection of the base of the incisor, slightly stronger than usual in living specimens of *cursor* examined, but the difference is not really very marked. The incisor is, as in *cursor,* well developed, and its depth is a little greater than that of the holotype and other specimens examined of *A. cursor montensis.*

The molar teeth are too worn to show many details of structure. However, the m_1 of specimen (a) is slightly less worn (fig. 10C) and shows a clear indication of an anteromedian flexid and an overall shape and development of the procingulum which matches perfectly with the procingulum of *cursor.* A well-marked anteroflexid is present in the holotype of *A. cursor montensis* (fig. 10D) and in each of 15 animals from Puerto Gisela, Misiones, in the British Museum referred to at subspecies. The shape of the enamel walls at the hypoflexid suggests that an ectostylid was present. This element is absent in the holotype of *A. c. montensis,* but is present in 7 of 14 specimens of the above-mentioned sample. From the enamel wall of the mesoflexid, no mesolophid remnant is evident in this fossil specimen. This structure is present in 80% of the modern sample, but it is almost completely absent in the type of *A. c. montensis.* In length and width, the m_1 falls within the limits of variation in a sample of living *cursor* (table 6), and there is also a complete correspondence in the length of the molar series between the fossil specimens and the living sample.

DISCUSSION—Apart from *Akodon cursor,* only *A. azarae,* now living in the same locality where the fossils were found (Reig, 1964), is a plausible relative. *Akodon azarae* agrees with the fossil specimens in being a medium-sized species with a rather elongated mandible. It is, however, significantly smaller than the fossil specimens (fig. 12) and has a stronger mandible. A sample of 55 *A. azarae* from Ezeiza, close to Buenos Aires, now in the Museum of Mar del Plata, shows an alveolar length of the lower series significantly smaller ($P < 0.001$) than the studied sample of *A. cursor* from Puerto Gisela (fig. 8). The alveolar molar length of the three fossil specimens here described have the same mean value as the *A. cursor* sample. Moreover, *azarae* is characterized by narrower molar teeth, as it is evident for the m_1 in the diagram of Figure 12. Thus, a close relationship of the fossil specimens with *azarae* must be ruled out. The morphological resemblance and the agreement in size with *cursor* is such that the fossils from Camet represent a form probably conspecific with living *cursor.*

Akodon cursor was first described by Winge (1887) from living and subfossil specimens from Lagoa Santa, Minas Gerais, Brazil, as a member of *Habrothrix* (Thomas [1884] had placed under *Habrothrix,* a misspelling of *Abrothrix* Waterhouse, all *Akodon*-like mice from South America). Afterward Thomas (1902) placed *cursor* in *Akodon,* and he later (1913) erected *A. arviculoides montensis,* which he compared with *cursor.* As already discussed (pages 354, 356) *arviculoides* Wagner is not an *Akodon,* but a synonym of *Bolomys lasiurus* (Lund); *montensis* Thomas is most likely a subspecies of *cursor* Winge, as proposed by Ximenez and Langguth (1970). Thus, two subspecies of *cursor* may be recognized: *A. c. cursor* Winge, which following Vieira (1955), extends over Minas Gerais, Espírito Santo, Guanabara (Rio de Janeiro), São Paulo, and Paraná; and *A. c. montensis,* which is known from Paraguay (Thomas, 1913), Misiones (Massoia & Fornes, 1962), and central Uruguay (Ximenez & Langguth, 1970). However, these subspecies are not well defined, and our fossil sample is too small and fragmentary to attempt a comparison with living subspecies or to place it in a subspecies of its own. The present evidence indicates only that they represent *Akodon* cf. *cursor* and that this species extended its range in Middle Pleistocene times at least 600 km south of the known limits of its present distribution.

Akodon cf. *cursor* from the Miramar Formation cannot be confused with *Akodon johannis,* found also in the same strata. The latter is much smaller, has a relatively stronger mandible and a relatively deeper incisor, a more upturned and higher coronoid process, and a much lower symphysis. Its m_1 is proportionally much wider, and its procingulum lacks a well-defined anteromedian flexid. The differences between these two species are as great as the differences between living *A. cursor* and *A. nigrita* in areas of sympatry.

Akodon (Akodon) lorenzinii, n. sp.

HOLOTYPE—MMP M-1081 (figs. 11I, 13H): the two lower jaws with incisors and molar teeth. Found by Mr. S. Lorenzini in the Atlantic slopes 5 km N of the city of Miramar customarily known as "Barranca Parodi," Partido de General Alvarado, SE of Buenos Aires Province, Argentina. The fossils were found in strata of San Andrés Formation (Kraglievich, 1952; see also Marshall et al., 1984), as confirmed in the field by G. J. Scaglia and the late J. Zetti.

HYPODIGM—The holotype; and MMP M-867 (figs. 11G; 13G): incomplete right maxilla with m_1, broken m_2, and m_3, and partially broken zygomatic plate. Found by G. J. Scaglia in the Vorohué Formation (Kraglievich, 1952; Marshall et al., 1984) at the Atlantic slopes of the Chapadmalal region, close to "Baliza San Andrés," Partido de General Pueyrredón, SE of Buenos Aires Province, Argentina. Found in association with next specimen and with MMP M-551, holotype of *Akodon (Abrothrix) magnus* (see above), and MMP M-869, holotype of *Cholomys pearsoni* Reig (see Reig, 1980).

MMP M-868 (fig. 11J): left lower jaw with incisor and all molar teeth, partially lacking coronoid, condyloid, and angular processes. Found in association with MMP M-867. M-867 and M-868 may belong to the same individual, but the lower molar teeth look more worn than the upper ones. Therefore, I prefer to treat them as belonging to two different individuals.

MLP 52-X-4-44 (a) (fig. 11M): most of right lower jaw broken in front of the middle of the symphysis, and at the posterior processes, with a broken incisor and all molar teeth, the second molar partially broken; fragment of right maxilla including the first molar and the posterior half of the zygomatic plate; left femur; portion of left tibia; right upper incisor. Found by the late J. Frenguelli in association with MLP 52-X-4-44 (b), a specimen referred to *Scapteromys* sp., in beds of San Andrés Formation (= "Prebelgranense" in Frenguelli's stratigraphic nomenclature), in the Atlantic slopes extending S of Punta Hermengo, Miramar, Partido de General Alvarado, SE of Buenos Aires Province, Argentina.

ETYMOLOGY—The species is named for Mr. Silvio Lorenzini, discoverer of the type specimen and other remarkable fossil cricetids, and an active collaborator of the Museum of Mar del Plata.

DIAGNOSIS—A small species of *Akodon* the size of *A. puer.* Mandible relatively short and high. Zygomatic plate moderately wide. Incisive foramina well behind the anterior border of first molar, but not reaching their protocones. M^1 with a well-marked anteromedian flexus, a shallow anteroflexus, and a projecting, narrow mesoloph remnant united to a mesostyle. Lower molars relatively narrow, without indication of mesolophid remnants, mesostylids, or ectolophids, and with somewhat oblique entolophids and posterolophids. Procingulum of m_1 narrow, with a shallow anteromedian flexid and a well-marked metaflexid. m_3 relatively small.

KNOWN DISTRIBUTION—Vorohuean and San Andresian subages, Lower Pleistocene (see Marshall et al., 1984), SE of Buenos Aires Province, Argentina.

DESCRIPTION—The skull fragments of specimens MMP M-867 and MLP 52-X-4-44 (a) afford only a few indications of the maxillary and palatal region. They show a rather robust zygomatic plate, probably wider than in *A. iniscatus* and as wide as in *A. puer.* The anterior border of the zygomatic plate is not preserved in either specimen, but its lower limit can be observed in MMP M-867, suggesting that it is slightly wider than the length of the M^1. The incisive foramina (fig. 11G) are expanded backward far more than in *johannis,* and they extend to the middle of the protoflexid of the M^1, not reaching to the protocone. In this respect, *lorenzinii* resembles *iniscatus* and *andinus* more than *puer,* in which the incisive foramina extend slightly behind the protocone of the M^1 in all 20 individuals I examined from different localities.

The mandible (fig. 11I–J) is much shorter than in *puer, iniscatus, andinus,* or *johannis.* It differs markedly from the slender and elongated mandible of *puer* (fig. 11C), and in proportions it is closer to the mandible of *iniscatus,* although noticeably smaller. The symphysis is fairly upturned, as the middle anterior point of the diastema reaches the level of the molar alveolar rows, differing from *puer,* and obviously from *johannis,* to approach more closely the condition found in *iniscatus.* The masseteric crests are located midway up the sides of the ramus, as in *iniscatus,* whereas in *puer* and *andinus,* they are placed higher. They

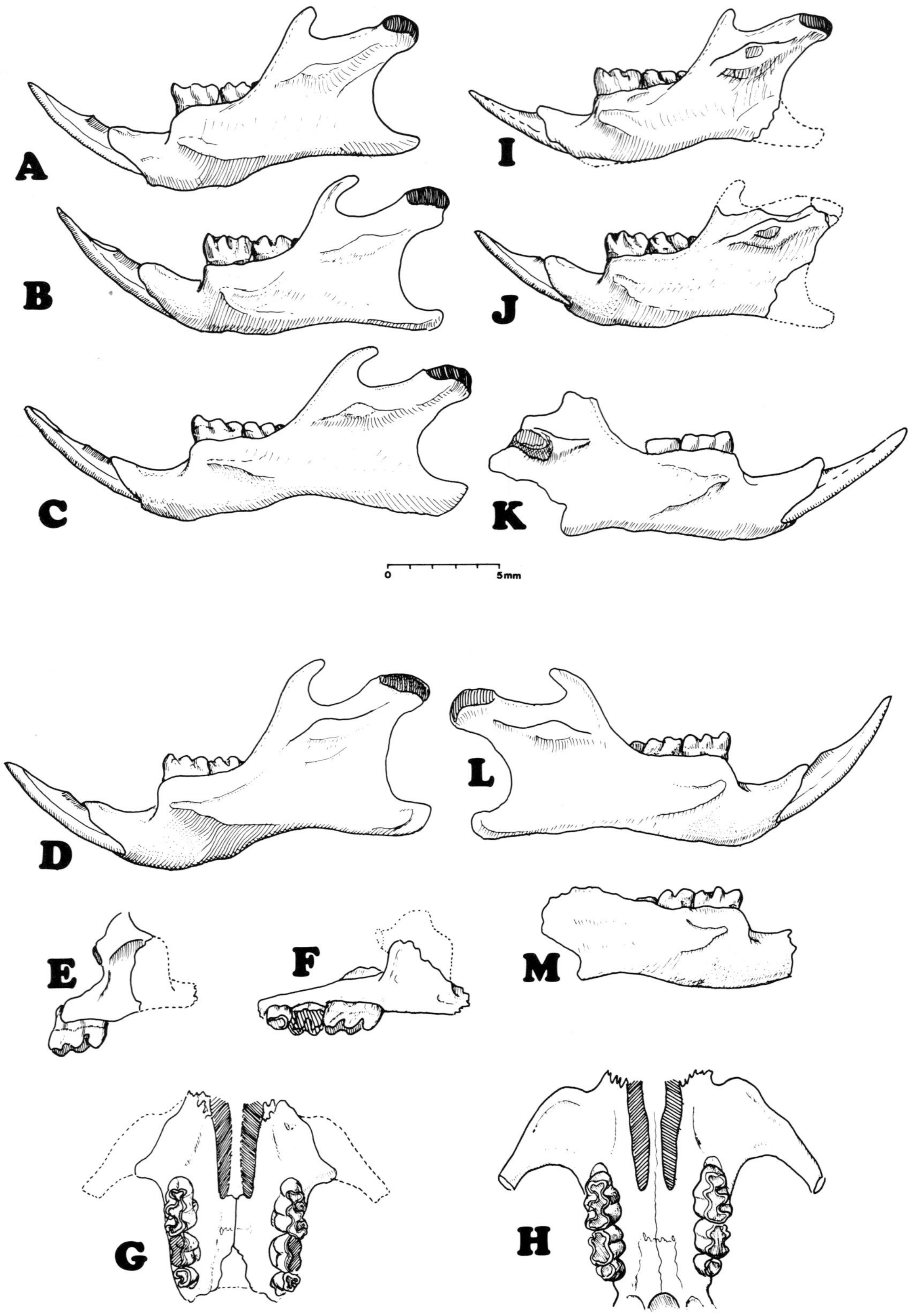

A
B
C
I
J
K
0
5mm
D
L
E
F
M
G
H

are less marked than in *iniscatus,* and the lower one is smooth, although stronger than the upper one. The depth of the mandible at the m_1 is greater than the diastema length, but it is less than the combined length of m_1–m_2. In *iniscatus,* the depth of the mandible is less than both the diastema length and the combined length of the first two molars, whereas in *puer* the diastema is longer than the depth of the mandible. The coronoid process slopes rather abruptly backward, more or less as in *iniscatus,* and more so than in *puer.* The process itself is short and low, so that the condyle is at a level higher than the tip of the coronoid process. The condyloid process is high, and it is not markedly projected backward. The capsular projection is moderately developed, but it is stronger than in *puer* and *andinus,* even a little stronger than in *iniscatus.*

The incisor is comparatively deep, clearly more so than in *puer,* and it is also slightly deeper than in *iniscatus.* Its depth equals or exceeds the length of the m_3 (fig. 14).

The upper molars of MMP M-867 show little wear, while the M^1 of MLP 52-X-4-44 (a) is moderately worn. They are very similar in morphology to upper molars of both *puer* and *iniscatus.* The only significant difference lies in the mesoloph remnant which unites to the mesostyle in the two available specimens and is projected further laterally than in *puer* or *iniscatus.* In this respect, *lorenzinii* resembles more closely *A. andinus,* but differs from it in the stronger procingulum of M^1, which has a well-marked anteromedian flexus and a less projecting parastyle. The M^2 is broken in its external half in the only specimens that show this organ (MMP M-867), but it is evident that it was relatively narrow, as in *puer* and *iniscatus.*

The lower molars are more distinctive in showing a little marked anteromedian flexid and a rather narrow procingulum in the m_1, without indication of a protostylid. There is, however, a well-developed anterolabial cingulum, but it does not contribute to the shape of the crown enamel pattern of the procingulum, as in other species. The metaflexid is also more re-entrant than in *puer* and *iniscatus,* but as in them, there is no trace of a mesolophid remnant or a mesostylid, and the simple entolophid is rather oblique in position, especially in the m_1. The posterolophid is even more oblique, so that the posteroflexid is noticeably wide. A protoflexid is well marked in the three lower molars, and the m_3, as in *puer,* is relatively small and has a sigmoid shape. No trace of ectolophid is shown in any of the three lower molars, but a tiny ectostylid is observed in the m_2 of the type specimen.

DISCUSSION—*Akodon lorenzinii* is a very small species of *Akodon* showing a distinctive combination of characters. It seems to be more closely related to *A. puer* and *A. iniscatus* than to any other species of the subgenus *Akodon,* and the balance of similarities would favor a closer relationship with *iniscatus.* This is also expected on biogeographic grounds. In fact, *puer* is a widespread species, but is restricted to the Andean and pampean mountains from Peru to Tucumán in Argentina. Thomas (1902) originally described *puer* from specimens of Chaquecamata in west-central Bolivia, and he subsequently identified as *puer* specimens I examined from south and central Peru.

Opposite Page:

FIG. 11. Lower jaws and maxillae of *Akodon* (*Akodon*) *lorenzinii* n. sp., and *Akodon* (*Akodon*) cf. *iniscatus* Thomas and of related living *Akodon* species. **A,** External view of right mandible of *A. iniscatus* Thomas; female; holotype, BMNH 3.7.9.64; Valle del Lago Blanco, Chubut, Argentina. **B,** External view of right mandible of *A. albiventer* (Thomas); male; BMNH 21.11.1.51; Sierra de Zenta, Jujuy, Argentina. **C,** External view of right mandible of *A. puer* Thomas; female; holotype, BMNH 2.1.1.78; Choquecamate, Bolivia. **D,** External view of right lower mandible of *A. andinus* (Philippi); female; holotype of *A. gossei* Thomas, BMNH 98.3.21.5; Puente del Inca, Mendoza, Argentina. **E,** External view of fragment of left maxilla with M^1 of *A. lorenzinii* n. sp.; MLP 52.X.4.44 (a); San Andrés Formation, Miramar, Buenos Aires Province, Argentina (San Andresian, uppermost Lower Pleistocene). **F,** External view of incomplete left maxilla with molar teeth of *A. lorenzinii* n. sp.; MMP M-867; Vorohué Formation, Chapadmalal region, Partido de General Pueyrredón, SE of Buenos Aires Province, Argentina/Vorohuéan (Lower Pleistocene). **G,** Reconstructed palatal view of *A. lorenzinii* n. sp., based on MMP M-867 (the right half is an inverted drawing of the original left half). **H,** Palatal view of the skull of *A. iniscatus* Thomas; male; BMNH 13.11.1.5; Pampa Central, Argentina. **I,** External view of left mandible of *A. lorenzinii* n. sp.; type specimen, MMP M-1081; San Andrés Formation, Barranca Parodi, Miramar, Buenos Aires Province, Argentina; San Andresian (Lower Pleistocene). **J,** External view of left mandible of *A. lorenzinii* n. sp.; MMP M-868; found in association with MMP M-867. **K,** External view of right mandible of *A.* cf. *iniscatus*: MMP S-640; Vorohué Formation, south of Arroyo Lobería, Chapadmalal region. **L,** External view of right mandible of *a. iniscatus* Thomas; male; BMNH 13.11.1.5. **M,** External view of incomplete mandible of *A. lorenzinii* n. sp.; MLP 52.X.4.44 (a).

TABLE 7. Measurements of *Akodon lorenzinii* n. sp., *Akodon johannis* n. sp., *Akodon* cf. *iniscatus*, and selected specimens of other small-sized living forms of the subgenus *Akodon*.

Variate	***Akodon lorenzinii*** Holotype, MMP M-1081	***Akodon lorenzinii*** MMP M-868	***Akodon lorenzinii*** MMP M-867	***Akodon lorenzinii*** MLP 52-X-4-44 (a)	***Akodon johannis***, type, MMP M-742	***Akodon* cf. *iniscatus***, MMP S-640	***Akodon iniscatus***, male, BMNH 13-11-1-5	***Akodon nigrita***, male, BMNH 3.7.1.74	***Akodon puer***, type, female, BMNH 2.1.1.78	***Akodon andinus*** (holotype of *A. gossei*), BMNH 98.3.21.5	***Akodon iniscatus***, type, BMNH 3.7.9.64	***Akodon puer*** (holotype of *A. caenosus*), BMNH 18.1.1.38
Mandibular condyle-diastema length	10.85	11.39	...	...	...	...	13.27	13.00	12.80	12.73	13.01	13.27
Mandibular condyle-m_1 length	8.90	...	...	...	...	...	10.05	9.91	9.91	9.65	...	9.78
Lower diastema length	2.37	2.49	...	...	3.26	3.26	3.07	3.00	2.75	2.94	...	2.94
Mandible depth at m_1	2.59	2.68	...	2.75	3.07	3.01	2.75	2.57	2.62	2.88	3.11	2.54
m_1–m_2 length	2.75	2.80	...	...	2.56	2.88	2.88	2.69	2.75	2.69	2.85	2.75
m_1–m_3 (alveolar) length	3.70	3.90	...	4.03	3.87	4.22	3.97	3.71	3.58	3.84	3.76	3.78
m_1–m_3 (coronal) length	3.66	3.90	...	3.90	...	...	3.97	3.71	3.66	3.71	3.73	3.78
m_1 length	1.55	1.59	...	1.58	1.49	1.64	1.67	1.53	1.58	1.55	1.64	1.56
m_1 width	0.99	1.07	...	0.99	1.08	1.12	1.02	0.94	0.90	1.05	0.99	0.89
m_2 length	1.24	1.36	...	1.18	1.15	1.24	1.21	1.18	1.15	1.21	1.21	1.18
m_2 width	0.97	1.02	...	0.97	1.12	1.12	0.93	0.93	0.87	1.02	0.99	0.90
m_3 length	0.91	0.90	...	1.08	...	1.08	1.05	1.02	0.90	0.90	0.99	0.96
m_3 width	0.71	0.78	...	0.83	...	...	0.74	0.83	0.72	0.77	0.77	0.77
Lower incisor depth	0.94	0.97	...	...	1.05	1.02	1.02	0.99	0.84	0.97	1.01	0.85
M^1–M^3 (alveolar) length	...	...	4.03	...	3.99	...	4.16	3.58	3.71	3.84	...	3.84
M^1–M^3 (coronal) length	...	...	3.88	...	3.78	...	3.84	3.39	3.45	3.52	3.61	3.65
M^1 length	...	...	2.01	1.83	1.92	...	2.01	1.74	1.74	1.78	1.80	1.89
M^1 width	...	...	1.08	1.08	1.21	...	1.08	0.96	1.05	1.05	0.90	1.00
M^2 length	...	...	1.21	...	1.21	...	1.21	1.16	1.05	1.08	1.12	1.12
M^2 width	...	...	...	...	1.15	...	1.08	0.99	1.02	0.99	0.87	0.93
M^3 length	...	...	0.77	...	0.71	...	0.65	0.62	0.68	0.71	0.71	0.65
M^3 width	...	...	0.87	...	0.87	...	0.77	0.84	0.74	0.87	0.68	0.81

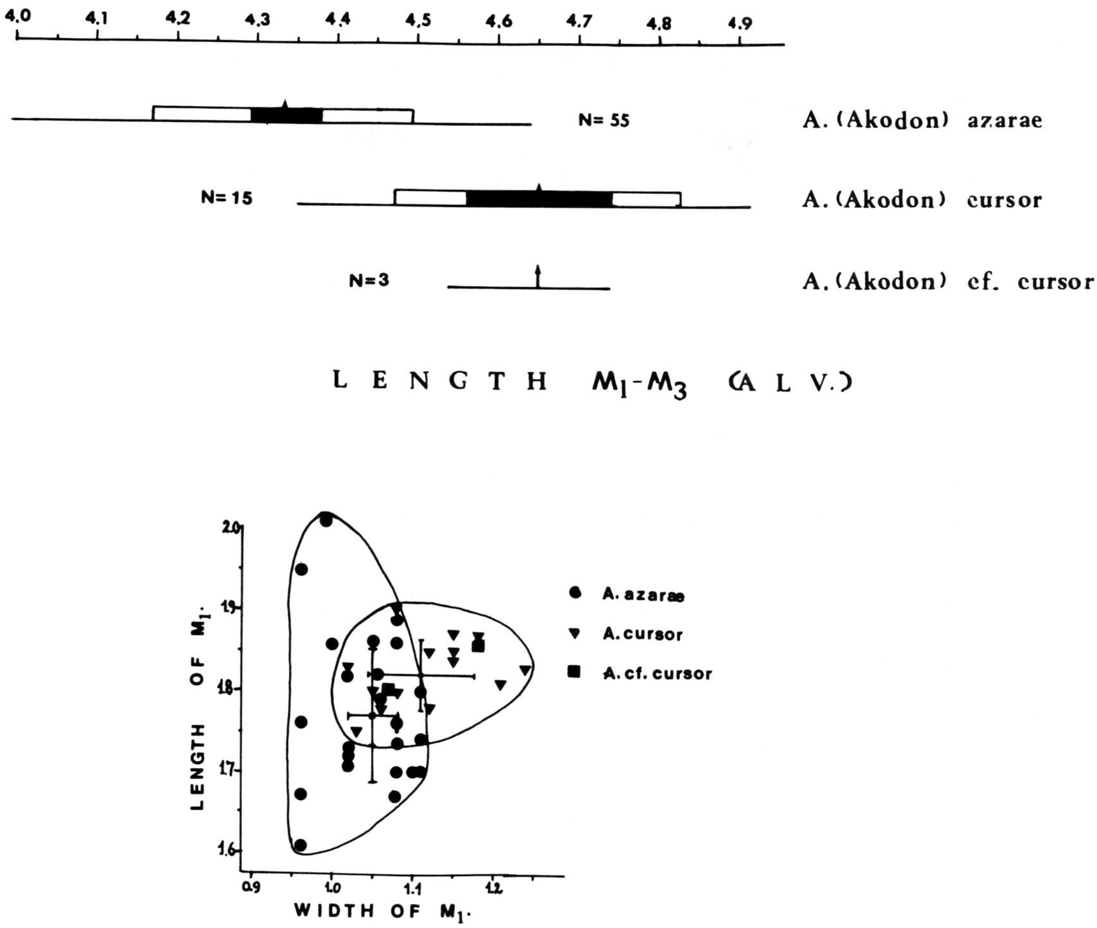

FIG. 12. Dice-grams and scattergram for measurements of dentition of species of *Akodon* (*Akodon*).

He later (Thomas, 1918) described *caenosus* from the mountains at León, Jujuy Province, in northwest Argentina, as a subspecies of *puer*. However, he later (Thomas, 1920) recorded more specimens from San Salvador de Jujuy, proposing specific status for *caenosus*. The extension of its range to Tucumán was recorded by Thomas (1926b) and Bárquez et al. (1980). On examination of the corresponding types and of all the specimens referred to *puer* and *caenosus* in the British Museum, I could not find any reasonable basis to accept specific or even subspecific recognition for *caenosus*, and I treat it as a junior synonym of *puer* (see also Vitullo et al., 1986).

Akodon iniscatus is based on an animal caught in the Andean region of Patagonia, from southwest of Chubut Province, but Thomas referred to it specimens from northern Patagonia to central La Pampa Province, believing (Thomas, 1919, p. 205) that it extended to the south of SE of Buenos Aires Province. *Akodon iniscatus collinus* was described (Thomas, 1919, p. 206) as a subspecies from northwestern Patagonia, and *A. nucus*, described as a full species (Thomas, 1926a) from specimens of western Neuquén and southern Mendoza, was considered as a subspecies of *iniscatus* by Cabrera (1961). Examination of the holotypes and fairly large series in the British Museum shows *A. nucus* is obviously different from the typical *iniscatus*. It is a much larger form, and I believe that it must be considered as a distinct species. The subspecific distinction of *collinus* from typical *iniscatus* is not at all evident, and I prefer to treat the former as a synonym of the latter. I examined specimens in the collection of the British Museum from central La Pampa Province (fig. 11H) which match per-

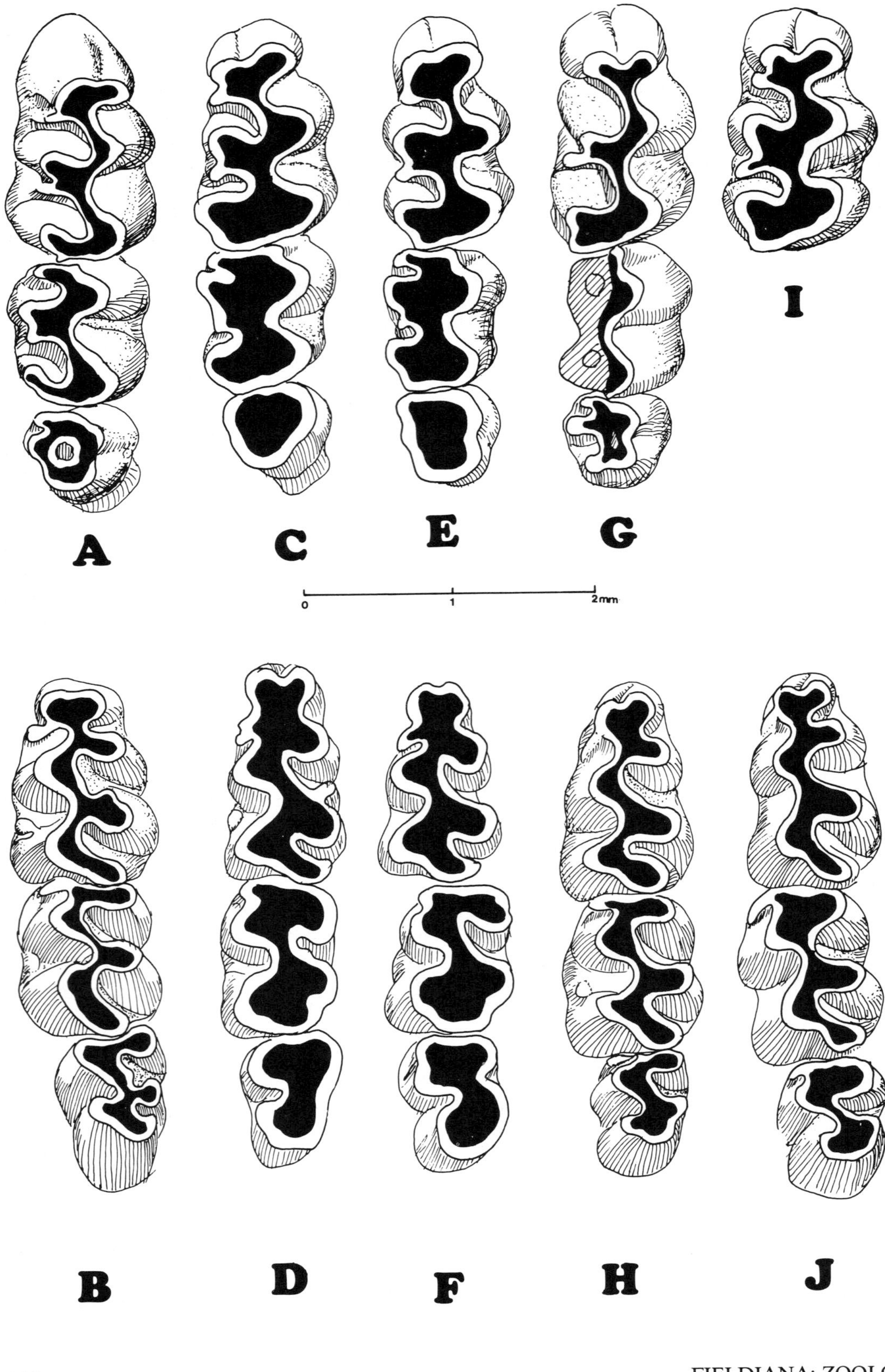
A
C
E
G
I
0
1
2 mm
B
D
F
H
J

fectly the type of *A. iniscatus* (fig. 11A), and I therefore agree with Thomas that *iniscatus* is widespread, reaching northeast to the south of Buenos Aires Province. This latter extension of its range seems to have been recently documented, as O. A. Scaglia and C. Velazquez (pers. comm.) reportedly found this species in a grassland field near Balcarce (Buenos Aires Province) where it occurs in sympatry with *A. azarae*.

Therefore, the presence of a form apparently related to *iniscatus* in the Lower Pleistocene of SE Buenos Aires Province is not surprising. Whether *lorenzinii* can be thought of as an ancestor of *iniscatus* or as a member of an independent, but related, lineage is a matter that cannot be settled now. The second alternative is more likely, as a form more closely related to the living *A. iniscatus* than *A. lorenzinii* was contemporaneous with the latter, as I shall present next.

Akodon (Akodon) cf. **iniscatus** Thomas

Referred Specimens—MMP S-640 (fig. 11K): right lower jaw with incisor and very worn m_1 and m_2; broken at the tip of the coronoid process and lacking condyloid and angular processes. Found by G. J. Scaglia in stratum II of Vorohué Formation (Kraglievich, 1952; Marshall et al., 1984) at the Atlantic slopes S of Arroyo Lobería, Chapadmalal region, Partido de General Pueyrredón, SE of Buenos Aires Province, Argentina.

Description and Discussion—This specimen cannot be included either in the coeval species *A. lorenzinii* or in *A.* cf. *cursor* or *A. johannis*, which immediately follow in the stratigraphic succession. It differs from the former mostly in size and in all the characters I discussed in comparing *lorenzinii* with *iniscatus*. It cannot be confused with *johannis* because of its somewhat greater size and more upturned symphysis. In all the observed characters, this specimen matches the states in studied mandibles of *iniscatus*, making plausible its reference to the living species. The alveolar length of the lower molar row is, however, somewhat greater (table 7) than in the type of *iniscatus*, but the difference obviously falls within the range of geographic variation of that species (fig. 12). Here again, *A. azarae* must be considered as a possible relative of the fossil specimen. Although a close relationship with *azarae* could be eventually demonstrated by new material, I believe that it is unlikely and unsupported by the present evidence. *Akodon azarae* shows a less marked capsular projection, a more elongated mandibular ramus, a less upturned symphysis, and a relatively deep incisor; and in all these characters, specimen MMP S-640 agrees more closely with *iniscatus* (fig. 11L). Unfortunately, the molar teeth are too worn to show details of the enamel pattern, which is quite different in *azarae* and *iniscatus*, as is shown by the high frequency in the former of ectolophids and ectostylids, mesolophid remnants, and mesostylids, which are almost completely absent in the latter. The m_2 of the fossil specimen, although very worn down, shows the external border of the mesoflexid fairly clearly, without trace of a mesolophid remnant or of a mesostylid, thus confirming a closer resemblance to *iniscatus*.

The Meaning of the Fossil Akodontini

One of the striking features of the fossil Akodontini known from the Pliocene and Lower and Middle Pleistocene of Argentina is that they represent diverse species closely related to some living ones or else advanced extinct taxa, such as *Dankomys*. The same conclusion emerged from the study of the Pliocene Phyllotini, described in another paper (Reig, 1978).

Table 8 summarizes our present knowledge of the fossil species of the tribe and their chrono-

---

Opposite Page:

Fig. 13. Occlusal crown views of right upper and left lower molar teeth of living species of *Akodon* (*Akodon*) and of the Lower Pleistocene *Akodon* (*Akodon*) *lorenzinii* n. sp. **A,** Upper molars and **B,** lower molars of living *A. andinus* (Phil.); type of *A. gossei* Thomas; female; BMNH 98.3.21.5; Puente del Inca, Mendoza, Argentina. **C,** Upper molar teeth and **D,** lower molar teeth of *A. puer* Thomas; female; holotype, BMNH 2.1.1.78; Choquecamata, Bolivia. **E,** Upper molar teeth and **F,** lower molar teeth of *A. iniscatus* Thomas; female; holotype, BMNH 3.7.9.64; Valle del Lago Planco, Chubut, Argentina. **G,** Upper molar teeth of *A. lorenzinii* n. sp.; MMP M-867; Vorohué Formation, Lower Pleistocene, Partido de General Pueyrredón, Buenos Aires Province, Argentina. **H,** Lower molar teeth of *A. lorenzinii* n. sp.; holotype, MMP M-1081; San Andrés Formation, Barranca Parodi, Miramar, Partido de General Alvarado, Buenos Aires Province, Argentina; Lower Pleistocene. **I,** Right M^1 of *A. lorenzinii* n. sp.; MLP 52.X.4.44 (a); San Andrés Formation, south of Punta Hermengo, Miramar, SE of Buenos Aires Province, Argentina; Lower Pleistocene. **J,** Lower molar teeth of *A. lorenzinii* n. sp., MMP M-868; found in association with MMP M-867.

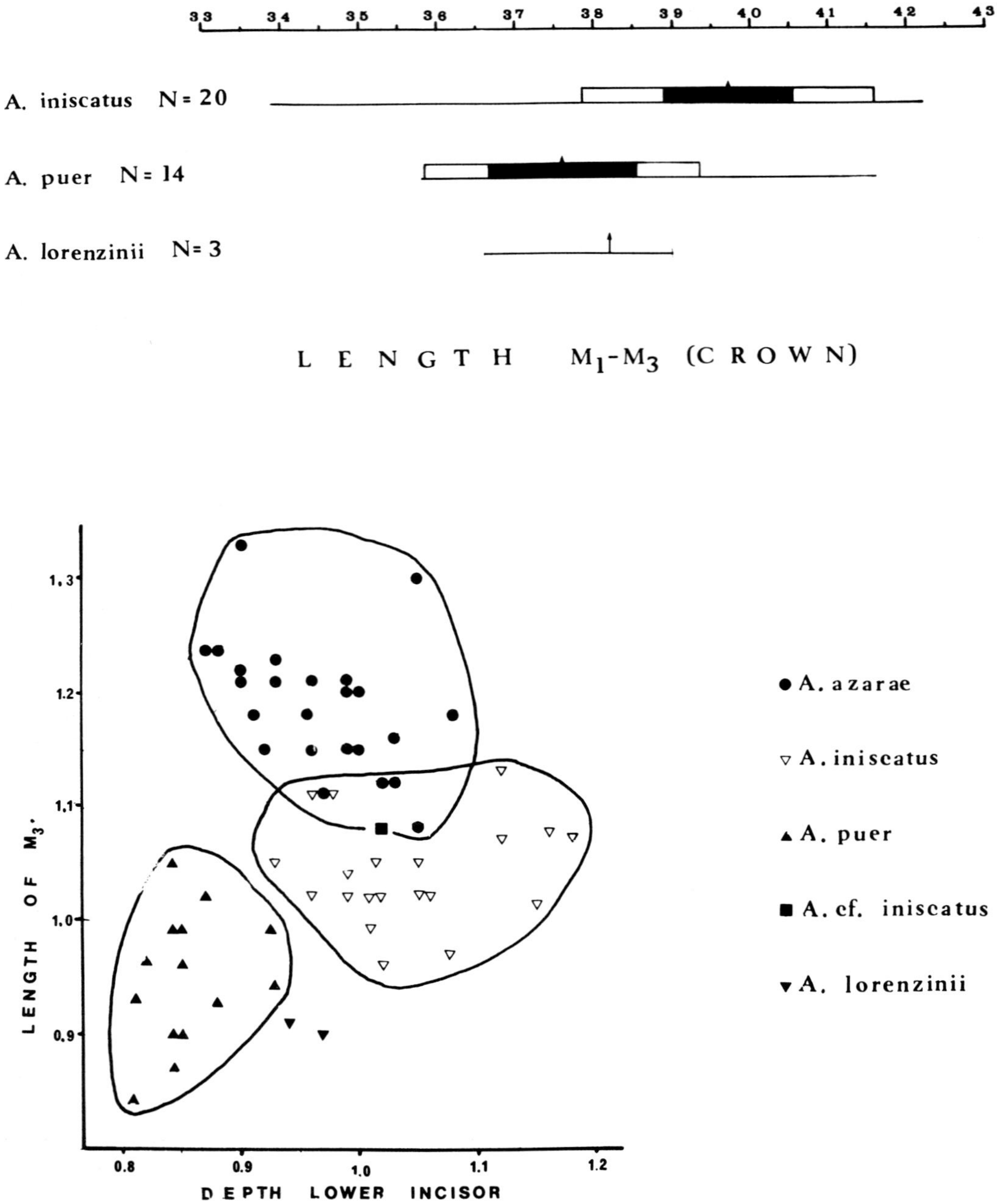

FIG. 14. Dice-grams and scattergrams of measurements of molar teeth in *Akodon* (*Akodon*) *azarae, Akodon* (*Akodon*) *cursor,* and a fossil sample referred to the latter.

stratigraphic distribution in the Plio-Pleistocene column of the south of Buenos Aires Province (Marshall et al., 1984). Included are representatives of *Bolomys* and *Dankomys* which are still undescribed, but which will be the subject of forthcoming papers. As discussed in the previous systematic part, the Upper Pliocene (*A. kermacki*) and Lower Pleistocene (*A. magnus*) species of the subgenus *Abrothrix* cannot be considered either ancestral to or more primitive than the living species of the subgenus. *Bolomys bonapartei* from the Lower Pliocene, far from representing primitive conditions in its character-states, is better interpreted as part of the spatiotemporal diversifi-

TABLE 8. Chronostratigraphic distribution of the known species of fossil Akodontini, as reported in this paper and in Reig (1978).

Species	Plio-Pleistocene ages and subages							
	Pliocene		Uquian (Lower Pleistocene)					
	Montehermosan	Chapadmalalan	Barrancaloban (= Uquian)	Vorohuean	San Andresian	Ensenadan (Miramar Formation)	Lujanian	Recent
Bolomys bonapartei	——							
Bolomys sp.*					——			
Bolomys sp.†						——		
Dankomys simpsoni		——						
Dankomys sp.‡				——	——			
Akodon (*Ab.*) *kermacki*		——	——					
Akodon (*Ab.*) *magnus*				——				
Akodon (*Ak.*) cf. *cursor*						——	- - -	——
Akodon (*Ak.*) *lorenzinii*				——	——			
Akodon (*Ak.*) *johannis*						——		
Akodon (*Ak.*) cf. *iniscatus*						——	- - -	——

* From MLP 52.X.4.30 (a), undescribed specimen from San Andrés Formation in the vicinity of Miramar, Partido de General Alvarado, SE of Buenos Aires Province.

† From MMP M-642 (b), undescribed M^2 from Miramar Formation at Sta. Helena, Partido de Mar Chiquita, SE of Buenos Aires Province.

‡ From MMP M-1064, undescribed jaws, partial skull and postcranial bones found in Vorohué Formation in the vicinity of Punta Lobería, Partido de General Pueyrredón, SE of Buenos Aires Province. This specimen is the basis of a new species which I shall describe in a forthcoming paper, and which is also represented by several other specimens from the Vorohué and San Andrés formations.

cation of a relatively advanced akodontine genus (Reig, 1978, p. 169). *Dankomys simpsoni* from the Upper Pliocene is related to *Bolomys,* but shows more advanced adaptations to an herbivorous diet; it is best thought of as a derivative of the latter. The genus is represented in the Lower Pleistocene by another, more advanced species (table 8). Fossil representatives of the subgenus *Akodon* from the Lower (*A. lorenzinii*) and the Middle (*A. johannis*) Pleistocene are neither ancestral nor more primitive than related living species, and two extant species (*A. cursor* and *A. iniscatus*) were present in the Middle Pleistocene. Thus, the fossil evidence from the pampean region does not indicate an early stage in akodontine evolution, but rather suggests that they had already attained a high degree of evolution and differentiation at the generic and subgeneric level in the Pliocene. It follows that the Akodontini started to differentiate and to diversify in times earlier than the Lower Pliocene, that is, during Miocene times. Several lines of reasoning support the parsimonious hypothesis that this differentiation took place in those times in South America from oryzomyine South American ancestors (Reig, 1984).

Fossil cricetids have not been found, however, in the rather rich deposits of the Upper Miocene (Chasicoan and Huayquerian) sediments of the pampean region. They are also absent in Miocene deposits elsewhere in Argentina or in South America. In view of the intrinsic incompleteness of the fossil record, the absence of akodontines (and other sigmodontines as well) in the known Miocene deposits may be just a matter of sampling, and they could eventually be discovered in those deposits after more careful collecting. It might be argued that the small size of cricetid remains makes their discovery less probable than those of larger rodents, which have actually been found in relative abundance in Miocene beds of Patagonia, the pampas and west of Argentina. In this sense, it may be meaningful that fossil cricetids have been mostly found in the Plio-Pleistocene outcrops of the Mar del Plata-Miramar region. This region has been exploited continuously and with careful scrutiny during the last 40 years by a collector, Galileo J. Scaglia, who was especially well trained in hunting tiny fossil remains.

Although these arguments are reasonable and do not discount the eventual discovery of fossil cricetids in the rich Miocene mammal-bearing formations of Argentina, I am inclined to the alternative hypothesis that the absence of cricetids in those formations is not a result of faulty sampling, but that it represents a real absence of mice in that time and place. This absence may reflect the biogeographic history of akodonts in particular and of the Sigmodontinae in general. There is suggestive evidence that the evolutionary history of sigmodontines was tightly linked with the history of the changing environments of the Andean region (Reig, 1984). As inferred from distributions of living species, South American sigmodontines are much more frequent in Andean highlands (71.6%) than in the remaining lowlands (28.4%; Reig, 1984). The pattern of species occurrences and the endemism of genera suggest the hypothesis that the Oryzomyini (which likely included the akodontine ancestors) had the northern Andes of Ecuador, Colombia, and Venezuela as its area of original differentiation (AOD). For the Akodontini, 71.2% of total reported occurrences are strictly Andean, and the south-central Andes and north-southern Andes between parallels 30°S and 40°S show the greatest diversity of genera and the greatest frequency of species occurrences, as defined and discussed in Reig (1984). Thus, it can be postulated that the AOD of the Akodontini was located in the general area of the present southern altiplano. The akodont rodents may have radiated from this area of original differentiation, expanding gradually to the northern Andes, the southern Andes of Argentina and Chile, and eventually to the lowlands of Bolivia, Paraguay, Brazil, Argentina, and Uruguay.

Thus, the colonization of the pampas by akodontines may have been preceded by a period of evolution in other geographic areas within South America, and the same may hold for the time of colonization of Patagonia. Therefore, the absence of cricetids in the Miocene deposits of the pampean region and Patagonia may reflect the fact that they had not yet colonized those areas. This being the case, the earliest occurrence of fossil akodontines in the Lower Pliocene deposits of Monte Hermoso would be interpreted as signaling the time of the first establishment of akodont rodents in the pampean plains.

It can be argued that this explanation remains pure speculation until fossils are found in Miocene formations of the Andes. However, the explanation is not mere guesswork, as it is based on strong empirical data coming from patterns of diversity and distribution of the living fauna. Nevertheless, I agree that the discovery of Miocene Andean fossil akodontines would be a critical additional corroboration to the hypothesis. Several Miocene fossil faunules from the altiplano have been discovered in the last 10 years by Dr. R. Hoffstetter and collaborators (see a review in Marshall et al., 1983). So far, these faunules yielded remains of large- to medium-sized mammals, but this may be a mere reflection of the exploratory stage of their study. It must also be recognized that preservation of tiny rodents demands special taphonomic conditions, which may or may not be found in the Andean Miocene deposits.

A Tentative Scenario of the Evolutionary Deployment of the Akodontini

My views on the origin, antiquity, and evolutionary history of the Sigmodontinae have been presented and discussed in previous papers (Reig, 1978, 1980, 1981, 1984) and are not the subject of further comment here. Instead, I shall base this argument on the main premises of my theory, namely: (1) The Sigmodontinae are a separate subfamily of the family Cricetidae that evolved in South America from North American ancestors; (2) their probable ancestors are the generalized cricetids of the North American Oligocene representing the subfamily Eucricetodontinae (*sensu* Martin, 1980); (3) the more primitive living sigmodontines belong to the tribe Oryzomyini, which represents the direct or indirect ancestral stock of the remaining tribes; (4) the Sigmodontinae started its evolutionary deployment in South America from a proto-oryzomyine ancestor which entered that continent by overwater dispersal by the early Miocene or late Oligocene and which became established in the northern Andes of Colombia; (5) the main episodes of the differentiation of the Sigmodontinae from the ancestral oryzomyine stock occurred within the Andes and were followed by successive invasions to the eastern lowlands; and (6) after the establishment of the Panamanian land bridge, several sigmodontine lineages that had differentiated in South America invaded Middle and North America in different dispersal episodes. To complete and update the picture, I now consider the tylomyines as a separate subfamily of the Cricetidae which independently evolved in Middle America from eucricetodontine ancestors (see Reig, 1984, commenting on Carleton, 1980).

Within this theory, the Akodontini are considered as a group directly descended from the Oryzomyini, which differentiated in the area of the present altiplano. The reasons supporting this conclusion are more extensively given in Reig (1984) and were alluded to in the previous sections. Let us examine the probable picture of their origin and further differentiation.

First, of the 64 extant species of akodontines recognized in this paper, 35 (54.7%) belong to a single genus, *Akodon,* of the 11 recognized living genera in the tribe. *Akodon, Bolomys,* and *Oxymycterus* comprise 81.3% of living species. The remaining eight genera are either monotypic or comprise two or three species. Thus, although an explanation of the evolution of the latter is important, a picture of the evolutionary history of the akodontines must necessarily focus on *Akodon, Bolomys,* and *Oxymycterus.*

Species of *Akodon* live in the puna, the paramos, in montane tropical and subtropical forests, in grassy pampas, dry montane Andean valleys, semidesert Patagonian tablelands, and cold southern Andean forests. The frequency of localities from which species of *Akodon* are reported shows that 82% of the occurrences belong to Andean environments. Seventy-two percent of species occurrences of the subgenus *Akodon* are Andean, while 100% of the living *Abrothrix, Chroeomys,* and *Hypsimys* are so distributed. *Bolomys* shows only 46% of Andean occurrences, inhabiting both the highest altitudes of the altiplano, and the Chacoan, pampean, and Brazilian lowlands. *Oxymycterus* also exhibits 46% Andean occurrences, represented there by three of the 10 tentatively recognized species. Its species live in puna localities, subtropical mountain and lowland forests, the Argentinian Mesopotamian region, and the grassy pampean steppes. The habitat versatility of species of *Akodon, Bolomys,* and *Oxymycterus* contrasts with the stenotopic nature of other genera: *Podoxymys* (restricted to the high *tepuis* of the Guaianan region), *Notiomys* and *Geoxus* (fossorial and restricted to the south temperate forests and neighboring areas), or *Microxus* (only inhabiting the Andean heights).

The evolutionary history of the akodonts is one of successful dispersal, which was surely fueled by the habitat and trophic versatility of the more speciose genera. In fact, they are broadly distributed in South America despite their overwhelming predominance in Andean and montane habitats.

The ancestral akodontine may have been a generalized *Akodon*-like form of North Andean origin which colonized the area of the puna from the north in Middle or Late Miocene times before the altiplano reached considerable heights. Elevation of the altiplano began in the Middle Pliocene (Ahlfeld, 1970). This ancestral form may have encountered adequate conditions in the southern proto-puna, and from here local differentiation may have developed as a response to heterogeneous environments in the changing Andes. Comparative cytogenetics suggests that this ancestral hypothetical akodontine might have possessed a karyotype of 58 pairs of telocentric autosomes plus the sexual pair (Vitullo et al., 1986). After a first branching which separated *Oxymycterus* and allies (see later) from the remaining akodontines, the earliest main radiation may have centered around what is now the genus *Akodon. Akodon* is the most diversified and the most generalized of the akodontines, and is therefore likely the original stock from which most of the remainder of the tribe radiated. A non-specialized member of the subgenus *Akodon* with 2n = 52 (autosomal FN = 58) chromosomes (which is most likely the primitive karyotype of *Akodon* and other related genera; see Bianchi & Merani, 1984; Vitullo et al., 1986; see also Gardner & Patton, 1976, and the discussion above), similar to *A. andinus,* may have been generalized enough to live in different habitats and to settle either in montane forests of the eastern slopes of the rising Andes or in the dry high mountain valleys and open semidesertic heights. *Akodon (Chroeomys) jelskii,* which retains the primitive 2n = 52 karyotype, may represent a well-differentiated *Akodon* offshoot which adapted to arid heights and remained endemic to the rising altiplano. An early main branch of the diversification within the subgenus *Akodon* acquired the derived karyotype of 2n = 40 (autosomal FN = 40), as found in the puna in *A. boliviensis* and *A. albiventer. Akodon puer* (= *A. caenosus,* see Vitullo et al., 1986, and above), which shows a derived 2n = 34, but keeps the same autosomal FN (Bárquez et al., 1980; Vitullo et al., 1986), must have originated in the puna from this branch and expanded to more southern Andean valleys and the lowlands of Tucumán. *Bolomys* may also have differentiated from the same branch early in the puna, as indicated by the generalized features of *B. amoenus* and, dating the time of its origination, by the occurrence of *B. bonapartei* in the Lower Pliocene. The derived 2n = 34, FN = 34 karyotype of *Bolomys* is more likely to have evolved from a 2n = 40 karyotype than from a 2n = 52 karyotype (Bianchi & Merani, 1984). *Akodon (Hypsimys) budini,* which shows a peculiar 2n = 38, FN = 42 karyotype with few arm-to-arm homologies in banding pattern with the 2n = 40 karyotypes (Vitullo et al., 1986), may represent either an isolated offshoot of the same main branch or a direct derivative of the earlier 2n = 52 branch which became endemic to the high Andean valleys. *Oxymycterus* is likely to be the result of the first dichotomy which occurred in the proto-puna from the hypothetical early akodontine ancestor. Its 2n = 54, FN = 58 karyotype seems to represent an independent derivation from the hypothetical 2n = 60 ancestor (Vitullo et al., 1986). *Oxymycterus* may have evolved in adaptation to an animal diet and more humid Andean slopes. Some of its forms became secondarily adapted to puna habitats (as represented by some subspecies of *O. paramensis*), but the main body of the genus eventually spread into the eastern lowlands. *Lenoxus* is certainly a well-differentiated *Oxymycterus* which is more likely to have evolved from the latter in the humid Andean slopes.

A complete picture of the evolutionary deployment of the Akodontini, however, must account for their high diversity in regions and habitats other than the original center of diversification. From what we have already said, it must be granted that much of their early diversification took place in what is now the southern puna region. From there, however, they must have migrated in different directions and radiated further in other areas. Three main directions of dispersal must be assumed: one to the north and another to the south, both following the Andean axis, and a third to the southeastern lowlands.

The southern dispersal took root in the generalized 2n = 52 *Akodon* and reached the southern Andes, which served as a secondary dispersal center. This branch may be based in *Akodon andinus,* which reaches south to the Andes of Mendoza and is now represented by species of the subgenus *Akodon* (*A. olivaceus, A. brachiotis,* and *A. markhami*) and species of the subgenus *Abrothrix* (including *A. xanthorhinus* and *A. hershkovitzi*). It is of interest to note here that *xanthorhinus* and *longipilis* share the same species of the parasitic louse *Hoplopleura andina* with *andinus* and *olivaceus* (Castro, 1981; pers. comm.), which might be interpreted as indicating persistence of an early host-parasite relation. The specialized genera *Chelemys, Notiomys,* and *Geoxus,* two of which are known to retain the 2n = 52 karyotype, the second being unknown in its chromosomes (Pearson, 1984), may easily be interpreted as independent offshoots of this southern branch which specialized in habits and diet. *Abrothrix* originated certainly in the Pliocene, as indicated by its presence in the Upper Pliocene (Chapadmalalan) of southern Buenos Aires Province. The presence of both fossil *Abrothrix* in the Upper Pliocene and Lower Pleistocene of the pampean region, and of the extant *A. longipilis* and *A. xanthorhinus*) in the Patagonian tablelands, indicates that this southern Andean dispersal subsequently spread into eastern steppes. Significantly, all southern akodonts so far known in their chromosomes retain the primitive 2n = 52 karyotype without meaningful modifications. It is unclear whether living Patagonian species as *A. iniscatus* and *A. nucus,* the chromosomes of which having not been described yet, derived from this branch or from the third dispersing branch described below. In view of the affinity of these species with *puer* and *azarae,* as discussed above, we are inclined to the second alternative.

A dispersal to the north from the original southern puna differentiation area is necessary to explain the distribution of *A. aerosus, A. orophilus, A. mollis, A. tolimae,* and *A. urichi,* as well as of species of *Microxus* and *Podoxymys.* Given the presence of most of these taxa in the northern Andes and connected mountain ranges, where they could have originated directly from the oryzomyines, it is necessary to postulate that the north Andean akodontines migrated there from a south-central Andean region in order to maintain the monophyly of the tribe. Chromosomal evidence is highly suggestive of the derived condition of the northern species of *Akodon,* with *A. orophilus* (2n = 26), *A. mollis* (2n = 22), and *A. urichi* (2n = 18) indicating a northward decrease in chromosome number (chromosomal data from Bianchi & Merani, 1984; Gardner & Patton, 1976; Reig et al., 1971). Unpublished chromosomal counts from Colombian *Akodon,* which may represent *A. tolimae* (from Santander; C. Ramírez, pers. comm.), with 2n = 24 chromosomes, and *A. urichi* (from Villavicencio; A. Gardner, pers. comm.), with 2n = 18 chromosomes, suggest that the pattern of decrease may not be regular. The less reduced karyotype of species of *Microxus* (as known in the polymorphic 2n = 35–37, FN = 48 of *Microxus bogotensis* from the Venezuelan paramos; see Barros & Reig, 1979) suggests a separate origin from the primitive 2n = 52 *Akodon* stock, consistent with its generic distinction. *Podoxymys* is an endemic genus from the *tepuis* that is likely to have been derived from the same main branch; its karyotype is unknown.

The third main direction of dispersal, from the southern puna area of original differentiation directly toward the eastern lowlands, is necessary to explain the present distribution of lowland representatives of *Akodon s.s.*, *Oxymycterus,* and *Bolomys,* as well of the exclusively lowland *Akodon* (*Deltamys*) of the genera *Blarinomys, Juscelinomys,* and fossil *Dankomys.* The itinerary of *Bolomys* may be inferred from the distribution of the living species, starting with a *B. amoenus*-like ancestral form in the puna, with *B. lactens* in pampean range valleys of northwestern Argentina, and *B. lenguarum* in the lowlands of Bolivia and Paraguay. Two diverging lines of dispersal can be inferred from the last region: one toward the pampean region (represented by *B. bonapartei, B. obscurus,* including *benefactus,* and a new undescribed species), the other toward southern and eastern Brazil (*B. lasiurus*), passing through the Argentinian Chaco and the northern Mesopotamian region (*B. temchuki*). A similar dispersal pattern may be postulated for *Oxymycterus,* which shows a similar pattern of species distribution in the lowlands of Paraguay, Argentina, Brazil, and Uruguay, as well as in the high valleys of northwestern Argentina. *Dankomys* may represent a local pampean derivative of *Bolomys,* and *Juscelinomys* may have been derived from a Brazilian branch of *Oxymycterus.* Regarding the lowland differentiation within the genus *Akodon,* it is interesting to realize that *A.* (*Ak.*) *boliviensis,* a typical inhabitant of the present puna, but also represented at lower altitudes in sites surrounding the puna and even in the lowlands of Tucumán, has 2n = 40 chromosomes. Most species of *Akodon* of the lowlands of central Argentina, the pampean region, the Mesopotamia, and Uruguay show either an identical 2n = 40 karyotype (*A. varius* complex) or a rather similar 2n = 42–43 (*A. molinae*), 2n = 38 (*A. azarae, A. dolores*), or 2n = 37 (*A. kempi*) karyotypes, sharing many resemblances to each other (Bianchi & Merani, 1984). Unfortunately, we still do not know the karyotype of *Akodon cursor,* which might belong to the same group. These Argentinian forms are then likely to represent an evolutionary group stemming from a puna ancestor probably closely related to *A. boliviensis. Akodon nigrita* poses an interesting problem, as its 2n = 52 karyotype precludes its membership in the same clade. The place within this picture of the unnamed Brazilian species with 2n = 14–16 and 2n = 24–25 chromosomes, as well as of the poorly known *A. serrensis, A. reinhardti,* and *Blarinomys breviceps,* is still unclear.

Needless to say, the above scenario is quite tentative and must be taken only as a set of working hypotheses open to partial or overall modification after the test of more detailed morphological, paleontological, cytogenetical, and biochemical studies. However, I believe that in its present version, it may represent a reasonable explanation of the most probable biogeographic and evolutionary events connected with the radiation of the akodontines and, therefore, a heuristic framework for further advances in the knowledge of the evolution of these rodents.

Acknowledgments

I thank S. Anderson, G. Corbet, K. Kermack, R. Pascual, G. J. Scaglia, and O. A. Scaglia for facilitating access to collections under their care. O. P. Pearson, A. Spotorno, N. O. Bianchi, and A. Langguth are also acknowledged for fruitful discussions. I thank Bruce Patterson and two anonymous referees for suggestions and a careful reading of the manuscript. This paper is also the outcome of a John Simon Guggenheim fellowship, and this Foundation is heartily acknowledged for its continuous support.

Literature Cited

AHLFELD, F. 1970. Zur Tektonik des andinen Bolivien. Geologischer Rundschau, **56:** 1124–1140.

ALSTON, W. F. 1964. Philosophy of Language. Prentice-Hall, Inc., Englewood Cliffs, N.J., 112 pp.

AMEGHINO, F. 1889. Contribución al conocimiento de los mamíferos fósiles de la República Argentina. Actas de la Academia Nacional de Ciencias de Córdoba, **6:** 1–1208 and Atlas.

ANTHONY, H. E. 1929. Two new genera of rodents from South America. American Museum Novitates, **388:** 1–9.

ARATA, A. 1967. Muroid, gliroid, and dipodoid rodents, pp. 226–253. *In* Anderson, S., and J. K. Jones, Jr., eds., Recent Mammals of the World. A Synopsis of Families. The Ronald Press Co., New York, viii + 453 pp.

BÁRQUEZ, R. M., D. F. WILLIAMS, M. A. MARES, AND H. H. GENOWAYS. 1980. Karyology and morphometrics of three species of *Akodon* (Mammalia: Muridae) from northwestern Argentina. Annals of Carnegie Museum, **49:** 379–403.

BARROS, M. A., AND O. A. REIG. 1979. Doble polimorfismo robertsoniano en *Microxus bogotensis* (Rodentia, Cricetidae) del Páramo de Mucubají (Mérida,

Venezuela). Acta Científica Venezolana, **30**(Suplemento N° 1): 106.

Beckner, M. 1959. The Biological Way of Thought. Columbia University Press, New York, 200 pp.

Bianchi, N. O., and J. R. Contreras. 1967. Mecanismo de determinación sexual en el ratón de campo, *Akodon azarae* (Rodentia, Cricetidae). Revista Argentina de Endocrinología y Metabolismo, **13:** 157–167.

Bianchi, N. O., J. R. Contreras, and F. N. Dulout. 1969. Intraspecies autosomal polymorphism and chromosomal replication in *Akodon azarae* (Rodentia, Cricetidae). Canadian Journal of Genetics and Cytology, **11:** 233–243.

Bianchi, N. O., and S. Merani. 1984. Cytogenetics of South American akodont rodents (Cricetidae). X. Karyological distances at generic and intrageneric levels. Journal of Mammalogy, **65:** 206–219.

Bianchi, N. O., S. Merani, and M. Lizarralde. 1979. Cytogenetics of South American akodont rodents (Cricetidae). V. Segregation of chromosome Nr. 1 polymorphism in *Akodon molinae.* Experientia, **35:** 1438–1439.

Bianchi, N. O., O. A. Reig, J. Molina, and F. N. Dulout. 1971. Cytogenetics of South American akodont rodents (Cricetidae). I. A progress record of Argentinian and Venezuelan forms. Evolution, **21:** 724–736.

Bianchi, N. O., L. Vidal Rioja, and M. S. A. Bianchi. 1973. Constitutive heterochromatin, G-bands and robertsonian rearrangements in the chromosomes of *Akodon molinae* (Rodentia, Cricetidae). Canadian Journal of Genetics and Cytology, **15:** 855–861.

———. 1976. Cytogenetics of South American akodont rodents (Cricetidae). II. Interspecific homologies of G-banding patterns. Cytologia, **41:** 139–144.

Cabrera, A. 1961. Catálogo de los mamíferos de América del Sur. II (Sirenia, Perissodactyla, Artiodactyla, Lagomorpha, Rodentia). Revista del Museo Argentino de Ciencias Naturales "Bernardino Rivadavia"; (Zoologia) **4:** 309–732.

Carleton, M. D. 1973. A survey of gross stomach morphology in New World Cricetinae (Rodentia, Muroidea), with comments on functional interpretations. Miscellaneous Publications Museum of Zoology, University of Michigan, **146:** 1–43.

———. 1980. Phylogenetic relationships in neotomine-peromyscine rodents (Muroidea) and a reappraisal of the dichotomy within New World Cricetinae. Miscellaneous Publications Museum of Zoology, University of Michigan, **157:** 1–146.

Castro, M. del C. 1981. Contribución al conocimiento de los Anoplura neotropicales. Revista Sociedad Entomológica Argentina, **40:** 231–236.

———. 1982. Las especies del grupo *travassosi* del género *Hoplopleura* Enderlein, 1904 en la República Argentina. Revista Sociedad Entomológica Argentina, **41:** 171–182.

Cestari, A. N., and J. Imada. 1968. Os cromossomos do roedor *Akodon arviculoides cursor* Winge, 1888 (Cricetidae, Rodentia). Ciencia e Cultura, **20:** 758–772.

Contreras, J. R. 1982. Notas acerca de *Bolomys temchuki* (Massoia 1982) en el noreste Argentino, con la descripción de dos nuevas subspecies (Rodentia, Cricetidae). Historia Natural (Corrientes), **2:** 174–176.

———, and M. I. Rosi. 1981. Notas sobre los Akodontini argentinos (Rodentia, Cricetidae). I. *Abrothrix longipilis moerens* Thomas, 1919, en el Parque Nacional Nahuel Huapi. Historia Natural (Corrientes), **1:** 209–212.

Copi, I. M. 1953. Introduction to Logic. The MacMillan Co., New York, 472 pp.

de Santis, L. J. M., and E. R. Justo. 1980. *Akodon* (*Abrothrix*) *mansoensis,* sp. nov., un nuevo "ratón lanoso" de la Provincia de Río Negro, Argentina (Rodentia, Cricetidae). Neotropica, **26:** 121–127.

Dulout, F. N., R. M. Ponssa, A. E. Schugurensky de Adris, and L. L. Dulout. 1976. Multiformidad y polimorfismo cromosómico en *Akodon* (*Abrothrix*) *illutea* Thomas (Rodentia, Cricetidae). Mendeliana, **1:** 49–52.

Echave Llanos, J. M., and C. A. Vilchez. 1964. Anatomía microscópica del estómago del ratón hocicudo (*Oxymycterus rutilans*). Revista de la Sociedad Argentina de Biología, **40:** 187–192.

Ellerman, J. R. 1941. The Families and Genera of Living Rodents. Vol. II. Family Muridae. British Museum (Natural History), London, xii + 690 pp.

Gallardo, M. H. 1982. Chromosomal homology in southern *Akodon.* Experientia, **38:** 1485–1487.

Gardner, A. L., and J. L. Patton. 1976. Karyotypic variation in oryzomyine rodents (Cricetinae) with comments on chromosomal evolution in the Neotropical cricetine complex. Occasional Papers, Museum of Zoology, Louisiana State University, **49:** 1–48.

Gentile de Fronza, T. 1970. Cariotipo de *Oxymycterus rutilans platensis* (Rodentia, Cricetidae). Physis, **30:** 343.

Gentile de Fronza, T., R. L. Wainberg, and G. E. Hurtado de Catalfo. 1979. Cromosomas sexuales múltiples en *Deltamys kempi* (Rodentia, Cricetidae): estadio preliminar a la instauración del sistema XY_1Y_2/XX. Mendeliana, **4:** 28–35.

Gyldenstolpe, N. 1932. A manual of Neotropical sigmodont rodents. Kungliga Svenska Vetenskapsakademiens Handliger, **11**(3): 1–164; 18 pl.

Hershkovitz, P. 1962. Evolution of Neotropical cricetine rodents (Muridae), with special reference to the phyllotine group. Fieldiana: Zoology, **46,** 1–524.

———. 1966. South American swamp and fossorial rats of the Scapteromyine group (Cricetinae, Muridae), with comments on the glans penis in murid taxonomy. Zeitschrift für Säugetierkunde, **31:** 81–149.

———. 1967. Dynamics of rodent molar evolution: A study based on New World Cricetinae, family Muridae. Journal of Dental Research, **46** (Supplement to No. 5): 829–842.

Honacki, J. H., K. E. Kinman, and J. W. Koeppl. 1982. Mammal Species of the World. Allen Press and Association of Systematic Collections, Lawrence, Kansas, 694 pp.

Hooper, E. T., and G. G. Musser. 1964. The glans penis in Neotropical cricetines (family Muridae) with comments on classification of muroid rodents. Miscellaneous Publications Museum of Zoology, University of Michigan, **123:** 1–57.

KIBLISKY, P., N. O. BIANCHI, AND M. S. MERANI. 1976. Citogenética de los roedores sudamericanos akodontinos (Cricetidae). III. Polimorfismos cromosómicos intraespecíficos en *Akodon dolores* Thomas. Mendeliana, **1:** 97–103.

KIBLISKY, P., I. LÖBIG, AND O. A. REIG. 1970. Estudio cromosómico en diversas poblaciones del ratón llanero, *Zygodontomys microtinus* Thomas. Acta Científica Venezolana, **21** (Suplemento 1): 35.

KRAGLIEVICH, J. L. 1952. El perfil geológico de Chapadmalal y Miramar, Provincia de Buenos Aires. Resumen preliminar. Revista del Museo Municipal de Ciencias Naturales y Tradicional de Mar del Plata, **1:** 8–37.

———. 1965. Spéciation phyletique dans les rongeurs fossiles du genre *Eumysops* (Echimyidae: Heteropsomyinae). Mammalia (Paris), **29:** 258–265.

KRAVETZ, F. O. 1973. Estudio de régimen alimentario, períodos de actividad y otros rasgos ecológicos en una población de "ratón hocicudo" (*Oxymycterus rufus platensis*) de Punta Lara. Acta Zoológica Lilloana, **29:** 201–212.

LOBATO, L., G. CANTOS, B. ARAUJO, N. O. BIANCHI, AND S. MERANI. 1982. Cytogenetics of South American akodont rodents (Cricetidae). VIII. *Akodon mollis*: A species with XY females and B-chromosomes. Genetica, **57:** 199–205.

LANGGUTH, A. 1975. La identidad de *Mus lasiotis* Lund y el status de género *Thalpomys* Thomas (Mammalia, Cricetidae). Papéis Avulsos de Zoologia (Sao Paulo), **29:** 45–54.

MAIA, V., AND A. LANGGUTH. 1981. New karyotypes of Brazilian akodont rodents, with notes on taxonomy. Zeitschrift für Säugetierkunde, **46:** 241–249.

MARSHALL, L. G., A. BERTA, R. HOFFSTETTER, R. PASCUAL, O. A. REIG, M. BOMBIN, AND A. MONES. 1984. Geochronology of the continental mammal-bearing Quaternary of South America. Palaeovertebrata, Mémoire Extraordinaire, **1984:** 1–80.

MARSHALL, L. G., R. HOFFSTETTER, AND R. PASCUAL. 1983. Geochronology of the continental mammal-bearing Tertiary of South America. Palaeovertebrata, Mémoire Extraordinaire, **1983:** 1–93.

MARTIN, L. D. 1980. The early evolution of the Cricetidae in North America. The University of Kansas Paleontological Contributions, **102:** 1–42.

MASSOIA, E. 1963a. Sobre la posición sistemática y distribución geográfica de *Akodon* (*Thaptomys*) *nigrita* (Rodentia, Cricetidae). Physis (Buenos Aires), **24**(67): 73–80.

———. 1963b. *Oxymycterus iheringi* (Rodentia, Cricetidae), nueva especie para la Argentina. Physis (Buenos Aires), **24**(67): 129–136.

———. 1981a. Nuevos datos sobre *Akodon, Deltamys* y *Cabreramys,* con la descripción de una especie y de una subespecie nuevas (Mammalia, Rodentia, Cricetidae). Historia Natural (Corrientes), **1:** 179–180.

———. 1981b. Notas sobre los cricétidos mendocinos. Historia Natural (Corrientes) **1:** 205–208.

———. 1982. Diagnosis previa de *Cabreramys temchuki,* nueva especie (Rodentia, Cricetidae). Historia Natural (Corrientes), **2:** 91–92.

MASSOIA, E., AND A. FORNES. 1962. Un cricétido nuevo para la Argentina: *Akodon arviculoides montensis* Thomas (Rodentia). Physis (Buenos Aires) **23**(65): 183–194.

———. 1967. El estado sistemático, distribución geográfica y datos etoecológicos de algunos mamíferos neotropicales (Marsupialia y Rodentia), con la descripción de *Cabreramys,* género nuevo (Cricetidae). Acta Zooloógica Lilloana, **23:** 407–430.

———. 1969. Caracteres comunes y distinitivos de *Oxymycterus nasutus* (Waterhouse) y *O. iheringi* Thomas (Rodentia, Cricetidae). Physis (Buenos Aires), **28**(77): 315–321.

MESERVE, P. L. 1981. Trophic relationships among small mammals in a Chilean semiarid thorn scrub community. Journal of Mammalogy, **62:** 304–314.

MESERVE, P. L., AND W. GLANZ. 1978. Geographical ecology of small mammals in the northern Chilean arid border. Journal of Biogeography, **5:** 135–148.

MOOJEN, J. 1965. Novo genero de Cricetidae do Brasil Central (Glires, Mammalia). Revista Brasileira de Biologia, **25:** 281–285.

OSGOOD, W. H. 1925. The long-clawed South American rodents of the genus *Notiomys.* Field Museum of Natural History, Zoological Series, **12:** 113–125.

———. 1943. The mammals of Chile. Field Museum of Natural History, Zoological Series, **30:** 1–268.

PATTERSON, B. D., M. H. GALLARDO, AND K. E. FREAS. 1984. Systematics of mice of the subgenus *Akodon* (Rodentia, Cricetidae) in southern South America, with the description of a new species. Fieldiana: Zoology, n.s., **23:** 1–16.

PEARSON, O. P. 1967. La estructura por edades y la dinámica reproductiva en una población de ratones de campo, *Akodon azarae.* Physis (Buenos Aires), **27:** 53–58.

———. 1983. Characteristics of mammalian faunas from forests in Patagonia, southern Argentina. Journal of Mammalogy, **64:** 476–492.

———. 1984. Taxonomy and natural history of some fossorial rodents of Patagonia, southern Argentina. Journal of Zoology (London), **202:** 225–237.

REIG, O. A. 1955. Un nuevo género y especie de cenolestinos del Plioceno de la Provincia de Buenos Aires (Rep. Argentina). Revista de la Asociación Geológica Argentina, **10:** 70–71.

———. 1959. Comunicación preliminar sobre nuevas especies del género *Thylatheridium* Reig (Mammalia, Didelphidae). Neotropica, **4:** 89–95.

———. 1964. Roedores y marsupiales del Partido de General Pueyrredón y regiones adyacentes (Provincia de Buenos Aires, República Argentina). Publicaciones del Museo Municipal de Ciencias Naturales de Mar del Plata, **1:** 203–224.

———. 1970. The Proterosuchia and the early evolution of the Archosauria; An essay about the origin of a major taxon. Bulletin of the Museum of Comparative Zoology at Harvard University, **139:** 229–292.

———. 1977. A proposed unified nomenclature for the enamelled components of the molar teeth of the Cricetidae (Rodentia). Journal of Zoology (London), **181:** 227–241.

———. 1978. Roedores cricétidos del Plioceno superior de la Provincia de Buenos Aires. Publicaciones del Museo Municipal de Ciencias Naturales de Mar del Plata, **2:** 164–190.

———. 1980. A new fossil genus of South American cricetid rodents allied to *Wiedemys*, with an assessment of the Sigmodontinae. Journal of Zoology (London), **192:** 257–281.

———. 1981. Teoría del origen y desarollo de la fauna de mamíferos de América del Sur. Monographiae Naturae, Museo Municipal de Ciencias Naturales "Lorenzo Scaglia," **1:** 1–161.

———. 1982. The reality of biological species: A conceptualistic and a systemic approach, pp. 479–499. *In* Los, J., and H. Pfeifer (eds.), Proceedings of the 6th International Congress of Logic, Methodology, and Philosophy of Science. D. Reidel Publishing Company, Dordrecht, Netherlands.

———. 1984. Distribução geográfica e historia evolutiva dos roedores muroideos sulamericanos (Cricetidae: Sigmodontinae). Revista Brasileira de Genética, **7:** 333–365.

Reig, O. A., P. Kiblisky, and O. J. Linares. 1971. Datos sobre *Akodon urichi venezuelensis* Allen (Rodentia: Cricetidae), un ratón de interés para la investigación citogenética. Acta Biológica Venezuélica, **7:** 157–189.

Reig, O. A., Kirsch, J. A. W., and Marshall, L. In press. Systematic relationships of the living and Neocenozoic opossum-like marsupials (suborder Didelphimorpha). *In* Archer, M., ed., Possums and Opossums. Royal Society of New South Wales, Sydney, Australia.

Reig, O. A., and O. J. Linares. 1969. The occurrence of *Akodon* in the Upper Pliocene of Argentina. Journal of Mammalogy, **50:** 643–647.

Reig, O. A., N. Olivo, and P. Kiblisky. 1971. The idiogram of the Venezuelan vole mouse, *Akodon urichi venezuelensis* Allen (Rodentia, Cricetidae). Cytogenetics, **10:** 99–114.

Reig, O. A., and G. G. Simpson. 1972. *Sparassocynus* (Marsupialia, Didelphidae), a peculiar mammal from the Late Cenozoic of Argentina. Journal of Zoology (London), **167:** 511–539.

Rodriquez, M., R. Montoya, and W. Venegas. 1983. Cytogenetic analysis of some Chilean species of the genus *Akodon* Meyen (Rodentia, Cricetidae). Caryologia, **36:** 129–138.

Sanborn, C. C. 1947. Geographic races of the rodent *Akodon jelskii*. Fieldiana: Zoology, **31:** 137–142.

Sbalqueiro, I. J., M. A. Mattevi, and L. F. B. Oliveira. 1984. An $X_1X_1X_2X_2/X_1X_2Y$ mechanism of sex determination in a South American rodent, *Deltamys kempi* (Rodentia, Cricetidae). Cytogenetics and Cell Genetics, **38:** 50–55.

Sokal, R. R., and P. H. A. Sneath. 1963. Principles of numerical taxonomy. W. H. Freeman, San Francisco.

Spotorno, A., and R. Fernández. 1976. Chromosome stability in southern *Akodon* (Rodentia, Cricetidae). Mammalian Chromosomes Newsletter, **17**(4): 13–14.

Tate, G. H. H. 1932a. The taxonomic history of the South and Central American oryzomine genera of rodents (excluding *Oryzomys*): *Nesoryzomys, Zygodontomys, Chilomys, Delomys, Phaenomys, Rhagomys, Rhipidomys, Nyctomys, Oecomys, Thomasomys, Inomys, Aepeomys, Neacomys*, and *Scolomys*. American Museum Novitates, **581:** 1–28.

———. 1932b. The taxonomic history of the South and Central American akodont rodent genera: *Thalpomys, Deltamys, Thaptomys, Hypsimys, Bolomys, Chroeomys, Abrothrix, Microxus, Podoxymys, Lenoxus, Oxymycterus, Notiomys* and *Blarinomys*. American Museum Novitates, **582:** 1–32.

Thomas, O. 1884. On a collection of Muridae from central Perú. Proceedings of the Zoological Society of London, **1884:** 447–458.

———. 1900. New Peruvian species of *Conepatus, Phyllotis*, and *Akodon*. Annals & Magazine of Natural History, (7)**6:** 466–469.

———. 1902. On mammals collected by Mr. Perry O. Simons in the Southern part of the Bolivian Plateau. Annals & Magazine of Natural History (7)**9:** 222–230.

———. 1909. Notes on some South American mammals, with description of new species. Annals & Magazine of Natural History (8)**4:** 230–242.

———. 1913. New forms of *Akodon* and *Phyllotis*, and a new genus for "*Akodon*" *teguina*. Annals & Magazine of Natural History, (8)**11:** 404–409.

———. 1916. The grouping of the South American Muridae commonly referred to *Akodon*. Annals & Magazine of Natural History, (8)**18:** 336–340.

———. 1918. On small mammals from Salta and Jujuy collected by Mr. E. Budin. Annals & Magazine of Natural History, (9)**1:** 186–193.

———. 1919. On small mammals collected by Sr. E. Budin in North Western Patagonia. Annals & Magazine of Natural History, (9)**3:** 199–212.

———. 1920. A further collection of mammals from Jujuy. Annals & Magazine of Natural History, (9)**5:** 188–198.

———. 1926a. Two new mammals from northern Argentina. Annals & Magazine of Natural History, (9)**17:** 311–312.

———. 1926b. The Spedan Lewis South American exploration. III. On mammals collected by Sr. Budin in the Province of Tucumán. Annals & Magazine of Natural History, (9)**17:** 602–609.

———. 1927. On a further collection of mammals made by Sr. E. Budin in Patagonia. Annals & Magazine of Natural History, (9)**19:** 545–554.

Tullberg, T. 1899. Ueber das System der Nagetiere; eine phylogenetische Studie. Nova Acta Regiae Societatis Scientiarum Upsalensis (3)**18:** 1–514.

Vieira, C. 1955. Roedores e lagomorfos do Estado de São Paulo. Arquivos de Zoologia do Estado de São Paulo, **8:** 129–167.

Vitullo, A. D., M. S. Merani, O. A. Reig, A. E. Kajon, O. Scaglia, M. B. Espinosa, and A. Pérez-Zapata. 1986. Cytogenetics of South American akodont rodents (Cricetidae): New karyotypes and chromosomal banding patterns of Argentinian and Uruguayan forms. Journal of Mammalogy, **67:** 69–80.

Vorontzov, N. N. 1959. Sistema khomiakov (Cricetinae) mirovoi fauny i ikh filogeneticheskie sviazi. Biuletin Moskovskogo Obshtschestva Ispitately Prirody, Otdel Biologia, **44:** 134–137.

———. 1967. Evoliutzia pishtshevaritelnoi sistemy gryzunov (mysheobraznye). Isdatelstvo "Nauka," Novosibirsk, 240 pp.

———. 1982. Nizshie khomiakoobraznye (Cricetidae) mirovoi fauny. Chast I. Morfologiia i Ekologiia. *In* Fauna S.S.S.R. (Nov. Seria No. 125), Mlekopitaiushtchie, T. III, Vyp. 6. Izdatielstvo "Nauka," Leningrad, 451 pp.

Voss, R. S., and A. V. Linzey. 1981. Comparative gross morphology of male accessory glands among Neotropical Muridae (Mammalia: Rodentia) with comments on systematic implications. Miscellaneous Publications Museum Zoology, University of Michigan, No. **159:** 1–41.

Williams, C. B. 1964. Patterns in the Balance of Nature. Academic Press, London and New York, 324 pp.

Winge, H. 1887. Jordfunde og nulevende Gnavere (Rodentia) fra Lagoa Santa, Minas Geraes, Brasilien. E Museo Lundii, **1**(3): 1–200.

Ximenez, A., and A. Langguth. 1970. *Akodon cursor montensis* en el Uruguay (Mammalia, Cricetinae). Comunicaciones Zoológicas del Museo de Historia Natural de Montevideo, **128:** 1–7.

Yañez, J., W. Siefeld, J. Valencia, and F. Jaksic. 1978. Relaciones entre la sistemática y la morfometría del subgénero *Abrothrix* (Rodentia: Cricetidae) en Chile. Anales del Instituto de la Patagonia, **9:** 185–197.

Yonenaga, Y. 1972. Chromosomal polymorphism in the rodent *Akodon arviculoides* ssp. (2n = 14) resulting from two pericentric inversions. Cytogenetics, **11:** 488–499.

———. 1975. Karyotypes and chromosome polymorphism in Brazilian rodents. Caryologia, **28:** 269–286.

———. 1979. New karyotypes and somatic and germ-cell banding in *Akodon arviculoides* (Rodentia, Cricetidae). Cytogenetics and Cell Genetics, **23:** 241–249.

Yonenaga, Y., S. Kasahara, E. J. C. Almeida, and A. L. Piracchi. 1975. Chromosomal banding patterns in *Akodon arviculoides* (2n = 14), *Akodon sp.* (2n = 24 and 25), and two male hybrids with 19 chromosomes. Cytogenetics and Cell Genetics, **15:** 388–399.

Yonenaga, Y., and F. Ricci. 1969. Estudos cromossomicos em especies de roedores. Ciença e Cultura, **21:** 249.

Biogeography of Octodontid Rodents: An Eco-Evolutionary Hypothesis

Luis C. Contreras, Juan C. Torres-Mura, and José L. Yáñez

ABSTRACTS

The family Octodontidae (Rodentia, Hystricognatha) is an old group of low diversity, currently found on both sides of the Andean mountains between 16°S and 41°S. Information on the geographic distribution of the octodontid genera is presented and discussed, and the systematic status of each species and subspecies is given.

An explanation is also proposed for the present distribution of the family, considering geological, climatic, floristic, faunistic, and ecological events that occurred after the first appearance of octodontids in the Deseadan age (early Oligocene) in Bolivia and Patagonia. The uplift of the Andes, the formation of Patagonian pampas, the disappearance of echimyids from the Patagonian Subregion, and the appearance of ctenomyids seem to be the most important factors determining the present distribution of octodontids.

La familia Octodontidae (Rodentia, Hystricognatha) es un antiguo grupo de roedores poco diversificados que se encuentra a ambos lados de la Cordillera de los Andes entre los 16° y los 41°S. En este trabajo presentamos y discutimos la información sobre la distribución de los géneros de octodóntidos y hacemos comentarios sobre el estatus sistemático de cada especie y subespecie.

Proponemos una explicación para la distribución actual de esta familia considerando eventos geológicos, climáticos, florísticos, faunísticos y ecológicos que ocurrieron después de la aparición de los octodóntidos en el registro fósil en el Deseadano (Oligoceno) de Bolivia y Patagonia.

El levantamiento de los Andes, la formación de las Pampas Patagónicas, la desaparición de los echímidos de la Subregión Patagónica y la aparición de los ctenómidos, parecen ser los factores más importantes en la estructuración de la distribución actual de la familia Octodontidae.

A família Octodontidae (Rodentia, Hystricognatha) é um antigo grupo de pouca diversidade, que se encontrado em ambos os lados das Cordilheiras dos Andes, entre 16° e 41°S. Apresentamos informações sôbre as distribuições geográficas dos gêneros octodontinos, e discuti-se as categorias sistemáticas de cada espécie ou subespécie.

Propõe-se uma explicação para a atual distribuição desta família, considerando-se os eventos geológicos, climáticos, florísticos e ecológicos que ocorreram após o aparecimento dos octodontinos no registro fóssil, durante a Era Deseadana (no Oligoceno inferior) da Bolívia e da Patagonia. A elevação dos Andes, a formação dos Pampas da Patagonia, o desaparecimento dos equimídeos da Subregião Patagônica, e o aparecimento dos ctenomídeos, parecem ser os fatores mais importantes determinando a atual distribuição dos octodontinos.

From the Departamento de Biología y Química, Facultad de Ciencias, Universidad de Talca, Casilla 747, Talca, Chile (Contreras and Torres-Murra); and Museo Nacional de Historia Natural, Casilla 787, Santiago, Chile (Yáñez).

Introduction

The order Rodentia is represented in South America by two main groups: the hystricognaths (*sensu* Woods, 1982) and the cricetids. The former had an independent evolutionary history lasting at least 35 million years while South America was an island continent. About 3.5 million years before present (MYBP), the Isthmus of Panama was formed, allowing the other main group of rodents, the cricetids, as well as other mammals, to invade from North America (Reig, 1981; Patterson & Wood, 1982; Webb & Marshall, 1982). South American hystricognaths are characterized by a relatively large body mass (80 g to 50 kg), low species diversity, long gestation period, and small litter size (Rowlands & Weir, 1974). The opposite is characteristic of South American cricetids (Pearson, 1958; Hershkovitz, 1962).

Octodontidae Waterhouse (excluding *Ctenomys*) is one of 11 families of hystricognath rodents found in South America. The family has a small number of lineages (Woods, 1982; Mares & Ojeda, 1982). It is distributed along the Andean mountains from 16°S to 41°S, especially on the western slopes. As a result of this, a large proportion of the species are endemic to or found mainly in Chile (seven of nine species). Despite its low phyletic diversity, this family is highly diverse in forms, ranging from the relatively generalized *Octodon degus*, capable of digging and climbing shrubs, to the specialized, fossorial *Spalacopus cyanus* (Glanz, 1977).

The octodontids are probably the least-studied family of South American hystricognaths. The phylogenetic relations of its genera are not clear, almost all species are poorly known from an ecological standpoint, and the geographic ranges of most species are uncertain (Mares & Ojeda, 1982). Here we analyze information published in the last few years and new data concerning the distribution and systematics of this family.

Geographic Distribution and Systematic Status

Genus *Octodon* Bennett, 1832

According to Osgood (1943) three monotypic species are recognized in this genus: *O. degus* Molina, 1782; *O. bridgesi* Waterhouse, 1844; and *O. lunatus* Osgood, 1943. Of these, the first two can be differentiated by external morphology and behavior (Osgood, 1943; Ipinza et al., 1971). *Octodon lunatus* was distinguished only by the absence of an indentation on the inner border of the last upper molars, all other features being the same as those in *O. bridgesi*. However, intrapopulational variability of this dental character is high, not only within *O. bridgesi* but also within *O. degus* (pers. obs.; Simonetti, pers. comm.), rendering it useless for taxonomic diagnosis. Although this does not invalidate *O. lunatus*, we believe it is a species of questionable status. Unfortunately, this uncertainty cannot be resolved due to scarcity of existing material.

Octodon degus is one of the best-studied small mammals of central Chile in terms of its ecology, behavior, and physiology (Woods & Boraker, 1975; Rosenmann, 1977; Yáñez & Jaksić, 1978; Contreras & Rosenmann, 1982; Meserve et al., 1984; and references therein). The degu is thought to be distributed south of Huasco Province (28°28′S) to Curicó (35°00′S) (Mann, 1978; Tamayo & Frassinetti, 1980). However, we are not aware of any specimens collected south of Santiago (fig. 1). The distributional limits of this abundant (Jaksić et al., 1981) and diurnal (Rosenmann et al., 1981) species seem correlated with those of the mediterranean shrubland formation known as *matorral* (Mann, 1978), whose northern and southern limits are determined by a scarcity and superabundance of water, respectively (Mann, 1964). The altitudinal limits of *O. degus* seem to be determined by its poor tolerance to low oxygen partial pressure (Rosenmann & Morrison, 1975) and also by altitudinal limits of its preferred habitat. This altitudinal limit is probably lower at higher latitudes because of temperature effects. Although *O. degus* can burrow, it feeds mainly on grasses and forbs (Meserve et al., 1983) above ground, which are covered by snow much of the year at high altitudes in the Andes. The highest capture record for this species is 2,000 m at 30°S. At 33°S (around Santiago) the altitudinal limit is probably below 1,200 m.

Octodon bridgesi is known to be distributed south from Cachapoal Province (34°15′S), along the foothills of the Andes to Malleco Province (38°40′S) (Greer, 1965; Tamayo & Frassinetti, 1980) (fig. 1). *Octodon lunatus* is reportedly found along the Cordillera Occidental (Cordillera de la Costa) from La Dormida (33°04′S) in Quillota Province to as far north as Illapel (31°30′S) (fig. 1). These limits

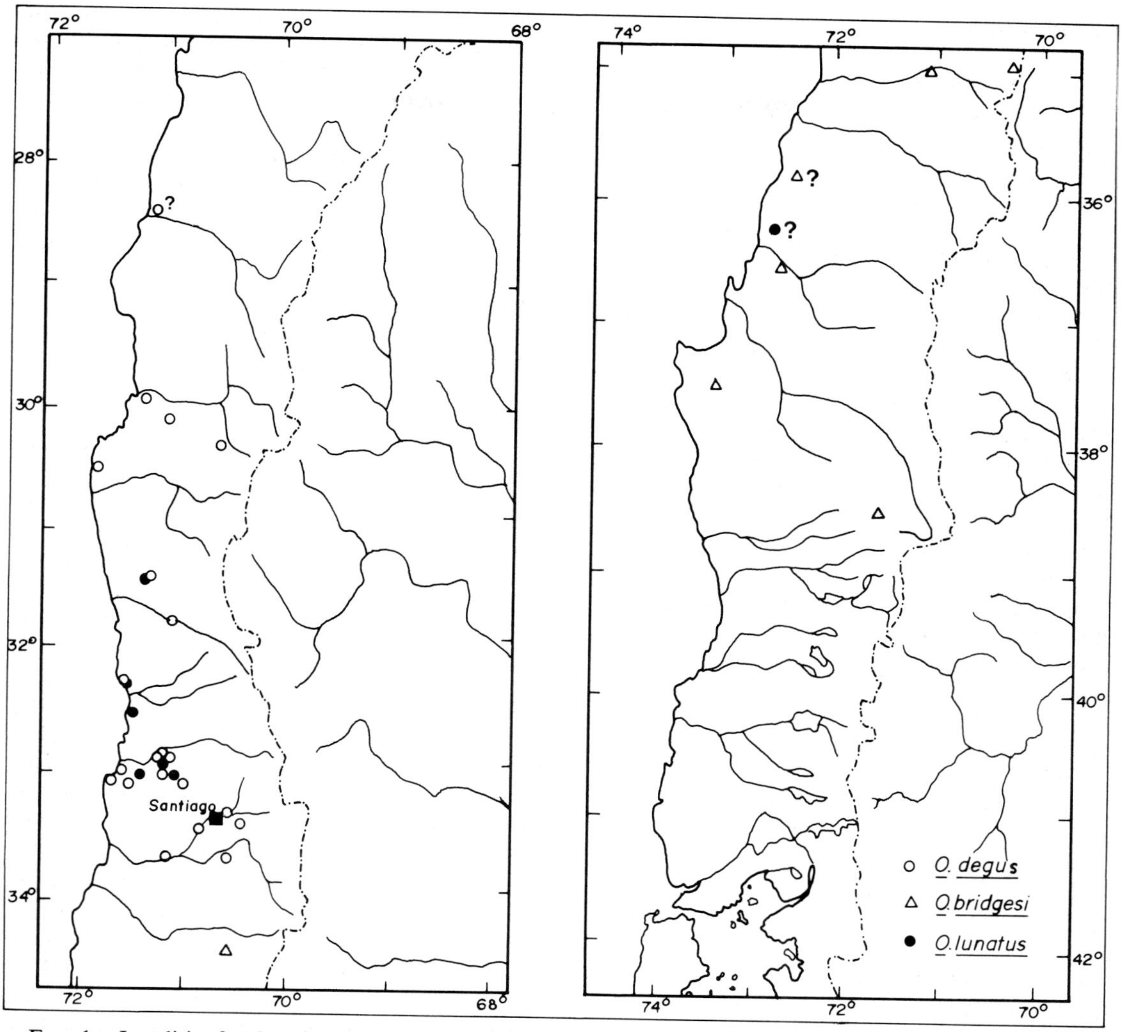

FIG. 1. Localities for *Octodon degus* (○), *O. bridgesi* (△), and *O. lunatus* (●). The "?" symbols represent uncertain and tentative identifications of captured specimens (see text).

clearly indicate a distributional overlap of at least *O. degus* and *O. lunatus*.

The distribution pattern of these three species has been obscured by the uncertain identification of *Octodon* specimens captured in Ñuble and Cauquenes provinces. These have been tentatively assigned to *O. lunatus* (cf. Tamayo & Frassinetti, 1980) and *O. bridgesi* (Rodríquez & Herrera, 1983), respectively (fig. 1). In our opinion this confusing situation reflects the lack of good diagnostic characters to separate *O. lunatus* from *O. bridgesi*. If the reported identifications prove correct, they would indicate that the distribution of *O. bridgesi* includes the Cordillera Occidental (Cordillera de la Costa), at least between 35°30′S and 37°00′S and/or that the southernmost limit of *O. lunatus* extends farther south along the coastal range than previously recognized. A third possibility is that *O. lunatus* and *O. bridgesi* are but a single species with a much larger geographic range.

Octodon bridgesi and *O. lunatus* seem not to burrow as much as *O. degus* (Greer, 1965; Ipinza et al., 1971), and their distributions are associated with denser, more humid scrub than that where *O. degus* is found. *Octodon bridgesi* is nocturnal, and preliminary data indicate that it has a greater evaporative water loss than *O. degus* (F. Bozinović, pers. comm.). Consequently, factors related to water availability may be important in determining the distribution of both *O. bridgesi* and *O. lunatus.*

Genus *Octodontomys* Palmer, 1903

The soco, *Octodontomys gliroides* (Gervais & D'Orbigny, 1844), is a monotypic species, found only in Andean and sub-Andean zones of southwestern Bolivia from La Paz to Potosí, in northwestern Argentina from Jujuy to La Rioja (Cabrera, 1961), and in northeastern Chile only in Tarapacá Province (Mann, 1945; Pine et al., 1979) (fig. 2). We believe that the apparently disjunct distribution of this species is due only to inadequate sampling. The soco has nocturnal habits according to Ipinza et al. (1971), but is diurnal according to Mann (1978). It inhabits very dry areas characterized by cacti and rock piles where it digs short burrows connected by superficial runways. This species eats succulent plants and the bark of resinous shrubs (Mann, 1945).

In many characteristics, such as a silky coat, a plantar surface with fine granulations, and an enlarged rostrum, *Octodontomys* is similar to *Abrocoma* (Abrocomidae), perhaps indicating adaptive convergence.

Genus *Octomys* Thomas, 1920

This is certainly the least-known octodontid genus. Woods (1982) included *Tympanoctomys barrerae* in the genus *Octomys,* so that the genus includes two species: *O. mimax* and *O. barrerae.*

Octomys inhabits mountainous regions in northwest Argentina in Catamarca, La Rioja, San Juan, and Mendoza provinces (Cabrera, 1961). The genus resembles *Octodontomys,* and the distributions of the two overlap in the northern provinces of Argentina. *Octomys* lives in desert scrub habitats and is nocturnal, a burrower, and an herbivore (Mares & Ojeda, 1982).

Genus *Aconaemys* Ameghino, 1891

According to Pearson (1984) two species are recognized in this genus: *A. fuscus* Waterhouse, 1841, and *A. sagei* Pearson, 1984. The former is considered to have two subspecies: *A. f. fuscus* Waterhouse, 1841, found in the slopes of the south-central Andes, and *A. f. porteri* Thomas, 1917, found in the southern extreme of the known geographic range of the genus (fig. 3). Unfortunately, type specimens of both forms lack reliable localities (Pearson, 1984). Available, geographically reliable representatives of these forms consist of two specimens of *fuscus* we captured at the confluence of Ríos Vergara and Nascimiento (35°08′S, 70°28′W) and eight specimens of *porteri* reported by Pearson (1984) from Ruca Malén. The southern form, *porteri,* is in part distinguished from *fuscus* by its bicolored tail. A word of caution on the validity of these is pertinent, because besides uncertain type localities (Osgood, 1943), the main diagnostic character has some variability not yet quantified. The tails of the specimens from Talca attributed to *fuscus* are bicolored, to some extent approaching the condition described for *A. f. porteri* (Pine et al., 1979), and the Ruca Malén specimens have moderately bicolored tails, not as the type of *A. f. porteri* (Pearson, 1984).

Aconaemys sagei, a smaller species than *A. fuscus,* was recently described from Neuquén Province, Argentina (Pearson, 1984). According to this author, some specimens from Chile, previously listed under *A. fuscus* by Osgood (1943), Greer (1965), and Pine et al. (1979), may also belong to this species. However, after examination of 21 specimens from Curicó, Ñuble, and Malleco (east and west) provinces, we concluded that none of the specimens can certainly be assigned to *A. sagei.* We found considerable overlap between specimens from these localities. Thus, we consider all known Chilean specimens as belonging to one species, *A. fuscus,* which is found along the Andes between 35°S and 41°S and in the coastal Cordillera de Nahuelbuta (fig. 3).

Aconaemys is fossorial, although to a lesser extent than *Spalacopus cyanus.* Its tunnel systems are more superficial, and its runways resemble somewhat those of voles (*Microtus*) of the northern hemisphere (Greer, 1965). This species seems to have a wide habitat tolerance, considering that it has been found in *Nothofagus* and *Araucaria* forests in regions with high rainfall (Osgood, 1943; Greer, 1965; Mann, 1978), to above timberline in close association with bunch grasses (*Festuca* and *Stipa* spp.) (Contreras & Torres-Mura, pers. comm.).

Competitive exclusion is thought to be common among fossorial mammals (Nevo, 1979). This might be the case with *Spalacopus* and *Aconaemys* (fig. 3), provided that the latter species replaces the former to the south. However, the range of *Aconaemys* seems to overlap that of other fossorial

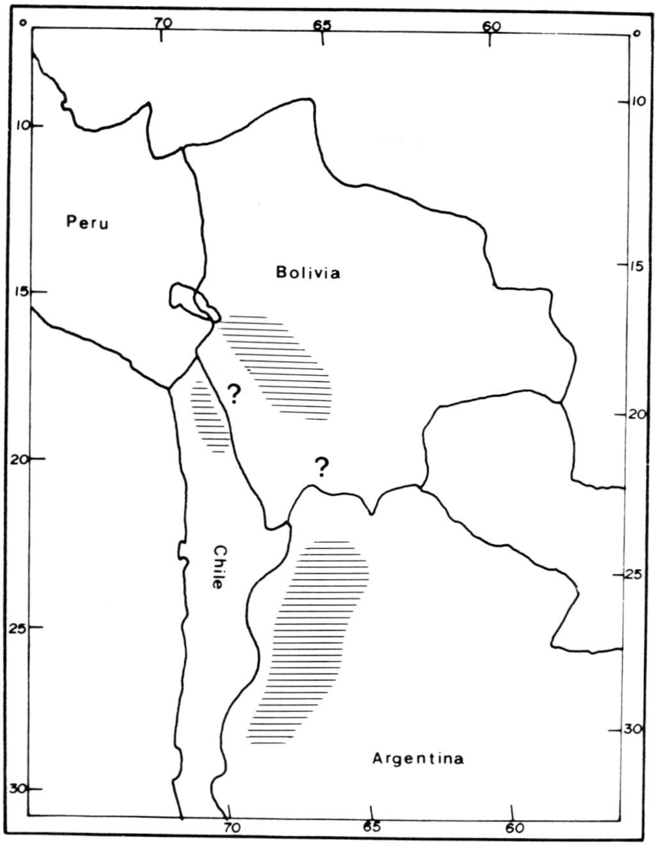

FIG. 2. Range of distribution (stippled) of *Octodontomys gliroides* in Bolivia, Chile, and Argentina.

herbivorous rodents, such as *Ctenomys maulinus* and *Ctenomys* sp. (see Gallardo, 1979; Pearson, 1984). At present the lack of data does not permit assessment of the actual fine-grain distributions of these two genera. They might be parapatric, but if so, it could be difficult to resolve whether their separation is caused by differences in microhabitat preferences, by competition, or by historical factors of colonization.

Genus *Spalacopus* Wagler, 1832

This genus is monotypic, and the taxonomic validity of its subspecies is debatable (Mann, 1978; Tamayo & Frassinetti, 1980). We agree with Mann (1978) that the distinction between *S. c. cyanus* (Molina, 1782) from the coast of central Chile and *S. c. maulinus* Osgood, 1943 from south-central Chile should be carefully reconsidered, because it is based on subtle cranial traits of a small number of specimens, and because individuals vary greatly. In contrast, the distinction between *S. c. cyanus* and *S. c. poeppigii* Wagler, 1832 from the Andes seems to be valid. Andean populations of *Spalacopus* are phenotypically distinct from coastal ones. Those from the Andes are larger (Reig et al., 1972; Yáñez & Zülch, 1981), probably in response to thermal factors and food availability (Contreras, 1983, 1986). These two forms also seem to differ in skull, tooth morphology, and color pattern (Reig et al., 1972). Because of the small number of *Spalacopus* in the Longitudinal Valley (Central Valley) and adjacent areas (see below), we believe that gene flow between the coastal and mountain populations is probably low, although these rodents do not show differences in G- and C-banding of chromosomes (Zülch et al., 1982) or in the electrophoretic patterns of six blood proteins (Woods & Kilpatrick, pers. comm.). *Spalacopus tabanus* Thomas, 1925 was treated as a subspecies of *S. cyanus* by Osgood (1943). Because the single known specimen has a provenance of "South of Chile" and is quite large, we agree with Reig et al.

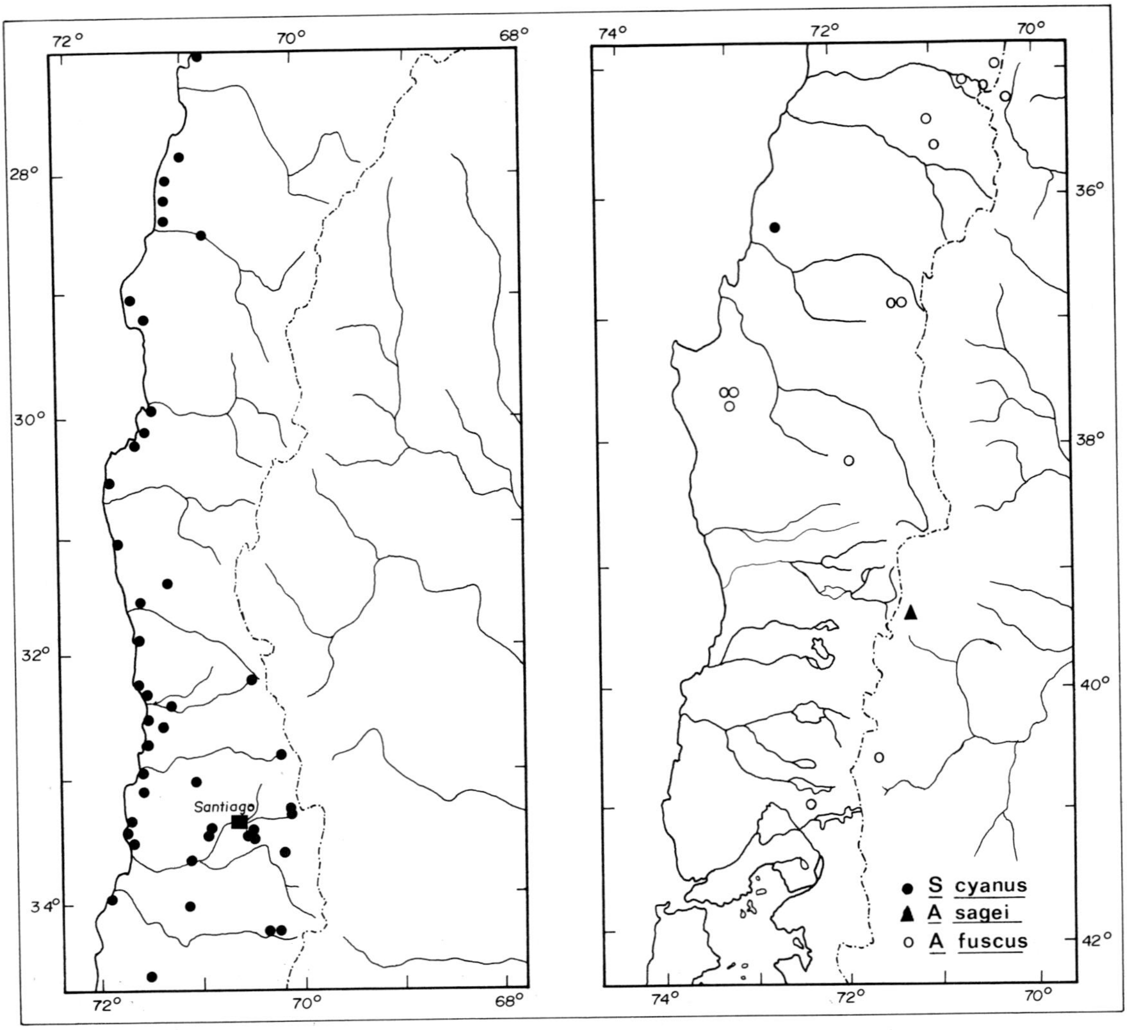

FIG. 3. Localities for *Aconaemys fuscus* (△), *A. sagei* (▲), and *Spalacopus cyanus* (●).

(1972) and Tamayo and Frassinetti (1980) that the animal probably represents *S. c. poeppigii* from the Andes, rather than *S. c. cyanus* as proposed by Mann (1978).

The low variability within *Spalacopus* has been attributed to the great mobility of its colonies (Reig, 1970). However, studies of its home ranges by radioactive tagging indicate that these are very stable areas (Torres-Mura & Contreras, 1983).

Taking the new data presented here into consideration, *Spalacopus cyanus* is probably the best known octodontid in terms of its geographic range (fig. 3). Its populations are found along the Pacific coast from Caldera (27°03′S) down to Quirihue (36°17′S) in Ñuble Province and also along the Andes from Alicahue (32°19′S, 70°39′W) to Los Cipreses (34°01′S, 70°29′W) up to above 3,000 m. Small populations are also found in ravines draining into the Longitudinal Valley (Central Valley) from the Cordillera Occidental (Cordillera de la Costa) and the Andes. The altitudinal limit of *S. cyanus* seems to be set by lack of food rather than by low oxygen critical pressure that allows it to tolerate severe hypoxic conditions resulting from the combination of high altitude and burrowing habits (Contreras, 1983). Within its geographic range, *S. cyanus* inhabits areas with shrub cover of no more than 60%, which allows the development of an herb stratum, containing the geophytes and hemicryptophytes that form its main food.

The northernmost population found along the coast is Quebrada Pajonales, located 22 km north of Caldera. In this locality we have found extensive unoccupied burrow systems, similar to those indicated by Osgood (1943). Although we failed to capture specimens, we found skull remains in burrowing owl pellets. The northern populations of *Spalacopus* are mainly restricted to the coast, but can also be found inland in the valley of some rivers (fig. 3). These populations can live there because the extreme desert conditions of the Atacama Desert are ameliorated along the coast by the formation of fog banks (Tricart, 1969), and also because extensive marine terraces with altitude lower than 300 m exist there, permitting the development of suitable vegetation for *Spalacopus.* This condition disappears along the coast north of 26°30′S where the land sharply rises from sea level to 1,000 m or more. Extreme desert-like conditions may also determine the northernmost limit of *S. cyanus* populations along the Andes, a limit which is farther south than along the coast (fig. 3). This may be due to the fact that in this area (about 29°S to 30°S) the desert penetrates the Andes and crosses to Argentina. There it continues southward along the eastern side of the Andes, producing sharp climatic and vegetational changes along the Andes (Arroyo et al., 1982; Villagrán et al., 1983) which are not suitable for *Spalacopus.*

The southernmost distributional limit of *S. cyanus* may be determined by two factors: (1) an increasing scarcity of open habitats along the coastal ranges approaching the south temperate rain forest and (2) the presence of *Aconaemys fuscus* on the southern Andes (fig. 3). *Spalacopus* seems absent from the valleys of Ríos Teno and Tinguiririca in the Andes, although the habitat there is not different from that farther north. However, *Aconaemys fuscus* occurs in these two valleys, therefore we postulate that the Andean southern limit of *S. cyanus* is somewhere between the basins of Ríos Cachapoal and Tinguiririca, which coincides with the northernmost limit of *A. fuscus.* Perhaps competitive exclusion is occurring between these species.

Discussion

The distribution of living organisms is determined by a combination of historical, evolutionary, and ecological factors. We propose that the present distribution of octodontids is related to the eco-evolutionary history of South America, especially to that of the Patagonian Subregion (Hershkovitz, 1972).

The earliest records of South American hystricognaths date from the Deseadan Oligocene (35 MYBP) of Patagonia and Bolivia (Wood & Patterson, 1959; Patterson & Wood, 1982). The earliest forms were represented by the families Octodontidae, Echimyidae, Eocardiidae, Dasyproctidae, Dinomyidae, Chinchillidae, and Erethizontidae. Of these, octodontids appear to be the most primitive (Patterson & Wood, 1982). Early octodontids and echimyids were probably similar to each other, at least in molar structure (Wood & Patterson, 1959). Since the postcranial skeleton of the early octodontids is similar to that of the present day *Octodon,* we infer that both octodontids and echimyids were generalized ground-dwelling forms living in the woodland habitats that covered much of southern South America. At that time the Andes had yet to rise, and the landscape was relatively flat and homogeneous from east to west (Webb, 1978), with a small thermal gradient from north to south (Simpson, 1983).

During the early Miocene, the Patagonian Subregion still had a humid climate supporting widespread woodlands (Menéndez, 1961). From the Miocene this pattern changed drastically, but not suddenly, with the Andean orogeny, the formation of the Humboldt Current and of the Pacific Anticyclone. The uplift of the Andes permitted the maintenance of the humid forest on the western slope, and the resulting rain shadow brought about the replacement of forest by savannas, grassland, and steppes to the east of the Andes, thus producing the present pampas. On the other hand, the Humboldt Current brought about the desertification of southern Peru and northern Chile, leading to the formation of the Atacama Desert, the southern retreat of forest, and the establishment of mediterranean scrub and sclerophyllous forest in central Chile. As a result of these events, two relatively xeric areas and two mesic areas were formed: (1) a desert scrub in the Atacama region, (2) a steppe vegetation in Patagonia, (3) a high-altitude steppe in the altiplano, and (4) an increasingly mesic gradient south of the Atacama Desert, with vegetation ranging from scrublands to temperate rain forests. Since that time, except for fluctuations due to Pleistocene glaciations, this pattern has remained basically the same.

The co-occurrence of octodontids and echimyids in the Patagonian Subregion persisted through the Miocene (Pascual et al., 1965). At this

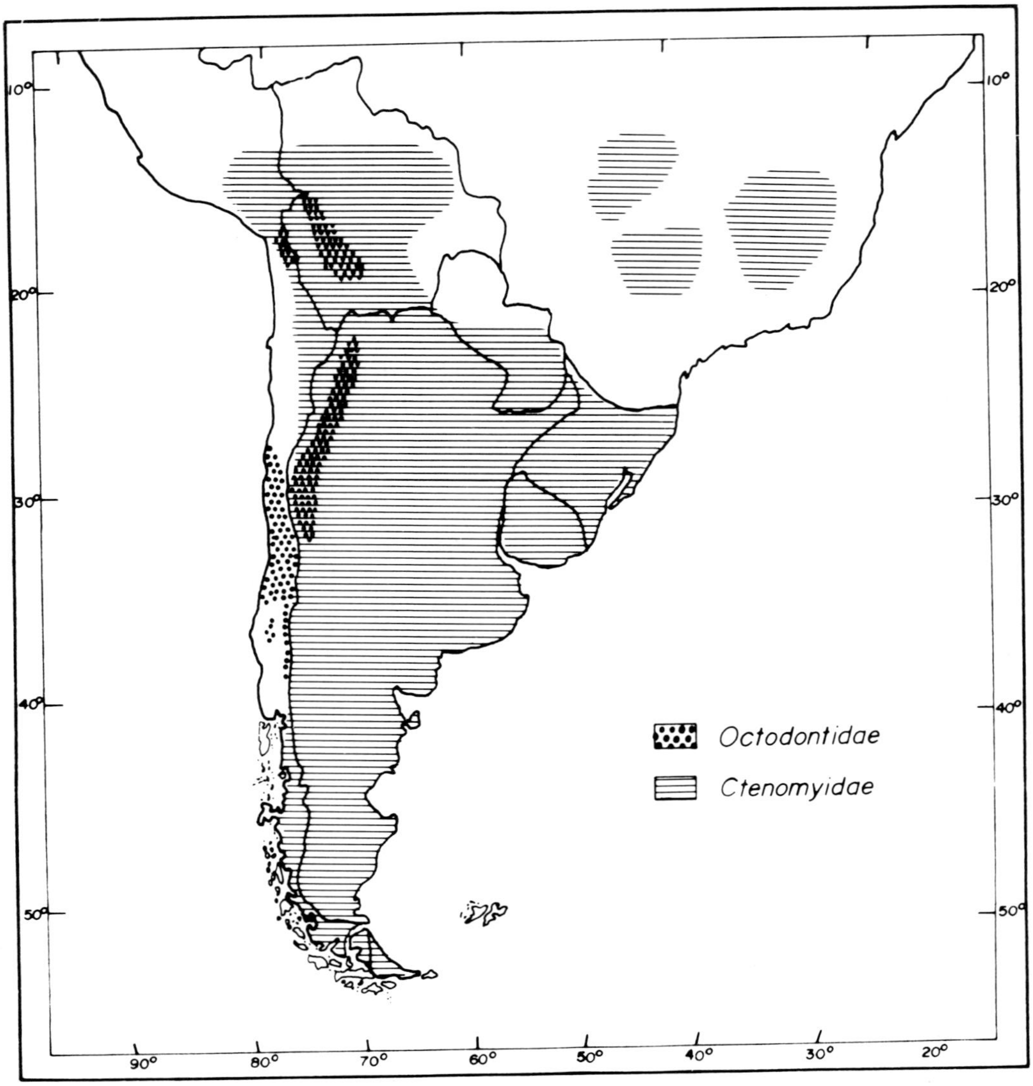

FIG. 4. Map of southern South America, showing the primarily eastern and western distributions of the Ctenomyidae and Octodontidae, respectively.

time octodontids were still similar to the earliest known forms in the family, but echimyids had clearly diverged from the primitive condition and had diversified into at least three groups (Wood & Patterson, 1959; Patterson & Wood, 1982). Unfortunately, echimyid fossil remains consist only of cranial fragments, with no postcranial elements. However, they were probably in a morphologically intermediate stage between the early generalized condition and the more scansorial-arboreal forms of today.

The fossil record indicates that echimyids disappeared from Patagonia during the early Pliocene. Subsequently, octodontid diversity increased markedly during the Pliocene in the pampas region (Patterson & Pascual, 1972). The reasons why echimyids disappeared, while octodontids remained in the Patagonian Subregion, are unclear. It is likely that echimyids had become more closely associated to tropical or subtropical forests and retreated to the north with them. This would explain the absence of echimyids from the temperate rain forest that became established in the south, west of the Andes. This forest developed from the most cold-adapted elements of the austral-antarctic flora (Van der Hammen & Cleef,

1983). A link between echimyids and tropical habitats is also suggested by their establishment in tropical North America after the Great American Interchange, but their failure to occupy the temperate zones of that continent (Patterson & Pascual, 1972).

During the Upper Pliocene, fossorial octodontid forms, not previously known, appear on the steppes. These fossorial forms belong to the family Ctenomyidae and were derived from octodontids (see Reig & Kiblisky, 1969; Reig, 1970).

The genus *Ctenomys* first appears in the early Pleistocene of Argentina (Chapadmalalian), but it more likely originated in the late Pliocene (Reig & Kiblisky, 1969). The several Pleistocene species of *Ctenomys* and the roughly 30 species at present denote a genus with a high rate of diversification and expansion from its probable center of origin (Chapadmalal?). *Ctenomys* now extends throughout the Patagonian Subregion and also to the southern part of the Brasilian Subregion of the Neotropics. During its expansion, *Ctenomys* entered Chile at the end of the Pleistocene (Tamayo & Frassinetti, 1980), mainly through mesic areas of the altiplano and the low xeric areas of Patagonia. They were thus able to colonize (1) the marginal and high-altitude zone of northern Chile and southern Peru (an extension of the Bolivian puna), (2) the steppes of Aysén and Magallanes, which are extensions of the Argentinian pampa into Chile, and (3) certain Andean areas in south-central Chile, where Patagonian vegetation interdigitates with montane forest between snow line and tree line. The distribution of *Ctenomys* thus essentially encloses the fossorial forms of the Octodontidae (see below) and partially overlaps with some nonfossorial octodontids (fig. 4).

After a Pliocene expansion, the area occupied by octodontids decreased; they disappeared from the pampas and persisted only along the slopes of the Andes. Unfortunately, there are almost no specimens from the Pleistocene, but some generalized ground-dwelling forms of the Holocene are very similar to those of the Oligocene and Miocene (Wood & Patterson, 1959). Forms present during the Pleistocene were also probably similar.

The low species diversity (monotypy) and the pronounced morphological and physiological specializations to fossorial life found in *Spalacopus* and *Aconaemys* indicate that these forms probably arose *in situ*. Thus, their origin would have been contemporaneous with the development of open areas of sclerophyllous scrub, savanna-like and grasslands vegetation in central Chile following Pleistocene glaciations (Troncoso et al., 1980). Their restricted geographic distribution may be a result of the rapid colonization of suitable habitat by *Ctenomys*.

In summary, the present distribution of octodontids seems to be the result of historical interactions between the Andean orogeny, the associated climatic and vegetational changes, and the various evolutionary trends of the echimyids, ctenomyids, and the octodontids themselves.

Acknowledgments

We would like to express our appreciation to F. Jaksić, M. T. Kalin-Arroyo, A. Troncoso, and C. Villagrán for valuable comments. We especially thank B. D. Patterson and R. H. Pine for their constructive comments as reviewers. Ms. V. Aguirre and M. L. Uribe provided competent technical assistance. Specimens of *Aconaemys* were made available to us by A. Spotorno, Universidad de Chile; T. Cekalović, Universidad de Concepción; and M. Gallardo, Universidad Austral de Chile. Partial support was provided by Facultad de Ciencias, Universidad de Talca. This work is in honor of P. Hershkovitz for his great contribution to the understanding of South American mammals.

Literature Cited

Arroyo, M. T. K., C. Villagrán, C. Marticorena, and J. Armesto. 1982. Flora y relaciones biogeográficas en los Andes del norte de Chile (18°19°S), pp. 71–92. *In* Veloso, A., and E. Bustos, eds., El ambiente natural y las poblaciones humanas de los Andes del Norte Grande de Chile (Arica, Lat. 18°28′S), Vol. 1. Roslac, Montevideo.

Cabrera, A. 1961. Catálogo de los mamíferos de América del Sur. Revista del Museo Argentino de Ciencias Naturales "Bernardino Rivadavia", Ciencias Zoológicas, **4**(2): 309–732.

Contreras, L. 1983. Ecophysiology of fossorial mammals: A comparative study. Ph.D. Diss., University of Florida, Gainesville, 185 pp.

———. 1986. Bioenergetics and distribution of fossorial *Spalacopus cyanus* (Rodentia): Thermal stress or cost of burrowing? Physiological Zoology, **59:** 20–28.

Contreras, L., and M. Rosenmann. 1982. Thermoregulation in the testes of *Octodon degus*: A non-scrotal rodent. Physiological Zoology, **55**(2): 144–147.

Gallardo, M. 1979. Las especies chilenas de *Cteno-*

mys (Rodentia, Octodontidae). I. Estabilidad cariotípica. Archivos Biología y Medicina Experimentales, **12:** 71–82.

GLANZ, W. E. 1977. Comparative ecology of small mammal communities in California and Chile. Ph.D. Diss., University of California, Berkeley, 300 pp.

GREER, J. K. 1965. Mammals of Malleco Province, Chile. Publications of The Museum, Michigan State University, Biological Series, **3:** 51–151.

HERSHKOVITZ, P. 1962. Evolution of Neotropical cricetine rodents (Muridae). Fieldiana: Zoology, **46:** 1–524.

———. 1972. The recent mammals of the Neotropical Region: A zoogeographic and ecological review, pp. 311–432. *In* Keast, A., F. C. Erk, and B. Glass, eds., Evolution, Mammals, and Southern Continents. State University of New York Press, Albany, 543 pp.

IPINZA, J., M. TAMAYO, AND J. ROTTMANN. 1971. Octodontidae en Chile. Noticiario Mensual Museo Nacional de Historia Natural, Santiago, Chile, **183:** 3–10.

JAKSIĆ, F. M., J. L. YÁÑEZ, AND E. R. FUENTES. 1981. Assessing a small mammal community in central Chile. Journal of Mammalogy, **62**(2): 391–396.

MANN, G. 1945. Mamíferos de Tarapacá. Observaciones realizadas durante una expedición al alto norte de Chile. Biológica, **2:** 23–134.

———. 1964. Compendio de Zoología. Vol. I. Ecología y Biogeografía. Santiago, Chile, 37 pp.

———. 1978. Los pequeños mamíferos de Chile. Gayana: Zoología, **40:** 1–342.

MARES, M. A., AND R. A. OJEDA. 1982. Patterns of diversity and adaptation in South American hystricognath rodents, pp. 393–432. *In* Mares, M. A., and H. H. Genoways, eds., Mammalian Biology in South America. A Symposium Held at the Pymatuning Laboratory of Ecology, May 10–14, 1981. Special Publication Series, Pymatuning Laboratory of Ecology, University of Pittsburgh, **6:** 1–539.

MENÉNDEZ, C. A. 1961. Estípite petrificado de una nueva Cyatheaceae del Terciario de Neuquén. Boletín de la Sociedad Argentina de Botánica, **9:** 331–358.

MESERVE, P. L., R. E. MARTIN, AND J. M. RODRÍGUEZ. 1983. Feeding ecology of two Chilean caviomorphs in a central mediterranean savanna. Journal of Mammalogy, **64**(2): 322–325.

———. 1984. Comparative ecology of the caviomorph rodent *Octodon degus* in two Chilean mediterranean type communities. Revista Chilena de Historia Natural, **57:** 79–89.

NEVO, E. 1979. Adaptive convergence and divergence of subterranean mammals. Annual Review of Ecology and Systematics, **10:** 269–308.

OSGOOD, W. H. 1943. The mammals of Chile. Field Museum of Natural History, Zoological Series, **30:** 1–268.

PASCUAL, R., J. PISANO, AND E. J. ORTEGA. 1965. Un nuevo Octodontidae (Rodentia, Caviomorpha) de la formación Epecuen (Plioceno Medio) de Hidalgo (Provincia de La Pampa). Consideraciones sobre los Ctenomyinae Reig, 1958, y la morfología de sus molariformes. Ameghiniana, **4:** 19–29.

PATTERSON, B., AND R. PASCUAL. 1972. The fossil mammal fauna of South America, pp. 247–309. *In* Keast, A., F. C. Erk, and B. Glass, eds., Evolution, Mammals, and Southern Continents. State University of New York Press, Albany, 543 pp.

PATTERSON, B., AND A. E. WOOD. 1982. Rodents from the Deseadan Oligocene of Bolivia and the relationships of the Caviomorpha. Bulletin of the Museum of Comparative Zoology, **149**(7): 371–543.

PEARSON, O. P. 1958. A taxonomic revision of the rodent genus *Phyllotis*. University of California Publications in Zoology, **56**(4): 391–496.

———. 1984. Taxonomy and natural history of some fossorial rodents of Patagonia, southern Argentina. Journal of Zoology (London), **202:** 225–237.

PINE, R. H., S. D. MILLER, AND M. L. SCHAMBERGER. 1979. Contributions to the mammalogy of Chile. Mammalia, **43:** 339–376.

REIG, O. A. 1970. Ecological notes on the fossorial octodont rodent *Spalacopus cyanus* (Molina). Journal of Mammalogy, **51**(3): 592–601.

———. 1981. Teoría del origen y desarrollo de la fauna de mamíferos de América del Sur. Monografía Naturae. Museo Municipal de Ciencias Naturales Lorenzo Scaglia, Mar del Plata, Argentina, 162 pp.

REIG, O. A., AND P. KIBLISKY. 1969. Chromosome multiformity in the genus *Ctenomys* (Rodentia, Octodontidae). Chromosoma (Berlin), **28:** 211–244.

REIG, O. A., A. SPOTORNO, AND R. FERNÁNDEZ. 1972. A preliminary survey of chromosomes in populations of the Chilean burrowing octodont rodent *Spalacopus cyanus* (Molina) (Caviomorpha, Octodontidae). Biological Journal of the Linnean Society, **4:** 29–38.

RODRÍGUEZ, J. A., AND L. A. HERRERA. 1983. Evaluación del daño causado por *Octodon bridgesi* en plantaciones de *Pinus radiata,* p. 163. Actas 9º Congreso Latinoamericano de Zoología, Arequipa, Perú.

ROSENMANN, M. 1977. Regulación térmica en *Octodon degus*. Medio Ambiente, **3:** 127–131.

ROSENMANN, M., AND P. R. MORRISON. 1975. Metabolic responses of highland and lowland rodents to simulated high altitude and cold. Comparative Biochemistry and Physiology, **51A**(4): 523–530.

ROSENMANN, M., G. RUIZ, AND A. CERDA. 1981. Relaciones de actividad espontánea de *Octodon degus* y *Abrocoma benneti* en Chile central. Medio Ambiente, **5**(1–2): 125–129.

ROWLANDS, I. W., AND B. J. WEIR (EDS.). 1974. The biology of hystricomorph rodents. Symposium of the Zoological Society of London, **34:** 1–482.

SIMPSON, B. B. 1983. An historical phytogeography of the high Andean flora. Revista Chilena de Historia Natural, **56:** 109–122.

TAMAYO, M., AND D. FRASSINETTI. 1980. Catálogo de los mamíferos fósiles y vivientes de Chile. Boletín Museo Nacional de Historia Natural, Chile, **37:** 323–399.

TORRES-MURA, J. C., AND L. C. CONTRERAS. 1983. Ambito de hogar en *Spalacopus cyanus* (Rodentia, Octodontidae). Archivos de Biología y Medicina Experimentales, **16**(2): R–223.

TRICART, J. 1969. Le modelé des région seches. *In*

Tricart, J., and A. Cailleux, eds., Traité de Géomorphologie, IV. SEDES, Paris, 472 pp.

TRONCOSO, A., C. VILLAGRÁN, AND M. MUÑOZ. 1980. Una nueva hipótesis acerca del origen y edad del bosque de Fray Jorge (Coquimbo, Chile). Boletín Museo Nacional de Historia Natural, Chile, **37:** 117–152.

VAN DER HAMMEN, T., AND A. M. CLEEF. 1983. Datos para la historia de la flora andina. Revista Chilena de Historia Natural, **56:** 97–107.

VILLAGRÁN, C., M. T. KALIN ARROYO, AND C. MARTICORENA. 1983. Efectos de la desertización en la distribución de la flora andina de Chile. Revista Chilena de Historia Natural, **56:** 137–157.

WEBB, S. D. 1978. A history of savanna vertebrates in the New World. Part II: South America and the Great Interchange. Annual Review of Ecology and Systematics, **9:** 393–426.

WEBB, S. D., AND L. G. MARSHALL. 1982. Historical biogeography of Recent South America land mammals, pp. 39–52. *In* Mares, M. A., and H. H. Genoways, eds., Mammalian Biology in South America. A Symposium Held at the Pymatuning Laboratory of Ecology, May 10–14, 1981. Volume 6, Special Publication Series, Pymatuning Laboratory of Ecology, University of Pittsburgh, Linesville, Pa., 539 pp.

WOOD, A. E., AND B. PATTERSON. 1959. The rodents of the Deseadan Oligocene of Patagonia and the beginning of South American rodent evolution. Bulletin of the Museum of Comparative Zoology at Harvard University, **120**(3): 279–428.

WOODS, C. A. 1982. The history and classification of South American hystricognath rodents: Reflections on the far away and long ago, pp. 377–392. *In* Mares, M. A., and H. H. Genoways, eds., Mammalian Biology in South America. A Symposium Held at the Pymatuning Laboratory of Ecology, May 10–14, 1981. Special Publication Series, Pymatuning Laboratory of Ecology, University of Pittsburgh, **6:** 1–539.

WOODS, C. A., AND D. K. BORAKER. 1975. Octodon degus. Mammalian Species, **67:** 1–5.

YÁÑEZ, J. L., AND F. M. JAKSIĆ. 1978. Historia Natural de *Octodon degus* (Molina) (Rodentia, Octodontidae). Publicación Ocasional Museo Nacional de Historia Natural, Santiago, Chile, **27:** 3–11.

YÁÑEZ, J., AND R. ZÜLCH. 1981. Morfometría de tres poblaciones de *Spalacopus cyanus* Rodentia. Archivos de Biología y Medicina Experimentales, **14**(3): R–304.

ZÜLCH, R., L. E. WALKER, AND J. PINCHEIRA. 1982. Comparación de los cariotipos bandeados G y C de las formas costera y andina de *Spalacopus cyanus* y *Octodon degus* (Rodentia, Octodontidae). Archivos de Biología y Medicina Experimentales, **15**(2): R–164.

Population Dynamics and Ecology of Small Mammals in the Northern Chilean Semiarid Region

Peter L. Meserve and Éric Le Boulengé

ABSTRACTS

Small mammal populations were studied in a northern Chilean semiarid thorn scrub community for 17 months during a period of sparse rainfall (27–76 mm annually, 1973–1975). A permanent 1.4-ha grid was trapped for four consecutive nights bimonthly with mark-and-recapture techniques, and standard data were taken for the estimation of population parameters. Snap-trapping in similar habitat provided reproductive information. Seven species were captured in the field (six rodents, one marsupial), but only four of them yielded sufficient captures for a quantitative study (three sigmodontines, one hystricomorph). Reproduction was strongly seasonal, starting and ending earlier in *Octodon degus* and *Akodon longipilis* (July–November) than in *Phyllotis darwini* and *Akodon olivaceus* (September–January for the former, September–November for the latter). The latter two species may produce more than one litter per reproduction period. Seasonal reproduction, generally high trappability, and clear distinction between the weight fluctuations of adults and juveniles—at least during the first two censuses after the latter appear—allowed for a classification of individuals into a cohort of adults, one of young of the year and a "cohort" of probable immigrants. Besides enumeration techniques which give an overall view of population trends, we used the Modified Calendar of Captures technique to describe the dynamics of cohorts and estimate the numbers of nontrappable juveniles from the date of their first capture back to their birth period. *Akodon olivaceus* had low survival in reproductive months and low recruitment in the second year of study; this resulted in a steady population decline through the study period. *Akodon longipilis* had a high survival rate and low recruitment in both years, and showed very constant population numbers. *Phyllotis darwini* and *O. degus* had a generally lower survival, but showed a high potential rate of reproduction, albeit with low juvenile survival. Dispersal is probably a crucial feature in the demography of these two species. Despite changes in the relative numbers and proportions of the component species, the fauna in this community was remarkably constant in overall numbers during four years of observation. Since the principal species studied here are widely distributed, physiologically and morphologically unspecialized for arid environments, and represent diverse trophic specializations, ecological adaptations and life history characteristics seem more important for their persistence and success than are biogeographical considerations or evolutionary preadaptations.

Poblaciones de pequeños mamíferos fueron estudiadas en una comunidad semiárida de matorrales espinosos en el norte de Chile por unos 17 meses durante un período de escasas

From the Department of Biological Sciences, Northern Illinois University, DeKalb, IL 60115-2861 (Meserve), and Unité d'Écologie, Université Catholique de Louvain, Place Croix du Sud, 5, 1348 Louvain-la-Neuve, Belgium (Le Boulengé).

lluvias (27–76 mm anuales, 1973–1975). Una red de trampas cubriendo 1.4 ha fue accionada por cuatro noches consecutivas bimensualmente. Los datos tipo para determinar los parámetros de población se obtuvieron por la técnica de captura, marquaje y recaptura. El uso de trampas de golpe en similar hábitat proveyo información reproductiva. Un total de siete especies fueron capturadas en el campo (seis roedores y un marsupial), pero solo cuatro de ellos produjeron suficientes capturas para un estudio cuantitativo (tres sigmodontines, un hystricomorfo). La reproducción fue marcadamente estacional, empezando y terminando más temprano en *Octodon degus* y *Akodon longipilis* (Julio–Noviembre) que en *Phyllotis darwini* y *Akodon olivaceus* (Septiembre–Enero para el primero, Septiembre–Noviembre para el último). Las últimas dos especies pueden producir más que una camada por período de reproducción. Estacionalidad de reproducción, generalmente alta probabilidad de captura y clara distinción entre las fluctuaciones de peso de los jovenes y de los adultos durante los dos primeros censos después de la apariencia de los primeros, permitieron una clasificación de individuos dentro de una cohorte de adultos, una de los jovenes del año y una "cohorte" de probables inmigrantes. Además de las técnicas de enumeración, las cuales dan una vista general de las tendencias de la población, usamos el Calendario Modificado de Capturas para describir las dinámicas de cohortes y estimar los números de jovenes no trampeables desde la fecha de su primera captura hacia atras al período de nacimiento. *Akodon olivaceus* tuvo una supervivencia baja en meses reproductivos y bajo reclutamiento en el segundo año de estudio, lo que resultó en un decrecimiento contínuo de sus números. *Akodon longipilis* tuvo una alta tasa de supervivencia y bajo reclutamiento en ambos años y mostró números muy constantes de población. *Phyllotis darwini* y *O. degus* mostraron una supervivencia generalmente baja, pero una alta tasa potencial de reproducción, sin embargo con alta tasa de mortalidad juvenil. Dispersión es probablemente una característica crucial en la demografía de estas dos especies. A pesar de los cambios en los números y proporciones relativas de las especies componentes, la fauna de esta comunidad fue notablemente constante en números globales durante cuatro años de observación. Ya que estas especies son ampliamente distribuidas, fisiológicamente y morfológicamente no especializadas para ambientes áridos, y tróficamente diversas, adaptaciones ecológicas y características de historia de vida parecen más importantes para su persistencia y éxito que consideraciones biogeográficas o preadaptaciones evolutivas.

Populações de pequenos mamíferos que habitam uma comunidade semi-árida de arbustos espinhosos no Chile, foram estudadas durante 17 meses, num período com pouca chuva (27–76 mm anuais entre 1973 e 1975). Durante quatro noites consecutivas, a cada dois meses, foram armadas armadilhas ao longo de uma grade permanente de 1,4 hectares. Foram usadas as técnicas de marcação e recaptura (mark and recapture), e todos dados convencionais foram tomados. Para informações reprodutivas, usaram-se alçapões (snap-traps) em um habitat semelhante. Um total de sete espécies foram capturadas (seis roedores e um marsupial), mas apenas quatro renderam números suficientemente altos para uma análise quantitativa (três cricetídeos e um histricomorfo). A época reprodutiva de cada espécie foi altamente relacionada às epocas do ano, começando e terminando mais cedo para *Octodon degus* e *Akodon longipilis* (julho a novembro), e mais tarde para *Phyllotis darwini* e *Akodon olivaceus* (setembro a janeiro e setembro a novembro, respectivamente). As últimas duas espécies podem produzir mais de uma cria por período reprodutivo. Dada a existencia de épocas reprodutivas bem-definidas, os pesos claramente diferentes dos jovens capturados nos seus primeiros dois censos, e, em geral, a frequente captura dos animais, foi possível classificar indivíduos em grupos de adultos, de jovens-do-ano, e de prováveis imigrantes. Além das técnicas de enumeração dos animais, que dão uma idéia dos padrões gerais das populações, usamos também a tecnica de Calendário Modificado de Capturas (Modified Calendar of Captures), para descrever dinamicas entre os grupos de idades diferentes, e para estimar o número de jovens não-capturáveis. Os resultados indicam que *A. olivaceus* sofreu baixa sobrevivencia durante os meses reprodutivos e baixa recruta durante o segundo ano de estudo, provocando assim um declíneo constante na sua população durante o período de observação. *Akodon longipilis* teve alta sobrevivencia mas baixa recruta durante os dois anos, mostrando então um número de população muito estável. *Phyllotis darwini* e *O. degus* tiveram um nível geral de sobrevivencia mais baixo, porem mostraram um alto potencial reprodutivo embora com baixa sobrevivencia juvenil. Dispersão é provàvelmente um fator crítico na demografia destas duas especies. Apesar das mudancas nos números relativos das populações e nas proporções das espécies componentes, a fauna desta comunidade semi-árida foi notàvelmente estável durante os quatro anos de observação. Como as espécies principais deste estudo representam animais de vasta distribuição, nãoespecializados fisiológica ou morfològicamente para ambientes áridos, e de níveis tróficos diversos, seu sucesso e sua persistencia parecem ser mais ligados a adaptações ecológicas, e a características biológicas vitais (life history), do que a biogeografia, ou a preadaptações evolutivas.

Introduction

The warm arid and semiarid regions of South America are of relatively recent origin. The Argentine Monte desert was formed in the Miocene-Pliocene period and was strongly influenced by the rise of the Andes (Solbrig, 1976). The western Pacific coastal desert and central mediterranean zone are even more recent, having been strongly influenced by the progressive cooling of the Pacific Ocean, the maximum rise of the Andes in the late Pliocene, and montane Pleistocene glaciations (Axelrod, 1973; Simpson, 1975a,b; Solbrig, 1976). The small mammal faunas of both the Monte and Pacific coastal deserts are typically depauperate in comparison to other desert faunas, are inhabited by few widely distributed species, and have few autochthonous forms (Osgood, 1943; Baker, 1967; Mares, 1975a, 1976, 1980). Even more indicative of a recent origin of these faunas, their component species are relatively unspecialized morphologically and few of them have developed the capacity to exist without free water, as is typical in other deserts (Mares, 1975a,b, 1980; Meserve, 1978). This has not prevented the development of convergence in ecologically related morphological traits for members of the Sonoran and Monte desert faunas (Mares, 1975a, 1976, 1980). The evidence for such convergence is less convincing for small mammals of the Chilean and Californian semiarid and mediterranean zones, perhaps because they may be more recent in origin and have a more generalized morphology and broader habitat ranges; also, there may be many alternative ecological strategies in newly formed ecosystems (Glanz, 1977; Glanz & Meserve, 1982).

Fulk's pioneering study (1975) of a Chilean semiarid small mammal community, involving qualitative comparisons with similar North American communities, emphasized the lower overall diversity and large annual fluctuations of the fauna there. He concluded that there was little evidence for intrinsic mechanisms of population regulation in the species studied. Fulk's study, however, was conducted during a period of exceptionally high rainfall; here, as well as elsewhere in Chile (Glanz, 1977; Péfaur et al., 1979), high population densities of many small mammal species were recorded. In addition, a ubiquitous caviomorph rodent, *Octodon degus* (family Octodontidae), frequently reported from here and elsewhere in north-central Chile, was virtually absent from his site.

Here we report the results of a subsequent study of the same small mammal community followed by Fulk (1975), for the years 1973–1975. We emphasize the patterns of population changes observed over some 17 months of both live- and snap-trapping, as well as the general implications of these results for viewing population dynamics of the principal species and faunal composition of Chilean semiarid small mammal communities.

Study Area and Methods

The study site was located in a semiarid thorn scrub community in Parque Nacional Fray Jorge, Coquimbo Province (IV Region), Chile (71°40′W, 30°38′S, 200 m). This area lies on the northern fringe of the Chilean mediterranean zone known as the "Norte Chico" and on the southern edge of the Pacific coastal desert. The site is situated in an interior valley ("Quebrada de las Vacas") on the east side of a coastal range (500 m) and about 5 km east of the Pacific Ocean. Descriptions of the vegetation have been given in Fulk (1975) and Meserve (1981a); briefly, the community is characterized by spiny drought-deciduous and evergreen shrubs, a sparse herbaceous understory, and open sandy areas between shrubs. The community has been termed the *Porlieria chilensis-Proustia pungens-Adesmia bedwellii* association for its most characteristic shrubs (Muñoz & Pisano, 1947). The flora of this region combines elements of the dry western Andean slopes and a few species from the Monte on the east side of the Andes (Sarmiento, 1975). The absolute cover of major plant categories as determined from line transect analysis of the live-trap grid conducted in November 1973 was 59.6% shrubs, 21.7% grasses and forbs, and 46.0% bare ground (total exceeds 100% due to overlapping projections of shrub canopies on other categories). A subsequent reanalysis of some stations in September 1974 showed no significant change in total shrub cover. Fulk (1975) reported a total shrub cover of 44% for his site, located approximately 500 m away across a dirt road.

A fundamental characteristic of this region is the presence of moderately warm temperatures, high insolation, and extremely variable precipitation which falls mostly (90%) between the months of May and September. Prior to 1965, the average annual precipitation was about 127 mm; in 1965, an exceptionally heavy rainfall of 326 mm was recorded (Kummerow, 1966). For the eight-year period of 1969–1976, during which continuous records have been kept in the park, precipitation averaged 68.9 ± 78.5 mm (SD); in the principal year of Fulk's study (1972–1973), precipitation was 255.3 mm. For the three subsequent years dealt with in this study (1973–1976), precipitation was 75.8, 26.5, and 37.0 mm. Kummerow (1966) showed that coastal fog can contribute up to ten times the amount of moisture available from precipitation on higher (500 m) west-facing mountain slopes; in the interior valley (200-m elevation), the impact of coastal fog is considerably less. Mean maximum temperature in the warmest month (January) is 24°C, and mean minimum temperature in the coolest month (July) is 4°C (all weather data are from the park headquarters located about 5 km southeast of the study site and collected by the Oficina Meterológica de Chile, Santiago).

Between November 1973 and January 1975, a 1.4-ha live-trapping grid with 48 stations 20 m apart (6 × 8 configuration) was censused at bimonthly intervals for four consecutive nights. Two medium Sherman live traps baited with rolled oats were placed within 1.5 m of each station and standard mark-and-recapture techniques employed, using toe-clipping or ear-tagging. Observations of species, body weight (to the nearest 0.5 g), sex, sexual condition (perforate or imperforate for females; scrotal or abdominal testes for males), evidence for reproduction (obviously pregnant or lactating females), number, and trap location were taken during handling. Traps were closed during the day and reopened in the late afternoon from November 1973 through March 1974; thereafter, traps were left open for at least one day each census, with twice-daily trap checks. Beginning in September 1973, adjacent snap-trapping was conducted during each census in similar habitat at least 400 m from the live-trap grid. Usually, a line of 66 Museum Specials and Victor 4-way rat traps was set in pairs approximately 10 m apart for two to four nights. Traps were checked twice daily and animals autopsied for presence of uterine scars and embryos, and relative testis size and position. Stomachs were also retained for subsequent dietary analysis (Meserve, 1981a).

Evidence for reproduction was based on the presence of scrotal males and pregnant and/or lactating females. Observations from both the live-trap grid and autopsy lines were utilized. Although the presence of perforate females may be considered to be evidence of sexual competence (as is the presence of scrotal testes in males), we utilized this mainly to indicate the minimum weight at which females are capable of reproduction and hence an approximate index to developmental age. For nonreproductive periods, developmental age was assessed by comparison with the last minimum weight at which evidence of sexual competence (i.e., perforate vagina in females, scrotal testes in males) was observed. This approach was deemed necessary because none of the species studied have distinct juvenile-subadult-adult molts, and there presently exists no satisfactory method for otherwise aging the species studied here in live-trapping studies. Even an index developed from snap-trapping results would still be dependent on some morphological character, such as weight taken in the field, in order to age live-trapped individuals. Tamarin (1984) and Dueser et al. (1984) have discussed the possible errors inherent in applying body weight criteria to live-trapped animals without independent verification.

Live-trap capture records were codified and recorded on computer for subsequent analysis with the CMR (capture-mark-recapture) package of Le Boulengé (1985a,b) at the University of Louvain computer facility, Louvain-la-Neuve, Belgium. This package produces data summaries such as the Calendar of Captures (Petrusewicz & Andrzejewski, 1962), sex ratios, reproductive condition,

Minimum Number Known Alive (Krebs, 1966) tabulations, plots of traps visited, and weights at first capture. In addition, higher level analyses are available for trappability analysis, density estimations (Manly-Parr, Hayne, and Zippin methods), home range parameters (using the bivariate normal method of area estimation, Mazurkiewicz, 1969; Jennrich & Turner, 1969), disappearance rates, and weight dynamics. Trappability is estimated either running "vertically" through the capture matrix to find the proportion of marked animals that are captured in a given time period (usually in a census; this is the Manly-Parr [1968] trappability estimator), or running "horizontally" through the recapture data to find the proportion of capture opportunities in which a given individual or group of individuals was effectively caught ("individual trappability" *sensu* Le Boulengé & Le Boulengé-Nguyen, 1981). The CMR package is available at cost from the second author upon request.

In small mammal studies, "classic" statistical methods very often yield unreliable estimates due to inadequate sample sizes or failure to fulfill underlying assumptions. Thus, in addition to the Minimum Number Known Alive method, we have calculated population sizes by the Modified Calendar of Captures technique whenever possible (Le Boulengé & Le Boulengé-Nguyen, 1981). Basically, the Modified Calendar of Captures technique is equivalent to the Calendar of Captures or Minimum Number Known Alive method, where an individual is counted as present from the first to last capture (Krebs, 1966); however, this technique includes two additional steps. The first consists of adjusting the observed residency time of each individual by adding the "mean section of time between successive captures" (*sensu* Andrzejewski, 1963, 1969) before its first and after its last capture (see details in Le Boulengé & Le Boulengé-Nguyen, 1981). The second step consists of counting young individuals as present since their approximate birthdate (i.e., in this case, the closest subsequent census) irrespective of the date of their first capture (a procedure called Age-Projected Number Alive by Ford & Pitelka, 1984). This backward projection of numbers of young is adjusted by the juvenile survival rate, assuming this to be constant from the date of first capture back to the date of birth (Le Boulengé & Le Boulengé-Nguyen, 1981). Direct survival estimates between successive sessions are biased by trappability fluctuations and are imprecise when, as often, sample sizes are small; whence survival values used in the above procedure were derived by modeling trappability and survival with the method of Clobert et al. (1985). This consists of finding the minimum sufficient multinomial trappability/survival model which adequately fits the observed data. Goodness-of-fit is assessed by comparing the candidate minimum model to a reference model in which survival and trappability parameters attached to each census are free from constraints. Comparison of two models is made using a log-likelihood ratio, distributed as a χ^2 variable with degrees of freedom equal to the difference in the number of free parameters of the two models.

The Modified Calendar of Captures method relies on the ability to distinguish age cohorts and to separate older cohorts from young born *in situ*. The major assumptions underlying the technique are that no individuals living on the plot escape capture entirely (as is also assumed by the Calendar of Captures/Minimum Number Known Alive methods) except for young ones prior to first capture, and that the mortality rate of marked and unmarked animals is similar (Le Boulengé & Le Boulengé-Nguyen, 1981). Definitions of the age cohorts that were utilized in this analysis are specified below for each species.

The objective of the analyses was to arrive at a more realistic view of changes occurring in the populations. In addition, insight into the nature of population dynamics may be gained. For example, when the backward projection procedure gave unrealistic estimates of numbers of young present before first capture (that is, far greater or less than the potential for known resident females to produce them), this could be interpreted as indicating a major role, respectively, of immigration or emigration.

Results

Trends of Minimum Numbers Known Alive

Figure 1 presents the Minimum Numbers Known Alive on the grid during the study period, for the four principal small mammal species as well as total. Numbers of *Akodon olivaceus* showed a constant decline throughout 15 months of observation; in contrast, those of *A. longipilis* were remarkably constant, from two to five individuals. *Phyllotis darwini* showed fairly regular increases during late spring–summer months and declines afterwards. Lacking diurnal trapping prior to May

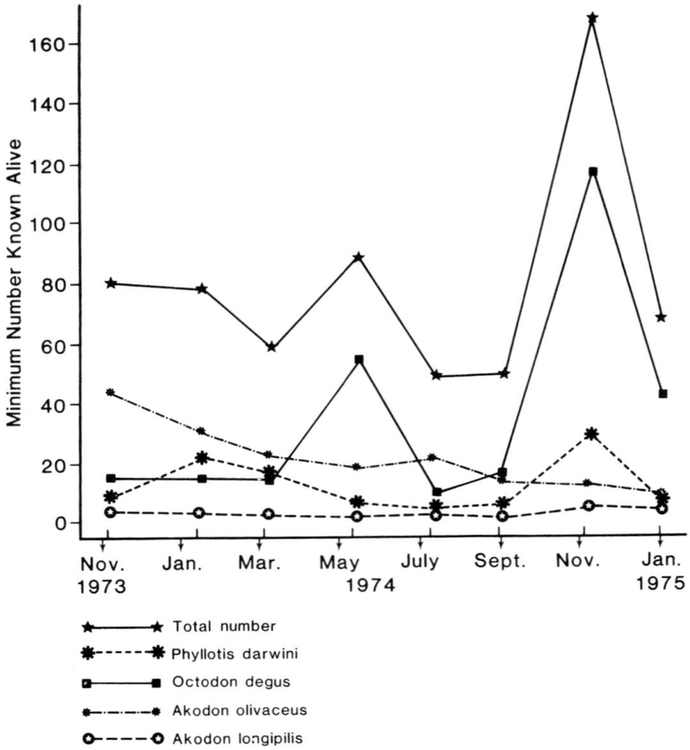

FIG. 1. Minimum Number Known Alive trends for *Akodon olivaceus, A. longipilis, Phyllotis darwini,* and *Octodon degus* and total numbers (all species including *Abrocoma bennetti, Oryzomys longicaudatus,* and *Marmosa elegans*) during November 1973–January 1975.

1974, earlier estimates of *Octodon degus* numbers are probably considerable underestimates; nevertheless, minimum numbers of degus were considerably greater than reported by Fulk (1975). In May 1974, a large number of new *O. degus* adults were recorded on the grid; most of these (64% of 39 individuals) were not recaptured in subsequent censuses. As this occurred during the beginning of the degu breeding season, when intense activity was observed on the grid following the first winter rains, these were likely peripheral nonresident animals (Meserve et al., 1984). Consequently, although these individuals were included in Minimum Numbers Known Alive figures, they have been excluded from the following demographic and age analysis.

TABLE 1. Observed sex-ratios of four rodent species given as percentages of females among the individuals captured in each season.

	Season				
Species	**Nov. '73–Jan. '74**	**March–May '74**	**July–Sept. '74**	**Nov. '74–Jan. '75**	**Total '73–'74**
Akodon olivaceus	0.31* (64)	0.44 (25)	0.38 (26)	0.44 (18)	0.34* (88)
Akodon longipilis	0.14 (7)	0.33 (3)	0.33 (3)	0.67 (6)	0.33 (12)
Phyllotis darwini	0.30* (30)	0.24 (17)	0.50 (8)	0.42 (36)	0.39 (75)
Octodon degus	0.36 (25)	0.53 (32)	0.73 (11)	0.49 (127)	0.50 (169)

Numbers in parentheses are numbers of individuals captured; total column includes all individuals caught during the study.
* Values differing significantly from a 0.5 ratio (chi-square test).

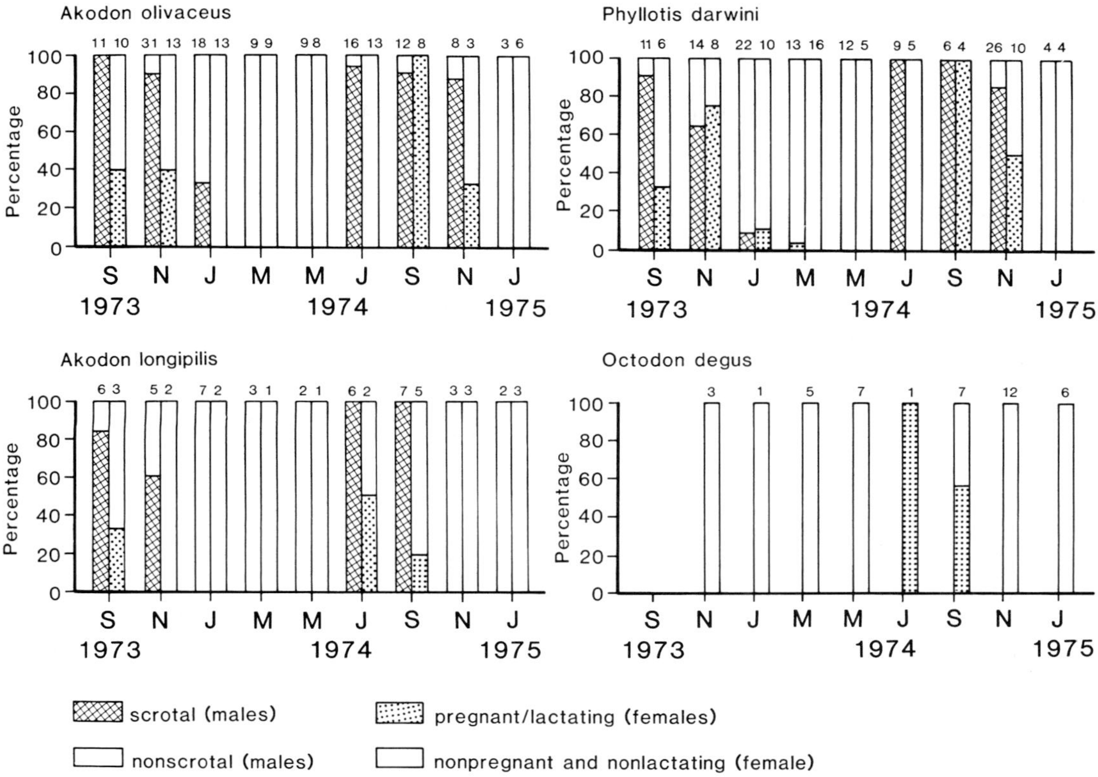

FIG. 2. Reproductive trends for *Akodon olivaceus, A. longipilis, Phyllotis darwini,* and *Octodon degus* during September 1973–January 1975, based on live- and snap-trap results. Percentages of reproductively active individuals are indicated by crosshatched areas (males) and dotted areas (females) in columns for each sample period; sample sizes are indicated over respective columns.

Other species captured on the grid include the caviomorph rodent *Abrocoma bennetti* (one to five individuals each census except in March and September 1974); *Oryzomys longicaudatus* (one to five individuals in five of eight censuses); and the mouse opossum *Marmosa elegans* (single individuals during three censuses).

Table 1 presents the sex-ratio for the four most common species in four seasons: late spring (November 1973–January 1974), summer–fall (March–May 1974), winter–early spring (July–September 1974), and late spring (November 1974–January 1975), based on the actual numbers of males and females caught in each season. Interestingly, the observed ratios are consistently biased in favor of males in the three sigmodontines, although significantly so only for *Akodon olivaceus* and *Phyllotis darwini* in late spring of the first year and in the total sample of the former species. Nonsignificant results for *Akodon longipilis,* which shows the most strongly biased ratio, are probably due to the low sample sizes in this species. *Octodon degus* in contrast seems to have a balanced sex-ratio.

Reproduction

Reproductive trends for all four species are summarized in Figure 2. For *Octodon degus* only information on females is presented, as relative testis size and position as determined by palpation is probably an inaccurate index of reproductive activity in this nonscrotal caviomorph rodent.

Pregnant or lactating females of *Akodon olivaceus* were briefly present in September–November, reproductively active males somewhat longer (fig. 2). A few scrotal males were present in January 1974 but none in January 1975, suggesting earlier termination of sexual competency. Embryo counts for 1974–1975 females (N = 11) averaged 5.6 ± 1.1 (SD), similar to that reported by Fulk (5.6; 1975) and Greer (5.5; 1965) in southern Chile, and somewhat higher than reported by Pearson (5.1; 1983) in the southern Argentine temperate rain forest. The maturation time for *A. olivaceus* is unknown, but Pearson (1983) and Murúa (pers. comm.) have reported breeding males and females about two months old. Field-caught male and female *A. olivaceus* reached sexual maturity at 22–24 g, but did not breed in the same season. One female born probably about November–December 1973 bred twice, in the following September and November 1974.

Data on *Akodon longipilis* is more limited due to small sample size. This species appeared to initiate and terminate reproduction earlier than either *Akodon olivaceus* or *Phyllotis darwini* (fig. 2). Fulk (1975), however, reported pregnant females in February 1973 and juveniles the following May. Similar to trends for other species, reproduction ceased earlier in 1974 than in 1973. Two females had six and four embryos; Greer (1965) and Pearson (1983) reported mean embryo counts of 3.7 and 3.8, respectively, for southern Chile and Argentina. Minimum weights for perforate females averaged 39.6 g, for scrotal males 47.5 g; no individual reached sexual maturity in less than five months. Pearson (1983), however, reported breeding in the same season as birth, and our observations are probably biased without data prior to September 1973.

In general, *Phyllotis darwini* had the longest period of potential and actual reproduction; scrotal males were recorded in seven out of nine periods and were only absent in May during the 1974 samples. Fulk (1975) reported no scrotal males later than January during a wet year in Fray Jorge, but they were present in the Santiago area in March of the same year. Pregnant or lactating females were present between September and January; captive females maintained in the Santiago area bred in virtually all months of the year. Embryo count for 11 autopsied females was 5.1 ± 1.0 (SD), close to that reported by Fulk (5.2; 1975) and Pearson (5.25; 1975) for an "outbreak" population in southern coastal Peru. Young born in captivity to field-pregnant females from Fray Jorge reached sexual maturity in 60.0 ± 10.2 days at a weight of 40.8 ± 6.5 g (females, $N = 3$), and 50.3 ± 7.2 days at 46.3 ± 6.4 g (males, $N = 6$). The lowest weight for a perforate female in the field was 39.2 g, but the next lowest weight was 47.0 g. Minimum weights for nine scrotal males in the field was somewhat lower than in captivity ($\bar{X} = 44.9 \pm 5.9$ g). Minimum time to first parturition for six captive females was 112.3 ± 3.3 days; more interestingly, minimum interval for second litters following birth of the first was 16–18 days and appeared to alternate between longer intervals of 30–60 days for continuously breeding *P. darwini* in captivity (captivity data from Le Boulengé, unpubl. data). Recently, this phenomenon was confirmed independently in pregnant field-caught females from Fray Jorge which gave birth in captivity to a second litter only 16 days after the first (M. H. Gallardo, pers. comm.). In the Fray Jorge population, males probably born in August–September reached sexual competency by November.

Octodon degus, as *Akodon longipilis* females, initiated and terminated reproduction earlier than those of *Phyllotis darwini* and *A. olivaceus*. Pregnant females were found only in June and July; thereafter only lactating ones were captured. With a gestation of 90 days (Weir, 1970, 1974), females probably conceived in May–June and gave birth in August–September. There was no evidence of a postpartum estrus or a second litter, as reported in Fray Jorge for 1972–1973 and in central Chile (Fulk, 1975; Rojas et al., 1977; Meserve et al., 1984). Embryo count for nine individuals was 5.7 ± 1.3, slightly greater than the 5.3 figure given for 1974 Fray Jorge females only, or for Santiago area degus (Meserve et al., 1984). Woods and Boraker (1975) reported litter sizes of 6.8 for degus in laboratory colonies. Young did not breed in the same season of birth in Fray Jorge; the minimum weight for perforate females was 127.5 g, and field-caught individuals did not reach this weight in less than 100 days of life (including a minimum estimated pre-weaning time of 28 days; Weir, 1970).

Cohort Definition

As live-trapped animals could not be aged directly using classic age-estimation techniques because of poor information on these species, we relied on several sources of evidence, including the seasonality of reproduction, body weight distributions, and sexual maturity of individuals at first capture. The study started at or near the end of the 1973–1974 reproductive season for most species; thus, young of the year had already reached adult size and/or sexual maturity by November 1973–January 1974 and could not be distinguished from older adults except in the case of *Akodon olivaceus*. For this species, there was a clear separation between sexually active, heavier individuals which characteristically lost weight (pregnant females excluded) through January 1974

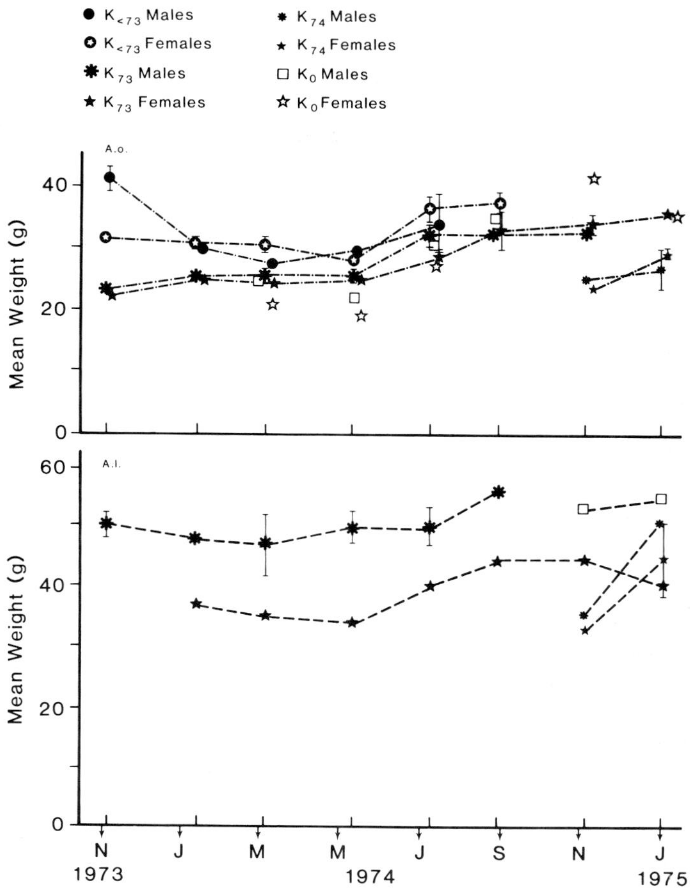

FIG. 3. Body weight trends for cohorts of *Akodon*: **top,** *A. olivaceus,* and **bottom,** *A. longipilis.* Meanings of symbols are defined in legend; see text for cohort definitions. Only two cohorts were defined for *A. longipilis.* Vertical line indicates interval of one standard error of the mean.

(fig. 3) and were significantly heavier than a group of smaller, mostly sexually inactive individuals which gained weight steadily; these were designated the $K_{<73}$ and K_{73} cohorts, respectively. Thus, $K_{<73}$ individuals were considered adults probably born early in the 1973 reproductive season (about September) and reaching weights of over 27.5 g by November 1973–January 1974. The K_{73} individuals were born late in the reproductive season and weighed less than 27.5 g then. After January 1974, weights of both cohorts converged, being virtually identical by July (fig. 3). The third cohort (K_{74}) consisted of individuals born in the 1974 reproductive season about September that entered the trappable population in November 1974–January 1975. Individuals caught for the first time at adult weights (> 27.5 g) after January 1974 were classified as unknown (K_0) individuals and probable immigrants; such a procedure is conservative, and particularly justified if trappability is high (see next section).

For the remaining species, only two cohorts were distinguished—a cohort of adults born before the beginning of the live-trapping study ($K_{\leq 73}$), and a cohort of younger, lower weight individuals appearing in September 1974–January 1975 (K_{74}). In addition, a cohort of probable immigrants (K_0) was distinguished. Figures 3 and 4 show the weight trends for the remaining three species; in all cases, a clear difference in body weight existed between $K_{\leq 73}$ and K_{74} individuals.

The K_0 group deserves some additional comments. Single new adults of *Akodon longipilis* appeared only in November 1974 and January 1975. On the other hand, new adults of *Phyllotis darwini* were captured in November 1973 and January 1974 which were far heavier than the average $K_{\leq 73}$ individuals (fig. 4); although these were probably $K_{\leq 73}$ individuals, they were classified in the K_0 cohort, as were all subsequent new adults. Finally, due to inadequate sampling prior to May 1974, all *Octodon degus* individuals marked through May

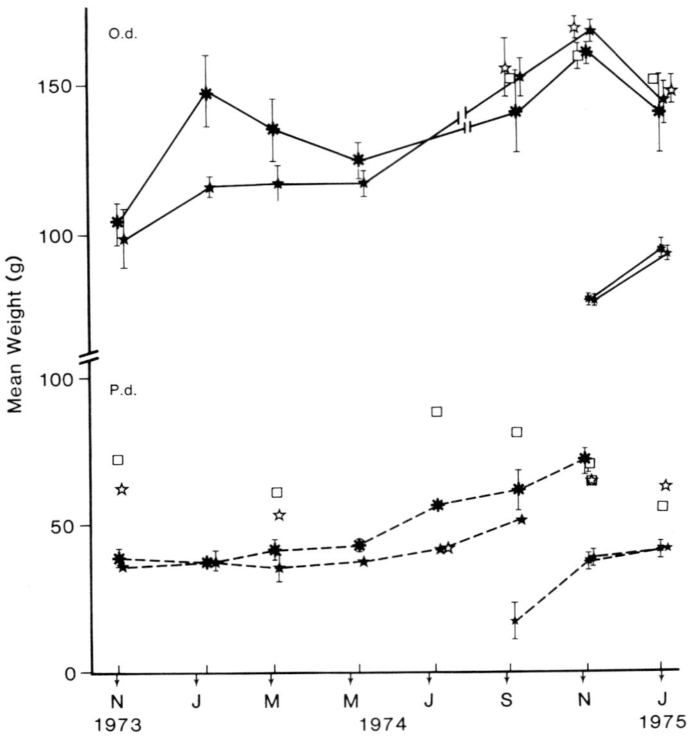

FIG. 4. Body weight trends: **top,** cohorts of *Octodon degus,* and **bottom,** cohorts of *Phyllotis darwini.* Symbols are as in Figure 3. Only two cohorts were defined for each of these species.

were considered $K_{\leq 73}$ cohort members, and all subsequent new adults (recorded from September on) as K_0 individuals and hence probable immigrants.

Trappability

Any evaluation of population dynamics in free-living animals requires an estimate of trappability. This may vary between species, sex, age group, and season, and hence should be estimated whenever possible. Enumeration methods such as the Minimum Number Known Alive have an inherent requirement of high trappability between sessions/censuses. Other estimation methods, such as the Jolly-Seber and derived models and the Modified Calendar of Captures, allow correction for variable trappabilities during the measurement period. Here, daily trappability and seasonal trappability during each of the four three-month periods outlined in Meserve (1981b) were determined: late spring (November 1973–January 1974), summer–fall (March–May 1974), winter–early spring (July–September 1974), and late spring (November 1974–January 1975). Daily "individual trappabilities" (see Study Area and Methods section) were estimated for each sex and cohort in each season; the results are presented in Table 2. Monthly or between-session trappabilities were determined by the method of Manly and Parr (1968). Finally, seasonal variations in daily trappability were evaluated using the modeling technique of Clobert et al. (1985).

Akodon olivaceus trappability might be considered homogeneous only in the summer–fall months (March–May) and in the winter months (July–September; $\chi^2 = 1.26$, 2 *d.f.*, NS); otherwise, monthly values were significantly heterogeneous ($\chi^2 = 10.9$, 3 *d.f.*, $P < 0.025$). Modeled daily trappability was highest in winter (0.75) and lowest in both January sessions (0.48 and 0.18, respectively). This means that the estimated percentage of animals missing a complete census ranged from 55% in January 1974 to 0.4% in winter censuses. The percentage of animals known alive and caught at least once in each census was generally high, however ($\bar{X} = 84.8 \pm 14.8\%$, 1 SD).

For *Akodon longipilis,* trappability could be pooled for all sessions except during January 1974 when it was significantly less than in other periods (modeled trappability of 0.27 vs. 0.73, $\chi^2 = 3.32$, 2 *d.f.*, NS for this model, vs. $\chi^2 = 9.4$, 3 *d.f.*, $P < 0.025$ when all sessions were pooled). Converted into probability of escaping capture in any given

TABLE 2. Daily trappability by season for *Akodon olivaceus, A. longipilis, Phyllotis darwini,* and *Octodon degus.*

Species, sex, cohort	Season: Nov. '73–Jan. '74	March–May '74	July–Sept. '74	Nov. '74–Jan. '75
Akodon olivaceus				
Adult males	0.67 (63)	0.48 (86)	0.63 (70)	0.50 (18)
Adult females	0.68 (41)	0.66 (56)	0.75 (57)	0.90 (10)
Young (both sexes)	...	...	...	0.45 (11)
Akodon longipilis				
Adult males	0.63 (8)	0.65 (20)	0.83 (12)	0.17 (6)
Adult females	...	0.64 (11)	0.88 (8)	1.00 (7)
Young (both sexes)	...	...	...	0.45 (11)
Phyllotis darwini				
Adult males	0.68 (28)	0.61 (66)	0.70 (23)	0.33 (30)
Adult females	1.00 (5)	0.58 (12)	0.88 (8)	...
Young (both sexes)	...	...	...	0.65 (20)
Octodon degus				
Adult males	0.40 (10)	0.31 (58)	0.11 (19)	0.21 (63)
Adult females	...	0.29 (35)	0.14 (35)	0.26 (53)
Young (both sexes)	...	...	...	0.60 (189)

Numbers in parentheses are numbers of capture occasions.

session, that for January was 0.47, and for the remaining months, 0.01. All individual *A. longipilis* known alive on the grid were captured at least once during each census when present.

Phyllotis darwini trappabilities were nonsignificantly lower in spring 1974–1975 than in the remaining sessions (modeled trappability of 0.36 vs. 0.64, $\chi^2 = 8.82$, 4 *d.f.*, $0.1 < P < 0.05$). This was equivalent to 17% of the animals escaping capture in the spring 1974–1975 censuses, and 2% in the remaining trap sessions. The mean proportion of *P. darwini* known alive and caught at least once in each session was $\bar{X} = 81.8 \pm 16.9\%$.

Daily trappabilities of *Octodon degus* were relatively homogeneous throughout the study ($\chi^2 = 7.67$, 5 *d.f.*, $P > 0.10$, NS) with a low mean value of 0.37. The percentage of individuals escaping capture during a session was estimated at 16%. The relative constancy of daily trappability despite the apparently inadequate sampling utilized prior to May 1974 is surprising. The percentages of animals known alive and caught at least once each census was quite low, however ($\bar{X} = 39.5 \pm 31.9\%$).

In general then, except for *Octodon degus* throughout the study, and isolated sessions for the remaining species (January 1974 for *Akodon longipilis,* January 1975 for *Akodon olivaceus,* November 1974–January1975 for *Phyllotis darwini*), we can assume that less than 2% of the individuals went uncaptured throughout a trapping session. The difference between this figure and that obtained only from observing occurrences of animals during consecutive censuses may be due to the temporary absence of individuals during specific (and relatively short) censuses with long between-census intervals (Clobert, pers. comm.). Thus, enumeration methods were fairly reliable for estimating minimum numbers of animals at risk of capture.

Demographic Trends

Figures 5–8 show the cohort dynamics as revealed by the Modified Calendar of Captures technique, as well as the observed survival rates for the defined cohorts and numbers of recruits into the populations of the four principal small mammal species. Due to the variable time intervals between bimonthly censuses (52 to 70 days), survival rates were standardized to monthly values. New adults were considered to have entered the population at the start of their corrected residency time and young recruits were estimated to have entered the population at the time of their birthdates, based on the presence of pregnant or lactating females in the population and approximate maturation times.

Akodon olivaceus showed low survival in the spring of 1973, but improved survival, especially during the winter, until the 1974–1975 reproductive period (fig. 5). No $K_{<73}$ individuals survived

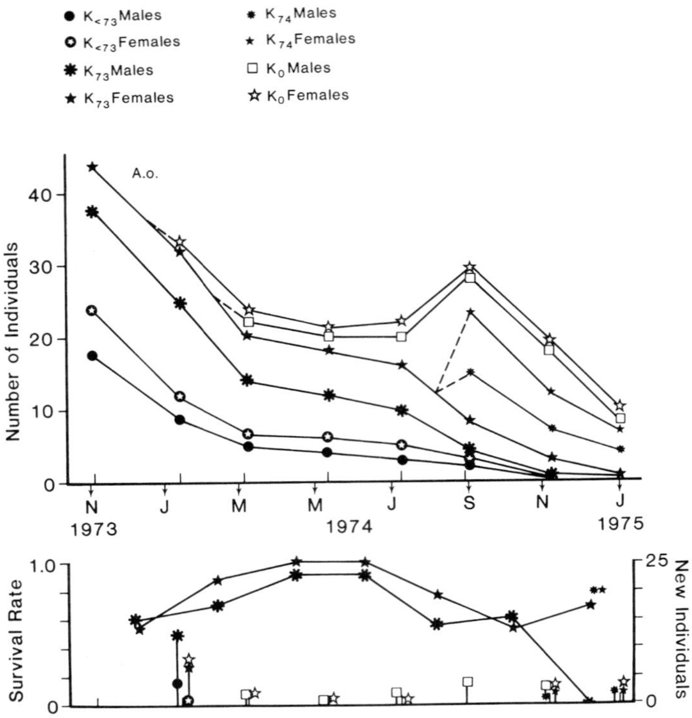

FIG. 5. **Top,** Estimated number of animals by cohort for *Akodon olivaceus* using the Modified Calendar of Captures technique during 1973–1975; and **bottom,** observed survival rates (line graphs) and recruitment (dots connected to base line, cumulated by cohort) during the same period. The $K_{<73}$ and K_{73} survival rates have been combined to simplify comparisons (see text for explanation of cohort definitions). Dashed lines indicate estimated entrance of newborn or immigrant individuals into the population.

to November 1974, and only one K_{73} survived to January 1975. The seasonality of survival was confirmed by the modeling technique of Clobert et al. (1985); four different survival values had to be distinguished—0.38 between consecutive censuses in both November–January periods and also in September–November 1974; 0.62 from January–March and March–May 1974; 0.31 from May to July; and 0.46 from July–September ($\chi^2 = 5.71$, $P > 0.10$, simplest nonsignificant model). Seven young were marked in November 1974 and January 1975, leading back to a projected estimate of 15 young born in September 1974. This is considerably less than the potential production of six pregnant females detected in September (one of which was pregnant again in November). As four of these females disappeared between September and November, a significant portion of the production of young for that season may have been lost close to birth. Recruitment in the second reproductive season was insufficient to compensate for disappearance rates of older individuals, and the population experienced a decline from 33 individuals the previous January to only 12 in January 1975.

Akodon longipilis presented a considerably different pattern (fig. 6). The $K_{\leq 73}$ cohort persisted into the last month and females especially had high survival rates. Modeled survival may be considered to have been constant throughout the study ($\chi^2 = 1.05$, $P > 0.05$). The projected production of six new individuals in August–September (possibly from only one female) was sufficient to entirely replace the population, especially with the high indicated survival rate for young *A. longipilis.* The overall effect was for the population number to be extremely stable in time.

The pattern for *Phyllotis darwini* was less consistent (fig. 7). Although observed survival rates seemed to vary from a low spring value to a winter peak and declined again to very low values in the spring of the following year, in fact they were nonsignificantly different from January through November 1974. Only the periods November 1973–January 1974 and November 1974–January 1975 had extraordinarily low survival rates (average of 13%). Due to this poor survival and low trappabilities in November–January of the second year, the number of young born in that reproductive period as estimated by the backward projection procedure is unrealistic—over 300 individuals in September. Thus, we have only indicated Minimum Numbers Known Alive for the K_{74} cohort in Figure 7. This discrepancy between observed number

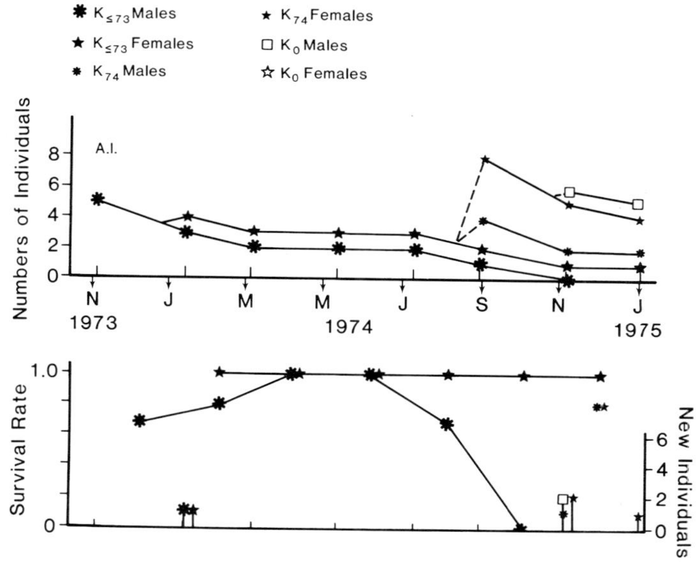

FIG. 6. **Top,** Estimated number of animals by cohort for *Akodon longipilis* using the Modified Calendar of Captures technique during 1973–1975; and **bottom,** observed survival rates and recruitment during the same period. Symbols and meaning of dashed lines are as in Figure 5. Only two cohorts were defined for this species (see text).

of young (which is a minimum figure) and potential production of young of the two reproductive females detected in September–November 1974 indicates a high immigration rate among younger individuals, a trend also observed in adults. In addition, immigrant females could have contributed significantly to reproduction, and the evidence for multiple short-interval pregnancies in captivity suggests a high potential production rate of litters by females in general. In view of the poor survival rates of younger individuals, immigration seems to play a major role in the ability of this species to reach high densities rapidly.

The results for *Octodon degus* (fig. 8) indicate a very large influx of juvenile animals occurred in November 1974, probably from a single episode of reproduction initiated in June with birth in early September and weaning in early October. The fact that 97 young were recorded in November when a maximum of 17 adult females were estimated to be present on the grid (including September animals) indicates that the degus were reproducing maximally. However, unless all of the young born on the plot survived until their first capture, immigration must be invoked to account for the large numbers of young actually observed. Survival of

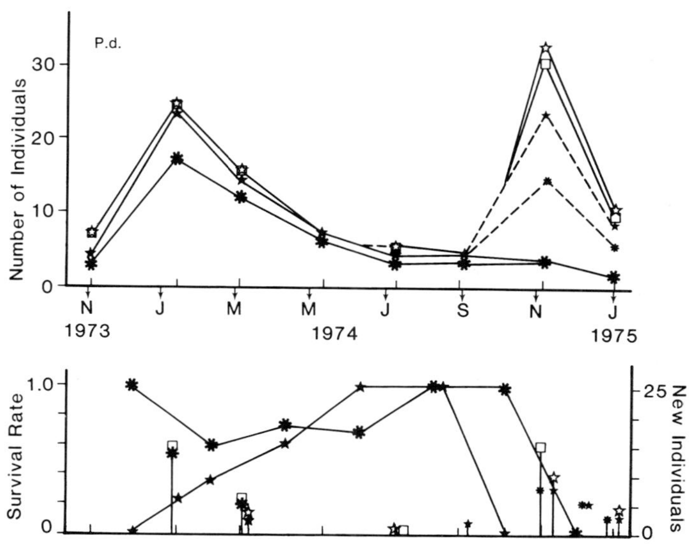

FIG. 7. Number of animals by cohort for *Phyllotis darwini* using the Modified Calendar of Captures technique during 1973–1975: **top,** Minimum Numbers Known Alive figures for the K_{74} cohort; no backward projection; and **bottom,** observed survival rates and recruitment during the same period. Symbols and meaning of dashed lines are as in Figure 5. Only two cohorts were defined for this species (see text).

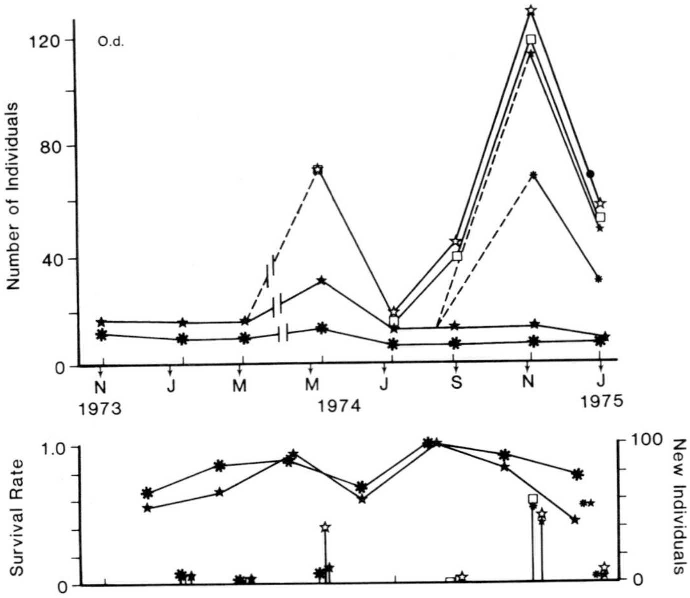

FIG. 8. Number of animals by cohort for *Octodon degus* using the Modified Calendar of Captures technique during 1973–1975: **top,** numbers in the K_{74} cohort as in Figure 7; and **bottom,** observed survival rates and recruitment during the same period. Symbols and meaning of dashed lines are as in Figure 5. The broken lines in March–May 1974 indicate uncertainty regarding previous population changes due to inadequate sampling. Only two cohorts were defined for this species (see text).

adults was fairly high except in late spring (November–January) and winter months (May–July). Disappearance rate of young once they reach trappable age was extremely high; this may reflect high dispersal and/or mortality rates. As in the case of *Phyllotis darwini,* both immigration and low survival of juveniles led to absurd estimates of total production of young by the backward projection procedure.

Discussion

Semiarid environments are often heterogeneous in space and time. Desert and semiarid mediterranean-type communities have frequent events of local extinction (e.g., MacMillen, 1964; M'Closkey, 1972; Glanz, 1977). Historically, the small mammals of the Chilean arid zone have not been isolated sufficiently long for speciation to occur, nor is there evidence for zoogeographic differentiation due to chance colonizations by members of the potential species pool elsewhere in South America. On a local scale, semiarid and arid communities in Chile are extremely heterogeneous in numbers of species encountered from year to year (Glanz, 1977; Meserve & Glanz, 1978). Overall, species number declines monotonically in a northerly direction due to the influence of declining precipitation and its concomitant effects on primary production and plant community structure (Meserve & Glanz, 1978). A similar situation exists along an altitudinal gradient on the western side of the southern Peruvian Andes where mammal species number increases with altitude, which in turn is related to vertical vegetation density (profile) and perhaps indirectly to precipitation (Pearson & Ralph, 1978). The frequent observations of irruptions of mice (= *ratadas*; Hershkovitz, 1962) following unusual rains and subsequent plant growth are further support for the role of extrinsic factors, such as climate, in this region (e.g., Pearson, 1975; Glanz, 1977; Péfaur et al., 1979). The 1972–1973 "outbreak" of *Phyllotis darwini* reported by Pearson (1975) in coastal southern Peru following a rainfall of 81 mm (in a locale averaging 32 mm annually) coincided with similar observations for this and other species in and near Fray Jorge by Fulk (1975) and Péfaur et al. (1979), and in the central to northern Chilean mediterranean zone by Glanz (1977).

In view of these well-documented outbreaks, it is significant that our results obtained in years subsequent to the above studies, during periods of meager rainfall, demonstrate the persistence of a permanent small mammal fauna after such outbreaks. As a way of viewing the significant changes that occurred in the fauna between the years of Fulk's study (1972–1973) and ours, we have presented data from four years of sampling using both live- and snap-trapping results and an index of numbers of animals caught per 100 trap-nights of effort (table 3). As a convention, to simplify com-

TABLE 3. Numbers of animals/100 trap-nights and relative proportions by species for small mammals trapped during four "rainfall" years (see text) during 1972–1976 in Fray Jorge.

		Species					
Year	Trap	*Akodon olivaceus*	*Akodon longipilis*	*Phyllotis darwini*	*Octodon degus*	Other	Total
July '72–June '73	Live-trap	12.92 (65.0%)	0.80 (4.0%)	5.75 (29.0%)	0.05 (0.3%)	0.33 (1.7%)	19.85 (100%)
July '73–June '74	Live-trap	7.62 (42.4%)	0.98 (5.5%)	3.65 (20.3%)	4.43 (24.6%)	1.30 (7.2%)	17.98 (100%)
July '73–June '74	Snap-trap	4.96 (23.8%)	1.68 (8.1%)	10.16 (48.8%)	3.36 (16.2%)	0.64 (3.1%)	20.80 (100%)
July '74–June '75	Live-trap	4.03 (18.9%)	1.04 (4.9%)	3.26 (15.3%)	12.36 (15.7%)	0.69 (3.2%)	21.38 (100%)
July '74–June '75	Snap-trap	2.45 (14.1%)	2.60 (14.9%)	6.88 (39.4%)	5.35 (30.7%)	0.15 (0.9%)	17.43 (100%)
July '75–June '76	Snap-trap	2.65 (14.8%)	3.53 (19.6%)	5.29 (29.4%)	3.24 (18.1%)	3.24 (18.1%)	17.95 (100%)

parisons, we have used periods of "rainfall years" which run from July of one year through June of the next, recognizing that virtually all precipitation falls during the May–September period, and that subsequent reproduction and recruitment occur in the population from July onward (Fulk, 1975; this study). While differences in trappabilities between species and between live- versus snap-traps would be expected, it is remarkable that total numbers of animals/100 trap-nights effort was constant from year to year. In the years most strictly comparable due to similar effort and methodology (1973–1974 and 1974–1975), mean total numbers of individuals/census (taken from fig. 1) were very similar despite large within-"rainfall year" variation (e.g., 76.5 ± 12.6 [1 SD] vs. 81.0 ± 58.5 individuals, respectively). This indicates that, despite within-year and between-year variations in species numbers, the major changes that occurred over time were in the relative numbers (and proportions) of faunal members. These changes included (1) a decline in the *Akodon olivaceus* population; (2) a relatively constant or even slightly increasing population of *Akodon longipilis*; (3) seasonal changes in the *Phyllotis darwini* population; and (4) a large increase in the *Octodon degus* population. The increase in numbers of degus is in part an artifact of trapping methodology, as Fulk (1975) did not often trap during the day for this bimodally diurnal to crepuscular species and used a small locally made trap which probably biased against capturing the larger degus. The effect of including degus is particularly significant in biomass estimates; for example, total small mammal biomass for the months of November 1974 and January 1975 was 9.2 and 3.1 kg/ha, respectively, of which 83.4% and 82.5% consisted of *O. degus* individuals. The former figure exceeds the maximum biomass observed by Fulk (1975) in any month by almost 100%.

The decline of *Akodon olivaceus* is intriguing, in view of the stable or even increasing number of *A. longipilis*. *Akodon olivaceus* has been shown to be a relatively omnivorous species preferring habitats with less shrub cover and greater herbaceous cover, while *A. longipilis* is more insectivorous and prefers habitats with higher shrub and litter cover (Fulk, 1975; Glanz, 1977, 1984; Meserve, 1981a,b). Geographically, *A. longipilis* is the more limited species in the arid zone, being restricted to lower elevations from La Serena southwards. In addition, it maintains a lower metabolic rate and has a lower energetic assimilation efficiency than *A. olivaceus* (Rau et al., 1981). Although little information is available on water balance relations, it appears to be a more mesic species than *A. olivaceus* (Meserve, 1978). Elsewhere, in moist primary and secondary growth temperate rain forests, *A. longipilis* maintains more constant and often numerically superior populations than *A. olivaceus* (Pearson & Pearson, 1982; Meserve et al., 1982; Murúa & González, 1985). Thus, the tendency to maintain more stable numbers is relatively independent of the community in which it is found. It is tempting to consider *A. longipilis* a more "K-selected" species in view of its lower density, relatively constant populations, long survivorship, slower maturation rate, and smaller litter sizes. In addition, we observed stable, nonoverlapping home ranges of adult females, as did Fulk (1975).

Phyllotis darwini populations showed, between years, fairly predictable patterns of population changes and reproduction which differed relatively little, except in magnitude, from the results of Fulk (1975). As litter size observed in our study was similar to that reported by Fulk (1975) and Pearson (1975) during an "outbreak" year, rapid increases in numbers are apparently achieved by rapid maturation of young to reproduce in the same season as birth, and by multiple short-interval pregnancies. With a minimum interval of only 16 days between consecutive litters, a postpartum estrus is indicated. Survival rate of older adults between reproductive seasons appears to be zero, so that the entire breeding population in a given year is made up of immigrant adults and of younger individuals born in the same season. Survival rates are low, but the species appears to colonize new habitats successfully and maintains high reproductive rates. *Phyllotis darwini* is the most ubiquitous species in the semiarid and arid regions of northern Chile and appears to be capable of seasonally adjusting its resistance to desiccation and to utilize seeds and succulents (Meserve, 1978; Meserve & Glanz, 1978). Interestingly, Fulk (pers. comm.) snap-trapped a dry overgrazed habitat dominated by *Baccaris* sp. and *Proustia pungens* immediately east of Fray Jorge in late May–early June 1973, simultaneously with trapping of irrigated bean fields near La Serena (about 100 km NE) by Péfaur et al. (1979). Whereas Fulk encountered 86.6% *P. darwini* in two nights of trapping (total of 82 animals, 51.3% trap success), Péfaur et al. (1979) reported 86.7% *Oryzomys longicaudatus* (135 animals, 44.4% success) in two nights. Mares (1977a,b) reported that both of these species were poor water conservers in the Argentine Monte desert. Percentages of the catch composed of *Akodon olivaceus* for the two Chilean sites mentioned above were 14.6 and 1.5, respectively, at a time when this species still made up a majority of the Fray Jorge live-trap grid population (Fulk, 1975).

The large numbers and relative importance of *Octodon degus* in this locality during two very dry years was surprising in view of the relatively stenothermic physiology and high evaporative water loss rates of this species (Rosenmann, 1977). While degus appear to have high population turnover and low juvenile survival rates under such conditions, a few individuals have very high reproductive potential after surviving to the next breeding season. There is evidence that precipitation directly triggers reproduction; in 1974 and 1976, reproduction immediately followed the first annual rains before the initiation of herbaceous growth (only 6.2 and 17 mm, respectively). With a gestation of 90 days (Woods & Boraker, 1975), such a strategy may be optimal in the northern arid zone. On the other hand, Rojas et al. (1977) have argued for the principal role of herbaceous growth in initiating reproduction in central Chile, paralleling the situation documented for heteromyid rodents by Beatley (1969, 1976) and Van De Graaff and Balda (1973).

Other species present in Fray Jorge included *Oryzomys longicaudatus,* which appears to be sporadic and irruptive and was transient in the Fray Jorge population. Interestingly, Fulk (1975) documented pregnant *O. longicaudatus* females in March 1973, beyond the reproductive period for other sigmodontines. Populations of *O. longicaudatus* might respond to increased seed production following unusual periods of rainfall, as this species is granivorous (Meserve, 1981a; Murúa & González, 1981). In the southern temperate rain forest, González and Murúa (pers. comm.) have suggested a relationship between the fluctuations in annual seed crops and those of *O. longicaudatus* populations. A second species, *Abrocoma bennetti,* was recorded in low numbers in most censuses but was never recaptured. It is the largest rodent found in this community (200–280 g) and appears to be a highly specialized herbivore with greatest similarity to *Octodon degus* in various aspects of its ecology (Meserve, 1981a; Meserve et al., 1983).

In conclusion then, we have documented the presence of a permanent and successful assemblage of small mammals living in what would be considered an extremely arid environment by virtually any standard. One might predict, without prior knowledge of the history and zoogeography of the area, that this community would be inhabited by highly specialized, granivorous, water-independent forms; instead one encounters an assemblage composed of relatively generalized, trophically diverse, and mostly water-dependent members. Such cases of "nonconvergence" perhaps only emphasize the importance of history, zoogeography, and alternative strategies for survival in harsh environments. While extrinsic environmental factors related to food and water availability may ultimately limit populations in this community, there is remarkable stability in overall numbers from year to year; most changes seem to involve relative numbers of component species. It may also be significant that Fray Jorge represents a relatively undisturbed native shrub community surrounded by vast areas of badly degraded and overgrazed shrublands and marginal cultivation. The extent to which environmental degradation elsewhere in arid Chile influences the patterns of population change, extinction, and community instability reported in previous studies can only be speculated on.

Acknowledgments

We wish to thank the Laboratório de Ecología, Pontífica Universidad Católica de Chile, for support provided both authors during the time that they were conducting fieldwork. The Laboratoire d'Écologie Théorique et de Biométrie, Université Catholique de Louvain-la-Neuve, Belgium, provided computing facilities and logistical support. Several reviewers made valuable comments on the manuscript. Drs. Eduardo R. Fuentes and Robert E. Martin assisted with aspects of fieldwork. Finally, we thank Janet L. Meserve and Paule Le Boulengé-Nguyen for unfailing assistance and forebearance during all aspects of the fieldwork and preparation of the manuscript.

Literature Cited

ANDRZEJEWSKI, R. 1963. Processes of incoming, settlement and disappearance of individuals and variations in the numbers of small rodents. Acta Theriologica, **7:** 169–213.

———. 1969. Analysis of results of catches of small mammals by the "Calendar of Catches" method. Zeszyty Naukowe, **2:** 1–104. (In Polish, English summary.)

AXELROD, D. I. 1973. History of the Mediterranean ecosystem in California, pp. 225–277. *In* di Castri, F., and H. A. Mooney, eds., Mediterranean Type Ecosystems. Springer-Verlag, Berlin, 395 pp.

BAKER, R. H. 1967. Distribution of Recent mammals along the Pacific coastal lowlands of the Western Hemisphere. Systematic Zoology, **16:** 28–37.

BEATLEY, J. C. 1969. Dependence of desert rodents on winter annuals and precipitation. Ecology, **50:** 721–724.

———. 1976. Rainfall and fluctuating plant populations in relation to distribution and numbers of desert rodents in southwest Nevada. Oecologia (Berlin), **24:** 21–42.

CLOBERT, J., J. D. LEBRETON, M. CLOBERT-GILLET, AND M. COQUILLART. 1985. Estimation of survival in bird populations by recaptures or sightings of marked individuals, pp. 197–213. *In* Morgan, B. J. T., and P. M. North, eds., Statistics in Ornithology. Springer-Verlag, Berlin, 418 pp.

DUESER, R. D., R. K. ROSE, AND J. H. PORTER. 1984. A body-weight criterion to identify dispersing small mammals. Journal of Mammalogy, **65:** 727–729.

FORD, R. G., AND F. A. PITELKA. 1984. Resource limitation in populations of the California vole. Ecology, **65:** 122–136.

FULK, G. W. 1975. Population ecology of rodents in the semiarid shrublands of Chile. Occasional Papers, The Museum, Texas Tech University, **33:** 1–40.

GLANZ, W. E. 1977. Comparative ecology of small mammal communities in California and Chile. Ph.D. diss., University of California, Berkeley, 300 pp.

———. 1984. Ecological relationships of two species of *Akodon* in central Chile. Journal of Mammalogy, **65:** 433–441.

GLANZ, W. E., AND P. L. MESERVE. 1982. An ecological comparison of small mammal communities in California and Chile, pp. 220–226. *In* Conrad, C. E., and W. C. Oechel, eds., Dynamics and Management of Mediterranean-type Ecosystems. U.S. Forest Service, Pacific Southwest Forest Range Experiment Station, General Technical Report PSW-58, 637 pp.

GREER, J. K. 1965. Mammals of Malleco Province, Chile. Publications of the Museum, Michigan State University, Biological Series, **3:** 59–152.

HERSHKOVITZ, P. 1962. Evolution of Neotropical cricetine rodents (Muridae), with special reference to the Phyllotine group. Fieldiana: Zoology, **46:** 1–524.

JENNRICH, R. I., AND F. B. TURNER. 1969. Measurement of non-circular home range. Journal of Theoretical Biology, **22:** 227–237.

KREBS, C. J. 1966. Demographic changes in fluctuating populations of *Microtus californicus*. Ecological Monographs, **36:** 239–273.

KUMMEROW, J. 1966. Aporte al conocimiento de las condiciones climáticas del bosque de Fray Jorge. Universidad de Chile, Facultad de Agronomía, Estación Experimental Agronómica, Boletin Técnico N° 24 (Santiago, Chile), pp. 21–24.

LE BOULENGÉ, É. 1985a. Package C.M.R., analysis of capture-recapture data. Package description and user's guide (version 4, January 1985). Mimeo report, Unité d'Écologie, Université Catholique de Louvain, Louvain, Belgium, IV + 50 pp.

———. 1985b. Computer package for the analysis of capture-recapture data. Acta Zoologica Fennica, **173:** 69–72.

LE BOULENGÉ, É., AND P. Y. LE BOULENGÉ-NGUYEN. 1981. Ecological study of a muskrat population. Acta Theriologica, **26:** 47–82.

MACMILLEN, R. E. 1964. Population ecology, water relations, and social behavior of a southern California semidesert rodent fauna. University of California Publications in Zoology, **7:** 1–59.

Manly, B. F. J., and M. J. Parr. 1968. A new method of estimating population size, survivorship, and birth rate from capture-recapture data. Transactions of the Society for British Entomology, **18:** 81–89.

Mares, M. A. 1975a. South American mammal zoogeography: Evidence from convergent evolution in desert rodents. Proceedings of the National Academy of Sciences (USA), **72:** 1702–1706.

———. 1975b. Observations of Argentine desert rodent ecology, with emphasis on water relations of *Eligmodontia typus,* pp. 155–175. *In* Prakash, I., and P. K. Ghosh, eds., Rodents in Desert Environments. Junk, The Hague, Netherlands.

———. 1976. Convergent evolution of desert rodents: Multivariate analysis and zoogeographic implications. Paleobiology, **2:** 39–63.

———. 1977a. The comparative water relations and habitat requirements of three South American phyllotine rodents. Journal of Mammalogy, **58:** 514–520.

———. 1977b. Aspects of water balance of *Oryzomys longicaudatus* from northwest Argentina. Comparative Biochemistry and Physiology, **57A:** 237–238.

———. 1980. Convergent evolution among desert rodents: A global perspective. Bulletin of the Carnegie Museum of Natural History, **16:** 1–51.

Mazurkiewicz, M. 1969. Elliptical modification of the home range pattern. Bulletin de l'Académie Polonaise des Sciences Classe II, **17:** 427–431.

M'Closkey, R. T. 1972. Temporal changes in populations and species diversity in a California rodent community. Journal of Mammalogy, **53:** 657–676.

Meserve, P. L. 1978. Water dependence in some Chilean arid zone rodents. Journal of Mammalogy, **59:** 217–219.

———. 1981a. Trophic relationships among small mammals in a Chilean semiarid thorn scrub community. Journal of Mammalogy, **62:** 304–314.

———. 1981b. Resource partitioning in a Chilean semiarid thorn scrub community. Journal of Animal Ecology, **50:** 745–757.

Meserve, P. L., and W. E. Glanz. 1978. Geographical ecology of small mammals in the northern Chilean arid zone. Journal of Biogeography, **5:** 135–148.

Meserve, P. L., R. E. Martin, and J. Rodriguez. 1983. Feeding ecology of two Chilean caviomorphs in a central mediterranean savanna. Journal of Mammalogy, **64:** 322–325.

———. 1984. Comparative ecology of the caviomorph rodent *Octodon degus* in two Chilean mediterranean type communities. Revista Chilena de Historia Natural (Santiago, Chile), **57:** 79–89.

Meserve, P. L., R. Murúa, O. Lopetegui, and J. R. Rau. 1982. Observations on the small mammal fauna of a primary temperate rain forest in southern Chile. Journal of Mammalogy, **63:** 315–317.

Muñoz, P. C., and E. Pisano V. 1947. Estudio de la vegetación y flora de los parques nacionales de Fray Jorge y Talinay. Agricultura Técnica (Santiago, Chile), **7:** 71–190.

Murúa, R., and L. A. González. 1981. Estudios de preferencias y hábitos alimentarios en dos especies de roedores cricétidos. Medio Ambiente (Valdivia, Chile), **5:** 115–124.

———. 1985. A cycling population of *Akodon olivaceus* (Rodentia: Cricetidae) in a Chilean temperate rain forest. Acta Zoologica Fennica, **173:** 77–79.

Osgood, W. H. 1943. The mammals of Chile. Field Museum of Natural History, Zoological Series, **30:** 1–268.

Pearson, O. P. 1975. An outbreak of mice in the coastal desert of Peru. Mammalia, **39:** 375–386.

———. 1983. Characteristics of a mammalian fauna from forests in Patagonia, southern Argentina. Journal of Mammalogy, **64:** 476–492.

Pearson, O. P., and A. K. Pearson. 1982. Ecology and biogeography of the southern rainforests of Argentina, pp. 129–142. *In* Mares, M. A., and H. H. Genoways, eds., Mammalian Biology in South America. Special Publication Series, Pymatuning Laboratory of Ecology, University of Pittsburgh, **6:** 1–539.

Pearson, O. P., and C. P. Ralph. 1978. The diversity and abundance of vertebrates along an altitudinal gradient in Peru. Memorias, Museo de Historia Natural "Javier Prado" (Lima, Peru), **18:** 1–97.

Péfaur, J. E., J. L. Yáñez, and F. M. Jaksić. 1979. Biological and environmental aspects of a mouse outbreak in the semi-arid region of Chile. Mammalia, **43:** 313–322.

Petrusewicz, K., and R. Andrzejewski. 1962. Natural history of a free-living population of house mice (*Mus musculus* Linnaeus) with particular reference to groupings within the population. Ekologia Polska, Seria A, **10:** 85–122.

Rau, J., R. Murúa, and M. Rosenmann. 1981. Bioenergetics and food preferences in sympatric southern Chilean rodents. Oecologia (Berlin), **50:** 205–209.

Rojas, M., O. Rivera, G. Montenegro, and C. Barros. 1977. Algunas observaciones de la reproducción de la hembra silvestre de *Octodon degus,* Molina y su posible relación con la fenología de la vegetación. Medio Ambiente (Valdivia, Chile), **3:** 78–82.

Rosenmann, M. 1977. Regulación térmica en *Octodon degus.* Medio Ambiente (Valdivia, Chile), **3:** 127–131.

Sarmiento, G. 1975. The dry plant formations of South America and their floristic connections. Journal of Biogeography, **2:** 233–251.

Simpson, B. B. 1975a. Glacial climates in the eastern tropical South Pacific. Nature, **253:** 34–36.

———. 1975b. Pleistocene changes in the flora of the high tropical Andes. Paleobiology, **1:** 273–294.

Solbrig, O. T. 1976. The origin and floristic affinities of the South American temperate desert and semidesert regions, pp. 7–49. *In* Goodall, D. W., ed., Evolution of Desert Biota. University of Texas Press, Austin, 250 pp.

Tamarin, R. H. 1984. Body mass as a criterion of dispersal in voles: A critique. Journal of Mammalogy, **65:** 691–692.

Van De Graaff, K. M., and R. P. Balda. 1973. Importance of green vegetation for reproduction in the kangaroo rat, *Dipodomys merriami merriami.* Journal of Mammalogy, **54:** 509–512.

Weir, B. J. 1970. The management and breeding of some more hystricomorph rodents. Laboratory Animals, **4:** 83–97.

———. 1974. Reproductive characteristics of hystricomorph rodents, pp. 269–299. *In* Rowlands, I. W., and B. J. Weir, eds., The Biology of Hystricomorph Rodents. Symposium, Zoological Society of London, **34:** 1–482.

Woods, C. A., and D. K. Boraker. 1975. *Octodon degus.* Mammalian Species, **67:** 1–5.

Demography and Reproduction of the Silky Desert Mouse (*Eligmodontia*) in Argentina

Oliver Pearson, Susana Martin, and Javier Bellati

ABSTRACTS

Specimens of *Eligmodontia typus* collected in semiarid steppe habitat were assigned to age classes according to tooth wear, and the age structure of the population was determined for different times of the year. The reproductive season extends from October to the end of April; average litter size is 5.9. Males and females can reach sexual maturity at approximately 1½ months of age and rarely live longer than 9 months. Adult females were 19% heavier than adult males and 6% longer. Abundance ranged from 0.4 per hectare in spring to 3.5 in autumn.

Although *Eligmodontia* is widely dispersed through many arid habitats, life history features such as small home range, large litters, and brief life span distinguish it from highly adapted desert rodents of other continents.

Especímenes de *Eligmodontia typus* colectados en hábitat de estepa semiárida fueron clasificados por edad según desgaste dentario; la estructura de edades de la población fué determinada para diferentes épocas del año. La estación reproductiva se extiende desde Octubre hasta fines de Abril y el tamaño promedio de camada es de 5.9. Machos y hembras pueden alcanzar la madurez sexual a los 1½ meses de edad y raramente ven más de 9 meses. Las hembras resultaron ser de mayor tamaño que los machos. La abundancia fluctua de 0.4 individuos por hectárea en primavera a 3.5 en otoño.

Aunque *Eligmodontia* es ampliamente distribuído en varios hábitats áridos, aspectos del ciclo vital tales como pequeños territorios, camadas numerosas y vida de corta duración la distinguen de roedores altamente adaptados de desiertos de otros continentes.

Espécimes de *Eligmodontia typus,* coletados nos habitat semi-áridos da estepe, foram designados a classes de idade de acôrdo com o gasto de seus dentes. A estrutura de idade da população foi assim determinada durante várias épocas do ano. A época reproductiva estende-se de outrubro ao fim de abril, e a ninhada média é de 5,9 crias. Machos e fêmeas podem atingir maturidade sexual por volta de 1,5 meses de idade, e raramente vivem mais do que 9 meses. As fêmeas são maiores que os machos. A densidade dos camundongos foi de 0,4 por hectare na primavera e de 3,5/ha. no verão.

Apesar da ampla distribuição dos Eligmodontes em vários hábitats áridos, aspectos da biolo-

From the Museum of Vertebrate Zoology, University of California, Berkeley, CA 94720 (Pearson); and Instituto Nacional de Tecnología Agropecuaria, Casilla de Correo 277, 8400 Bariloche, Argentina.

gia destes animais, como por exemplo o seu uso de área, o seu curto período de vida, e o seu grande número de crias, distinguem estes camundongos dos roedores de outros continentes que são altamente adaptados à vida árida do deserto.

Introduction

The silky desert mouse is abundant in arid or semiarid habitats from southern Argentina north to the altiplano of southern Peru. It occupies habitats that on other continents are filled by highly specialized rodents such as North American kangaroo rats, Asian dipodids, and African gerbils. Although *Eligmodontia* is probably as morphologically adapted for desert life as any other genus of mouse now living in South America, it is not highly specialized (Mares, 1980). Several tubercles on the soles of each hind foot are fused into a furry pad that may aid locomotion on sandy soil; the hind feet are somewhat elongate (about 20% longer than those of *Peromyscus maniculatus* of similar size), and many populations have pale silky fur. But, aside from these few presumed desert adaptations, *Eligmodontia* is in general appearance and proportions scarcely distinguishable from the wide-ranging *P. maniculatus* of North America. It lacks the large, inflated bullae, long tufted tail, and kangaroo-like hind limbs seen in many desert species on other continents.

In physiological properties as well, *Eligmodontia* does not appear to be highly adapted to desert living. It can live in captivity for weeks on a diet of dry seeds without drinking water, but this is true for mice from a variety of habitats. It will drink water if available, and also will utilize the water in succulent vegetation (Mares, 1975a, 1977). It has an unusual capability to drink salt water, which suggests that it could utilize succulent parts of desert plants with a high salt content (Mares, 1977). As measured by rate of loss of body weight by captives without water, the physiological capacity of *Eligmodontia* to resist dehydration is intermediate to that of desert-adapted pocket mice (*Perognathus*) and the relatively unspecialized *Peromyscus maniculatus* (Mares, 1975b).

Because *Eligmodontia* is the most abundant mammal in large areas of arid and semiarid habitat in South America and lacks a "desert" morphology and physiology, we thought that a study of its ecology and life history might reveal convergences with life history features of desert rodents elsewhere. Convergence of life history features of unrelated species living in different arid regions would suggest their adaptive significance.

Materials and Methods

The Argentine species of *Eligmodontia* is smaller than the species living in Peru and has a different karyotype (unpubl. obs.). Even within the province of Río Negro it shows considerable variation in color, size, and proportions. An isolated population 10 km west of Bariloche in an island of steppe vegetation consists of very small, dark individuals. Populations from the center and eastern parts of Río Negro are paler and have much longer tails. To minimize possible geographic variation of the demographic and reproductive features measured in our study, we have limited our samples to specimens collected within 25 km of Pilcaniyeu Viejo in Río Negro. They are referred to the subspecies *E. typus typus* pending resolution of nomenclatural complications pointed out by Hershkovitz (1962).

The study area (ca. 41°S and 70°30′W) is located within what has been described as the Patagonian Province by Cabrera (1971) or as the Western Mesetas and Sierras by Anchorena (1978). Average monthly temperatures range from a low of 1.4°C. in July to a high of 13.6°C. in January. No period is free of frosts. The mean annual temperature for the area is approximately 7.4°C. Precipitation averages 300 mm per year, almost three-quarters of it in the winter. Relative humidity for the area ranges from about 20% in January to 85% in July. Winds are especially strong in spring, although frequent and forceful at all seasons.

Land use is characterized by extensive sheep and cattle ranching, with seasonal use of summer and winter grazing areas. Sheep were introduced a century ago, and overgrazing has since modified the plant coverage and species composition of the area. The study area consists of steppe interrupted by small marshy areas referred to locally as *mallines*. The vegetation has a mean height of approximately 50 cm and is composed mainly of low shrubs and a variety of grasses. The main shrubs

FIG. 1. The census area in steppe at Campo Anexo Pilcaniyeu, December 1981.

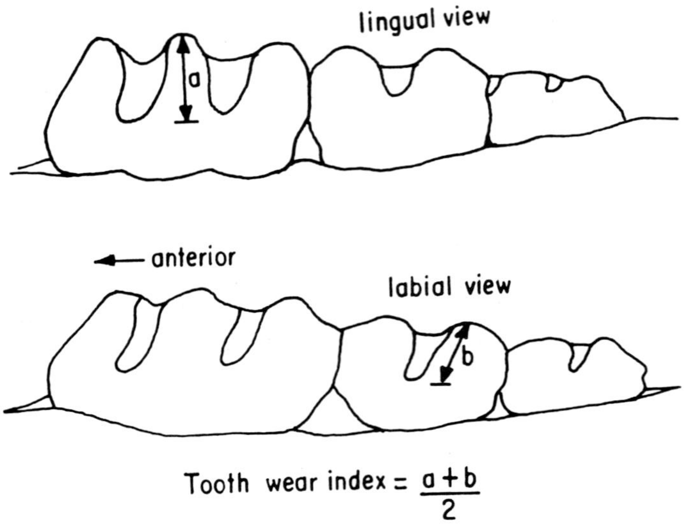

FIG. 2. Upper toothrows of *Eligmodontia* showing the two measurements taken for calculation of the tooth wear index.

are *neneo* (*Mulinum spinosum*), *mata torcida* (*Stillingia patagonica*), *charcao* (*Senecio bracteolatus*), *colapiche* (*Nassauvia glomerulosa*), *mamuel choique* (*Adesmia campestris*), *duraznillo* (*Coliguaya integerrima*), and *Senecio neaei*. The main grasses are various species of bunchgrass, especially *Stipa speciosa* var. *major, Poa lanuginosa, Festuca argentina, P. ligularis,* and *S. speciosa* var. *speciosa.*

The small-mammal fauna of the study area is dominated by *Eligmodontia* (nocturnal granivore-omnivore) and another mouse of about the same size, *Akodon xanthorhinus* (nocturnal and diurnal granivore-omnivore). Other small mammals present but caught much less frequently were *Phyllotis darwini* (nocturnal omnivore), *Reithrodon auritus* (nocturnal herbivore), *Ctenomys haigii* (fossorial herbivore), *Notiomys edwardsii* (semi-fossorial insectivore-omnivore), *Euneomys* sp. (herbivore), *Microcavia australis* (diurnal herbivore), and the introduced *Lepus capensis* (nocturnal herbivore). In the few patches of mesic or moist vegetation, *Akodon longipilis* and *Auliscomys micropus* may be abundant. Almost all of these small mammals range widely in southern Argentina. Some of them occur in much moister habitats.

Readers familiar with Mares's (1983) description of desert communities will note that the Argentine steppe community described above contains many of the elements found in desert communities on other continents. The guinea pig/ground squirrel, tuco-tuco/pocket gopher, and European hare/jackrabbit similarities are especially close. The small bipedal granivores are notably absent in South America.

Two hundred twenty *Eligmodontia* were collected in the study area between November 1981 and January 1984 using Sherman live traps baited with rolled oats, Museum Special kill-traps baited usually with cornmeal or rolled oats but occasionally with other baits, or smaller kill-traps baited with cornmeal and peanut butter. On several occasions traplines were set, alternating Sherman traps baited with rolled oats and Museum Specials baited with cornmeal. The Museum Specials were more effective. Almost all specimens were dissected immediately. Skulls were dried for later cleaning.

The reproductive condition of males was assessed from measurements of the length of one testis, the color and texture of the testes, and the length of the seminal vesicles. The seminal vesicles are J-shaped; the length was considered to be the distance from the base of the bladder to the curve of the J. The measurements were used to classify each male into one of three categories: sexually immature, mature, or postmature. Males in breeding condition had pale, firm testes more than 5.5 mm long and seminal vesicles more than 7.5 mm long; the diameter of the epididymis was great enough so that this contorted tube could be seen easily through the sheath of the cauda epididymis. For specimens of uncertain category, the presence or absence of abundant spermatozoa was confirmed by microscopic examination of a smear of an epididymis. Flabby or dark-colored testes, usually associated with submaximal seminal vesicles, were considered to be an indication of postbreeding condition.

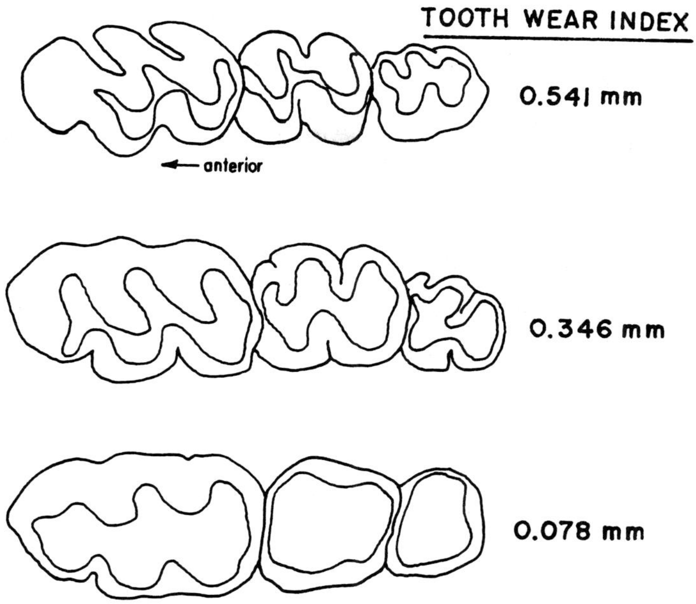

FIG. 3. View of the grinding surfaces of upper toothrows showing the degree of wear associated with three different tooth wear indices.

Females were placed in reproductive categories on the basis of examination of uteri, nipples, and the pubic symphysis. Thin, pale uterine horns without embryos or placental scars indicated nulliparous females. Females with swellings in the uteri were considered to be pregnant. Females with uterine scars, thick uteri, or large nipples were considered to be parous. If milk could be expressed from a nipple, a female was registered as lactating. Nonpregnant females with an open pubic symphysis were considered to be parous.

ESTIMATION OF AGE—The relative age of each individual was estimated by measuring the amount of wear on the cusps of two upper molar teeth. The tall, sharp cusps of young individuals wear down through life until, if the individual survives long enough, the tooth surface is smooth. Amount of wear was converted to approximate age by noting the progression of wear in successive collections of the first cohort of young each season. Similar techniques have been used by Pearson (1945, 1967, 1975), Happold (1967), French et al. (1974), and Feito et al. (1981).

In this study of *Eligmodontia,* the height of the middle cusp of the first upper molar was measured from the labial side, the depth of the groove between the two main cusps of the second molar in the opposite toothrow from the lingual side (fig. 2). Measurements were recorded to 0.001 mm using a microscope fitted with crosshairs and an electronic digital readout.

Although there was a high positive correlation between the two measurements for each specimen ($r = +0.88$, $n = 176$), we assumed that accuracy would be increased by averaging the two measurements. This average is the number presented as the "Tooth Wear Index" for each specimen: the larger the number, the younger the individual.

The cusps on the first molars are usually taller than those on the second molar (average 0.07 mm taller), but in young individuals the second is sometimes taller than the first. The M^2 wears more rapidly, however, and a regression of M^1 against M^2 shows that M^1 may be expected to have about 0.175 mm of cusp remaining at an age when the cusps on M^2 have worn away completely.

Measurement of the height of the cusps on two teeth is tedious and requires a special microscope, but the Tooth Wear Index is well correlated with the physical appearance of the teeth. Three examples of different ages are shown in Figure 3. It is possible that a subjective estimate of the amount of wear on the surface of all three molars is more reliable than actual measurement, because the observer can integrate information from all of the peaks and valleys of the entire toothrow.

To determine whether males and females were of equal size, it was desirable to establish a minimum age at which individuals could be considered adult. This was done by plotting length of head and body against tooth wear and estimating the point at which the growth curve levels off.

TABLE 1. Measurements of adult male and female *Eligmodontia* and of all male and female *Eligmodontia* from the population studied.

Measurement	Adult males	Adult females	All males	All females
Weight				
Mean ± SE	18.22 ± 0.52**	21.72 ± 0.65**	16.36 ± 0.44*	17.99 ± 0.61*
Range	10–25	13–31	7.2–25	7.5–31
No. of specimens	49	47	88	86
Coefficient of variation	19.9	20.5	...	...
Length of head and body				
Mean ± SE	82.12 ± 0.72**	87.08 ± 0.90**	79.35 ± 0.79*	82.17 ± 0.88*
Range	70–92	75–99	59–96	63–99
No. of specimens	49	47	88	86
Coefficient of variation	6.2	7.1	...	...
Length of tail				
Mean ± SE	84.29 ± 1.49*	88.51 ± 1.24*	80.45 ± 1.25	82.29 ± 1.29
Range	58–108	68–105	51–108	56–105
No. of specimens	49	47	88	86
Coefficient of variation	12.4	9.6	...	...
Greatest length of skull				
Mean ± SE	24.24 ± 0.12	24.43 ± 0.16	...	...
Range	23.17–25.97	22.40–26.40	...	...
No. of specimens	33	29	...	...
Coefficient of variation	2.9	3.5	...	...
Length of toothrow				
Mean ± SE	3.71 ± 0.02	3.68 ± 0.03	...	...
Range	3.51–4.05	3.35–4.05	...	...
No. of specimens	33	29	...	...
Coefficient of variation	3.6	4.0	...	...

* Difference significant at 0.05 level.
** Difference significant at 0.01 level.

Similarly, body weight was plotted against tooth wear. From these graphs, 0.460 mm was chosen as the separation between juveniles and adults. No adult defined in this way had a head and body shorter than 70 mm (although some juveniles were larger than this), and only two adults weighed less than 12 g (although numerous juveniles weighed more).

Results

SIZE—In general, females weigh more, have longer bodies, and have longer tails than males (table 1). When only adult individuals (Tooth Wear Index less than 0.460) are considered, females are statistically larger in all three of these dimensions. In length of skull and length of molar toothrow, however, males and females are of the same size. We have presented the size data in considerable detail because seldom are female mice, on average, demonstrably larger than males (Ralls, 1976). *Eligmodontia* and the desert dipodid *Jaculus jaculus* are rare exceptions (Happold, 1967).

The coefficient of variation was notably large for body weight (19.9, 20.5) and length of tail (12.4, 9.6). These coefficients are twice as large as those for tail length in one species of kangaroo rat (Lidicker, 1960).

SEX RATIO—The sex ratio of the entire sample was almost equal (table 1). When the sample is divided into age groups, no statistically significant difference in proportion of sexes appears, nor does the proportion of sexes depart appreciably from equality in spring or autumn. It appears, therefore, that the sexes are indeed equally abundant in the wild population, or else increased susceptibility to capture of one sex exactly compensates for its relative scarcity.

SEASON OF REPRODUCTION—The abundance of pregnant and lactating females early in November and the absence of young animals in the population at this time (fig. 4) demonstrate that the reproductive season begins in the Southern Hemisphere's spring (October). Reproduction continues

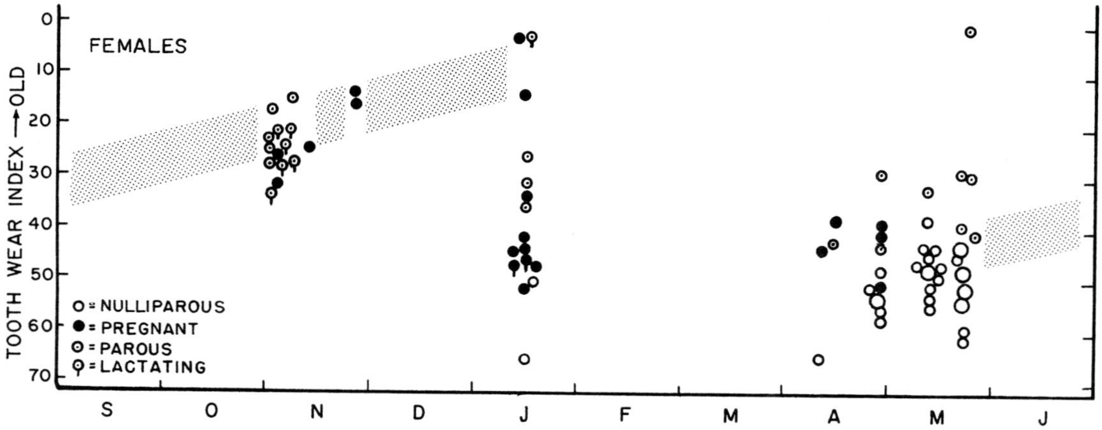

FIG. 4. Age and reproductive condition of female *Eligmodontia* during spring, summer, and autumn. Each double-sized symbol represents five individuals. Age is in tooth wear units, with the oldest animals at the top of the diagram. The stippled band suggests the rate of aging of one summer-born cohort.

during the spring and summer, during which time the proportion of juveniles in the population increases enormously (figs. 4–6). In the autumn pregnant females were caught until the end of April, but none of numerous females caught in the middle of May was pregnant (fig. 4).

The male reproductive season coincides with that of the females (fig. 5). Even old, sexually mature males pass out of breeding condition in May.

AGE OF SEXUAL MATURITY—Females born late in October or early in November may be visibly pregnant, or even lactating, with tooth wear of 0.50 mm in mid-January (fig. 4). They must have been inseminated at not more than six to eight weeks of age. Young males become sexually mature at six to eight weeks of age also (fig. 5).

LONGEVITY—The cluster of young breeding mice between tooth wear of 0.40 and 0.52 in January (figs. 4 and 5) must have been born late in October or early in November and therefore were about 10 weeks old. A different cluster of young mice of that age in May have aged to tooth wear of 0.20–0.30 in November, 22 weeks later. This indicates an age of 22 + 10 = 32 weeks for mice with tooth cusps 0.20 to 0.30 high. By January, when they have entered the oldest age category, they are about 41 weeks old. Since there is no accumulation of individuals in this old category, they must die soon after reaching that age. It is doubtful that any of the mice in our samples were as much as 12 months old. The oldest females and males were capable of breeding.

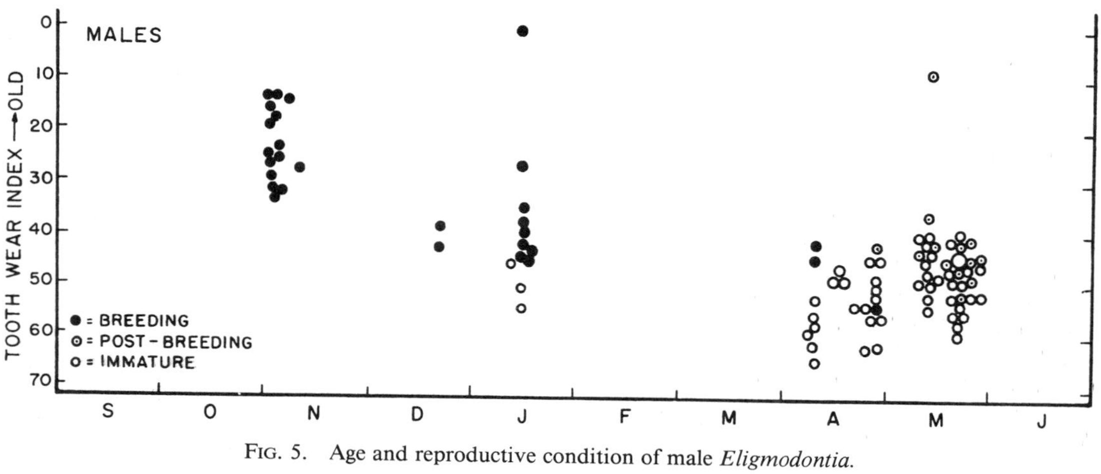

FIG. 5. Age and reproductive condition of male *Eligmodontia*.

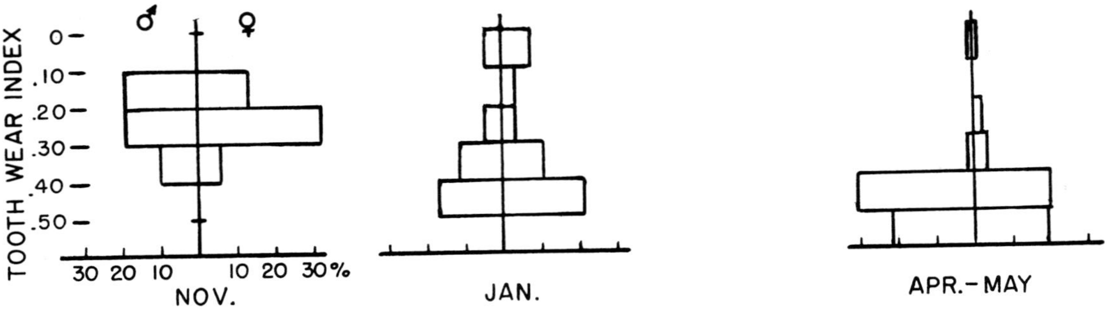

FIG. 6. Age pyramids and sex ratios of *Eligmodontia* in spring, summer, and autumn. Age is represented in tooth wear units.

The changes in age composition of the population can be followed easily in Figure 6. The entire population is middle-aged in November. This cohort is already much reduced in January and has probably disappeared entirely by April–May. Similarly, the young cohorts that have entered the population by January have almost disappeared before April–May, leaving the overwinter success of the population to the young born relatively late in the summer.

In summary, the reproductive activities of the population proceed as follows: In autumn (May) the population is made up almost entirely of animals born during the summer. The younger females are nulliparous and the middle-aged females parous. The young males have not matured sexually and the middle-aged and old males are no longer sexually competent. The overwinter survivors breed in October but die before the end of the summer. In fact, few individuals born early in the breeding season survive until autumn. The young born in November breed promptly, and some of their male and female offspring breed before the end of the reproductive season in April.

NUMBER OF FETUSES—Twenty females were visibly pregnant; each carried between three and nine fetuses (mean 5.90, SE ± 0.39). If number of fetuses is plotted against age of the mother, there is a loose but significant positive correlation ($r = +0.55$). The wearing away of 0.10 mm of molar cusps, which requires about six weeks, was accompanied by an average increase of 0.66 more fetuses.

ABUNDANCE—*Eligmodontia* is usually the most abundant species in its habitat, although in some places *Akodon xanthorhinus* outnumbers it. In the springtime, when populations are lowest, 17 traplines composed of a total of 1,723 trap-nights caught 35 *Eligmodontia,* giving a trap success of 2%. Nine of the lines caught no *Eligmodontia.* In the most successful line, 14% of the traps held *Eligmodontia.* In the autumn, 22 traplines adding up to 1,339 trap-nights produced a trap success of 13%. If trap success is directly proportional to abundance, this suggests that *Eligmodontia* was about six times as abundant in autumn as in spring. This ratio is confirmed by the census data. As many as 52% of the traps on one trapline in the autumn caught *Eligmodontia.*

Density of *Eligmodontia* was measured on a study area at the Campo Anexo Pilcaniyeu of the Instituto Nacional de Tecnología Agropecuaria. A 1-ha^2 plot was measured and marked as a grid with a stake at each 10-m intersection. A Sherman live-trap was set within 2 m of each stake (121 traps). At the time of each census, trapping was carried out for three or four nights; captured mice were marked with numbered metal ear tags and then released. The size of the population was estimated by Lincoln Index calculations, and the area occupied by this population was considered to be the 1-ha grid increased by a border strip the width of the three-night home range of *Eligmodontia* (measured by recapture of marked individuals). Trapping for supplementary animals for dissection was carried out at the same time a few kilometers away.

The census area was chosen to represent a moderately grazed sample of Patagonian steppe (fig. 1), but subsequent study revealed that it should probably be considered to be heavily grazed. It was also within the hunting range of several domestic cats. The vegetation in the census area was scrub with an abundance of *Poa lanuginosa* and *Stipa speciosa* var. *major,* with a high proportion of *Stillingia patagonica* (Lores et al., 1983).

An assessment of the vegetation at alternate stakes on the census grid is summarized in Table 2. The dominant three species at each stake were ranked as first, second, or third. The dominance ranking was based on an estimate of the biomass or of the area covered by each species growing within 1 m of each stake. The percentage of ground covered by vegetation at each stake was also estimated by imagining a 1-m hoop with the stake at its center. The lowest percentage of ground cover was 10%; the highest was 100% in a dense mat of *Berberis heterophylla*. The average coverage was 43%.

Table 2 shows that *Stillingia*, bunchgrass (*Poa ligularis* and *Festuca argentina*), *Mulinum*, and *Senecio bracteolatus*, in that sequence, were the species most frequently listed as one of the dominant three species. *Stillingia* was dominant at 25 of the stakes, *Mulinum* at 17, bunchgrass at 11, *Senecio bracteolatus* at four, and *Nassauvia* at two. Other species that were among the dominant three species at a few stakes were *Cerastium arvense*, *Adesmia campestris*, and *Berberis heterophylla*.

The area was censused for mice on 24–26 November 1981, 8–11 April 1982, and 10–12 November 1982. On each of the spring censuses, only a single *Eligmodontia* was captured, giving a population size of one. On the April census, eight *Eligmodontia* were captured on the study grid, and Lincoln Index calculations indicate a population of 9.3 individuals. Seven recaptures of five of the mice indicate an average distance moved of 31 m. If a border strip of 31 m is added to the four sides of the 1-ha trapping grid to allow for the area utilized by the population during three nights, then the density of *Eligmodontia* was 3.5/ha. The average weight of these *Eligmodontia* was 15.9 g, which gives a biomass of 56 g/ha.

If 31 m is assumed to be an appropriate width of a border strip in spring as well as in autumn, then the single captures in November in 1981 and in 1982 represent a density of *Eligmodontia* of 0.4/ha. This is only one-ninth as abundant as in autumn, a ratio that is in fairly good agreement with the difference in percentage trap success in spring and autumn.

The only other mammalian species captured on the census grid was *Akodon xanthorhinus*; but feces of *Reithrodon auritus* (rata conejo) were seen at several places on the grid, armadillo burrows were present, and a hare (*Lepus capensis*) was seen.

DIET—*Eligmodontia* is primarily granivorous. Stomach contents of individuals captured in two different areas (Pichileufú and Los Menucos) were examined. Each sample was made up of contents from 23 individuals captured in the month of May. The samples consisted mainly of seeds which were identified and separated from the remaining contents. The latter consisted of green vegetation which was identified microscopically using epidermal characteristics. The most frequent seeds in the sample from Pichileufú were of *Berberis* and, in the sample from Los Menucos, *Prosopis*. The green vegetation was made up exclusively of dicots, including *Mulinum spinosum* and *Acaena* sp. in the sample from Pichileufú and predominantly *Lycium* sp. in the sample from Los Menucos. Grasses were completely absent in the contents examined. Mares (1977) reported that captive individuals ate moths, a grasshopper, and a locust. Our captives ate, among other things, the acrid seeds of *Stillingia* and the legs of the common black desert beetles.

TABLE 2. Dominant vegetation at 61 evenly distributed sites on the census grid. Dominance was estimated from the apparent biomass of each species.

Vegetation	Dominant at sites (N)	Among dominant 3 species at sites (N)
Mata torcida (*Stillingia patagonica*)	25	57
Neneo (*Mulinum spinosum*)	17	37
Bunchgrass (*Poa ligularis, Festuca argentina*)	11	45
Charcao (*Senecio bracteolatus*)	4	30
Colapiche (*Nassauvia glomerulosa*)	2	6

Discussion

We have noted above that the anatomy and physiology of *Eligmodontia typus* do not seem to be highly adapted for desert life. Mares (1983) has listed various behavioral and ecological traits that one would expect to find in a desert rodent, such as nocturnality, nesting in burrows, strong territorial behavior, relatively large home range, reproduction associated with a rainy season, low fecundity, and long survivorship. He points out that these traits have become associated in our minds with desert existence because a number of well-studied desert species display them, but he cautions that confirmation is needed from many more desert communities before we can safely assume that they are indeed adaptive traits fixed by the arid environment. We shall show that *Eligmodontia* displays some of them but not others, and that it shares more traits with its relatives living nearby in South Temperate forest (Pearson, 1983) than with desert forms elsewhere.

Eligmodontia is strictly nocturnal. Its nests and retreats are underground, sometimes in burrows made by tuco-tucos or armadillos. We know nothing of the exclusiveness of its territories, but its home range, unlike the expectation, is relatively small (see table 3; compare French et al., 1975). Curiously, *E. puerulus* in rather similar habitat in Peru and measured by the same methods has a much larger home range (table 3).

A survey of the number of fetuses carried by desert mice indicates that the average for *Eligmodontia* (5.9) is unusually large compared with the highly adapted desert rodents belonging to the Dipodidae, Heteromyidae, and Gerbillinae (table 3; Smith & Jorgensen, 1975; Naumov & Lobachev, 1975; French et al., 1975; Conley et al., 1977). Layne (1968) listed litter sizes for 21 species and subspecies of *Peromyscus,* all between 2.8 and 5.0 and with no obvious relationship between litter size and aridity of habitat. Eisenberg and Isaac (1963) listed average litter sizes between 2.2 and 4.0 for several desert species and pointed out that in arid regions small litters would conserve water during lactation. *Phyllotis gerbillus* is a species restricted to the Sechura Desert of Peru. Ten pregnant females of that species in the collection of the Museum of Vertebrate Zoology at Berkeley, California, were carrying an average of only 2.8 fetuses (range 1–4). In this feature, therefore, *P. gerbillus* conforms more to the expected, small-litter pattern of desert mice than does *Eligmodontia.*

Male and female *Eligmodontia typus* breed in the same season in which they are born, and females are capable of producing more than one litter in a season, even becoming pregnant at a postpartum estrus. These features are found also in a variety of highly adapted desert rodents (McCulloch & Inglis, 1961; Speth et al., 1968; French et al., 1975; Smith & Jorgensen, 1975; Conley et al., 1977) and contribute to a population's potential to respond rapidly to favorable (but rare and unpredictable) climatic departures from a stressful norm. The population increase of many highly adapted desert species, however, is impeded by the small size of their litters.

Initiation of reproduction in many species of desert mice seems to depend upon the sprouting and growth of seedlings after rains (Chew & Butterworth, 1964; Beatley, 1969; Van De Graaff & Balda, 1973; Reichman & Van De Graaff, 1975). *Eligmodontia* breeds during the plant growing season and, although it is granivorous and insectivorous, its stomach during the breeding season frequently contains green vegetation. It is probable, therefore, that the timing and success of reproduction of *Eligmodontia* also is controlled by the response of vegetation to temperature and rainfall. Such a response is clearly adaptive because keying the timing and the intensity of reproduction to the sprouting of seeds of annual plants maximizes the probability that there will be an adequate seed crop to support offspring. If rainfall should be unusually abundant or should be repeated for two or more years, populations of mice may increase to very high densities. This has occurred in North America (Soholt, 1973; Pearson, 1963), South America (Hershkovitz, 1962; Fulk, 1975; Pearson, 1975), and Africa (Christian, 1977; Poulet, 1978).

Populations of *Eligmodontia* did not reach high densities during the three years of this study. On the most productive traplines, as many as 52% of the traps captured *Eligmodontia,* but other traplines at the same time in similar habitat nearby were much less productive. It seems, therefore, that populations were very local and that we encountered no regional outbreak of either *Eligmodontia* or *Akodon xanthorhinus.* Trap success on the grid was usually lower than on other traplines. This suggests that our measurements of density (table 3) are lower than would have been found in the most favorable habitats.

Extreme fluctuations of population densities make it difficult to compare small mammal communities of different deserts. Only in studies lasting many years can one be certain whether one is studying a sparse, an average, or an abundant population. Nevertheless, we have listed data in Table 3 that permit a crude comparison of densities and biomasses of *Eligmodontia* with those of other desert rodents. Extreme variation is immediately apparent. Fulk's (1975) data from Chilean deserts were from very abundant populations following two rainy years. The highly adapted desert genera such as *Dipodomys, Gerbillurus, Desmodillus,* and *Taterillus* were found at greater densities than *Eligmodontia.* Of special interest is *Eligmodontia puerulus* in the high desert of Peru. This habitat is quite similar to the steppe of our study area, and *Eligmodontia* was living in Peru at densities and biomasses similar to those of *Eligmodontia typus.* It is also of interest that two species of the North American cricetid *Peromyscus,* relatively

TABLE 3. A comparison of some characteristics of small rodents in various deserts. The studies cited were chosen because the field data were gathered using techniques comparable with our own.

Location	Ground cover (%)	Species	Diam. of home range (m)	Density (no./ha)	Biomass (g/ha)	Average litter size	Reference
Steppe, Argentina	43	*Eligmodontia typus*	31	(0.4–3.5)	(6–56)	5.9	This study
Tola, Peru	33	*Eligmodontia puerulus*	92	1.0	25	...	Pearson & Ralph (1978)
Loma, Peru	37	*Phyllotis darwini*	36	6.5	206	...	Pearson & Ralph (1978); Pearson (1975)
Desert scrub, Peru	15	*Phyllotis darwini*	60	2.6	100	...	Pearson & Ralph (1978)
Mountain scrub, Peru	36	*Phyllotis darwini*	54	2.5	95	3.7	Pearson & Ralph (1978); Pearson (1975)
Semiarid shrub, La Rinconada, Chile	54	*Phyllotis darwini*	77	(3.2–4.4)	(184–191)	...	Fulk (1975)
Semiarid shrub, La Rinconada, Chile	54	*Akodon olivaceus*	54	(6–16)	(158–502)	...	Fulk (1975)
Semiarid shrub, Fray Jorge, Chile	44	*Phyllotis darwini*	(36–41)	(29–46)	(1,798–2,463)	5.2	Fulk (1975)
Semiarid shrub, Fray Jorge, Chile	44	*Akodon olivaceus*	(30–45)	(30–97)	(992–2,707)	5.5	Fulk (1975)
Larrea, USA	23	*Dipodomys merriami*	70	11.5 (7.8–15.6)	453 (370–590)	...	Chew & Chew (1970)
Larrea, USA	23	*Onychomys torridus*	118	1.83 (0.6–3.3)	40.3 (15–71)	...	Chew & Chew (1970)
Larrea, USA	23	*Perognathus flavus*	62	1.1 (0–3.9)	7.3 (0–25)	...	Chew & Chew (1970)
Larrea, USA	23	*Peromyscus eremicus*	87	1.1 (0.5–3.3)	21.9 (4–67)	...	Chew & Chew (1970)
Larrea-yucca, USA	+27	*Dipodomys merriami*	61	0.7–3.7	...	...	Chew & Butterworth (1964)
Larrea-yucca, USA	+27	*Perognathus longimembris*	64	0.8–1.7	...	...	Chew & Butterworth (1964)
Larrea-cassia, USA	6.6	*Dipodomys merriami*	41	16.2 (12.3–19.5)	(459–741)	2.2	Soholt (1973)
Coastal sage, USA	...	*Dipodomys agilis*	...	(1.2–4.8)	(74–295)	2.6 (2–4)	MacMillen (1964)
Coastal sage, USA	...	*Perognathus fallax*	...	(0.2–1.3)	(4.2–24.8)	...	MacMillen (1964)
Coastal sage, USA	...	*Peromyscus eremicus*	...	(0.5–4.0)	(8.5–75.6)	2.9 (2–6)	MacMillen (1964)
Coastal sage, USA	...	*Peromyscus maniculatus*	...	(0–2.5)	(0–51)	4.3 (2–7)	MacMillen (1964)
Desert grassland, South-West Africa	...	*Gerbillurus paeba*	27	(7–12)	...	...	Christian (1977)
Desert grassland, South-West Africa	...	*Desmodillus auricularis*	23	(12–16)	...	...	Christian (1977)
Dry bush, dunes, Senegal	...	*Taterillus pygargus*	25	3.0, 8.0	108,288	±4	Poulet (1972)

unspecialized and in general appearance similar to *Eligmodontia,* were living at densities similar to those of *Eligmodontia* (table 3). Our conclusion after comparing the densities of *Eligmodontia* with those of other species in Table 3 and with the densities reported by French et al. (1975), Smith and Jorgensen (1975), Naumov and Lobachev (1975), and Conley et al. (1977) is that the highly adapted desert mice on the other continents maintain higher densities than does *Eligmodontia* and other genera inhabiting arid zones in South America. No long-term studies have been carried out, however, in South America.

One of the most impressive demographic differences between *Eligmodontia* and the North American genera *Dipodomys* and *Perognathus* is in longevity. Our estimate of longevity for *Eligmodontia* is made by comparing age pyramids of samples trapped at different seasons of the year. These samples indicated such a rapid turnover of the population that very few individuals could be expected to survive nine months after birth and almost none could be expected to reach one year. All studies of *Dipodomys* and *Perognathus* show that these genera survive much longer than this. French et al. (1967) recaptured 25 *Perognathus longimembris* that had been marked three to five years earlier. Chew and Butterworth (1964) recaptured up to 19% of their marked *Dipodomys merriami* and 31% of their *Perognathus longimembris* a year later. Chew and Chew (1970) found that at least 50% of their marked *Dipodomys merriami* were still alive six months later, but 50% of the *Peromyscus eremicus* haa disappeared in two months. M'Closkey (1972) reported survivals per month of 79%, 79%, and 54% for *Dipodomys agilis, Peromyscus eremicus,* and *Peromyscus maniculatus,* respectively. In Soholt's (1973) study of *D. merriami,* at least 42% of the marked animals remained 50 weeks later.

It is not known how *Dipodomys* achieves such longevity. The long survival of *Perognathus* is usually attributed to the fact that it is capable of hibernation and thereby avoids both metabolic degradation and above-ground dangers. We have no evidence that *Eligmodontia* hibernates and doubt that it does so because it does not accumulate appreciable quantities of fat in the autumn.

In the Namib Desert of South-West Africa, nearly half the *Gerbillurus* and *Desmodillus* disappeared from the population each month (Christian, 1977). This suggests that their longevities and their age structure would be more like those of *Eligmodontia* than of *Dipodomys* and *Perognathus.* Many jerboas in the Sudan, however, survive more than one year (Happold, 1967). *Phyllotis darwini* in the coastal deserts of Peru and Chile is almost as short-lived as *Eligmodontia* (Pearson, 1975; Fulk, 1975).

The fact that *Eligmodontia* has been able to colonize an enormous expanse of arid and semiarid habitat without having many of the anatomical, physiological, reproductive, behavioral, and ecological features traditionally associated with desert rodents indicates either 1) that survival in deserts is not so stressful as to require extreme specializations, or 2) that *Eligmodontia* possesses very effective, undetected desert adaptations. From our study of *Eligmodontia* in the Patagonian steppe, we conclude that the first alternative is closer to the truth. Other species of relatively unspecialized mice are able, in the absence of highly-adapted competitors, to survive in much harsher deserts, such as *Mus domesticus* in Peruvian and Australian deserts or *Peromyscus maniculatus* in North American deserts. The key consideration is one of population density. Densities of highly adapted desert species are held down by even the usual or "normal" climatic regime of the region that they occupy. This is demonstrated by the population increase that occurs following unusual sequences of rain. Relatively unadapted species respond to rains also, but start their increase from lower densities and rely on their greater reproductive potential to carry them to high densities. The highly adapted species rely on their greater longevity and their other adaptations to maintain relatively high densities during the "normal" unfavorable years.

At high densities (much higher than *Eligmodontia* reached in our study), desert rodents eat as much as 70% (Chew & Chew, 1970) or 95% (Soholt, 1973) of the seed production of some plant species, yet this impact seems supportable by the plant populations. Therefore, unless the mouse population increases to much higher densities than we measured, *Eligmodontia* is not likely to be a threat to the steppe vegetation.

Literature Cited

Anchorena, J. de. 1978. Regiones ecologicas de la Patagonia. Primer Curso de Manejo de Pastizales. Internal Report of Instituto Nacional de Tecnología Agropecuaria, Bariloche.

Beatley, J. 1969. Dependence of desert rodents on winter annuals and precipitation. Ecology, **50:** 721–724.

CABRERA, A. 1971. Fitogeografía de la Republica Argentina. Boletin Sociedad Argentina Botanica, **14:** 1–42.

CHEW, R., AND B. BUTTERWORTH. 1964. Ecology of rodents in Indian Cove (Mojave Desert), Joshua Tree National Monument, California. Journal of Mammalogy, **45:** 203–225.

CHEW, R., AND A. CHEW. 1970. Energy relationships of the mammals of a desert shrub (*Larrea tridentata*) community. Ecological Monographs, **40:** 1–21.

CHRISTIAN, D. 1977. Effects of fire on small mammal populations in a desert grassland. Journal of Mammalogy, **58:** 423–427.

CONLEY, W., J. NICHOLS, AND A. TIPTON. 1977. Reproductive strategies in desert rodents, pp. 193–215. *In* Wauer, R., and D. Riskind, eds., Transactions of the Symposium on the Biological Resources of the Chihuahuan Desert Region. United States and Mexico. National Park Service Transactions and Proceedings Series, No. 3. U.S. Department of the Interior, Washington, D.C.

EISENBERG, J., AND D. ISAAC. 1963. The reproduction of heteromyid rodents in captivity. Journal of Mammalogy, **44:** 61–67.

FEITO, R., L. GONZALES, AND R. MONDION. 1981. Determination of age and age classes of *Akodon olivaceus* in Southern Chile. Neotropical Fauna and Environment, **16:** 201–207.

FRENCH, N., B. MAZA, AND A. ASCHWANDEN. 1967. Life spans of *Dipodomys* and *Perognathus* in the Mojave Desert. Journal of Mammalogy, **48:** 537–548.

FRENCH, N., B. MAZA, H. HILL, A. ASCHWANDEN, AND H. KAAZ. 1974. A population study of irradiated desert rodents. Ecological Monographs, **44:** 45–72.

FRENCH, N., D. STODDART, AND B. BOBEK. 1975. Patterns of demography in small mammal populations, pp. 73–102. *In* Golley, F., K. Petrusewicz, and L. Ryszkowski, eds., Small Mammals: Their Productivity and Population Dynamics. International Biological Programme 5. Cambridge University Press, London.

FULK, G. 1975. Population ecology of rodents in the semiarid shrublands of Chile. Occasional Papers of The Museum, Texas Tech University, No. 33, 40 pp.

HAPPOLD, D. 1967. Biology of the jerboa, *Jaculus jaculus butleri* (Rodentia, Dipodidae), in the Sudan. Journal of Zoology, **151:** 257–275.

HERSHKOVITZ, P. 1962. Evolution of Neotropical cricetine rodents (Muridae), with special reference to the phyllotine group. Fieldiana: Zoology, **46:** 1–524.

LAYNE, J. 1968. Ontogeny, pp. 148–253. *In* King, J., ed., Biology of *Peromyscus* (Rodentia). Special Publication No. 2, American Society of Mammalogists, xiii + 593 pp.

LIDICKER, W., JR. 1960. An analysis of intraspecific variation in the kangaroo rat Dipodomys merriami. University of California Publications in Zoology, **67:** 125–218.

LORES, R., C. FERREIRA, J. DE ANCHORENA, AND A. MARCOLIN. 1983. Las unidades ecologicas del campo experimental Pilcaniyeu (Provincia de Rio Negro). Su importancia regional. Gaceta Agronomica, **4:** 661–691.

MACMILLEN, R. 1964. Population ecology, water relations, and social behavior of a southern California semidesert rodent fauna. University of California Publications in Zoology, **71:** 1–59.

MARES, M. 1975a. Observations of Argentine desert rodent ecology, with emphasis on water relations of *Eligmodontia typus,* pp. 155–175. *In* Prakash, I., and P. Ghosh, eds., Rodents in Desert Environments. Monographiae Biologicae, vol. 28. The Hague, Netherlands.

———. 1975b. South American mammal zoogeography: Evidence from convergent evolution in desert rodents. Proceedings of the National Academy of Sciences of the United States of America, **72:** 1702–1706.

———. 1977. Water economy and salt balance in a South American desert rodent, *Eligmodontia typus.* Comparative Biochemistry and Physiology, **56A:** 325–332.

———. 1980. Convergent evolution among desert rodents: A global perspective. Bulletin of Carnegie Museum of Natural History, No. 16, 51 pp.

———. 1983. Desert rodent adaptation and community structure, pp. 30–43. *In* Biology of Desert Rodents. Great Basin Memoirs No. 7, Brigham Young University, Provo, Utah.

MCCULLOCH, C., AND J. INGLIS. 1961. Breeding periods of the Ord kangaroo rat. Journal of Mammalogy, **42:** 337–344.

M'CLOSKEY, R. 1972. Temporal changes in populations and species diversity in a California rodent community. Journal of Mammalogy, **53:** 657–676.

NAUMOV, N., AND V. LOBACHEV. 1975. Ecology of desert rodents of the U.S.S.R. (jerboas and gerbils), pp. 465–598. *In* Prakash, I., and P. Ghosh, eds., Rodents in Desert Environments. Monographiae Biologicae, vol. 28. The Hague, Netherlands.

PEARSON, O. 1945. Longevity of the short-tailed shrew. American Midland Naturalist, **34:** 531–546.

———. 1963. History of two local outbreaks of feral house mice. Ecology, **44:** 540–549.

———. 1967. La estructura por edades y la dinamica reproductiva en una poblacion de ratones de campo, *Akodon azarae.* Physis, **27:** 53–58.

———. 1975. An outbreak of mice in the coastal desert of Peru. Mammalia, **39:** 375–386.

———. 1983. Characteristics of a mammalian fauna from forests in Patagonia, southern Argentina. Journal of Mammalogy, **64:** 476–492.

PEARSON, O., AND C. RALPH. 1978. The diversity and abundance of vertebrates along an altitudinal gradient in Peru. Memorias del Museo de Historia Natural "Javier Prado," Lima, No. 18, 97 pp.

POULET, A. 1972. Recherches ecologiques sur une savane sahelienne du Ferlo Septentrional, Senegal: les mammiferes. Terre et Vie, **26:** 440–472.

———. 1978. Evolution of the rodent population of a dry bush savanna in the Senegalese Sahel from 1969 to 1977, pp. 113–117. *In* Schlitter, D., ed., Ecology and Taxonomy of African Small Mammals. Bulletin of the Carnegie Museum of Natural History, No. 6.

RALLS, K. 1976. Mammals in which females are larger than males. Quarterly Review of Biology, **51:** 245–276.

Reichman, O., and K. Van De Graaff. 1975. Association between ingestion of green vegetation and desert rodent reproduction. Journal of Mammalogy, **56:** 503–506.

Smith, D., and C. Jorgensen. 1975. Reproductive biology of North American desert rodents, pp. 305–330. *In* Prakash, I., and P. Ghosh, eds., Rodents in Desert Environments. Monographiae Biologicae, vol. 28. The Hague, Netherlands.

Soholt, L. 1973. Consumption of primary production by a population of kangaroo rats (*Dipodomys merriami*) in the Mojave Desert. Ecological Monographs, **43:** 357–376.

Speth, R., C. Pritchett, and C. Jorgensen. 1968. Reproductive activity of *Perognathus parvus.* Journal of Mammalogy, **49:** 336–337.

Van De Graaff, K., and R. Balda. 1973. Importance of green vegetation for reproduction in the kangaroo rat, *Dipodomys merriami merriami.* Journal of Mammalogy, **54:** 509–512.

Baculum of the Lesser Andean Coati, *Nasuella olivacea* (Gray), and of the Larger Grison, *Galictis vittata* (Schreber)

Edgardo Mondolfi

ABSTRACTS

The bacula of *Nasuella olivacea* (Gray) and *Galictis vittata* (Schreber) are described for the first time. Despite its smaller size, *Nasuella* has a larger and more robust baculum than *Nasua*; the bifid distal tip resembles that of *Procyon cancrivorus.* The baculum of *G. vittata* is larger and stouter than that of the smaller *Galictis cuja,* with a more expanded spatulate tip vaguely resembling that seen in *Eira barbara.* Mustelid specimens previously described by Didier (1947) are reidentified on the basis of these descriptions.

Los báculos de *Nasuella olivacea* (Gray) y *Galictis vittata* (Schreber) son descritos por primera vez. A pesar de su talla más pequeña, *Nasuella* tiene un báculo más largo y robusto que *Nasua*; la punta distal bifida se parece a la de *Procyon cancrivorus.* El báculo de *G. vittata* es más grande y fuerte el del pequeño *Galictis cuja,* con una punta espatulada más expandida, vagamente similar a la observada en *Eira barbara.* Especímenes mustélidos previamente descritos por Didier (1947) son reidentificados en base a estas descripciones.

Descreve-se, pela primeira vez, os bácula de *Nasuella olivacea* (Gray) e de *Galictis vittata* (Schreber). Apesar de seu tamanho menor, *Nasuella* possue um baculum maior e mais robusto do que *Nasua,* e a ponta distal bífida assemelha-se à de *Procyon cancrivorus.* O baculum de *G. vittata* é maior e mais robusto do que o da espécie menor, *Galictis cuja,* possuindo uma ponta larga e espatulada que lembra a forma da de *Eira barbara.* Espécimes mustelídeos prèviamente descritos por Didier (1947), são aqui reidentificados, baseado nas descrições dos bácula.

Introduction

The baculum or os penis of different species of Carnivora has distinctive morphological characteristics, with possible taxonomic value. However, as noted by Ewer (1973), the differences do not always reflect taxonomic relationships in any simple way.

The bacula of several species of Procyonidae and Mustelidae have been described by Pohl (1909, 1911), Pocock (1918, 1921), Chaine (1925), Didier (1947, 1950), and Burt (1960). However, none of these studies refers to the bacula of *Nasuella olivacea* (Gray) or *Galictis vittata* (Schreber).

The purpose of this paper is to describe the bacula of these species, with comments on their taxonomy.

Materials and Methods

Five bacula of the larger grison (*Galictis vittata*) and two of the lesser Andean coati (*Nasuella oli-*

From the Embassy of Venezuela, P. O. Box 34477, Nairobi, Kenya.

vacea) were analyzed for shape and measurements. Four bacula of the South American coati, *Nasua nasua* (Linnaeus), two of the crab-eating raccoon, *Procyon cancrivorus* (G. Cuvier), one of the lesser grison, *Galictis* (*Grisonella*) *cuja* (Molina), and two of the tayra, *Eira barbara* (Linnaeus) were available for comparison. All these specimens are in my private collection.

Results and Discussion

Lesser Andean Coati, *Nasuella olivacea* (Gray)

One of the two bacula is from a skeleton of an old individual salvaged on 18 March 1975 by Omar Linares at the Páramo de Piedras Blancas, 30 km NE of Mérida City, state of Mérida, in western Venezuela. It was found in *Polylepis sericea* Wedd. brush at an altitude of 4200 m. This baculum is astonishingly robust for the size of the animal, whose skull has a condylobasal length of 106 mm. It is larger and stouter than the baculum of *Nasua nasua* and about the same size as that of *Procyon cancrivorus*, but thicker (figs. 1–2). The shaft is nearly straight, slightly curved dorsally in the distal third and ventrally in the proximal third. The proximal end is very wide, rugose, blunt at the tip, with a deep groove on the dorsal face. On the ventral face of the base, there is a wider but shallower groove. The shaft tapers gradually toward the distal end. The dorsal surface has a prominent crest, extending from the end of the groove at the base to almost the distal tip. The ventral face is nearly flat on about two-thirds of the shaft, with a very slight indication of a urethral groove. The shaft is compressed laterally, appearing triangular in cross section. The tip is widened and bifid. It is forked in two thick blunt branches or knobs, one slightly longer than the other, separated by a notch. Measurements: total length, 97.2 mm; maximum height at base, 6.5 mm; maximum width at base, 8.7 mm; width at middle of shaft, 4.1 mm; width across tips of distal knobs, 9.0 mm; length of distal knobs, 5.4–6.5 mm.

The other baculum comes from a specimen, preserved in formalin, found killed on the road on the Páramo de Piedras Negras, altitude 4,250 m. The total length of this animal is 620 mm. It is similar in shape, but much smaller than the former specimen and considerably thinner. Measurements: total length, 65.5 mm; maximum width at base, 3.5 mm; width at middle of shaft, 2.7 mm; width across tips of distal knobs, 6.1 mm.

The differences in size and massiveness of the two specimens may be due to age. The first baculum is from an old animal, the second probably belongs to a juvenile. In this respect it could be mentioned that the figure given by Powell (1981) of four bacula of fisher (*Martes pennanti*) shows progressive changes with age, with bone deposition at the basal end and a more massive appearance in the adult.

The baculum of *Nasua,* as described by Pohl (1911), Pocock (1921), Chaine (1925), Didier (1950), and Burt (1960) and as shown by four specimens at hand, has an expanded, flattened, subspatulate, indistinctly bifid distal end (figs. 1–2). The distal end of the baculum of *Procyon cancrivorus* is very similar to that of *Nasuella olivacea,* that is, forked in two divergent condyle-like knobs separated by a medial notch. The largest baculum of *Procyon cancrivorus* at hand shows two tiny dorsal knobs at the base of the larger condyle-like ones (figs. 1–2).

The monotypic genus *Nasuella* was erected by Hollister (1915) for the little-known lesser Andean coati, *Nasuella olivacea.* Hollister designated as genotype *Nasuella olivacea meridensis* (Thomas) from the Andes of Mérida, Venezuela. *Nasuella* resembles *Nasua* in general features, but is smaller in size, head and body length, 420–478 mm. The tail is much shorter, 228–270 mm. Weight: 1,072–1,500 g. Hollister (1915) described a series of cranial characteristics that clearly distinguishes *Nasuella* from *Nasua.* He pointed out that the teeth of *Nasuella* are similar in many respects to those of *Bassaricyon.*

In regard to the postcranial skeleton of *Nasuella,* the following features could be pointed out: *Nasua nasua* is unusual among carnivores in that the slender fibula is fused proximally with the tibia, but articulates distally by a synovial joint (Ewer, 1973). The same condition is present in *Nasuella olivacea.* Stains (1973) described the calcaneum of *Nasua nasua* and *Nasua narica,* but made no reference to that of *Nasuella olivacea.* Regarding a calcaneum I sent him, he gave (in litt.) the following comments: "It appears much smaller than that of *Nasua nasua.* When compared to other procyonids, it appears to be most similar to *Bassariscus astutus* and is about the same size." Information regarding this aberrant coati is very scanty. It inhabits the Páramo Life Belt of western Venezuela, throughout Colombia to Ecuador, at

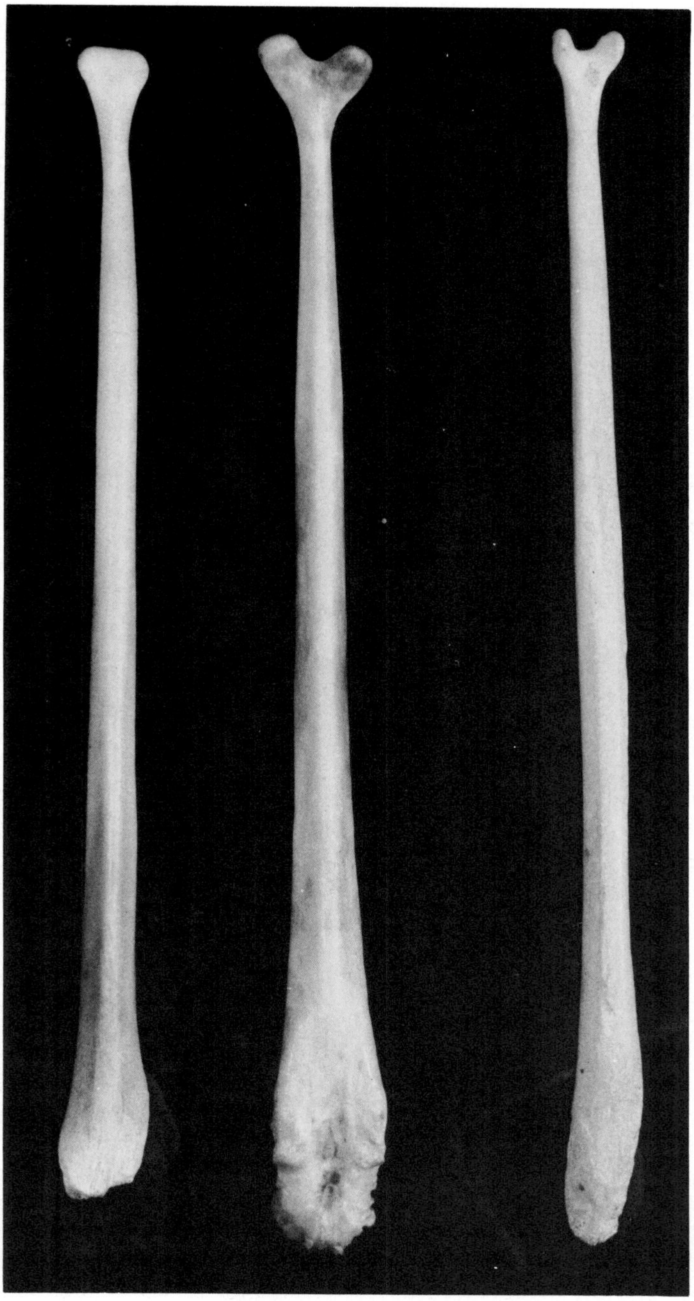

FIG. 1. **Left,** Dorsal view of baculum of *Nasua nasua*; **middle,** dorsal view of baculum of *Nasuella olivacea*; **right,** ventral view of baculum of *Procyon cancrivorus.*

altitudes of 2300 to 4250 m. In the páramos of the state of Mérida, it is called "*guache,*" and there is a small lagoon that bears the name "Laguna de los Guaches." The habitat preference of the lesser Andean coati seems to be páramo brush vegetation, and it is often seen on the ground. On the label of a specimen (USNM-143658) collected in the Montes de la Culata, state of Mérida, by Briceño-Gabaldón, is written "nido en tierra, pare cuatro hijos" (nest on ground, gave birth to four

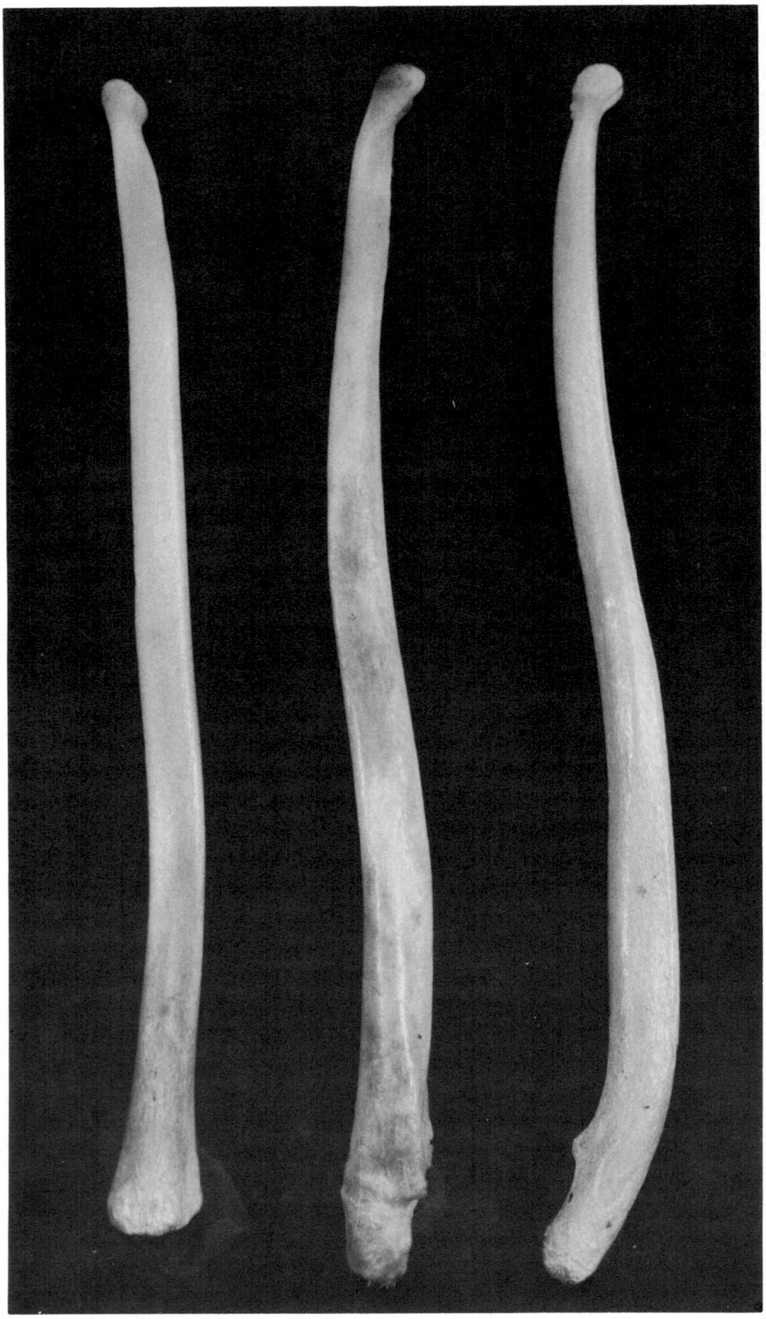

FIG. 2. Lateral views of bacula: **top,** *Nasua nasua;* **middle,** *Nasuella olivacea;* **bottom,** *Procyon cancrivorus.*

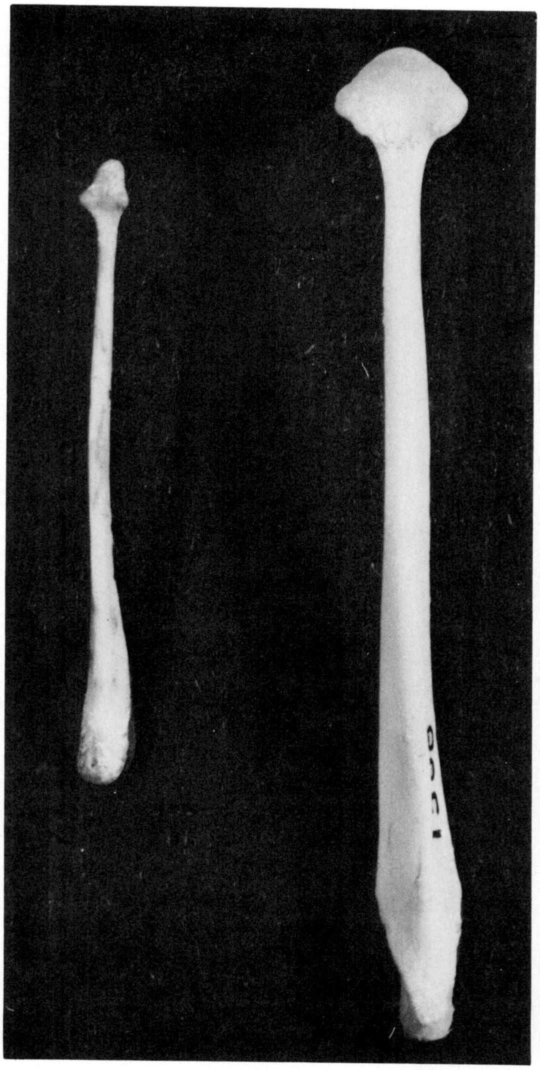

FIG. 3. Dorsal views of bacula: **left,** *Galictis* (*Grisonella*) *cuja*; **right,** *Galictis* (*Galictis*) *vittata.*

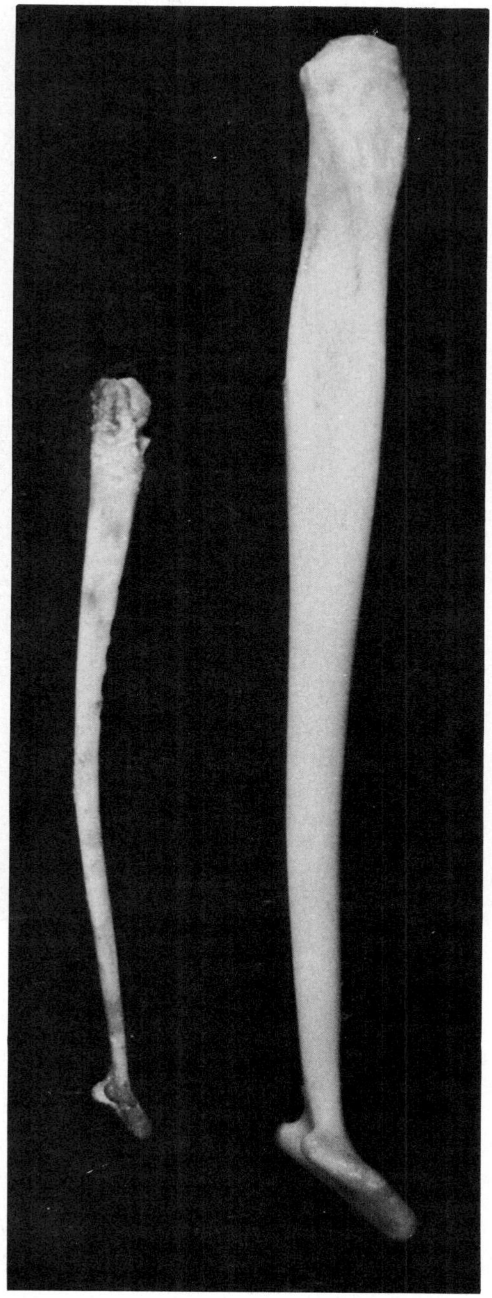

FIG. 4. Lateral views of bacula: **top,** *Galictis* (*Grisonella*) *cuja*; **bottom,** *Galictis* (*Galictis*) *vittata.*

young). The stomach of the male collected at Páramo de Piedras Negras, which I examined, contained insect remains.

Three weakly distinguished subspecies have been described: *Nasuella olivacea olivacea* (Gray) from Santa Fé de Bogotá, Colombia; *Nasuella olivacea meridensis* (Thomas) from the Andes of Mérida; and *Nasuella olivacea quitensis* (Lönnberg) from "más abajo de Lloa, ladera meridional del Pichincha, Ecuador." According to the descriptions, it seems that the separation of these subspecies on details of pelage coloration is not justified.

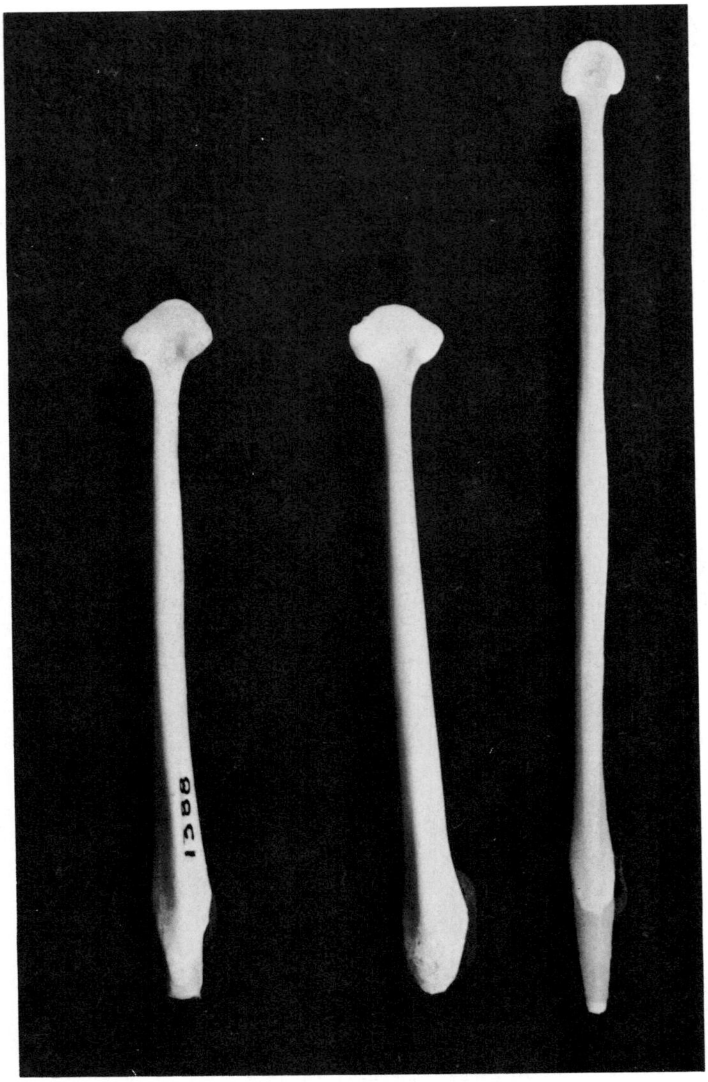

FIG. 5. Dorsal views of bacula: **left and middle,** *Galictis vittata*; **right,** *Eira barbara.*

Larger Grison, Galictis (Galictis) vittata (Schreber)

Didier (1947) described a baculum of the lesser grison, *Galictis* (*Grisonella*) *cuja* (Molina), under the name *Grison* (*Galictis*) *vittata,* which corresponds to the larger grison. Ewer (1973, p. 30, fig. 2.15) reproduced Didier's drawing of the baculum. A detailed description of the baculum of the lesser grison is given by Pocock (1918) under the name of *Grison furax.*

The description that follows of the baculum of the larger grison is based on five adult specimens. Three of these were collected in the state of Guárico, both north and south of Calabozo, Central Llanos of Venezuela; another comes from south of Ciudad Bolívar, state of Bolívar, eastern Venezuela; a fifth is from a captive animal that died at the El Pinar Zoo in Caracas. For comparison, I used a baculum of an adult lesser grison from the state of Minas Gerais, Brazil, as well as the descriptions given by Pocock and Didier.

The baculum of the larger grison is larger and much stouter than that of *Galictis cuja* and shows some differences in shape (figs. 3–4). The shaft is nearly straight. The proximal half is wider, thicker,

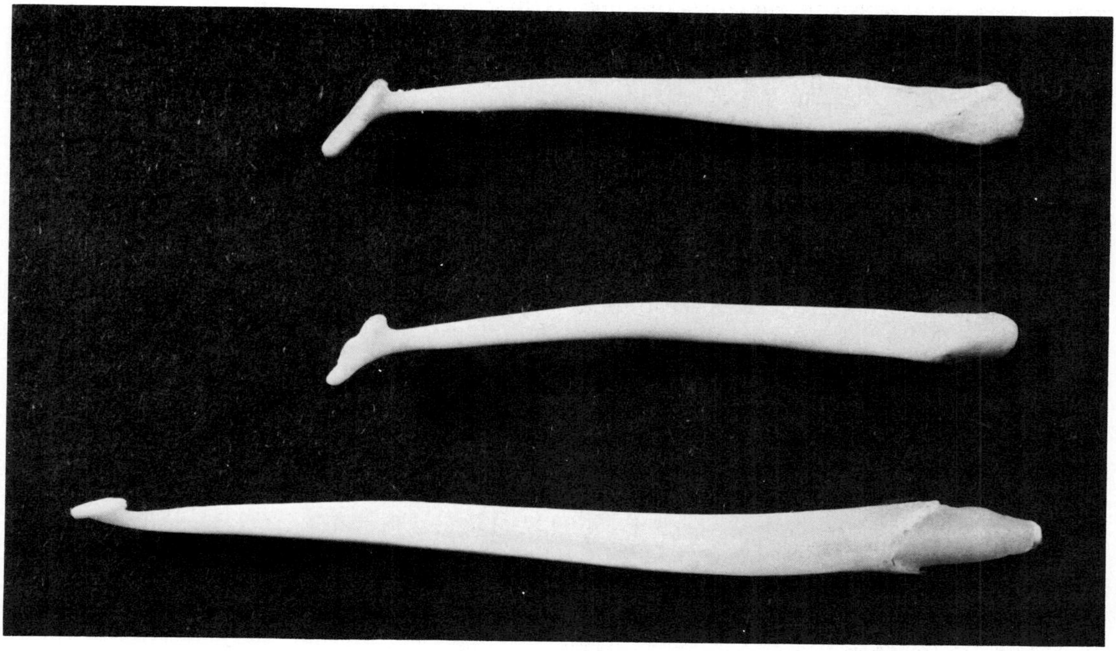

FIG. 6. Lateral views of bacula: **top and middle,** *Galictis vittata*; **bottom,** *Eira barbara.*

and compressed laterally, with a blunt end. The ventral surface is flatter, the dorsal prominently ridged. In cross section the basal half presents a triangular outline. The shaft tapers distally. The distal half is somewhat rounded on its ventral face and is slightly bent ventrally. The distal end is depressed, much widened, and bent ventrally, recalling a golf club. This spatulate distal end is shaped like a heart or an arrowhead, is flattened on its ventral surface, and is slightly concave on its dorsal one. Close to the neck of the shaft, there is a pair of erect, apically rounded excrescences or knobs pointing backward. There is some variation in the curvature of the shaft: in the specimen from the state of Bolívar, the shaft is bent to the right at its distal third; the one from the zoo specimen has a pronounced downward curvature. Extremes and mean measurements for five bacula are: total length, 54.6–56.9 (55.74) mm; maximum width at proximal extremity, 4.3–5.2 (4.72) mm; length of depressed tip, 6.2–7.5 (6.86) mm; width of depressed tip, 6.2–7.4 (7.05) mm; least width of shaft just behind the widened tip, 2.2–3.1 (2.82) mm.

The baculum of *G. cuja* is much smaller and thinner than that of *G. vittata,* but its shape is similar. The shaft is nearly straight, its distal half slightly bent ventrally. The ventral and dorsal surfaces are similar to the baculum of the larger grison. The widened distal end has the shape of an arrowhead; it is narrower than that in *G. vittata*; the pair of prominent excrescences pointing upward close to the neck, called "horns" by Pocock (1918), are prominent. Measurements: total length, 36.5 mm; maximum width at base, 2.7 mm; length of depressed tip, 3.4 mm; width of depressed tip, 2.7 mm; least width of shaft just behind the horns, 1.3 mm.

The baculum of the tayra (*Eira barbara*) is longer (total length, 75–83 mm) than that of the larger grison (figs. 5–6). It has a horseshoe-shaped distal end, formed by an arched concavity in the middle of its dorsal surface, surrounded by a thick rim. The shaft is straight. The enlarged proximal half is laterally flattened, forming a prominent ridge on its dorsal face. Burt (1960) pointed out in his description and figure of a baculum of *Eira barbara* that it is quite different from the one illustrated by Didier (1947, pp. 139–140). It is noticeable that the description and the three drawings (Didier, 1947, fig. 1) of the baculum of a mustelid Didier called "Le Grison Taíra (*Galera barbara* (L.), Brésil" matches the Venezuelan specimens of the larger grison, *G. vittata,* very closely in size and shape, particularly the one with the shaft curved downward on the distal half. It can be deduced that the specimen studied by Didier was incorrectly identified and probably belongs to a larger grison.

Galictis vittata is a larger animal than *G. cuja*. Adult males have an average body and head length of 450–626 (577.6) mm; average tail length, 150–182 (168.4) mm; average weight (based on seven adult males), 2,600–3,665 (3,076) g. According to Walker (1964), *G. cuja* has a head and body length of 400–450 mm; tail length, 150–190 mm; weight, about 1 kg. Nehring (1886), Krumbiegel (1942), and Cabrera (1958) considered the presence of a metaconid on the first lower molar an important characteristic which distinguishes *G. vittata* from *G. cuja*. Thomas (1912) erected the subgenus *Grisonella* for the latter species, based on the absence of the metaconid.

The distribution of the larger grison ranges from southeastern Mexico through Central America southward to central Peru and southeastern Brazil. Cabrera (1958) listed four subspecies: *Galictis vittata andina* Thomas from Pozuzo, Peru; *G. vittata vittata* (Schreber) from Surinam; *G. vittata brasiliensis* (Thunberg) from Rio de Janeiro; and *G. vittata canaster* Nelson from Cercanias de Tunkas, Yucatan, Mexico. My preliminary work suggests that a thorough revision may reduce the number of subspecies to only two: *G. vittata vittata* and *G. vittata andina*.

Acknowledgments

I am grateful to Dr. Bruce Patterson for reviewing the manuscript and to Hilary M. Airey for typing the manuscript.

Literature Cited

Burt, W. H. 1960. Bacula of North American mammals. Miscellaneous Publications, Museum of Zoology, University of Michigan, **113:** 1–76.

Cabrera, A. 1958. Catálogo de los mamíferos de América del Sur. Revista del Museo Argentino de Ciencias Naturales "Bernardino Rivadavia" e Instituto Nacional de Investigación de las Ciencias Naturales. Ciencias Zoológicas, **4:** i–xvi; 1–307.

Chaine, J. 1925. L'os pénien, étude descriptive et comparative. Actes de la Société Linnéenne de Bordeaux, **78:** 5–195.

Didier, R. 1947. Etude systématique de l'os pénien des mammifères. Carnivores. 2. Famille des Mustélides (suite). Mammalia, **11:** 139–152.

———. 1950. Etude systématique de l'os pénien des mammifères (suite). Famille de Procyonides et Ursides. Mammalia, **14:** 78–94.

Ewer, R. F. 1973. The Carnivores. Weidenfeld and Nicolson, London.

Hollister, N. 1915. The genera and subgenera of raccoons and their allies. Proceedings of the United States National Museum, **49:** 143–150.

Krumbiegel, I. 1942. Hyrare und Grisons (Tayra und Grison). Die Saügetiere der Südamerika—Expeditionen Prof. Dr. Kriegs. 17. Zoologischer Anzeiger, **139:** 81–108.

Nehring, A. 1886. Beiträge zur Kenntniss der *Galictis*-Arten. Zoologische Jahrbücher Zeitschrift für Systematik, Geographie und Biologie der Thiere, **1:** 177–221.

Pocock, R. I. 1918. The baculum or os penis of some genera of Mustelidae. Annals and Magazine of Natural History, ser. 9, **1:** 307–312.

———. 1921. The external characters and classification of the Procyonidae. Proceedings of the Zoological Society of London, **1921:** 389–422.

Pohl, L. 1909. Ueber das os penis der Musteliden. Jenaische Zeitschrift für Naturwissenschaft, **45:** 381–394.

———. 1911. Das os penis der Carnivoren einschliesslich der Pinnipedier. Jenaische Zeitschrift für Naturwissenschaft, **47:** 115–160.

Powell, R. A. 1981. Martes pennanti. Mammalian Species, **156:** 1–6.

Stains, H. J. 1973. Comparative study of the calcanea of members of the Ursidae and Procyonidae. Bulletin Southern California Academy of Sciences, **72**(3): 137–148.

Thomas, O. 1912. Small mammals from South America. Annals and Magazine of Natural History, ser. 8, **10:** 44–48.

Walker, E. P. 1964. Mammals of the World, vol. II. Johns Hopkins Press, Baltimore, Md.

Origin, Diversification, and Zoogeography of the South American Canidae

Annalisa Berta

ABSTRACTS

Members of the Canidae are known from the late Pliocene and early Pleistocene (Uquian) through the Recent in South America. Ten genera and 28 species of wolves and foxes are represented. Cladistic analysis supports recognition of four monophyletic groups: (1) *Urocyon* (including *Vulpes, Urocyon,* and *Otocyon*; (2) *Dusicyon* (including *Pseudalopex, Dusicyon, Protocyon,* and *Theriodictis*); (3) *Cerdocyon* (including *Nyctereutes, Cerdocyon, Atelocynus,* and *Speothos*); and (4) *Chrysocyon* (including *Chrysocyon* and *Canis*).

Zoogeographic implications of the cladistic hypotheses presented here are supported by the fossil record, suggesting the origin of canids in North America and their subsequent dispersal and extensive radiation in South America. The extinction of large canids in South America at the end of the Pleistocene is a consequence of extinction of their specialized large herbivorous prey. The current high diversity of South American foxes is, at least in part, the result of an opportunistic feeding strategy that utilizes small prey as well as fruits and grains.

Miembros de los Canidae son conocidos del Plioceno tardío y Pleistoceno temprano (Uquian) hasta el Reciente en Sudamérica. Diez géneros y 28 especies de lobos y zorros estan representados. El análisis cladístico soporta el reconocimiento de cuatro grupos mayores: (1) *Urocyon* (incluyendo *Vulpes, Urocyon,* y *Otocyon*); (2) *Dusicyon* (incluyendo *Pseudalopex, Dusicyon, Protocyon,* y *Theriodictis*); (3) *Cerdocyon* (incluyendo *Nyctereutes, Cerdocyon, Atelocynus,* y *Speothos*); y (4) *Chrysocyon* (incluyendo *Chrysocyon* y *Canis*).

Implicaciones zoogeográficas de la hipótesis cladística de parentesco presentadas aquí son soportadas por el registro fósil y ellas sugieren el origen de los cánidos en Norteamérica y su subsecuente dispersion y extensiva radiacion en Sudamérica. La extinción de cánidos grandes en el sur del continente a fines del Pleistoceno es una consecuencia de la pérdida de sus presas, los especializados herbívoros grandes. La presente alta diversidad de zorros sudamericanos es, al menos, en parte el resultado de una estrategia oportunista de alimentacion que utiliza pequeñas presas como también frutos y cereales.

Membros dos cânidos são conhecidos do Pleioceno superior, e desde o Pleistoceno inferior (Uquiano) à época Recente, na América do Sul. Dez gêneros e 28 espécies de lobos e raposas estão representados. As análises cladísticas fundamentam o reconhecimento de quatro grupos principais: (1) *Urocyon* (incluindo *Vulpes, Urocyon,* e *Otocyon*); (2) *Dusicyon* (incluindo *Pseu-*

From the Department of Biology, San Diego State University, San Diego, CA 92182.

dalopex, Dusicyon, Protocyon, e *Theriodictis*); (3) *Cerdocyon* (incluindo *Nyctereutes, Cerdocyon, Atelocynus,* e *Speothos*); e (4) *Chrysocyon* (incluindo *Chrysocyon* e *Canis*).

Implicações zoogeográficas, provenientes das hipóteses cladísticas apresentadas neste trabalho, são sustentadas também pelo registro fóssil. O cenário, então, é dos canídeos terem originado na América do Norte, dispersando-se, subsequentemente, à América do Sul, onde ter-se-iam radiado amplamente. A extinção dos canídeos grandes sulamericanos, ao final do Pleistoceno, ocorrera devido à extinção de sua especializada rapina—a dos grandes animais herbívoros. O alto grau de diversificação atual das raposas sulamericanas devese, ao menos em parte, aos seus hábitos alimentares oportunistas, que utilizam tanto animais pequenos como também diversos grãos e frutos.

Introduction

Although South American canids are more diverse than those of any other continent, they are less well known. Living South American foxes or wild dogs include seven genera and eleven species. Most are small to medium in size and predominantly omnivorous. Except for the mostly North American gray fox, *Urocyon,* South American foxes occur from Panama to Tierra del Fuego in a wide range of habitats that include rain forests, tree-covered steppes, grasslands, and deserts. In the past, canids attained a much greater diversity in South America and included wolves and wolf-like forms. In addition to foxes, they comprised a major component of the large carnivorous adaptive zone on that continent (Berta, 1981, in press).

The fossil record suggests that the origin and initial diversification of modern canids took place in North America and possibly Middle America during the late Miocene and early Pliocene. Subsequently they entered South America after emergence of the Panamanian land bridge, approximately 3 MYBP. Thereafter an impressive Plio-Pleistocene radiation of canids is documented in Uquian (late Pliocene and early Pleistocene) through Recent faunas. This paper summarizes current knowledge of the evolution and radiation of South American canids. Cladistic hypotheses of relationship presented provide a test of proposals regarding origin, patterns of dispersal, and changes in diversity through time.

Previous Systematic Studies

The first reference to canids in South America was Kerr's (1792) description of a large wild dog, *Canis australis,* from the Falkland Islands, off the eastern coast of Argentina. Since that time the nomenclatural and taxonomic history of living and fossil South American canids has been confusing and problematic.

Important contributors to the classification of South American canids prior to 1945 include Thomas (1914), Kraglievich (1930), Cabrera (1931), and Osgood (1934). Kraglievich (1930) provided the first comprehensive classification of living and fossil taxa, basing his systematic arrangement on traditionally emphasized characters of the skull and teeth. Since Simpson's (1945) monograph, classification of the South American canids has taken a variety of different approaches. Langguth (1969, 1970) proposed systematic arrangements based on morphology and ecology. He recognized a forest fox group ("*zorros de monte*") which included the genus *Cerdocyon* and the subgenera *Cerdocyon, Atelocynus,* and *Speothos.* Characters uniting this group included dark-colored pelage, short, robust skull and limb proportions, and a short, uncoiled cecum. A second group, the grassland foxes ("*zorros de campo*") is comprised of the genus *Dusicyon,* with *Dusicyon* and *Pseudalopex* as subgenera. Members of this group were characterized by a light-colored pelage, skull and limbs of average proportion, and a long, coiled cecum. In 1975, Langguth revised his classification and gave generic rank to members of the "differentiated group," *Cerdocyon, Speothos, Lycalopex,* and *Atelocynus.* The genus *Canis* comprised his "generalized group," including "dogs possessing features common to the majority of canid species" (1975, p. 193). Within this genus he recognized the subgenus *Dusicyon* for the Falkland Island Wolf, *D. australis,* and placed four other species (*culpaeus, gymnocercus, griseus,* and *sechurae*) in a second subgenus, *Pseudalopex. Urocyon* was categorized as a "*Vulpes*-like" fox and its generic distinction maintained.

Clutton-Brock et al. (1976) used numerical taxonomy to assess relationships within members of the family Canidae based on anatomical and behavioral characters. They concluded that subfamilial level separations suggested by Hough (1948) and Thenius (1954) were not warranted. *Chrysocyon* and *Speothos* were recognized as monotypic, *Urocyon* was synonymized with *Vulpes*, and *Pseudalopex, Atelocynus, Cerdocyon,* and *Lycalopex* were included in the polytypic genus *Dusicyon*. This taxonomic arrangement is similar to Simpson's (1945) but without subgeneric designations.

Van Gelder (1978) proposed a classification based on the nature and extent of hybridization between taxa. He recognized the genus *Canis* as polytypic. *Dusicyon, Pseudalopex, Lycalopex, Atelocynus,* and *Cerdocyon* were arranged as subgenera of *Canis,* and *Urocyon* and *Vulpes* were considered congeners. *Chrysocyon* and *Speothos* were identified as monotypic. Unfortunately, as acknowledged by Van Gelder (1978), too few data from molecular, immunological, or karyological studies are available to establish clear relationships among members of this group.

The arrangement prescribed here is based on cladistic hypotheses of relationship (see fig. 2, table 3). Seven genera of living South American canids are recognized: *Chrysocyon* (maned wolf), *Speothos* (bush dog), *Cerdocyon* (crab-eating fox), *Atelocynus* (small-eared dog), *Dusicyon* (Falkland Island wolf), *Pseudalopex* (South American fox), and *Urocyon* (gray fox). These genera include eleven species. The extinct large canids, *Theriodictis, Protocyon,* and *Canis,* and fossil species of both foxes and wolves bring the total number of recognized living and fossil South American canids to 10 genera and 28 species (table 1).

Despite numerous systematic treatments, relatively little is known about the behavior and ecology of canids in South America (reviewed by Langguth, 1975). The recent field studies of Brady (1978, 1979) on *Cerdocyon thous,* Jaksíc et al. (1980) on *Pseudalopex culpaeus* and *P. griseus,* Crespo (1975) on *P. culpaeus* and *P. gymnocercus,* and Dietz (1984) on *Chrysocyon brachyurus* are notable exceptions and hopefully indicative of a new trend.

Fossil Record

Canids have been recovered from the fossil localities shown in Figure 1. The chronology and usage of South American Land Mammal Ages is as follows: Uquian (2.5–1.5 MYBP), Ensenadan (1.5–0.3 MYBP), and Lujanian (300,000–10,000 YBP), which represent the late Pliocene and early Pleistocene, middle Pleistocene, and late Pleistocene, respectively (Marshall et al., 1982, 1984). The earliest record of canids in South America is from the Vorohué Formation, Buenos Aires Province, Argentina, which has been considered Uquian in age (Pascual et al., 1966). Canids recovered from these deposits include the highly derived dhole-like species, *Protocyon scagliarium,* and the generalized fox, *Pseudalopex gymnocercus* (Kraglievich, 1952).

Large wolves and wolflike canids were especially well represented in South America during the Ensenadan mammal age. Large canids that make their first appearance during this interval include *Canis, Theriodictis,* and *Chrysocyon* from diverse localities in Argentina and the classic Tarija Basin in Bolivia (fig. 1). The foxes *Pseudalopex* and *Cerdocyon* are also recorded from Argentina during the Ensenadan (table 2). Maximum diversity of both large and small canids was attained during the succeeding mammal age, the Lujanian. The best known Lujanian faunas containing canids include those from Talara, Peru; Andean Ecuador; Lagoa Santa Caves, Brazil; and Muaco, Venezuela (fig. 1). Two derived species of the true wolf *Canis* and two species of the highly specialized large canid *Protocyon* are known. The majority of living South American foxes are also first recorded during this interval: *Dusicyon australis, Pseudalopex culpaeus, P. sechurae, P. vetulus, Cerdocyon thous,* and *Speothos* (table 2). At the end of the Pleistocene the large canids except *Chrysocyon* became extinct. The modern forms *Atelocynus microtis, Urocyon cinereoargenteus,* and *Pseudalopex griseus* are unknown in the South American fossil record.

Phylogeny

Cladistic analysis yields the phylogenetic relationships presented in Figure 2. Derived characters that support this arrangement are listed in Table 3 and were based on Tedford and Taylor's (in prep.) comprehensive study of North American canids. A more detailed discussion of these characters is provided in the Appendix and elsewhere (Tedford & Taylor, North American fossil Canidae [Mammalia: Carnivora]: Tribe Canini [Caninae], unpubl. data; Berta, in press). The polarity of characters for all South American canids,

TABLE 1. Taxonomic arrangement of genera and species of South American Canidae.

Genera	Stratigraphic occurrence*
FAMILY CANIDAE	
Atelocynus Cabrera, 1940	
A. microtis (Sclater, 1882)	Rec.
Canis Linnaeus, 1758	
C. dirus† Leidy, 1858	L. Pleist. (Lujanian)
C. gezi† L. Kraglievich, 1928	M.-L. Pleist. (Ensenadan-Lujanian)
C. nehringi† (F. Ameghino, 1902)	L. Pleist. (Lujanian)
Cerdocyon Hamilton Smith, 1839	
C. avius† Torres and Ferrusquia, 1981	L. Plio. E. Pleist. (Blancan)
C. ensenadensis† (Ameghino, 1885)	M. Pleist. (Ensenadan)
C. thous (Linnaeus, 1766)	L. Pleist.-Rec. (Lujanian-Rec.)
?*C.* new sp.† (Tedford, pers. comm.)	E.-M. Plio. (L. Hemphillian)
Chrysocyon Hamilton Smith, 1839	
C. brachyurus Illiger, 1815	M. Pleist.-Rec. (Ensenadan-Rec.)
C. new sp.† (Tedford, pers. comm.)	M.-L. Plio. (E.-M. Blancan)
Dusicyon Hamilton Smith, 1839	
D. australis† (Kerr, 1792)	Rec.
D. avus† (Burmeister, 1864)	L. Pleist.-Rec. (Lujanian-Rec.)
Protocyon Giebel, 1855	
P. orcesi† Hoffstetter, 1952	L. Pleist. (Lujanian)
P. scagliarum† J. Kraglievich, 1952	L. Plio. & E. Pleist. (Uquian)
P. troglodytes† Lund, 1839b	L. Pleist.-Rec. (Lujanian-Rec.)
Pseudalopex Burmeister, 1856	
P. culpaeus (Molina, 1782)	L. Pleist.-Rec. (Lujanian-Rec.)
P. griseus (Gray, 1837)	Rec.
P. gymnocercus (Fischer, 1814)	L. Plio. & E. Pleist.-Rec. (Uquian-Rec.)
P. peruanus† (Nordenskiold, 1908)	L. Pleist. (Lujanian)
P. sechurae (Thomas, 1900)	L. Pleist.-Rec. (Lujanian-Rec.)
P. vetulus Lund, 1842	L. Pleist.-Rec. (Lujanian-Rec.)
Speothos Lund, 1839a	
S. pacivorus† Lund, 1839a	L. Pleist.-Rec. (Lujanian-Rec.)
S. venaticus (Lund, 1842)	L. Pleist.-Rec. (Lujanian-Rec.)
Theriodictis	
T. platensis† Mercerat, 1891	M.-L. Pleist. (Ensenadan-Lujanian)
T. tarijensis† (F. Ameghino, 1902)	M.-L. Pleist. (late Ensenadan and/or early Lujanian)
Urocyon	
U. cinereoargenteus (Schreber, 1775)	Rec.
U. progressus† Stevens, 1965.	M.-L. Plio. (Blancan)

* E. = early, M. = middle, L. = late, Plio. = Pliocene, Pleist. = Pleistocene, and Rec. = Recent.
† Extinct.

with the exception of *Urocyon,* was assessed by outgroup comparison with "*Canis*" *davisi,* a primitive North American canid. The bat-eared fox, *Otocyon,* was the outgroup for *Urocyon.* With the exception of the gray fox, *Urocyon,* and the maned wolf, *Chrysocyon,* the South American canids are a monophyletic group. *Urocyon* is a sister group of *Otocyon,* and *Chrysocyon* is the sister taxon of *Canis.* Based on cladistic analysis, four groups of South American canids are recognized: (1) *Urocyon,* (2) *Dusicyon,* (3) *Cerdocyon,* and (4) *Chrysocyon* (fig. 2).

Urocyon Group

This group includes the foxes *Vulpes, Otocyon,* and *Urocyon* and their fossil relatives. It is united by the derived characters (5, 6) listed in Table 3. *Urocyon* is distinguished from other members of this group by canines small relative to cheekteeth, and mandibular condyle above the level of the alveolar border of the cheekteeth. However, both characters also occur in the *Dusicyon* group, interpreted as parallel acquisitions. The *Urocyon* group is more primitive in lacking characters 7–9

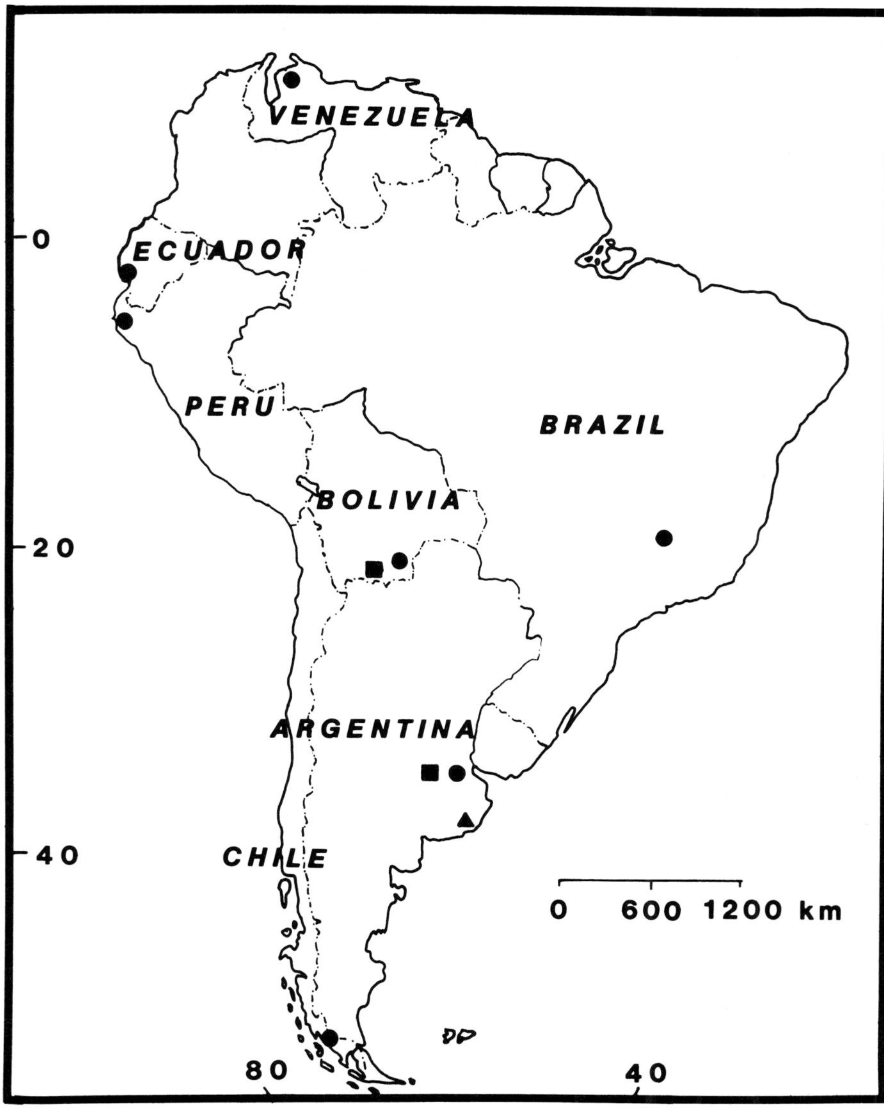

FIG. 1. Geographic and stratigraphic distribution of South American fossil canids. ● = Lujanian: Brazil (Lagoa Santa Caves), Bolivia (Nuapua), Chile (Eberhart Cave), Ecuador (Salinas, La Carolina), Peru (Talara), Venezuela (Muaco); ■ = Ensenadan: Argentina (Buenos Aires, La Plata, and environs), Bolivia (Tarija); ▲ = Uquian: Argentina (near Mar del Plata).

which unite the *Dusicyon* group with more advanced canids. The only living South American representative of this group, the gray fox, *Urocyon cinereoargenteus,* extends its range from Canada through the eastern and western United States, Mexico, and Central America, into the northern part of Colombia and Venezuela. This species lacks a South American fossil record, although it has been recorded from nearly 40 North American Pleistocene localities, the oldest of which are late Irvingtonian or possibly early Irvingtonian (Kurtén & Anderson, 1980). The earliest record of the genus is *Urocyon progressus* described by Stevens (1965) from the Blancan of Texas, Kansas, and possibly Nebraska.

Dusicyon Group

The *Dusicyon* lineage can be subdivided into two modern genera, the more generalized *Pseudalopex* foxes and the more derived *Dusicyon*

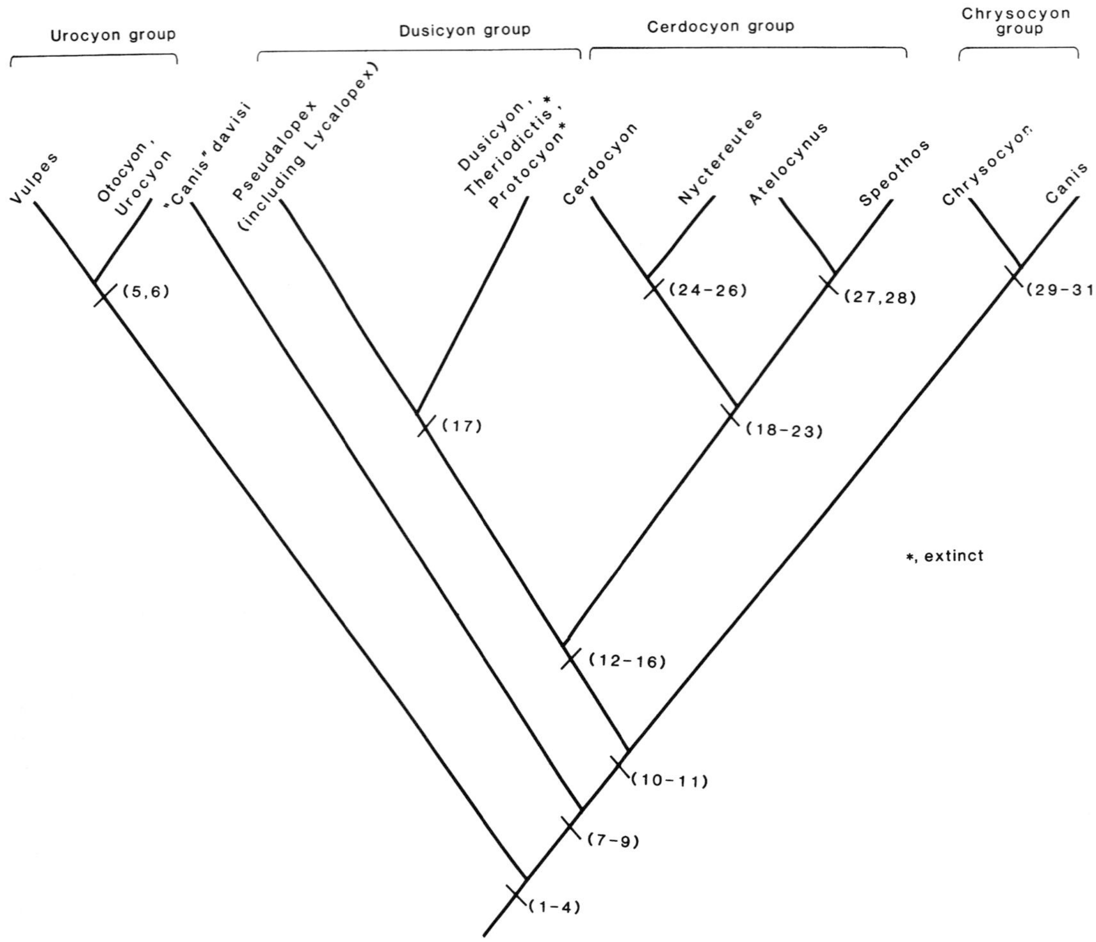

FIG. 2. Cladogram of proposed relationships among South American canids and related taxa.

"wolves." Monophyly of this group is supported by the derived characters 12–16 (table 3). The *Pseudalopex* foxes include six species: the modern culpeo fox, *P. culpaeus,* and its fossil relative, *P. peruanus*; the Argentine chilla fox, *P. griseus*; the Sechura Desert fox, *P. sechurae*; Azara's fox, *P. gymnocercus*; and the hoary fox, *P. vetulus. Pseudalopex culpaeus* (fig. 3) and *P. peruanus* are coy-

TABLE 2. Stratigraphic ranges of South American canid genera.

Genera	Uquian	Ensenadan	Lujanian	Recent
Atelocynus				——→
Canis		——	——	
Cerdocyon ←——		——	——	——→
Chrysocyon ←——		——	——	——→
Dusicyon			——	——
Protocyon			——	——
Pseudalopex	——	——	——	——→
Speothos			——	——→
Theriodictis		——	——	
Urocyon ←——				——→

TABLE 3. Derived characters that support phylogeny (see fig. 3).

No.	Character
1.	m_2 anterolabial cingulum enlarged.
2.	m_2 posterior cingulum present.
3.	m_2 metaconid enlarged, taller than protoconid.
4.	Metatarsal I reduced to proximal rudiment.
5.	Upper incisors simple, I^{1-2} cusplets weak or absent.
6.	Paroccipital process broad, tip does not extend below body of process.
7.	Frontal sinus large, penetrating postorbital process.
8.	I^3 enlarged, extending markedly below level of I^{1-2}.
9.	Humerus lacks entepicondylar foramen.
10.	Strongly arched zygoma with inverted jugals.
11.	Angular process large with expanded fossa for inferior/superior branch of medial pterygoideus muscle or expanded pterygoid fossa.
12.	Long palatines extending to or beyond toothrow.
13.	Coronoid process anteroposteriorly broad and dorsoventrally low.
14.	m_1 with protostylid.
15.	m_{1-2} with mesoconid (fig. 3).
16.	m_2 with strong paracristid (fig. 3).
17.	Frontal sinus large, does not penetrate postorbital process.
18.	Canines small relative to cheekteeth.
19.	Angular process with pterygoid fossa greatly expanded.
20.	External auditory meatus very short and of small diameter.
21.	Cecum short and straight.
22.	Ears short.
23.	Limbs short.
24.	Carnassials small relative to cheekteeth.
25.	Mandibular condyle above level of alveolar border of cheekteeth.
26.	Subangular lobe.
27.	Nasals short.
28.	Frontal sinus small, does not penetrate postorbital process.
29.	Frontal sinus large, penetrating postorbital process, extending anteriorly and particularly posteriorly, ultimately to the frontal-parietal suture.
30.	Angular process with large fossa for superior branch of medial pterygoideus muscle.
31.	I^3 enlarged with accessory cusps and a strong posteromedial cingulum.

ote-like in size, proportionately larger and broader than other South American foxes, with a reduced m_2 metaconid. *Pseudalopex sechurae* and *P. vetulus* share small size, short rostrum, very small carnassials relative to the cheekteeth, and M^{1-2} very narrow for their length.

The earliest recorded fox, *Pseudalopex gymnocercus,* presumably differentiated into two lineages. The large, robust, coyote-like lineage represented by *P. culpaeus* in Ecuador and *P. peruanus* in Peru is known from the Lujanian. *Pseudalopex culpaeus* is now widely distributed throughout the

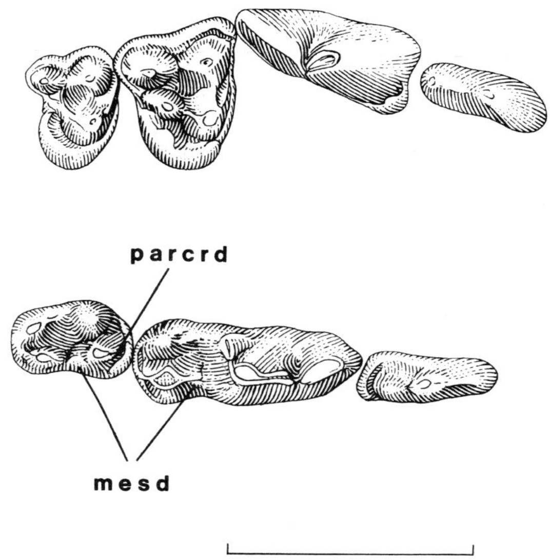

FIG. 3. Left upper (top) and lower (bottom) dentition of *Pseudalopex culpaeus,* illustrating dental features characteristic of the *Dusicyon* lineage. parcrd = paracristid; mesd = mesoconid; scale = 2 cm.

Andes from Colombia to the southern tip of Chile, and is found throughout the Patagonian plateau. A second, smaller lineage comprised of *P. sechurae* and *P. vetulus* is also known from the late Pleistocene. The past and present distribution of *P. sechurae* is restricted to northwestern Peru and southwestern Ecuador. *Pseudalopex vetulus* was recorded as a fossil in Argentina but survives today in southeastern Brazil (Minas Gerais and Matto Grosso states).

Members of the more derived *Dusicyon* "wolves," the recently extinct Falkland Island wolf, *D. australis,* and its fossil relative *D. avus* share short, high-crowned premolars and small $M\frac{2}{2}$ relative to $M\frac{1}{1}$. These taxa are related to the large fossil canids, *Theriodictis* and *Protocyon,* in possessing the following derived characters: broad palate, deep zygoma with wide masseteric scar, m_1 metaconid reduced or absent, and m_2 metaconid relatively unreduced (Berta, 1981, in press). *Protocyon* and *D. australis* share extreme reduction of P^4 protocone and have a posterior tilt to the p_3 crown. Although *D. australis* is unknown as a fossil and *D. avus* is only known from the late Pleistocene, their closest relatives, *Protocyon* and *Theriodictis,* are known from the late Pliocene and early Pleistocene through the late Pleistocene and possiby into the Recent.

Theriodictis, a medium-sized, short-faced canid represented by *T. platensis* and *T. tarijensis,* is

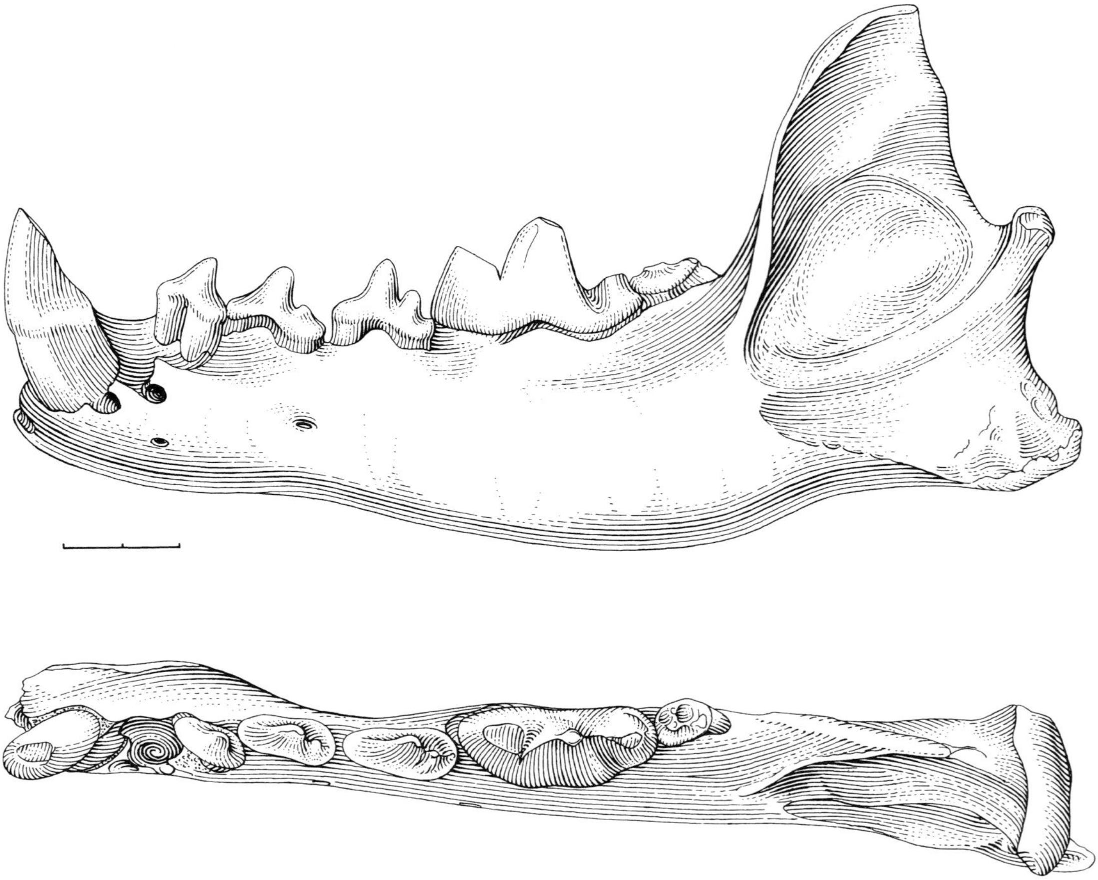

FIG. 4. Lateral (top) and occlusal (bottom) views of left mandibular ramus of *Protocyon orcesi.* Escuela Politechnica Nacional, Quito, EPN V2871. Scale = 2 cm.

recorded from Ensenadan- and Lujanian-aged deposits in Argentina, Bolivia, and Ecuador. The type species, *T. platensis,* with its simplification of the m_1 talonid (metaconid lost and the entoconid retained as a distinct cusp), is a likely ancestor for the more derived genus *Protocyon.* Among *Protocyon* species these distinctive dental characters and others are further modified: both the metaconid and entoconid are lost on m_1, as are the hypocones on M^{1-2}. Three species of this well-known canid are recognized, *P. orcesi* (fig. 4), *P. scagliarum,* and *P. troglodytes* (fig. 5), from Uquian through Lujanian and possibly Recent deposits in Argentina, Bolivia, Brazil, and Ecuador. Derived members of the *Dusicyon* group, with their high-crowned premolars and trenchant m_1 with very reduced or absent talonid cusps, show a trend toward hypercarnivory—that is, increased specialization of the shearing mechanism.

Cerdocyon Group

The crab-eating fox, *Cerdocyon,* and the Asian raccoon dog, *Nyctereutes,* are primitive members of the *Cerdocyon* clade. They form a sister group defined by derived characters 24–26 (table 3). *Cerdocyon* is known from late Miocene-early Pliocene (6–3 MYBP) deposits in North America; *Nyctereutes* has been recorded from coeval deposits in Europe (R. H. Tedford, pers. comm.). *Cerdocyon avius* is reported from the Blancan of Baja California, Mexico (Torres & Ferrusquia, 1981), and an undescribed lower jaw is questionably assigned to this genus from the late Hemphillian of the Texas Panhandle (R. H. Tedford, pers. comm.). Its South American fossil record includes two species, *Cerdocyon ensenadensis* from the Ensenadan of Argentina, and the closely related living species *C. thous* (including *lydekkeri*; see Berta,

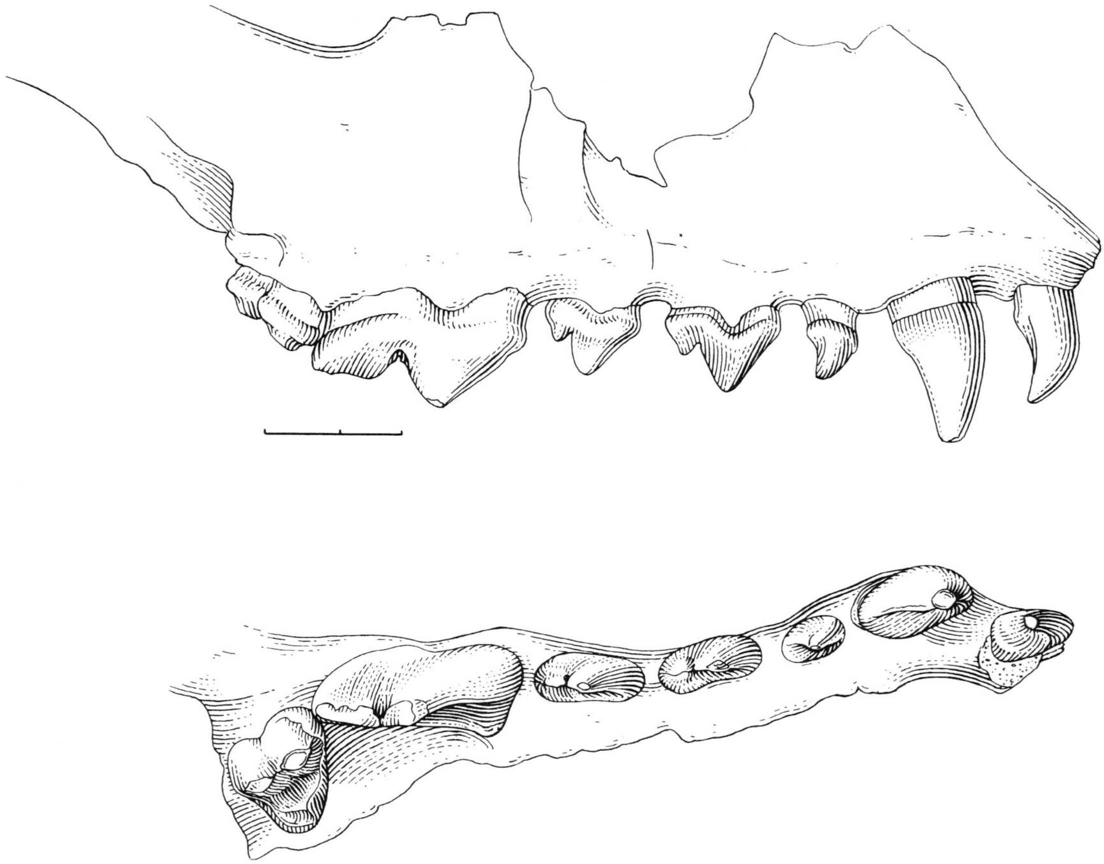

FIG. 5. Lateral (top) and occlusal (bottom) views of right maxillary of *Protocyon troglodytes*. Universitets Zoologiske Museum (Peter Wilhelm Lund Collection), Copenhagen, UZML 5697–5698. Scale = 2 cm.

1982) (fig. 6), from Lujanian to Recent deposits in Brazil. Today crab-eating foxes inhabit the savannah and woodland areas of northeastern South America, with a range extending from Colombia, northern Argentina, and Uruguay.

Derived members of the *Cerdocyon* clade, the small-eared dog, *Atelocynus*, and the bush dog, *Speothos*, are distinguished by shared derived characters 27–28 (table 3). *Speothos*, with two known species, is the most derived member of the *Cerdocyon* clade. *Speothos pacivorus*, from the late Pleistocene-Recent Lagoa Santa Caves of Brazil, is characterized by large size, presence of a metaconule and hypocone on M^1, and double-rooted m_2. The living species *S. venaticus*, also recorded from these cave deposits, is distinguished by its reduced size, loss of metaconule and hypocone on M^1, and absence of M^2 and m_3 (Berta, 1984). The current range of the bush dog extends from Panama throughout the Amazonian basin.

No fossils of *Atelocynus* are known. The living species, *A. microtis*, occurs in tropical rain forests in the Amazonian basin in Brazil, Peru, Ecuador, and Colombia; the upper Río Orinoco basin in Colombia and Venezuela; and the upper Rio Paraná basin in Matto Grosso, Brazil (Hershkovitz, 1961).

Chrysocyon Group

This group includes the maned wolf, *Chrysocyon brachyurus*, and three extinct species of South American "true" wolves, *Canis gezi, C. nehringi*, and *C. dirus*. Derived characters shared by these taxa include characters 29–31 (table 3).

The maned wolf, *Chrysocyon*, is the most distinctive South American canid. It is distinguished from *Canis* in having small carnassials relative to the cheekteeth, a short, straight cecum, and straight and greatly elongate limbs. In addition to the living *C. brachyurus*, reported from the Ensenadan of

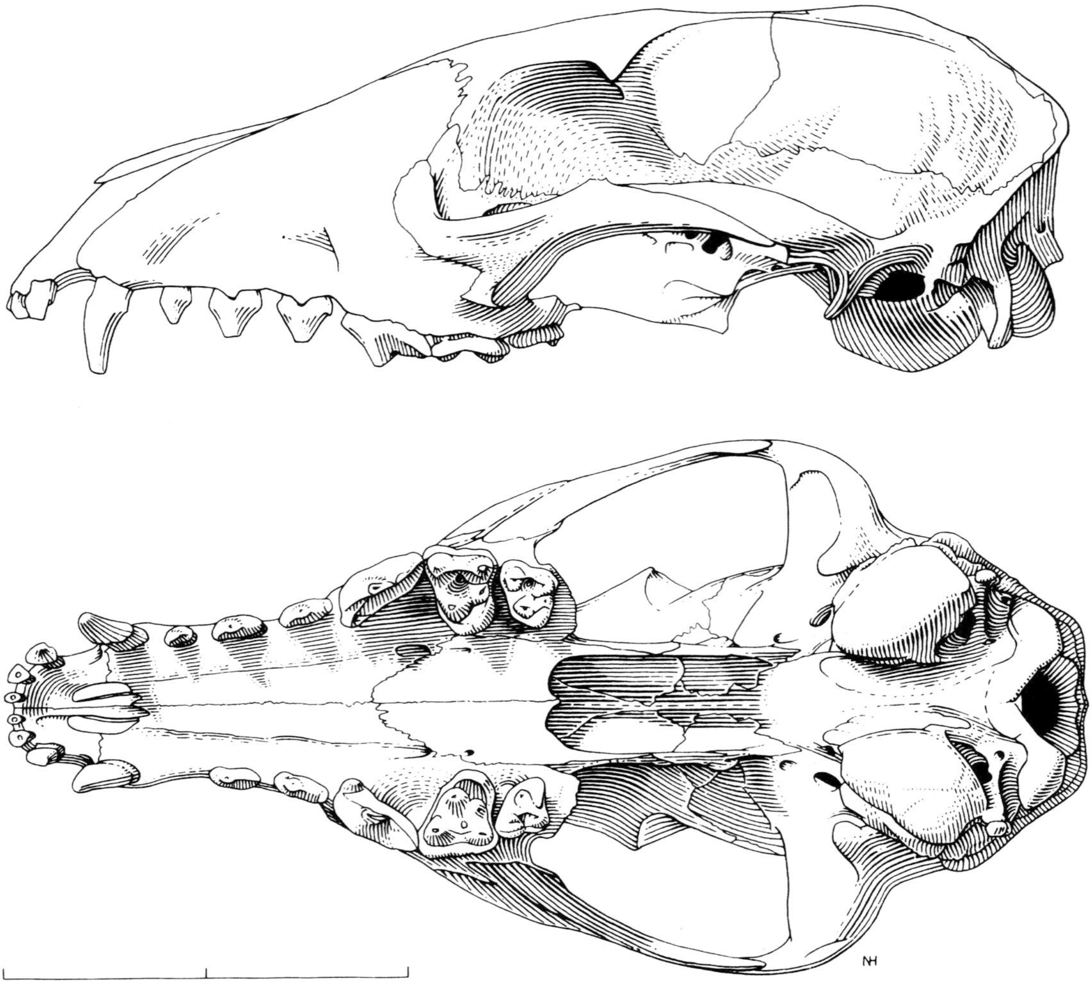

FIG. 6. Lateral (top) and ventral (bottom) views of skull of *Cerdocyon thous thous*. American Museum of Natural History, New York, AMNH 130475. Scale = 5 cm.

Bolivia and Lujanian and possibly Recent deposits in Brazil (Berta, 1981, in press), an undescribed new species is known from the early and middle Blancan of Arizona and Mexico (R. H. Tedford, pers. comm.). The present range of the maned wolf indicates its subsequent southern dispersal into northern Argentina and Paraguay.

North American Pleistocene wolves (including *C. armbrusteri, C.* cf. *C. dirus,* and *C. lupus*) were ancestors of South American species. The earliest recorded species, *Canis gezi* from the Ensenadan of Argentina, is a good structural ancestor for the Lujanian species, *C.nehringi. Canis nehringi* differs from *C. gezi* in the continued trend toward larger size, development of a narrow, triangular supraoccipital shield, and greater complication of cusps on the upper and lower molars. The best-represented and most derived species, *C. dirus* (fig. 7), is distinguished from *C. nehringi* in its larger size, more massive proportions, and more complex construction of the lower molars (Berta, 1981, in press). *Canis dirus,* widely distributed in North America, has a more limited South American distribution, occurring only in Bolivia, Peru, and Venezuela.

Zoogeographic History: Problems of Origin, Dispersal, and Ecology

Inherent in any cladistic analysis is the hypothesis that two taxa shared a closest common ances-

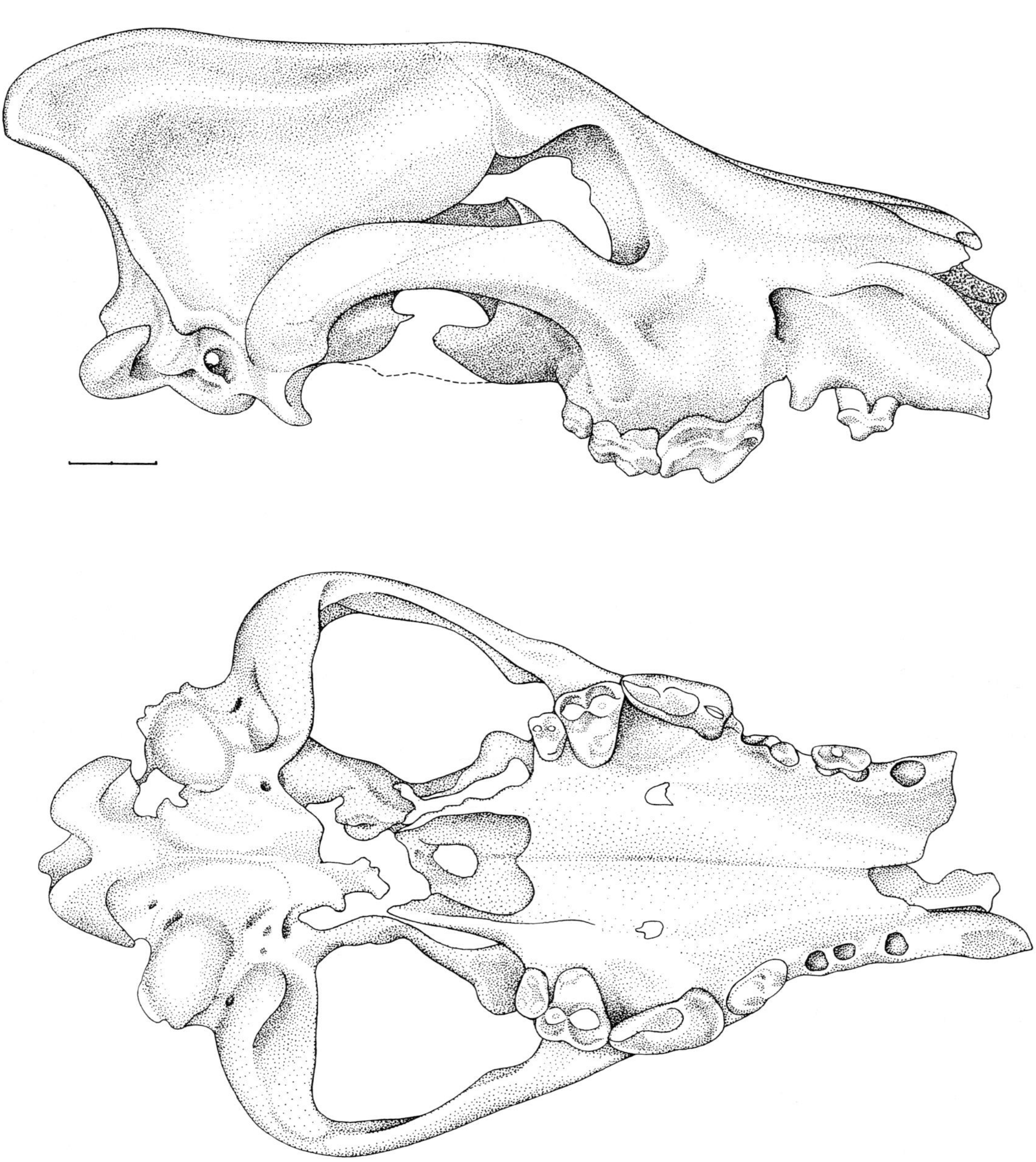

FIG. 7. Lateral (top) and ventral (bottom) views of skull of *Canis dirus*. Royal Ontario Museum, Toronto, ROM 4303. Scale = 2 cm.

tor. For this reason every cladistic hypothesis entails zoogeographic implications about the temporal and spatial distribution of the animals studied. Early consideration of the stratigraphic position of the fossils can bias a cladistic analysis, which initially should be based on morphological information. However, once a cladogram has been constructed, the stratigraphic and spatial occurrence of fossil forms can provide additional valuable information regarding patterns of distribution that can be tested.

A North American origin for the canids of South America is supported by the fossil record. The earlier record of *Canis, Chrysocyon, Cerdocyon,* and *Urocyon* in North America suggests that this group originated as early as the late Miocene and early Pliocene (6–3 MYBP, table 4). This same phylogenetic pattern is seen among other "endem-

TABLE 4. First appearance of South American canid genera in North and South America.

Genera	Record in North America	Record in South America
Atelocynus	...	Recent
Canis	Blancan	Ensenadan
Cerdocyon	Hemphillian	Ensenadan
Chrysocyon	Blancan	Ensenadan
Dusicyon	...	Lujanian
Protocyon	...	Uquian
Pseudalopex	...	Uquian
Speothos	...	Lujanian
Theriodictis	...	Ensenadan
Urocyon	Hemphillian	Recent

ic" groups that later dispersed to South America, including the sigmodontine rodents (Jacobs & Lindsay, 1981), equine horses (MacFadden, 1979), and llamas (Webb, 1974). Although four of the 10 recognized South American canid genera have more ancient records in North America, canids probably did not actually arrive on the southern continent until much later. It is suggested that canids originated and initially diversified in North America (and possibly Middle America, but data there are lacking). After emergence of the Panamanian Land Bridge, during the late Pliocene and early Pleistocene (beginning about 3 MYBP), canids entered South America during the "Great American Interchange" and then radiated to achieve their present diversity. Support for this proposal again comes from the fossil record. Canids are conspicuously absent from late Tertiary deposits in South America, and their first record of occurrence on the southern continent is from Uquian (2.5–1.5 MYBP) deposits. There is no known physical evidence which places canids or their possible ancestors in South America earlier than the late Pliocene.

Our knowledge of patterns of diversification and dispersal of canids, once in South America, is limited by their more common occurrence in fossil localities in the southern portion of the continent (fig. 1). The known record suggests Argentina as a major center of evolution for this group, but this is at least partly due to the fact that both Quaternary and Tertiary faunas from Argentina are the best known and most intensively studied (for recent reviews see Marshall et al., 1984). Undoubtedly, the Brazilian highlands were also an important area of canid differentiation (Langguth, 1975). The cave faunas of Lagoa Santa and environs which range from the late Pleistocene to the Recent provide some measure of this rich sample of past diversity. Among the nearly 40 mammalian genera known (excluding bats) are four genera of canids—*Cerdocyon, Atelocynus, Speothos,* and *Protocyon.*

Pleistocene canids and other vertebrates recorded from the Lagoa Santa Caves and Muaco, Venezuela, presumably followed Webb's (1978) "Eastern Savanna Route," a major dispersal corridor that extended from the Caribbean perimeter of Central America into eastern Venezuela and there into the Amazon Basin. Savanna habitats predominated along this route, the major North-South corridor for amphibians and reptiles. In Argentina, this route converged with a second major dispersal corridor, the "Andean Route," which extended from the isthmian region along the Andean chain (Webb, 1978). The cool, dry unforested habitats of this route were occupied by most bird and mammal groups. Fossil canid faunas along this pathway were the Pacific coast fauna of Talara, Peru, and those of coastal and Andean Ecuador and Bolivia.

If one considers the current distributions of *Cerdocyon* in South America and *Nyctereutes* in Asia, their hypothesized sister group relationship is puzzling. Additional stratigraphic and geographic information clarifies this systematic proposal. The fossil record suggests that the common ancestor of those taxa ranged over Eurasia and North America between 10 and 4 MYBP. *Cerdocyon* is known from late Miocene and early Pliocene (6–3 MYBP) deposits in North America, and *Nyctereutes* has been reported from coeval deposits in Eurasia (Torres & Ferrusquia, 1981; Soria & Aguirre, 1976). The disappearance of the Bering Land Bridge between 4.5 and 4.0 MYBP can be seen as a vicariant event, isolating *Nyctereutes* in Eurasia and *Cerdocyon* in North America. After establishment of the Panamanian Land Bridge, approximately 3 MYBP, *Cerdocyon* dispersed from North America to South America, where it is recorded in late Pleistocene through Recent faunas.

One of the more interesting biogeographic problems concerns the origin of the Falkland Island wolf, *Dusicyon australis,* which has been extinct since 1880. According to Clutton-Brock (1977), the white tail tip, enlarged frontal sinuses, and wide muzzle of *D. australis* are characters signifying domestication; the same traits frequently occur in the Australian dingo. Clutton-Brock argues that *D. australis* was domesticated by early man and then introduced to the Falklands. However, the geological history of this area and phylogenetic

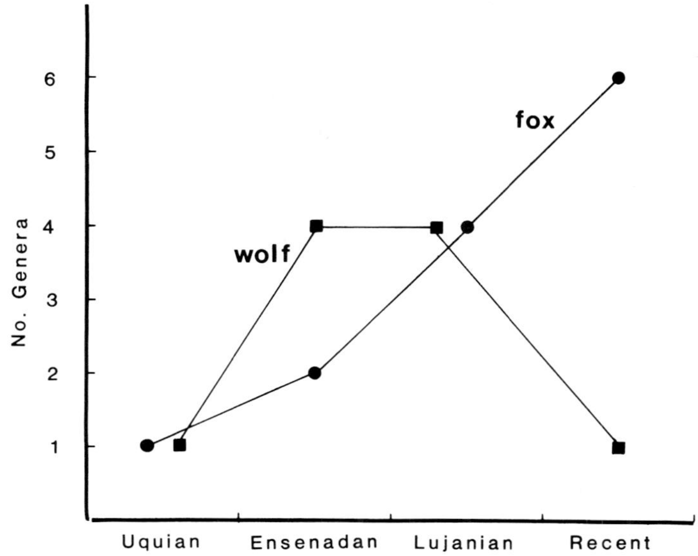

FIG. 8. South American canid diversity through time. Genera considered as either "fox" or "wolf" morphotypes.

information offer an alternative hypothesis. Cladistic analysis indicates a close sister group relationship between *D. australis* and *D. avus*. Remains of *D. avus* are known from Eberhart Cave in southern Chile. Sloth dung, hide, hair, and bone from this cave have yielded ^{14}C dates ranging from 10,200 ± 200 to 13,500 ± 190 YBP (Long & Martin, 1974), suggesting a minimum time of divergence for these species. However, evidence that *Dusicyon avus* survived even longer is indicated by isolated teeth recovered from the southern coast of Argentina which were referred to this species by Tonni and Politis (1981). In any event, lowered sea level during glacial times would have brought the mainland and the Falkland Islands closer together and facilitated dispersal of terrestrial mammals, including *D. australis* (and possibly *D. avus*). The distinctiveness of the Falkland Island wolf, with its high-crowned teeth, reduced molar cusps, and highly modified shearing carnassials, is more likely the result of its isolation as the island's only indigenous carnivore rather than its domestication.

It has been suggested that during the Pleistocene placental carnivores replaced the doglike marsupial family, Borhyaenidae, which occupied the carnivorous adaptive zones during the late Tertiary (Marshall, 1978). As Marshall (1978, p. 82) noted,

> Whether this "relay" of various carnivorous groups through time was due to *active* competition between the successive groups filling these roles, or to the disappearance of one carnivorous group (possibly linked with concurrent environmental changes) with subsequent *passive* replacement by another group which came to fill a similar role in a later fauna, (or whether the faunal changes were the result of a combination of these possibilities or others) is unknown.

However, it is clear that before we can understand interactions of competition, extinction, and/or replacement, the pattern must be known. Figure 8 summarizes the pattern of canid diversity through the Plio-Pleistocene. It is apparent that beginning in the late Pliocene and early Pleistocene, when canids first appear in South America, both adaptive types are represented, the small-medium fox and the large wolf. By the end of the Pleistocene these two adaptive types have shifted dramatically in diversity. The large wolves declined rapidly, with only the maned wolf, *Chrysocyon,* surviving into the present.

It should be noted that morphologically and ecologically *Chrysocyon* is a large omnivore. The fox morphotype with its generalized omnivorous habits seems to have been favored. Today, among North American, African, and Eurasian canids, foxes are the most numerous and display the greatest diversity. It seems that animals with more flexible food habits have been able to adapt more easily to environmental changes, especially those at the end of the Pleistocene.

Such an opportunistic feeding strategy has been documented for South American foxes. Field studies of *Pseudalopex culpaeus* and *P. griseus* have shown that increasing amounts of fruit are eaten from spring to winter as rodent densities decrease towards the winter (Yáñez & Jaksíc, 1978; Jaksíc et al., 1980). Brady (1979) observed that *Cerdocyon thous* on the llanos of central Venezuela showed seasonal food shifts. When insects and fruit (primary food sources during the wet season) become scarce during the dry season, crab-eating foxes hunt crabs and vertebrates (lizards, snakes, and rodents).

The extinction of large canids in South America at the end of the Pleistocene has been related to the extinction of their large herbivorous prey (Berta, 1981, in press). Today, the large cats—the puma, *Felis concolor,* and the jaguar, *Felis onca*—survive as sole occupants of the large carnivorous adaptive zone. Reasons for the success of large felids at the expense of large canids is not known, although their differing feeding and hunting strategies and social behavior probably played a part. Fox (1975) classified the wolf, the dhole, and the African hunting dog "social hunters" which obtain large ungulate prey by pack hunting. A second group, "solitary-social hunters" typified by the coyote and jackal, hunt alone or in pairs depending on the size, abudance, and distribution of prey. A third group, "solitary hunters" which hunt small prey and may also be omnivorous, is exemplified by foxes, mustelids, procyonids, and felids (lions). The large South American canids of the Plio-Pleistocene *Canis, Theriodictis,* and *Protocyon* were likely both "social hunters" and "solitary-social hunters"; thus they would have been profoundly affected by the extinction of large ungulates between 15,000 and 8,000 YBP. During the late Pleistocene, the herbivorous megafauna (including ground sloths, glyptodonts, proboscidians, horses, notoungulates, and litopterns) reached its acme and numbered nearly 50 genera. This stands in sharp contrast to the 11 genera represented today. The present-day high diversity seen among the foxes (especially the *Pseudalopex* complex), procyonids, mustelids, and felids is a consequence of an opportunistic feeding strategy and the greater availability of small to medium-sized prey (e.g., rodents, birds) and fruits and grains.

Acknowledgments

I am particularly grateful to Dr. Richard H. Tedford, whose continuing encouragement, guidance, and extensive knowledge of the Canidae greatly facilitated this study. J. David Archibald and Roger Carpenter are acknowledged for critically reading the manuscript. Specimen illustrations were skillfully prepared by Nancy Halliday (fig. 6) and Patricia Lufkin (figs. 3–5, 7).

Literature Cited

AMEGHINO, F. 1885. Nuevos restos de mamíferos fosiles oligocenos, recognidos por el professor Pedro Sclabrin: y pertenecientes al Museo provincial de la ciudad de Parana. Boletín de la Academia Nacional de Ciencias, Córdoba, **8:** 3–207.

———. 1902. Notas sobre algunos mamíferos fosiles nuevos o poco conocidos del valle de Tarija. Anales del Museo Nacional de Buenos Aires, **3:** 225–261.

BERTA, A. 1981. Evolution of large canids in South America. Anais II Congresso Latino Americano de Paleontologia, Porto Alegre, **2:** 835–845.

———. 1982. *Cerdocyon thous.* Mammalian Species, **186:** 1–4.

———. 1984. The Pleistocene bush dog *Speothos pacivorus* from the Lagoa Santa Caves, Brazil. Journal of Mammalogy **65**(4): 549–559.

———. In press. Quaternary evolution and biogeography of the large South American Canidae (Mammalia: Carnivora). University of California Publications in the Geological Sciences.

BRADY, C. A. 1978. Reproduction, growth and parental care in crab-eating foxes (*Cerdocyon thous*) at the National Zoological Park, Washington. International Zoo Yearbook **18:** 130–134.

———. 1979. Observations on the behavior and ecology of the crab-eating fox (*Cerdocyon thous*), pp. 161–171. *In* Eisenberg, J. F., ed., Vertebrate Ecology in the Northern Neotropics. Smithsonian Institution Press, Washington, D.C., 271 pp.

BURMEISTER, H. 1856. Erläuterungen zur Fauna Brasiliens, enthaltend Abbildungen und ausführliche Beschreibungen neuer oder ungenügend bekannter Thier-Arten. Georg Reimer, Berlin, viii + 115 pp.

———. 1864. Fauna Argentina. I. Mamiferos fosiles. Anales del Museo Nacional de Buenos Aires, **1:** 123–142.

CABRERA, A. 1931. On some South American canine genera. Journal of Mammalogy, **12:** 54–67.

———. 1940. Notas sobre carnivoros sudamericanos. Notas Museo de La Plata, Zoology, **29:** 1–22.

CLUTTON-BROCK, J. 1977. Man-made dogs. Science, **197:** 1340–1342.

CLUTTON-BROCK, J., G. B. CORBET, AND M. HILLS. 1976. A review of the family Canidae, with a classification by numerical methods. Bulletin of the British Museum (Natural History), **29**(3): 120–199.

CRESPO, J. A. 1975. Ecology of the Pampas gray fox and the large fox (culpeo), pp. 179–191. *In* Fox, M. W., ed., The Wild Canids. Van Nostrand Reinhold Co., New York.

DIETZ, J. 1984. Ecology and social organization of the maned wolf (*Chrysocyon brachyurus*). Smithsonian Contributions to Zoology, **392:** 1–55.

EWER, R. F. 1973. The Carnivores. Cornell University Press, New York.

FISCHER, G. VON WALDHEIM. 1814. Zoognosia Tabulis Synopticis Illustrata. Vol. 3, Quadrupedum reliquorum, cetorûm et monotrymatum descriptionem continens. Typis Nicolai S. Vsevolozsky, Mosquae, 605 pp.

FOX, M. W. 1975. Evolution of social behavior in canids, pp. 429–460. *In* Fox, M. W., ed., The Wild Canids. Van Nostrand Reinhold Co., New York.

GIEBEL, C. G. 1855. Die Säugethiere in zoologischer, anatomischer, und palæontologischer Beziehung umfassend dargestellt. A. Abel, Leipzig, 1,108 pp.

GRAY, J. E. 1837. Descriptions of some new or little known Mammalia principally in the British Museum Collection. Magazine of Natural History (Charlesworth's New Series), **1:** 577–587.

HERSHKOVITZ, P. 1961. On the South American small-eared zorro *Atelocynus microtis* Sclater (Canidae). Fieldiana: Zoology, **39**(44): 505–523.

HOFFSTETTER, R. 1952. Les Mammifèrès Pléistocènes de la République de l'Equateur. Mémoires de la Société Géologie, France (nouvelle série), **31**(66): 1–391.

HOUGH, J. R. 1948. The auditory region in some members of the Procyonidae, Canidae and Ursidae. Its significance in the phylogeny of the Carnivora. Bulletin of the American Museum of Natural History, **92**(2): 70–118.

ILLIGER, K. 1815. Uberlick der Saugethiere nach ihrer Verteilung uber die Weittheile. Abhandlungen der (Königlichen preussischen) Akademie der Wissenschaften, Berlin, **1804-1811:** 39–159.

JACOBS, L. L., AND E. H. LINDSAY. 1981. *Prosigmodon oroscoi,* a new sigmodont rodent from the late Tertiary of Mexico. Journal of Paleontology, **55**(2): 425–430.

JAKSÍC, F. M., R. P. SCHLATTER, AND J. L. YÁÑEZ. 1980. Feeding ecology of central Chilean foxes, *Dusicyon culpaeus* and *Dusicyon griseus.* Journal of Mammalogy, **61**(2): 254–260.

KERR, R. 1792. The animal kingdom or zoological system, of . . . Linnaeus: Class I. Mammalia [Class II. Birds] . . . being a translation of that part of the Systema Naturae, as lately published, by . . . Professor Gmelin. . . . 2 pts. in 1 vol. J. Murray and R. Faulder, London, xii [xviii] + 644 pp.

KRAGLIEVICH, J. L. 1952. Un canido del Eocuartario de Mar del Plata y sus relaciones con otras formas brasileñas y norteamericanas. Revista del Museo Municipal de Ciencias Naturales y Tradicional de Mar del Plata, **1:** 53–70.

KRAGLIEVICH, L. 1928. Contribución al conocimiento de los grandes cánidos extinguidos de Sud América. Anales de la Sociedad Argentina de Ciencias Naturales, **106:** 35–66.

———. 1930. Craneometría y clasificación de los canidos sudamericanos especialmente los argentinos actuales y fósiles. Physis (Revista de la Sociedad Argentina de Ciencias Naturales), **10:** 35–73.

KURTÉN, B. J., AND E. ANDERSON. 1980. Pleistocene Mammals of North America. Columbia University Press, New York.

LANGGUTH, A. 1969. Die sudamerikanischen Canidae unter besonderer Berücksichtigung des Mähnenwolfes *Chrysocyon brachyurus* Illiger. Zeitschrift fur wissenschaftliche Zoologie **179**(1, 2): 1–88.

———. 1970. Una nueva clasificación de los canidos sudamericanos, pp. 129–143. *In* Klappenbach, M., et al., eds., IV Congresso Latinoamericano de Zoologia, Actas. Vol. 1. Caracas, Venezuela.

———. 1975. Ecology and evolution in the South American canids, pp. 192–206. *In* Fox, M. W., ed., The Wild Canids. Van Nostrand Reinhold Co., New York.

LEIDY, J. 1858. Notice of remains of extinct vertebrata from the valley of the Niobrara River, collected during the exploring expedition of 1857, in Nebraska, under the command of Lieut. G. K. Warren, U.S. Top Eng., by Dr. F. V. Hayden, Geologist to the Expedition. Proceedings of the Academy of Natural Sciences of Philadelphia, **10:** 20–29.

LINNAEUS, C. 1766. Systema Naturae, vol. 1, 12th ed. Laurentii Salvii, Uppsala, Sweden, 532 pp.

LONG, A., AND P. S. MARTIN. 1974. Death of American ground sloths. Science, **186:** 638–640.

LUND, P. W. 1839a. Blik paa Brasiliens Dyreverden för sidste jordomvaeltning. Anden Afhandlinger Pattedyrene. Danske Videnskabernes Selskabs Naturvidenskabelig og Mathematisk Afdeling, **8:** 61–144.

———. 1839b. Coup d'oeil sur les espèces eteintes de mammiferès du Brésil; extrait de quelques mémoires présentés a l'Academie royale de Copenhaguen. Annales des Sciences Naturelles (Paris), **11:** 214–234.

———. 1842. Blik paa Brasiliens Dyreverden för sidste jordomvaeltning. Fjerde Afhandling: Fortsaettelse af Pattedyrene. Danske Videnskabernes Selskabs Naturvidenskabelig og Mathematisk Afdeling, **9:** 137–208.

MACFADDEN, B. J. 1979. Diversification and biogeography of the one-toed horses *Onohippidium* and *Hippidion.* Postilla, **175:** 1–10.

MARSHALL, L. G. 1978. Evolution of the Borhyaenidae, extinct South American predaceous marsupials. University of California Publications in the Geological Sciences, **117:** 1–89.

MARSHALL, L. G., A. BERTA, R. H. HOFFSTETTER, R. PASCUAL, O. A. REIG, M. BOMBIN, AND A. MONES. 1984. Geochronology of the continental mammal-bearing Quaternary of South America. Palaeovertebrata, Mémoire Extraordinaire: 1–76.

MARSHALL, L. G., R. F. BUTLER, R. E. DRAKE, AND G. H. CURTIS. 1982. Geochronology of type Uquian (Late Cenozoic) land mammal age, Argentina. Science, **216:** 986–989.

MERCERAT, A. 1891. Caracteres diagnósticos de algunas especies de Creodonta conservados en el Museo de La Plata. Revista Museo de La Plata, **2:** 51–56.

MOLINA, G. I. 1782. Saggio sulla Storia Naturale del Chile, 367 pp.

NORDENSKIÖLD, E. 1908. Ein neuer Fundort für Säugetier fossilien in Peru. Arkiv für Zoologi, **4**(11): 1–22.

OSGOOD, W. H. 1934. The genera and subgenera of South American Canidae. Journal of Mammalogy, **15:** 45–50.

PASCUAL, R., ET AL. 1966. Paleontografía Bonaerense. Fascículo IV: Vertebrata. Comision de Investigacion Cientifica, Provincia de Buenos Aires, La Plata, Argentina.

SCHREBER, J. C. D. 1775. Die Säugethiere in Abbildungen nach der Nature mit Beschreibungen. W. Walther, Leipzig, **3:** 281–590.

SCLATER, P. L. 1882. Reports on the additions to the society's menagerie in June, July, August, September, and October. Proceedings of the Zoological Society of London, **1882:** 630–631.

SIMPSON, G. G. 1945. The principles of classification and a classification of mammals. Bulletin of the American Museum of Natural History, **85:** 1–350.

SMITH, C. H. 1839. The natural history of dogs, pp. 1–267. *In* Jardine, W., ed., The Naturalist's Library. Mammalia, vol. 9. W. H. Lizars, Edinburgh.

SORIA, D., AND E. AGUIRRE. 1976. El canido de Layna: revision de los *Nyctereutes* fosiles. Trabajos sobre Neogeno-Cuaternario, Seccion de Paleontologia de Vertebrados y Humana (I.L.C., C. S.I.C., Madrid), No. 5, Miscelanea Neogena Alberdi and Aguirre, eds., pp. 83–115.

STEVENS, M. S. 1965. A new species of *Urocyon* from the Upper Pliocene of Kansas. Journal of Mammalogy, **46**(2): 265–269.

THENIUS, E. 1954. Zur Abstammung der Rotwolfe (Gattung *Cuon* Hodgson). Osterreichische Zoologische Zeitschrift, **5**(3): 378–387.

THOMAS, O. 1900. New South American mammals. Annals and Magazine of Natural History, (7)**5:** 148–153.

———. 1914. The generic and subgeneric names of the South American Canidae. Annals and Magazine of Natural History, (8)**13:** 350–352.

TONNI, E. P., AND G. P. POLITIS. 1981. Un gran canido del Holoceno de la Provincia de Buenos Aires y el registro prehispanico de *Canis* (*Canis*) *familiaris* en las areas pampeana y patagonica. Ameghiniana, **18**(3–4): 251–265.

TORRES, R. V., AND I. FERRUSQUIA, V. 1981. *Cerdocyon* sp. nov. A (Mammalia, Carnivora) en Mexico y su significacion evolutiva y zoogeografia en relacion a los canidos sud americanos. Anais II Congresso Latino-Americano de Paleontologia, Porto Alegre, Brazil, **2:** 709–719.

VAN GELDER, R. G. 1978. A review of canid classification. American Museum of Natural History Novitates, **2646:** 1–10.

WEBB, S. D. 1974. Pleistocene llamas of Florida, with a brief review of the Lamini, pp. 170–213. *In* Webb, S. D., ed., Pleistocene Mammals of Florida. University of Florida Press, Gainesville.

———. 1978. A history of savanna vertebrates in the New World. Part II: South America and the Great Interchange. Annual Review of Ecology and Systematics, **9:** 393–426.

YÁÑEZ, J., AND F. JAKSÍC. 1978. Rol ecologico de los zorros (*Dusicyon*) en Chile central. Anales del Museo de Historia Natural, **11:** 105–112.

Appendix

The following are notes on the evaluations of characters listed in Table 3.

CHARACTERS 1–4—These derived characters distinguish the living Canini from the extinct Borophaghini, a Miocene-Pliocene lineage of hyaena-like dogs.

CHARACTERS 5–6—The simple construction of the upper incisors lacking accessory cusps distinguishes *Vulpes, Urocyon,* and *Otocyon* from more advanced canids. Another diagnostic feature of the *Urocyon* group is the development of a broad paroccipital process with a tip that does not extend below the body of the process.

CHARACTER 7—The presence and expansion of a frontal sinus is the derived condition among advanced canids (Tedford & Taylor, in prep.; Berta, in press; fig. 9).

CHARACTER 8—The increase in size of I^3 relative to I^{1-2} is recognized as a derived state, diagnostic of advanced canids.

CHARACTER 9—The absence of an entepicondylar foramen on the humerus distinguishes all advanced canids from the *Urocyon* group.

CHARACTER 10—Primitive canids including "*Canis*" *davisi* possess a moderately arched or nearly flat zygoma with an everted jugal. The derived state, a strongly arched zygoma with an inverted jugal, unites advanced canids.

CHARACTER 11—The development of a large angular process with expansion of various fossa for branches of the medial pterygoideus muscle is the derived condition among advanced canids (Tedford & Taylor, in prep.; Berta, in press; fig. 10).

CHARACTER 12—The primitive condition, short palatines extending to or just anterior to the toothrow, is observed in most advanced canids. Long palatines extending to or beyond the toothrow is recognized as one of two derived states of this character and it unites the *Dusicyon* group.

CHARACTER 13—The broad, low coronoid process diagnostic of the *Dusicyon* and *Cerdocyon* groups is a marked departure from the narrow, high coronoid process of other advanced canids (e.g., *Canis* and *Chrysocyon*).

CHARACTERS 14–16—These dental features indicate a trend toward greater complication of cusps on the lower molars, characteristic of the *Dusicyon* group (Berta, in press; fig. 8).

CHARACTER 17—The frontal sinus, although large, does not penetrate the postorbital process, recognized as the least derived sinus condition. This character unites the *Dusicyon* group. See comments under character 7.

CHARACTER 18—Canines that are proportionally small relative to the cheekteeth is a derived character which distinguishes the *Cerdocyon* clade from the *Dusicyon* clade.

CHARACTER 19—Further expansion of the pterygoid fossa is recognized as continued modification of the trend established with character 11.

CHARACTERS 20–23—Short extremities are derived features of the *Cerdocyon* clade.

CHARACTERS 24–26—These derived dental and mandibular characters distinguish *Cerdocyon* and *Nyctereutes* from *Atelocynus* and *Speothos.* Development of a subangular lobe has been correlated with an insectivorous diet (Ewer, 1973).

CHARACTER 27—Short nasals which rarely extend beyond the maxillary-frontal suture is a derived character which unites *Atelocynus* and *Speothos.*

CHARACTER 28—The marked reduction in size of the frontal sinus in *Speothos, Atelocynus,* and *Nyctereutes* is a reversal of the derived condition. See comments under character 7.

CHARACTER 29—Posterior expansion of the frontal sinus is regarded as a further modification of the trend established with characters 7 and 17.

CHARACTER 30—Expansion of the fossa for the superior branch of the medial pterygoideus muscle is regarded as a further modification of the trend established with character 11.

CHARACTER 31—The further increase in size and complication of I^3 relative to I^{1-2} observed in *Canis* and *Chrysocyon* is recognized as a more derived state of character 8.

Comparative Cytogenetics of South American Deer

Angel E. Spotorno, Nadir Brum, and Mariela Di Tomaso

ABSTRACTS

Karyotypes of a male *Hippocamelus bisulcus* from Chile had 2n = 70, and those of a female *Blastoceros bezoarticus* from Uruguay, 2n = 68, both with NF = 74. Most chromosomes were similar, and at least pairs 1 (also NOR-bearing), the X long arms, the small metacentric pair, and *Hippocamelus* 2 and 4 with *Blastoceros* 32 p and q were identical in G-band patterns. C-bands were large and paracentromeric, but absent in the small metacentric pair that is shared with other deer. Similarity relationships based on X shapes and lengths compared in a karyo-idiogram are (*Pudu-Blastoceros, Hippocamelus-Mazama*; *Odocoileus*) Cervinae. All the other large metacentric chromosome (*Blastoceros* 32 and *Mazama* 1 through 9) are inferred to be unique, favoring the hypothesis of centric and tandem fusions. Such metacentrics and the distribution of 2n and NF suggest the occurrence of many parallel and independent fusions throughout the three phyletic lines of living Cervidae.

Los cariotipos de un macho *Hippocamelus bisulcus* de Chile tienen 2n = 70 y los de una hembra *Blastoceros bezoarticus* de Uruguay 2n = 68; ambos tenían un NF = 74. La mayor parte de los cromosomas fueron similares en ambas especies, y por lo menos los pares 1 (que es portador del sector NOR), los brazos largos de los cromosomas X. El pequeño par metacéntrico, e *Hippocamelus* 2 y 4 con *Blastoceros* 32 p y q eran idénticos en sus patrones de bandas G. Las bandas C eran grandes y paracentroméricas pero estaban absentes en el pequeño par metacéntrico que es compartido con otros ciervos. Las relaciones de similitud basadas en las longitudes y formas comparadas del X en un cario-idiograma son (*Pudu-Blastoceros, Hippocamelus-Mazama*; *Odocoileus*) Cervinae. Todos los otros cromosomas metacéntricos grandes (*Blastoceros* 32 y *Mazama* 1 al 9) se infiere que son únicos, lo que favorece la hipótesis de fusiones céntricas y en tandem. Tales metacéntricos y la distribución de 2n y NF sugieren la ocurrencia de varias fusiones paralelas e independientes en las tres líneas filéticas de Cervidae vivientes.

Cariótipos de um macho *Hippocamelus bisulcus* do Chile contam com 2n = 70, e os de uma fêmea *Blastoceros bezoarticus* do Uruguai, com 2n = 68, ambos com NF = 74. A maior parte dos cromossomos são parecidos. Os pares 1 (também contendo NOR), os braços longos do X, o pequeno par metacêntrico; *Hippocamelus* 2 e 4, e *Blastoceros* 32 p e q, todos possuem padrões idênticos na faixa G (G-band). As faixas C (C-band) são grandes e paracentroméricas porém não ocorrem no par metacêntrico pequeno, comuns nas outras espécies de veados. Baseando-

From the Departamento Biología Celular y Genética, Facultad de Medicina, Universidad de Chile, Casilla 70061, Santiago 7, Chile (Spotorno); and División Citogenética, Instituto de Investigaciones Biológicas Clemente Estable, Av. Italia 3318, Montevideo, Uruguay (Brum and Di Tomaso).

se nas formas e comprimentos dos seus cromossomos X (comparados através de cário-idiogramas) as semelhanças são (*Pudu-Blastoceros, Hippocamelus-Mazama*; *Odocoileus*) Cervinae. Infere-se que todos outros cromossomos grandes metacêntricos (*Blastoceros* 32 e *Mazama* 1–9) são exclusivos, o que sugere ocorrências de fusões alinhadas (tandem) e centrais. Tais metacêntricos, junto com a distribuição dos 2n e NF, sugerem a ocorrência de várias fusões paralelas e independentes nas três linhas filéticas dos Cervidae agora existentes.

Introduction

Of the 11 species of deer now living in South America (Cabrera, 1961), only three species are cytologically known: *Odocoileus virginianus* (Wurster & Benirschke, 1967), *Mazama americana* (Taylor et al., 1969) and *Pudu puda* (Koulischer et al., 1972; Spotorno & Fernández, 1975). Furthermore, no detailed chromosome analysis or G- and C-band descriptions have been published yet. We report here cytogenetic descriptions and comparisons of material from *Hippocamelus bisulcus* and *Blastoceros bezoarticus,* which complete representative karyotypic descriptions of all South American genera of deer recognized by most authorities (Koopman, 1967).

Chromosomes can provide reliable taxonomic characters when they are examined in detail, particularly the specific patterns of G- or R-bands or certain kinds of DNA (C-bands and NOR). Moreover, mechanism of change in character-states can eventually be inferred. These analyses have been rarely done on cervid chromosomes. With newly reported data, we think there is now enough data to compare the chromosomes of American deer systematically.

Materials and Methods

An adult male *Hippocamelus bisulcus* (fig. 1a), captured in Región de Aisén, Chile, and kept alive by CONAF (Chilean National Forestry Service), was the source of a 10-ml blood sample. Conventional 72-hour blood cultures were done in TC-199 with phytohemagglutinin. Cells were then treated with Colcemid, 0.075 M KCl hypotonic solution, fixed in Carnoy, dropped on clean slides, and air-dried.

A female *Blastoceros bezoarticus* (sister of the male shown in fig. 1b), captured in Departamento de Salto, Uruguay, died by accidential injury during the trip to Montevideo. Immediately, bone marrow samples were obtained. Cells were incubated in TC-199 with 0.04% colchicine plus 3 drops of Liquemin for 4 hours and then treated as above.

Slides were Giemsa-stained. G- and C-bands and NOR were induced by treating slides with the methods of Chiarelli et al. (1972), Sumner (1972), and Rufas et al. (1982), respectively. A selected number of metaphases were photographed with fine-grain high-contrast copy-film.

Measurements were made on enlargements us-

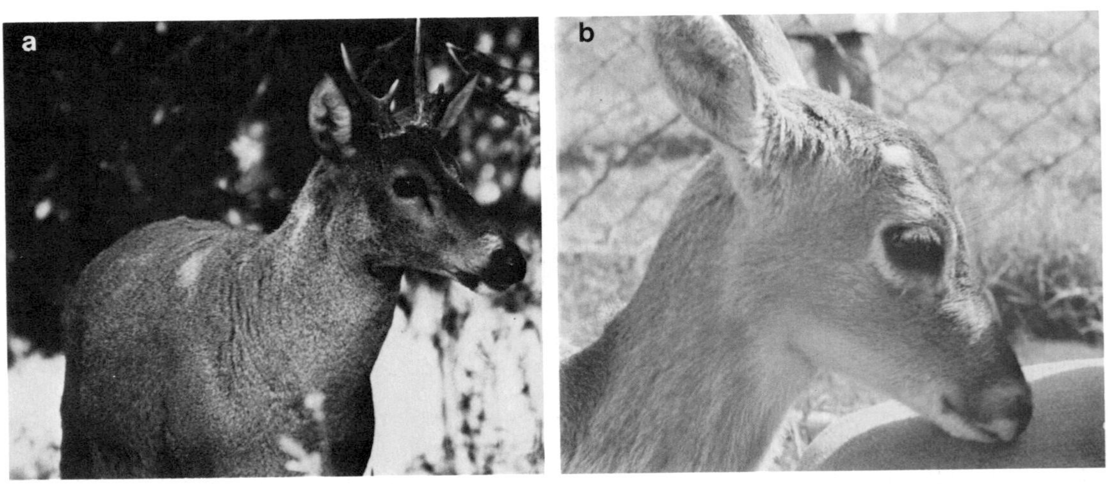

FIG. 1. Specimens examined: **a,** *Hippocamelus bisulcus* male in Reserva de Peñuelas (photograph by A. Spotorno); **b,** *Blastoceros bezoarticus* male in Reserva Maldonado (photograph by H. Cardoso).

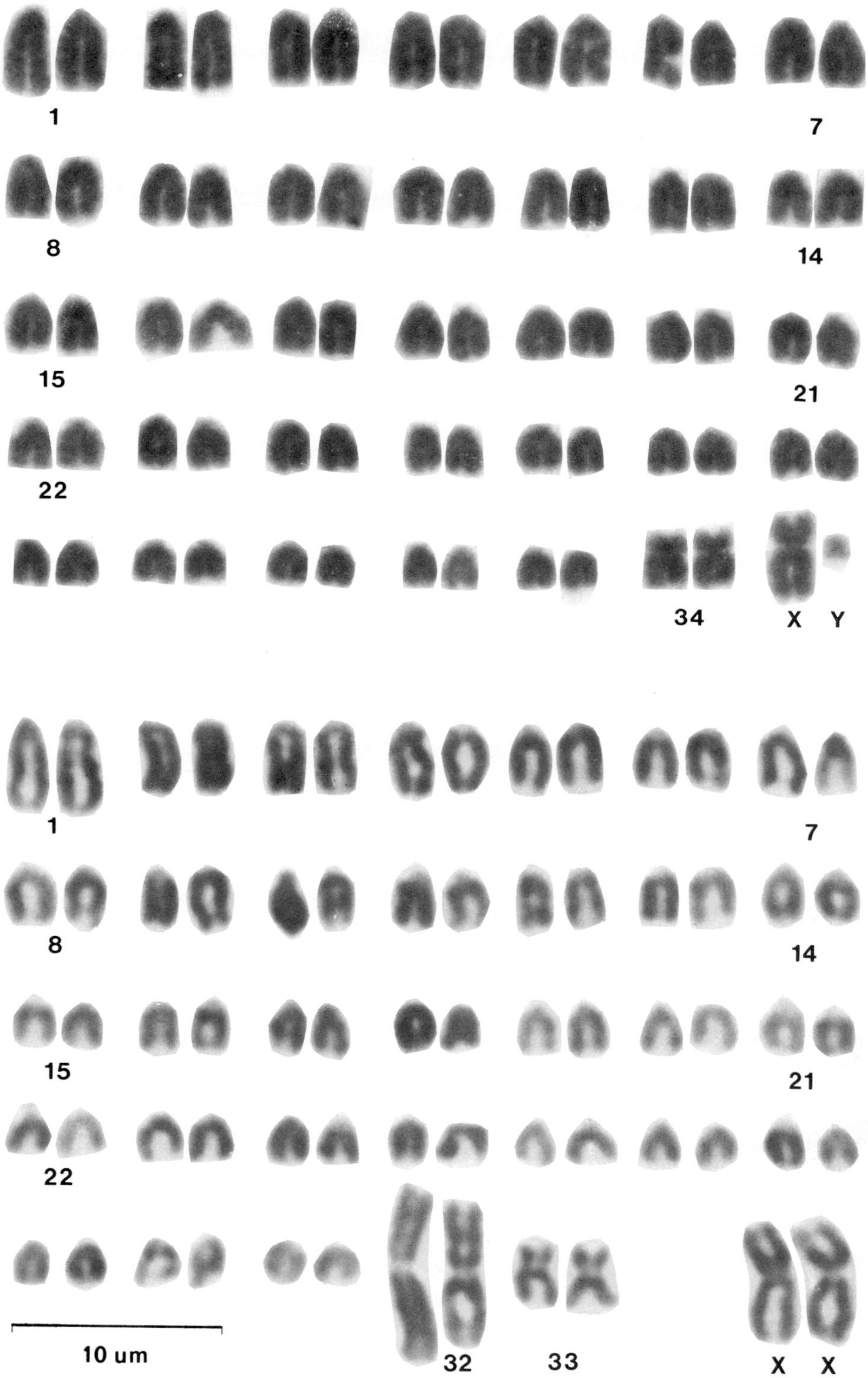

FIG. 2. Karyotype of single cells from: **top,** *Hippocamelus bisulcus* male; and **bottom,** *Blastoceros bezoarticus* female.

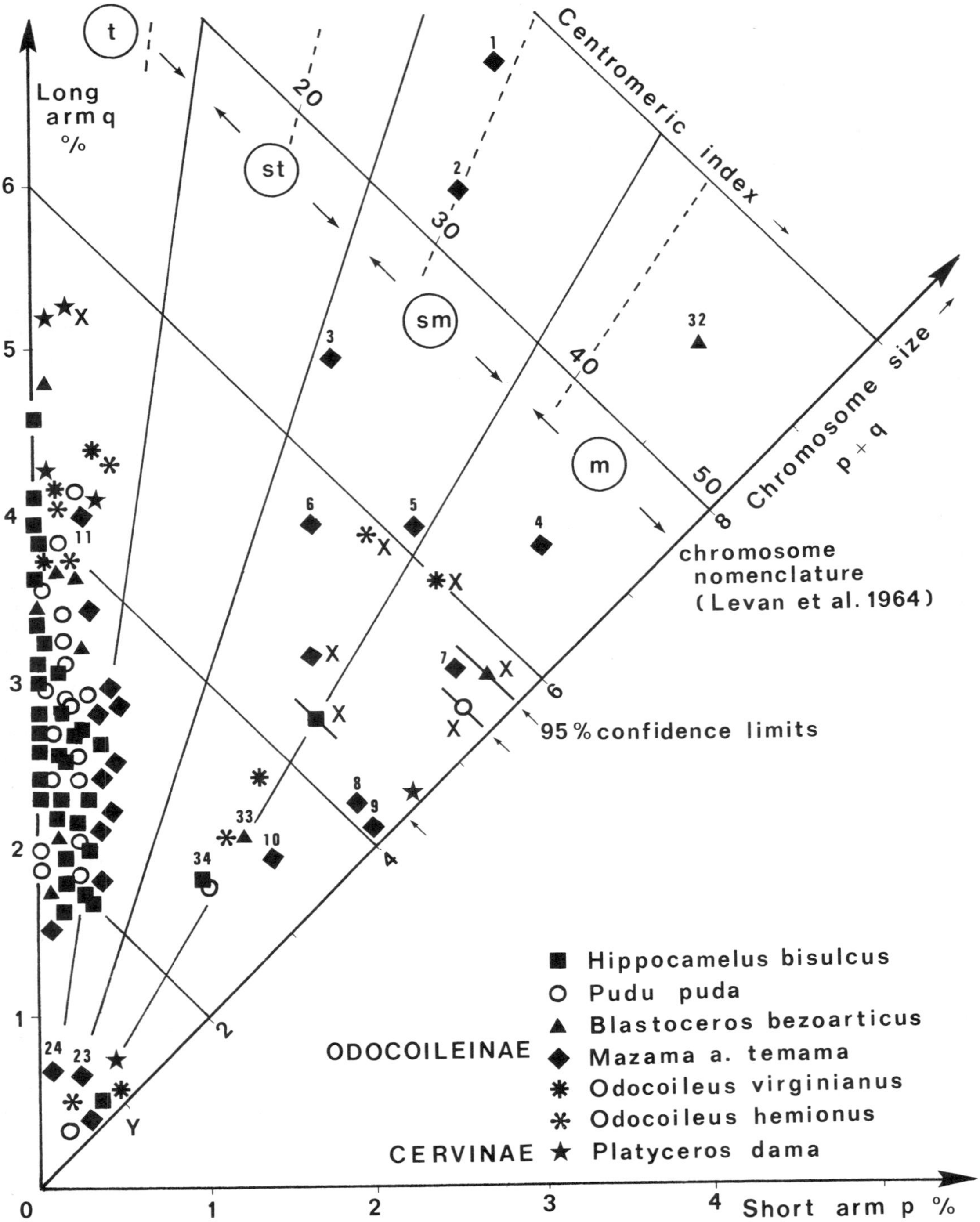

FIG. 3. Karyo-idiogram displaying relative chromosome lengths of some deer; chromosome size and shape can be read on diagonals. Numbers and letters are from original karyotype descriptions. Some overlapping chromosomes are not displayed.

ing the best single chromatid per pair, and values were transformed into percentages of the total haploid plus X set. Such relative lengths are displayed in a scatter diagram called a karyo-idiogram (Spotorno et al., 1979; Spotorno, 1985), a useful device that allows eventual chromosome identification and comparison through two independent variables, total chromosome size (short arm length plus long arm length), and centromeric index (100 times short arm length divided by total chromosome length). This procedure assumes the conservation of total nuclear material. Although this assumption is generally true for mammals, it can be validated by C-banding techniques which detect heterochromatin-containing satellite DNA or by marker chromosomes (Spotorno, 1977). A karyo-idiogram can also portray confidence intervals (Spotorno et al., 1979); we use them here to compare some centromeric indices calculated without measuring all chromosomes from a karyotype. Chromosomes of other deer used in comparisons were measured from the following sources: *Mazama americana temama* from Jorge and Benirschke (1977, p. 712), *Odocoileus virginiana* and *O. hemionus* from Wurster and Benirschke (1967, p. 275), and *Platyceros dama* from Wurster and Benirschke, 1967, p. 277).

Results

The diploid number of *Hippocamelus bisulcus* was 70 in almost all of the 30 cells analyzed, with 33 telocentric pairs of decreasing size (fig. 2), a submetacentric (here labeled 34), and a heteromorphic pair of metacentrics. The latter are the largest and the smallest elements of the karyotype, having relative sizes of 4.3% and 0.9% of the total haploid set (means from eight cells measured; see also fig. 4); they probably are the X and Y chromosomes, respectively. The total number of chromosome arms per cell (FN or fundamental number) was 74.

The diploid number of *Blastoceros bezoarticus* was 68 in six cells analyzed, with 31 telocentric pairs of decreasing size (fig. 2) and three metacentrics, here labeled 32, 33, and a presumptive X. These had relative sizes of 6.9%, 3.3%, and 5.6% of total haploid set (means from six cells). The FN was also 74.

Many similarities and some differences in the length and shape of chromosomes were immediately detected when both karyotypes were compared, particularly with the aid of a karyo-idiogram constructed from their measurements (fig. 3). Thus, 31 telocentrics, ranging from 1.6% to 4.6% of haploid set, and one metacentric having a 3%, tended to cluster at the same positions in the karyo-idiogram, indicating that they share similar size and shape.

In contrast, the large chromosome 32 of *Blastoceros* was not found in *Hippocamelus* (fig. 3), but the latter had two additional telocentrics whose sizes corresponded to the short and long arms of this chromosome 32. Another difference was detected in the X chromosome. The presumptive X of *Blastoceros* seemed to be larger than that of *Hippocamelus,* mainly in the short arm. This is reflected in their statistically different centromeric indices, 46.49 $\pm$ 3.89 and 38.15 $\pm$ 2.60, respectively (confidence intervals do not overlap in fig. 3; $N = 8$ for *Blastoceros,* $N = 15$ for *Hippocamelus*).

G-banded karyotypes of the two species (illustrated in fig. 4) allowed a better identification of chromosome pairs than that displayed in Figure 2. Many gross similarities were found when banded chromosomes of the species were compared side by side. This was a difficult task, because cells with similar states of contraction and treatment must be used. Nevertheless, approximate banding patterns were found in *Hippocamelus* for the following *Blastoceros* chromosomes: 1, 32, 33, and the X long arm (illustrated in fig. 5a).

The position of the nucleolar organizing region (NOR) was detected at the largest telocentric pair. Ag-stained plates of *Blastoceros* exhibited clear terminal dark spots over chromosome 1 only (fig. 5c). The same technique was unsuccessful in *Hippocamelus,* but in the G-banded plates the homologous chromosomes of pair No. 1 were usually associated by faint terminal material (fig. 5b), probably nucleolar material. This suggests the probable position of the NOR.

C-bands in almost all chromosomes of both species were paracentric (fig. 5d,e), and one of the largest autosomes (probably the 1) and the X chromosomes had rather large C-bands near the centromere. This was also shown in the G-bands, where large centromeric light bands were observed at the same positions. The only exception to this pattern was the minute amount of the C-bands in pairs 34 of *Hippocamelus* and 33 of *Blastoceros* (fig. 5e).

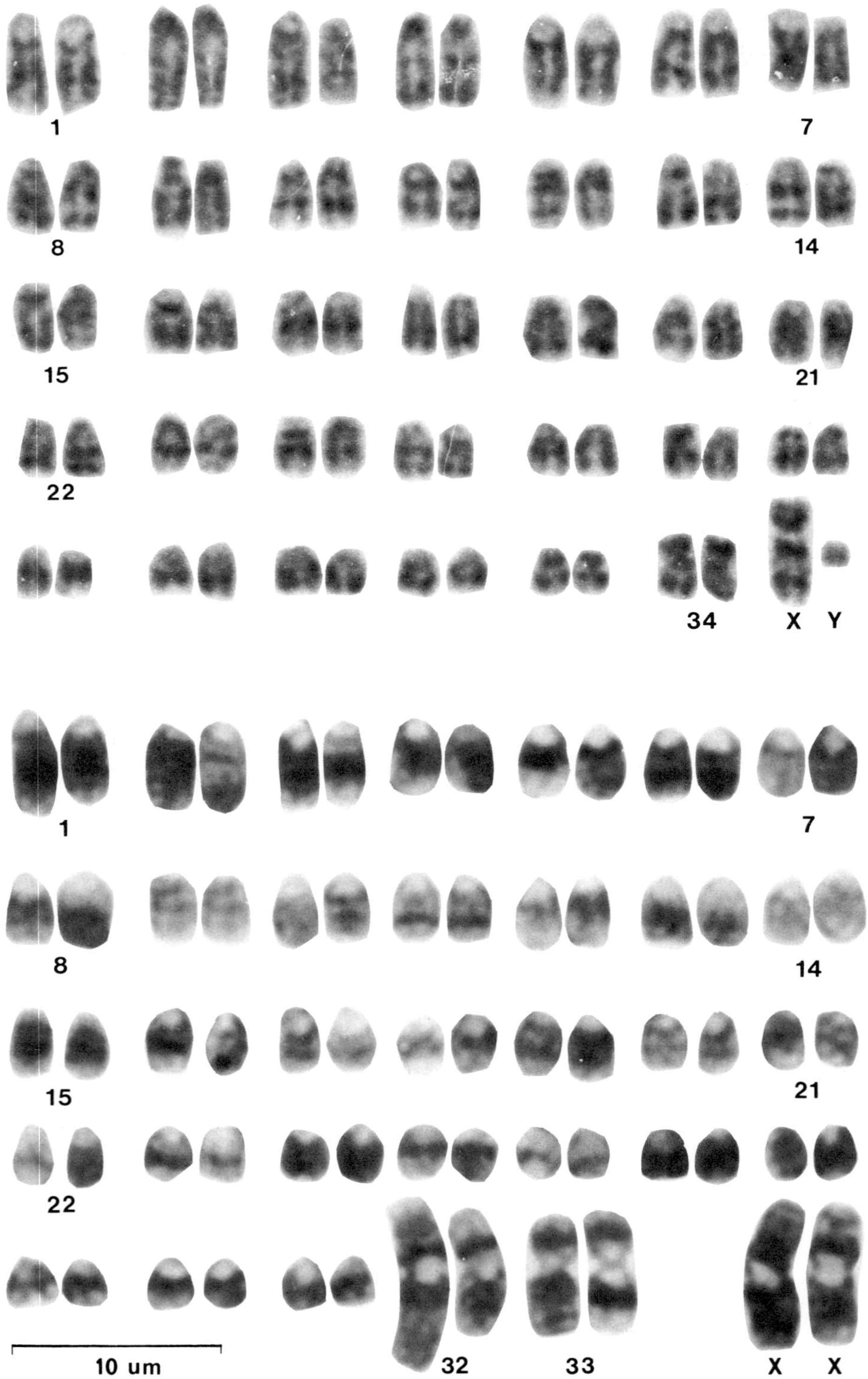

FIG. 4. G-banded karyotypes from: **top,** *Hippocamelus bisulcus* male; and **bottom,** *Blastoceros bezoarticus* female.

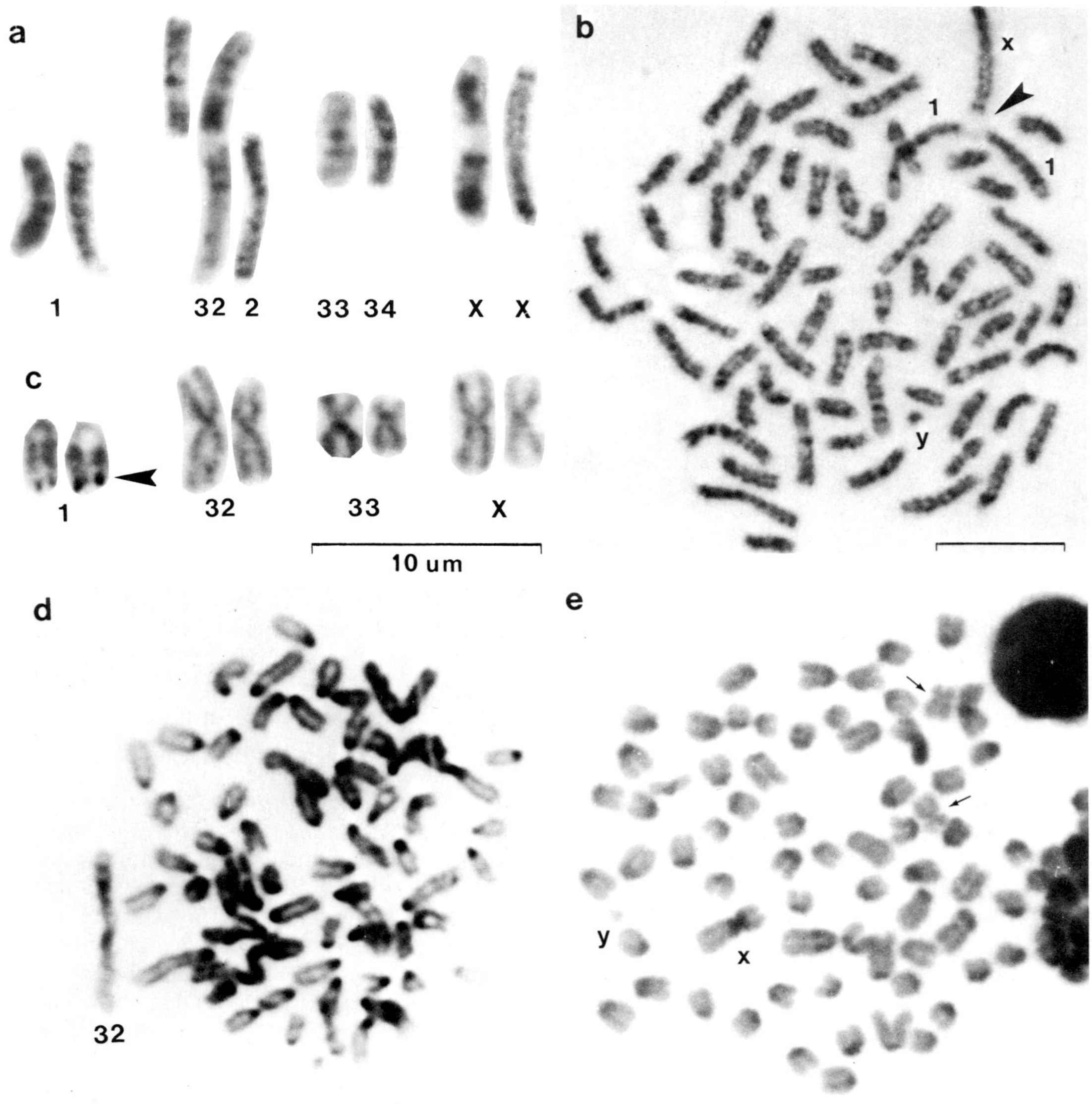

FIG. 5. Banding patterns in *Hippocamelus* and *Blastoceros* chromosomes. **a,** G-banded chromosomes of *Blastoceros* (left) compared with those of *Hippocamelus* (right). **b,** G-banded metaphase of *Hippocamelus* showing terminal nucleolar association (arrow). **c,** Ag-NOR bands (arrow) from a *Blastoceros* metaphase. **d,** C-banding in a *Blastoceros* cell. **e,** C-bands in a *Hippocamelus* cell; note absence of C-bands (arrows). Numbers and letters from original karyotype descriptions.

Discussion

Detailed banded karyotypes of only three species of Odocoileinae are known, but conventional non-banded karyotypes can still be used in comparative studies, particularly when few changes have occurred. Thus, we will discuss our results, taking into account the chromosomes of other related species, which have been also displayed on the karyo-idiogram of Figure 3. The assumption of conservation in the total amount of chromatin required by this method is verified by the absence of heterochromatic arms and the small quantities of C material detected in at least three species.

Conservation of many chromosomes is the rule within these cervids. The tight clustering of most telocentric and Y chromosomes from many species at the left of Figure 3 is also exhibited by the small

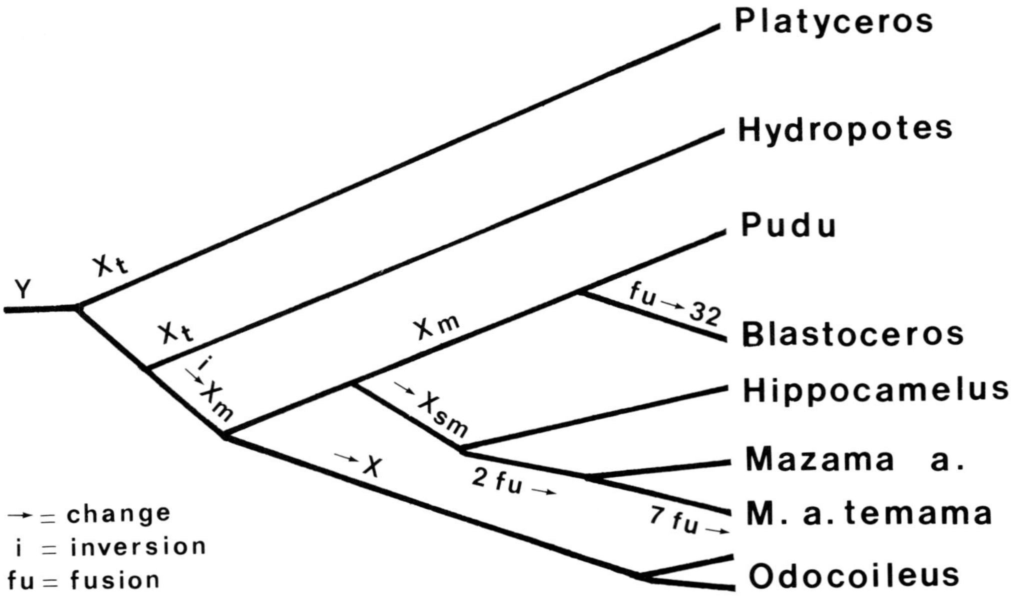

FIG. 6. Diagram of chromosome similarities and inferred changes in South American deer, mainly based on X chromosomes. X and Y are sex chromosomes. Small letters are from chromosome nomenclature of Figure 3.

submetacentric or metacentric autosome. In this case, we have also documented the conservation of its G-bands (fig. 5a) and of its characteristically small C-bands. The metacentric 10 of *Mazama a. temama* had the same characteristics (Jorge & Benirschke, 1977, fig. 3) as the sika deer (*Cervus nippon*; Cervinae) (Van Tuinen et al., 1983). In *Platyceros* (= *Dama*) *dama* there is also a small metacentric element somewhat larger than the above (fig. 3). If all these chromosomes have been inherited intact from a common ancestor, as it is indicated by their metacentric shape, total size, unique small C-band, and correspondence of G-bands, this is a clearly documented case of chromosome conservation in both Cervinae and Odocoileinae.

The NOR-bearing chromosomes of South American deer appear to be the largest telocentric pair. It was clearly identified as No. 1 in *Blastoceros bezoarticus, Hippocamelus bisulcus* (this report), *Pudu puda* (Spotorno & Fernández, 1975), and *Odocoileus virginianus* and *Odocoileus hemionus* (Wurster & Benirschke, 1967). This condition is also displayed by the karyotypes of *Alces alces, Cervus elaphus, Capreolus capreolus, Platyceros dama,* and *Cervus nippon nippon* (Gustavsson & Sundt, 1968). Nevertheless, not one but two chromosome pairs carrying terminal NOR were identified with sequential Q-banding and Ag-NOR techniques in introduced *Cervus nippon nippon* (Van Tuinen et al., 1983). In any case, uniNOR seems to be a conservative and extended condition among American Cervidae.

The X chromosome seems to have changed the most in these deer. The submetacentric X of *Hippocamelus* is very similar to the one of *Mazama,* but these are clearly different in shape and bands from those of *Blastoceros-Pudu* ones (see fig. 3; unpublished measurements for *Pudu* based on 12 metaphases). The X chromosomes of the two *Odocoileus* species appear to be larger than those above, but they seem similar in shape to those of *Hippocamelus* and *Mazama.* All these chromosomes can be inferred to be uniquely derived in the Odocoileinae, not only because of their widespread condition within the group but because of the consistent telocentric shape of the X within Cervinae (all species) and Muntiacinae material reported until now (Jorge & Benirschke, 1977, and references therein). Given that total size has been grossly conserved despite shape changes, we conclude that an ancestral telocentric X probably had a pericentric inversion during the initial evolution of the Odocoileinae. Our view of all these changes is included in Figure 6.

There is a single exception to the metacentric X condition within the Odocoileinae, namely the telocentric X of *Hydropotes inermis* (karyotype not analyzed by present authors; data from Jorge & Benirschke, 1977). Such observations on *Hydrop-*

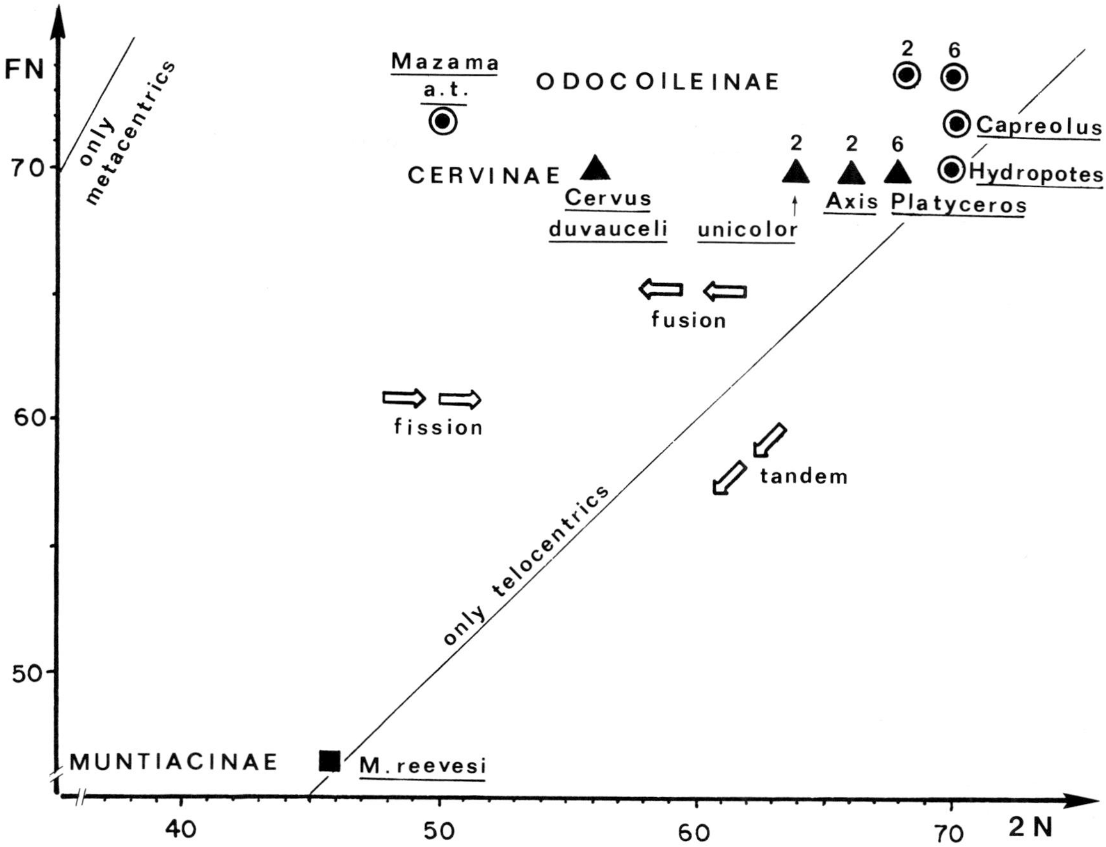

FIG. 7. Karyograph of known Cervidae species. Each symbol represents a karyotype whose position depends on its FN and diploid number. Karyotypes with only telocentric chromosomes fall on the line to the right. Numbers over symbols are numbers of species having same values. Large arrows are processes of change.

otes can be explained by the following: It is not an Odocoileinae, it is an early offshoot of Odocoileinae, or its metacentric X has reverted to the ancestral telocentric condition. The fact that *Hydropotes* FN is also 70 (Jorge & Benirschke, 1977), like all Cervinae and unlike almost all Odocoileinae, favors the first two possibilities.

The presence of large metacentric chromosomes in *Blastoceros* and *Mazama* deserves close study because those also predominate in the karyotype of *Muntiacus muntjak* (2n = 6 and 7; FN = 12). Two alternative hypotheses about their origin have been proposed: (1) the classic fusion processes, including centromeric and tandem fusions (Matthey, 1973), where large and diverse metacentric chromosomes are final products; and (2) the fission hypothesis (Imai & Crozier, 1980), where unaltered metacentric elements are the remains of an ancestral karyotype. In the latter case, ancestral metacentric chromosomes would be identical in interspecies comparisons. A simple inspection of the karyo-idiogram in Figure 3 shows that chromosomes 32 of *Blastoceros* and 1 through 9 of *Mazama* are unique elements. This is also the case with the chromosomes of *M. muntjak,* whose sizes are too large to be included in Figure 3. A fusion of two telocentric elements is the simplest explanation of the origin of *Blastoceros* 32, given that banding patterns of its short and long arms correspond to those of two telocentric pairs from *Hippocamelus* (fig. 5) and that arms of similar sizes are present in the telocentrics of many other Odocoileinae species.

The hypothesis about deer chromosome evolution can be also examined from the distribution of diploid numbers and FN of all deer karyotypes known. This can be easily done through the karyograph (Imai & Crozier, 1980) shown in Figure 7 with deer data. Most karyotypes within each subfamily have mainly telocentric elements, that is, they tend to fall to the extreme right in Figure 7. This suggests that such is the ancestral karyotype not only for each subfamily, but also for the Cervidae as a whole. Therefore, it is probably that many fusions throughout the Cervidae (Jorge & Benirschke, 1977; Shi et al., 1980) have occurred with differing degrees in parallel and independent ways.

There are two submetacentric chromosomes, *Mazama* 1 and 2, whose long arms are exceptionally long (fig. 3). This suggests the possible occurrence of tandem fusions or translocations. The simultaneous appearance of two unique tiny telocentric pairs in *Mazama* with both FN = 74 and 72 (chromosomes 23 and 24 of the latter in fig. 3) suggests past translocations from one to the other. Interstitial C-bands in *Mazama* 1 and 3 from a heterozygous doe with a presumed translocation *t* (4,24) (Jorge & Benirschke, 1977) clearly indicates that additional translocations have occurred in the *Mazama* line.

Conclusions

Despite the conservation of many chromosomes and chromosome arms, the karyotypes of South American deer seem to have undergone drastic changes in some phyletic lines. The development of large metacentrics is a process that seems to have occurred at least three times in a parallel form among the three subfamilies of deer. It is possible that factors acting on rates of chromosomal evolution in horses, for example (e.g., Bush et al., 1977), are also acting within deer species. In the case of South American deer, the conjunction of small population demes (for review, see Hershkovitz, 1972, 1982) and new ecological opportunities (Keast, 1972) might be such factors. We probably will never fully understand the biology of these vanishing gracile mammals.

Acknowledgments

We are grateful to the officers of the Reserva Peñuelas (CONAF), Chile and Reserva Salto, Uruguay, to Lic. J. Simonetti and Sr. Blengini, who helped the accession and handling of specimens, as well as to P. Valenzuela, who typed the final manuscript. This work was funded in part by Proyecto B-1979-8523, DIB, Universidad de Chile.

Literature Cited

Bush, G. L., S. M. Case, A. C. Wilson, and J. L. Patton. 1977. Rapid speciation and chromosomal evolution in mammals. Proceedings of the National Academy of Sciences, U.S.A., **74:** 3942–3946.

Cabrera, A. 1961 (1958–1961). Catálogo de los Mamíferos de América del Sur. Revista del Museo Argentino de Ciencias Naturales "Bernardino Rivadavia," **4:** 309–732.

Chiarelli, B. A., M. Sarti-Chiarelli, and D. A. Shafer. 1972. Chromosome banding with trypsin. Mammalian Chromosomes Newsletter, **13:** 44–45.

Gustavsson, I., and C. O. Sundt. 1968. Karyotypes in five species of deer. Hereditas, **60:** 233–248.

Hershkovitz, P. 1972. The recent mammals of the neotropical region: A zoogeographic and ecological review, pp. 311–431. *In* Keast, A., F. G. Erk, and B. Glass, eds., Evolution, Mammals and Southern Continents. State University of New York Press, Albany, 543 pp.

———. 1982. Neotropical deer (Cervidae). Part I. Pudus, genus *Pudu* Gray. Fieldiana: Zoology, n.s., **11:** 1–86.

Imai, H. I., and R. H. Crozier. 1980. Quantitative analysis of directionality in mammalian karyotype evolution. American Naturalist, **116:** 537–569.

Jorge, W., and K. Benirschke. 1977. Centromeric heterochromatin and G-banding of the red brocket deer, *Mazama americana temama* (Cervoidea, Artiodactyla) with a probable non-Robertsonian translocation. Cytologia, **42:** 711–721.

Keast, A. 1972. Comparisons of contemporary mammal faunas of southern continents, pp. 433–501. *In* Keast, A., F. G. Erk, and B. Glass, eds., Evolution, Mammals and Southern Continents. State University of New York Press, Albany, 543 pp.

Koopman, K. F. 1967. Artiodactyls, pp. 385–406. *In* Anderson, S., and J. K. Jones, Jr., eds., Recent Mammals of the World. The Ronald Press Company, New York, 453 pp.

Koulischer, L., J. Tyskens, and J. Mortlemans. 1972. Mammalian cytogenetics. VII. The chromosomes of *Cervus canadensis, Elaphurus davidianus, Cervus nippon* (Temminck) and *Pudu pudu.* Acta Zoologica et Pathologia Antverpiensia, no. **56:** 25–30; 3 figs.

Levan, A., K. Fregda, and A. Sandberg. 1964. Nomenclature for centromeric position on chromosomes. Hereditas, **52:** 201–220.

Matthey, R. 1973. The chromosome formulae of eutherian mammals, pp. 531–616. *In* Chiarelli, A. B., and E. Capanna, eds., Cytotaxonomy and Vertebrate Evolution. Academic Press, London, 783 pp.

Rufas, J. S., G. Gimenes-Martin, and P. Esponda. 1982. Presence of a chromatid core in mitotic and meiotic chromosomes of grasshoppers. Cell Biology International Reports, **6:** 261–267.

Shi, L. M., Y. Ye, and X. Duan. 1980. Comparative cytogenetic studies on the red muntjak, Chinese muntjak, and their F_1 hybrids. Cytogenetics and Cell Genetics, **26:** 22–27.

SPOTORNO, A. 1977. El número fundamental (NF) como descriptor citotaxonómico en grupos de mamíferos, pp. 50–51. *In* Resúmenes del VII Congreso Latinoamericano de Zoología, Tucumán, Argentina.

———. 1985. Concepts and methods in descriptive and comparative karyology, pp. 135–165. *In* Fernández-Donoso, R., ed., El núcleo, los cromosomas y la evolución. (In Spanish, with English summary.)

SPOTORNO, O., AND R. FERNÁNDEZ-DONOSO. 1975. The chromosomes of the Chilean dwarf-deer "pudu," *Pudu pudu* (Molina). Mammalian Chromosomes Newsletter, **16:** 17–18, fig. 10.

SPOTORNO, A., R. FERNÁNDEZ-DONOSO, AND J. PINCHEIRA. 1979. Similitud cromosómica: Un nuevo método cuantitativo de descripción y comparación. Archivos de Biología y Medicina Experimentales, **12:** 223 (abstract).

SUMNER, A. T. 1972. A simple technique for demonstrating centromeric heterochromatin. Experimental Cell Research, **75:** 304–306.

TAYLOR, K. M., D. A. HUNGERFORD, AND R. L. SNYDER. 1969. Artiodactyl mammals: Their chromosome cytology in relation to patterns of evolution, pp. 346–356. *In* Benirschke, K., ed., Comparative Mammalian Cytogenetics. Springer Verlag, New York.

VAN TUINEN, P., T. J. ROBINSON, AND G. A. FELDHAMER. 1983. Chromosome banding and NOR location in sika deer. The Journal of Heredity, **74:** 473–474.

WURSTER, D. H., AND K. BENIRSCHKE. 1967. Chromosome studies in some deer, the springbok, and the pronghorn, with notes on placentation in deer. Cytologia, **32:** 273–285.

Faunal Representation in Museum Collections of Mammals: Osgood's Mammals of Chile

Bruce D. Patterson and Clare E. Feigl

ABSTRACTS

Possible bias in faunal representation of museum mammal collections is evaluated using Chilean mammals and their representation in the collections of Field Museum of Natural History. Rank correlation analyses were used to relate the number of Chilean species and the number of specimens of these species to several independent variables: order, body size, trophic level and habitat affinity, life zone, and current status in Chile. Results indicate significant positive relations between specimen number and natural abundance, as this was variously inferred. For the Chilean fauna at least, museum collections provide an adequate basis for faunistic inference.

La posible desviación en la representación faunística de colecciones en museos es evaluada usando mamíferos chilenos y su representación en colecciones del Field Museum of Natural History. Análisis de rangos de correlación fueron usados para relacionar el número de especímenes de estas especies con diversas variables independientes: orden, talla del cuerpo, nivel trófico y afinidad de hábitat, zona de vida y estado actual en Chile. Los resultados indican significativas relaciones positivas entre el número de especímenes y abundancia natural, tal como esto fue variadamente inferido. Al menos para la fauna Chilena las colecciones museológicas proveen una base adecuada de inferencia faunística.

Possíveis inexatidões nas representações faunísticas em coleções de mamíferos em Museus, foram avaliadas através de uma comparação entre a composição de mamíferos no Chile e a sua representação no Field Museu de História Natural. Análises correlacionais de ordem foram usadas para investigar relações entre o número de espécies Chilenas e o número de representantes destas espécies no Museu, com variáveis independentes, como ordem, tamãnho, nível trófico e afinidade de habitat, zona de vida, e status atual no Chile. Os resultados indicam relações positivas significantes entre o número de espécimes no Museu e a abundancia atual no campo (deduzida de várias maneiras diferentes). Conclue-se assim que, pelo menos quanto à fauna Chilena, coleções em museus oferecem uma base adequada para deduções sobre a fauna de um local.

Introduction

Much of what is known of the biology of Neotropical mammals is based on museum specimens. Of course museum specimens form the basis of scientific nomenclature, but further they figure prominently in studies of evolution, descriptive and functional morphology, genetics, cytology, and even ecology. Given this essential role, especially in initial studies, it is important to understand the limitations of museum collections. Some limitations are well known, as the nature of preservation

From Division of Mammals, Field Museum of Natural History, Chicago, IL 60605-2496 (Patterson); and Department of Evolutionary Biology, Northwestern University, Evanston, IL 60201 (Feigl).

(skulls, fluid-preserved) obviously determines the kinds of questions that can be asked. However, the limitation imposed by collection bias, by collections which represent only part of a native fauna, remains unexplored. How representative are museum collections of the natural Neotropical mammal fauna?

As examples of the sort of bias we consider here, one can point to the monotypic sigmodontine genera *Abrawayaomys* Cunha and Cruz, 1979, *Scolomys* Anthony, 1929, *Juscelinomys* Moojen, 1965, *Podoxymys* Anthony, 1929, *Galenomys* Thomas, 1916, and *Anotomys* Thomas, 1906, all known from one or a few localities and represented in the world's museum collections by a mere handful of specimens (Nowak & Paradiso, 1983). At a finer level, the ichthyomyine genus *Daptomys* Anthony, 1929, is comprised of three species, *D. venezuelae, D. peruviensis,* and *D. oyapocki,* represented in world collections by three, one, and one specimens, respectively (Nowak & Paradiso, 1983). Finally, of the 47 recognized species of the murine opossum *Marmosa* Gray, 1821 (*sensu lato*; Honacki et al., 1982), four are known only from the type localities (*M. agricolai, M. andersoni, M. scapulata,* and *M. tatei*) and two others from the vicinity of their type localities (*M. handleyi* and *M. cracens*). For this short list at least, how different our understanding of these faunas might be if museum collectors had camped at different localities, used different varieties of traps or baits, placed traps in different positions, or had delayed their travel plans by as little as a week. Fieldworkers are aware firsthand of the serendipity involved in certain captures, and many have spent days and even weeks at a locality, trying in vain to secure additional representatives of a new or poorly represented form. However, little attention (beyond the routine caveats which often accompany faunal listings) and no systematic scrutiny have been given to the question these data raise: again, how representative *are* museum mammal collections?

The question is a frustrating one, for it immediately leads full circle. If our knowledge of Neotropical mammals depends even in part on museum collections, how can the representation of museum collections be independently assessed? The question as posed is tautologous. However, introducing certain assumptions about the native fauna permits an investigation of the question of representation. Other studies have shown that the population density of a species in a given area is related to several morphological and ecological characteristics, namely body size, trophic habits, and habitat affinities. A given area can support fewer animals of large size, other factors being equal, because of the greater energy demands of larger biomass (e.g., Peters & Raelson, 1984). Similarly, that area supports fewer carnivores than omnivores, and fewer omnivores than herbivores, because of the lower available biomass for consumption. Finally, species having catholic habitat affinities can sustain a greater number of individuals than species which are highly restricted in the habitats that they can exploit, other factors being equal (e.g., Patterson, 1982). While many other factors are known to affect population density, knowledge of only these three may permit refined inference of patterns of spatial and temporal distribution (e.g., Brown, 1971; Patterson, 1984). Applying these assumptions to a faunal list, we can hypothetically specify the relative number of individuals in nature expected for each species, and use this hypothetical expectation to independently assess faunal representation in museum mammal collections.

We elected to study the representation of Field Museum's mammals of Chile for several reasons: (1) the Chilean mammal fauna is thoroughly studied and is one of the best known faunas in the Neotropics (Pine, 1982); its faunal list can be considered virtually complete; (2) this fauna was the subject of systematic and expressly faunistic studies by FMNH curator W. H. Osgood, who led two expeditions there in 1922–1923 and 1939–1940; as a result, Field Museum maintains the world's premier Chilean mammal collection (pers. obs.; Yáñez, 1982); and (3) a listing of Chilean mammals that are endangered, vulnerable, rare, or inadequately known has recently appeared (Miller et al., 1983), yielding an independent criterion for determining natural population numbers.

Materials and Methods

Species of mammals occurring in Chile were assembled from the primary and secondary literature. Osgood's (1943) treatise summarizes the earlier literature. Discrepancies between Osgood's faunal list and that presented in Table 1 reflect additions to the fauna from more recent collecting, as well as taxonomic reappraisals of previously known forms. Many additions were made shortly after Osgood's work with the publication of Mann's (1945) mammals of Tarapacá, and records from this region were augmented by Spotorno

(1976) and Pine et al. (1979). Several additional species were added by taxonomic revisions of forms Osgood reported (e.g., *Phyllotis xanthopygus, Myotis atacamensis*). Documentation for the extended list in Table 1 may be found in Mann (1978), Pine et al. (1979), Pine (1973), Spotorno (1976), Honacki et al. (1982), Patterson et al. (1984), Pearson (1984), Walker et al. (1984), and literature cited therein. Introduced species (e.g., beaver, rabbits, hares, European rats) were excluded from the analysis. Finally, a native species reported to occur in Chile, *Felis geoffroyi,* was deleted from the Chilean fauna on the authority of Honacki et al. (1982).

For each terrestrial species occurring in Chile, we determined body size, trophic habits, and habitat affinities. Body size, as measured by head and body lengths (most taxa) or shoulder heights (ungulates), was taken from Osgood (1943), the FMNH collection, or primary sources, using means or midpoints of ranges. An average value was used for sexually dimorphic taxa (e.g., mustelids, pinnipeds).

We used only four categories of trophic-habitat characteristics to reduce subjectivity and to gain suitable sample sizes for statistical analyses (cf. Eisenberg, 1981): carnivores, insectivorous carnivores, sanguinivores, piscivores, and omnivores were grouped together as carnivores; insectivores included both insect-eating and insectivorous-frugivorous taxa; animals subsisting on granivorous, graminivorous, or herbivorous diets were considered herbivores. This category included many species and was further subdivided into generalized (marked by dietary and habitat breadth) and specialized (restricted in dietary and habitat breadth) components. Information on diet and habitat was taken from all available sources, but especially Osgood (1943), Hershkovitz (1962), Mann (1978), and Pearson (1983, 1984). The continuum between generalized and specialized habits of herbivores was assessed relative to the distribution of resources in Chile; thus, *Auliscomys boliviensis,* which has broad habitat requirements in Peru (Pearson, 1951) but occurs in only a minute portion of Chile, was considered a specialized herbivore. The four trophic-habitat categories were coded 1–4 as carnivores, insectivores, specialized herbivores, and generalized herbivores, respectively.

The current status of these species in Chile was taken directly from Miller et al. (1983). Taxa were assigned numerical codes 1–5 according to the status given by Miller et al.: endangered, vulnerable, rare, inadequately known, and secure, respectively. Where Miller et al. gave different status assessments to subspecies of polytypic forms, the subspecies were assigned numerical codes and an average determined, rounding down to the nearest integer. Species or subspecies not explicitly mentioned in the account of Miller et al. were presumed to be secure.

To determine specimen representation for these taxa in the FMNH mammal collections, we used those numbers reported by Osgood (1943) in the text and "Specimens examined" sections of his species accounts. In a few cases, this number is less than the number collected by Osgood and associates and currently in the collection; in these cases we used the number of Osgood-era specimens in the collection.

We used specimen numbers reported by Osgood rather than the number currently in the collection for two principal reasons. First, FMNH now has fewer specimens of some taxa than Osgood and associates collected, due to specimen exchange programs and attrition of material on loan. Our primary goal is to assess bias in museum collecting techniques and collections, not the bias in curatorial exchange programs or collection users. Second, subsequent fieldwork by R. E. Martin on *Octodon* and B. D. Patterson on *Akodon, Oryzomys,* and marsupials has greatly augmented FMNH holdings of these taxa; inclusion of specimens obtained in these highly focused studies could strongly confound the results.

The data were analyzed using the Statistical Analysis System (SAS) procedures at the University of Chicago Computation Center. Correlation procedures used variable ranks in nonparametric analyses. Because of the lesser efficiency of nonparametric procedures in comparison to parametric ones, and hence the greater probability of Type II errors, we report values having probability levels between 0.05 and 0.10 as being "marginally significant." This assumes that larger sample size (i.e., a richer fauna or faunal category) would suffice to generate a traditionally significant value ($P < 0.05$).

Results

The native Chilean mammal fauna includes six orders, 55 genera, and 93 species (table 1). The taxonomic distribution is as follows: Marsupialia with three genera (5%) and three species (3%); Chiroptera with seven genera (13%) and 10 species

TABLE 1. Species of mammals known to occur in Chile, order to which they belong, body lengths (or shoulder heights for ungulates) in mm, trophic and habitat characterizations, current status in Chile (after Miller et al., 1983), the number of specimens at Field Museum obtained by Osgood, and the life zone they inhabit.

Taxon	Order	Size (mm)	Trophic habits	Current status	No. of specimens	Life zone
Marmosa elegans	Marsupialia	133	Insectivore	Secure	34	Terrestrial
Dromiciops australis	Marsupialia	100	Insectivore	Secure	15	Terrestrial
Rhyncholestes raphanurus	Marsupialia	119	Insectivore	Incompletely known	3	Terrestrial
Desmodus rotundus	Chiroptera	85	Carnivore	Secure	1	Aerial
Amorphochilus schnablii	Chiroptera	58	Insectivore	Secure	0	Aerial
Myotis atacamensis	Chiroptera	52	Insectivore	Secure	36	Aerial
Myotis chiloensis	Chiroptera	53	Insectivore	Secure	36	Aerial
Histiotis macrotus	Chiroptera	70	Insectivore	Secure	0	Aerial
Histiotis montanus	Chiroptera	65	Insectivore	Secure	7	Aerial
Lasiurus cinereus	Chiroptera	85	Insectivore	Secure	3	Aerial
Lasiurus borealis	Chiroptera	55	Insectivore	Secure	10	Aerial
Tadarida brasiliensis	Chiroptera	55	Insectivore	Secure	7	Aerial
Mormopterus kalinowskii	Chiroptera	51	Insectivore	Secure	0	Aerial
Chaetophractus nationi (?)	Edentata	370	Insectivore	Incompletely known	0	Terrestrial
Euphractus sexcinctus	Edentata	440	Insectivore	Secure	0	Terrestrial
Zaedyus pichiy	Edentata	230	Insectivore	Secure	0	Terrestrial
Dusicyon culpaeus	Carnivora	620	Carnivore	Secure	11	Terrestrial
Dusicyon griseus	Carnivora	520	Carnivore	Secure	38	Terrestrial
Dusicyon (*griseus*?) *fulvipes*	Carnivora	542	Carnivore	Rare	2	Terrestrial
Galictis cuja	Carnivora	425	Carnivore	Secure	12	Terrestrial
Lynchodon patagonicus	Carnivora	350	Carnivore	Rare	0	Terrestrial
Conepatus chinga	Carnivora	335	Carnivore	Secure	3	Terrestrial
Conepatus (*chinga*?) *rex*	Carnivora	419	Carnivore	Incompletely known	2	Terrestrial
Conepatus humboldtii	Carnivora	355	Carnivore	Secure	10	Terrestrial
Lutra felina	Carnivora	570	Carnivore	Endangered	8	Aquatic
Lutra provocax	Carnivora	590	Carnivore	Endangered	2	Aquatic
Felis colocolo	Carnivora	605	Carnivore	Rare	3	Terrestrial
Felis concolor	Carnivora	1,330	Carnivore	Incompletely known	7	Terrestrial
Felis guigna	Carnivora	457	Carnivore	Vulnerable	4	Terrestrial
Felis jacobita	Carnivora	600	Carnivore	Vulnerable	0	Terrestrial

TABLE 1. *Continued.*

Taxon	Order	Size (mm)	Trophic habits	Current status	No. of specimens	Life zone
Arctocephalus australis	Carnivora	2,500	Carnivore	Vulnerable	0	Aquatic
Otaria byronia	Carnivora	3,000	Carnivore	Secure	0	Aquatic
Hydrurga leptonyx	Carnivora	4,000	Carnivore	Secure	0	Aquatic
Leptonychotes weddelli	Carnivora	3,000	Carnivore	Secure	0	Aquatic
Mirounga leonina	Carnivora	4,500	Carnivore	Endangered	0	Aquatic
Lama guanicoe	Artiodactyla	1,140	Generalized herbivore	Vulnerable	1	Terrestrial
Vicugna vicugna	Artiodactyla	910	Specialized herbivore	Vulnerable	0	Terrestrial
Hippocamelus antisensis	Artiodactyla	770	Specialized herbivore	Endangered	0	Terrestrial
Hippocamelus bisulcus	Artiodactyla	800	Generalized herbivore	Endangered	3	Terrestrial
Pudu puda	Artiodactyla	410	Specialized herbivore	Incompletely known	14	Terrestrial
Oryzomys longicaudatus	Rodentia	106	Generalized herbivore	Secure	180	Terrestrial
Akodon andinus	Rodentia	88	Specialized herbivore	Secure	22	Terrestrial
Akodon berlepschii	Rodentia	100	Specialized herbivore	Secure	0	Terrestrial
Akodon hershkovitzi	Rodentia	104	Specialized herbivore	Secure	0	Terrestrial
Akodon markhami	Rodentia	104	Specialized herbivore	Secure	0	Terrestrial
Akodon olivaceus	Rodentia	99	Generalized herbivore	Secure	323	Terrestrial
Akodon xanthorhinus	Rodentia	103	Generalized herbivore	Secure	116	Terrestrial
Akodon lanosus	Rodentia	104	Specialized herbivore	Secure	24	Terrestrial
Akodon longipilis	Rodentia	118	Generalized herbivore	Secure	200	Terrestrial
Akodon sanborni	Rodentia	104	Specialized herbivore	Secure	37	Terrestrial
Geoxus valdivianus	Rodentia	111	Insectivore	Secure	30	Terrestrial
Chelemys delfini	Rodentia	106	Specialized herbivore	Incompletely known	0	Terrestrial
Chelemys macronyx	Rodentia	129	Specialized herbivore	Secure	35	Terrestrial
Chelemys megalonyx	Rodentia	120	Specialized herbivore	Secure	3	Terrestrial
Calomys lepidus	Rodentia	80	Specialized herbivore	Secure	0	Terrestrial
Eligmodontia puerulus	Rodentia	92	Specialized herbivore	Secure	5	Terrestrial
Eligmodontia typus	Rodentia	80	Specialized herbivore	Secure	1	Terrestrial
Phyllotis darwini	Rodentia	120	Generalized herbivore	Secure	37	Terrestrial
Phyllotis (darwini?) osgoodi	Rodentia	117	Specialized herbivore	Secure	0	Terrestrial
Phyllotis xanthopygus	Rodentia	135	Generalized herbivore	Secure	77	Terrestrial
Phyllotis magister	Rodentia	152	Specialized herbivore	Secure	0	Terrestrial
Auliscomys boliviensis	Rodentia	125	Specialized herbivore	Secure	7	Terrestrial
Auliscomys micropus	Rodentia	136	Generalized herbivore	Secure	78	Terrestrial

TABLE 1. *Continued.*

Taxon	Order	Size (mm)	Trophic habits	Current status	No. of specimens	Life zone
Auliscomys sublimis	Rodentia	115	Specialized herbivore	Secure	0	Terrestrial
Galenomys garleppi	Rodentia	105	Specialized herbivore	Secure	0	Terrestrial
Irenomys tarsalis	Rodentia	128	Specialized herbivore	Secure	23	Terrestrial
Chinchillula sahamae	Rodentia	171	Specialized herbivore	Rare	0	Terrestrial
Neotomys ebriosus	Rodentia	125	Specialized herbivore	Secure	0	Terrestrial
Reithrodon physodes	Rodentia	147	Specialized herbivore	Secure	24	Terrestrial
Euneomys chinchilloides	Rodentia	145	Specialized herbivore	Secure	19	Terrestrial
Andinomys edax	Rodentia	185	Specialized herbivore	Secure	0	Terrestrial
Cavia tschudii	Rodentia	220	Specialized herbivore	Secure	0	Terrestrial
Galea musteloides	Rodentia	220	Specialized herbivore	Secure	0	Terrestrial
Microcavia australis	Rodentia	220	Specialized herbivore	Secure	0	Terrestrial
Lagidium viscacia	Rodentia	397	Specialized herbivore	Incompletely known	11	Terrestrial
Lagidium wolffsohni	Rodentia	470	Specialized herbivore	Secure	0	Terrestrial
Chinchilla brevicaudata	Rodentia	370	Specialized herbivore	Endangered	0	Terrestrial
Chinchilla lanigera	Rodentia	254	Specialized herbivore	Endangered	4	Terrestrial
Myocastor coypus	Rodentia	495	Generalized herbivore	Secure	15	Aquatic
Octodon bridgesi	Rodentia	194	Generalized herbivore	Vulnerable	1	Terrestrial
Octodon degus	Rodentia	178	Generalized herbivore	Secure	38	Terrestrial
Octodon lunatus	Rodentia	203	Generalized herbivore	Secure	8	Terrestrial
Octodontomys gliroides	Rodentia	158	Specialized herbivore	Secure	0	Terrestrial
Spalacopus cyanus	Rodentia	150	Specialized herbivore	Secure	8	Terrestrial
Aconaemys fuscus	Rodentia	178	Generalized herbivore	Secure	0	Terrestrial
Aconaemys sagei	Rodentia	170	Generalized herbivore	Secure	11	Terrestrial
Ctenomys fulvus	Rodentia	218	Generalized herbivore	Secure	9	Terrestrial
Ctenomys (fulvus?) robustus	Rodentia	249	Generalized herbivore	Rare	4	Terrestrial
Ctenomys magellanicus	Rodentia	208	Generalized herbivore	Rare	5	Terrestrial
Ctenomys maulinus	Rodentia	204	Generalized herbivore	Secure	4	Terrestrial
Ctenomys opimus	Rodentia	252	Generalized herbivore	Secure	1	Terrestrial
Abrocoma bennetti	Rodentia	217	Generalized herbivore	Secure	29	Terrestrial
Abrocoma cinerea	Rodentia	192	Specialized herbivore	Rare	0	Terrestrial

(11%); Edentata with three genera (5%) and three species (3%); Carnivora, including pinnipeds, with 11 genera (20%) and 19 species (20%); Artiodactyla with four genera (7%) and five species (5%); and Rodentia with 27 genera (49%) and 53 species (57%). Four orders (80%), 34 genera (62%), and 56 species (60%) were represented in the 1943 FMNH collections, totaling 1,661 specimens. The taxonomic distribution of specimens is as follows: Marsupialia (3%), Chiroptera (6%), Carnivora (6%), Artiodactyla (1%), and Rodentia (84%).

Taxa missing from the 1943 Chilean collections at FMNH have the following distribution. No marsupial genus or species is missing. Two genera and three of 10 species of bats are missing. All three genera and species of edentates are absent. Six of 11 genera of carnivores (including all 5 pinnipeds) and seven of 19 species are unrepresented. One of four artiodactyl genera and two of five species are lacking. Finally, nine of 27 rodent genera and 20 of 53 species of rodents are not represented. Three of these groups, the edentates, pinniped carnivores, and artiodactyls, appear by these figures to be grossly underrepresented in the collections, but whether this is due to bias in collecting effort and efficiency or to their morphological and ecological characteristics remains to be determined.

We consider the evident features of this data set in greater detail below. The first section focuses on the taxonomic, trophic-habitat, and morphological composition of the Chilean mammal fauna, the second on the representation of this fauna in the FMNH Chilean mammal collections.

Faunal Analyses

Across all taxa, there is the expected correlation between trophic-habitat affinities and order, and a negative correlation between trophic-habitat affinities and size ($r_s = -0.22$; $P \cong 0.03$). Current status in Chile is strongly correlated with body size ($r_s = -0.56$; $P < 0.001$), but not with trophic-habitat affinities ($r_s = 0.17$; NS). Contingency table analysis of current status by life zone characteristics indicates a significantly larger proportion of endangered aquatic species than would be expected by chance alone. Results of the contingency table analysis should be considered tentative, however, because of small expected numbers in certain cells.

When the taxa are stratified by life zone and correlations made within groups, neither the aquatic or volant groups present interesting relationships, probably because both lack the diversity of habits, habitats, and behaviors requisite for such patterns. However, patterns are evident in the terrestrial group that support those apparent in the fauna as a whole. Trophic habits and size are strongly correlated ($r_s = -0.29$; $P < 0.02$). As in the fauna as a whole, current status in Chile is significantly related to body size ($r_s = -0.56$; $P < 0.001$); this correlation excludes the large pinnipeds of conservation concern and the small bats with presumably secure populations. As before, current status is not correlated with the trophic-habitat affinities of any group ($r_s = 0.14$; NS).

When the taxa are stratified by order, several of these patterns disappear. Only the Carnivora and Rodentia are considered here because of sample size. Neither group shows a significant correlation between trophic habits and size (carnivores, $r_s = 0$; rodents, $r_s = 0.25$; $P < 0.08$). However, both groups exhibit significant or marginally significant correlations between life zone and size (carnivores, $r_s = -0.68$; $P < 0.002$; rodents, $r_s = -0.24$; $P < 0.09$). Neither group shows a relation between status and life zone, but rodents (and not carnivores) show a significant correlation between current status and body size ($r_s = -0.40$; $P < 0.005$), larger rodents being progressively more vulnerable than smaller ones.

Collection Analyses

For the Chilean fauna as a whole, the number of specimens in the FMNH mammal collection is correlated with trophic-habitat affinities ($r_s = 0.24$; $P < 0.02$), there being fewer specimens of carnivores and insectivores than herbivores of both categories. Also, there are significantly fewer specimens of large taxa, as shown by the inverse relation of number and body size ($r_s = -0.23$; $P < 0.03$). Additionally, there is a highly significant relationship between the number of specimens and the current status of taxa in Chile ($r_s = 0.25$; $P < 0.02$).

Similar results emerge when the fauna is stratified by life zone. There is a significant relationship between body size and number of specimens only among aquatic forms—aquatic, $r_s = -0.88$ ($P < 0.005$); terrestrial, $r_s = -0.21$ ($P < 0.08$); and volant, $r_s = -0.36$ (NS)—although the two groups showing this relationship have small sample sizes. For the terrestrial group (N = 75), number of specimens and trophic-habitat affinities are related (r_s =

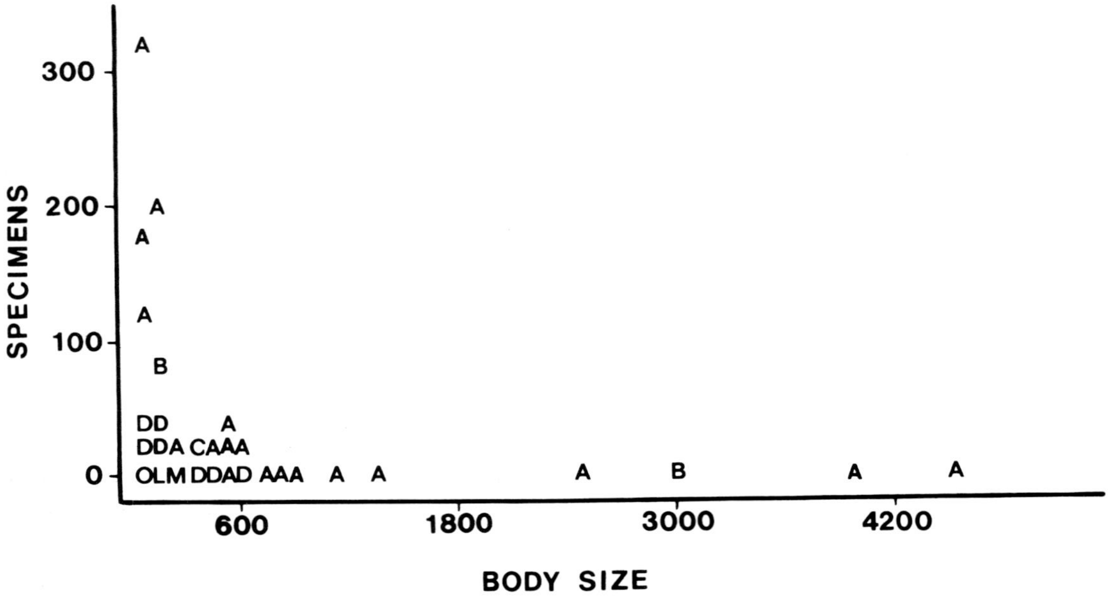

FIG. 1. Plot of number of Field Museum specimens of 58 Chilean mammals against body size. Plotting symbols represent number of species: A = one species; B = two species, etc. The rank correlation of these variables is −0.32 ($P < 0.02$). See text for discussion.

0.23; $P < 0.05$), and number of specimens and current status in Chile are strongly and positively related ($r_s = 0.29$; $P < 0.02$).

When orders are considered separately, nearly all show the inverse relationship between specimen number and body size, although small samples preclude significance in some cases: marsupials, $r_s = 0.50$ (NS); bats, $r_s = -0.36$ (NS); carnivores, $r_s = -0.47$ ($P < 0.05$); rodents, $r_s = -0.17$ (NS); edentates, $r_s = 0$; and artiodactyls, $r_s = -0.41$ (NS). Only the rodents (N = 53) exhibit a significant range of trophic-habitat affinities, and these are correlated with specimen numbers ($r_s = 0.47$; $P < 0.001$). Finally, in all represented orders showing variation in current status, there is a positive correlation between status and specimen numbers: marsupials, $r_s = 0.87$ (NS); carnivores, $r_s = 0.28$ (NS); rodents, $r_s = 0.22$ (NS); and artiodactyls, $r_s = 0.43$ (NS); however, in none of these cases is the relationship significant.

It is noteworthy that these correlations remain if species that are altogether lacking in the FMNH Chilean mammal collection are excluded. Using the 58 species having nonzero numbers of specimens in Table 1, specimen number is positively correlated with trophic-habitat affinities ($r_s = 0.26$; $P = 0.05$) and negatively correlated with body size ($r_s = -0.32$; $P < 0.02$). As before, number of specimens is highly correlated with current status in Chile ($r_s = 0.53$; $P < 0.001$) (figs. 1 and 2).

Interestingly, the correlation between specimen number and trophic-habitat affinities disappears when both unrepresented and exceedingly well represented taxa (N > 100) are excluded. Basing correlations on 54 species, this correlation equals 0.15 ($P = 0.27$). However, correlations of specimen number with body size ($r_s = -0.26$; $P < 0.06$) and with current status in Chile ($r_s = 0.52$; $P < 0.001$) still hold.

Discussion

In *The Mammals of Chile,* Osgood (1943) described the itinerary of the two FMNH Chilean expeditions, undertaken "with the intention of making a survey of the vertebrate fauna of that country" (p. 9). His own appraisal of the resulting collection is as follows:

> The mammals obtained by these two expeditions form a collection vastly larger and more varied than anything previously existing. . . . This collection is still deficient in many respects, but it covers the principal

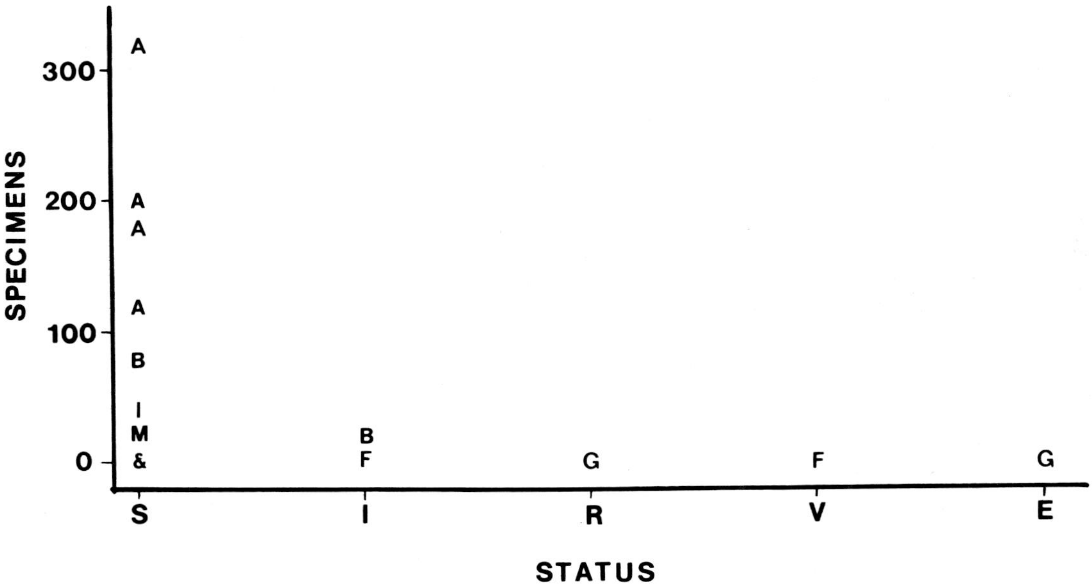

FIG. 2. Plot of number of Field Museum specimens of 58 Chilean mammals against current status in Chile. Status categories are: S = secure; I = incompletely known; R = rare; V = vulnerable; E = endangered. Plotting symbols represent numbers of species: A = one species; B = two species, etc.; & = 38 species. The rank correlation of these variables is 0.53 ($P < 0.001$). See text for discussion.

> faunal areas of Chile and probably furnishes a fairly accurate and nearly complete picture of the whole mammal fauna. This may seem to be a rash statement, not justified by experience in other fields . . . but the main facts seem to be already in hand (pp. 9–10).

Results of this analysis suggest that Osgood was correct in his inferences. The two expeditions secured most taxa that are reported from the country, and they secured these in numbers broadly indicative of natural abundances. Well represented taxa in the collections tend to represent smaller species at lower trophic levels; these species tend also to have generalized habitat requirements and secure population status. In contrast, unrepresented or poorly represented taxa tend to be larger, are restricted to higher trophic levels or specialized habitats, and are often characterized by less secure status.

The diversity of collecting techniques and procedures used by Osgood and associates was clearly adequate to sample the Chilean mammal fauna as a whole. Osgood's field notes indicate that he used mouse traps, carnivore traps, shotguns, and rifles in making his collections, and he employed field parties of as many as four coworkers. In addition, he reports having bought *Pudu* captured by native hunters with dogs and purchased *Lutra* from *gateros* who hunted them for their pelts. His use of commercial hunters and salvage operations to sample marine mammals accounts for the striking absence of seals and sea lions in the collection. The Chilean pinnipeds had been decimated by sealing operations in the 18th and 19th centuries, and now occur in highly restricted portions of their former ranges.

No taxonomic group presents stronger evidence of collecting bias than do the five pinnipeds missing from the FMNH collections. These indicate the insufficiency of Osgood's sampling of the littoral zone. However, even this omission from the FMNH collection pales in comparison to that involving forms restricted to northernmost Chile. Most of these are *puna* species of the central Andes, occurring in southern Peru, Bolivia, and northwestern Argentina, extending into Chile only in the northernmost Tarapacá region. Their absence in the 1943 collections reflects Osgood's failure to devote sufficient time to sampling these marginal localities, not to biases in the collecting techniques themselves. Lacking such material in his collections, Osgood (1943) did not recognize an important tropical and subtropical component to the Chilean mammal fauna, although Hellmayr (1932) had earlier recognized such components to the Chilean avifauna. The omission enhanced the apparent endemicity of the Chilean mammal fauna and did little to aid identification of relationships with other regions.

Two of the remaining species, *Akodon markhami* and *A. hershkovitzi,* are island derivatives of mainland species (Pine, 1973; Patterson et al., 1984) and exhibit restricted geographic ranges that FMNH collecting parties did not sample. A mainland taxon, *Aconaemys fuscus,* was recently restricted by Pearson's revision (1984), so that its known range in Chile is quite small. Similarly, *Microcavia australis* is a pampas species barely crossing the Chilean frontier in southern Chile, while *Chelemys delfini* is known only from the holotype. [We dissent with the opinion of Miller et al. (1983), who considered *delfini* of doubtful validity. The disjunct *Geoxus* taxon *michaelseni* is the only other long-clawed rodent occurring near Punta Arenas, and it seems only reasonable to believe that Cabrera could distinguish the substantially different skulls of these forms (cf. figures of *Geoxus* and *Chelemys* in accounts of "*Notiomys*" in Osgood (1943).] Osgood's failure to secure other taxa, for example *Histiotius macrotus,* is perplexing, given the number of localities reported for these species by Mann (1978). For the bats at least, modern records may be attributed to the use of mist nets rather than shotguns to collect bats.

The use of trophic-habitat affinities and body size to estimate naturally occurring population densities seems justifiable in view of their contributions to other studies. This justification is reinforced by the strong positive correlations between current status in Chile and the numbers of specimens Osgood and associates collected; these correlations hold for the fauna as a whole, and also for selected subsets (e.g., stratifications by life zone and by order). The correlations hold in spite of uncontrolled variation in other factors known to affect the current status of Chilean species. For example, Miller et al. (1983) note regional variation in endangerment; species living in central Chile have been most severely impacted by human activities. The correlation analyses assumed that, during the first half of this century, Osgood and associates should have collected mammals proportional to their 1983 abundances. However, the assumption seems approximated in view of the demonstrated causes of species decline, principally exploitation through hunting and habitat deterioration. Chile was settled by Europeans in the 16th century, and much of its fauna and flora were known by the end of the 18th century (e.g., Molina, 1782). Although economic exploitation and habitat destruction are accelerating with population growth in many parts of Chile, patterns of species densities are probably broadly comparable to those Osgood encountered.

This analysis suggests that it should be possible to predict the species *currently lacking* from the Chilean mammal inventory. If unbiased collecting techniques sample animal populations proportional to their densities, then one could expect that species of carnivores or specialized herbivores might thus far have been overlooked. However, a critically confounding variable in making this inference is the geographic range size of species. While carnivores may exist in nature at lower densities than herbivores and thus are sampled less often at a given locality, they tend to have larger ranges (Rapoport, 1982), which means they can be sampled at more localities. Averaged across an entire country or across an entire fauna, the expected correlation between numbers of specimens and expected population density holds. But when a fauna has been largely sampled, as is certainly true for Chile's mammal fauna, further additions to the faunal list are apt to be those species with highly restricted geographic ranges, especially those in remote areas (e.g., *Akodon hershkovitzi,* Patterson et al., 1984).

What are the implications of this study for other mammal collections made in other faunas? Chile's mammal fauna is highly peculiar, stemming from its isolation from the remainder of South America by the Atacama Desert to the north and the Andes to the east. Excluding bats, pinnipeds, and forms that barely cross Chile's borders, fully one-third of the genera of Chilean mammals are endemic (Osgood, 1943, p. 36), but the fauna does share higher-level affinities with adjacent areas. The fauna is also a comparatively depauperate one: the 93 mammal species recorded from Chile stand in contrast to the 141 species of bats alone that are thought to occur in Colombia (Koopman, 1982). Finally, the Chilean fauna occurs in habitats that are structurally simple by tropical standards and similar in many respects to those of the north temperate zone. Thus trapping procedures and experiences of north temperate scientists might enable them to sample Chilean habitats more effectively than tropical ones. For these reasons, this study should be repeated in a strictly Neotropical area (*sensu* Hershkovitz, 1972). Both Suriname and Venezuela have diverse Neotropical mammal faunas that are comparatively well studied (see Husson, 1978; Genoways & Williams, 1979; Handley, 1976, and continuing reports, respectively). Both countries would represent prospective areas for repetition of this study.

Acknowledgments

This study was an outgrowth of a Senior Linkage Seminar entitled "The biology and conservation of South American mammals" at Northwestern University (EEB C-94) during Spring 1984; we thank classmates for helpful discussion. The senior author thanks Milton Gallardo for fostering his appreciation of the mammals of Chile, and we both thank Ronald H. Pine, Guy G. Musser, and an anonymous reviewer for criticisms of the manuscript.

Literature Cited

Brown, J. H. 1971. Mammals on mountaintops: Nonequilibrium insular biogeography. American Naturalist, **105:** 467–478.

Eisenberg, J. F. 1981. The Mammalian Radiations: An Analysis of Trends in Evolution, Adaptation, and Behavior. University of Chicago Press, Chicago, 610 pp.

Genoways, H. H., and S. L. Williams. 1979. Records of bats (Mammalia: Chiroptera) from Suriname. Annals of Carnegie Museum, **48:** 323–335.

Handley, C. O., Jr. 1976. Mammals of the Smithsonian Venezuelan Project. Brigham Young University Science Bulletin, Biological Series, **20**(5): 1–89.

Hellmayr, C. 1932. The birds of Chile. Field Museum of Natural History, Zoological Series, **19:** 1–472.

Hershkovitz, P. 1962. Evolution of Neotropical cricetine rodents (Muridae), with special reference to the phyllotine group. Fieldiana: Zoology, **46:** 1–524.

———. 1972. The Recent mammals of the Neotropical region: A zoogeographic and ecological review, pp. 311–431. *In* Keast, A., F. C. Erk, and B. Glass, eds., Evolution, Mammals, and Southern Continents. State University of New York Press, Albany, 543 pp.

Honacki, J. H., K. E. Kinman, and J. W. Koeppl. 1982. Mammal Species of the World: A Taxonomic and Geographic Reference. Allen Press and Association of Systematic Collections, Lawrence, Kansas, 694 pp.

Husson, A. M. 1978. The mammals of Suriname. Zoölogische Monographieën van het Rijksmuseum van Natuurlijke Historie, **2:** xv + 1–569.

Koopman, K. F. 1982. Biogeography of the bats of South America, pp. 273–302. *In* Mares, M. A., and H. H. Genoways, eds., Mammalian biology in South America. Special Publication Series, Pymatuning Laboratory of Ecology, University of Pittsburgh, **6:** 1–539.

Mann F., G. 1945. Mamíferos de Tarapacá. Observaciones realizadas durante una expedición al Alto Norte de Chile. Biologica, **2:** 23–134.

———. 1978. Los pequeños mamíferos de Chile. Gayana: Zoología, **40:** 1–342.

Miller, S. D., J. Rottmann, K. J. Raedeke, and R. D. Taber. 1983. Endangered mammals of Chile: Status and conservation. Biological Conservation, **25:** 335–352.

Molina, G. I. 1782. Saggio sulla storia naturale del Chili. Bologna, 368 pp.

Nowak, R. M., and J. L. Paradiso. 1983. Walker's Mammals of the World, vol. II, pp. 569–1362. Johns Hopkins University Press, Baltimore.

Osgood, W. H. 1943. The mammals of Chile. Field Museum of Natural History, Zoological Series, **30:** 1–268.

Patterson, B. D. 1982. Pleistocene vicariance, montane islands, and the evolutionary divergence of some chipmunks (genus *Eutamias*). Journal of Mammalogy, **63:** 387–398.

———. 1984. Mammalian extinction and biogeography in the southern Rocky Mountains, pp. 247–293. *In* Nitecki, M. H., ed., Extinctions. University of Chicago Press, Chicago, 354 pp.

Patterson, B. D., M. H. Gallardo, and K. E. Freas. 1984. Systematics of mice of the subgenus *Akodon* (Rodentia: Cricetidae) in southern South America, with the description of a new species. Fieldiana: Zoology, n.s., **23:** 1–16.

Pearson, O. P. 1951. Mammals in the highlands of southern Peru. Bulletin of the Museum of Comparative Zoology, **106:** 117–174.

———. 1983. Characteristics of a mammalian fauna from forests in Patagonia, southern Argentina. Journal of Mammalogy, **64:** 476–492.

———. 1984. Taxonomy and natural history of some fossorial rodents of Patagonia, southern Argentina. Journal of Zoology, London, **202:** 225–237.

Peters, R. H., and J. V. Raelson. 1984. Relations between individual size and mammalian population density. American Naturalist, **124:** 498–517.

Pine, R. H. 1973. Una nueva especie de Akodon (Mammalia: Rodentia: Muridae) de la isla de Wellington, Magallanes, Chile. Anales del Instituto de la Patagonia, **IV**(1–3): 423–426.

———. 1982. Current status of South American mammalogy, pp. 27–37. *In* Mares, M. A., and H. H. Genoways, eds., Mammalian biology in South America. Special Publication Series, Pymatuning Laboratory of Ecology, University of Pittsburgh, **6:** 1–539.

Pine, R. H., S. D. Miller, and M. L. Schamberger. 1979. Contributions to the mammalogy of Chile. Mammalia, **43:** 339–376.

Rapoport, E. H. 1982. Areography: Geographical Strategies of Species. Pergamon Press, New York, 269 pp.

Spotorno O., A. E. 1976. Analisis taxonomico de tres

especies altiplanicas del genero *Phyllotis* (Rodentia, Cricetidae). Anales del Museo de Historia Natural, Valparaiso, **9**: 141–161.

WALKER, L. I., A. E. SPOTORNO, AND J. ARRAU. 1984. Cytogenetic and reproductive studies of two nominal subspecies of *Phyllotis darwini* and their experimental hybrids. Journal of Mammalogy, **65**: 220–230.

YÁÑEZ V., J. 1982. Estado de las colecciones mastozoológicas nacionales. Publicación Ocasional, Museo Nacional de Historia Natural, Santiago, **38**: 13–28.

Taxonomic Index

Subject Index